Laser-Induced Breakdown Spectroscopy

Reinhard Noll

Laser-Induced Breakdown Spectroscopy

Fundamentals and Applications

Springer

Dr. Reinhard Noll
Fraunhofer-Institut für Lasertechnik (ILT)
Steinbachstr. 15
52074 Aachen
Germany
reinhard.noll@ilt.fraunhofer.de

ISBN 978-3-642-20667-2 e-ISBN 978-3-642-20668-9
DOI 10.1007/978-3-642-20668-9
Springer Heidelberg Dordrecht London New York

Library of Congress Control Number: 2011940321

Printed on acid-free paper

Springer is part of Springer Science+Business Media (www.springer.com)

In loving memory of my mother Maria,
who passed away November 12th, 2011

Preface

This book is meant as comprehensive source of the fundamentals and applications of laser-induced breakdown spectroscopy (LIBS). It provides a systematic introduction to the principles, the dominant process parameters, and the instrumental components for LIBS. The effect of multiple pulses on material ablation, plasma dynamics, and plasma emission is presented in detail. The double-pulse approach is described, and the underlying physics by integral as well as temporally and spatially resolved plasma diagnostics is clarified. A heuristic plasma modeling allows to simulate complex experimental plasma spectra. These methods and findings form the basis for a variety of applications to perform quantitative multielement analysis with LIBS.

These application potentials of LIBS have really boosted in the last years ranging from bulk analysis of metallic alloys and nonconducting materials, via spatially resolved analysis and depth profiling covering measuring objects in all physical states: gaseous, liquid, and solid. Dedicated chapters present LIBS investigations for these tasks with special emphasis on the methodical and instrumental concepts, as well as the optimization strategies for a quantitative analysis.

Requirements, concepts, design, and characteristic features of LIBS instruments are described covering laboratory systems, inspection systems for inline process control, mobile systems, and remote systems.

Based on this applied research and development, LIBS was pushed forward significantly achieving limits of detection for quantitative trace element determination and measuring frequencies not achieved so far. Industrial applications of LIBS systems are presented demonstrating the benefits of inline process control for improved process guiding and quality assurance purposes. Among these the identification testing of pipe fittings with LIBS impressively shows the first routine application established where more than five million products were tested automatically in a production line within a period of less than 2 years.

The author moved into this challenging subject in the early nineties and the present book has grown out of work performed at the Fraunhofer-Institut für Lasertechnik (ILT) in Aachen, Germany. Thanks are due to many colleagues and students at ILT as well as former coworkers who continued their career in industry and universities for their contributions and discussions (in alphabetic order): Ü. Aydin, H. Balzer, H. Bette, R. Bleich, A. Brysch, D. Eilers, R. Fleige, C. Fricke-Begemann, C. Gehlen, M. Gröbner, C. Haas, J. Hertzberg, F. Hilbk-Kortenbruck, M. Höhne, S. Hölters, P. Jander, C. Janzen, R. Klauke, O. Klein, H. Krause, M. Kraushaar, T. Kuhlen, H. Kunze, A. Lamott, F. Legewie, A. Löbe, J. Makowe, I. Mönch, L. Peter, S. Pflüger, R. Quay, P. Roth, R. Sattmann, M. Sellhorst, M. Stepputat, N. Strauss, V. Sturm, J. Vrenegor, Q. Wang, P. Werheit, R. Wester, E. Wiens, S. Winkelmann, and M. Wypadlo.

The author likes to express his special thanks to Mrs. Anna Mosna, who drew many of the figures and illustrations shown in this book with great diligence.

Very special thanks go to the author's wife for her help, patience, and encouragement.

Aachen,
August 2011

Reinhard Noll

Contents

Chapter 1
Introduction

Laser radiation is a high-quality form of electromagnetic energy enabling a multitude of new methods and applications, such as material processing, biomedical and communication technologies, and measuring methods. The advantages of lasers in measuring technologies are the noncontacting measurement, high flexibility, and high measuring speeds. Due to these features, laser measuring methods and applications have encountered a dynamic development during the last years. Laser measuring methods were introduced successfully in production technology, process engineering, quality assurance, environmental technology, and life sciences. The most important measuring quantities are:

- Geometrical quantities, such as distance, contour, shape, roughness, strain
- Dynamic quantities, such as velocities, vibration modes
- Thermodynamic and chemical quantities, such as temperature, density, concentration, chemical bonds

The above-mentioned quantities are determined preferably with the use of spectroscopic methods. From the very first the invention of laser was tightly linked with spectroscopic questions [1.1]. The special properties of laser radiation enable a variety of new spectroscopic methods to analyze the chemical constituents of a substance or to determine their physical state. Examples are laser absorption spectroscopy, light detection and ranging (LIDAR), laser-induced fluorescence (LIF), and coherent antistokes Raman spectroscopy (CARS) [1.2, 1.3]. On the one hand, the tunability of the laser wavelength to atomic or molecular transitions allows for a high selectivity, and on the other hand the high spectral brightness of laser radiation enhances the detection sensitivity enabling the determination of traces.

Laser spectroscopic methods are a powerful tool for fundamental investigations, such as high-resolution spectroscopy within the Doppler width to study the fine structure of excited states [1.4]. Femtosecond laser pulses allow observing directly the dynamics of chemical reactions. The Nobel prize for physics in the year 1999 was awarded to A. Zerwail for his pioneering work in this field [1.5]. In an increasing degree laser spectroscopy enters into new application fields. Among these are the remote investigation of harmful substances in the atmosphere, the

R. Noll, *Laser-Induced Breakdown Spectroscopy*, DOI 10.1007/978-3-642-20668-9_1,

monitoring of combustion processes or material-dependent production processes, and the quality assurance of semi-finished products [1.6, 1.7, 1.8, 1.9].

For technical applications, laser spectroscopic methods are of special interest as they are able to determine several species simultaneously with minimum equipment. Laser-induced breakdown spectroscopy (LIBS) belongs to these methods [1.10]. Fundamentals and applications of LIBS are the subject of this book. LIBS is able to analyze solid, liquid, and gaseous substances. In principle, laser absorption spectroscopy and LIF are also able to determine several species, but in this case different laser wavelengths are necessary, which cause a corresponding high instrumental effort for technical applications.

Focusability and temporal modulation of the power of the laser beam are the two main features of laser radiation which are exploited for LIBS. At irradiances above $10^9\,\mathrm{W/cm^2}$ a solid substance to be studied evaporates locally within a short time forming a transient plasma. In the plasma, the ablated atoms and ions are excited and emit their characteristic line radiation [1.11]. Emission wavelengths cover a broad spectral range from the deep ultraviolet to the infrared. A short time after the invention of the laser the emission features of the laser-induced plasma were investigated [1.12]. However, the used ruby lasers could only be operated at low repetition rates. Additionally the emission of the plasma was observed time integrated, so only a limited analytical quantification was achievable.

The dynamic technological developments in the field of solid state lasers, electro-optic detectors, and signal processing during the last decade were successfully utilized for LIBS. The analytical performance for a multielement analysis achieves a level, which equals or is even better than that of classical methods [1.13]. The vital point is that due to optical excitation of the measuring radiation over a distance, new fields of application were opened up, which are not accessible for conventional physical methods, such as X-ray fluorescence analysis or spark emission spectrometry. Such fields are, e.g., the inline analysis of liquid steel or the material identification for sorting and quality assurance tasks [1.14, 1.15]. Quality assurance close to the process on the basis of an automated inline measurement is the first step to fast and efficient feedback actions which result in significant cost savings and improved competitiveness.

Pulsed laser radiation can be used not only for analysis, but also for the preparation of substances to be investigated. Surface layers, as, e.g., corrosion layers, oxides, scale, contaminations, whose chemical composition differs from one of the subjacent substances, can be ablated locally by the laser beam. A mechanical preparation step, by milling, turning, or grinding, is not necessary in many cases. This dual use of the laser beam as ablation and analyzing tool has a high technological potential in terms of automation capabilities for inline inspection systems.

With diode-pumped solid state lasers, compact radiation sources of high beam quality are available offering electro-optical Q-switching with pulse repetition rates of 1 kHz and beyond, average powers of several watts, and excellent beam qualities [1.16]. The operating life time of these lasers amounts to more than 20,000 h. This opens for LIBS the area of spatially resolved determinations of element distributions on the surface of macroscopic samples. An example of this approach is the analysis of the cleanliness of steel samples, where microscopic inclusions in a steel matrix are determined, consisting of hard materials as, e.g., carbides and nitrides. These inclusions influence the processability and the material properties of steel grades. While competing methods such as scanning electron microscopy (SEM) for a spatially resolved energy-dispersive X-ray fluorescence analysis (SEM-EDX) or the electron microprobe require high vacuum conditions and an extensive sample preparation such as diamond polishing, the laser-based method measures under an inert gas atmosphere and a simple preparation of the sample surface by milling or grinding is sufficient.

The features of LIBS compared with those of other laser spectroscopic methods are given in Table 1.1. Laser-induced breakdown spectroscopy is the only method, which allows for a simultaneous multispecies analysis in all states of aggregation of a substance.

Table 1.1 Comparative survey of laser spectroscopic methods with exemplary data of the used laser sources and measuring quantities

Method	LIBS	LIF	Raman	CARS	LAS
λ_{exc} (nm)	266, 532, 1,064	200–800; 337	347, 694; 488, 515, 532; 266	532, 607–609	760, 1,570, 1,800; 3,160–3,670; 10,600
$\Delta\tilde{\nu}(cm^{-1})$	1	0.5; 1.2	<1	0.8; 180	<0.05
t_{exc} (ns)	5–100	10–cw	10–cw	10	6–cw
$I(W/cm^2)$	10^8–10^{11}	10^3–10^6	10^3–10^6	10^9–10^{11}	
Measuring species	g, l, s, atoms	g, l, atoms, molecules	g, l, s, molecules	g, l, s, molecules	g, molecules
Measuring quantity	c_i	c_i	c_i, T	T, c_i	c_i
Measuring range	>1 μg/g	>0.01 μg/g	>10 μg/g, 300–3,500 K	300–3,500 K, 1 mbar–100 bar	>1 ng/g
Measuring distance (mm)	10–1,500	10–1,000	10–1,000	50–1,000	500–4,000
Spatial resolution	>1 μm	>1 μm	>1 μm	$>1\ mm^3$	./.
Minimum measuring time	~10 μs	<10 μs	~10 ns	~10 ns	~100 ms
References	[1.13], [1.17], [1.18]	[1.19], [1.20], [1.21]	[1.22], [1.23], [1.24]	[1.25], [1.26]	[1.7], [1.27], [1.28]

LIBS laser-induced breakdown spectroscopy, *LIF* laser-induced fluorescence, *CARS* coherent anti-Stokes Raman spectroscopy, *LAS* laser absorption spectroscopy, λ_{exc} wavelength of the laser used for excitation, $\Delta\tilde{\nu}$ bandwidth of the laser, t_{exc} pulse duration of the laser beam, I irradiance at the location of interaction, g gaseous, l liquid, s solid, c_i concentration of species i, T temperature, *cw* continuous wave

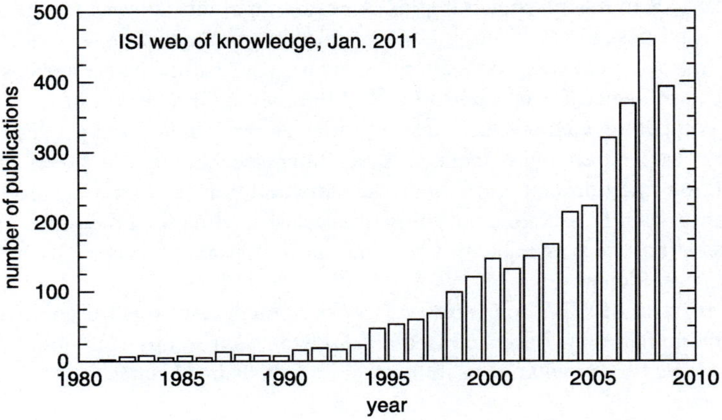

Fig. 1.1 Number of publications in the field of laser-induced breakdown spectroscopy or LIBS in the years 1980–2010 from a database search in ISI Web of Knowledge

The number of publications in the field of LIBS shows a strong growth in the last years [1.29]. Figure 1.1 shows the result of a database search from January 2011 using the search term "laser-induced breakdown spectroscopy or LIBS" in the ISI Web of Knowledge showing the nearly exponential growth of publications in the last years reaching more than 450 publications dealing with LIBS in the year 2009.

The purpose of this book is to give a comprehensive introduction to the basic methodological approaches, the relevant process parameters, and the instrumental components being subject of Chaps. 2–4. The interaction of the laser beam with the measuring object leading to evaporation and plasma formation is presented in Chap. 5. The use of multiple pulses for LIBS has led to a significant progress in the analytical performance, and Chap. 6 gives an overview of the various concepts and achieved improvements. Their effect on the mass ablation and plasma dynamics is discussed in Chaps. 7 and 8. Chapter 9 deals with the various emission processes of the laser-induced plasma such as continuum emission and line emission. A calculation of the plasma emission based on a heuristic model is presented in Chap. 10. The procedures to be used for quantitative measurements with LIBS are given in Chap. 11. The combination of the two laser spectroscopic methods LIBS and LIF is the topic of Chap. 12. Chapters 13–16 present various analytical investigations performed with LIBS ranging from the bulk analysis of metallic matrices via nonconducting materials to spatially resolved analysis and depth profiling. LIBS instruments are described in Chap. 17. Industrial applications of LIBS are the subject of Chap. 18. Finally, the Annex comprises an estimate of the necessary energy to displace the ambient atmosphere by the expanding plasma, lists of abbreviations and symbols, a compilation of Stark data important for LIBS, and finally a description of a spectral line selection method for Boltzmann plots.

References

1.1. Ch. Townes, The birth of the laser. Opto Laser Eur. **69**, 20–30 (1999)
1.2. W. Demtröder, *Laserspektroskopie* (Springer-Verlag, Berlin, 1999)
1.3. C. Weitkamp, *LIDAR – Range-Resolved Optical Remote Sensing of the Atmosphere* (Springer-Verlag, Berlin, 2005), ISBN 0-387-40075-3
1.4. C. Duke, H. Fischer, H.J. Kluge, H. Kremmling, Th. Kühl, E.W. Otten, Determination of isotope shift of ^{190}Hg by on-line laser spectroscopy. Phys. Lett. **60A**, 303–306 (1977)
1.5. G. Gerber, Die Momentaufnahme der Molekülspaltung. Phys. Blätter **55**, 23–25 (1999)
1.6. M. Proffitt, A. Langford, Ground based differential LIDAR system for day or night measurement of ozone throughout the free troposphere. Appl. Optics **36**, 2568–2585 (1997)
1.7. R. Mihalcea, D. Baer, R. Hanson, A diode-laser absorption sensor system for combustion emission measurements. Meas. Sci. Technol. **9**, 327–338 (1998)
1.8. L. Paksy, B. Nemet, A. Lengyel, L. Kozma, J. Czekkel, Production control of metal alloys by laser spectroscopy of molten metals. Part 1. Preliminary investigation. Spectrochim. Acta B **51** 279–290 (1996)
1.9. R. Noll, L. Peter, I. Mönch, V. Sturm, Automatic laser-based identification and marking of high-grade steel qualities, in *Progress in Analytical Chemistry in the Steel and Metals Industries*, ed. by R. Tomellni (European Communities, Luxembourg, 1999), pp. 345–351
1.10. D. Cremers, The analysis of metals at a distance using laser-induced breakdown spectroscopy. Appl. Spectrosc. **41**, 572–579 (1987)
1.11. R. Adrain, J. Watson, Laser microspectral analysis: a review of principles and applications. J. Phys. D Appl. Phys. **17**, 1915–1940 (1984)
1.12. E. Runge, R. Minck, F. Bryan, Spectrochemical analysis using a pulsed laser source. Spectrochim. Acta **20**, 733–736 (1964)
1.13. V. Sturm, L. Peter, R. Noll, Steel analysis with laser-induced breakdown spectrometry in the vacuum ultraviolet. Appl. Spectrosc. **54**, 1275–1278 (2000)
1.14. C. Carlhoff, Laserinduzierte Emissionsspektroskopie für die Direktanalyse von flüssigem Stahl im Konverter. Laser Optoelektr. **23**, 50–52 (1991)
1.15. H. Sattler, Metallschrotte nach Analyse automatisch sortieren. Materialprüfung **35**, 312–315 (1993)
1.16. P. Peuser, N. Schmitt, *Dioden-gepumpte Festkörperlaser* (Springer-Verlag, New York, 1995)
1.17. Final report of the Brite-Euram project: study of emission spectroscopy on laser produced plasma for localised multielemental analysis in solids with surface imaging, Coordinator Saclay, France, Contract MAT 1-CT-93–0029, 1996
1.18. L. Cabalin, D. Romero, J. Baena, J. Laserna, Effect of surface topography in the characterization of stainless steel using laser-induced breakdown spectrometry. Surf. Interface Anal. **27**, 805–810 (1999)
1.19. W. Sdorra, A. Quentmeier, K. Niemax, Basic investigations for laser microanalysis. II. Laser-induced fluorescence in laser-produced sample plumes. Mikrochim. Acta II **2**, 201–218 (1989)
1.20. R. Wamsley, T. O'Brian, K. Mitsuhashi, J. Lawler, Laser-induced fluorescence on Hg^{+} in Hg-Ar discharges. Appl. Phys. Lett. **59**, 2947–2949 (1991)
1.21. J. Bublitz, M. Dickenhausen, M. Grätz, S. Todt, W. Schade, Fiber-optic laser-induced fluorescence probe for the detection of environmental pollutants. Appl. Optics **34**, 3223–3233 (1995)
1.22. P. Kasal, B. Mewes, D. Brüggemann, Lasergestützte Verbrennungsdiagnostik, VDI-Broschüre zur Abschlußpräsentation von BMBF-Verbundprojekten, Düsseldorf, 1997, S. 1–13
1.23. B. Schrader (ed.), *Infrared and Raman Spectroscopy* (VCH Verlagsgesellschaft, Weinheim, 1995)
1.24. P. Jander, R. Noll, Automated detection of fingerprint traces of high explosives using ultraviolet Raman spectroscopy. Appl. Spectrosc. **63**, 559–563 (2009)

1.25. A.C. Eckbreth, *Laser Diagnostics for Combustion Temperature and Species* (Abacus, Tunbridge Wells, 1988)
1.26. R. Clark, R. Hester (eds.), *Advances in Non-linear Spectroscopy* (Wiley, Chichester, 1988)
1.27. M. Seiter, M. Sigrist, On-line multicomponent trace-gas analysis with a broadly tunable pulsed difference-frequency laser spectrometer. Appl. Optics **38**, 4691–4698 (1999)
1.28. T. Fink, S. Büscher, R. Gäbler, Q. Yu, A. Dax, W. Urban, Laser intracavity photoacoustic spectrometer for trace gas analysis. Rev. Sci. Instrum. **67**, 4000–4004 (1996)
1.29. R. Noll, Terms and notations for laser-induced breakdown spectroscopy. Anal. Bioanal. Chem. **385**, 214–218 (2006)

Chapter 2
Laser-Induced Breakdown Spectroscopy

In this chapter, the principle of laser-induced breakdown spectroscopy is presented. The intention is to give an overview of the main parameters of the method, which will be discussed in more detail in Chap 3. By means of a typical setup of an analyzing device based on LIBS, the important instrumental components are introduced. These components are the subject of Chap 4. In Sect. 2.3, the single steps of an analysis are introduced.

2.1 Principle and Measuring Parameters

The principle of laser-induced breakdown spectroscopy is shown schematically in Fig. 2.1.

A pulsed laser beam is focused onto the surface of a substance to be analyzed, see (1) in Fig. 2.1. Radiation energy is locally coupled into the material (2) and the material starts to evaporate (3). Within this material vapor and the surrounding gas atmosphere a plasma is generated (4), leading to the excitation of the material constituents and their spontaneous emission of radiation. The plasma decays and emits element-specific radiation (5)–(7). This emission is resolved spectrally and is detected by a spectrometer. For solid substances, a crater is formed finally (8). The evaporated material is removed partially from the interaction zone driven by the intrinsic dynamics of the plasma expansion and by an externally impressed gas flow.

The process denoted in phase (3) as "evaporation" is a simplified description and refers to solid inorganic substances. In general, there is no pure sublimation from the solid phase to the gaseous phase. Depending on the laser parameters – e.g., the pulse duration – and the material properties a transient liquid phase may also exist. Besides the evaporation and dissociation processes also particles are ablated, e.g., small particulates or droplets, which are ejected due to the impact of the pressure exerted by the plasma and the accompanying shock waves on a melt layer [2.1]. For organic substances, the initial material is disintegrated, fragmented, and dissociated.

R. Noll, *Laser-Induced Breakdown Spectroscopy*, DOI 10.1007/978-3-642-20668-9_2,

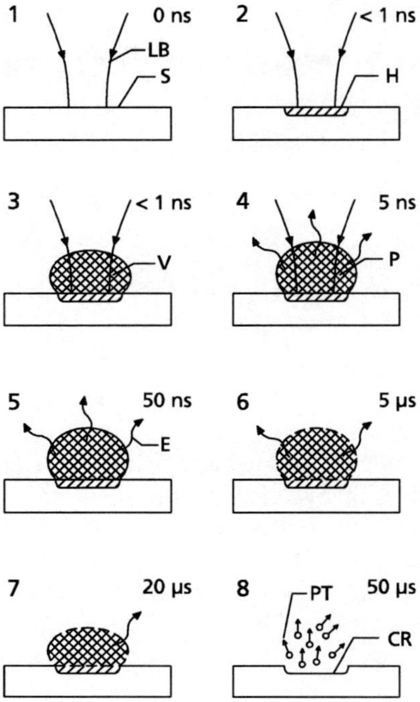

Fig. 2.1 Principle of laser-induced breakdown spectroscopy shown in phases 1–8; LB = incoming laser beam, S = sample, H = region of energy deposition, V = material vapor, P = plasma, E = element-specific emission, CR = crater, PT = particles . The times given depict the temporal evolution after start of irradiation of the laser pulse

tHE life time of the plasma depends on the laser beam parameters chosen, the conditions of the surrounding gas atmosphere, and the substance to be analyzed. The life time lies typically in the range of 0.5–10 μs. The whole process depicted in Fig. 2.1 can be repeated with frequencies of 10 Hz up to 1 kHz (see Chap 15, Sect. 15.3).

Figure 2.2 shows schematically some of the measuring parameters and the spectrum emitted from the plasma. The measuring parameters will be described in more detail in Chap 3. The incident laser irradiance $I^{\mathrm{i}}_{\lambda_{\mathrm{L}}}(t)$ at the wavelength λ_{L} is a function of time (the spatial dependence is not discussed here). The temporal structure of the laser pulse has a decisive influence on the generated plasma states and hence the emitted spectrum, see also Sect. 3.2 and Chap. 6. The quantity Δs describes the position of the beam waist of the laser beam in relation to the sample surface for a solid or a liquid measuring object. A part of the incident irradiance of the laser beam is reflected by the sample and the plasma in the backward direction, and the other part is scattered in various directions. These intensities are denoted with $I^{\mathrm{r}}_{\lambda_{\mathrm{L}}}$ and $I^{\mathrm{s}}_{\lambda_{\mathrm{L}}}$. Besides the irradiance $I^{\mathrm{i}}_{\lambda_{\mathrm{L}}}(t)$, the ambient gas pressure p_{g} and the type of the ambient gas have an influence on the plasma dynamics (see Sect. 3.5

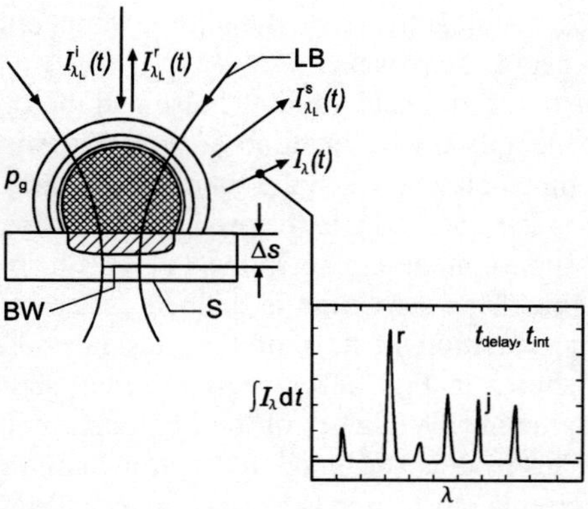

Fig. 2.2 Schematic illustration of some measuring parameters and the emitted spectrum; LB = incident laser beam, S = sample, BW = beam waist, $I^{\mathrm{i}}_{\lambda_\mathrm{L}}(t)$ = irradiance of the incident laser beam, $I^{\mathrm{r}}_{\lambda_\mathrm{L}}(t)$ = reflected laser irradiance, $I^{\mathrm{s}}_{\lambda_\mathrm{L}}(t)$ = scattered laser irradiance, p_g = ambient gas pressure, Δs = beam waist position, j = emission line of an analyte, r = emission line of a reference line, t_{delay} = delay time between the laser pulse and the start of the integration window to record the spectrum, t_{int} = width of the integration window

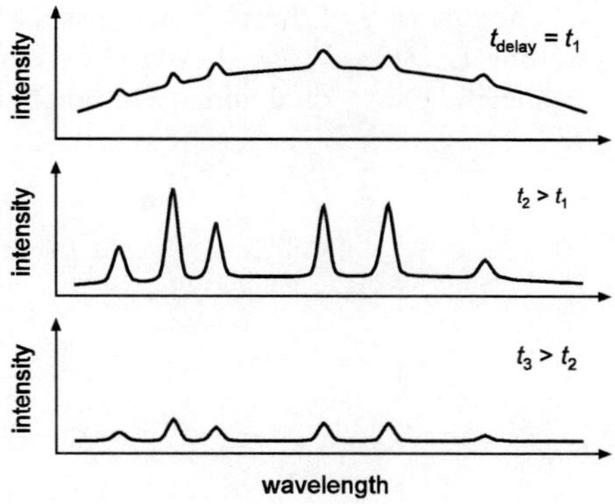

Fig. 2.3 Schematic illustration of the emission spectra of the laser-induced plasma for different time delays with respect to the irradiation of the laser pulse

and Chap 7). The emitted spectrum of the plasma $I_\lambda(t)$ is also a function of time. Figure 2.2, bottom right, shows schematically a spectrum detected at a time t_{delay} with respect to the irradiation time of the laser pulse and integrated for a duration of t_{int}.

During the life time of the plasma the emission spectrum changes. Figure 2.3 illustrates schematically emission spectra of the laser-induced plasma for three different time delays after the irradiation of the laser pulse. At the time t_1, the plasma emits predominantly a continuous spectrum, caused by free–free transitions of electrons. Only small peaks of the line intensities of atoms and ions are visible, and the ratio of the peak intensity of an emission line to the neighboring intensity of the spectral continuum emission is low. At the time t_2, the plasma has cooled down and the intensity of the line emission, as well as the ratio peak intensity to continuous background, increases significantly. At t_3, the plasma temperature decreases further and the emission intensities decrease.

Figure 2.4a, b illustrates the incident laser pulse and the plasma emission at an element-specific wavelength λ_i as a function of time. Experimental investigations have shown that the time delay between the rising of the laser pulse and the rising of the plasma emission is small being in the order of <1 ns. Hence, in the following paragraphs we can assume that the signals shown in Fig. 2.4a, b start simultaneously, see dashed vertical line. This start time is defined as the reference time for the delay time t_{delay}. The variation in time of the plasma emission at an element-specific wavelength shown in Fig. 2.4b reveals different signal components. The full width at half maximum (FWHM) τ_1 of the first maximum lies typically in the range of 40–500 ns. This peak is attributed to the strong continuum emission of the plasma in the early phases (corresponds to $t_{delay} = t_1$ in Fig. 2.3). After this first generally distinct maximum, a further temporal variation is observed characterized by a significantly greater time scale. In many cases, e.g., for atomic lines with upper energy levels around 3 eV, a second maximum occurs, whose FWHM τ_2 is between 2 and 10 μs. This signal component represents the element-specific line emission to be evaluated for LIBS.

The following measurands are used for LIBS:

(a) The intensity of the plasma emission at discrete wavelengths as a function of time is: $I(\lambda = \lambda_i, t)$, where λ_i denotes a set of wavelengths of element-specific lines (for simplicity of notation the index λ at the quantity I to denote

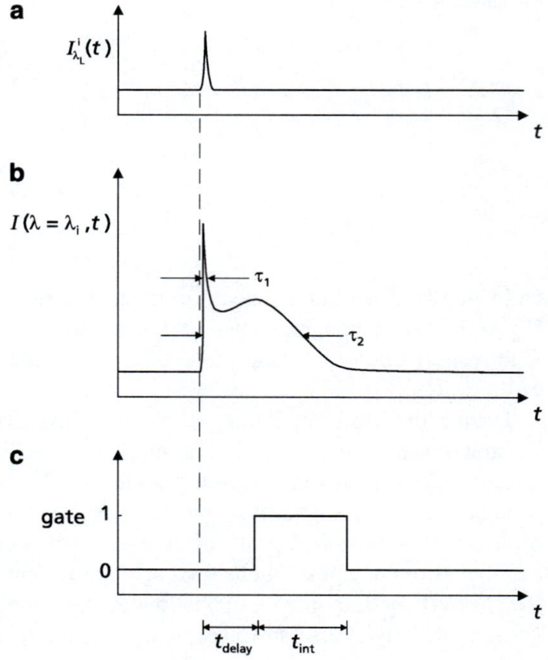

Fig. 2.4 Schematic illustration of the temporal variation of: (**a**) laser pulse, (**b**) plasma emission at an element-specific wavelength λ_i. $I(\lambda = \lambda_i, t)$ = intensity of the plasma emission at the wavelength λ_i; τ_1, τ_2 = FWHM times of the two signal components. (**c**) Gate signal defining the position and duration of the integration window; t_{delay}, t_{int} = cf. Fig. 2.2

the spectral intensity is omitted). Usually this signal is integrated over a time interval $[t_{delay}, t_{delay} + t_{int}]$.

(b) A spectrum, which is integrated over a time interval $[t_{delay}, t_{delay} + t_{int}]$: $S(\lambda) = \int_{t_{delay}}^{t_{delay}+t_{int}} I(\lambda, t)dt$ with t_{delay} = start of integration interval with respect to the start of the laser pulse and t_{int} = temporal width of the integration window.

The measurands (a) and (b) are used for the quantitative determination of the chemical composition of a substance. As a rule this composition is described by the concentrations of the chemical elements expressed as the mass of the analyte in relation to the total mass using the unit g/g or $\mu g/g$ for traces. For higher concentrations, the unit m.-% is used in the following expressing the mass percentage of an analyte. For quantitative measurements, the method has to be calibrated using reference samples to determine the functional dependence between the measurands and resultant quantities on the known concentrations of a set of reference samples, cf. Sect. 11.2.

Figure 2.5 illustrates the procedure for a quantitative measurement. The left diagram shows a spectrum $S(\lambda)$ with a number of emission lines depending on the composition of the sample to be analyzed. The spectral position of the lines has to be allocated to the respective elements on the basis of literature data as, e.g., those given in [2.2, 2.3]. The height of a line is a measure of the concentration of the respective element in the sample. However, this line intensity depends also on a number of other factors, e.g., the laser pulse energy, plasma temperature, plasma size, atomic parameters of the line transition, sample surface, detector response function. Generally, the influence of these factors can be reduced by taking the ratio of the intensity of an analyte line j to the line intensity of a dominant element r of the sample, a so-called matrix element cf. (11.2), (11.3). This line acts as an internal standard or reference. To gain quantitative results, the spectral signals have to be calibrated using a set of samples with known chemical composition, e.g., certified reference samples (CRM). Figure 2.5, right, shows a calibration curve where the intensity ratio I_j/I_r is plotted as a function of the concentration c_j of these reference samples. In general, the calibration curve shows a nonlinear behavior. For an unknown sample, the intensity ratio I_j/I_r is measured, yielding the concentration of the analyte via the analysis function (also called working curve) which is the

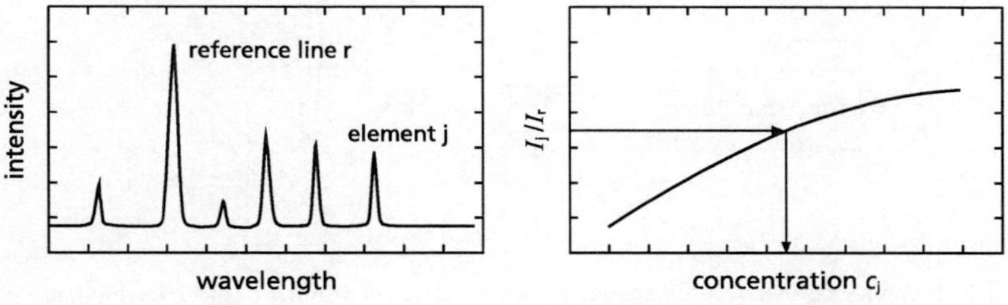

Fig. 2.5 *Left*: emission spectrum of laser-induced plasma with analyte and reference line. *Right*: calibration curve

inverse function of the calibration curve (for simplicity, this step is illustrated by the arrows in Fig. 2.5, right). The procedure for quantitative measurements will be described in more detail in Chap 11.

2.2 Setup for Laser-Induced Breakdown Spectroscopy

Figure 2.6 shows the principle setup for laser-induced breakdown spectroscopy. A mirror guides the pulsed laser radiation to a focusing lens. The sample to be analyzed is placed in a measuring chamber. As a rule the incident direction of the laser radiation is oriented perpendicularly to the sample surface. The focused radiation generates a plasma at the sample surface. The emission of this plasma is observed in a direction, which includes an angle α_O to the incident direction of the laser radiation. In Fig. 2.6, the measuring radiation is transmitted via a fiber optics to a spectrometer, where it is spectrally dispersed and converted to electrical signals.

The measuring chamber is gas tight. Laser radiation and measuring radiation are transmitted via built-in windows. Via gas fittings the type of gas filling as well as the gas pressure and gas exchange rate can be adjusted in a defined manner. A translation stage moves the sample in relation to the incident laser beam to measure at different locations on the sample surface.

The use of a measuring chamber is not a necessary precondition for laser-induced breakdown spectroscopy. For inline analyzing tasks, the measuring chamber is often set aside and the measurement is performed under atmospheric conditions. In this case, the usable measuring radiation is limited to wavelength greater than 190 nm, since air absorbs shorter wavelength strongly (cf. Sect. 3.5). For quantitative measurements with high requirements on measuring precision and uncertainty, a measuring chamber is used to adjust the ambient gas conditions and the gas exchange at the interaction region in a defined way.

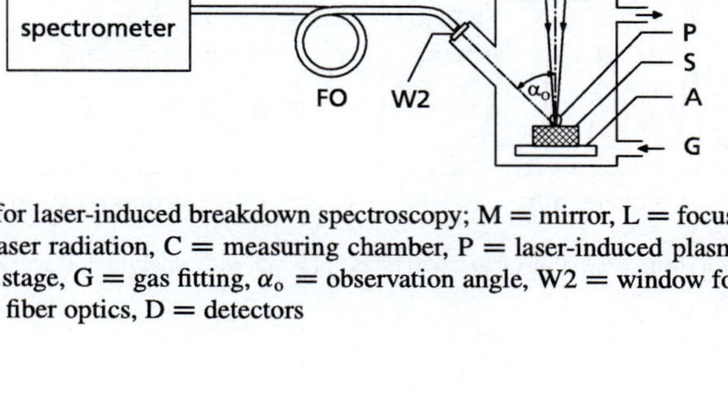

Fig. 2.6 Setup for laser-induced breakdown spectroscopy; M = mirror, L = focusing lens, W1 = window of the laser radiation, C = measuring chamber, P = laser-induced plasma, S = sample, A = translation stage, G = gas fitting, α_o = observation angle, W2 = window for the measuring radiation, FO = fiber optics, D = detectors

A control unit triggers the laser and reads the signals of the detectors. To improve the signal-to-noise ratio the plasma radiation is recorded only during the life time of the plasma. For this purpose, the spectrally dispersed radiation is detected time resolved and integrated over a time gate within the life time of the plasma. The control unit adjusts the position and duration of that time gate (cf. t_{delay} and t_{int} in Fig. 2.4).

Figure 2.7 shows schematically the arrangement of components in an analyzing system based on LIBS. The sample is positioned on a sample table. The laser beam is focused onto the bottom side of the sample. The sample table can be set into rotation, whereby the rotational axes and the optical axes of the incoming laser beam are parallel and shifted laterally. Due to this eccentricity, a relative motion between sample and laser beam occurs, where the locations of laser irradiation lie on a circle. By this a spatial averaging is realized (cf. Sect. 3.6). Sample table and measuring chamber are parts of the so-called sample stand (cf. Sect. 4.6).

Figure 2.8 shows a view of an analyzing system. The sample stand can be seen on the left side at the top of the system.

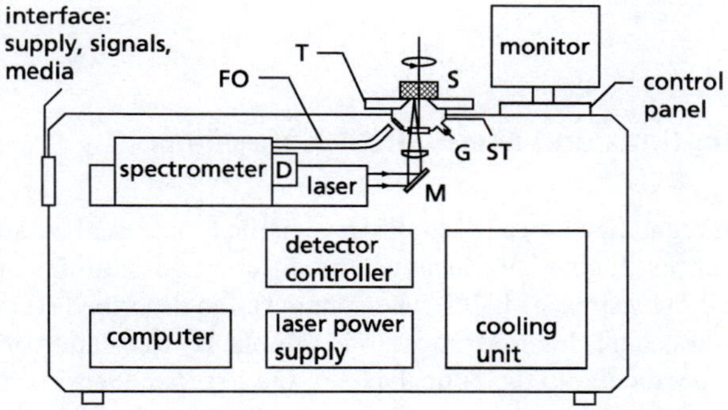

Fig. 2.7 Arrangement of components in a LIBS-based analyzing system; T = sample table, S = sample, ST = sample stand, FO = fiber optics, G = gas supply, D = detector, M = mirror

Fig. 2.8 Analyzing system based on LIBS. *Left*: sample stand

2.3 Measuring Procedure

A measuring procedure comprises the following steps:

1. Definition of a measuring method or retrieval of a measuring method which was already defined. Putting the sample on the sample stand
2. Start of the measurement
3. Evaluation of the spectral signals
4. Display of the measuring results

A measuring method is defined in particular by the selection of measuring parameters and their temporal sequence, cf. Sect. 11.1. Measuring parameters are, e.g., the laser pulse energy, the number of prepulses and measuring pulses, and the gas flow in the measuring chamber. These various parameters influencing the measuring process will be discussed in more detail in Chap. 3.

A typical measuring sequence for a quantitative analysis of a sample including data evaluation takes about 30 s–2 min. For an identification testing of work pieces, inspection times of a few seconds or even fractions of a second are achievable, cf. Chap. 18.

2.4 Applications and Methodical Extensions

Laser-induced breakdown spectroscopy is a versatile tool to analyze solid, liquid, or gaseous substances. Figure 2.9 illustrates the different possibilities of applications and methodical extensions of LIBS. The configuration described so far is shown in item 1. The laser beam is focused onto a solid sample. By translation of the sample in a direction perpendicular to the optical axis of the irradiated laser beam the sample surface is scanned to obtain a spatially resolved information about the chemical composition. By this approach maps of element distributions can be gained, cf. Sect. 18.3. The laser beam can be guided via fiber optics to a measuring head and then be focused to analyze a liquid (item 2 in Fig. 2.9). Analysis of, e.g., liquid steel by LIBS will be described in more detail in Sect. 13.1.3. The laser beam can also be focused through a window into a closed tube to analyze aerosoles or gases, see item 3 (cf. Sect. 14.6.). A methodical extension is the combination of LIBS with separation techniques such as capillary electrophoresis (CE) or high-pressure liquid chromatography (HPLC), see item 4. In this case, a molecule-specific separation is followed by an element-specific analysis to improve speciation analytics of smallest amounts of complex samples, cf. Sect. 14.5. LIBS can be combined with other laser spectroscopic methods such as laser-induced fluorescence (LIF), see item 5. In a first step, a pulsed laser having a fixed wavelength λ_1 ablates material and induces a plasma, thus performing LIBS as described in the previous sections. After the plasma plume has evolved, a second laser with a tunable wavelength λ_2 irradiates the material cloud inducing element-specific transitions to generate fluorescence

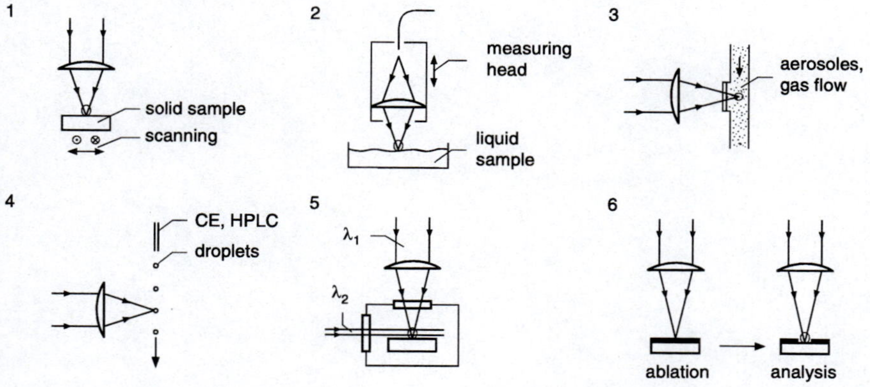

Fig. 2.9 Applications and methodical extensions of LIBS: *1* = element mapping of sample surfaces by scanning, *2* = analysis of liquids, *3* = analysis of aerosoles and gases excited through an optical window, *4* = combination of molecule-specific separation techniques with LIBS (CE = capillary electrophoresis, HPLC = high-pressure liquid chromatography), *5* = LIBS combined with laser-induced fluorescence (LIF), *6* = laser ablation of surface layers and subsequent analysis of the bulk material

signals to detect selected species with high sensitivity, see Chap. 12. Finally, the laser is a multipurpose tool allowing to locally ablate surface layers and to perform chemical analysis of the underlying bulk material, see item 6 and Sect. 13.1.2.

References

2.1. A. Pakhomov, M. Thompson, D. Gregory, Laser-induced phase explosions in lead, tin and other elements: microsecond regime and UV-emission. J. Phys. D Appl. Phys. **36**, 2067–2075 (2003)
2.2. A. Zaidel, V. Prokofev, S. Raiskii, V. Slavnyi, E. Shreider, *Tables of Spectral Lines* (IFI/Plenum, New York, 1970)
2.3. http://cfa-www.harvard.edu/amdata/ampdata/kurucz23/

Chapter 3
Process Parameters

This chapter describes the main process parameters and their influence on LIBS. The effect on the plasma state and the emission spectrum will be described in more detail in Chaps. 4, 6–8.

3.1 Laser Pulse Energy and Repetition Rate

The energy of the laser pulse constitutes the main part of the total energy available for evaporation, plasma formation, and generation of spontaneous emission. Exothermic chemical reactions of the evaporated material with the ambient gas or within the ambient gas may provide additional process energy.

Typical laser pulse energies lie in the range from 1 to 1,000 mJ depending on the specific requirements. As a rule pulse energies below 50 mJ are used for small focal length of the focusing optics (<100 mm), high spatial lateral resolution (<20 μm), high measuring frequencies (>100 Hz), or if the laser pulses are transmitted via fiber optics. Pulse energies greater than 50 mJ are used for focal lengths ranging from 200 mm to several meters, low spatial lateral resolution (>200 μm), and measuring frequencies in the range of typically 10–100 Hz.

The repetition rate of the laser pulses defines the maximum measuring frequency for LIBS. For flashlamp-pumped Nd:YAG lasers with electro-optical Q-switching, this rate amounts to 10–100 Hz yielding pulses with a temporal width of typically 5–20 ns. Continuously pumped solid state lasers with acousto-optical Q-switching achieve repetition rates beyond 10 kHz; however, the pulse energies are significantly smaller and the pulse width is greater (>100 ns). Diode-pumped solid state lasers with electro-optic Q-switching run with repetition rates of up to 2 kHz and beyond, delivering pulses of about 1 mJ with pulse widths <10 ns. The average optical power is typically in the range between 1 and 50 W.

Figure 3.1 shows the parameter range of repetition rates and laser pulse energies used for LIBS to analyze metals, steel alloys, slags, and oxidic materials. The laser parameters given in most of the scientific publications can be grouped around

R. Noll, *Laser-Induced Breakdown Spectroscopy*, DOI 10.1007/978-3-642-20668-9_3,

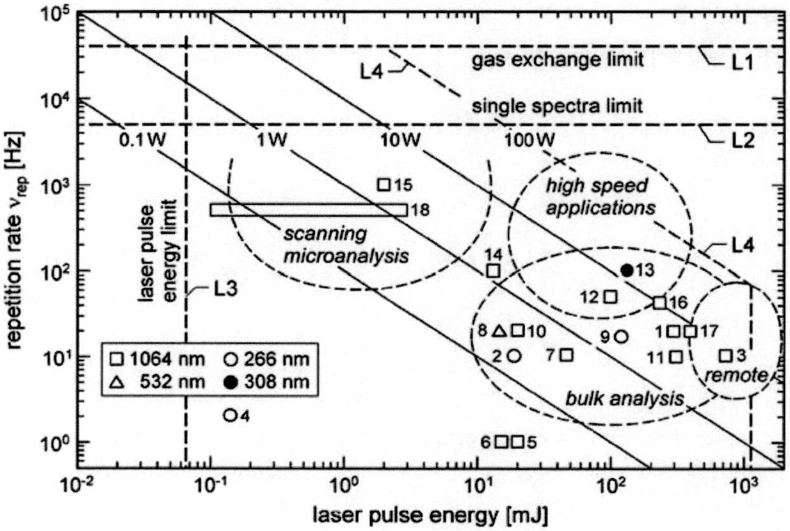

Fig. 3.1 Repetition rate versus laser pulse energy of Q-switched laser sources used for LIBS to analyze metallic species, aluminum, steel, slag, and oxides (the numbers close to the data points in the graph refer to the following references: 1 = [3.1], 2 = [3.2], 3 = [3.3], 4 = [3.4], 5 = [3.5], 6 = [3.6], 7 = [3.7], 8 = [3.8], 9 = [3.9], 10 = [3.10], 11 = [3.11], 12 = [3.12], 13 = [3.13], 14 = [3.14], 15 = see Chap. 15, 16 = see Sect. 18.4, 17 = see Sect. 17.4, 18 = see Sect. 16.1. *L*1–*L*4 = limits, see text)

repetition rates of 20 Hz and pulse energies of 200 mJ. The dominant field of application in this parameter range is bulk analysis. The diagonals plotted in Fig. 3.1 depict the average laser power. For inline applications of LIBS to directly analyze or identify the chemical composition of a substance in a process line, typical average laser powers are 5–10 W, see data points 1, 3, 12, and 13 in Fig. 3.1. This is a consequence of the respective requirements: (a) fast measurement, (b) sufficiently large working distance (distance between the last optical element of the measuring system and the laser-induced plasma) in the range from 150 to 1,800 mm to assure a flexible access to the measuring object.

High repetition rates (point 15 in Fig. 3.1) are advantageous for spatially resolved analysis to scan macroscopic samples within a short time. In this case, the typical requirements are: sample surface to be scanned 50 mm × 100 mm, spatial lateral resolution <20 μm. The spatially resolved analysis will be discussed in detail in Chap. 15.

Figure 3.1 shows as dashed lines various limits denoted with L1 to L4. The limit L1 describes an upper value of the repetition rate of the laser given by restrictions of the gas exchange in the interaction volume, see relation (3.1). L2 denotes a limit for the repetition rate of the laser for single spectra evaluation, i.e., each spectrum generated by each laser pulse is processed separately. This estimated limit is given by the detector response and the processing speed of the subsequent

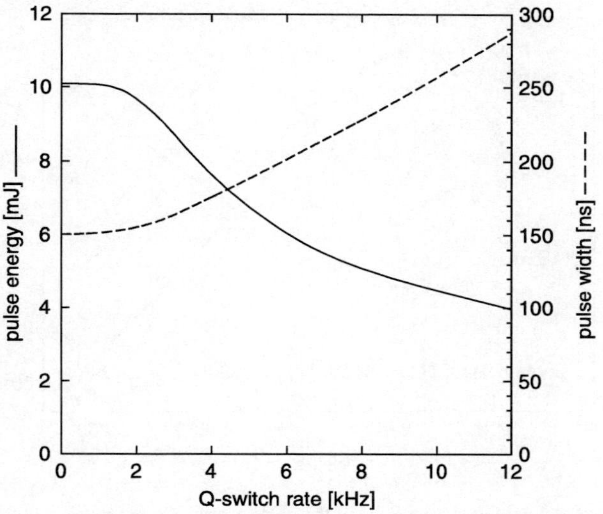

Fig. 3.2 Pulse energy and pulse width as a function of the Q-switch rate for a continuously pumped Nd:YAG laser [3.15]

signal electronics. Maximum achieved repetition rates with single spectra evaluation amount to 1 kHz, cf. Chap. 15. The limit L3 presents a lower limit for the laser pulse energy of less than 100 μJ, where the intensity of the plasma emission becomes so weak that noise levels of the detectors are approached. The limit L4 is given on the one hand by pulse energies of about 1 J and on the other hand by an average laser power of 100 W, both involving bulky Q-switched laser systems, which are at the limit of technical feasibility and commercial availability.

Generally, a change in the repetition rate of a given laser source changes also the beam parameters pulse energy, pulse width, and beam profile. As an example, Fig. 3.2 shows energy and width of the pulses (FWHM) generated by a continuously pumped Nd:YAG laser as a function of the repetition rate. With increasing repetition rate, the level of inversion and hence the achievable amplification of the laser medium between two consecutive pulses decreases with the consequence of a reduced pulse energy and enlarged pulse width. Due to this behavior, the laser power within a pulse is reduced from 66 kW for repetition rates <1 kHz to 14 kW for repetition rates of 12 kHz. pulse energy and pulse width of the laser source itself cannot be chosen independently, as they are linked with each other due to the properties of the laser medium and the pumping source.

Nd:YAG lasers pumped by pulsed diode lasers (DPSSL = diode-pumped solid state laser) can generate pulse energies of 2 mJ at 2 kHz repetition rate with electro-optic Q-switching. The pulse width amounts to about 5 ns. Although the pulse energy is only one-fifth of the one delivered by the continuously pumped Nd:YAG

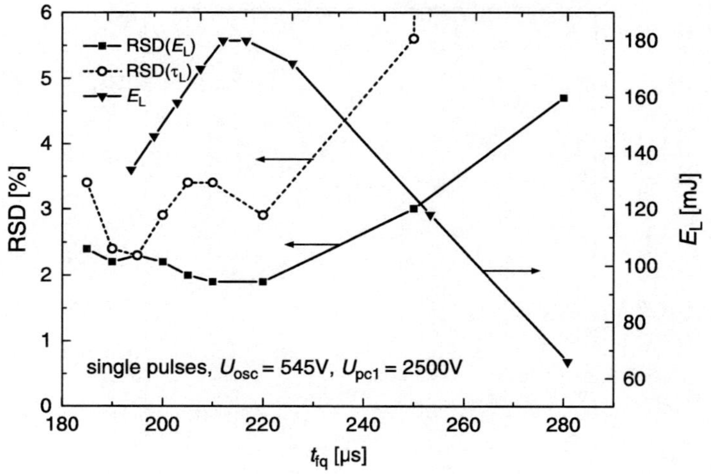

Fig. 3.3 Relative standard deviation RSD of the pulse energy E_L and pulse width τ_L as a function of the delay time between flashlamp and Pockels cell triggering t_{fq}. On the *right scale*, the corresponding pulse energy is shown. U_{osc} = charging voltage of the capacitors driving the flashlamp discharge, U_{pc1} = bias voltage at Pockels cell

laser from Fig. 3.2 at 2 kHz, the pulse power is greater by a factor of 6, due to the reduced pulse width being shorter by a factor of 30.

The generation of plasma for LIBS requires a sufficient high irradiance (unit W/cm^2) at the surface of the material to be analyzed. Typical values are $>10^7\ W/cm^2$. An estimation of the necessary irradiances will be given in Chap. 5. The irradiance depends on the pulse energy, the pulse width, the beam diameter, and the beam quality at the location of interaction with the sample, see Sect. 3.4.

Figure 3.3 illustrates the relative standard deviation (RSD) of the pulse energy and pulse width of a flashlamp-pumped Nd:YAG laser as a function of the time between the ignition of the flashlamp and the Q-switching t_{fq} (cf. Sect. 4.1). There is an optimum time t_{fq}, where the highest pulse energies are achieved and at the same time the RSD of the pulse energy describing the pulse-to-pulse fluctuations of the pulse energy attains a minimum at about 2%. The RSD of the pulse width amounts to about 3% at $t_{fq} = 220\ \mu s$ and increases strongly for greater t_{fq}. These features of a pulsed pumped laser demonstrate that a change in the laser operating parameters – here shown for t_{fq} – has an effect not only on the level of pulse energy generated but also on the stability of the laser beam parameters.

Plasmas generated by two subsequent laser pulses at one sample location can be considered as independent events, if the following conditions are fulfilled: (a) the state of the ambient gas prior to the irradiance of the second pulse is identical to the one persisting prior to the first pulse, (b) the surface of the sample to be analyzed is not modified in a way, which influences the interaction of the laser radiation with the material. The first condition is fulfilled, if the material ablated

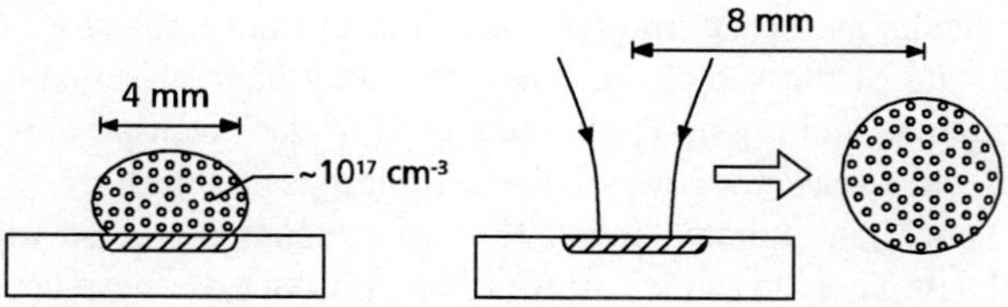

Fig. 3.4 The ablated material generated by a previous laser pulse has to be removed from the interaction region prior to the irradiance of the next laser pulse, see text

by the laser pulse irradiation and/or recondensated material is transported off the interaction volume by diffusion or gas flow. Furthermore, the ambient gas flows back into this interaction volume and establishes a stationary and homogeneous gas atmosphere prior to the arrival of the next pulse.

The time for a complete gas exchange in the interaction volume is denoted as local gas exchange time τ_{ge}. Figure 3.4 illustrates the situation. As an example we examine the interaction of a collinear double pulse (cf. Chap. 6) with a steel sample. For a burst energy of 2×140 mJ, the ablated mass of iron amounts to 300 ng, which corresponds to 3.22×10^{15} iron atoms. After the plasma ignition, the material cloud expands and attains a volume approximated here by a sphere having a diameter of 4 mm. The average iron species density is then $2 \times 10^{17}\,\mathrm{cm}^{-3}$. This number density is by about two orders of magnitude smaller than the ambient gas density at 1 bar of $2.5 \times 10^{19}\,\mathrm{cm}^{-3}$. Hence, there is a chance to push this material cloud away from the interaction zone by a flow of the ambient gas. A practical limit will be the sound velocity, which is for argon at room temperature 300 m/s. Thus, the condition (a) described above can be expressed as follows:

$$t_{rep} > \tau_{ge} = \frac{2d_{mc}}{v_{sound}}, \tag{3.1}$$

where $t_{rep} = 1/\nu_{rep}$ is the time between two subsequent laser pulses, ν_{rep} is the laser pulse repetition rate, τ_{ge} is the local gas exchange time, d_{mc} is the diameter of material cloud, and v_{sound} is the velocity of sound of the ambient gas.

For (3.1), it is assumed that the material cloud should be shifted by twice its diameter to assure a sufficient lateral translation. With $d_{mc} = 4$ mm (cf. Fig. 3.4), we obtain $\tau_{ge} = 26\,\mu$s and the limit for the repetition frequency is $\nu_{rep} < 38$ kHz. The highest repetition frequency realized so far for LIBS amounts to 1 kHz, cf. Chap. 15. This is still about a factor of 40 below the principal limit estimated using relation (3.1).

To which extent the second condition (b) holds depends on the selected parameters: pulse energy, temporal structure of the laser pulse, focusing, ambient gas, as well as on the properties of the sample to be analyzed. Qualitatively the situation can be described as follows. The laser pulse causes a material ablation, i.e., the local topology of the sample surface is modified. Depending on the selected parameters and the studied sample material, these modifications can be described phenomenologically as formation of a crater. Furthermore, the structural, physical, and chemical state of the sample in this crater may have changed due to a series of processes such as melting and resolidification of the material, ablation of oxidic surface layers, chemical reactions of the sample species with constituents of the ambient atmosphere and others. Hence, the subsequent laser pulse will encounter a different local sample state, which in turn has an effect on the interaction process and the state and dynamics of the generated plasma (cf. Chap. 8).

3.2 Temporal Characteristics of the Laser Pulse

A key parameter of the temporal characteristics of a laser pulse is the pulse width. Figure 3.5 illustrates the main difference in the interaction of nanosecond and femtosecond pulses with a sample for LIBS. For nanosecond pulses, the plasma is generated while the laser irradiation still persists (cf. phase 2 in the first row of Fig. 3.5). Hence, there is the possibility of an additional interaction between the laser beam and the plasma, i.e., the plasma can be heated by the laser radiation. For femtosecond pulses, the energy is deposited in the target within a very short time before the material ablation and plasma formation starts. The laser beam plasma interaction does not exist in this case (cf. phase 2 in the second row of Fig. 3.5).

A broad range of pulse widths was studied for laser ablation and LIBS varying over nearly seven orders of magnitude, see Fig. 3.6. Looking at this diagram, it is obvious that there is a cluster of data points at pulse widths in the nanosecond range. In most cases, nanosecond pulses are generated by Q-switching of solid state lasers. The pulse width depends on the Q-switching of the laser source and the amplification of the laser medium. For solid state lasers with electro-optic Q-switching, the typical pulse widths amount to 10–20 ns. Depending on the pulse energy, this results in powers of the laser pulses ranging from 50 kW to 25 MW.

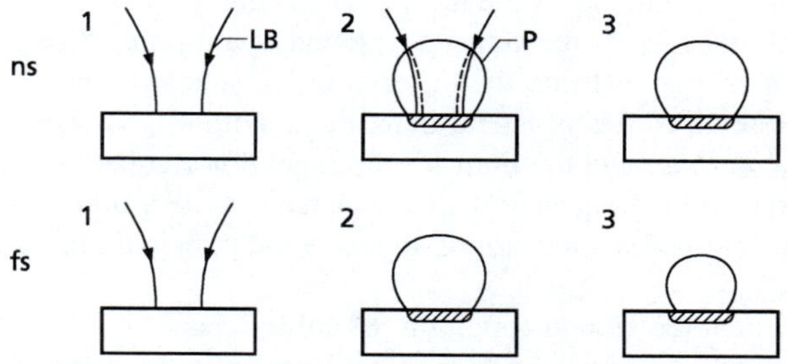

Fig. 3.5 Schematic illustration of the interaction of nanosecond (ns, *top row*) and femtosecond laser pulses (fs, *bottom row*) with a sample for LIBS, for abbreviations see Fig. 2.1

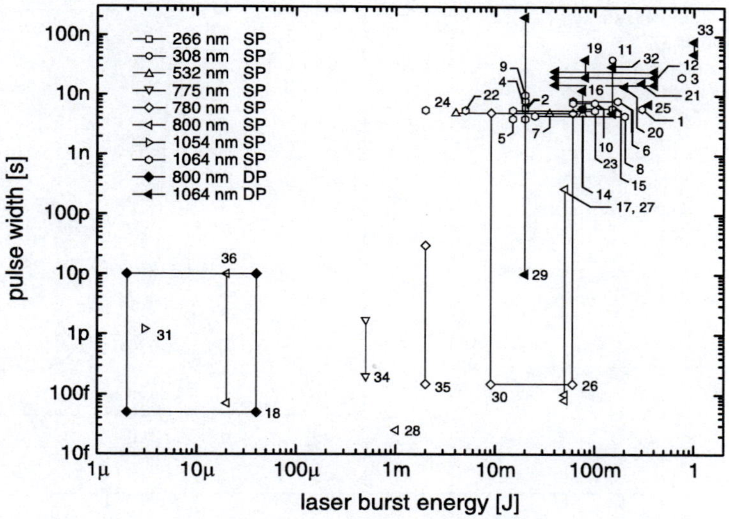

Fig. 3.6 Pulse width versus laser burst energy studied for laser ablation and LIBS. *SP* = single pulses, *DP* = double pulses (cf. Chap. 6). The numbers refer to the references as follows: 1 = [3.1], 2 = [3.2], 3 = [3.3], 4–6 = [3.5–3.7], 7–9 = [3.8–3.10], 10–12 = [3.12–3.14], 13–36 = [3.16–3.39]

The pulse duration influences the interaction of the laser beam with the measuring object. For solid targets, the diameter of the crater is reduced for shorter pulse durations, cf. Sect. 3.6, Fig. 3.18.

Figure 3.7 shows as an example the temporal shape of a laser pulse generated by a Q-switched Nd:YAG laser [3.36]. The pulse width influences the relative significance of the various interaction processes with the target material. For pulse durations in the sub-nanosecond regime, i.e., picoseconds and femtoseconds, the laser energy is dominantly coupled into the material causing a fast transition from the solid state to the vapor phase prior to the formation of the plasma. By this well-defined ablation geometries can be produced. However, the energy content in the plasma is small which in turn results in a weaker and shorter line emission [3.30]. For pulse durations beyond 50 ns and metallic targets, the melt phase becomes increasingly important, i.e., the ratio between particles transported off the interaction region in the melt phase in relation to those transported off as vapor increases. For pulse durations between 1 and 50 ns, the laser energy partitioning for material evaporation and energy coupling into the radiating plasma seem to be in a favorable range for LIBS in terms of the total number of analytes excited into a state, where a significant line emission persists for a few microseconds.

During a single flashlamp discharge cycle, which pumps the laser rod, multiple laser pulses can be generated with a step-by-step control of the Pockels cell [3.36]. An example for a double pulse is shown in Fig. 3.8. A group of pulses generated during a single flashlamp discharge is called a burst. The selected voltage levels for

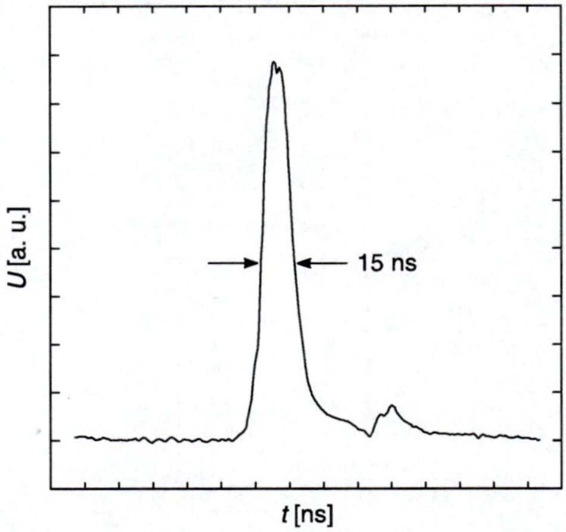

Fig. 3.7 Temporal shape of a laser pulse of a flashlamp-pumped Q-switched Nd:YAG laser

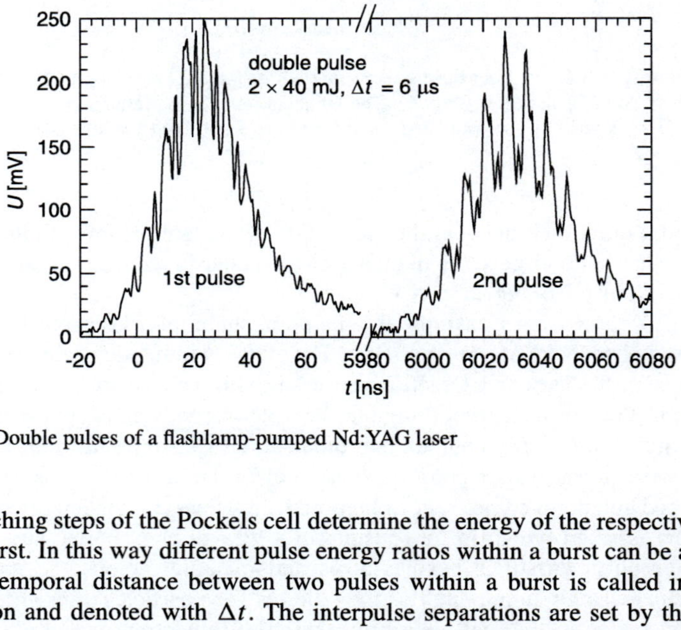

Fig. 3.8 Double pulses of a flashlamp-pumped Nd:YAG laser

the switching steps of the Pockels cell determine the energy of the respective pulses in the burst. In this way different pulse energy ratios within a burst can be adjusted.

The temporal distance between two pulses within a burst is called interpulse separation and denoted with Δt. The interpulse separations are set by the trigger delays for the respective switching steps. The minimum interpulse separation Δt amounts to $2\,\mu s$ with this approach. The maximum duration of a burst amounts to $100\,\mu s$ and is limited by the variation of the amplification as a function of time of the flashlamp-pumped laser. The bursts are generated with the repetition frequency of the flashlamp of, e.g., 10 Hz.

Figure 3.9 illustrates the different pulse modes: single pulses (SP), double pulses (DP), and triple pulses (TP) and the describing parameters interpulse separation Δt, cycle duration T_{rep}, pulse energy E_j, pulse duration τ_{L_j} [3.40]. Generating DP

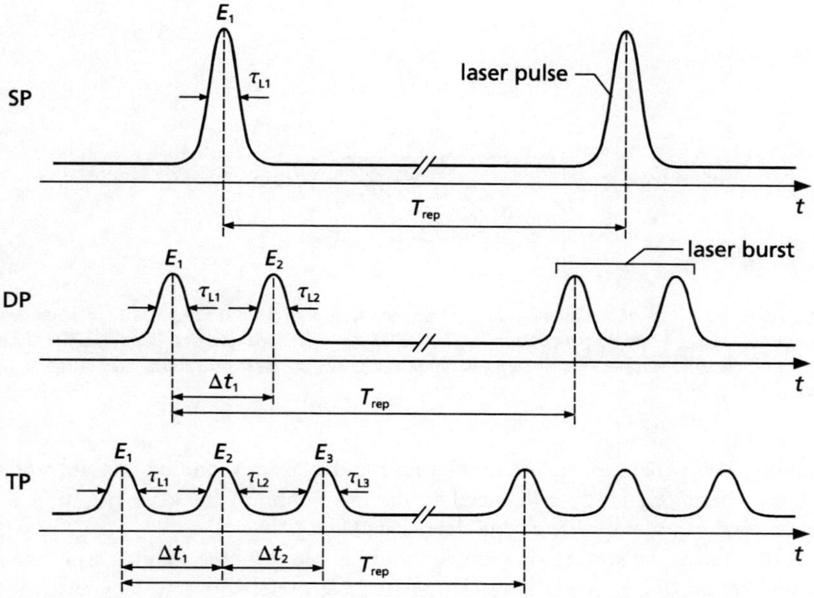

Fig. 3.9 Illustration of different pulse modes: single pulses (SP), double pulses (DP), and triple pulses (TP) and the describing parameters

or TP within a single flashlamp discharge implies that the duration of the pulses within a burst changes such that $\tau_{L_j}(\text{TP}) > \tau_{L_j}(\text{DP}) > \tau_L(\text{SP})$. An alternative for the generation of pulse bursts is the superposition of the laser beams of two laser systems by, e.g., a polarizing beam combiner. The sum of pulse energies within a burst is called burst energy $E_b = \sum_j E_j$.

Typical Δt values are in the range of a few microseconds. A pulse structure with time scales in the microsecond range has an effect on the transport processes in the interaction region as well as on the laser-induced plasma, whose characteristic time scale is of the same order of magnitude, cf. Sect. 8.5. Double or multiple pulses may generate plasma states with more intense and longer lasting line emission which is advantageous for LIBS. The effect of double and multiple pulses will be discussed in more detail in Chaps. 6–8.

3.3 Laser Wavelength

The absorption properties of a substance depend on the wavelength of the laser radiation. This natural absorption determines the portion of the laser pulse energy, which is coupled into the material, especially in the phases (1) and (2) as shown in Fig. 2.1 just before the initial state of aggregation of the substance to be analyzed

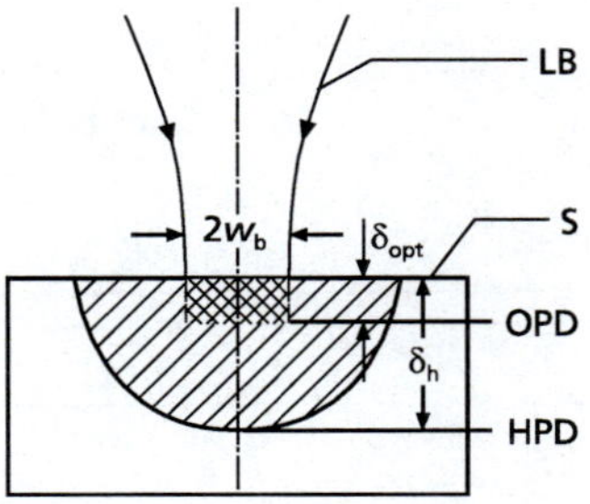

Fig. 3.10 Parameters of the interaction of laser radiation with a solid sample, LB = laser beam, S = sample, δ_{opt} = optical penetration depth (OPD), δ_{h} = heat penetration depth (HPD) of laser radiation in a solid sample (the drawing is not scaled), $2w_{\text{b}}$ = laser beam diameter at the sample surface

changes. The generation of material vapor and plasma modifies this absorption, which is then dominantly influenced by the interaction of the laser radiation with the material vapor, the plasma, and the particulates.

Furthermore, the surface properties of the sample, e.g., the roughness, influence the absorption. For a specular reflecting surface, the absorptivity decreases from 0.64 at 200 nm to 0.36 at 1 μm [3.41].

Figure 3.10 illustrates the optical penetration depth and the heat penetration depth of laser radiation for a solid sample. The optical penetration depth δ_{opt} is the reciprocal of the absorption coefficient and describes that length along which the incoming laser irradiance decreases to 37% of its initial value. The heat penetration depth δ_{h} is a measure of the propagation length of the thermal energy into the sample after start of laser irradiation. The spatial and temporal run of the temperature inside the sample depends on the temporal structure of the laser irradiance, as well as on the laser beam profile and hence the geometry of the heat source. The heat penetration depth as a function of time can be estimated as follows [3.42]:

$$\delta_{\text{h}} = 2\sqrt{\kappa t}, \tag{3.2}$$

where κ is the thermal diffusivity, $[\kappa] = \text{cm}^2/\text{s}$.

The heating of the sample can be described by the heat conduction equation, which will be considered in more detail in Chap. 5.

Figure 3.11 shows δ_{opt} and δ_{h} as a function of the wavelength between 200 nm and 1.2 μm for aluminum and iron. The values for δ_{opt} are calculated from tabulated data of the complex index of refraction [3.43, 3.44]. For a first calculation of δ_{h}, we have taken $t = 10$ ns, which is the typical order of magnitude of the full width at half maximum of a Q-switched laser pulse. Thus, the values for δ_{h} represent an upper limit since only a part of the laser pulse energy is absorbed by the sample via natural absorption. This is due to the induced plasma, whose emission can be observed already after time delays of <1 ns after start of irradiation of the laser pulse. On the other hand, the laser-induced plasma itself interacts with the sample

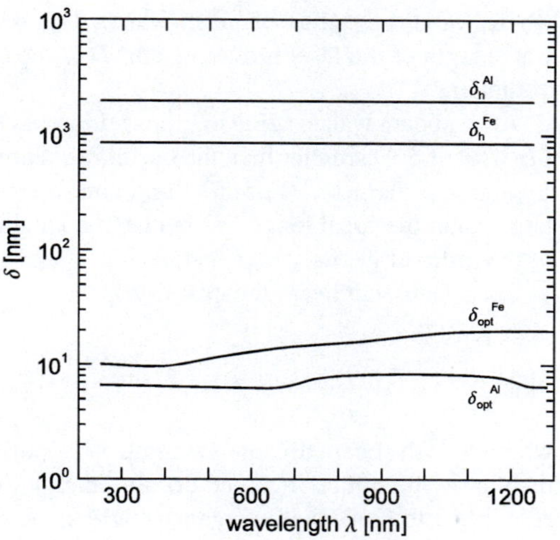

Fig. 3.11 Optical penetration depth δ_{opt} and heat penetration depth δ_h as a function of the wavelength for aluminum and iron

surface and acts as a local heat source transferring energy to the sample by radiation and heat conduction. This secondary heat source is characterized by the size of the plasma of typically a few millimeters, its life time being of the order of some microseconds and energy density as a result of the temperature inside the plasma, which are in the range of several thousand kelvin.

The absorption of polymer materials in the near infrared is determined by electronic as well as vibrational excitations. For Nd:YAG laser radiation at a wavelength of 1.06 μm, the optical penetration depth was determined experimentally [3.45]. For polyamide (PA 6), the optical penetration depths amounts to about 3.5 mm in a temperature range of the polymer between 20 and 220°C. For polycarbonate (PC), δ_{opt} is about 0.25 mm in this temperature range. These values are by orders of magnitude greater than those for metals.

The absorption of laser radiation by the plasma as a function of the laser wavelength will be discussed in more detail in Sect. 5.3.

3.4 Beam Quality and Focusing

The laser radiation is focused to increase the irradiance at the location of interaction to values greater than 10^7 W/cm^2. Each substance exposed to these irradiances will be transferred to the plasma state within a few nanoseconds.

The focusability of laser radiation depends on the beam quality and the wavelength. In the ideal case of a so-called Gaussian beam, the laser beam can be focused by an aberration free lens to the following beam waist radius:

$$w_0 = \frac{2f\lambda}{\pi D}, \tag{3.3}$$

where w_0 is the radius of beam waist, f is the focal length of the lens, λ is the wavelength of the laser radiation, and D is the diameter of the illuminated aperture of the lens.

At the beam waist radius w_0, the irradiance attains a level which is by a factor of $1/e^2(\approx 13.5\%)$ smaller than the one at the beam axis. Due to physical and technical reasons, the diameter D of the illuminated aperture of a lens cannot be significantly larger than the focal length f; hence, the smallest possible beam waist diameter is of the order of the used wavelength λ.

For a Gaussian beam the quantity:

$$\frac{w_0 \theta_0}{\lambda} = \frac{1}{\pi}, \tag{3.4}$$

where θ_0 is the beam divergence angle of a Gaussian beam, is a propagation invariant insofar as the optics involved do not change the stigmatic character of the beam. According to (3.4), the divergence angle increases for smaller beam waists.

The Gaussian beam is an idealization of a laser beam with the smallest possible beam divergence. A real laser beam deviates from this ideal case, which is described by the standardized beam propagation ratio M^2 [3.46]:

$$M^2 = \frac{\pi}{\lambda} \frac{d\Theta}{4}, \tag{3.5}$$

where M^2 is the beam propagation ratio, d is the diameter of laser beam, and Θ is the beam divergence angle (full angle) of the laser beam.

The smallest possible value of M^2 amounts to 1. For each real laser beam holds $M^2>1$.

For a typical flashlamp-pumped Nd:YAG solid state laser follows with $d = 10.8$ mm and $\Theta = 0.35$ mrad: $M^2 = 2.8$. Figure 3.12 shows the experimentally determined beam radius of a Nd:YAG laser beam as a function of the distance from the focal plane, i.e., the location of the beam waist. The quantity z_R describes the Rayleigh length and defines the distance from the focal plane at which the beam radius increases to $\sqrt{2}w_0$.

For the average irradiance (average within a single laser pulse) at the location of the beam waist holds:

$$I_f = \frac{E_L}{\tau_L \pi w_0^2} = \frac{\pi E_L D^2}{4\tau_L f^2 \lambda^2 M^4}, \tag{3.6}$$

where E_L is the energy of the laser pulse, D is the diameter of the illuminated aperture of the focusing lens, τ_L is the temporal width (FWHM) of the laser pulse, f is the focal length of the lens; λ is the wavelength of the laser radiation, and M^2 is the beam propagation ratio.

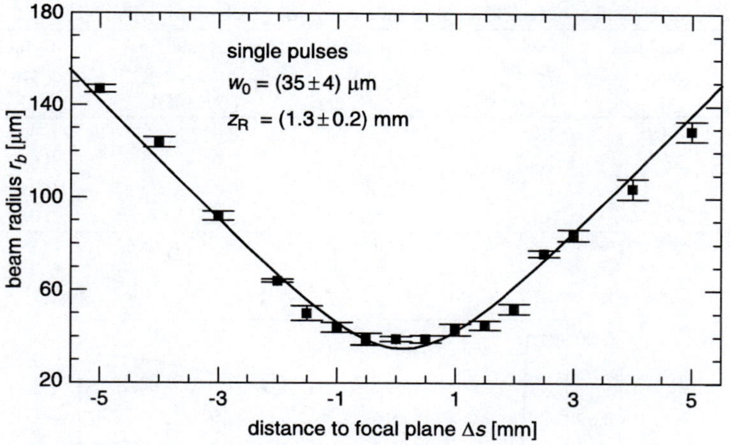

Fig. 3.12 Beam radius of single pulses of a flashlamp-pumped Nd:YAG laser as a function of the distance from the focal plane (pulse energy $E_b = 80$ mJ, focal length $f = 200$ mm)

For typical values ($E_L = 80$ mJ, $D = 10.8$ mm, $\tau_L = 20$ ns, $f = 200$ mm, $\lambda = 1{,}064$ nm, $M^2 = 2.8$), relation (3.6) yields an average irradiance of $I_f = 10^{11}$ W/cm^2.

The diameter of the beam waist influences the achievable spatial resolution for LIBS. In Fig. 3.12, this diameter amounts to 70 µm. Due to the interaction of the laser beam with a solid target, a crater is generated having a diameter greater than the beam waist caused by heat conduction and material transport processes.

For a solid or liquid measuring object, the quantity Δs describes the distance between the beam waist location and the surface of the target, cf. Fig. 2.2. A beam waist lying at the surface corresponds to $\Delta s = 0$. For $\Delta s > 0$, the beam waist is within the sample and for $\Delta s < 0$ the beam waist lies between the focusing optics and the sample surface. In many cases, $\Delta s > 0$ is chosen to avoid a gas breakdown in front of the sample surface, which would lead to a plasma formation in the ambient atmosphere and prevents an efficient ablation of the material to be investigated.

3.5 Ambient Gas Atmosphere

The ambient gas atmosphere influences the properties and the dynamics of the laser-induced plasma, as well as the transmission of the generated line radiation along the beam path to the spectrometer. Typical ambient gases applied for LIBS to analyze solid or liquid substances are argon, nitrogen, and atmospheric air. Table 3.1 shows physical data for these gases and helium.

Studies of the mass ablation efficiency show that the material ablated per laser pulse (or laser burst) increases with rising ambient gas density at a pressure of

Table 3.1 Physical data of typical ambient gases used for LIBS

Gas	Molar mass (g)	Gas density (kg/m^3)	Ionization energy (eV)	Thermal conductivity (W/(mK))	Dynamic viscosity (10^{-3} Ns/m^2)
Ar	39.95	1.784	15.76	0.0178	0.022
O_2	32	1.429	12.06	0.0266	0.019
N_2	28	1.25	15.58	0.026	0.018
He	4	0.179	24.59	0.15	0.022

The gas density refers to a pressure of 1 bar and a temperature of 293 K [3.41]

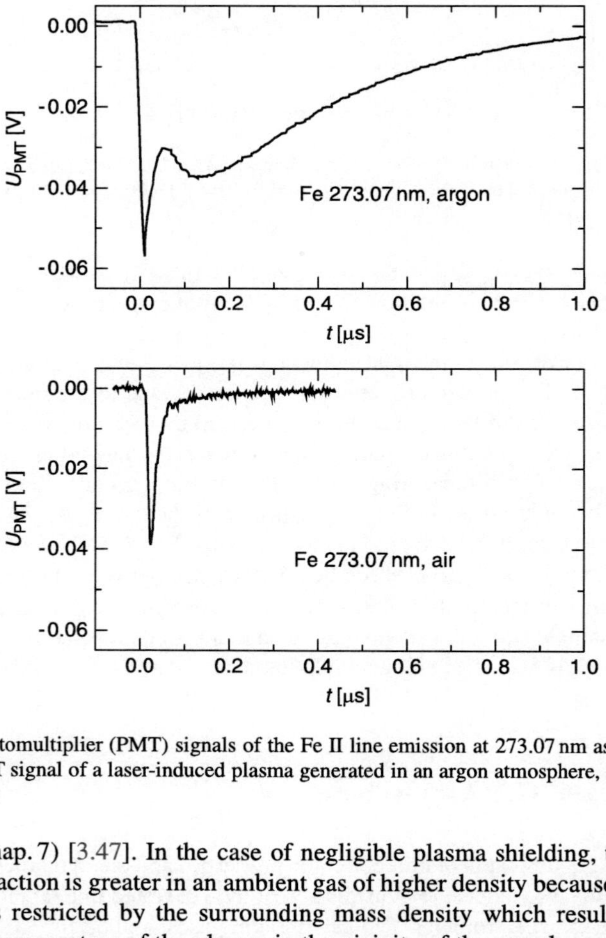

Fig. 3.13 Photomultiplier (PMT) signals of the Fe II line emission at 273.07 nm as a function of time, *top:* PMT signal of a laser-induced plasma generated in an argon atmosphere, *bottom:* for an air atmosphere

1 bar (cf. Chap. 7) [3.47]. In the case of negligible plasma shielding, the plasma–sample interaction is greater in an ambient gas of higher density because the plasma expansion is restricted by the surrounding mass density which results in higher density and temperature of the plasma in the vicinity of the sample surface leading to enhanced material ablation.

Figure 3.13 shows photomultiplier (PMT) signals as a function of time detected at a wavelength of 273.07 nm corresponding to a singly ionized iron line (Fe II) generated by focusing a 1-mJ laser pulse onto an iron sample under two different atmospheres: argon and air (the experimental setup is shown in Sect. 15.1). In air the short living continuum peak (cf. Sect. 2.1) dominates the signal course and only a small signal level of the element-specific emission is detected (for times >70 ns in Fig. 3.13, bottom), whereas the situation in argon is completely different. A distinct second maximum appears followed by a decay lasting several 100 ns. This second maximum is attributed to the line emission, which is strongly enhanced for argon in comparison with air. The line emission increases due to enhanced ablation and higher plasma temperatures (see previous paragraph).

For a given ambient gas, e.g., air, the ablation rate increases for reduced ambient pressures [3.48]. A reduction of the pressure from 1 bar to 0.1 mbar increases the ablation rate by a factor of 5. This behavior is explained by the reduced plasma shielding for lower ambient pressures and the reduced recondensation of the ablated material in the interaction region due to an increased plasma volume whose center is shifted to larger distances from the target surface. Further details are described in Chap. 7.

Besides the mass effect of the ambient gas, the electronic configurations of the gas species influence the absorption features for electromagnetic radiation. Figure 3.14 shows the mass absorption coefficient of the gases oxygen, water vapor, and ozone in the wavelength region from 100 nm to 30 μm [3.49]. In the ultraviolet region below 400 nm, the absorption of oxygen and water vapor increases strongly.

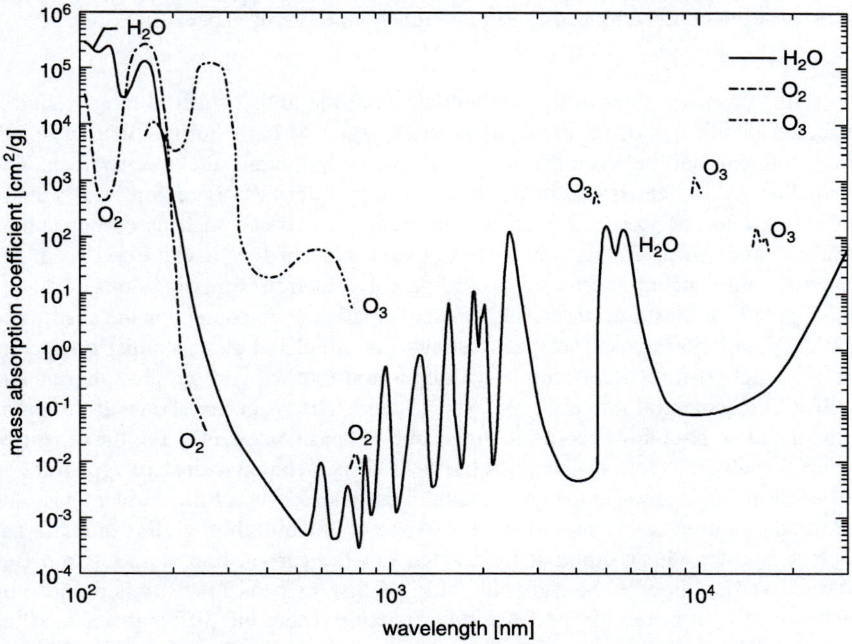

Fig. 3.14 Mass absorption coefficient of oxygen, ozone, and water vapor as a function of the wavelength

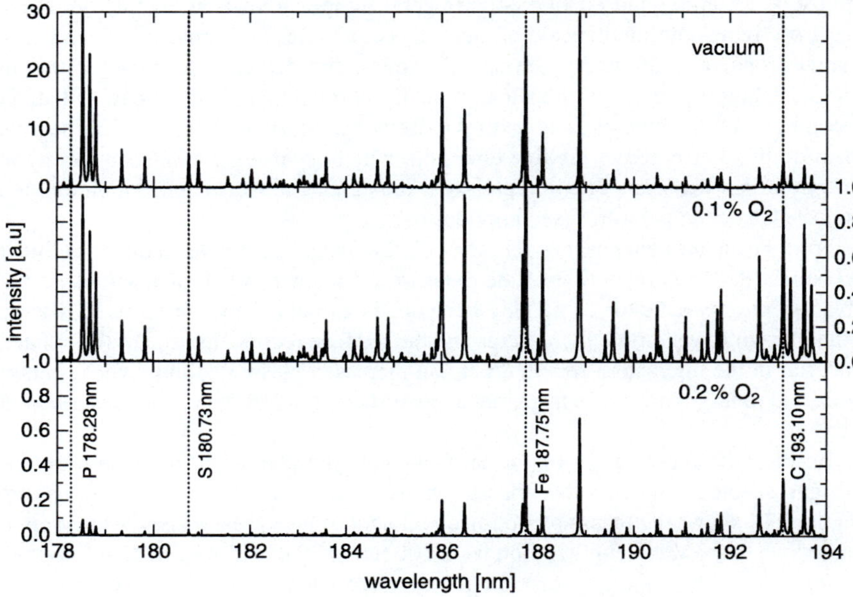

Fig. 3.15 Simulated emission spectra in the VUV range for a steel sample with emission lines of carbon, phosphorus, and sulfur. *Top*: no absorption in the ambient atmosphere, *middle/bottom*: 0.1% and 0.2% oxygen content in an inert gas causing absorption of VUV emission lines

Using emission lines of the laser-induced plasma in this spectral range requires vacuum conditions or an inert gas such as argon of high purity for a sufficient high transmission between the source of line emission and the spectrometer. The detection of line emissions in the vacuum ultraviolet (VUV) region is especially of interest to analyze light elements in metallic matrices such as carbon, phosphorus, and sulfur [3.24]. This spectral range offers low background from the Bremsstrahlung continuum and less interferences by other emission lines.

Figure 3.15 illustrates the strong effect of small oxygen content on the absorption of VUV emission lines. The spectra shown are simulated plasma emission spectra (cf. Sect. 10.3) for a steel sample with emission lines of carbon, phosphorus, and sulfur in the spectral range from 178 to 194 nm. The spectrum shown at the top is calculated without any absorption in the ambient gas ("vacuum"). For the other two spectra, an inert gas atmosphere of 1 bar at 293 K is assumed with an oxygen fraction of 0.1% and 0.2%, respectively, of the total particle density of the ambient gas. The emitted radiation propagates along a distance of 180 mm through this ambient gas before it enters the vacuum spectrometer. For the carbon line at 193.10 nm (see dashed vertical line on the right side in Fig. 3.15) the peak intensity is reduced by a factor of 5 from the top to the middle spectrum (note the different scales at the y-axes of the three spectra), whereas this factor is more than 20 for the Fe 187.75-nm line, 28 for S 180.73-nm line, and >40 for the P 178.28-nm line.

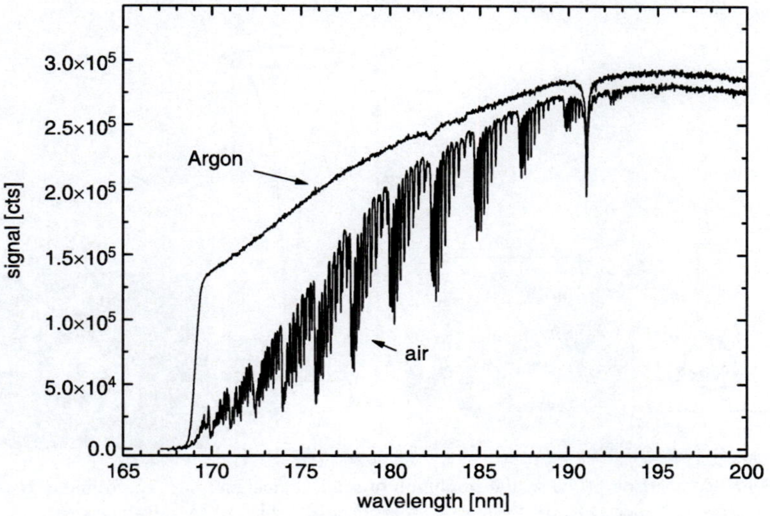

Fig. 3.16 Emission spectrum of a deuterium lamp for wavelengths below 200 nm transmitted through a thin gas layer flushed with argon or air

Figure 3.16 illustrates the absorption properties of air looking at the emission spectrum of a deuterium lamp in the near UV region, where the short distance between the lamp and the spectrometer is flushed with argon or with air.

3.6 Spatial Resolution and Spatial Averaging

For LIBS, the laser beam is focused to induce a plasma. This intrinsically allows for a high lateral spatial resolution. The point of analysis can be defined by the location of the laser focus, i.e., the position of the waist of the converging laser beam. Two kinds of spatial resolution have to be distinguished: the lateral resolution and the longitudinal or depth resolution. Lateral and longitudinal resolutions are primarily influenced by the laser beam parameters: focus diameter and Rayleigh length. Both parameters cannot be selected independently but are linked to each other by the well-known relation given for a Gaussian beam: $z_R = \pi w_0^2/\lambda$, where z_R is the Rayleigh length, w_0 is the waist radius of the beam, and λ is the wavelength. In the following paragraphs, we will first focus the discussion on the lateral spatial resolution.

Figure 3.17 illustrates the lateral spatial resolution of LIBS compared with conventional spark optical emission spectrometry, where an electric spark is ignited between an electrode and a metallic sample surface. While the spark jumps to any position within a circle of about 5 mm diameter depending on the respective local peaks of the electric field on the surface, the laser beam irradiates a well-defined

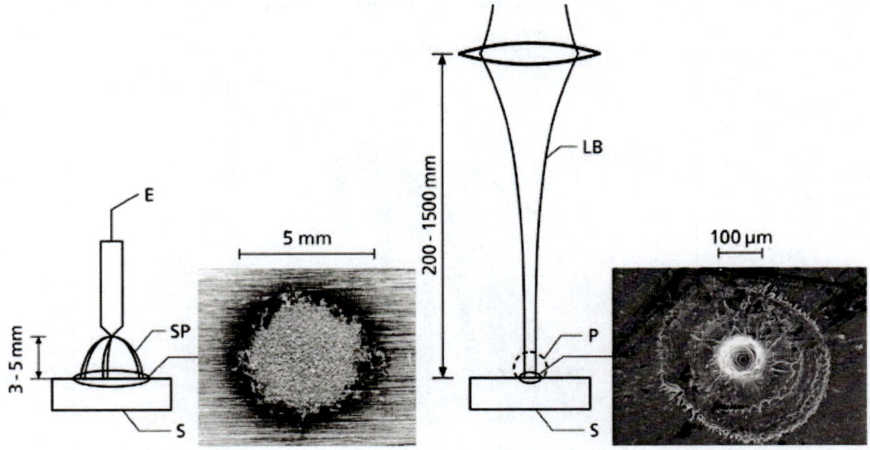

Fig. 3.17 Comparison of the spatial resolution of spark optical emission spectrometry (*left*) and LIBS (*right*). E = electrode, SP = spark, S = sample, LB = laser beam, P = plasma

position. For a series of several hundreds of sparks – this is the case for the photograph shown on the left side of Fig. 3.17 – the interrogated surface is a circle with a diameter of 5 mm. Hence, the individual location of each spark is uncertain within this 5 mm.

For the laser beam, the lateral extension of the interaction region is significantly smaller. In the example shown in Fig. 3.17, the central part of the crater has a diameter of less than 100 μm. Of course this diameter depends among others on the focusing conditions, cf. Sect. 3.4. With a tightly focused laser beam of high beam quality, much smaller crater diameters can be achieved. High spatial resolution is especially of interest for scanning microanalysis with LIBS, see Chap. 15.

The achievable lateral spatial resolution with LIBS is determined by the size of the produced laser craters within the material. The crater diameter depends not only on the laser focus diameter but also on the heat affected zone and therewith on the thermal diffusivity and on the laser pulse duration, cf. Sect. 3.3. The dependence of the crater radius produced by the laser pulse on the radius of the beam waist and the heat penetration depth is described in a simplified approach for constant laser pulse energy and beam profile:

$$r_c \propto w_b + \delta_h, \tag{3.7}$$

where r_c is the radius of the produced laser crater, w_b is the radius of the beam waist on the surface of the specimen, and δ_h is the heat penetration depth, see (3.2).

A shorter laser pulse duration will lead to a decreased heat penetration depth and hence to a smaller crater radius.

Smaller laser craters can also be achieved by further reducing the applied laser pulse energy while keeping the beam waist radius constant. However, this will lead to smaller irradiances within the focus and therefore elements with high excitation energies could not be excited sufficiently with the consequence of lower sensitivity and higher detection limits.

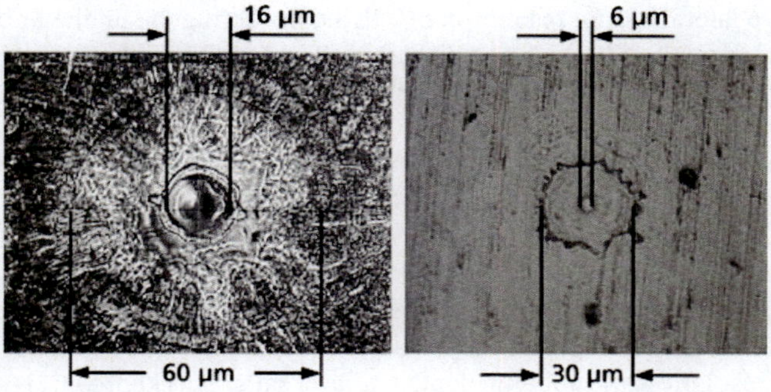

Fig. 3.18 Microscopic photographs of craters produced in a steel sample by single laser pulses generated with DPSSL systems. *Left*: nanosecond pulse with an irradiance of 1.2 TW/cm^2, *right*: picosecond pulse, 25 TW/cm^2

Figure 3.18 shows two photographs of laser craters produced by single laser pulses of different pulse durations. The nanosecond pulse is produced by the DPSSL described in Sect. 4.1 (see Table 4.3; Figs. 4.2 and 4.3). The picosecond pulse is generated also by a DPSSL. The crater shown on the left side was generated with a laser pulse of 5.5-ns duration, 2 mJ pulse energy, and a focus irradiance of 1.2 TW/cm^2, whereas the crater shown on the right side was generated with a laser pulse of 40-ps duration, 0.4 mJ pulse energy, and a focus irradiance of 25 TW/cm^2. The laser wavelength in both cases is 1,064 nm. The crater diameter is reduced from 16 µm diameter for the nanosecond laser to 6 µm for the picosecond laser, and the diameter of the debris area is reduced from 60 to 30 µm. Although the irradiance within the laser focus is approximately 21 times higher for the picosecond laser used in the right picture, the crater and the debris diameter are smaller than those for the nanosecond laser in the left picture of Fig. 3.18. Hence, picosecond lasers offer the potential to further improve the spatial resolution for scanning LIBS. An even further increase in resolution would be gained with the use of femtosecond lasers. However, an application of such lasers has to cope with the fact that these lasers are at present more expensive by a factor of two than picosecond lasers.

While subnanosecond laser pulses allow to reduce the crater diameters, their efficiency for plasma excitation for LIBS is expected to decrease. A compromise has to be found to optimize the ablation geometry and the plasma excitation. There is no reason to assume that the laser pulse durations applied so far for LIBS are already at this optimum. The temporal pulse structure has a strong influence on the ablated mass and the efficiency of plasma excitation (cf. Sect. 3.2 and Chap. 7). Different approaches were studied based on combinations of nanosecond pulses, which can easily be generated by conventional lasers. In the near future, further progress in the development of DPSSL systems will open up new perspectives to generate temporally modulated laser pulses within time scales from picosecond to nanosecond. This will give the chance to find the optimum tailored pulse structure for the various LIBS applications.

So far the lateral spatial resolution of LIBS considering the diameter of the laser-induced crater in a solid target was discussed. The longitudinal spatial resolution or depth resolution is determined by the depth of the crater.

Depth resolution is a key parameter for depth profiling of coated samples by LIBS, which was studied by different research groups mostly with the aim to achieve high depth resolution [3.47, 3.50–3.55]. The experiments were performed on stationary samples by irradiating a single spot or a microline on the sample with a series of laser pulses [3.53]. Each pulse ablates an amount of material, thus penetrating step by step into the sample. From the spectra of the successive laser pulses, information about the depth profile of the elements was obtained. To achieve high depth resolution, the ablation depth per laser pulse was kept low (1–100 nm) by using single pulses and large spot diameters or a line focus on the sample surface. Figure 3.19 shows an overview of the achieved ablation rates with the corresponding number of pulses per sample position. For depth profiling of moving coated samples, a different approach is chosen, where collinear double pulses can penetrate with a single burst coating thicknesses of up to 10 μm and more (see data points with filled triangles at $n = 1$ in Fig. 3.19, cf. Chap. 16 [3.56]).

For a representative analysis of an inhomogeneous sample, spatial averaging is necessary. Spatial averaging with LIBS is possible via two principal approaches: sequential and simultaneous. For sequential averaging, the laser focus is shifted laterally relative to the measuring object, by moving the object or by moving of the

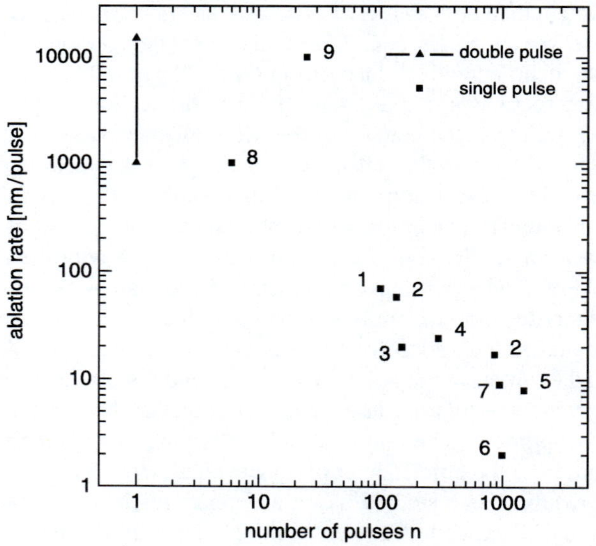

Fig. 3.19 Comparison of ablation rates as a function of the number of pulses per sample position used in LIBS depth profiling experiments. Studies on static samples are plotted as *filled squares* (the numbers close to the data points in the graph refer to the following references: 1–9 = [3.47, 3.50–3.55])

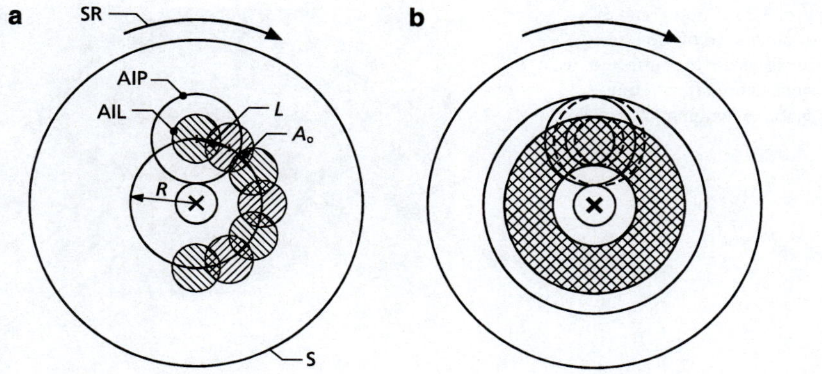

Fig. 3.20 Principle of spatial averaging by scanning the laser beam across a sample surface by rotation of the sample. S = sample, SR = sample rotation, AIL = area irradiated by the laser beam, AIP = area where interaction of the laser-induced plasma with the sample surface takes place; A_o, L = see text. (**a**) Overlapping areas after a few laser pulses and (**b**) laser track after several hundred laser pulses

laser beam. In the first simple approach, the laser focus is scanned in a circle across the surface of the measuring object, where the diameter of that circle is chosen to be sufficiently large to cover the respective inhomogeneity scale of the object under study. Figure 3.20 illustrates this approach, where the area irradiated by the laser beam runs along a circle having a radius R on the surface of the sample. Depending on the repetition rate of the laser, the radius of the beam focus and the angular velocity of the sample rotation different degrees of area overlap can be adjusted described by the following relation:

$$\eta = \frac{A_o}{\pi w_b^2} = \frac{2}{\pi}\arccos\left(\frac{L}{2w_b}\right) - \frac{L}{\pi w_b}\sqrt{1-\left(\frac{L}{2w_b}\right)^2}, \tag{3.8}$$

where η is the degree of overlap, A_o is the overlapping area, see Fig. 3.20, w_b is the radius of laser beam on the sample surface, and L is the distance between the centers of two adjacent laser beam positions.

The distance L is approximately given by the relation $L \approx \omega R/\nu_{rep}$, where ω is the angular frequency of the sample rotation, R is the radius shown in Fig. 3.20, and ν_{rep} the repetition frequency of the laser.

Figure 3.21 shows a photograph of such a scanned circle on a steel sample with a series of LIBS craters along the circumference (for the setup see Chap. 2, Fig. 2.7). Such an approach is especially of interest to perform a reliable bulk analysis of inhomogeneous samples, e.g., slag samples, cf. Sect. 14.3 [3.57].

Besides moving the sample, the laser beam itself can be translated very fast using a galvanometric scanner. By this different areas of a sample are irradiated and moreover tailored crater geometries can be generated, cf. Sect. 13.1.2.

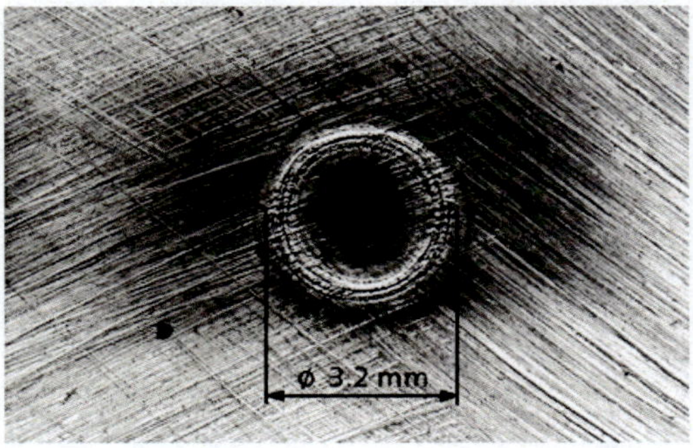

Fig. 3.21 Photograph of a circular scan of the laser focus across the surface of a steel sample illustrating spatial averaging, cf. Fig. 3.20

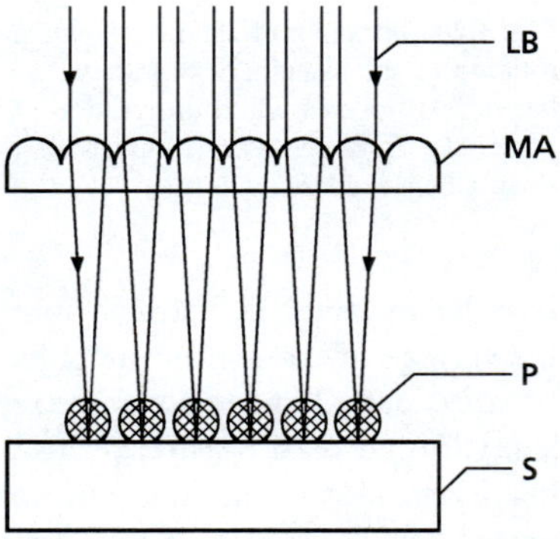

Fig. 3.22 Optical microlens array to generate simultaneously an array of plasmas. LB = laser beam, MA = microlens array, P = plasma, and S = sample

With a microlens array a simultaneous generation of up to $N_p = 40$ plasmas is possible spread over an area of about 6 mm diameter, see Fig. 3.22 [3.58]. The sampling area A_s is defined by the high-intensity spots of the laser beam at the sample surface. With 40 spots and a focal length of the microlens array of 37 mm, the single spot diameter for a laser beam with $M^2 = 1$ amounts to $d_s = 58\,\mu\text{m}$. The sampling area A_s is therefore: $A_s = 0.25 N_p \pi d_s^2 \approx 0.11\,\text{mm}^2$. For comparison, the corresponding values for a single lens ($f = 200$ mm) with a beam diameter of 6 mm are: $d_s = 45\,\mu\text{m}$ and $A_s = 0.25 \pi d_s^2 \approx 0.0016\,\text{mm}^2$. Therefore, the sampling area with the microlens array is approximately a factor of 66 higher than for the single lens. The advantage of the microlens array is that it spatially averages over a distance of approximately the diameter of the input laser beam.

3.7 Direction of Incidence of the Laser Beam and Direction of Observation of the Plasma Emission

In many cases, the laser beam axis is oriented perpendicularly to the sample surface to be analyzed. Figure 3.23 illustrates the more general case of a laser beam irradiated under an angle of incidence of α_L. For flat solid samples $\alpha_L = 0°$ is

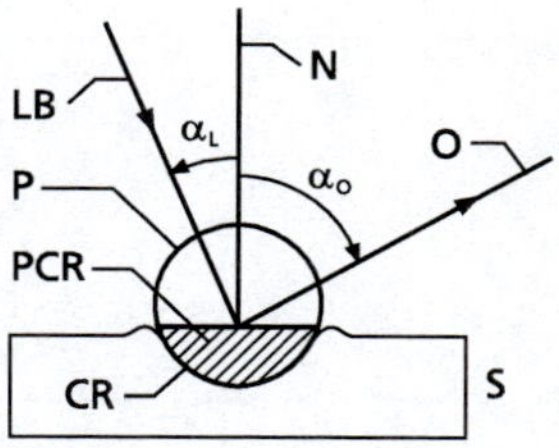

Fig. 3.23 Incident laser beam and observation direction relative to the normal of the sample surface. LB = laser beam, O = observation direction, N = normal to local sample surface, S = sample, P = plasma, CR = crater, PCR = part of the plasma inside the crater

chosen offering the highest irradiances on the sample surface for a given laser pulse energy, whereas $\alpha_L \neq 0°$ has to be considered for nonflat measuring objects, e.g., scrap pieces in high-speed sorting applications, cf. Sect. 18.4. For a rotationally symmetric beam, this leads to an elliptic cross section of the irradiance distribution on the sample surface.

The observation angle α_O determines the part of the plasma contributing to the detected spectrum. For α_O close to 90°, the plasma is observed in side-on direction and only those parts of the plasma contribute to the spectrum which are above the sample surface and therefore not shielded by the generated crater (plasma part PCR in Fig. 3.23 cannot be observed for $\alpha_O = 90°$). If only a few laser pulses are applied, this effect of shielding plays only a minor role. However, in the case of a series of laser pulses irradiated on the same sample location (in some applications several hundreds up to thousands of pulses are applied), the shielding of the plasma by the crater walls becomes more and more important. The realization of different observation angles in sample stands is described in Sect. 4.6.

For many inline applications where only one direction of access is technically feasible, α_O is chosen close to $-\alpha_L$, i.e., the plasma emission is observed antiparallel to the incident laser beam, cf. Sects. 13.1.3, 14.1, and 14.6.

3.8 Warming-up Pulses, Prepulses, and Measuring Pulses

If a series of laser pulses are irradiated at one location on a sample to perform a LIBS measurement, then we have to distinguish at least three groups of pulses. Before irradiating the sample the laser has to be in a stabile operational mode. Besides a sufficiently long switch-on time to assure a thermal stationary state of power supply and laser head, the laser system has to generate a certain number of laser pulses to achieve a stationary state of the laser medium, the cavity and the cooling system as well. The number of pulses to achieve this state is called warming-up pulses N_{wu} [3.40]. The laser pulses are blocked outside the laser by a shutter or beam dump and do not irradiate the sample. The warming-up pulses assure a sufficient degree

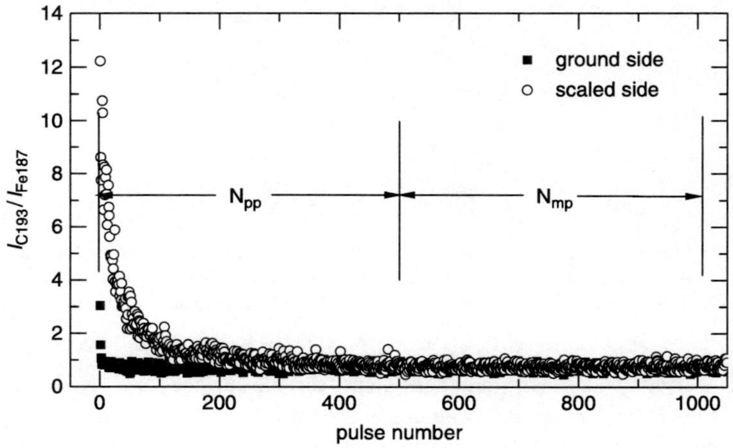

Fig. 3.24 Line ratio of the carbon line 193.1 nm to the matrix line of iron 187.7 nm as a function of the number of laser pulses irradiated on a steel sample

of thermal stability of the laser operation with respect to beam profile, pulse energy, and pointing stability.

The first series of laser pulses irradiating the sample under study is called prepulses. The number of prepulses N_{pp} causes a local ablation to remove surface contaminations or surface layers which are not representative for the bulk composition of the sample. The spectral emission of the plasmas generated with the prepulses is not used for the bulk analysis. After the prepulses, a series of measuring pulses N_{mp} is irradiated to generate plasmas, whose spectral emission is evaluated to determine the composition of the sample.

Figure 3.24 shows a series of spectral signals as a function of the pulse number irradiated onto one location on a scaled steel sample (measuring parameters: double pulses, $E_b = 2 \times 80$ mJ, interpulse separation $\Delta t_{12} = 30\,\mu$s, beam waist position $\Delta s = 8$ mm, $t_{del} = 2\,\mu$s, $t_{int} = 10\,\mu$s, for further details see Sect. 13.1.2). The scale layer has a composition which is different from the bulk composition. The spectral signal shown is the ratio of the carbon line 193.1 nm to the matrix line of iron at 187.7 nm. For the scaled side, the spectral ratio shows a strong decrease for pulse numbers 0–200, and beyond 500 pulses the ratio converges to a stationary value. If the scale layer is removed by grinding, then the stationary value is achieved much faster. Hence, in the case shown in Fig. 3.24, the number of prepulses is chosen as, e.g., $N_{pp} = 500$ to assure that the scale layer is penetrated completely and a representative value for the bulk composition is gained.

Table 3.2 Examples for moving measuring objects in LIBS setups

Velocity of measuring object	Configuration	Application, remark, references
0 mm/s	Fixed sample on sample stand	Bulk analysis of homogeneous samples, gas exchange has to be regarded; cf. Sects. 3.1 and 4.6 [3.24]
10–50 mm/s	Sample on linear translational stages	Scanning microanalysis, cf. Chap. 15 [3.27]
12–75 mm/s	Sample moved on sample stand by excentric mechanism	Bulk analysis of inhomogeneous samples, cf. Sect.14.3 [3.57]
Up to 3 m/s	Measuring object on moving belt conveyor	Identification of scrap pieces or raw materials, relative wind forwards efficient gas exchange [3.59]

3.9 State of the Measuring Object

Besides its elemental composition, the state of the measuring object can be described by the following categories: (a) state of movement, (b) geometry and surface properties (e.g., roughness), (c) state of aggregation, and (d) homogeneity.

Since a LIBS measurement can be performed within a few 10 μs, there is an intrinsic potential to measure objects which are moving. Table 3.2 gives an overview of different states of movement of the measuring object studied with LIBS. In the predominant number of LIBS studies published, the measuring object is fixed. For scanning microanalysis, the sample is shifted relative to the laser beam to spatially resolve element distributions. Circular sample movements are applied to perform a spatial averaging for inhomogeneous samples, see Sect. 3.6. Highest velocities of measuring objects were achieved for the identification of scrap pieces transported on a belt conveyor with up to 3 m/s. For the stationary measuring objects or those moving at slow velocities <100 mm/s the relative wind is negligible, whereas this is not the case for measuring objects moving with speeds of several meters per second.

The effect of relative wind becomes obvious in the analysis functions shown in Fig. 3.25 for static and moving aluminum reference samples, measured with the setup described in Sect. 18.4 [3.59]. The coefficient of determination is better for the moving samples than for the static samples, which is attributed to an improved gas exchange in the interaction region by the relative wind (cf. Sect. 3.1, Fig. 3.4).

The geometry of the measuring object influences the measuring process. In many cases, flat measuring objects are analyzed and the laser beam is irradiated perpendicularly to the sample surface, i.e., parallel to the normal of the surface. However, the general case, especially with respect to industrial applications of LIBS, is measuring objects where the local surface element is inclined against the propagation axis of the laser. In the following we consider the situation already shown in Fig. 3.23, where the laser beam propagation axis encloses an angle α_L

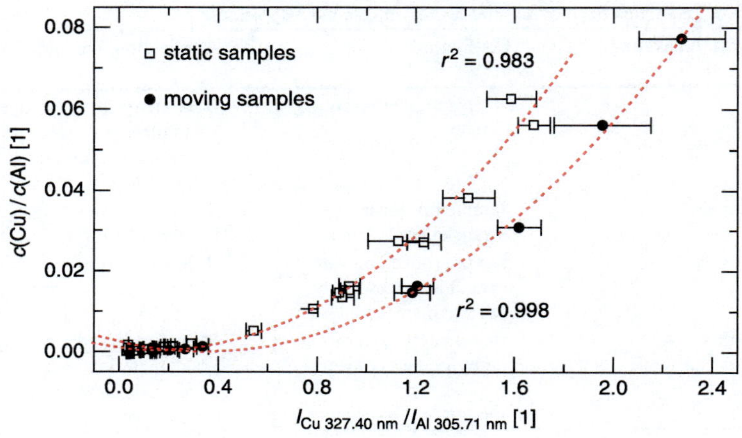

Fig. 3.25 Analysis functions of static and moving aluminum samples at 3 m/s for copper

Fig. 3.26 Laser beam irradiating a surface element where the propagation axis of the laser encloses an angle of α_L with the normal of the surface element. PI = point of intersection of optical axis and surface element, FL = focusing lens, SE = surface element, PB = projection of beam profile onto the surface element, OA = optical axis, BW = beam waist, N = normal of surface element, d = distance between *FL* and PI, Δs = beam waist position, cf. Fig. 2.2

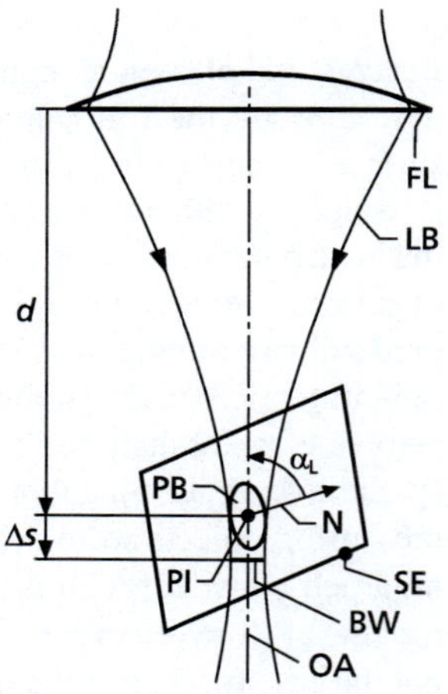

to the normal of the irradiated surface element. Figure 3.26 shows this situation in a perspective view. If $\alpha_L \neq 0°$, the projection of the laser beam profile onto the surface element has an elliptic shape and the irradiance is reduced in relation to the case $\alpha_L = 0°$. Hence, the orientation of the surface element irradiated influences the local irradiance and therefore the interaction process with the object. This effect has to be taken into account especially for samples with nonflat surfaces or measuring objects with a 3D geometry, e.g., scrap pieces, cf. Sect. 18.4.

LIBS is able to analyze matter in each state of aggregation: solid, liquid, and gaseous. The relevance of various measuring parameters depends on the state of aggregation of the measuring object. For example, the distance to the focal plane Δs (cf. Figs. 2.2 and 3.11) determines the irradiance at the surface of a solid or liquid metal sample causing the ablation and plasma ignition. For a gas, no surface can be defined and the plasma ignition will start in those regions where the irradiance exceeds the threshold for ionization and plasma formation.

After laser pulse irradiation at the surface of a free liquid surface or gaseous measuring objects, the matter flows back into the interaction region and after a sufficient time interval the same initial conditions of the measuring object are achieved. The laser pulse liquid interaction at a free liquid surface can cause splashes and droplets ejected from the liquid into the adjacent half space. These droplets may reach the focusing objective and can cause damage. An approach to prevent such splashes from reaching the optics is a gas flow to push these particles away [3.60–3.62].

Typical threshold irradiances in air leading to a breakdown initialized via free electrons from multiphoton absorption and ionization and subsequent cascade processes are in the order of 10^{10}–$10^{13}\,\mathrm{W/cm^2}$ depending on the wavelength of the laser radiation and the experimental conditions [3.63].

For solid samples, the homogeneity of the sample is an important issue. The extension of spatial variations of element concentrations has to be considered carefully in order to decide which way of spatial averaging shall be applied to perform a representative bulk analysis, cf. Sect. 3.6. In some cases, the sample has to be prepared before analysis with LIBS. Slag samples are, e.g., crushed, milled, and pressed into pellets to get a homogenization, cf. Sect. 14.2.

For the spatially resolved LIBS analysis of, e.g., metallic samples with inclusions, the sample surface has to be prepared by grinding or milling, cf. Chap. 15. However, the selection of the grinding paper has to be chosen carefully, since there is a danger to contaminate the surface of the sample by species originating from the grinding paper so that the measuring results are disturbed. For example, SiC paper leads to a change in the histogram of carbon signals, whereas for corundum and zirconium paper no such change is observed [3.64]. A similar effect occurs for Ca histograms indicating that the resin used as binder in the SiC paper may lead to a contamination of the sample surface. Milling of the sample surface to be scanned by LIBS is an approach to avoid such effects.

References

3.1. E. Muller, A. Lauritzen, P. Eggimann, Laser spark optical emission spectrometry: A revolutionary method for the iron and steel industry. Steel Times **6**, 224–225 (1998)

3.2. GmbH & Co KG, *Product Information of Spectro Analytical Instruments* (GmbH & Co KG, Kleve, 2000)

3.3. I. Whiteside, R. Jowitt, Laser liquid metal analysis, in *Progress of Analytical Chemistry in the Iron and Steel Industry* (Commission of the European Communities, Brussels, 1992), pp. 135–141

3.4. Final report of the Brite-Euram project: study of emission spectroscopy on laser produced plasma for localised multielemental analysis in solids with surface imaging, Coordinator Saclay, France, Contract MAT 1-CT-93–0029, 1996

3.5. K. Yamamoto, D. Cremers, M. Ferris, L. Foster, Detection of metals in the environment using a portable laser-induced breakdown spectroscopy instrument. Appl. Spectrosc. **50**, 222–233 (1996)

3.6. L. Paksy, B. Nemet, Production control of metal alloys by laser spectroscopy of the molten metals. Part 1. Preliminary investigations. Spectrochim. Acta B **51**, 279–290 (1996)
3.7. M. Sabsadi, P. Cielo, Quantitative analysis of aluminium alloys by laser-induced breakdown spectroscopy and plasma characterization. Appl. Spectrosc. **49**, 499–507 (1995)
3.8. L. Cabalin, D. Romero, J. Baena, J. Laserna, Effect of surface topography in the characterization of stainless steel using laser-induced breakdown spectrometry. Surf. Interface Anal. **27**, 805–810 (1999)
3.9. C. Aragon, J. Aguilera, F. Penalba, Improvements in quantitative analysis of steel composition by laser-induced breakdown spectroscopy at atmospheric pressure using an infrared Nd:YAG laser. Appl. Spectrosc. **53**, 1259–1267 (1999)
3.10. M. Hatcher, Breakdown spectroscopy finds nuclear application. Opto Laser Eur. April, 20–22 (2000)
3.11. V. Sturm, L. Peter, R. Noll, J. Viirret, R. Hakala, L. Ernenputsch, K. Mavrommatis, H. Gudenau, P. Koke, B. Overkamp, Elemental analysis of liquid steel by means of laser technology. International Meeting on Chemical Engineering, Environmental Protection and Biotechnology, ACHEMA 2000, Materials Technology and Testing, 9–11 (2000)
3.12. I. Mönch, R. Noll, R. Buchholz, J. Worringer, Laser identifies steel grades. Stainless Steel World **12**(4), 25–29 (2000)
3.13. H. Sattler, Metallschrotte nach Analyse automatisch sortieren. Materialprüfung **35**, 312–315 (1993)
3.14. R. Noll, V. Sturm, L. Peter, I. Whiteside, Analysis using lasers. Proceedings of 49th Chemist's Conference, Scarborough, 22–27 (1997)
3.15. cw-Nd:YAG Laser BLS 600, Manufacturer Carl Baasel Lasertechnik GmbH, Postfach 1453, 82304 Starnberg
3.16. R. Sattmann, V. Sturm, R. Noll, Laser-induced breakdown spectroscopy of steel samples using multiple Q-switch Nd:YAG laser pulses. J. Phys. D Appl. Phys. **28**, 2181 (1995)
3.17. L.M. Cabalin, J.J. Laserna, Atomic emission spectroscopy of laser-induced plasmas generated with an annular-shaped laser beam. J. Anal. At. Spectrom. **19**, 445–450 (2004)
3.18. L. St-Onge, M. Sabsabi, P. Cielo, Quantitative analysis of additives in solid zinc alloys by laser-induced plasma spectrometry. J. Anal. At. Spectrom. **12**, 997–1004 (1997)
3.19. L. St-Onge, M. Sabsabi, P. Cielo, Analysis of solids using laser-induced plasma spectroscopy in double-pulse mode. Spectrochim. Acta B **53**, 407–415 (1998)
3.20. B. Le Drogoff, J. Margot, S. Laville, M. Chaker, M. Sabsabi, T.W. Johnston, O. Barthelemy, Influence of the laser pulse duration on laser-produced plasma properties. Plasma Sources Sci. Technol. **13**, 223–230 (2004)
3.21. A. Semerok, C. Dutouquet, Ultrashort double pulse laser ablation of metals. Thin Solid Films **453–454**, 501–505 (2004)
3.22. R. Noll, R. Sattmann, V. Sturm, Multi-Element-Analyse von Stahllegierungen mit Laser-Emissionsspektroskopie, in *Proceedings of 12 International Congress Laser 95* (Springer-Verlag, Berlin, 1996), pp. 626–629
3.23. L. Peter, V. Sturm, R. Noll, R. Hakala, J. Viirret, B. Overkamp, P. Koke, Multi-elemental analysis of low-alloyed steel by laser-induced breakdown spectrometry. SPIE **3823**, 256–265 (1999)
3.24. V. Sturm, L. Peter, R. Noll, Steel analysis with laser-induced breakdown spectrometry in the vacuum ultraviolet. Appl. Spectrosc. **54**, 1275–1278 (2000)
3.25. L. St-Onge, M. Sabsabi, P. Cielo, Towards quantitative depth-profile analysis using laser-induced plasma spectroscopy: investigation of galvannealed coatings on steel. Spectrochim. Acta B **55**, 299–308 (2000)
3.26. L. St-Onge, R. Sing, S. Bechard, M. Sabsabi, Carbon emissions following 1.064 μm laser ablation of graphite and organic samples in ambient air. Appl. Phys. A **69**, S913–S916 (1999)
3.27. H. Bette, R. Noll, High-speed laser-induced breakdown spectrometry for scanning microanalysis. J. Phys. D Appl. Phys. **37**, 1281–1288 (2004)
3.28. M. Stepputat, R. Noll, On-line detection of heavy metals and brominated flame retardants in technical polymers with laser-induced breakdown spectrometry. Appl. Optics **42**, 6210–6220 (2003)

3.29. B. Le Drogoff, J. Margot, M. Chaker, M. Sabsabi, O. Barthelemy, T.W. Johnston, S. Laville, F. Vidal, Y. von Kaenel, Temporal characterization of femtosecond laser pulses induced plasma for spectrochemical analysis of aluminium alloys. Spectrochim. Acta B **56**, 987–1002 (2001)
3.30. B. Le Drogoff, M. Chaker, J. Margot, M. Sabsabi, O. Barthelemy, T.W. Johnston, S. Laville, F. Vidal, Influence of the laser pulse duration on spectrochemical analysis of solids by laser-induced plasma spectroscopy. Appl. Spectrosc. **58**, 122–128 (2004)
3.31. J. Kutzner, M. Silies, T. Witting, G. Tsilimis, H. Zacharias, Efficient high-repetition-rate fs-laser based X-ray source. Appl. Phys. B **78**, 949–955 (2004)
3.32. S. Klimentov, S. Garnov, T. Kononenko, V. Konov, P. Pivovarov, E. Dausinger, High rate deep channel ablative formation by picosecond-nanosecond combined laser pulses. Appl. Phys. A **69**, S633–S636 (1999)
3.33. C. Momma, B.N. Chichkov, S. Nolte, F. von Alvensleben, A. Tünnermann, H. Welling, B. Wellegehausen, Short-pulse laser ablation of solid targets. Opt. Commun. **129**, 134–142 (1996)
3.34. P. Herman, A. Oettl, K. Chen, R. Marjoribanks, Laser micromachining of "transparent" fused silica with 1-ps pulses and pulse trains. Proc. SPIE **3616**, 148–154 (1999)
3.35. H.-U. Schmitz, R. Noll, M. Kraushaar, Laser-OES: A universal technique for slag analysis, *50th Chemist's Conference*, Ashorne Hill, 1999, 11–15
3.36. V. Sturm, R. Sattmann, R. Noll, Optical fiber transmission of multiple Q-Switch Nd:YAG laser pulses with microsecond interpulse separations. Appl. Phys. B **63**, 363–370 (1996)
3.37. V. Margetic, M. Bolshov, A. Stockhaus, K. Niemax, R. Hergenröder, Depth profiling of multi-layer samples using femtosecond laser ablation. J. Anal. At. Spectrom. **16**, 616–621 (2001)
3.38. S. Nolte, C. Momma, G. Kamlage, A. Ostendorf, C. Fallnich, F. von Alvensleben, H. Welling, Polarization effects in ultrashort-pulse laser drilling. Appl. Phys. A **68**, 563–567 (1999)
3.39. B. Salle, O. Gobert, P. Meynadier, M. Perdrix, G. Petite, A. Semerok, Femtosecond and picosecond laser microablation: ablation efficiency and laser microplasma expansion. Appl. Phys. A **69**, S381–S383 (1999)
3.40. R. Noll, Terms and notations for laser-induced breakdown spectroscopy. Anal. Bioanal. Chem. **385**, 214–218 (2006)
3.41. D.R. Lide, *Handbook of Chemistry and Physics*, 67th edn. (CRC Press, Boca Raton, 1986–1987)
3.42. H. Carslaw, J. Jaeger, *Conduction of Heat in Solids*, 2nd edn. (Oxford University Press, Oxford, 1959)
3.43. E. Shiles, T. Sasaki, M. Inokuti, D. Smith, Self-consistency and sum-rule tests in the Kramers-Kronig analysis of optical data: applications to aluminum. Phys. Rev. B **22**, 1612–1628 (1989)
3.44. J. Weaver, E. Colavita, D. Lynch, R. Rosei, Low-energy interband absorption in bcc Fe and hcp Co. Phys. Rev. B **19**, 3850–3856 (1979)
3.45. R. Klein, Bearbeitung von Polymerwerkstoffen mit infraroter Laserstrahlung, D 82, Dissertation, Aachen, 1990
3.46. European Standard DIN EN ISO 11146–1, Lasers and laser-related equipment – test methods for laser beam widths, divergence angles and beam propagation ratios – part 1: stigmatic and simple astigmatic beams, 2005
3.47. J. Vadillo, C. Garcia, S. Palanco, J. Laserna, Nanometric range depth-resolved analysis of coated-steels using laser-induced breakdown spectrometry with a 308 nm collimated beam. J. Anal. At. Spectrom. **13**, 793–797 (1998)
3.48. L. Peter, R. Noll, Material ablation and plasma state for single and collinear double pulses interacting with iron samples at ambient gas pressures below 1 bar. Appl. Phys. B **86**, 159–167 (2007)
3.49. J. Pecker, *Space Observatories* (D. Reidel Publishing, Dordrecht, 1970)

3.50. D. Anderson, C. McLeod, T. English, A. Smith, Depth profile studies using laser-induced plasma emission spectrometry. Appl. Spectrosc. **49**, 691–701 (1995)
3.51. M. Mateo, J. Vadillo, J. Laserna, Irradiance-dependent depth profiling of layered materials using laser-induced plasma spectrometry. J. Anal. At. Spectrom. **16**, 1317–1321 (2001)
3.52. C. Garcia, M. Corral, J. Vadillo, J. Laserna, Angle-resolved laser-induced breakdown spectrometry for depth profiling of coated materials. Appl. Spectrosc. **54**, 1027–1031 (2000)
3.53. M. Mateo, L. Cabalín, J. Laserna, Line-focused laser ablation for depth-profiling analysis of coated and layered materials. Appl. Optics **42**, 6057–6062 (2003)
3.54. D. Papazoglou, V. Papadakis, D. Anglos, In situ depth and topography monitoring in LIBS elemental profiling of multi-layer structures. J. Anal. At. Spectrom. **19**, 483–488 (2004)
3.55. M. Mowery, R. Sing, J. Kirsch, A. Razaghi, S. Bechard, R. Reed, Rapid at-line analysis of coating thickness and uniformity on tablets using laser induced breakdown spectroscopy. J. Pharm. Biomed. Anal. **28**, 935–943 (2002)
3.56. H. Balzer, M. Hoehne, V. Sturm, R. Noll, Online coating thickness measurement and depth profiling of zinc coated sheet steel by laser-induced breakdown spectroscopy. Spectrochim. Acta B **60**, 1172–1178 (2005)
3.57. M. Kraushaar, R. Noll, H.-U. Schmitz, Slag analysis with laser-induced breakdown spectrometry. Appl. Spectrosc. **57**, 1282–1287 (2003)
3.58. V. Sturm, Optical microlens array for plasma generation in spectrochemical analysis. J. Anal. At. Spectrom. **22**, 1495–1500 (2007)
3.59. Ü. Aydin, R. Noll, J. Makowe, Automatic sorting of aluminium alloys by fast LIBS identification, in *Seventh International Workshop Progress in Analytical Chemistry in the Steel and Metal Industries*, ed. by J. Angeli (Glückauf GmbH, Essen, 2006), pp. 309–314
3.60. L. Peter, V. Sturm, R. Noll, Liquid steel analysis with laser-induced breakdown spectrometry in the vacuum ultraviolet. Appl. Optics **42**, 6199–6204 (2003)
3.61. L. Barrette, S. Turmel, On-line iron-ore slurry monitoring for real-time process control of pellet making processes using laser-induced breakdown spectroscopy: graphitic vs. total carbon detection, Spectrochim. Acta B **56**, 715–723 (2001)
3.62. S. Oh, T. Miller, F. Yueh, J.P. Singh, Comparative study of laser-induced breakdown spectroscopy measurement using two slurry circulation systems. Appl. Optics **46**, 4020–4025 (2007)
3.63. A. Miziolek, V. Palleschi, I. Schechter (eds.), *Laser-Induced Breakdown Spectroscopy* (Cambridge University Press, Cambridge, 2006), p. 212
3.64. Improved production control through rapid characterisation of non-metallic inclusions in steel, Report, EUR 21627, 2005
3.65. A. Löbe, J. Vrenegor, R. Fleige, V. Sturm, R. Noll, Laser-induced ablation of a steel sample in different ambient gases by use of collinear multiple laser pulses. Anal. Bioanal. Chem. **385**, 326–332 (2006)

Chapter 4
Instrumental Components

4.1 Laser

Various laser types are applied for LIBS and laser ablation studies. The most wide spread are flashlamp-pumped solid-state lasers with Nd:YAG as laser medium operated in the Q-switch mode to generate high-energy laser pulses with pulse durations in the nanosecond range. Table 4.1 gives an overview of the lasers and laser parameters used for LIBS.

For further details concerning characteristics, design, construction, and performance of solid-state lasers, see [4.29, 4.30]. In the following paragraphs, the main features of solid-state lasers and here especially the Nd:YAG lasers are presented, since these are of predominant importance for LIBS. The laser medium is a doped insulator; dopants are rare earths or transition metals. The laser medium is pumped optically with broadband noble gas or halogen lamps (emission range 200–1,000 nm) or narrow band semiconductor lasers (806 nm).

Figure 4.1 shows the principal setup of a flashlamp-pumped solid-state laser. The laser medium is usually a rod cooled with water. The flashlamps are arranged parallel to this rod. The pump light is reflected and concentrated by the pump cavity. The inner surface of the pump cavity has a highly reflecting coating or a scattering surface to homogenize the illumination of the laser rod. For pulsed Nd:YAG lasers – as typically applied for LIBS – Xe flashlamps are used with typical life times of 20–50 million flashes depending on the total energy dissipated during the flashlamp discharge. The emission spectrum of the Xe flashlamp only partly matches the excitation spectrum of Nd:YAG; hence, only a small fraction of the absorbed pump power is transferred to the upper laser level. As a consequence less than 3% of the electrical input power is converted into laser radiation. Diode-pumped systems offer a better matching to the absorption band and reach 10–20% overall efficiency (laser beam power in relation to electrical power input). Therefore, diode-pumped solid-state lasers (DPSSL) systems save energy costs and secondary costs like cooling engines. The reduced heat load of the laser crystal leads to smaller internal stresses

R. Noll, *Laser-Induced Breakdown Spectroscopy*, DOI 10.1007/978-3-642-20668-9_4,

Table 4.1 Overview of laser types used for LIBS

Laser type	λ (nm)	Pulse energy, pulse duration	Repetition rate (Hz)	References and remarks
Nd:YAG	1,064	1–1,000 mJ, 3–50 ns	1–100	Standard laser for LIBS, laser medium is pumped by flashlamps, typical beam quality is in the range of $M^2 = 3$–6. Multiple laser pulses can be generated from a single laser unit [4.1–4.10]
Nd:YAG, SHG	532	1–90 mJ, 5–14 ns	5–10	[4.11–4.16]
Nd:YAG, THG	355	10 mJ, 6 ns	5–30	[4.11,4.17]
Nd:YAG, FHG	266	3–70 mJ, 4–8 ns	2–30	[4.18–4.21]
Nd:YAG	1,064	2.7 mJ, 25 ps		[4.22]
Nd:YAG, SHG	532	1.5 mJ, 18 ps		[4.22]
Nd:YAG, FHG	266	0.7 mJ, 13 ps		[4.22]
Diode-pumped Nd:YAG	1,064	2 mJ, 5 ns	1,000	Beam quality $M^2 = 1.3$ [4.23]
Ti–sapphire, SHG	400	150 fs		[4.22]
Excimer XeCl	308	70–150 mJ, 20–30 ns	5–50	[4.24,4.25]
Excimer KrF	248	40–70 mJ, 30 ns		[4.26]
Excimer ArF	193	250 mJ, 20 ns	1–50	[4.27]
Dye	500	22 mJ, 30 ns	10	Pumped by a XeCl excimer laser [4.28]

The range of parameters given in the columns – pulse energy, pulse duration, and repetition rate – refer to data of different laser systems for the given laser type
SHG = second harmonic generation, *THG* = third harmonic generation, *FHG* = fourth harmonic generation

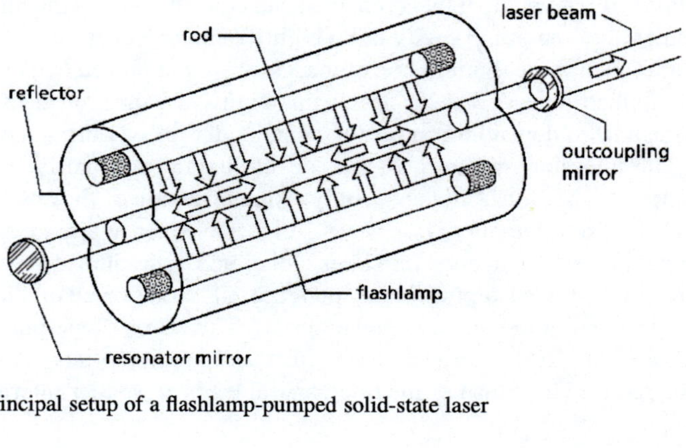

Fig. 4.1 Principal setup of a flashlamp-pumped solid-state laser

Table 4.2 Typical diode-pumpable laser crystals, their strongest laser emission wavelengths and suitable pumping wavelengths to be generated by laser diodes

Laser crystal	Laser wavelength (nm)	Pumping wavelength (nm)
Nd:YAG	1,064, 1,319	808
Nd:YVO_4	1,064, 1,340	809
Nd:YLF	1,047, 1,053	792, 798
Yb:YAG	1,023–1,052	941
Cr:LiSAF	800–920	670, 760

and changes in the optical properties which improves the mode quality of the laser beam (cf. Table 4.1, row 8).

The specific features of solid-state lasers in comparison with gas, liquid, or semiconductor lasers are high power, short pulses, efficient frequency conversion, and the capability to transmit the laser radiation via fiber optics (cf. Sect. 4.3).

The active ion in a Nd:YAG laser rod is Nd^{3+} embedded in an yttrium aluminum granate (YAG, $Y_3Al_5O_{12}$) crystal. The Nd:YAG laser is a four-level system offering a high amplification and suitable mechanical and thermal properties. The excitation by optical pumping occurs into broad absorption bands and subsequent radiationless transitions to the upper laser level. The lifetime of this level ($^4F_{3/2}$) is 230 μs, whereas the lower level has a lifetime of 30 ns only. The lower level lies 0.24 eV above the ground level, and hence at room temperature the population of this level is negligible. As a four-level system, the laser threshold of Nd:YAG is small. The laser lines are broadened homogeneously by thermal lattice vibrations. The line width is typically about 100 GHz at room temperature.

During the last years, the availability of high-power semiconductor laser diodes has stimulated the development of all-solid-state laser systems [4.29, 4.31–4.33]. Instead of the spectrally broad emitting flashlamp, laser diodes pump the laser crystal in a narrow spectral band selected in such a way to increase the efficiency of the pumping by minimizing energy dissipation. These DPSSL in Q-switch operation offer high pulse repetition rates of up to 1 kHz and more (cf. Chap. 15). Although the pulse energy is still limited to a few millijoule, high intensities in the focal spot can be achieved due to the excellent beam quality of DPSSL systems. The potential benefits of DPSSL compared with flashlamp-pumped lasers for LIBS are faster acquisition of spectral information, higher spatial resolution, improved precision on the basis of significantly enlarged data ensembles (mean values, pulse discrimination analysis), lower energy consumption, and increased life time. Laser diodes have typical lifetimes of 10,000 h in continuous wave mode and 10^9–10^{10} pulses in pulsed mode. In contrast to this typical flashlamp lifetimes amount to about 5×10^7 pulses. Typical diode-pumped laser materials and their important pumping wavelengths are listed in Table 4.2 [4.29, 4.31].

The main advantage of diode-pumped solid-state laser systems is the high beam quality in combination with the achievable pumping efficiency. This means that a small laser focus diameter is easier to achieve with a DPSSL than with a flashlamp-pumped system. To get a small focus with a flashlamp-pumped system requires the use of apertures within the oscillator for mode selection, which in turn limits the efficiency factor optical-to-optical down to 1–3%.

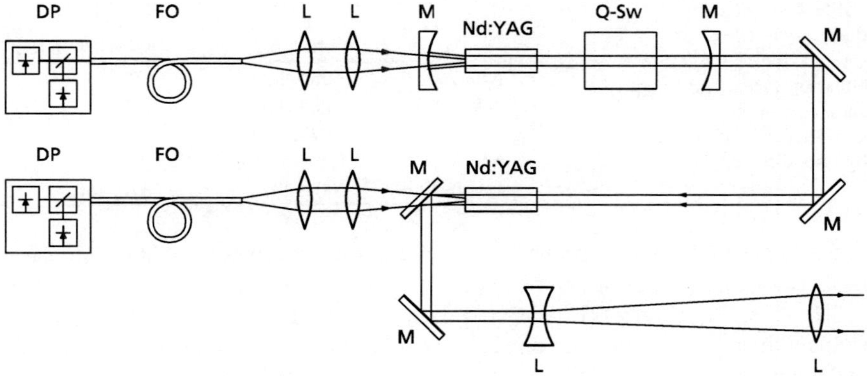

Fig. 4.2 Typical layout of a diode-pumped solid-state laser (DPSSL) in an end-pumped configuration. DP = diode pump module, FO = fiber optic, L = lens, M = mirror, Q-Sw = Pockels cell for Q-switch operation

Fig. 4.3 Photograph of the Nd:YAG DPSSL developed at ILT, data is given in Table 4.3

DPSSL systems can be pumped in a transversal configuration as it is the common arrangement for flashlamp-pumped systems, as well as in a longitudinal (called "end-pumped") configuration, see Fig. 4.2. High beam quality and efficiency in an end-pumped configuration are achieved by gain guiding. In this case, the pumped volume and laser mode have a high degree of overlap. Figure 4.3 shows a photograph of a DPSSL based on this principle developed at Fraunhofer ILT, Aachen. Table 4.3 lists the data of that laser.

Q-switched DPSSL have tunable repetition rates from below 1 Hz to several kilohertz, favoring them for the generation of high repetitive laser pulses in the

Table 4.3 Data of the Nd:YAG diode-pumped laser shown in Fig. 4.3

Wavelength	1,064 nm
Repetition frequency	1 kHz
Pulse energy	2 mJ
Pulse duration	5.5 ns
Beam quality M^2	≤1.3

millijoule range. Preflashing of the laser to achieve thermal stability of the system and by that to generate reproducible laser pulses is strongly reduced or even not necessary, thus eliminating dead times.

4.2 Spectrometer

Spectrometers disperse the emitted radiation of the laser-induced plasma to get a spectrum in terms of intensity as a function of the wavelength. The dominant spectrometer types used for LIBS are: (a) Czerny–Turner, (b) echelle, and (c) Paschen–Runge. For a detailed description of the various spectrometer types, the reader is referred to the literature [4.34, 4.35].

Figure 4.4 shows the setup of a Czerny–Turner spectrometer. The incoming light passing the entrance slit is collimated by a first concave mirror and directed onto a plane grating. The dispersed radiation is imaged by a second concave mirror to a detector plane.

The dispersion is described by the grating equation:

$$d_g(\sin\alpha + \sin\beta) = \mathrm{n}\lambda, \tag{4.1}$$

where d_g is the grating period, α is the angle of incidence, the sign of α is negative if the incident beam is on the opposite side of the grating normal with respect to the diffracted beam, β is the angle of diffraction, n is the diffraction order, and λ is the wavelength.

The linear dispersion in the detector plane at a central wavelength λ along an axis x is given by:

$$\frac{\mathrm{d}\lambda}{\mathrm{d}x} = \frac{d_g\cos\beta}{\mathrm{n}f}, \tag{4.2}$$

where x is the coordinate in the detector plane, parallel to the dispersion direction, and f is the focal length of the exit mirror.

For a spectrometer with a grating having 1,200 grooves/mm ($d = 0.83\,\mu\mathrm{m}$), a deviation angle ($\beta - \alpha$) of 27.5° wavelength, a central wavelength of 500 nm, first order (n = 1), and a focal length of the second mirror of 640 mm, we obtain: $\mathrm{d}\lambda/\mathrm{d}x = 1.2$ nm/mm.

The resolving power of a spectrometer is given by:

$$R = \frac{\lambda}{\Delta\lambda} = \mathrm{nN}, \tag{4.3}$$

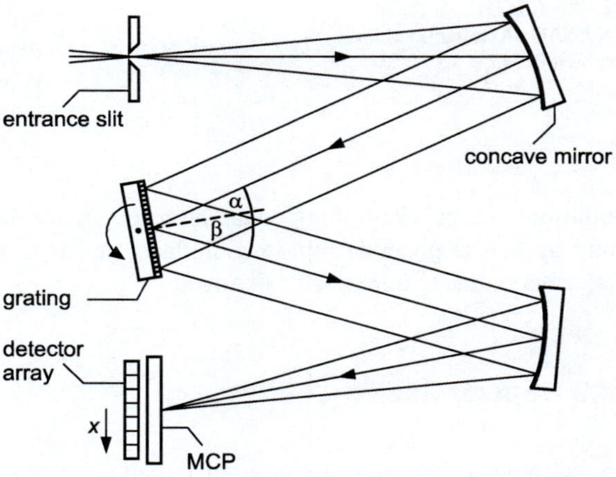

Fig. 4.4 Setup of a Czerny–Turner spectrometer. MCP = microchannel plate

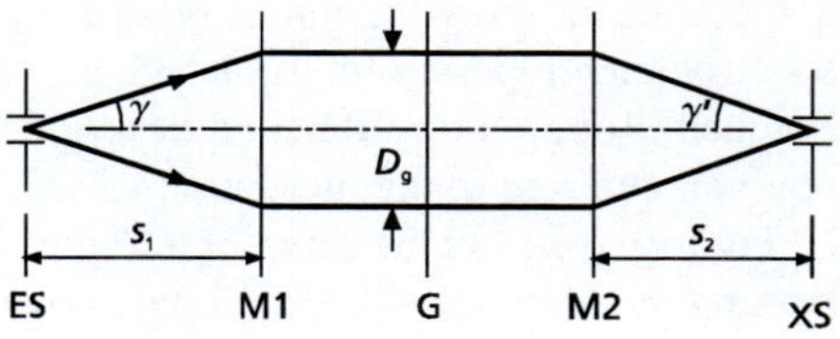

Fig. 4.5 Schematic optical setup of a Czerny–Turner spectrometer. ES = entrance slit, M1, M2 = mirrors, G = grating, XS = exit slit

where $\Delta\lambda$ is the spectral separation of two lines and N is the total number of illuminated grooves of the grating.

The quantity $\Delta\lambda$ describes the difference between two neighboring spectral lines of equal intensity. These two lines can be resolved, if the Rayleigh criterion is fulfilled, i.e., the maximum of one line falls on the first minimum of the other. Taking as an example the grating from above with 1,200 grooves/mm, an illuminated grating size of 100 mm, 500 nm, first order, we obtain from (4.3): $R = 1.2 \times 10^5$ and $\Delta\lambda = 4.2$ pm.

Figure 4.5 illustrates schematically the beam propagation in a Czerny–Turner spectrometer from the entrance slit to the exit slit. The optical setup is shown here in an equivalent linear arrangement. The numerical aperture of the spectrometer seen from the entrance slit is defined as follows:

$$\mathrm{NA_{ES}} = n_{\mathrm{ref}} \sin\gamma, \tag{4.4}$$

where n_{ref} is the refractive index and γ is the angle shown in Fig. 4.5.

An analogous expression holds for the numerical aperture seen from the exit slit. The f-number or f-value for the spectrometer entrance is given by:

$$f\text{-value} = \frac{s_1}{D_g}, \tag{4.5}$$

where s_1 is the distance between entrance slit and mirror M1, cf. Fig. 4.5, and D_g is the effective diameter of the grating G, cf. Fig. 4.5.

For a circular grating, the effective diameter seen by the entrance slit is $D_g = W_g \cos\alpha$, where W_g is the diameter of the grating. For a rectangular grating, an area equivalent diameter is calculated. For f-values greater than 2, the following approximation holds:

$$f\text{-value} \approx \frac{1}{2\mathrm{NA}_{\mathrm{ES}}}. \tag{4.6}$$

As an example, the NA and f-value are for a Czerny–Turner system with 640 mm focal length of the first concave mirror, a grating of 110 mm × 110 mm, and an angle of incidence of $\alpha = 0.14$ rad: $\mathrm{NA}_{\mathrm{ES}} = 0.095$, f-value = 5.3.

The Czerny–Turner configuration is the most widely used spectrometer type for LIBS. It offers a high resolution at compact dimensions. In most cases, a combination of a microchannel plate (MCP) and a photodiode array (PDA) is used to detect the spectrum, cf. Sect. 4.4. However, the spectral range, which can be detected simultaneously, is limited typically to 10–30 nm.

A much broader spectral range can be dispersed with an echelle spectrometer, whose principle setup is shown in Fig. 4.6 [4.36]. Behind the entrance slit, the incoming light is collimated by a first mirror and directed to a prism. The refracted light irradiates the echelle grating under a flat angle. The echelle grating has a low groove density of typically 80 grooves/mm. The diffracted light passes again the prism and is imaged by a camera mirror to the detector plane (typical dimensions 25 mm × 25 mm). The dispersion direction of the prism is perpendicular to the dispersion direction of the echelle grating, and by this the high diffraction orders are separated in the detector plane. A set of spectra are obtained for different orders (see wavelength and order axis in Fig. 4.6). The orders are in the range of n = 30–120. The spectral range detected simultaneously is 200–780 nm. The resolving power including the MCP–PDA is about 20,000 [4.37].

Due to the dense packaging of lines and orders in the detector plane, there is a danger of a cross talking between a strong spectral line and adjacent weak lines. Since the sensitivity of the MCP–PDA is nearly equal for the whole detector plane, there is no possibility for an individual adaption to these different line intensities. Also the gating of the MCP with the parameters t_{delay} and t_{int} (cf. Sect. 2.1) is the same for the whole detector plane.

The Paschen–Runge spectrometer shown in Fig. 4.7 offers a wide spectral range, a high spectral resolution, and the possibility for an individual adaption of the detector sensitivity and gating parameters for each spectral line. Entrance slit, grating, and a series of exit slits are mounted along the Rowland circle. The grating has a spherical curvature with a radius equal to the diameter of the Rowland circle. Behind the exit slits, photomultiplier tubes (PMT) are mounted detecting the light passing through the respective exit slit. Typical diameters of the Rowland circle are 500, 750, and 1,000 mm. The groove density of the applied gratings is, e.g., 1,080, 1,200, 2,400, and 3,600 grooves/mm. Depending on the spectral lines,

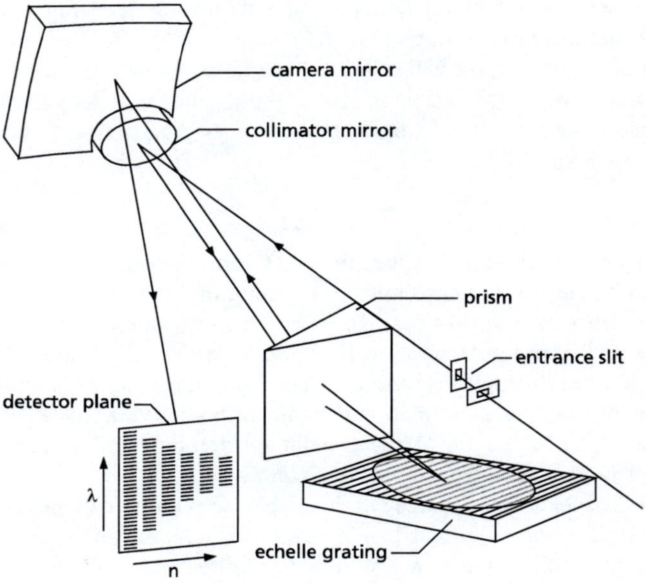

Fig. 4.6 Setup of an echelle spectrometer

Fig. 4.7 Setup of a Paschen–Runge spectrometer. PMT = photomultiplier tube

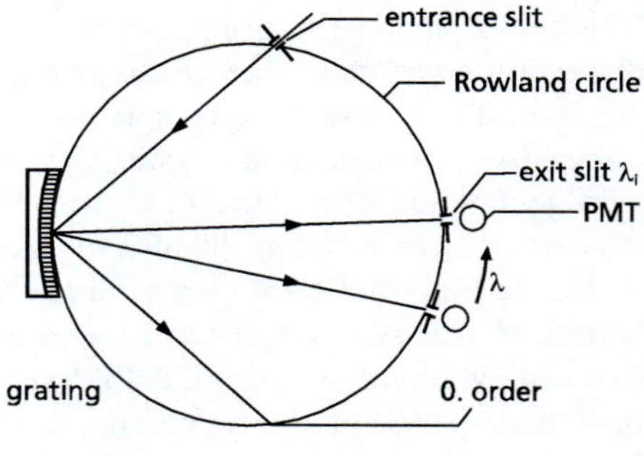

different PMTs are selected with optimized cathode sensitivity especially in the deep ultraviolet and near infrared region (cf. Sect. 4.4).

Table 4.4 gives an overview of typical data of the spectrometer types used for LIBS.

Figure 4.8 illustrates the wavelength–time domains for different LIBS applications. The vertical axis shows the time delay with respect to the laser pulse irradiation (at $t = 0\,\mu s$) and the duration of the integration of the spectral signals. Czerny–Turner–MCP systems generally cover a rather narrow spectral range of about 10–20 nm. The integration time is the same for the whole registered spectrum, and hence these domains are drawn as closed boxes in Fig. 4.8, cf. numbers 1 and 2. The echelle–MCP systems detect a broader wavelength range, where the integration

Table 4.4 Typical data of the spectrometer types Czerny–Turner, echelle, and Paschen–Runge

Spectrometer type	Czerny–Turner	Echelle	Paschen–Runge
Focal length (mm)	640	250	500
Grooves (mm^{-1})	1,200	75	2,400
Diffraction order	1–2	30–120	1–3
$f-$ value	5	10	12
Width of spectral range detected simultaneously (nm)	30	580[a]	460[b]
Spectral resolution per pixel, exit slit width (pm)	<20	5–19	<20
Detectors	MCP–PDA	MCP–CCD	PMT
Size (mm × mm × mm)	740 × 350 × 350	320 × 140 × 160	940 × 790 × 345

MCP = microchannel plate, PDA = photodiode array, CCD = charge-coupled device, PMT = photomultiplier tube
[a]From 200 to 780 nm
[b]From 130 to 590 nm

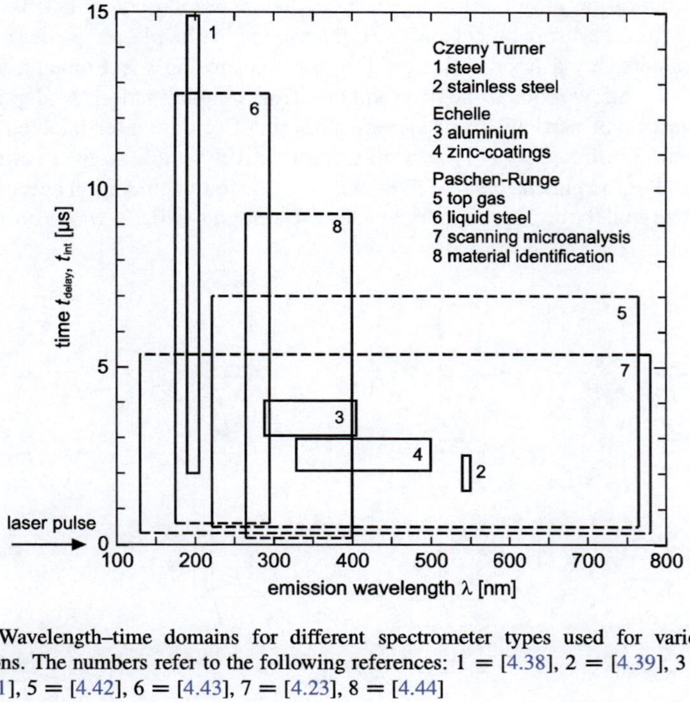

Fig. 4.8 Wavelength–time domains for different spectrometer types used for various LIBS applications. The numbers refer to the following references: 1 = [4.38], 2 = [4.39], 3 = [4.40], 4 = [4.41], 5 = [4.42], 6 = [4.43], 7 = [4.23], 8 = [4.44]

time is also fixed for all wavelengths, see closed boxes 3 and 4 in Fig. 4.8. By contrast Paschen–Runge systems comprise a broad spectral range and the delay time and the integration time can be adjusted selectively for the different element emission lines, see boxes with dashed horizontal lines for the numbers 5–8 in

Fig. 4.8. Furthermore, the sensitivity of each detector channel can be adapted according to the intensity range of the respective emission line. In this way, the dynamic range can be expanded by many orders of magnitude.

For some application, the observation of selected small spectral regions is sufficient to identify a material [4.45]. In these cases, the plasma emission is filtered spectrally by using band-pass filters which are centered at the interesting wavelength and have a width of typically 10 nm. The radiation passing the filter is detected, e.g., with a photomultiplier.

4.3 Beam Guiding Optics

The beam guiding optics comprises all optical components used to guide the laser radiation to the measuring object and to collect and guide the measuring radiation to the spectrometer. Figure 4.9 gives an overview of the different variants of optical setups applied for LIBS.

In configuration (1) of Fig. 4.9, the laser beam is focused by a lens onto the surface of the measuring object and the light emitted by the plasma passes a window and is collected by a fiber optics guiding the light to the spectrometer, see, e.g., [4.46]. The window protects the fiber end face from contaminations by deposition or recondensation of particulates or vapor originating from the laser-induced material ablation and plasma. Setup (1) is simple and relatively stable against variations of the position of the plasma and the fiber end face due to the numerical aperture of the fiber. For a typical fiber, the numerical aperture amounts to 0.22, which corresponds

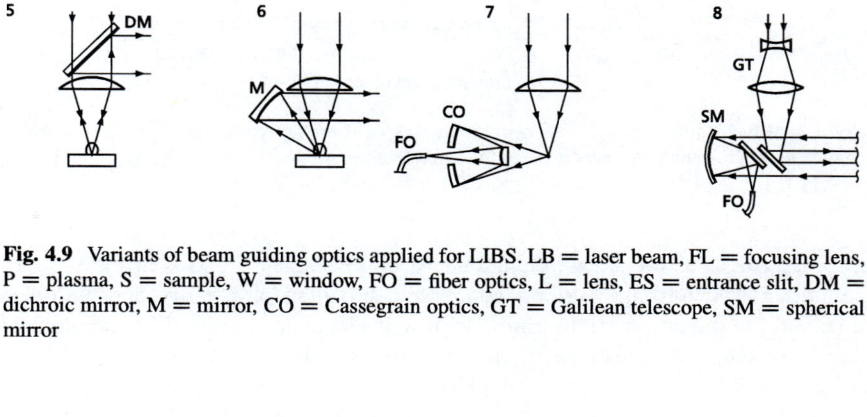

Fig. 4.9 Variants of beam guiding optics applied for LIBS. LB = laser beam, FL = focusing lens, P = plasma, S = sample, W = window, FO = fiber optics, L = lens, ES = entrance slit, DM = dichroic mirror, M = mirror, CO = Cassegrain optics, GT = Galilean telescope, SM = spherical mirror

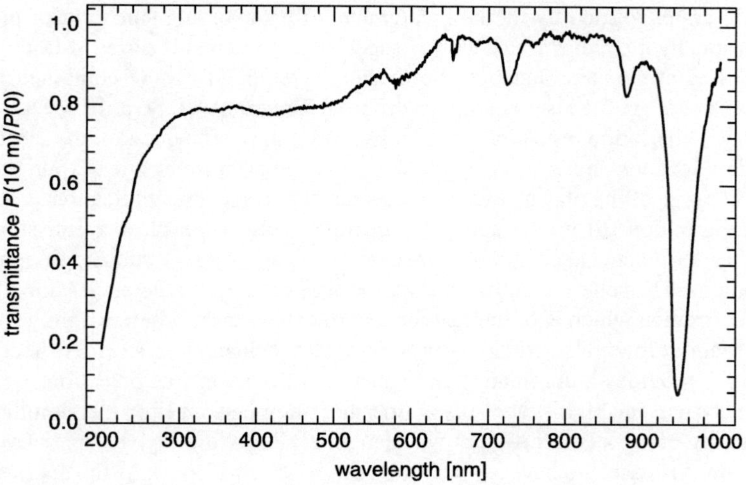

Fig. 4.10 Transmittance of an all-silica fiber optic (core and cladding are made of fused silica) with a core diameter of 600 μm and a length of 10 m as a function of the wavelength

to a full angle of the acceptance cone of 25°. However, the solid angle collected from the plasma light emission is very small, so the sensitivity achievable is reduced. In setup (2), a lens is used to collect the light emitted by the plasma and to couple the light into the fiber optics. In this way, the detected solid angle can be increased by orders of magnitude. However, the lens has a chromatic aberration, i.e., the focusing features are different for the various wavelengths emitted by the plasma. Hence, the imaging conditions have to be selected carefully in order not to cut-off parts of the emitting plasma volume. For this setup plasma position, lens, and fiber end face have to be aligned precisely in order to achieve a high coupling efficiency.

Using fiber optics in the configurations (1) and (2) allows to position the spectrometer up to several meters away from the measuring object. Especially for LIBS applications under harsh environments such as the inline analysis of measuring objects at high temperatures, this is a significant advantage. However, the fiber optics limits the transmittable spectral range for wavelengths below 200 nm, cf. Fig. 4.10. At configuration (3), the plasma emission is directly coupled into the entrance slit of the spectrometer. Again the solid angle received is small and on the other hand UV lines can be observed, if the surrounding atmosphere is an inert gas having no absorption bands in the UV, e.g., argon. With this setup emission lines down to 120 nm can be detected, which is of special interest for the analysis of light elements in steel, see Sect. 13.1. Configuration (4) is similar to (2). If a magnesium fluoride (MgF_2) lens is used, wavelengths in the deep UV can also be detected.

In setup (5), the measuring radiation propagating antiparallel to the incoming laser radiation is collected by the focusing lens and reflected by a dichroic mirror guiding the radiation to the spectrometer. For this configuration holds $\alpha_L = -\alpha_O$, cf. Fig. 3.23. Observing the plasma emission along the same axis allows to look into the crater formed by the laser interaction thus avoiding shielding of parts of the plasma by the crater walls. This is especially of interest if series of laser pulses are applied at the same location. Furthermore, setup (5) can be combined with a scanner to deflect the laser beam into different directions, cf. Sect. 18.4. Due to the chromatic aberration of the focusing lens, the coupling efficiency for the measuring radiation depends on the wavelength. Hence, a compromise has to be found for the spectral range of the plasma radiation of interest in a specific application.

Configuration (6) uses a spherical mirror to collect the plasma emission thus avoiding chromatic aberrations. However, such a mirror is subject to spherical aberrations. The spherical mirror allows for high collection efficiencies down to the deep UV region which is of interest for microanalysis tasks where a tightly focused laser beam at low pulse energy is used to detect inclusions in steel, cf. Sect. 18.3. Setup (7) provides a chromatic- and spherical-aberration-free detection, which is of importance for high-spatial-resolution measurements and for the simultaneous monitoring of signals in different wavelength regions from a given spatial volume [4.47]. In this case, we have $\alpha_L = 0°$ and $\alpha_O = 90°$, which limits the applicability to the analysis of gaseous media.

Finally, configuration (8) is used for remote LIBS analyzing systems, where the laser beam is focused at distances of several meters by a Galilean telescope and the plasma radiation is collected by a Newtonian telescope consisting of a spherical primary mirror and a flat secondary mirror guiding the collected radiation to a fiber optics [4.48].

A key feature of the refractive optical components used for the various configurations shown in Fig. 4.9 to collect and guide the measuring radiation to the spectrometer is the transmittance as a function of the wavelength. Figure 4.10 shows the transmittance curve for a silica fiber optics with a length of 10 m. For wavelength below 300 nm, the transmittance decreases significantly attaining values of about 0.2 at 200 nm. In the visible and near infrared regions, high transmittance values greater than 0.8 exist. For wavelengths beyond 900 nm, absorption bands of OH are effective.

For wavelength below 190 nm, fluorides are the only materials offering significant transmittance. Figure 4.11 shows the transmittance of CaF_2 and MgF_2 windows for wavelength down to 120 nm. At short wavelength, the cleanliness of the surfaces plays an important role. A comparative study of the transmittance of CaF_2 windows with thicknesses of 2 and 20 mm shows that for wavelength greater than 140 nm the volume absorption is negligible in relation to the surface absorption.

Fiber optics can also be used to transmit the laser radiation from the laser source to the focusing lens. However, the transmittable laser pulse energy is limited by laser-induced damage of the fiber core material and stimulated backscattering processes. Table 4.5 summarizes threshold data for laser-induced damage of bulk fused silica, which present an upper limit for the fiber damage thresholds. The energy thresholds for laser-induced damage $\varepsilon_{d,th}$ depend on the pulse width as $\varepsilon_{d,th} \propto t_p^{0.4,\ldots,0.5}$ for bulk samples [4.53, 4.55]. The threshold values decrease with increasing focal spot size which is explained by the increasing probability to irradiate surface defects having lower damage thresholds.

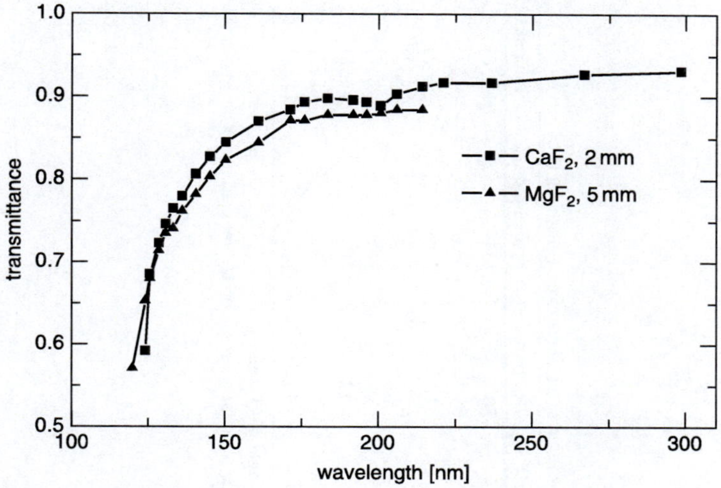

Fig. 4.11 Transmittance of CaF_2 (thickness 2 mm) and MgF_2 (thickness 5 mm) in the ultraviolet spectral range [4.49–4.51]

Table 4.5 Damage threshold radiant exposure $\varepsilon_{\mathrm{d,th}}$ and corresponding irradiance $I_{\mathrm{d,th}}$ for bulk fused silica samples at 1.06 μm wavelength

T_p(ns)	$2r$ (mm)	$\varepsilon_{\mathrm{d,th}}$ (J/cm^2)	$I_{\mathrm{d,th}}$ (GW/cm^2)	References
18	0.0072	3,960	220	[4.52]
27	0.0090	3,240	120	[4.52]
15	≈1	60	4	[4.53]
30	≈1 − 2	210	7	[4.54]
80	≈1	130	1.5	[4.53]

t_p = pulse width (FWHM), $2r$ = focal spot diameter

Table 4.6 shows the maximum fiber output energy per pulse for single, double, and multiple pulses. In the rows 1–4, maximum transmitted energies are reported for single Q-switch pulses when stimulated backscattering is not observed due to the short fiber lengths, short pulse widths, and/or spectral width and spatial profile of the laser. Measured values for single and double pulses at a fiber length of 5 m are shown in rows 5 and 6. The damage threshold increases for a multimode spatial profile of the laser and for a higher input beam divergence. The references [4.56, 4.57] only achieved the higher transmitted energy levels with a highly multimode spatial profile of a Nd:glass laser and high input divergence $2\theta_{\mathrm{in}}$ resulting in a high output divergence of the transmitted beam. If a near-Gaussian spatial profile is used, then much lower thresholds are achieved [4.57], see row 4. These radiant exposure thresholds $\varepsilon_{\max}$ are about a factor of 1.7 higher than the values measured in [4.58] for a Gaussian beam, see row 7, but the input divergence was five times smaller. From this, no essential decrease of the fiber damage radiant exposure thresholds

Table 4.6 Maximum fiber output energy per pulse E_{max}, radiant exposure ε_{max}, and irradiance I_{max} for fused silica fibers of core diameter d

No.	t_p(ns)	d (mm)	L (m)	N	E_{max} (mJ)	ε_{max} (J/cm^2)	I_{max} (GW/cm^2)	$2\theta_{in}$ (mrad)	Spatial profile	References
1	16	0.4	1	1	100	80	5.0	50	MM	[4.56]
2	25	0.4	≈1	1	97	77	3.1	64	MM	[4.57]
3	25	1	≈1	1	>225[a]	>29	>1.1	64	MM	[4.57]
4	16	1	≈1	1	70	8.9	0.6	64	NG	[4.57]
5	15	1	5	1	>38[a]	>4.8	>0.32	<12	NG	[4.58]
6	25	1	5	2	>60[a]	>7.6	>0.31	<12	NG	[4.58]
7	50–80	1	0.5	6	40	5.1	0.1	12	G	[4.58]
8	50–80	1	0.5	6	>50	>6.4	>0.13	<12	NG	[4.58]

t_p = pulse width (FWHM), L = fiber length, N = number of pulses per burst, $2\theta_{in}$ = fiber input full-angle divergence

Spatial profile of laser output: MM = multimode, NG = near-Gaussian, G = Gaussian. Interpulse separations used for the measurements in row 6 are $\Delta t = 6\,\mu s$ and for rows 7, 8: 15 μs

[a]Limited by air breakdown ahead of fiber input face

Table 4.7 SBS threshold energies $E_{p,th}$ (in mJ) dependent on the spectral width $\Delta\tilde{\nu}$ of the laser radiation[a]

Type	$\Delta\tilde{\nu} < 0.002\ cm^{-1}$	$\Delta\tilde{\nu} = 0.1\ cm^{-1}$	$\Delta\tilde{\nu} = 0.65\ cm^{-1}$
S, WF	5.8	45	>90 (SP)
S, UV	3.4	26	>90 (SP), 300 (FPB)
G	0.75	8	>30 (SP)

S, WF = step-index fiber, water-free with enhanced near infrared transmission, UV = fiber glass with enhanced short wavelength transmission, G = gradient-index fiber, SP = single pulse, FPB = five-pulse burst with 10-μs interpulse separations

[a]For the broad spectral width, the fiber damage thresholds are given as a lower limit estimation of the actually higher SBS thresholds. Fiber core diameter 0.6 mm, fiber length 5 m, input NA ≈ 0.07

for the individual pulses is obvious, due to the microsecond-spaced multiple pulse irradiation, i.e., the transmissible energy of a N-pulse burst increases to about N times the maximum energy of a single pulse.

For laser radiation with narrow spectral width stimulated Brillouin scattering (SBS) limits the transmission of Q-switched laser pulses in large-core optical fibers well below the threshold energy of fiber damage [4.59]. Table 4.7 summarizes the SBS thresholds measured for different spectral widths of the laser radiation. The threshold increases by a factor of ≈7.5 for the step-index fibers and ≈ 10 for the gradient-index fiber, respectively, when the spectral width of the laser is changed from <0.002 to 0.1 cm^{-1} (i.e., by at least a factor of 50). For a laser with a spectral width of 0.65 cm^{-1}, the SBS thresholds exceed the damage thresholds of the fiber input surface for all samples when single laser pulses are used.

Table 4.7 shows that energies of >90mJ for single pulses and 300 mJ for five-pulse bursts are transmitted through 0.6-mm-diameter fibers of 5 m length, when a broadband laser with a spectral width of 0.65 cm^{-1} is used.

Fiber optic laser beam delivery was applied for remote material analysis by LIBS using all-silica fibers with a core diameter of 550 μm and a length of 75 m [4.60]. Via this fiber Q-switched Nd:YAG laser pulses at 1,064 nm with 20-Hz pulse repetition frequency and a pulse width of 6 ns were transmitted with pulse energies of 5.5 mJ. At pulse energies greater than 9 mJ catastrophic damages occurred either at the input face of the fiber, or more often, approximately 75 mm into the fiber.

4.4 Detectors

Various detector types are applied to register the spectrally dispersed radiation. Table 4.8 lists the most import detectors and their typical data used in spectrometers for LIBS measurements.

The PMT is a highly sensitive detector offering a large current amplification, dynamic range, and measuring frequency. These detectors are used mainly in combination with Paschen–Runge spectrometers, cf. Sect. 4.2, Fig. 4.7. Depending on the respective LIBS application up to 45 and more PMTs are installed along the Rowland circle of a Paschen–Runge spectrometer. PMTs allow for a time-resolved

Table 4.8 Detector types applied in spectrometers used for LIBS and their typical data and features

Detector	PMT	CCD	MCP–CCD
Number of sensitive elements	1	1,024	1,024 × 1,024(CCD)
Size of detector element	4 mm × 13 mm	14 μm × 14 μm, length of sensing area 14.4 mm	Ø25 mm(MCP), 25.4 mm × 25.4 mm(CCD array)
Spatial resolution		14 μm	36 lp/mm
Window materials	U, Q, MgF_2	Glass, UV-enhanced fused silica	Synthetic silica
Quantum efficiency (%)	<65	<90	<17
Sensitivity	50–60 mA/W at peak wavelength	$40\,V/\mu J/cm^2$ at 740 nm	45 mA/W at 400 nm
Gain	Current amplification: 3×10^5		Radiant emittance gain: 9×10^3
Rise time	1.5 ns	Integrating	Integrating
Exposure control	Possible	>1 ms, 10 μs	10 ns–10 μs
Dynamic range	$>10^4$	2,500:1	Limited by CCD
Measuring frequency	>1 kHz	<500 Hz	<20 Hz
Size	Ø13.5 mm×40 mm	38.1 mm × 10.03 mm × 7.62 mm	Ø53 mm × 19 mm(MCP)
Spectrometer	Paschen–Runge	Czerny–Turner, Paschen–Runge	Czerny–Turner, echelle

PMT = photomultiplier tube, CCD = charge-coupled device, MCP = microchannel plate, U = UV glass, Q = synthetic silica

detection of the line emission; typical signals are shown in Fig. 3.13. The PMT output signals are processed by a subsequent electronics where parts of the signal are integrated as illustrated in Fig. 2.4. The high measuring frequency is deployed for high-speed LIBS applications, see Chap. 15. The spectral PMT sensitivity depends on the type of the cathode and window material. Figure 4.12 shows photocathode sensitivity and quantum efficiencies as a function of the wavelength for typical cathode and window materials. Spectral signals down to the deep vacuum ultraviolet region with wavelength of, e.g., 130.22 nm for oxygen can be detected. For the major part of industrial LIBS applications where the objects to be analyzed are metals, PMTs are the detector type used preferentially.

CCD detectors consist of an array of single photosensitive elements thus they are capable to detect a spectral range of, e.g., several tens of nanometers depending on the dispersion of the spectrometer, see, e.g., Fig. 4.8. Their dynamic range is much smaller than the one of PMTs. For best signal-to-noise ratios, CCD detectors are cooled by, e.g., thermoelectric peltier elements. The CCD is an intrinsically

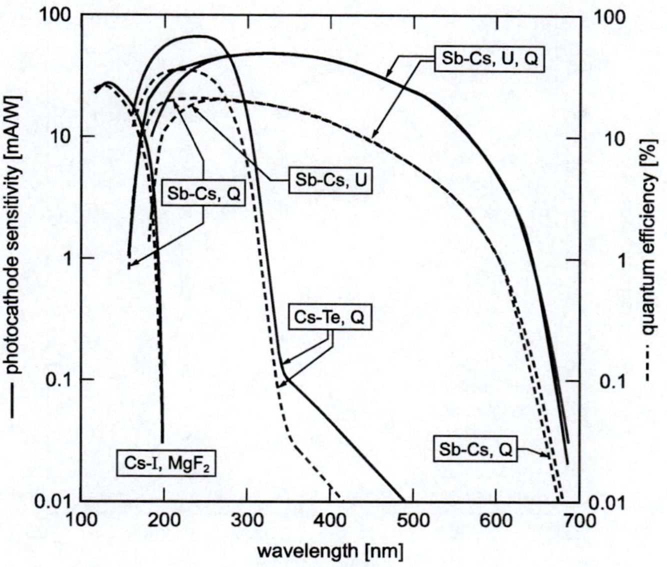

Fig. 4.12 Photocathode sensitivity and quantum efficiency of various photocathode and window materials. U = UV glass, Q = synthetic silica [4.61]

integrating detector type with integration times of typically 1 ms and more. Some detector types allow for an exposure control with integration times down to about 10 μs.

Figure 4.13 shows the quantum efficiency as a function of the wavelength depending on the light input direction. Front-illuminated CCDs ("front-sided" in Fig. 4.13) have smaller quantum efficiencies than back-illuminated CCDs ("back-thinned" in Fig. 4.13). Polysilicon and insulating layers usually deposited on the photosensitive surface of a front-illuminated CCD limit the sensitivity for short wavelengths because of the high absorption of the radiation, which leads to short penetration depth (<2 nm), and absorption of the radiation in the gate material rather than within the channel of the CCD [4.62, 4.63].

To detect UV light with front-illuminated CCDs, their photosensitive surface is coated with a phosphor. The phosphor coating is prepared by sedimentation of phosphor grains on the CCD with grain sizes in the range of 1–3 μm. A typical phosphor used consists of Gd_2O_2S:Tb having the maximum light emission at 545 nm with a decay time of 3.5 ms [4.64].

The MCP–CCD listed in the last column of Table 4.8 is a widely used detector for LIBS studies. This combination of a microchannel plate and a CCD is also called intensified CCD (= ICCD, MCP acts as an intensifier). The principle setup is shown in Fig. 4.14. The irradiated photons generate free electrons at the photocathode, which are accelerated toward the microchannel plate due to the applied electric field. The MCP is a thin disc consisting of millions of small channels where each channel has a diameter of 6 μm. The channels serve as independent electron multipliers. The

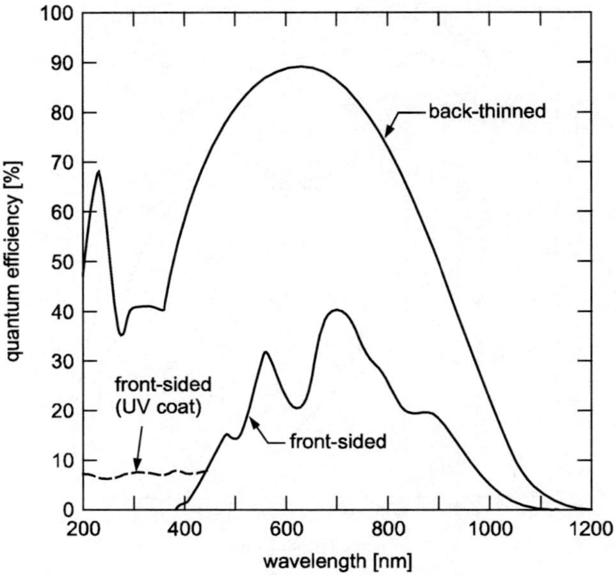

Fig. 4.13 Quantum efficiency of CCD detectors without window. The spectral sensitivity of front-illuminated CCDs can be extended to the UV region by coating of the photosensitive surface (UV coat)

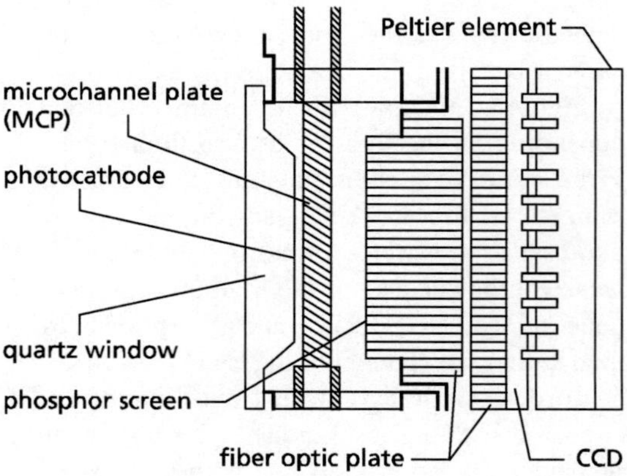

Fig. 4.14 Schematic setup of a MCP-CCD. The gap between the fiber optic plates is shown here to illustrate the different components. This gap is zero in practice

input electrons impinging on the channel wall produce secondary electrons. This process is repeated hundreds of times by the potential gradient across both ends of the MCP and a large number of electrons are in this way released from the output end. They are further accelerated onto the phosphor screen that reconverts electrons into light. Finally, a fiber optic plate guides the light to the CCD.

Figure 4.15 shows an example of quantum efficiencies for image intensifiers (MCP).

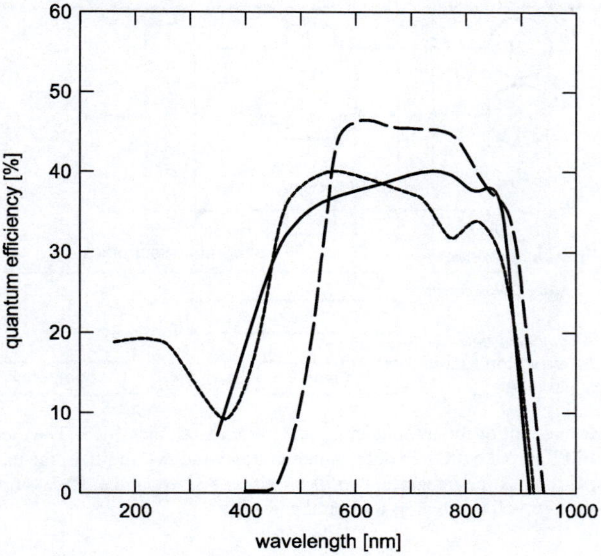

Fig. 4.15 Quantum efficiency of various image intensifiers [4.65]

4.5 Signal Electronics

For the detection and processing of the photomultiplier signals, a multichannel integrator electronics MCI was developed [4.66]. Figure 4.16 shows the block diagram, and the data is given in Table 4.9. The electronics is capable to process up to 64 PMT signals in a pulse-by-pulse way. The anode currents of the PMT are transmitted via coaxial cables to a gated multichannel integrator electronics. Each channel is equipped with two different integration capacitors allowing to adapt the sensitivity to different signal levels of the PMT output.

The dynamic range with respect to the ratio of the integration capacities (cf. Table 4.9) and the digitalization resolution amounts to: $(C_B/C_A) \times 2^{15} = 1.5 \times 10^6$. The experimentally achieved dynamic range covers a range of integrated currents, i.e., electric charges between 20 fC and 10 nC, corresponding to a ratio of 5×10^5. Taking into account the possibility to adapt the integration time t_{int} and the voltage supply of the PMT, dynamic ranges being greater than $>10^7$ are achievable.

The integrator electronics MCI is used to process PMT signals in a broad range of industrial applications (cf. Chap. 18).

The development of the next generation of MCI, called MCI-2, was completed in 2011 achieving maximum measuring frequencies of 3 kHz and beyond.

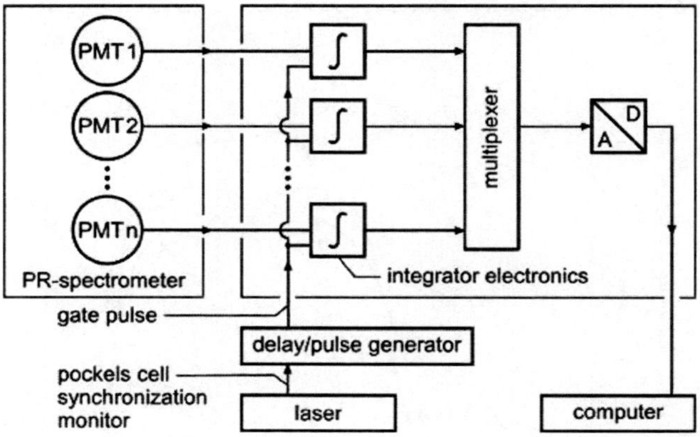

Fig. 4.16 Block diagram of the multichannel integrator electronics MCI. The electronics is able to integrate the PMT output current in defined integration windows, digitizes the integrated electric charges, and synchronizes the integration window with the laser source. PMT = photomultiplier, PR = Paschen–Runge, A/D = analog-to-digital converter

Table 4.9 Data of the integrator electronics developed at Fraunhofer ILT

Number of channels	up to 64
Integration capacitors	22 pF(C_A), 1 nF(C_B)
Delay time for the integration window, t_{delay}	240 ns–300 μs
Width of the integration window, t_{int}	50 ns–200 μs
Digitalization resolution	15 bit
Maximum measuring frequency	1,000 Hz
Charge detection limit	20×10^{-15} C
Sensitivity, delay time, and width of integration	Programmable per pair of channels

4.6 Sample Stand and Measuring Chamber

LIBS is a spectroscopic method which can be applied under ambient atmospheric conditions, which is especially of interest for inline applications such as high-speed identification of scrap pieces for material specific sorting, cf. Sect. 18.4. If the analytical requirements with respect to sensitivity and precision are more stringent, then well-defined ambient atmospheric conditions are necessary (cf. Sect. 3.5). In those cases, a sample stand or a measuring chamber is used to establish a controllable gas atmosphere in the interaction region. In the case of a measuring chamber, the sample to be analyzed is brought into a gas tight vessel, see, e.g., Fig. 2.6. A sample stand is a measuring chamber with an opening in a sample table.

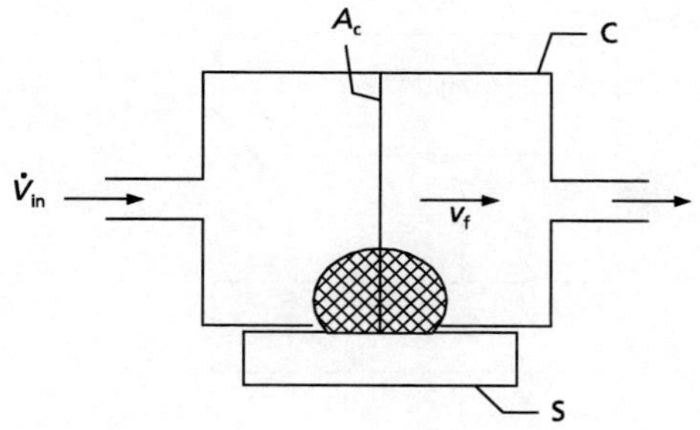

Fig. 4.17 Schematic drawing of a sample stand and measuring chamber with respect to the parameters defining the gas exchange. C = measuring chamber, S = sample, for the other quantities see text

A solid sample having a plane surface is put on the sample table in such a way that the opening is closed, see Fig. 2.7. The sample is then irradiated by the laser beam within the area of the opening in the sample table.

A sample stand or a measuring chamber allows to control the following parameters: gas type, gas flow rate, transport of the ablated material out of the interaction region, transmission characteristics between the laser-induced plasma and the spectrometer entrance window. Figure 4.17 illustrates schematically a measuring chamber with respect to the parameters defining the gas exchange. The time for a complete exchange of the gas volume inside the chamber – assuming that the inlet gas flow rate is equal to the outlet gas flow rate – is simply given by:

$$T_{ge} = \frac{V_C}{\dot{V}_{in}}, \tag{4.7}$$

where T_{ge} is the gas exchange time of the measuring chamber, V_C is the volume of measuring chamber, and $\dot{V}_{in}$ is the inlet gas flow rate.

For typical values ($V_C = 140$ mm^3, $\dot{V}_{in} = 20$ l/ min), we obtain: $T_{ge} = 420\,\mu$s. This is the minimum necessary time required to completely exchange the gas volume inside the measuring chamber. This time is greater than the local exchange time estimated with relation (3.1) in Chap. 3. With an effective cross section of the chamber of A_C, the average gas flow velocity amounts to $v_f = \dot{V}_{in}/A_C$. With $A_C = 30$ mm^2, this velocity is 11 m/s. A design goal of the sample stands described in the following was to minimize V_C and A_C for short gas exchange times and high gas flow velocities in the interaction region.

Figure 4.18 shows a sectional view of a sample stand integrated in a LIBS setup for the analysis of light elements in steel samples requiring the observation of emission lines in the vacuum ultraviolet spectral range (<200 nm). The plasma is observed in side-on direction, and the detection angle α_D amounts to 10°. The sample stand and the tube between the sample stand and the entrance window of the spectrometer are purged with argon gas of high purity. Typical gas flow rates are in the range of 10–20 l/min.

Due to the small detection angle radiation originating from the plasma region being in contact with the surface of the sample is partly shielded by the crater formed

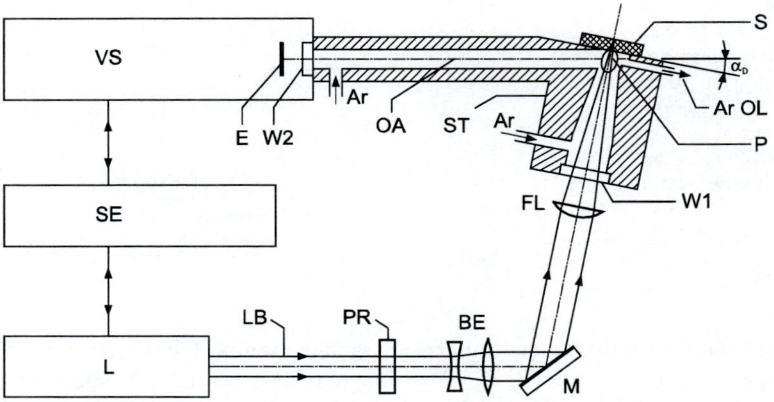

Fig. 4.18 Sectional view of a sample stand with side-on observation of the emitted radiation integrated in a LIBS setup. VS = vacuum spectrometer, SE = signal electronics, L = laser, LB = laser beam, PR = Glan-laser prism, BE = beam expansion, M = mirror, FL = focusing lens, ST = argon-flushed sample stand, Ar = argon inlet, W1 = window of ST, P = laser-induced plasma, Ar OL = argon outlet, α_D = detection angle, OA = optical axis of VS, W2 = entrance window of VS, E = entrance slit of VS

Fig. 4.19 Sample stand of Fig. 4.18 mounted at a Paschen–Runge spectrometer

thus reducing the portion of continuum radiation reaching the spectrometer. With such a configuration, detection limits for carbon, phosphorus, and sulfur in low-alloyed steel samples below $10\,\mu g/g$ were achieved, cf. Sect. 13.1 [4.67].

Figure 4.19 shows the sample stand of Fig. 4.18 mounted at a Paschen–Runge spectrometer. A relative motion between the laser beam and the sample can be achieved by a displacement of the sample table (upper part of the sample stand in contact with the sample) in the plane perpendicular to the laser beam. By this a

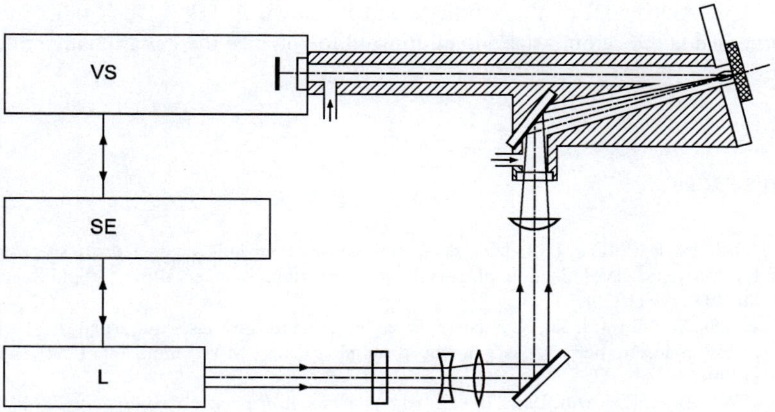

Fig. 4.20 Sectional view of a sample stand with end-on observation of the emitted radiation integrated in a LIBS setup. Other components as in Fig. 4.18

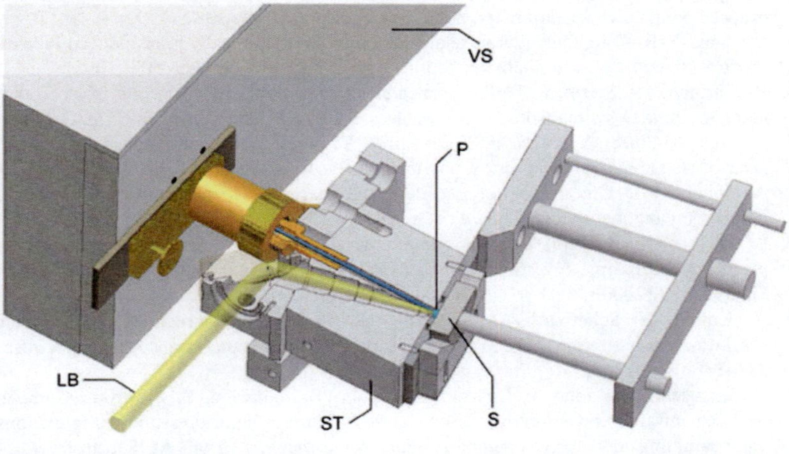

Fig. 4.21 Sample stand of Fig. 4.20 mounted at a Paschen–Runge spectrometer. For abbreviations see Fig. 4.18

spatial averaging is realized, which is of interest for the analysis of inhomogeneous samples, e.g., slag samples, see Sects. 3.6 and 14.2.

Figure 4.20 shows a sectional view of a sample stand with end-on observation of the laser-induced plasma. Such a configuration is advantageous if samples have to be analyzed having an oxide layer at the surface which is not representative for the bulk composition, cf. Sect. 13.1.2. In this case, deeper craters have to be formed and hence a greater part of the plasma lies inside the crater volume. In order to detect a sufficient part of the line emission, an end-on observation is chosen. In this case, the detection angle (cf. Fig. 4.18) amounts to $\alpha_D = 78°$.

A 3D-CAD model of the sample stand is shown in Fig. 4.21 [4.68]. The gas volume inside the sample stand is minimized to enhance the gas exchange rate, cf. Sect. 3.1.

References

4.1. J. Millard, R. Dalling, L. Radziemski, Time-resolved laser-induced breakdown spectrometry for the rapid determination of beryllium in beryllium-copper alloys. Appl. Spectrosc. **40**, 491–494 (1986)
4.2. F. Leis, W. Sdorra, J. Ko, K. Niemax, Basic investigations for laser microanalysis: I. optical emission spectrometry of laser-produced sample plumes. Mikrochim. Acta II **98**, 185–199 (1989)
4.3. I. Whiteside, R. Jowitt, Laser liquid metal analysis, in *Progress of Analytical Chemistry in the Iron and Steel Industry*, ed. by R. Nauche (Luxembourg Office for Official Publications of the European Communities, Luxembourg, 1992), pp. 135–141, ISBN 92-826-3877-4, EUR 14113
4.4. J. Aguilera, C. Aragon, J. Campos, Determination of carbon content in steel using laser-induced breakdown spectrometry. Appl. Spectrosc. **46**, 1382–1387 (1992)
4.5. M. Sabsadi, P. Cielo, Quantitative analysis of aluminum alloys by laser-induced breakdown spectroscopy and plasma characterization. Appl. Spectrosc. **49**, 499–507 (1995)
4.6. R. Sattmann, V. Sturm, R. Noll, Laser-induced breakdown spectroscopy of steel samples using multiple Q-switch Nd:YAG laser pulses. J. Phys. D Appl. Phys. **28**, 2181–2187 (1995)
4.7. G. Arca, A. Ciucci, V. Palleschi, S. Rastelli, E. Tognoni, Trace element analysis in water by laser-induced breakdown spectroscopy techniques. Appl. Spectrosc. **51**, 1102–1105 (1997)
4.8. A. Pichahchy, D. Cremers, M. Ferris, Elemental analysis of metals under water using laser-induced breakdown spectroscopy. Spectrochim. Acta B **52**, 25–39 (1997)
4.9. L. St-Onge, M. Sabsabi, Towards quantitative depth-profile analysis using laser-induced plasma spectroscopy: investigation of galvanealed coatings on steel. Spectrochim. Acta B **55**, 299–308 (2000)
4.10. A. Knight, N. Scherbarth, D. Cremers, M. Ferris, Characterization of laser-induced breakdown spectroscopy (LIBS) for application to space exploration. Appl. Spectrosc. **54**, 331–340 (2000)
4.11. C. Geertsen, A. Briand, F. Chartier, J. Lacour, P. Mauchien, S. Sjöström, Comparison between infrared and ultraviolet laser ablation at atmospheric pressure – implications for solid sampling inductively coupled plasma spectrometry. J. Anal. At. Spectrom. **9**, 17–22 (1994)
4.12. W. Ho, C. Ng, N. Cheung, Spectrochemical analysis of liquids using laser-induced plasma emissions: effects of laser wavelength. Appl. Spectrosc. **51**, 87–91 (1997)
4.13. H. Zhang, F. Yueh, J. Singh, Laser-induced breakdown spectrometry as a multimetal continuous-emission monitor. Appl. Optics **38**, 1459–1466 (1999)
4.14. D. Romero, J. Romero, J. Laserna, Distribution of metal impurities in silicon wafers using imaging-mode multi-elemental laser-induced breakdown spectrometry. J. Anal. At. Spectrom. **14**, 199–204 (1999)
4.15. R. Neuhauser, U. Panne, R. Niessner, Utilization of fiber optics for remote sensing by laser-induced plasma spectroscopy (LIPS). Appl. Spectrosc. **54**, 923–927 (2000)
4.16. P. Fichet, P. Mauchien, J. Wagner, C. Moulin, Quantitative elemental determination in water and oil by laser-induced breakdown spectroscopy. Anal. Chim. Acta **429**, 269–278 (2001)
4.17. L. Berman, P. Wolf, Laser-induced breakdown spectroscopy of liquids: aqueous solutions of nickel and chlorinated hydrocarbons. Appl. Spectrosc. **52**, 438–443 (1998)

4.18. J. Simeonsson, A. Miziolek, Spectroscopic studies of laser-produced plasmas formed in CO and CO_2 using 193, 266, 355, 532 and 1064 nm laser radiation. Appl. Phys. B **59**, 1–9 (1994)
4.19. C. Geertsen, J. Lacour, P. Mauchien, L. Pierrard, Evaluation of laser ablation optical emission spectrometry for microanalysis in aluminium samples. Spectrochim. Acta B **51**, 1403–1416 (1996)
4.20. M. Guillong, I. Horn, D. Günther, Capabilities of a homogenized 266 nm Nd:YAG laser ablation system for LA-ICP-MS. J. Anal. At. Spectrom. **17**, 8–14 (2002)
4.21. L. St-Onge, V. Detalle, M. Sabsabi, Enhanced laser-induced breakdown spectroscopy using the combination of fourth-harmonic and fundamental Nd:YAG laser pulses. Spectrochim. Acta B **57**, 121–135 (2002)
4.22. A. Semerok, C. Chaléard, V. Detalle, J. Lacour, P. Mauchien, P. Meynadier, C. Nouvellon, B. Sallé, P. Palianov, M. Perdrix, G. Petite, Experimental investigations of laser ablation efficiency of pure metals with femto, pico and nanosecond pulses. Appl. Surf. Sci. **138–139**, 311–314 (1999)
4.23. H. Bette, R. Noll, High-speed laser-induced breakdown spectrometry for scanning microanalysis. J. Phys. D Appl. Phys. **37**, 1281–1288 (2004)
4.24. K. Kagawa, K. Kawai, M. Tani, T. Kobayashi, XeCl excimer laser-induced shock wave plasma and its application to emission spectrochemical analysis. Appl. Spectrosc. **48**, 198–205 (1994)
4.25. N. André, C. Geertsen, J. Lacour, P. Mauchien, S. Sjöström, UV laser ablation optical emission spectrometry on aluminium alloys in air at atmospheric pressure. Spectrochim. Acta **49B**, 1363–1372 (1994)
4.26. N. Arnold, J. Gruber, J. Heitz, Spherical expansion of the vapor plume into ambient gas: an analytical model. Appl. Phys. A **69**, S87–S93 (1999)
4.27. Z. Hwang, Y. Teng, K. Li, J. Sneddon, Interaction of a laser beam with metals, Part I: Quantitative studies of plasma emission. Appl. Spectrosc. **45**, 435–441 (1991)
4.28. R. Knopp, F. Scherbaum, J. Kim, Laser-induced breakdown spectroscopy (LIBS) as an analytical tool for the detection of metal ions in aqueous solutions. Fresenius J. Anal. Chem. **355**, 16–20 (1996)
4.29. W. Koechner, *Solid-State Laser Engineering*, 6th ed., Springer Series in Optical Sciences, vol. 1 (Springer, Berlin, 2006), XVI, p. 748
4.30. R. Diehl (ed.), *High-Power Diode Lasers – Fundamentals, Technology, Applications* (Springer, Berlin, 2000), p. 416
4.31. P. Peuser, N. Schmitt, *Diodengepumpte Festkörperlaser* (Springer-Verlag, Berlin, 1995)
4.32. D. Hoffmann, G. Bonati, P. Kayser, P. Loosen, R. Poprawe, R. Dinger, C. Haas, H. Martinen, Modular, fiber coupled, diode-pumped solid state laser with up to 5 kW average output power, in *Proceedings of Advanced Solid-State Lasers*, OSA, Seattle, 2001, 33–35
4.33. K. Du, D. Li, H. Zhang, P. Shi, X. Wei, R. Diart, Electro-optically Q-switched Nd:YVO_4 slab laser with high repetition rate and short pulse width. Opt. Lett. **28**, 87–89 (2003)
4.34. A. Thorne, U. Litzen, S. Johansson, *Spectrophysics* (Springer-Verlag, Berlin, 1999)
4.35. J. Samson, *Techniques of Vacuum Ultraviolet Spectroscopy* (Pied, Lincoln, 1967)
4.36. H. Becker-Ross, S. Florek, Echelle spectrometers and charged-coupled devices. Spectrochim. Acta B **52**, 1367–1375 (1997)
4.37. Echelle spectrometer 4000 EV/I from LLA Instruments GmbH, Berlin. http://www.lla.de
4.38. C. Aragon, J. Aguilera, F. Penalba, Improvements in quantitative analysis of steel composition by laser-induced breakdown spectroscopy at atmospheric pressure using an infrared Nd:YAG laser. Appl. Spectrosc. **53**, 1259–1267 (1999)
4.39. L. Cabalin, D. Romero, J. Baena, J. Laserna, Effect of surface topography in the characterization of stainless steel using laser-induced breakdown spectrometry. Surf. Interface Anal. **27**, 805–810 (1999)
4.40. M. Sabsabi, V. Detalle, M. Harith, W. Tawfik, H. Imam, Comparative study of two new commercial echelle spectrometers equipped with intensified CCD for analysis of laser-induced breakdown spectroscopy. Appl. Optics **42**, 6094–6098 (2003)

4.41. H. Balzer, S. Hölters, V. Sturm, R. Noll, Systematic line selection for online coating thickness measurements of galvanised sheet steel using LIBS. Anal. Bioanal. Chem. **385**, 234–239 (2006)
4.42. V. Sturm, A. Brysch, R. Noll, H. Brinkmann, R. Schwalbe, K. Mülheims, P. Luoto, P. Mannila, K. Heinänen, D. Carrascal, L. Sancho, A. Opfermann, K. Mavrommatis, H.W. Gudenau, A. Hatziapostolou, S. Couris, Online multi-element analysis of the top gas of a blast furnace, in *Seventh International Workshop Progress in Analytica Chemistry in the Steel and Metal Industries*, ed. by J. Angeli (Glückauf GmbH, Essen, 2006), pp. 183–188
4.43. L. Peter, V. Sturm, R. Noll, Liquid steel analysis with laser-induced breakdown spectrometry in the vacuum ultraviolet. Appl. Optics **42**, 6199–6204 (2003)
4.44. R. Noll, I. Mönch, O. Klein, A. Lamott, Concept and performance of inspection machines for industrial use based on LIBS. Spectrochim. Acta B **60**, 1070–1075 (2005)
4.45. C. Bohling, K. Hohmann, D. Scheel, W. Schade, M. Reuter, G. Holl, Anti-personnel-mine detection by laser-induced breakdown spectroscopy, in *Conf. on Lasers and Electro-Optics/Quantum Electronics and Laser Science and Photonic Applications, Systems and Technologies*, Technical Digest (CD), Optical Society of America, 2005, paper PWD3
4.46. K. Yamamoto, D. Cremers, M. Ferris, L. Foster, Detection of metals in the environment using a portable laser-induced breakdown spectroscopy instrument. Appl. Spectrosc. **50**, 222–233 (1996)
4.47. F. Akamatsu, T. Wakabayashi, S. Tsushima, M. Katsuki, Y. Mizutani, Y. Ikeda, N. Kawahara, T. Nakaiima, The development of a light-collection probe with high spatial resolution applicable to randomly fluctuating combustion fields. Meas. Sci. Technol. **10**, 1240–1246 (1999)
4.48. S. Palanco, J. Laserna, Remote sensing instrument for solid samples based on open-path atomic emission spectrometry. Rev. Sci. Instrum. **75**, 2068–2074 (2004)
4.49. P. Shaw, R. Gupta, T. Germer, U. Arp, T. Lucatorto, K. Lykke, Characterization of materials using an ultraviolet radiometric beamline at SURF III. Metrologia **37**, 551–554 (2000)
4.50. K. Knorr, Fluoridkristalle – alte Materialien mit glänzender Zukunft, Photonik 1/2002, p. 62–64
4.51. Transmittance data for MgF_2 from Korth Kristalle GmbH, Am Jägersberg 3, D-24161 Altenholz (Kiel), http://www.korth.de
4.52. L. Merkle, N. Koumvakalis, M. Bass, Laser-induced bulk damage in SiO_2 at 1.064, 0.532, and 0.355 μm. J. Appl. Phys. **55**, 772–775 (1984)
4.53. F. Rainer, R. Gonzales, A. Morgan, in *Laser-Induced Damage in Optical Materials*, ed. by H. Bennett, L. Chase, A. Guenther, B. Newnam, M. Soileau, Proc. SPIE **1438**, 58–73 (1989)
4.54. N. Boling, G. Dubé, Laser-induced inclusion damage at surfaces of transparent dielectrics. Appl. Phys. Lett. **23**, 658–660 (1973)
4.55. R. Taylor, K. Leopold, S. Mihailov, R. Brimacombe, Damage measurements of fused silica fibres using long optical pulse XeCl lasers. Opt. Commun. **63**, 26–31 (1987)
4.56. R Setchell, K. Meeks, W. Trott, P. Klingsporn, D. Berry, High-power transmission through step-index multimode fibers, in *Laser-Induced Damage in Optical Materials*: 1990, ed. by H. Bennett, L. Chase, A. Guenther, B. Newnam, M. Soileau, Proc. SPIE 1441 (1991) 61–70
4.57. W. Trott, K. Meeks, High-power Nd:glass laser transmission through optical fibers and its use in acceleration of thin foil targets. J. Appl. Phys. 67 (1990) 3297–3301
4.58. V. Sturm, R. Sattmann, R. Noll, Optical fiber transmission of multiple Q-switch Nd:YAG laser pulses with microsecond interpulse separations. Appl. Phys. B **63**, 363–370 (1996)
4.59. V. Pashinin, V. Sturm, V. Tumorin, R. Noll, Stimulated Brillouin scattering of Q-switched laser pulses in large-core optical fibers. Opt. Laser Technol. **33**, 617–622 (2001)
4.60. A. Whitehouse, J. Young, I. Botheroyd, S. Lawson, C. Evans, J. Wright, Remote material analysis of nuclear power station steam generator tubes by laser-induced breakdown spectroscopy. Spectrochim. Acta B **56**, 821–830 (2001)
4.61. Hamamatsu data sheets for the PMT types R 6350, R 6351, R 6354, R 7511, 1998
4.62. W. Franks, M. Kiik, A. Nathan, UV-responsive CCD image sensors with enhanced inorganic phosphor coatings. IEEE Trans. Electron Dev. **20**, 352–358 (2003)

4.63. L. Poletto, A. Boscolo, G. Tondello, Characterization of a charge-coupled-device detector in the 1100 – 0.14 nm (1-eV to 9-keV) spectral region. Appl. Optics **38**, 29–36 (1999)
4.64. T. Jüstel, C. Feldmann, C. Ronda, Leuchtstoffe für aktive Displays. Phys. Blätter **56**, 55–58 (2000)
4.65. Princeton instruments data sheets for intensified spectroscopy cameras types HB GEN III, UNIGEN and HQ GEN III, 2007
4.66. H. Bette, R. Noll, High-speed, high-resolution LIBS using diode-pumped solid-state lasers, in *Laser-Induced Breakdown Spectroscopy*, ed. by A.W. Miziolek, V. Palleschi, I. Schechter (Cambridge University Press, Cambridge, 2006), Chapter 14, pp. 490–515
4.67. V. Sturm, L. Peter, R. Noll, Steel analysis with laser-induced breakdown spectrometry in the vacuum ultraviolet. Appl. Spectrosc. **54**, 1275–1278 (2000)
4.68. R. Noll, V. Sturm, Ü. Aydin, D. Eilers, C. Gehlen, M. Höhne, A. Lamott, J. Makowe, J. Vrenegor, LIBS – from research to industry, new frontiers for process control. Spectrochim. Acta B **63**, 1159–1166 (2008)

Chapter 5
Evaporation and Plasma Generation

5.1 Evaporation

For LIBS of solid samples, the laser radiation evaporates the material and excites spontaneous emission of the material species (cf. Chap. 2). The material is heated locally to boiling or decomposition temperature to achieve an efficient evaporation. The fraction of laser energy absorbed by the material contributes to this process. The amount of temperature rise is influenced by the heat conduction of the material, which transports the absorbed energy out of the region, where the laser energy is coupled into the material.

The characteristics of the interaction between laser radiation and solid sample are illustrated in Fig. 5.1. In the following paragraphs, two cases are considered.

(a) *Surface absorption* ($\delta_{\mathrm{opt}} \ll \delta_{\mathrm{h}}$)

The optical penetration depth is much smaller than the heat penetration depth. For metals and laser wavelengths in the visible and near infrared spectral region, δ_{opt} amounts to 10^{-6}–10^{-5}cm typically. The condition $\delta_{\mathrm{opt}} \ll \delta_{\mathrm{h}}$ is, e.g., fulfilled for steel and times $t > 10^{-9}$s after start of the laser irradiation (the thermal diffusivity of steel at room temperature is $\kappa_{\mathrm{steel}} = 0.04\,\mathrm{cm}^2/\mathrm{s}$). In this case, the source term in the heat equation can be described as a surface source. For a Gaussian intensity profile of the laser beam, the source term is:

$$Q(r,t) = AI_0 \mathrm{e}^{-\frac{2r^2}{w_{\mathrm{b}}^2}}, \tag{5.1}$$

where A is the absorptance, I_0 is the irradiance of the laser beam in the beam center at $r = 0$, and w_{b} is the beam radius, cf. Fig. 3.10.

In this case, a solution of the heat equation yields the temperature at the surface and the beam center as a function of time [5.1]:

$$T = \frac{AI_0 w_{\mathrm{b}}}{\sqrt{2\pi}\kappa\rho c} \arctan \frac{\sqrt{8\kappa t}}{w_{\mathrm{b}}}, \tag{5.2}$$

R. Noll, *Laser-Induced Breakdown Spectroscopy*, DOI 10.1007/978-3-642-20668-9_5,

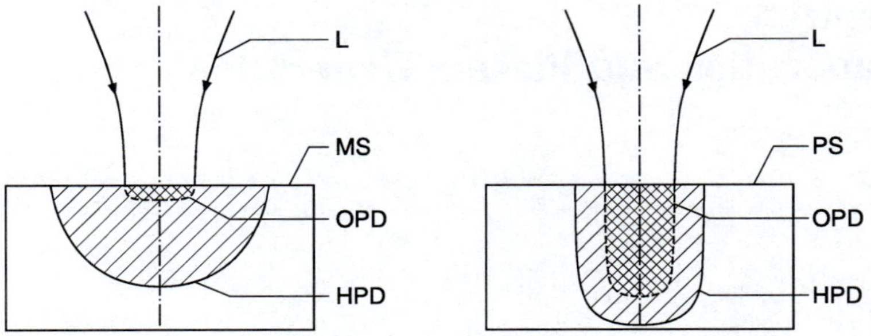

Fig. 5.1 *Left:* the optical penetration depth is smaller than the heat penetration depth; *right*: the optical penetration depth is greater than the heat penetration depth. L = laser beam, MS = metal sample, OPD = optical penetration depth, HPD = heat penetration depth, PS = polymer sample

where T is the temperature at the sample surface in the center of the laser beam, κ is the thermal diffusivity, ρ is the mass density, c is the specific heat capacity, and t is the time.

For a given absorbed irradiance, (5.2) yields the time at which the temperature at the sample surface attains, e.g., the boiling temperature and efficient evaporation sets in. Figure 5.2 shows the absorbed irradiance as a function of time for steel and aluminum according to (5.2). The heat conductivity $\kappa\rho c$ and the thermal diffusivity κ are also functions of the temperature. To simplify the calculation of these curves, average values were taken for the conductivity and diffusivity in the temperature range from 300 K to the evaporation temperature T_v (the used parameters are: (a) steel, average heat conductivity $\langle\rho\kappa c\rangle = 0.27\,\mathrm{W/(cm\,K)}$, average thermal diffusivity $\langle\kappa\rangle = 0.06\,\mathrm{cm^2/s}$, $T = T_v = 2{,}900\,\mathrm{K}$; (b) aluminum, $\langle\rho\kappa c\rangle = 1.79\,\mathrm{W/cm\,K}$, $\langle\kappa\rangle = 0.64\,\mathrm{cm^2/s}$, $T_v = 2{,}700\,\mathrm{K}$; the beam radius amounts in both cases $w_b = 100\,\mu\mathrm{m}$). The absorbed irradiance necessary to reach evaporation temperature decreases with longer irradiation times. Depending on the absorption coefficient, the laser beam irradiance at the sample surface has to be chosen sufficiently high to reach the evaporation threshold. For steel and aluminum, the absorption coefficient has typically values between 0.03 and 0.1. For Q-switched lasers with pulse durations of 10–100 ns, the necessary irradiances are greater than $10^8\,\mathrm{W/cm^2}$ to evaporate steel or aluminum according to Fig. 5.2. The area of irradiances accessible with commercially available Q-switched Nd:YAG lasers is shown as hatched box in Fig. 5.2.

(b) *Volume absorption* ($\delta_{\mathrm{opt}} \gg \delta_{\mathrm{h}}$)

The optical penetration depth is much greater than the heat penetration depth. This is, e.g., the case for polymers. The optical penetration depth for polymers as polyamide amounts to about 3.5 mm for the wavelength of the Nd:YAG laser [5.2]. With a thermal diffusivity of $\kappa_{\mathrm{polyamide}} = 1.3 \times 10^{-3}\,\mathrm{cm^2/s}$ [5.3], (3.2) yields that the optical penetration depth is greater than the heat penetration depth for times $t < 20\,\mathrm{s}$.

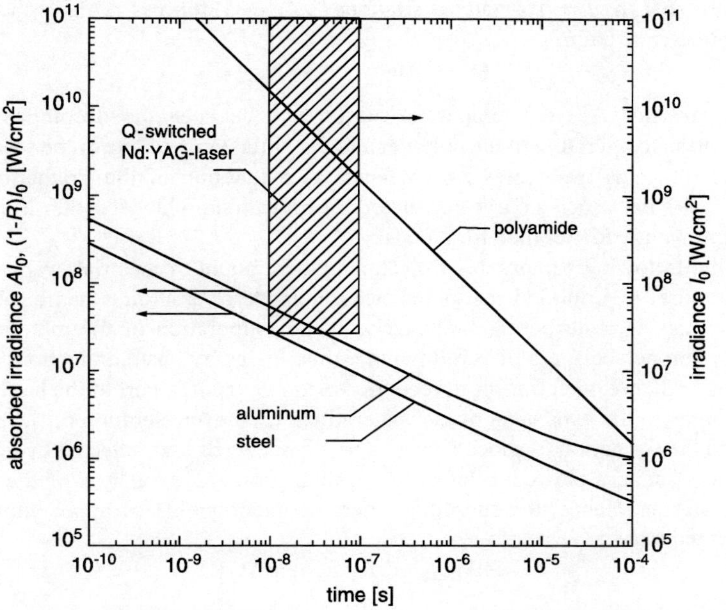

Fig. 5.2 Absorbed irradiance as a function of time necessary to reach evaporation or decomposition temperature for various materials. The *hatched box* marks the area of irradiances accessible with commercial Nd:YAG lasers. The parameters used for calculation of the curves are given in the text

If the heat conduction is negligible, then the temperature rise can be calculated directly for the locally absorbed laser irradiance. For the irradiance in the absorbing medium holds:

$$I = (1 - R) I_0 \mathrm{e}^{-\frac{z}{\delta_{\mathrm{opt}}}}, \tag{5.3}$$

where R is the reflectivity, I_0 is the irradiance at the sample surface, and z is the coordinate in the propagation direction of the laser beam with the sample surface at $z = 0$.

The factor $(1 - R)$ in (5.3) describes that part of the incoming irradiance, which enters the sample. In contrast to the case (a) described above for the surface absorption, the absorption A cannot be taken here, since in the considered case of volume absorption the absorption is not limited to the surface alone. For the absorbed energy per unit volume after a time t, we obtain from (5.3):

$$-(\mathrm{d}I/\mathrm{d}z)t = (1 - R)\frac{I_0}{\delta_{\mathrm{opt}}}\mathrm{e}^{-\frac{z}{\delta_{\mathrm{opt}}}} t. \tag{5.4}$$

This energy heats up the volume element leading to a temperature rise of ΔT:

$$\Delta q = \rho c \Delta T, \tag{5.5}$$

where Δq is the heat energy per unit volume and ΔT is the temperature rise.

With $-(dI/dz)t = \Delta q$ and the relations (5.4), (5.5) follows at the surface of the workpiece ($z = 0$ mm):

$$(1 - R)I_0 = \rho c \Delta T \delta_{\text{opt}}/t. \tag{5.6}$$

If we take for ΔT the temperature difference between the decomposition or evaporation temperature of the polymer and the initial temperature of the workpiece, then (5.6) yields the necessary irradiance as a function of the irradiation time. Figure 5.2 shows such a curve calculated for polyamide. The necessary irradiances are significantly higher than for metals.

In the following paragraphs, we estimate the amount of material evaporated by a laser pulse. A simplified energy balance is considered assuming that the absorbed laser energy is contributing exclusively to the evaporation of the material. This assumption neglects, e.g., the following terms: losses by heat conduction; specific melting enthalpy; melt transport; recondensation of vapor; a part of the laser energy contributes to the expansion of the material cloud, the acceleration of the ambient gas and the formation of shock waves; a part is scattered and reflected by the vapor and the plasma; a part contributes to the ionization and excitation of the species in the plasma. Hence, the simplified energy balance yields an upper limit of the evaporated mass:

$$AW_{\text{Laser}} = \rho V(E_{\text{v}} + c\Delta T), \tag{5.7}$$

where A is the absorptance, W_{Laser} is the irradiated energy of the laser pulse, V is the ablated volume, E_{v} is the specific evaporation enthalpy, and ΔT is the temperature difference between boiling temperature and room temperature.

For simplification, we furthermore neglect that the quantities A, ρ, c are generally not constant during the heating and evaporation phase. The values of the initial state were taken instead. As numerical example (5.7) yields as upper limit of the mass ablated by a laser pulse with $AW_{\text{Laser}} = 10$ mJ: $\rho V = 1.6 \times 10^{-6}$ g ($E_{\text{v}} = 6 \times 10^3$ J/g, $c = 0.51$ J/g K, $\Delta T = 2{,}600$ K). Experimentally determined mass ablations lie in the range of micrograms per pulse depending on the pulse energy, pulse width, burst mode, and target material, cf. Chap. 7.

5.2 Plasma Generation

The density of the evolving vapor can be estimated by use of the Clausius–Clapeyron equation [5.4]. For orientation solely some typical numbers will be indicated. Absorbed laser irradiances of 10^6–10^7 W/cm^2 result in vapor densities in the range of about 10^{18}–10^{20} cm^{-3}. Due to the high temperature of the vapor, there always exist some free electrons, which can gain energy from the radiation field and transfer this energy via collisions to the atoms of the vapor. Collisions of electrons with atoms are a precondition for a net energy gain of the electrons from the oscillating electric field of the laser beam. Collision-free electrons would take up and loose energy periodically and hence on average would not gain any kinetic energy. Within a short time, the electrons get high average energies, which are sufficient to partly ionize the atoms of the material vapor. In this way, the plasma is induced, in which the atoms and ions are excited and emit their specific radiation.

The state of the plasma can be described with the use of balance equations for particle densities, momenta, and energies. Model calculations show that within a few nanoseconds after the start of the laser beam irradiation, the electron energies reach values in the range of 0.25–1 eV [5.5]. Due to collisional ionization processes, the electron density rises to a level of 10^{16}–10^{18} cm^{-3}.

5.3 Absorption of Laser Radiation in the Plasma

The absorption of the laser radiation in the evolving material vapor and plasma depends on the wavelength, which will be discussed in the following paragraphs in more detail. The absorption coefficient can be calculated by means of the Maxwell equations and the equation of motion for free nonrelativistic electrons [5.6]:

$$\alpha = 2k \sqrt{\frac{1}{2}\left\{\sqrt{\left[1-\left(\frac{\omega_\mathrm{p}}{\omega}\right)^2 \frac{1}{1+(\nu/\omega)^2}\right]^2 + \left[\frac{\nu}{\omega}\left(\frac{\omega_\mathrm{p}}{\omega}\right)^2 \frac{1}{1+(\nu/\omega)^2}\right]^2} - \left[1-\left(\frac{\omega_\mathrm{p}}{\omega}\right)^2 \frac{1}{1+(\nu/\omega)^2}\right]\right\}}\,, \tag{5.8}$$

where $k = \omega/c$ is the wave vector in vacuum, ω is the angular frequency of the laser radiation, c is the vacuum speed of light, ω_p is the plasma frequency, and ν is the collision frequency of electrons with ions and atoms.

The plasma frequency is the characteristic eigenfrequency of the plasma describing the oscillation of the electrons with respect to the ions [5.7]:

$$\omega_\mathrm{p} = \sqrt{\frac{n_\mathrm{e} e^2}{\varepsilon_0 m_\mathrm{e}}}\,, \tag{5.9}$$

where n_e is the electron density, e is the elementary charge, ε_0 is the vacuum permittivity, and m_e electron mass.

Taking the typical conditions of a laser-induced plasma for LIBS allows to simplify the relation (5.8) under the following assumptions:

$$\nu \ll \omega, \quad \omega_\mathrm{p} < \omega. \tag{5.10}$$

For the absorption coefficient holds:

$$\alpha = \frac{\nu}{c} \frac{\omega_\mathrm{p}^2}{\omega^2 \sqrt{1-(\omega_\mathrm{p}/\omega)^2}}. \tag{5.11}$$

The absorption coefficient is proportional to the collision frequency and approximately inversely proportional to the square of the frequency of the laser light. The collision frequency comprises two terms:

$$\nu = \nu_{ei} + \nu_{en}, \tag{5.12}$$

where ν_{ei} is the collision frequency of electrons with ions, and ν_{en} is the collision frequency of electrons with atoms.

In the following paragraphs, the LIBS plasma of iron samples is taken as a numerical example. The electron atom collision frequency for iron at a temperature of 0.8 eV can be estimated to be $\nu_{en} = (5 \times 10^{-14}\,\mathrm{m^3/s}) n_{\mathrm{Fe\,I}}$ [5.8]. Typical values of the iron atom density determined experimentally at a time of 1 μs after the irradiation of a double pulse are: $n_{\mathrm{Fe\,I}} \leq 10^{16}\,\mathrm{cm^{-3}}$ [5.9]. Then we obtain for the electron iron atom collision frequency: $\nu_{en} \leq 5 \times 10^8\,\mathrm{s^{-1}}$. The electron ion collision frequency is given by [5.10]:

$$\nu_{ei} = \frac{1}{(4\pi\varepsilon_0)^2} \frac{4\sqrt{2\pi} Z^2 e^4 n_i \ln \Lambda}{3\sqrt{m_e}(k_B T_e)^{3/2}}, \tag{5.13}$$

where ε_0, e, m_e see (5.9), Z is the ion charge, n_i is the density of ions, $\ln \Lambda$ is the Coulomb logarithm, k_B is the Boltzmann constant, and T_e is the electron temperature.

For the Coulomb logarithm holds [5.10]:

$$\ln \Lambda = \ln \left[12\pi \varepsilon_0^{3/2} \frac{(k_B T_e)^{3/2}}{e^3 \sqrt{n_e}} \right]. \tag{5.14}$$

With the experimental values (1 μs after irradiation of a double pulse, cf. Chap. 8): $n_e = n_i = n_{\mathrm{Fe\,II}} = 6 \times 10^{16}\,\mathrm{cm^{-3}}$, $T_e = 9{,}200\,\mathrm{K}$ follows: $\ln \Lambda = 3.8$ and $\nu_{ei} = 9.4 \times 10^{11}\,\mathrm{s^{-1}}$. Hence, the collision frequency of the electrons with the ions is significantly greater than the collision frequency between electrons and neutral particles. For the plasma frequency follows in this case using relation (5.9): $\omega_p = 1.4 \times 10^{13}\,\mathrm{rad/s}$. For a Nd:YAG laser with a wavelength of 1.06 μm, the angular frequency is $\omega = 1.78 \times 10^{15}\,\mathrm{rad/s}$. Hence, the conditions (5.10) are fulfilled to apply the approximation (5.11) to calculate the absorption coefficient (This estimation holds for $t_{delay} = 1\,\mu s$; at earlier times being closer to the irradiation time of the laser pulse the situation is completely different).

Figure 5.3 shows the absorption coefficient as a function of the electron density. The temperatures given cover the range typically observed for LIBS plasmas where the temperature is measured by use of Boltzmann plots (cf. Chap. 8). The range of measured plasma parameters n_e and T_e is illustrated as hatched box (cf. Chap. 8 and, e.g., [5.11]). Assuming as an upper limit a plasma dimension of 5 mm, the right axis shows the corresponding transmission of the laser radiation through the plasma. Hence, for plasma conditions prevailing typically some 100 ns after the laser

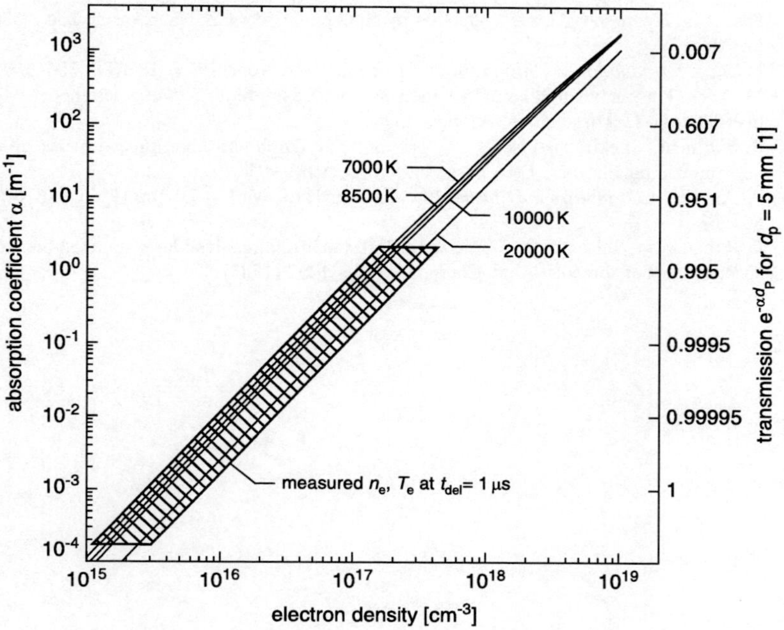

Fig. 5.3 Absorption coefficient for laser radiation with a wavelength of 1.06 μm in a plasma as a function of the electron density. Parameter is the electron temperature of the plasma. The *right axis* shows the transmission of the laser radiation through a plasma having a thickness of 5 mm. The *hatched box* illustrates the typical range of electron densities and temperatures measured for LIBS plasmas at a delay time of 1 μs after irradiation of the laser pulse

pulse irradiation, the plasma is already highly transparent for the laser radiation of a Nd:YAG laser. Therefore, a second collinear pulse of a double pulse irradiated with an interpulse separation of $\Delta t > 1\,\mu s$ reaches the surface of the measuring object again with negligible absorption by the plasma generated by the first pulse (cf. Chap. 8).

References

5.1. H. Carslaw, J. Jaeger, *Conduction of Heat in Solids*, 2nd edn. (Oxford University Press, Oxford, 1959)
5.2. R. Klein, Bearbeitung von Polymerwerkstoffen mit infraroter Laserstrahlung, D82, Dissertation, Aachen, 1990
5.3. VDI-Wärmeatlas, 4. Auflage, Düsseldorf, 1984
5.4. R. Wester, Laser-induced evaporation as basic process of drilling, caving and cutting, Laser und Optoelektronik **23**, 60–63 (1991)
5.5. E. Beyer, Einfluß des laserinduzierten Plasmas beim Schweißen mit CO_2-Lasern, Schweißtechnische Forschungsberichte, Band 2, Düsseldorf, Deutscher Verlag für Schweißtechnik, 1985

5.6. P. Mulser, Hydrogen plasma production by giant pulse lasers. Z. Naturforsch. **25a**, 282–295 (1970)
5.7. F. Chen, *Introduction to Plasma Physics* (Plenum, New York, 1974), ISBN 0–306–30755–3
5.8. M. Aden, Plasmadynamik beim laserinduzierten Verdampfungsprozeß einer ebenen Metalloberfläche, D82, Dissertation, Aachen, 1994
5.9. R. Sattmann, Spektrochemische Analyse von Stahl durch optische Emissionsspektrometrie laserinduzierter Plasmen, D82, Dissertation, Aachen, 1997
5.10. A. Anders, *A Formulary for Plasma Physics* (Akademie-Verlag, Berlin, 1990), ISBN 3–05–501263–1
5.11. L. Radziemski, T. Loree, D. Cremers, N. Hoffman, Time-resolved laser-induced breakdown spectrometry of aerosols. Anal. Chem. **55**, 1246–1252 (1983)

Chapter 6
Multiple Pulses for LIBS

Laser-induced breakdown spectroscopy was investigated already for a broad variety of chemical analytical purposes in the 1970s to 1990s [6.1–6.6]. However, a transfer to applications was often prevented because of insufficient analytical performance compared with other atomic emission spectrometric methods such as spark discharge or inductively coupled plasma (ICP) atomic emission spectroscopy. Thus, at that time LIBS was used in a limited number of applications only, in which the analytical requirements are low and the advantages of LIBS as a rapid, noncontact, in situ method are essential [6.7].

To use LIBS in a broader field of applications, the analytical performance has to be raised considerably. A crucial point is the laser parameters and the interaction of the laser beam with the material to be analyzed. Most of the work reported is done with pulsed pumped Q-switch Nd:YAG lasers, although other lasers have also been used for LIBS (cf. Sect. 4.1). Single pulses, i.e., one laser pulse per pump pulse, are the conventional approach to ablate and vaporize material and to induce the plasma. Typical laser parameters are 10–100 Hz repetition rate, 8–50 ns pulse width, and 5–200 mJ pulse energy.

If single pulses are used for LIBS, then ablation and plasma excitation cannot be optimized separately, in contrast to two-step techniques, such as laser ablation with subsequent excitation by electrical sparks, microwaves, or inductively coupled plasmas [6.8–6.10]. These two-step techniques require additional equipment close to the interaction area. Spark excitation of the laser ablated material, for example, requires electrodes located close to the samples. These reduce the flexibility and may interfere with the analysis.

Uebbing et al. used a second Q-switch Nd:YAG laser to re-heat the vaporized material [6.11]. The second laser was fired with controlled delay respective to the first laser. The direction of the first laser beam was perpendicular to the sample surface, whereas the second laser beam propagated parallel to the surface. Hence, the second laser beam interacted only with the vapor above the sample. This arrangement may be adequate for fundamental studies, but it is hardly suited for a practical application of LIBS as an analytical method.

R. Noll, *Laser-Induced Breakdown Spectroscopy*, DOI 10.1007/978-3-642-20668-9_6,

The use of controlled multiple laser pulses combines the advantages of a two-step techniques without losing the flexibility of one-step LIBS. Multiple pulses have been used in former approaches as well, but the separate pulses could not be controlled actively [6.12, 6.13]. For that reason and due to worse pulse-to-pulse stability compared with single pulses, these authors recommended single pulses for LIBS. In the mid-nineties, a pioneering work demonstrated clearly the benefits of actively controlled collinear multiple pulses for LIBS [6.14]. Since then double pulses and multiple pulses were studied to improve the analytical performance of LIBS – e.g., for the determination of traces in metallic matrices or to enable special applications, e.g., the analysis of samples under water – or to increase the ablation rate in laser material processing [6.15–6.19].

6.1 Double Pulses

The term "double pulse" is defined here as follows: two laser pulses having each a width in the range of about 5–30 ns, which are separated in time by several 100 ns up to several microseconds (cf. Sect. 3.2). This interpulse separation is denoted by Δt (Fig. 3.9). Several methods are available to generate these double pulses: (a) a single-pulsed laser with a modified electronic control of the Q-switch to generate two pulses within a single flashlamp discharge, (b) superposition of the beams of two pulsed lasers. For case (a), only one laser is necessary and the two pulses are emitted collinearly. However, the interpulse separation is limited to about $0.5\,\mu s < \Delta t < 180\,\mu s$. Case (b) requires two lasers and a precise adjustment of the spatial superposition of the two pulses. This approach offers a high flexibility in terms of the wavelengths of the pulses, the pulse widths, and the sequence of pulses.

Figure 6.1 illustrates schematically the various configurations studied so far for LIBS with double pulses. The arrows depict the direction of propagation of the laser pulses, and the numbers show the temporal sequence. In the collinear case, both pulses have the same axis of propagation and are directed orthogonally to the sample

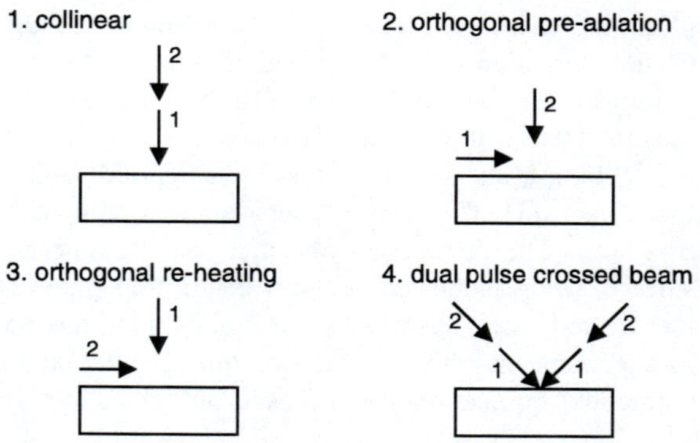

Fig. 6.1 Double-pulse configurations studied for LIBS. The *arrows* depict the laser pulses and their direction of propagation and the *numbers 1, 2* their temporal sequence

surface [6.14–6.16]. Various pulse combinations in terms of different wavelengths and different pulse widths have been studied [6.18, 6.19]. In the case of orthogonal preablation, a first laser pulse is irradiated parallel to the sample surface generating a laser-induced plasma above the sample surface in the ambient atmosphere and the second pulse is directed orthogonally to the sample surface [6.20]. For orthogonal re-heating, the sequence of the pulses is exchanged, i.e., the laser pulses irradiates the sample at first and then the re-heating pulse propagating parallel to the sample surface is applied [6.21]. In the dual pulse crossed beam approach, the first laser pulse is splitted into two beams irradiating the sample under an angle of incidence of 45°, and the second pulse follows in the same directions [6.22].

Investigations of collinear double pulses and triple pulses generated within a single flashlamp discharge of a Nd:YAG laser for plasma excitation showed a significant improvement in the detection limits [6.14,6.16,6.23]. With this approach, detection limits for the elements C, P, S, Mn, Ni, Cr in low-alloyed solid steel samples of less than 10 μg/g were achieved for the first time using LIBS. Collinear double and multiple pulses led to a significant increase in the ablated mass [6.14]. Comparing the ablated mass of a steel target using double pulses with 2×40 mJ and $\Delta t = 6\,\mu$s with a single pulse of 1×80 mJ yields an increase by about a factor of 2.5. While for single pulses the ablated mass strongly saturates with increasing pulse energy, no such saturation behavior is observed for double and multiple pulses up to burst energies (denoting the sum of pulse energies within a double or multiple pulse) of 320 mJ. Electron temperatures in the laser-induced plasma determined spectroscopically are higher by 1,000–1,500 K using double pulses with respect to those measured in the plasmas generated by single pulses of the same energy (i.e., the energy of the single pulse is equal to the total energy of the double pulse). Hence, the increased line emission observed using double-pulse excitation is linked with the greater mass ablated and the higher temperatures achieved in the laser-induced plasma [6.14].

For collinear double pulses studied for the analysis of solid metallic samples under water, an increase in the ablated mass of aluminum samples by a factor of 3.8 was reported [6.15]. The observed increase in line emissions is attributed to the formation of a bubble in the water layer above the sample surface generated by the first pulse into which the emitting atoms can expand. Due to the absence of atomic emissions from the plasma produced by the single pulse, it was not possible to determine the temperature and electron density for underwater excitation. The authors conclude that in this case the temperature is less than 3,000 K. For double-pulse excitation, an electron temperature of 8,900 K was determined [6.15].

Studies of collinear double pulses interacting with aluminum samples in air showed higher line intensities lasting longer than corresponding emissions generated by single laser pulses [6.17]. The time evolution of the line emission follows an exponential decay, where the decay time constant for the Al II line 624.34 nm and double-pulse excitation is more than twice that determined for single-pulse excitation. Looking at the ionic nitrogen emission N II at 500.1 nm revealed greater line intensities in the case of single-pulse excitation than for double pulses indicating that the preplasma formation induced by the first pulse reduces the air present nearby the surface and hence the concentration of nitrogen available for emission [6.17].

In the orthogonal preablation configuration, an enhancement of Fe lines of about a factor of 30 was observed comparing preablation double pulses with single pulses [6.20]. A temperature increase of up to 5,000 K was observed for preablation double pulses. The signal enhancement is correlated with an increased crater volume, i.e., the ablated mass is greater for preablation double pulses than for single pulses [6.20].

From a practical point of view, the collinear configuration is the most application relevant approach for double pulses offering an easy access to the sample to be analyzed. Hence, the investigations presented in the following are focused on the collinear double pulse configuration.

For a qualitative assessment of the influence of multiple pulses on the spatial structure of the laser-induced plasma, the visible plasma emission was observed by time-integrated photographs. For the experiments, the multiple Q-switch pulses are generated by a Nd:YAG laser with one oscillator (3″ × 4 mm rod) and two amplifiers (3″×1/4″ and 4″×″3/8″ rods). The high voltage of the Pockels cell can be increased in up to six steps with variable step height and time delay. Each voltage step results in the emission of a Q-switch pulse. The time delay and the energy of the pulses are adjustable. The minimum delay between consecutive pulses is about 2 μs; the maximum duration of a burst of pulses is given by the flashlamp pump pulse and amounts to about 100 μs. The laser emits the fundamental wavelength of 1,064 nm with a burst repetition rate of 30 Hz. The pulse duration and maximum energy per burst depend on the number of pulses per burst and the delays (which determine the interpulse separations Δt, cf. Fig. 3.9). The following notation is used: E^{s}, $E_{\mathrm{b}}^{\mathrm{d}}$, or $E_{\mathrm{b}}^{\mathrm{m}}$ denotes the total energy of a single-, double- or multiple-pulse burst, and E_1^{d} or E_2^{d} denotes the energy of the first or second pulse of a double pulse, and so on. The maximum burst energy is measured with a volume-absorbing disc calorimeter (Scientech model 38-0101). The energy distribution of the separate pulses of a burst is determined by integration of a photodiode signal of the burst. Figure 3.8 shows the photodiode signal of the laser output for a double pulse with $E_1^{\mathrm{d}} = E_2^{\mathrm{d}}$, $E_{\mathrm{b}}^{\mathrm{d}} =$ 80 mJ and $\Delta t = 6\,\mu$s interpulse separation. For 6 μs interpulse separation, the pulse duration (FWHM) of each pulse of a double burst amounts to about 25 ns. The pulse duration for single pulses is about 15 ns.

The laser beam is focused on plane samples by a planoconvex lens with a focal length of $f = 200$ mm. The diameter of the laser beam in the plane of the lens amounts to about 12 mm. The sample surface is adjusted perpendicularly to the laser beam. All experiments are performed under air and a pressure of 10^5 Pa. For each measurement, the laser beam is focused on a new part of a ground surface of the sample. A pure iron sample and samples with certified composition from the Bureau of Analysed Samples, Ltd., UK (BAS), are used for the experiments.

For the photographs of the laser-induced plasma, the direction of observation is oriented perpendicularly to the propagation direction of the laser beam, see Fig. 6.2. A constant exposure time of 1 s is taken for all photographs, so the plasma radiation induced by 30 laser bursts is accumulated. The photographs indicate the extent of the light-emitting plasma.

Figure 6.2 shows photographs of the plasmas generated by single laser pulses of $E^s = 80\,\text{mJ}$ and double pulses of the same total energy $E_b^d = 80\,\text{mJ}$ with an interpulse separation of $\Delta t = 6\,\mu\text{s}$. The photographs in the left-hand column are taken without sample. The middle of the air breakdown plasma is taken as the zero point of Δs. Positive Δs corresponds to a smaller distance between focusing lens and sample, that is, the focal region moves into the sample. In the experiments, different Δs are realized by a shift of the focusing lens. The photographs taken for $\Delta s = 5$ and $\pm 2\,\text{mm}$ show radiating volumes, which are approximately a factor of two greater for double pulses than for single pulses.

Another important fact is obvious from the photographs with $\Delta s = -5\,\text{mm}$: for the 80 mJ single pulse, the plasma induced in the surrounding atmosphere shields off a major part of the laser energy. With double pulses, the plasma induced in the atmosphere does not prevent a considerable part of the laser energy from reaching the sample, and an intense plasma is also generated on the sample surface.

The plasma radiation is collected by an optical fiber bundle (4 mm diameter, about 160 all-silica fibers, 200 μm diameter per fiber, bundle length 4 m) without additional optical elements. The distance of the entrance face of the fiber bundle to the sample surface is about 67 mm, and the angle to the laser beam about 39°. The numerical aperture of the fiber amounts to 0.22, so the received radiation is spatially integrated over the emitting plasma volume. The plasma light is guided to a Paschen–Runge spectrometer with 750 mm focal length, 2,400 lines per millimeter

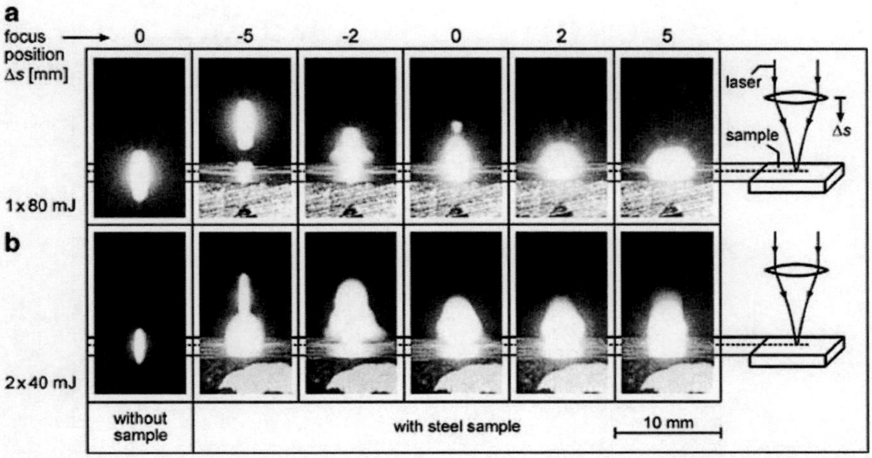

Fig. 6.2 Time-integrated photographs of the plasmas in the visible spectral range induced in the atmosphere and on a steel surface by single and double laser pulses of 80 mJ burst energy. The distance between the focal plane of the laser beam and the surface of the sample is denoted by Δs, see text for details

grating, and photomultiplier tubes as detectors. The spectral resolution amounts to 0.03 nm in first order.

Figure 6.3 shows the output signal of a photomultiplier tube mounted in the Paschen–Runge spectrometer at the Fe II 273.07 nm line for single and double pulses of 80 mJ (interpulse separation $\Delta t = 6\,\mu s$, $\Delta s = 5$ mm). Each curve is averaged over 16 laser bursts on a pure iron sample after 84 prebursts. An analysis of the signal induced by the single pulses (Fig. 6.3, top) shows that the decay following the maximum can be described by an exponential function for times up to 0.3 μs after the maximum of the signal. The $1/e$ time constant amounts to about 90 ns. This section of the signal is attributed to the continuum emission of the laser-induced plasma (cf. Sect. 2.1). The subsequent signal behavior is characterized by decay times greater than 800 ns. Line radiation is the dominant contribution to this part of the signal.

The temporal behavior of the plasma emissionindexContinuum!emission of the plasma induced by the first pulse of the double pulse is similar to that of the single pulse, whereas the intensity is significantly smaller due to the reduced laser pulse energy. In contrast to this, the plasma emission induced by the second pulse of the

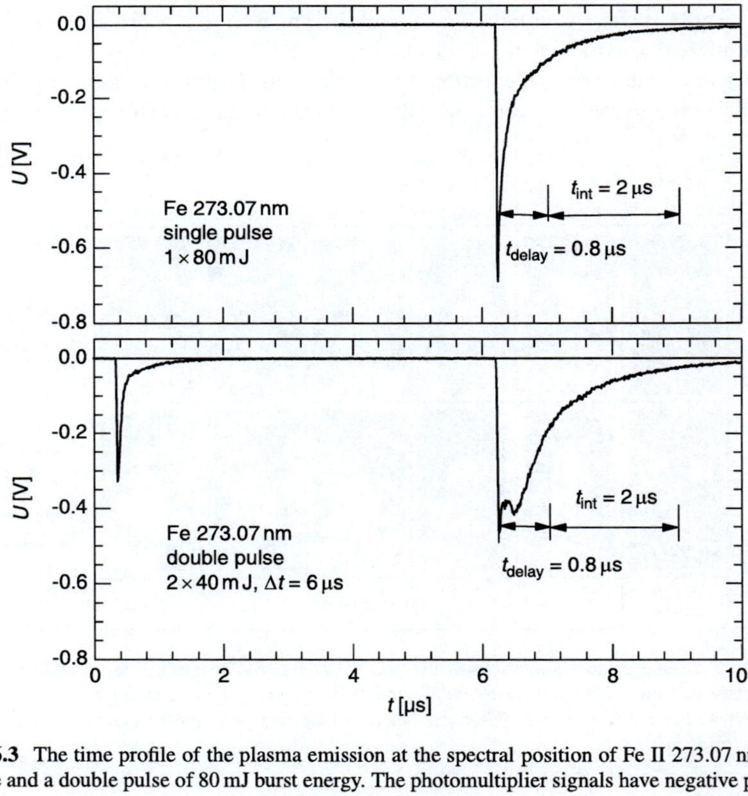

Fig. 6.3 The time profile of the plasma emission at the spectral position of Fe II 273.07 nm for a single and a double pulse of 80 mJ burst energy. The photomultiplier signals have negative polarity

double pulse reveals a completely different behavior. About 0.4 μs after the first maximum, the time profile of the photomultiplier signal attains a second maximum. At this time, the number of singly ionized species and their excitation achieves a maximum. For times greater than 0.4 μs (relative to the start of the emission induced by the single pulse or to the start of the emission induced by the second pulse of the double pulses), the line intensity is approximately a factor of two greater for the double pulse than for the single pulse, although the laser pulse energy of the second pulse is only half of the single pulse energy.

For analytical measurements, the ratio of the analyte signal, i.e., the element specific emission, to the background caused by the continuum emission is a key issue. Figure 6.4 shows the photomultiplier signals of the Mn II line at 293.31 nm for a pure iron sample (Hu 85) and a low-alloyed steel sample with a Mn content of 0.62% (BAS 451). All experimental parameters are the same as for Fig. 6.3.

The emission maxima which coincide with the laser pulses are the same for both samples. Obviously the origin of this emission is the element-independent

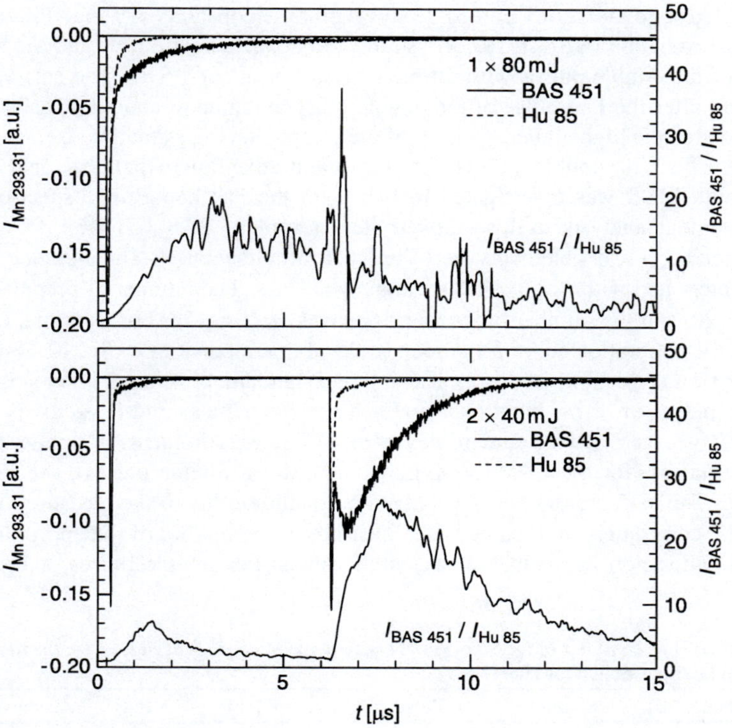

Fig. 6.4 Emission of the Mn II line at 293.31 nm for single (*top*) and double pulses (*bottom*) as a function of time and two samples, where BAS 451 (*solid line*) has a Mn concentration of 0.62% and Hu 85 (*dashed line*) is a pure iron sample. The *right axis* shows the ratio of the PMT signals of the two samples

continuum radiation. After about 100 ns, differences in the PMT curves occur. This difference is due to the manganese line radiation. Again significantly higher Mn line intensities show up after the second pulse of the double pulse.

The ratio of line to background emission is also shown in Fig. 6.4. For a single pulse after 1 μs, this ratio amounts to ≈ 7, and 1 μs after the second pulse of the double pulse it achieves a value of ≈ 20. Besides an increase in the line intensity with double pulses, the ratio of line to background radiation improves significantly.

Table 6.1 shows the time constants of the PMT curves shown in Fig. 6.4. All curves can be approximated by an exponential decay with two time constants. For the fitting of the measuring values obtained for the BAS 451 samples after the second pulse of the double pulse, the time regime after the second emission maximum was evaluated.

In the time interval up to 200–500 ns after the laser pulse, all four curves (solid and dashed lines of the Mn intensity shown in Fig. 6.4) show an exponential decay with time constants less than 100 ns. After the 40 mJ pulses, the time constant is in the range of 30–40 ns, and for the 80 mJ pulse 60 ns. The higher value is attributed to an enlarged hot core of the plasma and hence smaller energy losses. At later times, the background radiation (Hu 85) decays with time constants of about 400–850 ns. The line radiation (BAS 451) decays after the 80 mJ single pulses and the second pulse of the double pulses with a time constant of about 1.5 μs. The smaller time constant after the first pulse of the double pulse amounts to about 900 ns which is attributed to the higher energy losses of the plasma having a smaller size.

The effect of double pulses for the quantitative microchemical analysis of steel with LIBS was investigated [6.14]. With the Paschen–Runge spectrometer, multielement analysis of low-alloyed steel samples (BAS 451–460, 097) were performed. Up to 14 elements were analyzed simultaneously. The element signals were integrated with $t_{delay} = 0.8\,\mu s$ and $t_{int} = 2\,\mu s$. The number of prepulses was $N_{pp} = 20$, and the number of measuring pulses $N_{mp} = 128$. For example, Fig. 6.5 shows the calibration curve for silicon (ratio of line intensities of Si 212.41 nm and Fe 273.07 nm, this line ratio is used as internal standard, cf. Chap. 11) for single and double pulses of 80 mJ burst energy. Each sample is measured three to six times. On average, the relative standard deviation of the measurements is smaller for the double pulses (for Si, 1.8% for single and 1.5% for double pulses), see Fig. 6.5, middle. The calibration curves are given by quadratic fits to the measured values, with the correlation coefficient r as a measure of the quality of the approximation of the calibration curve to the measured values. For double pulses, r increases

Table 6.1 Time constants of the exponential decay of the Mn signals of Fig. 6.4 for the first 500 ns after the laser pulses and for later times

Pulse mode (mJ)	Sample	1. Pulse, time constant (ns)		2. Pulse, time constant (ns)	
1 × 80	Hu 85	60	≈ 400	–	–
	BAS 451	60	1,500	–	–
2 × 40	Hu 85	33	360	40	850
	BAS 451	32	880	≈ 30	1,450

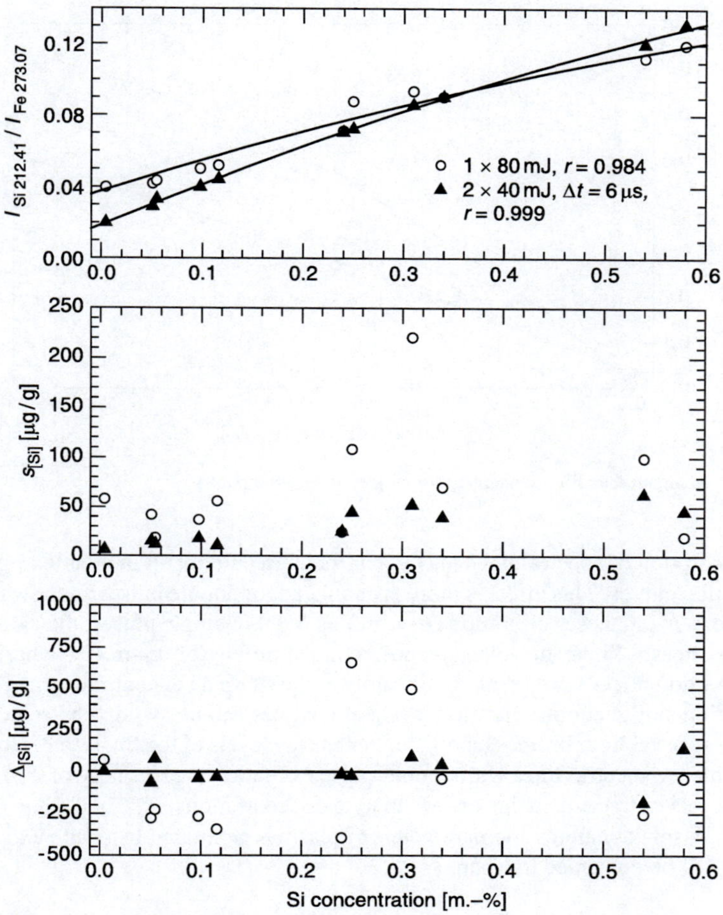

Fig. 6.5 *Top*: Silicon calibration curve for single and double pulses; *middle*: standard deviation of measured concentrations; *bottom*: difference between measured concentration and certified concentration of the reference samples

to 0.999 in comparison with 0.984 for single pulses. The measuring values lie closer to the calibration curve, and the correctness of the analysis is improved, see Fig. 6.5, bottom. Finally, for double pulses, the calibration curve has a factor of two lower background equivalent concentration (BEC, cf. Sect. 11.2, (11.19), Fig. 11.3, reduction from 2,100 to 790 μg/g) and a higher gain, which further improve the analytical performance. The limit of detection using the 3s-criterion is improved from 174 to 19 μg/g [see (11.16)].

As shown in Fig. 6.2, the plasma induced by double pulses has a greater extension than that generated by single pulses using the same total laser energy. Hence,

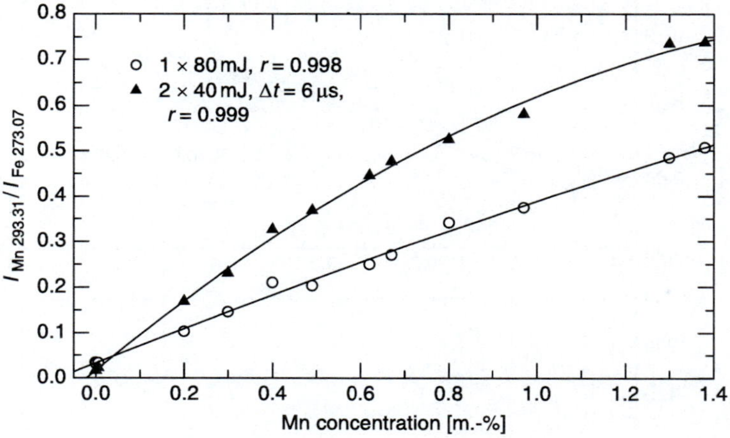

Fig. 6.6 Manganese calibration curves for single and double pulses

self-absorption of spectral lines may occur which would result in a flattening of the calibration curves. This effect is more pronounced for double pulses as shown exemplary for a manganese calibration curve in Fig. 6.6. For single pulses, the calibration curve is nearly linear up to manganese concentrations of 1.4 m.-%, whereas the curve for double pulses flattens significantly in the studied concentration range. Self-absorption at higher concentration levels of analytes can be avoided by selection of weaker spectral lines or lines having higher energy levels of the final state. However, using intense spectral lines with double-pulse excitation is especially of interest for the analysis of trace elements or low analyte concentrations, cf. Chap. 13.

The plasma dynamics and parameters of plasmas generated by single and double pulses will be presented in Chap. 8.

6.2 Triple and Multiple Pulses

Collinear triple pulses can be generated within a single flashlamp discharge of a Nd:YAG laser [6.16]. The laser emits three equal-energy pulses of 16 ns duration (FWHM) with an interpulse separation of 25 μs between the first and second pulse, and 40 μs between the second and the third pulse, respectively. The burst energy amounts to 300 mJ. The analytical performances achieved with this approach are described in detail in Sect. 13.1.1.

The effect of multiple pulses on the line emission of Fe I 532.80 nm is shown in Fig. 6.7. The multiple pulse comprises four pulses with a total burst energy of 160 mJ. The intensity levels of the element-specific signal part after the third and fourth pulse are about the same and they are greater than after the second pulse. Hence, for an n-fold multiple pulse ($n > 2$) and use of several time integration

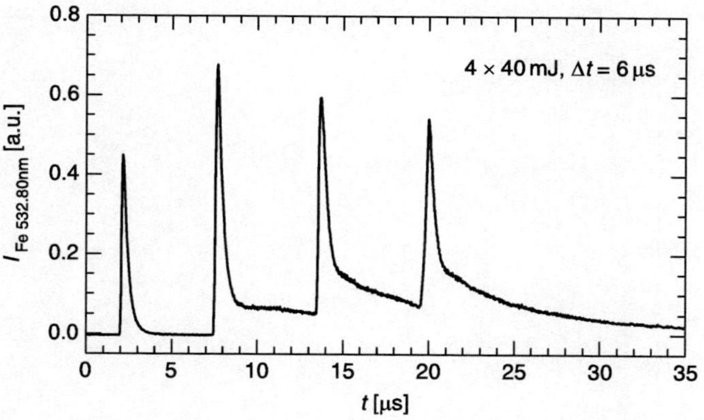

Fig. 6.7 Intensity of the line Fe I 532.80 nm as a function of time for a fourfold multiple pulse with 4×40 mJ burst energy and interpulse separations of 6 μs (the PMT output is shown as positive signal in this diagram)

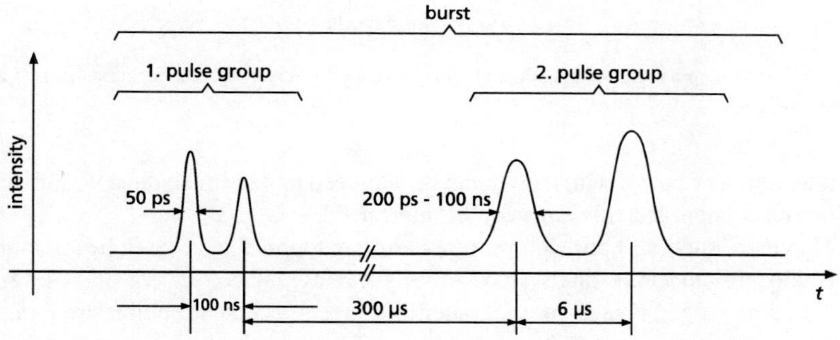

Fig. 6.8 Temporal structure of a tailored pulse train as a general approach for bursts to optimize element-specific emission

gates during such a burst, the detected element-specific intensity can be increased approximately by a factor of $(n - 1)$ with respect to the analyte intensity of the second pulse of a double pulse.

6.3 Tailored Pulse Trains

So far we have considered bursts comprising a series of two to four Q-switch pulses with typical pulse widths of a few nanoseconds and interpulse separations in the microsecond regime. In a more general approach, a burst can consist of a series of pulses having different pulse widths and interpulse separations as illustrated exemplarily in Fig. 6.8. By such an approach, a further optimization of

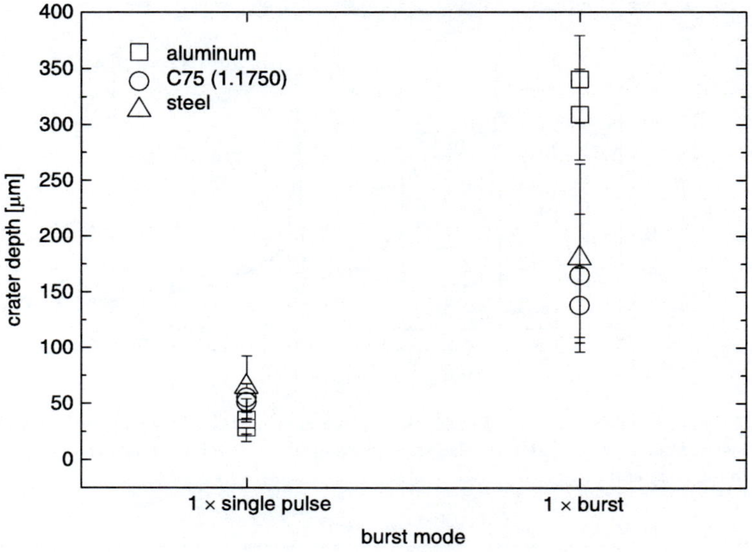

Fig. 6.9 Crater depths achieved with a single pulse and a single optimized burst for different materials

the interaction of bursts with matter can be achieved to maximize the ablation depth and/or the element-specific emission of interest.

Of course, such an approach requires a more sophisticated laser beam source. Especially, for the bulk analysis of objects having surface layers of a different composition – e.g., lacquers, oxides, anodized layers – burst structures are required being capable to penetrate this surface layers effectively and to ablate and excite the material underneath (cf. Sect. 18.4).

Figure 6.9 shows the results of an optimized burst structure in terms of the crater depth achieved with a single Q-switch pulse compared with an optimized burst with total laser burst energy in the 100 mJ range. By temporal pulse tailoring the crater depth can be increased with a single burst by a factor of 3 for steel to 7 for aluminum [6.24]. Such potentials of tailored pulse trains are, e.g., of interest for high-speed applications of LIBS for material recycling tasks, cf. Sect. 18.4.

References

6.1. C. Aragon, J. Aguilera, J. Campos, Determination of carbon content in molten steel using laser-induced breakdown spectroscopy. Appl. Spectrosc. **47**, 606–608 (1993)

6.2. J. Belliveau, L. Cadwell, K. Coleman, L. Hüwel, H. Griffin, Laser-induced breakdown spectroscopy of steels at atmospheric pressure and in air. Appl. Spectrosc. **39**, 727–729 (1985)

6.3. K. Grant, G. Paul, J. O'Neill, Quantitative elemental analysis of iron ore by laser-induced breakdown spectroscopy. Appl. Spectrosc. **45**, 701–705 (1991)
6.4. R. Kirchheim, U. Nagorny, K. Maier, G. Tölg, Laser microprobe spectrometry of single-crystal metals and alloys. Anal. Chem. **48**, 1505–1508 (1976)
6.5. J. Millard, R. Dalling, L. Radziemski, Time-resolved laser-induced breakdown spectrometry for the rapid determination of beryllium in beryllium-copper alloys. Appl. Spectrosc. **40**, 491–494 (1986)
6.6. R. Wisbrun, R. Niessner, H. Schröder, Laser-induced breakdown spectrometry as a fast screening for environmental analysis of trace amounts of heavy metals in soils. Anal. Meth. Instrum. **1**, 17–22 (1993)
6.7. D. Cremers, D. Romero, An evaluation of the factors affecting the analysis of metals using laser-induced breakdown spectroscopy (LIBS). Proc. SPIE **644**, 7–12 (1986)
6.8. W. van Deijck, J. Balke, F. Maessen, An assessment of the laser microprobe analyzer as a tool for quantitative analysis in atomic emission spectrometry. Spectrochim. Acta **34B**, 359–369 (1979)
6.9. T. Ishizuka, Y. Uwamino, Atomic emission spectrometry of solid samples with laser vaporization-microwave induced plasma system. Anal. Chem. **52**, 125–129 (1980)
6.10. M. Thompson, J. Goulter, F. Sieper, Laser ablation for the introduction of solid samples into an inductively coupled plasma for atomic emission spectrometry. Analyst **106**, 32–39 (1981)
6.11. J. Uebbing, J. Brust, W. Sdorra, F. Leis, K. Niemax, Reheating of a laser-produced plasma by a second pulse laser. Appl. Spectrosc. **45**, 1419–1423 (1991)
6.12. E. Piepmeier, H. Malmstadt, Q-switched laser energy absorption in the plume of an aluminium alloy. Anal. Chem. **41**, 700–707 (1969)
6.13. R. Scott, A. Strasheim, Laser induced plasmas for analytical spectroscopy. Spectrochim. Acta **25B**, 311–332 (1970)
6.14. R. Sattmann, V. Sturm, R. Noll, Laser-induced breakdown spectroscopy of steel samples using multiple Q-switch Nd:YAG laser pulses. J. Phys. D. Appl. Phys. **28**, 2181–2187 (1995)
6.15. A. Pichahchy, D. Cremers, M. Ferris, Elemental analysis of metals under water using laser-induced breakdown spectroscopy. Spectrochim. Acta B **52**, 25–39 (1997)
6.16. V. Sturm, L. Peter, R. Noll, Steel analysis with laser-induced breakdown spectrometry in the vacuum ultraviolet. Appl. Spectrosc. **54**, 1275–1278 (2000)
6.17. F. Colao, V. Lazic, R. Fantoni, S. Pershin, A comparison of single and double pulse laser-induced breakdown spectroscopy of aluminum samples. Spectrochim. Acta B **57**, 1167–1179 (2002)
6.18. L. St-Onge, V. Detalle, M. Sabsabi, Enhanced laser-induced breakdown spectroscopy using the combination of fourth-harmonic and fundamental Nd:YAG laser pulses. Spectrochim. Acta B **57**, 121–135 (2002)
6.19. S. Klimentov, S. Garnov, T. Kononenko, V. Konov, P. Pivovarov, F. Dausinger, High rate deep channel ablative formation by picosecond-nanosecond combined laser pulses. Appl. Phys. A **69**, S633–S636 (1999)
6.20. D. Stratis, K. Eland, S. Angel, Effect of pulse delay time on a pre-ablation dual-pulse LIBS plasma. Appl. Spectrosc. **55**, 1297–1303 (2001)
6.21. D. Stratis, K. Eland, S. Angel, Dual-pulse LIBS using a pre-ablation spark for enhanced ablation and emission. Appl. Spectrosc. **54**, 1270–1274 (2000)
6.22. A. Kuwako, Y. Uchida, K. Maeda, Supersensitive detection of sodium in water with use of dual-pulse laser-induced breakdown spectroscopy. Appl. Optics **42**, 6052–6056 (2003)
6.23. R. Noll, R. Sattmann, V. Sturm, S. Lüngen, H.-J. von Wachtendonk, Schnelle Multielement-analyse in der Stahlschmelze mit laserinduzierter Emissionsspektrometrie. Stahl u. Eisen **117**, 57–62 (1997)
6.24. DE 10 2008 032 532, Verfahren und Vorrichtung zum präparierenden Lasermaterialabtrag, Patent application, Publication date 11.2.2010, 13 p.

Chapter 7
Material Ablation

The interaction of the pulsed laser beam with the material causes an ablation of mass from a liquid or solid measuring object. This mass is partially evaporated and transferred into the plasma state, where the species are excited to emit element-specific radiation used for LIBS. However, a part of the ablated mass leaves the measuring object as material vapor, particle, or liquid. At a later stage, this vapor condenses, forms particulates, or precipitates in the neighborhood of the interaction region. The particles ejected may be carried away by a gas flow or they propagate back to the surface of the measuring object forming a deposition. If a melt phase occurs – e.g., in case of the interaction of pulsed laser radiation with metals – a part of the mass is ejected as a liquid forming splashes or droplets which resolidify in the surrounding atmosphere or on the surface of the measuring object. Due to the complexity of processes, it is generally a difficult task to determine quantitatively the amount of ablated mass for single laser pulses or laser bursts.

In Sect. 7.1, the formation of a crater in a solid object will be described. Section 7.2 presents measurements of the ablated mass for single and multiple pulses at different ambient pressures and gas types.

7.1 Crater Formation

Figure 7.1 illustrates schematically a cross section of a crater formed in a metallic sample after the interaction with a series of laser pulses. The position of the original surface RP can be used to define the upper border of the actual crater volume. The material above RP, e.g., the collar formed around the crater rim (CL in Fig. 7.1) and the splashes and droplets (SP in Fig. 7.1), is due to resolidifcation and recondensation processes occurring during and after the laser–matter interaction. In case of a rotationally symmetrical geometry, the diameter d_1 describes the size of the crater at the plane RP corresponding to the area of high irradiance of the laser beam and d_2 is the diameter of the area affected by splashes, recondensed particles,

R. Noll, *Laser-Induced Breakdown Spectroscopy*, DOI 10.1007/978-3-642-20668-9_7,

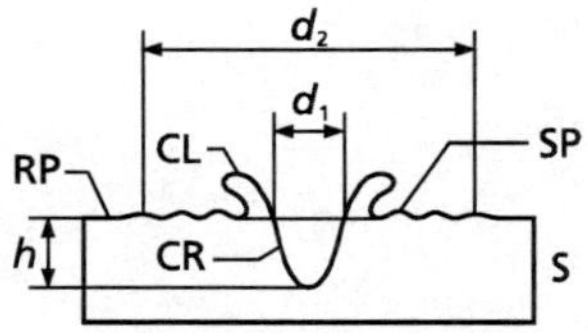

Fig. 7.1 Schematic illustration of the cross section of a crater formed in a metallic sample after interaction with a series of laser pulses. S = sample, RP = original surface of the sample serving as reference plane, CR = crater underneath RP, CL = collar formed by solidified melt ejected along the crater walls during the interaction with the laser beam, SP = splashes and particles deposited on the surface, d_1, d_2 = diameters, see text; h = depth of the crater

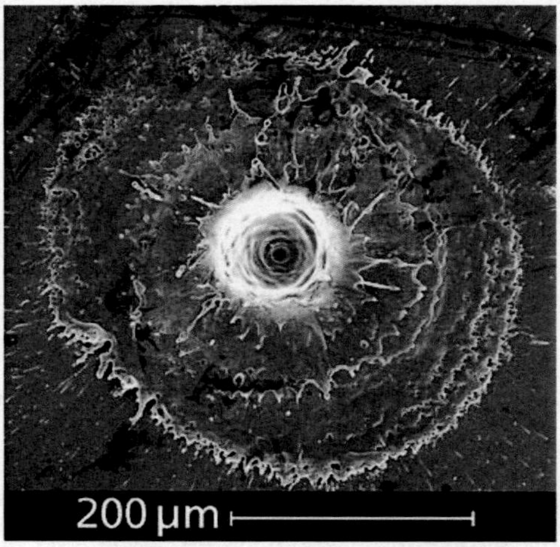

Fig. 7.2 SEM image of a crater formed in a steel sample after interaction with ten bursts, 2 × 40 mJ, $\Delta s = 0$ mm

depositions, and secondary interactions between the formed plasma and the sample surface.

Figure 7.2 shows an image of a crater formed by a series of ten double pulses in a steel sample taken with a scanning electron microscope (SEM). Evaluating this image yields for the diameters described above: $d_1 = 73\,\mu\text{m}$ and $d_2 = 390\,\mu\text{m}$. The diameter d_1 is approximately equal to the determined diameter of the laser beam at the plane RP of 70 μm (cf. Fig. 3.12). While SEM images offer a high lateral resolution, it is not possible to determine the depth of deep craters (the depth of shallow craters can be determined with stereo SEM).

Figure 7.3 shows a metallographic cross section of a crater taken with an optical microscope. The collar of the crater is clearly visible. Such cross sections enable the determination of the crater dimensions diameter and depth. However the necessary grinding process is time consuming and the achievable accuracy depends on the accuracy to which extent the middle of the crater is met. The results of the determination of the crater volume by light microscopy and stereo SEM are shown exemplarily in Table 7.1 for single and double pulses irradiated onto a steel sample.

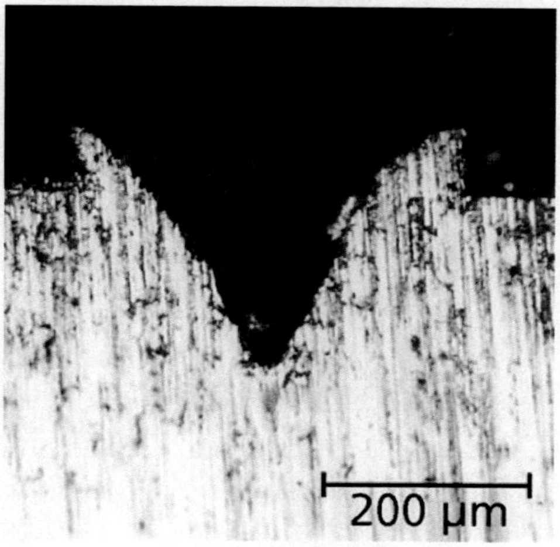

Fig. 7.3 Optical microscope image of a cross section of a crater formed after 200 bursts, 1×80 mJ, $\Delta s = 0$ mm

Table 7.1 Determination of the crater volume per burst by light microscopy and stereo SEM for single and double pulses

N_b	E_b (mJ)	V/N_b (μm^3) light microscopy	V/N_b (μm^3) stereo SEM
50	1×80	3,400	5,300
1	2×40	53,000	48,000
10	2×40	34,600	77,900

N_b = number of laser bursts, E_b = burst energy, V/N_b = crater volume per burst, SEM = scanning electron microscope, $\Delta s = 0$ mm

Differences in the determined crater volume per burst of up to a factor of two occur comparing the two methods. Yet the data show clearly the increase in the crater volume for double pulses by a factor of 9 to 16.

The shape of the crater depends among others on the laser beam profile. A spatial beam profiling of the laser allows to modify the crater shape. A Gaussian beam profile of a laser source can be formed to a top hat beam profile by, e.g., slightly focusing the beam on a 600-μm pinhole, which cuts off the outer part of the Gaussian beam, and leads to a near flat top beam. A disadvantage of this method is the loss of 95% of the burst energy. The aperture is then imaged onto the surface of the specimen with a demagnification factor of 10. Figure 7.4 shows an image of a crater generated with such a top hat laser beam profile on a galvanized sheet steel. A crater with a flat crater bottom is formed.

7.2 Ablated Mass

Various methods have been applied to determine the ablated mass. Table 7.2 gives an overview of results for the ablated masses for various materials, laser pulse parameters, and ambient conditions. The masses ablated per pulse range from a few 100 pg to about 10 μg depending on the experimental conditions.

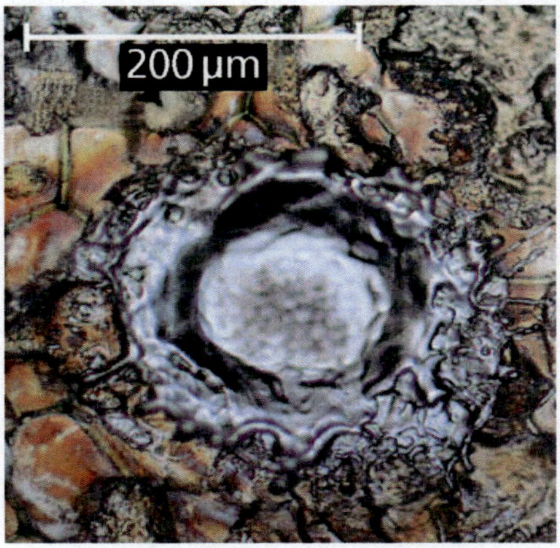

Fig. 7.4 LIBS crater generated with a top-hat beam profile using two double pulses, $E_b = 8.5$ mJ, crater diameter $d_1 \approx 155\,\mu$m

Piepmeier et al. showed that the amount of ablated material at atmospheric pressure was relatively constant over the range of energies used [7.3]. The stated uncertainty of the mass determination is about 30%. At 1.3 mbar, the amount of copper vaporized increased significantly with laser energy. With a 100-mJ laser pulse, the amount of ablated copper is about an order of magnitude greater at 1.3 mbar, 300 ng, than at 1,000 mbar. Sattmann et al. studied the influence of single and double pulses on the ablated mass [7.7]. The number of bursts which are necessary to drill through a foil of stainless steel of 100-μm thickness is determined experimentally. This number is approximately inversely proportional to the material removal per burst. The mean drilling depth per burst is the foil thickness divided by the mean value of the number of laser bursts until penetration. It should be noted that the material removal for a foil is different from that of a semi-infinite sample because of restricted heat transfer conditions, but is expected to behave similarly. Semerok et al. used polished samples and measured the craters formed by means of a phase-stepping profilometer with lateral resolution of $\approx 0.5\,\mu$m and depth resolution of $\approx 0.01\,\mu$m [7.8]. Crater volumes were measured with reference to the initial surface of the sample (without "corona" around the crater). The crater depth as a function of the fluence and the crater volume as a function of the pulse energy were studied. Ablation efficiency dependence on pulse duration and wavelength was discussed.

In the following paragraphs, the material ablation studied for single and multiple pulses will be presented in more detail. Figure 7.5 shows the drilling depth d per burst as a function of the laser burst energy at an ambient air pressure of 60 and 1,013 mbar for single-, double-, and four-pulse bursts ($\Delta s = 0$ mm for these measurements). The interpulse separation for double- and four-pulse bursts is 6 μs. The mass ablated per burst (right-hand ordinate) is determined under the assumption of a bore hole diameter of 200 μm, which is the mean diameter of the holes, measured with a microscope. For an ambient pressure of 1,013 mbar, the material ablation of single pulses cannot be raised by increasing the laser burst energy as

Table 7.2 Overview of determinations of ablated mass per laser pulse (or per laser burst) for solid samples

Reference	Material	E_L (mJ)	τ_L (ns)	m (μg)	Remarks
Lorenzen [7.1]	Steel	25–300	6–9	0.1–7	1,064 nm, ablated area 0.1–6 mm^2, n.s.
De Young [7.2]	Al, Cu	400	8	0.4–1.5	High vacuum, 10^9 W/cm^2,weighing
Piepmeier [7.3]	Cu	3–120	60	0.04	Pressure dependence, $p = 1{,}000$ mbar
	Cu	3–120	60	0.04–0.31	$p = 1.3$ mbar, dissolution of condensed matter and flame atomic absorption spectroscopy
Felske [7.4]	Al Cu Steel			1.9 2.7 01.4	Mechanical Q-switching, 1.8×10^{10} W/cm^2, mass calculated from crater volume
Leis [7.5]	Steel	5	8	0.03	1,064 nm, crater diameter 100 μm, p(Ar) = 140 mbar, n.s.
Jowitt [7.6]	Steel	750	20	0.03	1,064 nm, weighing, weight of ablated mass is approximately proportional to the laser pulse energy in the range of 500–800 mJ
Sattmann [7.7]	Steel	40–320	25	SP: 0.1–0.14 DP: 0.3–0.7	1,064 nm, drilling of steel foil, first demonstration of enhanced mass ablation with DP, $\Delta t = 6\,\mu$s
Semerok [7.8]	Fe	0.06	4	532 nm, $d = 8\,\mu$m: 710 pg	1,064, 532, 266 nm, laser spot diameter d
			25 ps	532 nm, $d = 9.5\,\mu$m: 188 pg	Phase-stepping profilometer to determine crater volume

E_L = laser pulse or laser burst energy, τ_L = pulse width of the laser pulses; in case of double pulses, the interpulse separation Δt is stated as well. If not stated otherwise the experiments were performed at normal atmospheric conditions and single pulses were applied. SP, DP = single/double pulses, $n.s.$ = method of determination of the ablated mass is not specified

observed by Piepmeier et al., see [7.3]. This behavior is attributed to a shielding of the target by a plasma formed in the atmosphere in front of the sample. The material removal of double pulses is higher than that for single pulses with the same energy, and it can be further raised by using four-pulse bursts. No saturation of the removal is observed for increasing burst energy of double pulses up to $E_b^d = 320$ mJ. Due to the lower peak intensity of the double pulse, shielding effects are reduced which favors – among other effects to be discussed later – an increase of the mass ablation.

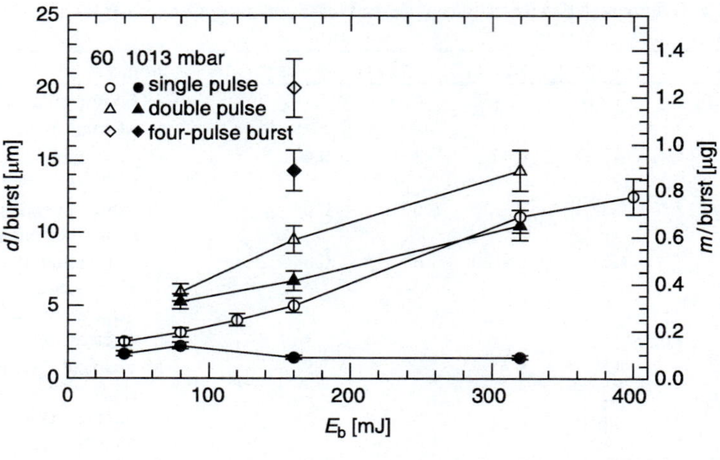

Fig. 7.5 Drilling depth d per burst in a steel foil as a function of the burst energy for single-, double-, and four-pulse bursts at 60 and 1,013 mbar ambient atmospheric pressure. The interpulse separation of double- and four-pulse bursts is 6 μs. The right axis shows the corresponding ablated mass per burst. $\Delta s = 0$ mm

In the following paragraphs, we compare the ablated mass of a double pulse with energy E_b^d with the ablated material for two single pulses with half of the energy of the double pulse, $E^s = E_b^d/2$. Figure 7.5 shows that, e.g., a $2 \times 40-$mJ double pulse removes about 0.32 μg (at 1,013 mbar ambient pressure), whereas two 40-mJ single pulses ablate only about 2×0.1 μg. The significant difference between the ablation of a double pulse with energy E_b^d and the ablation of two single pulses with energy $E^s = E_b^d/2$ shows that the material ablation is strongly affected by the combined effect of the subsequent pulses of a double pulse at interpulse separations of 6 μs.

Figure 7.5 also shows results for a pressure of 60 mbar. Due to the reduced ambient pressure, the material vapor can escape faster from the interaction region, which results in higher specific drill depths for all studied laser parameters. Furthermore, shielding effects by a plasma in the ambient atmosphere are diminished.

7.3 Material Ablation in Different Ambient Gases by Collinear Multiple Laser Pulses

Typically the interaction region is flushed with an ambient gas to avoid a breakdown at residual particulates above the sample surface which are generated by preceding laser pulses [7.9,7.10,7.11,7.12,7.13,7.14]. Furthermore the plasma conditions like electron density and electron temperature are strongly dependent on the ambient gas [7.13,7.15]. Often inert gases such as argon are used because they offer a higher transmittance for wavelengths below 200 nm [7.9, 7.10, 7.11] compared with air.

The oxygen content of 20% absorbs significantly wavelengths below 200 nm (cf. Fig. 3.16). This wavelength region is interesting especially for the elements C, P, S, N, and O, which are important, for example, for the analysis of steel samples which are used for process monitoring during steelmaking.

The repetition rates of Q-switched solid-state lasers with pulse energies between 100 and 500 mJ and a pulse length of 10 ns are typically in the range of 10–50 Hz. Many studies about laser ablation were performed using these parameters [7.7,7.12, 7.13,7.14,7.16,7.17]. Sattmann et al. describe the influence of interpulse separation and burst energy of collinear double pulses on laser ablation [7.7]. Russo studied the influence of the pressure of helium and argon atmospheres [7.16]. The lower ionization energy of argon leads to a more severe plasma shielding and thus smaller crater depths compared with those obtained in a helium atmosphere. Mass ablation rate depends on the laser pulse duration and the power density on the sample surface [7.17]. The type of ambient gases like helium, argon, and neon have an effect on the particle size distribution showing that in helium significantly smaller particles occur compared with Ar [7.14]. Aguilera et al. looked for the effect of ambient gases on the temporal evolution of electron temperatures and electron densities of the laser-induced plasmas [7.13]. The decay times are shorter in helium than in air and argon.

In the following paragraphs, the influence of single, collinear double, and triple laser pulse bursts in different ambient gases consisting of mixtures of argon, nitrogen, and oxygen on the ablation of a steel sheet is described in more detail [7.18].

The experimental setup is shown in Fig. 4.18, Chap. 4. The laser beam generated by a Q-switched Nd:YAG laser working at the fundamental wavelength 1,064 nm is guided via a Glan-laser prism, a twofold beam expansion, and a mirror to a lens focusing the laser beam into a gas-flushed sample stand onto the sample. The Glan-laser prism is used to attenuate the laser pulse energy while keeping the operational conditions of the laser source constant. Two lasers are used for the experiments. Laser 1: Surelite SLI-10 (Continuum, Santa Clara, CA), repetition rate of 10 Hz, with double pulse option, interpulse separation $\geq$ 30 μs. Laser 2: YM/R-201 (Innolas, Krailling, Germany), repetition rate of 50 Hz, with double and triple pulse option, interpulse separations 5–50 μs.

The plasma light is detected via a direct light channel with a vacuum Paschen–Runge spectrometer without further optical imaging. In the Paschen–Runge spectrometer, the plasma light is dispersed spectrally covering a spectral range from 178 to 406 nm (grating, 2,700 l/mm) with a spectral resolution of ~20 pm. The distance of the sample surface to the entrance slit of the spectrometer is ~300 mm. The plasma signals are detected at a detection angle of $\alpha_D = 10°$ between the optical axis of the spectrometer and the sample surface. Twenty photomultipliers are located on the Rowland circle (diameter 0.5 m) to detect 19 element lines and the zeroth order signal. A multichannel-integrator electronics (see Sect. 4.5) integrates the signals in programmable time windows for each generated plasma (SE in Fig. 4.18).

The flow rate of the ambient gases through the sample stand amounts to about 17 l/min. The gas mixtures studied were provided in gas bottles from Linde AG with specified composition. A series of laser pulses are irradiated to one sample position.

The focal length of the plano-convex lens is $f = 150\,\text{mm}$ and the focal position is $\Delta s = 8\,\text{mm}$ within the sample to avoid a gas breakdown in front of the sample surface. The experiments are carried out with the above-mentioned two laser sources in different pulse modes as illustrated in Fig. 3.9, Chap. 3.

The pulse ratio and the interpulse separation within a burst can be varied. Double- and triple-pulse bursts (DP/TP) with equidistant interpulse separations, uniform energy distribution (i.e., pulse ratios = 1), and equal total energy (i.e., burst energy) are used. The pulse width for laser 1 in single-pulse mode (SP) is $\tau_{SP} = 6\,\text{ns}$ and in DP mode 10 ns for both pulses. The interpulse separation was set to $\Delta t_1 = 50\,\mu\text{s}$, where laser 1 is able to generate stable DP with uniform energy distribution (burst energy 160 mJ = 80 + 80 mJ). For laser 2, the pulse widths for SP is $\tau_{SP} = 7\,\text{ns}$ and for DP, TP $\tau \sim 14$–40 ns for each pulse. Comparing the different burst modes, it is not possible to keep the pulse widths constant because the double- and triple-pulse bursts are generated with one laser system within a single flashlamp discharge. The pulse widths depend on the flashlamp voltage, the delay between the start of the flashlamp discharge and the Q-switch trigger as well as on the interpulse separations. Generally the material ablation will be influenced by both the pulse structure (SP, DP, TP) as well as by the individual pulse widths. However, with the setup used here, there is no possibility to vary the pulse widths independent from the pulse structure. Nevertheless, the use of a single laser system to generate collinear multiple pulses is of great practical interest since it simplifies the setup of industrial LIBS systems.

The line emissions are integrated with a time delay of $t_{delay} = 4\,\mu\text{s}$ and an integration width of $t_{int} = 10\,\mu\text{s}$ after the last pulse of a burst. Five measurements at different sample locations are carried out for each parameter set.

The efficiency of the laser ablation process is described by the ablation burst number $N_{abl,\,3s}$, which is defined as the number of laser bursts needed to penetrate a steel sheet of a given thickness. In order to quantify $N_{abl,\,3s}$ experimentally in a precise and reproducible way, a block of copper is put onto the rear side of the steel sheet, see Fig. 7.6, and during the ablation sequence LIBS signals of the copper line 324.8 nm are detected.

The steel sheet has a low copper content and therefore only a background Cu signal is detected during the ablation of the steel sheet. Once the steel is penetrated, the Cu signal rises strongly, see Fig. 7.7.

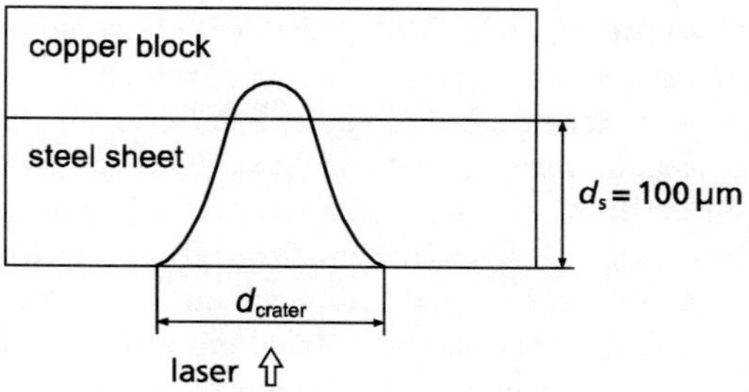

Fig. 7.6 Schematic drawing of steel sheet and copper block with laser-generated crater. If the depth of the crater is greater than 100 μm a copper signal is detected indicating the penetration of the steel sheet

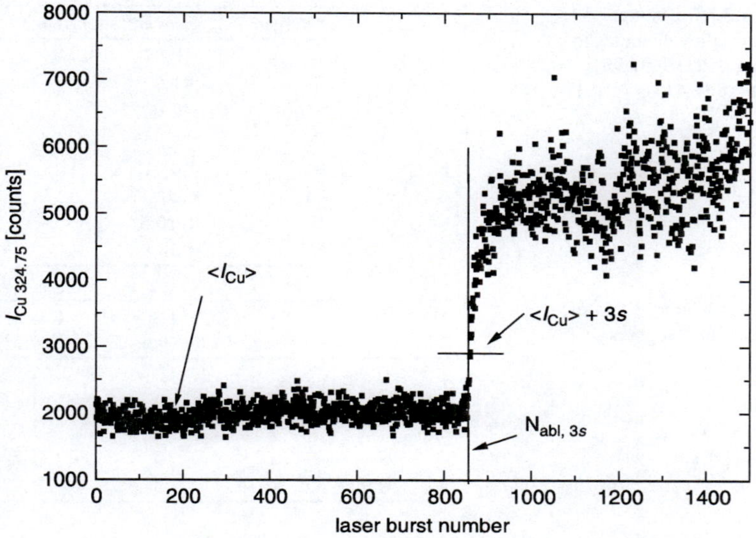

Fig. 7.7 Intensity of the copper line 324.75 nm as a function of the laser burst number. At a burst number of $N_{abl,\,3s}$, the intensity increases to $\langle I_{Cu}\rangle + 3s$, where $\langle I_{Cu}\rangle$ is the average copper intensity prior to penetration and s the corresponding standard deviation

First, we determine the average background Cu signal $\langle I_{Cu}\rangle$ prior to penetration of the steel sheet and the corresponding standard deviation s (in the example shown in Fig. 7.7 from pulse 101–700). We define $N_{abl,\,3s}$ as the lowest ablation burst number where the Cu signal exceeds the sum $(\langle I_{Cu}\rangle + 3s)$. Hence, the probability of a wrong result due to statistical fluctuations in the background Cu signal is less than 1%.

Steel sheets with known thickness are used as samples for the experiments to determine the ablation burst number $N_{abl,\,3s}$. The steel sheets are made of cold-rolled, spring hard metal sheet X5CrNi18 (1.4301) with a well-defined thickness. The sheets are 13 mm wide with a thickness of $100 \pm 1.2\,\mu m$. Their density amounts to $7.9\,g/cm^3$. The concentration ranges of the elements are listed in Table 7.3 [7.19]. From the composition given in Table 7.3, the minimum iron content is greater than $c_{Fe} = 66.8$ m.-%.

Gas mixtures with varying proportions of argon are chosen to examine a range of chemical reactivity and gas densities. An idea to enhance the ablation process is to use a mixture of argon and oxygen as ambient gas. An exothermic reaction between ablated material in the outer plasma region may contribute an additional energy supply for the plasma and thus enhance the plasma–matter interaction. Therefore, five ambient gases are studied, see Table 7.4, to determine the influence of gas composition and density on the ablation efficiency.

Figure 7.8 shows the ablation burst number $N_{abl,\,3s}$ as a function of the gas density for steel sheets with a thickness of $100\,\mu m$, single- and double-pulse bursts of

Table 7.3 Concentration ranges of alloy elements in the steel sheet of grade X5CrNi18 (1.4301) [7.19]

Element	Concentration range (m.-%)
C	≤0.07
Si	≤1.00
Mn	≤2.00
P	≤0.045
S	≤0.015
Cr	17.0–19.5
Ni	8–10.5
N	≤0.11

Table 7.4 Composition of used mixtures of gases

No.	Ar (vol.-%)	O_2 (vol.-%)	N_2 (vol.-%)	Gas density (kg/m^3)
1	0	20	80	1.29
2	50	50	0	1.61
3	92	8	0	1.76
4	99	1	0	1.78
5	99.999	0	0	1.784

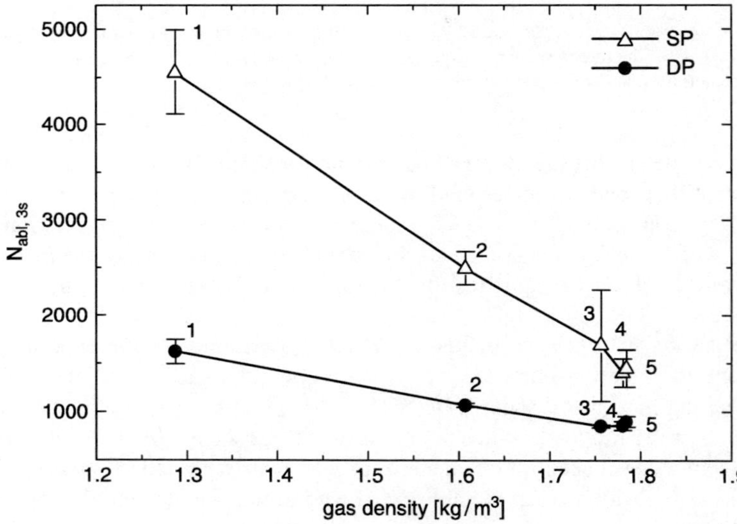

Fig. 7.8 Ablation burst number $N_{abl,3s}$ as a function of the gas density. Laser 1, $E_b = 160$ mJ, $\Delta t_1 = 50\,\mu$s, $\Delta s = 8$ mm. Errors bars mark 3s standard deviation. The numbers 1–5 at the data points refer to the gas mixtures listed in Table 7.4

$E_b = 160$ mJ generated with laser 1. The interpulse separation of the DP amounts to $\Delta t_1 = 50\,\mu$s.

A nearly linear relationship between the ablation burst number $N_{abl,3s}$ and the gas density is observed for both SP and DP. The ablation burst number for DP is always smaller than for SP and can be reduced by a factor of 2.8 in ambient gas no. 1 (cf. Table 7.4) using DP instead of SP. For gas no. 5, the factor of improvement is 1.6. The influence of the burst mode on the ablation process decreases with rising gas density. The ablation burst number plotted as a function of the oxygen content shows no clear correlation. There is certainly a reaction of the oxygen with the plasma constituents but the effect might only take place in the outer region of the plasma and has no significant influence on the steel ablation.

The crater diameters are in single-pulse mode $d_{\text{crater,SP}} \approx 1{,}300\,\mu\text{m}$ and in double-pulse mode $d_{\text{crater,DP}} \approx 750\,\mu\text{m}$ for all mixtures of gases. In the studied parameter regime, the gas density of the ambient gases has no effect on the crater diameters. Therefore, the ablation burst number is a measure for the steel ablation and approximately proportional to the ablated depth for both SP and DP.

The smaller crater diameters for double-pulse bursts are an effect of the reduced plasma–sample interaction. Also St-Onge et al. observed smaller crater diameters using DP instead of SP [7.20]. An explanation of the reduced plasma–sample interaction for DP is given in [7.21]. For double pulses, the first pulse leads to a local dilution of the ambient gas in the interaction region between laser radiation and sample. The second pulse of the double-pulse burst interacts again with the sample surface and generates a plasma which detaches from the sample surface providing an efficient way to transport material away from the interaction region. For this the thermal influence of the plasma on the sample decreases compared with single pulses. The plasma generated by single pulses still stays in contact with the sample surface causing an enhanced thermal interaction with the sample.

The material ablated per burst increases with rising gas density (a reduction of the oxygen and an increase of the argon content enlarge the gas density of the ambient gas). In the case of negligible plasma shielding, the plasma–sample interaction is greater in an ambient gas of higher density because the plasma expansion is restricted by the surrounding mass density which results in higher density and temperature of the plasma in the vicinity of the sample surface leading to enhanced steel ablation.

On the contrary, investigations on copper samples showed that the crater depth and the ablated mass are larger in air than in argon atmosphere [7.15]. This behavior was attributed to the high electron density in case of an argon atmosphere which results in an optical shielding of the sample surface. In [7.15] the laser beam is focused on the sample surface (i.e., $\Delta s = 0\,\text{mm}$, instead of 8 mm as applied here), resulting in laser radiant exposures of about $150\,\text{J/cm}^2$, which are more than a factor of three higher than those used in this study. The shielding effect may be effective in [7.15], whereas it plays a minor role here. Furthermore, the gas flow rate applied in [7.15] amounts to 100 ml/min, which is more than two orders of magnitude less the one we used (17 l/min) while the laser repetition rate is comparable (20 Hz in [7.15] and 10 Hz here). The high flow rate allows for an effective transport of the ablated or condensated particulates out of the interaction region, thus avoiding any accumulation or redeposition effects. This is a precondition for reproducible measurements of the ablation as manifested in the small error bars in Fig. 7.8. Vadillo et al. studied also the ablation behavior in different ambient gases [7.22]. Focusing a laser of 581 nm on a Fe foil (i.e., $\Delta s = 0\,\text{mm}$), higher ablation rates were found in an air atmosphere than in an argon atmosphere as well as higher ablation rates for low pressure than at atmospheric pressure. The results obtained were explained as an effect of the redeposition of rejected material at atmospheric pressure. However, no flow rate of the gas inside the measuring chamber is stated in [7.22]. If the flow rate was zero, then an accumulation of particulates in the vicinity of the interaction region originating from previous laser pulses will occur, causing a scattering and absorption of the laser radiation of subsequent pulses thus leading to a reduced interaction of the laser radiation with the sample surface. Due to the higher gas density, this effect is expected to be stronger for argon than for air (additionally in air a chemical reaction between the fine iron particulates and the oxygen will take place). This may explain the difference in the observed ablation behavior in air and argon to the one found here, where a high flow rate was used. Also the experiments described in [7.23] were performed with a very low flow rate of 50 ml/min and fluences being a factor of two greater than those used here. Mass ablation was quantified by an electronic microbalance, which implies that the result is to be interpreted as a net mass loss, i.e., the mass vaporized minus the mass redeposited.

A further difference of the setup used here (cf. Fig. 4.18) in contrast to those described in [7.15, 7.22, 7.23] is the orientation of the normal of the sample surface in relation to the axis of gravity. In [7.15, 7.22] the laser is irradiated onto the sample from above (i.e., surface normal and gravity vector are antiparallel), whereas the laser propagates from the bottom to the sample in the setup shown in Fig. 4.18 (i.e., the surface normal and the gravity vector are parallel). The latter configuration favors transport of the ablated material or condensated particulates away from the sample surface, thus reducing a potential redeposition.

Besides the gas density also other physical properties of the gas mixtures used were considered, which may show a correlation with the ablation burst number, such as ionization potential, chemical reactivity (via the oxygen content), and thermal conductivity. Only for the latter quantity, a clear correlation was found in terms of a decrease of the ablation burst number with decreasing thermal conductivity (gas mixture no. 1 has the highest thermal conductivity of about 0.026 W/m K, and gas mixture 5 has 0.018 W/m K). However, the interpretation given above discussing the effect of the higher gas density is congruent to the effect of a lower thermal conductivity which reduces the heat flow from the plasma to the ambient gas thus favoring higher temperatures in the plasma.

Furthermore, the influence of the burst mode and interpulse separation on laser ablation was studied. In these measurements, SP, DP, and TP are applied with symmetric burst energy partition and equal interpulse separation for triple-pulse bursts generated with laser 2. The ablation burst numbers $N_{\mathrm{abl,\,3s}}$ are measured on steel sheets with a thickness of $d_s = 100\,\mu\mathrm{m}$ and a total burst energy of $E_b = 105\,\mathrm{mJ}$. (This burst energy is the maximum possible burst energy for TP and interpulse separations of $\Delta t_1 = \Delta t_2 = 5\,\mu\mathrm{s}$. For all other SP, DP, and TP configurations, laser 2 is able to generate higher burst energies, which are then attenuated externally down to 105 mJ to keep the burst energy for the different pulse structures studied constant.) The ambient gases no. 4 and 5, see Table 7.4, are used to examine the influence of an oxygen content on the ablation process. Figure 7.9

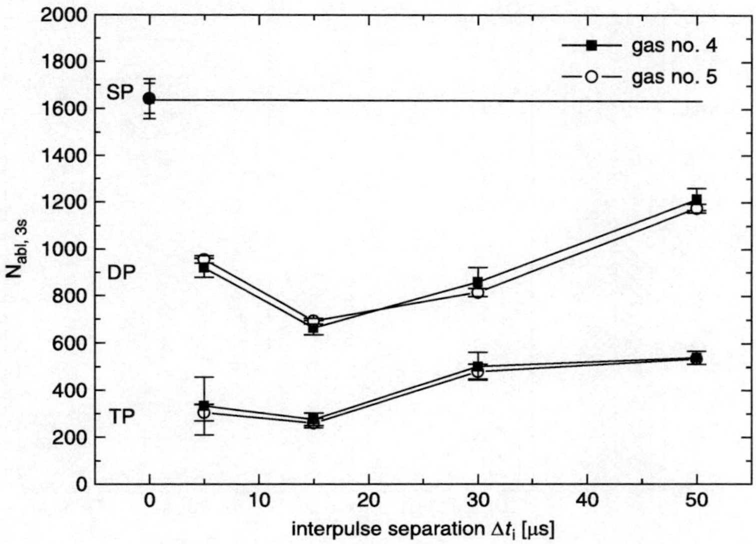

Fig. 7.9 Ablation burst number $N_{abl, 3s}$ as a function of the interpulse separation and burst mode for ambient gases no. 4 (99 vol.-% argon, 1 vol.-% O_2) and 5 (99.999 vol.-% argon) of Table 7.4. Laser 2, $E_b = 105\,mJ$, $\Delta s = 8\,mm$

shows the ablation burst numbers as a function of the interpulse separation Δt_i for SP, DP, and TP. The ablation burst number for SP is plotted at $\Delta t = 0$ and marked with a horizontal line.

No significant difference for gas no. 4 and 5 is observed. Hence, the gas density seems to have the dominant influence on depth penetration, whereas the oxygen content plays only a minor role for the studied gas mixtures.

For all interpulse separations Δt_i, the ablation burst numbers are significantly smaller for DP and TP than for SP. Using DP with an interpulse separation of $\Delta t_1 = 50\,\mu s$ – like in the measurement with laser 1 – the ablation burst number $N_{abl, 3s}$ for DP is reduced by a factor of 1.4. With DP and an interpulse separation of $\Delta t_1 = 15\,\mu s$, the depth penetration increases by a factor of 2.4 in comparison with SP. Using TP with interpulse separations of $\Delta t_1 = \Delta t_2 = 50\,\mu s$, the ablation burst number decreases by a factor of 3.1. With TP and interpulse separations of $\Delta t_1 = \Delta t_2 = 15\,\mu s$, the ablation burst number decreases in ambient gases no. 4 and 5 by a factor of 6.3 in comparison with single pulses. Surprising are the measurements with interpulse separations of $\Delta t_1 = \Delta t_2 = 5\,\mu s$. A further lowering of the ablation burst number was expected. However, the ablation burst number $N_{abl, 3s}$ increases for DP and stays about constant for TP.

In Fig. 7.10, ablation burst numbers $N_{abl, 3s}$ of measurements in ambient gas no. 5 performed on different days are illustrated to show the reproducibility of the results. The days are labeled with the letters A, B, and C. "A" marks the first day

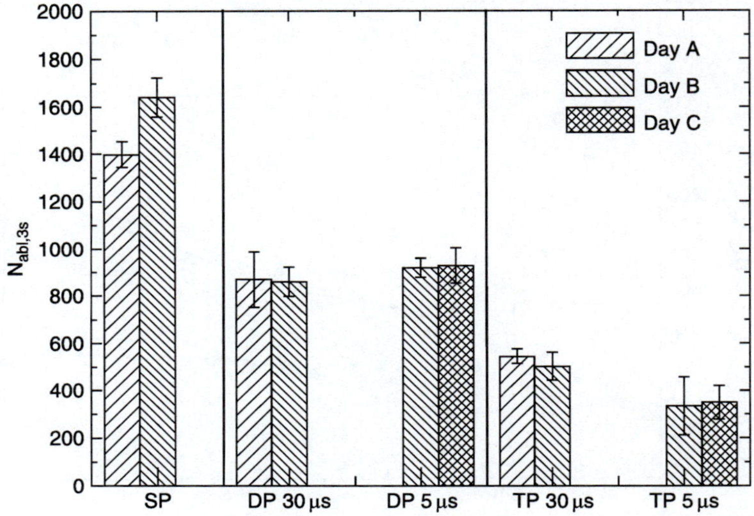

Fig. 7.10 Ablation burst numbers $N_{abl, 3s}$ for SP, DP, and TP measured at different days. Laser 2, $E_b = 105$ mJ, SP/DP/TP = single/double/triple pulse

of measurements, "B" is 8 days after "A," and "C" is 11 days after day "A." The error bars represent the standard deviation of five measurements on the samples (the same sample type is used as before). The labeling "DP 30 μs" means that the ablation burst numbers $N_{abl, 3s}$ are measured in double-pulse mode with an interpulse separation of $\Delta t_1 = 30\,\mu s$. The other labels are given in an analogous way. In the first part of the figure, the SP measurements are presented showing a difference in the ablation burst numbers for SP on day A and B which is about a factor 2 greater than the standard deviation. In the second part of Fig. 7.10, the ablation burst numbers using DP are shown. There is no significant difference between the results using interpulse separations for $\Delta t_1 = 30\,\mu s$ or $\Delta t_1 = 5\,\mu s$. For TP again, no significant difference for bursts with interpulse separations of $\Delta t_1 = \Delta t_2 = 30\,\mu s$ and $\Delta t_1 = \Delta t_2 = 5\,\mu s$ was found. The error bars for $\Delta t_1 = \Delta t_2 = 5\,\mu s$ are significantly greater. The reason is that for TP, the LIBS signals used to determine the ablation burst number $N_{abl, 3s}$ show a higher scattering attributed to a reduced stability of laser 2 in the TP mode.

7.4 Material Ablation by Collinear Double Laser Pulses at Reduced Ambient Gas Pressures

Several authors have studied the influence of ambient gases and different gas pressures on laser ablation with single pulses [7.5, 7.24]. Argon as ambient gas in the pressure range from 0.1 to 1,000 hPa was studied for LIBS using single Nd:YAG laser pulses having an energy of 5 mJ [7.5]. The target materials were low-alloyed steel samples. For the silicon line Si I 288.2 nm, the authors found a pronounced maximum at an ambient pressure of 140 hPa attributed to an optimum atomization process of the ablated and expanding sample material interacting with the argon buffer gas. The expansion dynamics of the plasma generated by laser ablation of an aluminum target was investigated using frequency-doubled radiation from a Q-switched Nd:YAG laser [7.24]. The ambient gas studied was air at pressures ranging from 10^{-6} to 100 torr. In the pressure region between 10^{-6} and 2×10^{-2} torr the plasma expands freely. At pressures of 10^{-1} torr, the plasma breaks away from the target surface and a contact boundary is formed between plasma and a shock wave in the gas. At 1.3 torr, the onset of an instability is observed perturbing the interface between plasma and ambient gas. At pressures between 10 and 100 torr a plume confinement is observed. The dynamics of plume boundaries generated by irradiation of steel targets using a KrF laser at 248 nm were studied at ambient argon pressures between 3×10^{-5} and 20 mbar [7.25]. The authors showed that over a wide range of parameters the dynamics can be described by a generalized shock wave expansion model. Ambient gas effects were also studied for resonance-enhanced laser-induced plasma spectroscopy irradiating potassium iodate pellets with frequency-doubled laser pulses from a Nd:YAG laser followed after 27 ns by a dye laser pulse at 404.4 nm to resonantly excite the K atoms [7.26]. By using a rarefied atmosphere of heavy inert gas, the hot plume could be thermally insulated and confined.

Most studies dealing with the effect of ambient gas pressure were restricted to the interaction of single laser pulses. However, Sattmann et al. studied the material ablation for single and collinear double pulses as a function of the burst energy for air pressures of 60 and 1,013 mbar giving a first indication that the saturation effect observed for the material ablation as a function of the irradiated laser pulse energy using single pulses at 1,013 mbar does not occur at pressures of 60 mbar and can be overcome by double pulses [7.7].

In the following studies are presented to clarify the influence of the ambient gas pressure for collinear double pulses interacting with a solid iron target [7.27]. Material ablation is determined as a function of the ambient gas pressure being varied between 0.1 and 1,000 mbar.

The experimental setup is shown in Fig. 7.11. A flashlamp-pumped Q-switched Nd:YAG laser emits at 1,064 nm and has a repetition rate of 10 Hz. The Pockels cell voltage allows for a stepwise increase of the voltage up to the $\lambda/4$-voltage to generate a burst of laser pulses with adjustable interpulse separations and pulse heights during a single flashlamp discharge. The laser beam is guided via two mirrors and two beam splitters to attenuate the burst energy. The resulting burst energy is measured with a volume absorbing disk calorimeter (Scientech 38–0101). A lens of 200-mm focal length focuses the laser beam through a window onto a sample positioned inside a measuring chamber. This chamber can be evacuated by a vacuum pump. Via a gas inlet the chamber is filled with pressures ranging from 0.1 mbar to 1 bar. The pressure is measured with two pressure gauges with measuring ranges of 0.1–27 mbar, 1–1,060 mbar, respectively (Wallace & Tiernan). The light emitted by the

Fig. 7.11 Experimental setup. LS = laser system, M1, M2 = mirrors, BS1, BS2 = beam splitters, CM = calorimeter, L = lens, W1, W2 = windows, S = sample, P = plasma, MC = measuring chamber, F = filter, FO = fiber optics, G = gas inlet, VP = vacuum pump, PG = pressure gauge

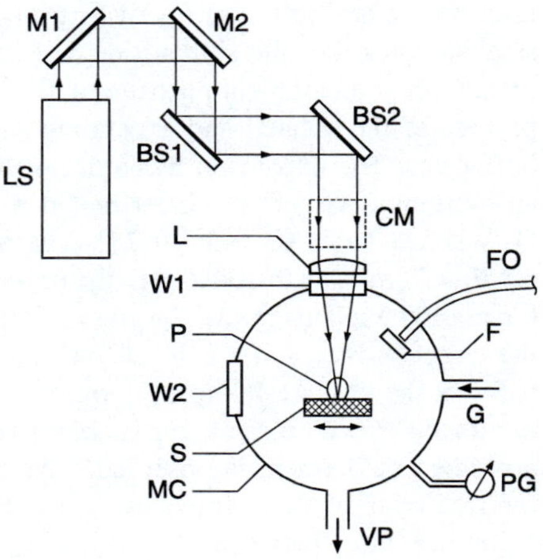

plasma is guided to the spectrometer with a fiber optics positioned at a distance of 75 mm from the plasma center and oriented at an angle of 45° to the direction of the incoming laser beam. The fiber optics consists of a bundle of quartz fibers each with a core diameter of 200 μm. The diameter of the bundle amounts to 3 mm. In front of the fiber, an edge filter is inserted to assure that with the spectrometer only emission lines in first diffraction order are observed (Schott type GG 395). With the numerical aperture of the fibers and the distance of 75 mm, the diameter of the receiving cone at the plasma location amounts to about 30 mm; hence, it is expected that the emission received is spatially integrated over the whole plasma volume.

The laser beam profile was measured with a CCD camera being shifted axially along the beam propagation axis to determine the Rayleigh length and the beam radius. The converging beam behind the focusing lens is further attenuated by reflection at a wedge prism and a stack of gray filters in front of the CCD camera. Table 7.5 shows the results for the beam waist radius and the Rayleigh length for single and double pulses of the same total energy of 80 mJ and an interpulse separation of $\Delta t = 6\,\mu s$. No significant differences were found.

For the measurements, the following parameters were used: single pulses 80, 160 mJ, pulse width 20 ns; double pulses $2 \times 40, 2 \times 80$ mJ, width 40 ns; interpulse separation $\Delta t = 6\,\mu s$, energy ratio 1:1; position of the beam waist relative to the sample surface $\Delta s = 10$ mm (i.e., the focus lies underneath the sample surface), detection after the single pulse or the second pulse of a double pulse with $t_{delay} = 1\,\mu s$, $t_{int} = 1\,\mu s$. Time-resolved studies are out of the scope of this study.

The following samples were used: (a) high-alloyed steel foils of the grade 1.4301 (18% Cr, 9% Ni) with a thickness ranging from 25 to 300 μm and (b) pure iron of Thyssen Krupp Stahl AG, TRP 1043.

The material ablation was determined by counting the number of bursts necessary to drill through high-alloyed steel foils. Figure 7.12a, b shows the number of bursts as a function of the foil thickness for air pressures of 1 bar and 0.1 mbar. Each data

Table 7.5 Beam waist radius w_0 and Rayleigh length z_R determined for single and double pulses focused by a 200-mm lens

	w_0 (μm)	z_R (mm)
Double pulse 2 × 40 mJ	32 ± 4	1.3 ± 0.2
Single pulse 1 × 80 mJ	35 ± 4	1.3 ± 0.2

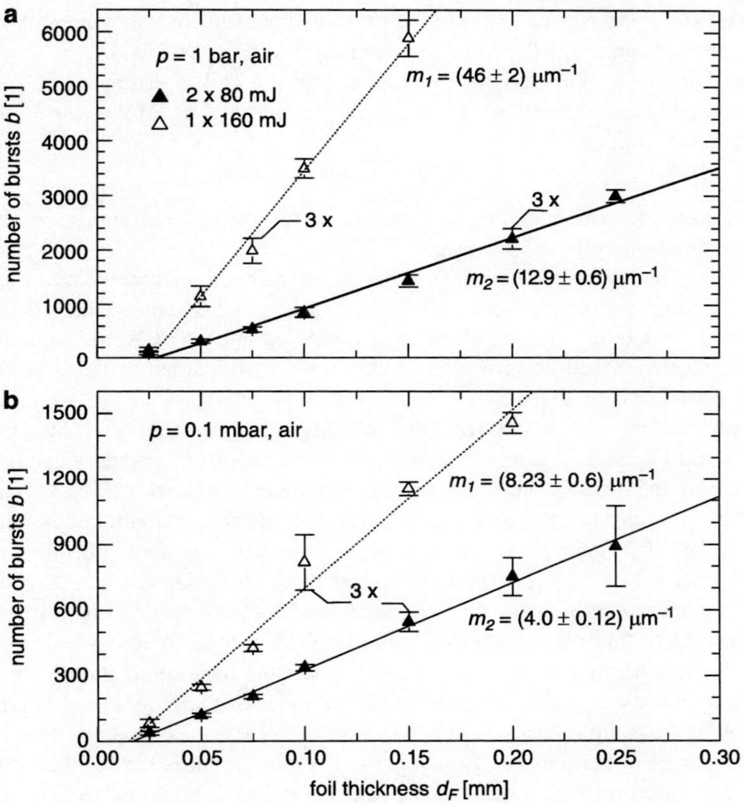

Fig. 7.12 Number of bursts needed to drill through high-alloyed steel foils as a function of the foil thickness for air pressures of **(a)** 1 bar and **(b)** 0.1 mbar for single and double pulses. Burst energy $E_b = 160\,\mathrm{mJ}$

point is the average of seven measurements. The error bars correspond to 99% confidence intervals and are plotted enlarged by a factor of 3 for better visibility. Linear relationships are found revealing significant differences in the slopes between single and double pulses and for the different pressures. For double pulses, the number of bursts necessary to drill through the steel foils is in general significantly smaller compared with single pulses. These differences cannot be explained by different beam profiles, since these were nearly identical for single and double pulses (see above). Furthermore, during drilling the effective laser irradiance is not changing significantly with increasing drilling depth. Taking the determined Rayleigh length $z_R = 1.3$ mm and $\Delta s = 10$ mm the reduction of the beam diameter at the maximum interaction depth of 300 μm is 3% with respect to the beam diameter at the surface of the foil. This corresponds to an increase of the irradiance by 6% while penetrating through the foil from the front to the backside. This increase has obviously no influence on the observed linear relationship between the number of bursts and the foil thickness shown in Fig. 7.12.

The potentially restricted heat conduction in the foil is estimated as follows. The foil can be considered to be a semi-infinite material as long as the following condition is fulfilled:

$$d_f \geq 4\sqrt{\kappa \tau_L}, \tag{7.1}$$

where d_f is the foil thickness, κ is the thermal diffusivity of the sample material, and τ_L is the duration of the laser pulse.

The right-hand side of relation (7.1) describes the depth in a semi-infinite sample where the temperature rise is 3×10^{-3} of the value achieved at the surface. With $\kappa = 0.04\,\text{cm}^2/\text{s}$ and $\tau_L = 40$ ns, relation (7.1) yields: $d_f \geq 1.6$ μm. At the beginning of the drilling this is fulfilled for all foil thicknesses used down to the thinnest foil having a thickness of 25 μm.

During drilling, the successive ablation leads to a local stepwise decrease of the residual foil thickness and hence after a certain number of laser pulses irradiated at one location the residual thickness may approach the estimated limit of 1.6 μm and condition (7.1) no longer holds. Figure 7.12b shows that for double pulses and the thinnest foil of 25 μm, the number of bursts determined is about 50. For a rough estimate, we calculate the ablated depths per burst by 25 μm/50 = 0.5 μm/burst. From this we expect that for the last three bursts, condition (7.1) is not fulfilled. With respect to the total number of bursts applied, this is obviously only a minor effect. Hence, no significant influence of a restricted heat conduction is expected which is confirmed by the observed linear behavior of the numbers of bursts as a function of the foil thickness as shown in Fig. 7.12.

A reduction of the pressure from 1 bar to 0.1 mbar reduces the number of bursts per ablated micrometer (i.e., the slope of the regression lines plotted in Fig. 7.12) by a factor of 5 for single pulses and a factor of 3 for double pulses. The best fit straight lines do not go through the origin but intersect the abscissa at a final value of the foil thickness d_{f0}. For a given pressure the intersection point has for single and double pulses about the same value. From the curves of Fig. 7.12, we determine for 1 bar $d_{f0} = 27 \pm 4$ μm and 16 ± 5 μm for 0.1 mbar. These values are much greater than the material thickness ablated per burst (reciprocal of the slopes in Fig. 7.12) which lies in the range of 20–250 nm. The quantity d_{f0} is interpreted as the extrapolated foil thickness where the induced plasma pressure of the first laser burst is sufficiently high to break the foil. The thickness d_{f0} is greater for 1 bar than for 0.1 mbar, since the ambient gas acts as a mass load for the expanding plasma leading to a local confinement and thus increasing the pressure exerted onto the foil.

Figure 7.13 shows the number of bursts needed to drill through a foil of 50 μm as a function of the pressure for two different burst energies. For double pulses, only a weak dependence is observed becoming manifest in an increase of the burst number by about a factor of two while approaching 1,013 mbar. Considering the error bars, the number of burstsindexBurst!number of bursts does not change significantly while increasing the burst energy from 80 to 160 mJ in the pressure range <50 mbar and at 1 bar for single pulses and double pulses. For a burst energy of 80 mJ, the number of bursts for double pulses is always less than the number of bursts for single pulses, see Fig. 7.13b. Figure 7.13a (E_b = 160 mJ) shows a different behavior. At a pressure of 500 mbar, the number of bursts is greater for double pulses than for single pulses. For pressures <100 mbar, the number of bursts is independent of the pressure. For E_b = 80 mJ, the number of bursts does not change for pressures in the range <50 mbar.

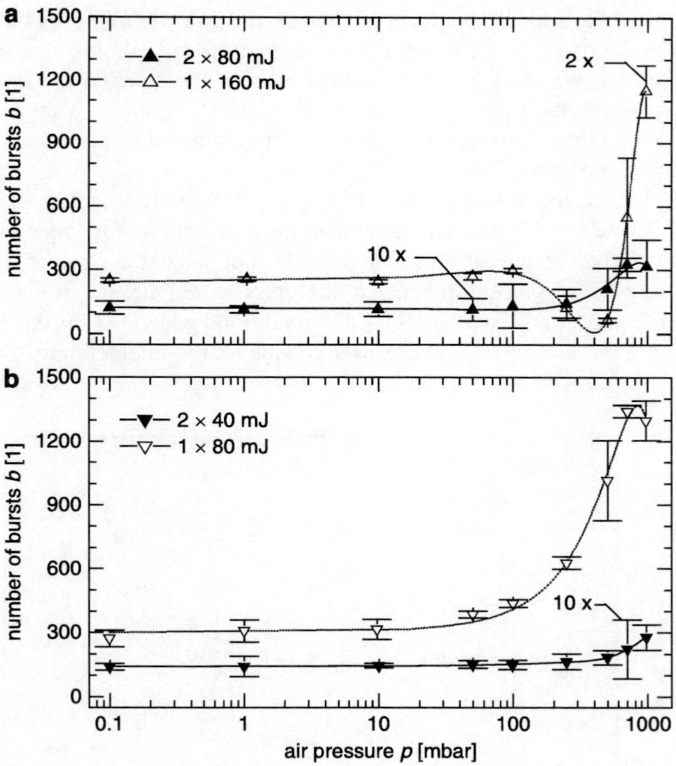

Fig. 7.13 Number of bursts needed to drill through a high-alloyed steel foil with a thickness of 50 μm as a function of the ambient air pressure for single and double pulses of burst energies **(a)** 160 mJ and **(b)** 80 mJ

Figure 7.13a shows that the number of double pulses to drill through the 50-μm foil at 1,013 mbar is about 300. The same number is obtained for single pulses and pressures ≤100 mbar. This means at 1 bar, the material ablation of double pulses is similar to the one observed in the case of single pulses (of the same total energy) irradiated at a reduced ambient pressure of ≤100 mbar. This suggests the interpretation that double pulses act as if the ambient pressure is reduced. This is consistent with the findings of a paper studying the space- and time-resolved dynamics of plasmas generated by collinear double pulses [7.21].

The regression lines of Fig. 7.12 can be described by the following relation:

$$b_i(p) = m_i(p)[d_f - d_{f\,0}(p)], \tag{7.2}$$

where $b_i(p)$ is the number of bursts; $i = 1$ for single pulses, $i = 2$ for double pulses; $m_i(p)$ is the slope of regression line; d_f is the foil thickness; and $d_{f\,0}(p)$ is the extrapolated foil thickness for $b_i(p) = 0$.

While the slope depends on the burst mode i, the quantity $d_{f0}(p)$ is independent of i, cf. Fig. 7.12 for E_b = 160 mJ. Assuming that the latter holds also for a burst energy of 80 mJ, we see from (7.2) that the ratio of the ablation rates defined by $a_2/a_1 = m_1/m_2$ equals the ratio b_1/b_2 for d_f = const. Figure 7.14 shows the ratio of the ablation rates a_2/a_1 as a function of the pressure determined for a steel foil with a thickness of $d_f = 50\,\mu$m.

At 1,013 mbar, the ablation rate for double pulses is by a factor of 3.6 (burst energy of 160 mJ) to 4.7 (80 mJ) higher than the ablation rate for single pulses. For pressures <100 mbar, the ratio of ablation rates approaches a value of 2 which is equal to the ratio of pulse numbers. In this pressure regime, the effect of double pulses is merely an additive one, i.e., the conditions generated by the first pulse of the double pulse have no significant influence on the interaction of the second

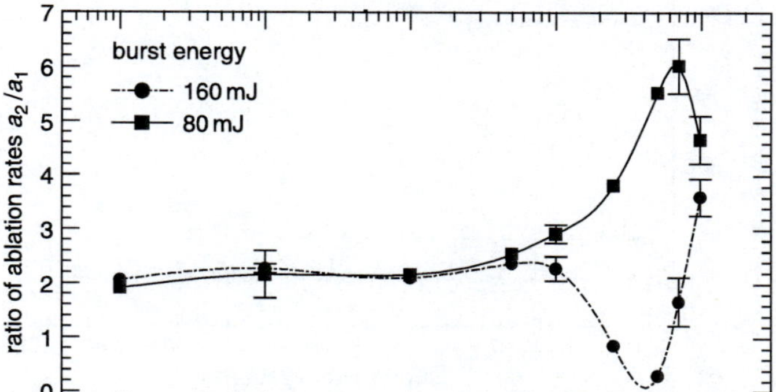

Fig. 7.14 Ratio of ablation rates of double pulses to single pulses as a function of ambient air pressure for a foil thickness of 50 μm and two burst energies

pulse irradiated 6 μs later. Hence, the two pulses of a double pulse burst can be considered to cause independent interaction processes. However, in the range 100 mbar $< p <$ 1,013 mbar, the situation is completely different; here the ratio of ablation rates shows a strong variation with changing ambient pressure. For the majority of data points in Fig. 7.14, we find $a_2/a_1 > 1$. The only exception occurs for a burst energy of 160 mJ and pressures of 250 and 500 mbar. In this case, the ratio a_2/a_1 is less than 1, i.e., the ablation rate for single pulses is higher than the one for double pulses. At a pressure of 700 mbar and a burst energy of 80 mJ, the ratio of ablation rates achieves a maximum value of 6, i.e., the efficiency of ablation is increased by a factor of 6 using double pulses instead of single pulses of the same energy. The position of the maximum is assumed to depend on the interpulse separation, the burst energy, and the pulse height ratio within a double-pulse burst.

References

7.1. C. Lorenzen, C. Carlhoff, U. Hahn, M. Jogwich, Applications of laser-induced emission spectral analysis for industrial process and quality control. J. Anal. At. Spectrom. **7**, 1029–1035 (1992)
7.2. R. De Young, W. Situ, Elemental mass spectroscopy of remote surfaces from laser-induced plasmas. Appl. Spectrosc. **48**, 1297–1306 (1994)
7.3. E. Piepmeier, D. Osten, Atmospheric influences on Q-switched laser sampling and resulting plumes. Appl. Spectrosc. **25**, 642–652 (1971)
7.4. A. Felske, Über einige Erfahrungen bei der Makrospektralanalyse mit Laserlichtquellen. Spectrochim. Acta **27B**, 1–21 (1972)
7.5. F. Leis, W. Sdorra, J. Ko, K. Niemax, Basic investigations for laser microanalysis: I. Optical emission spectrometry of laser-produced sample plumes. Mikrochim. Acta II **98**, 185–199 (1989)
7.6. R. Jowitt, Direct analysis of liquid steel by laser, Proceedings of 38. Chemistry Conf., BSC Teesside Laboratories, 1985, 19–29
7.7. R. Sattmann, V. Sturm, R. Noll, Laser-induced breakdown spectroscopy of steel samples using multiple Q-switch Nd:YAG laser pulses. J. Phys. D Appl. Phys. **28**, 2181–2187 (1995)
7.8. A. Semerok, C. Chaléard, V. Detalle, J. Lacour, P. Mauchien, P. Meynadier, C. Nouvellon, B. Sallé, P. Palianov, M. Perdrix, G. Petite, Experimental investigations of laser ablation efficiency of pure metals with femto, pico and nanosecond pulses. Appl. Surf. Sci. **138–139**, 311–314 (1999)
7.9. V. Sturm, L. Peter, R. Noll, Steel analysis with laser-induced breakdown spectrometry in the vacuum ultraviolet. Appl. Spectrosc. **54**, 1275–1278 (2000)
7.10. L. Peter, V. Sturm, R. Noll, Liquid steel analysis with laser-induced breakdown spectrometry in the vacuum ultraviolet. Appl. Optics **42**, 6199–6203 (2003)
7.11. J. Vrenegor, R. Noll, V. Sturm, Investigation of matrix effects in LIBS plasmas of high-alloy steel for matrix and minor elements. Spectrochim. Acta B **60**, 1083–1091 (2005)
7.12. J. Aguilera, C. Aragón, F. Penalba, Plasma shielding effect in laser ablation of metallic samples and its influence on LIBS analysis. Appl. Surf. Sci. **127–129**, 309–314 (1998)
7.13. J. Aguilera, C. Aragón, A comparison of the temperatures and electron densities of laser-produced plasmas obtained in air, argon, and helium at atmospheric pressure. Appl. Phys. A **69**, S475–S478 (1999)
7.14. I. Horn, D. Günther, The influence of ablation carrier gasses Ar, He and Ne on the particle size distribution and transport efficiencies of laser ablation-induced aerosols: implications for LA–ICP–MS. Appl. Surf. Sci. **207**, 144–157 (2003)

7.15. W. Sdorra, K. Niemax, Basic investigations for laser microanalysis: III. Application of different buffer gases for laser-produced sample plumes. Mikrochim. Acta **107**, 319–327 (1992)
7.16. R. Russo, Laser ablation. Appl. Spectrosc. **49**, 14A–28A (1995)
7.17. M. Guillong, I. Horn, D. Günther, Description and characterization of a homogenized high power 266 nm Nd:YAG laser ablation systems for LA-ICP-MS. J. Anal. At. Spectrom. **17**, 8–14 (2002)
7.18. A. Löbe, J. Vrenegor, R. Fleige, V. Sturm, R. Noll, Laser-induced ablation of a steel sample in different ambient gases by use of collinear multiple laser pulses. Anal. Bioanal. Chem. **385**, 326–332 (2006)
7.19. C. Wegst, Stahlschlüssel 2001, Marbach, Stahlschlüssel Wegst GmbH, 19. Auflage
7.20. L. St-Onge, V. Detalle, M. Sabsabi, Enhanced laser-induced breakdown spectroscopy using the combination of fourth-harmonic and fundamental Nd:YAG laser pulses. Spectrochim. Acta B **57**, 121–135 (2002)
7.21. R. Noll, R. Sattmann, V. Sturm, S. Winkelmann, Space- and time-resolved dynamics of plasmas generated by laser double pulses interacting with metallic samples. J. Anal. At. Spectrom. **19**, 419–428 (2004)
7.22. J. Vadillo, J. Romero, C. Rodriguez, J. Laserna, Depth-resolved analysis by laser-induced breakdown spectrometry at reduced pressure. Surf. Interface Anal. **26**, 995–1000 (1998)
7.23. Y. Iida, Effect of atmosphere on laser vaporization and excitation processes of solid samples. Spectrochim. Acta B **45**, 1353–1367 (1990)
7.24. S. Harilal, C. Bindhu, M. Tillack, F. Najmabadi, A. Gaeris, Internal structure and expansion dynamics of laser ablation plumes into ambient gases. J. Appl. Phys. **93**, 2380–2388 (2003)
7.25. J. Heitz, J. Gruber, N. Arnold, D. Bäuerle, N. Ramaseder, W. Meyer, J. Hochörtler, In-situ analysis of steel under reduced ambient pressure by laser-induced break-down spectroscopy. Proc. SPIE **5120**, 588–595 (2003)
7.26. S. Lui, N. Cheung, Resonance-enhanced laser-induced plasma spectroscopy: ambient gas effects. Spectrochim. Acta B **58**, 1613–1623 (2003)
7.27. L. Peter, R. Noll, Material ablation and plasma state for single and collinear double pulses interacting with iron samples at ambient gas pressures below 1 bar. Appl. Phys. B **86**, 159–167 (2007)

Chapter 8
Plasma Dynamics and Plasma Parameters

In this chapter, investigations are presented to clarify the underlying physical processes of the observed improvements achieved with collinear double pulses studying the space- and time-resolved dynamics of the plasmas generated by laser double pulses interacting with metallic samples, as well as the plasma state in terms of electron density and temperature.

8.1 Expansion and Decay of the Plasma

The experimental methodology is based on high-speed framing and streak images to observe the spatial and temporal development of the plasma using the optical emission of the plasma [8.1]. Furthermore, the spatiotemporal changes of the refractive index of the plasma are investigated using a Mach–Zehnder interferometer, where the shifts of the interference fringes are detected with a high-speed electro-optic camera.

The experimental setup for the time-resolved photography of the luminous laser-induced plasma is shown in Fig. 8.1. A Nd:YAG laser (Lumonics, model HY 1200, modified) with multipulse option, operated at 10 Hz and 1,064 nm emission wavelength is used to generate the plasmas on the surface of a plane metallic sample in ambient air. Details of the laser are described in [13.4]. For the experiments presented in this chapter, the plasma dynamics resulting from the laser–sample interaction with single and double Q-switched Nd:YAG laser pulses is investigated. The pulse width of a single pulse amounts to ~15 ns, whereas each pulse of a double pulse has a width of ~25 ns. In the case of double pulses, the energy ratio adjusted is 1:1. The interpulse separation of the double pulses amounts to $\Delta t = 6\,\mu s$. The pulses are focused by a planoconvex lens of 200 mm focal length on pure iron targets (Fe concentration > 99.99%). The beam waist of the laser focus lies inside the sample, 5 mm beneath its surface ($\Delta s = 5$ mm). The diameter of the laser beam on the surface of the sample is 280 μm. The pulse energies are adjusted by use of a Glan–Thompson polarizer and measured with

R. Noll, *Laser-Induced Breakdown Spectroscopy*, DOI 10.1007/978-3-642-20668-9_8,

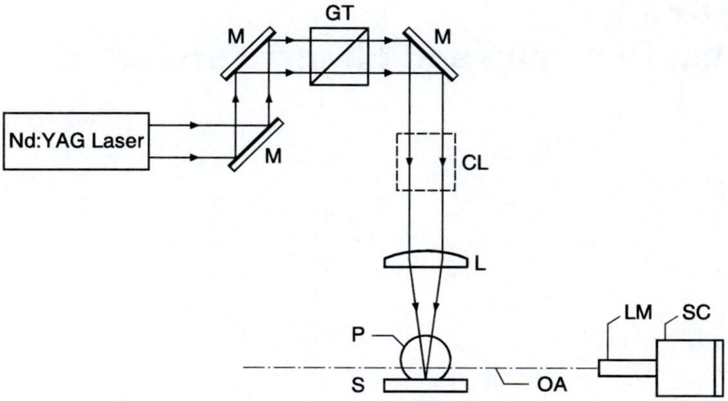

Fig. 8.1 Experimental setup for high-speed photography of the luminous laser-induced plasma. M = mirror, GT = Glan–Thompson polarizer, CL = volume absorbing disc calorimeter, L = lens, P = laser-induced plasma, S = sample, OA = optical axis of streak camera, LM = long distance microscope, SC = streak camera

a volume absorbing disc calorimeter (Scientech 38–0101). The optical axis of the streak camera is oriented perpendicularly to the propagation direction of the laser beam irradiating the sample. The laser-induced plasma is imaged by a long-distance microscope on the streak camera (Hadland, model Imacon 790) with a S 20 photocathode having the dimensions $9 \times 8\,\mathrm{mm}^2$. The distance between the laser-induced plasma and the front end of the microscope is about 160 mm. The streak camera can be operated in framing mode and in streak mode. In streak mode, the slit of the streak camera is oriented collinearly to the propagation axis of the irradiated laser beam. The magnification object-to-image was determined using a line graticule placed at the position of the laser propagation axis just above the sample and observed with the streak camera in the focus mode. The experiments are carried out under normal atmospheric conditions.

The experimental setup used for the interferometric investigations is shown schematically in Fig. 8.2. The components to generate the laser-induced plasma are identical to Fig. 8.1, but the sample is now located in a gas-tight measuring chamber allowing to perform experiments with air pressures of $p_0 = 100$ and 1,013 mbar. The Mach–Zehnder interferometer is oriented in such a way that one arm passes through the interaction volume above the sample surface perpendicular to the irradiated Nd:YAG laser beam. The radiation source for the interferometry is a cw argon ion laser (Coherent, model Innova 100–10) operated at an emission wavelength of 514.5 nm. The interference fringes are observed with the streak camera. To avoid damage of the streak camera due to continuous illumination by the argon laser beam, a mechanical shutter and an acousto-optic modulator (AOM) are used to extract pulses of $\sim 10\,\mu\mathrm{s}$ duration of the argon laser beam synchronized to the triggering of the Nd:YAG laser. The first-order diffracted beam of the AOM

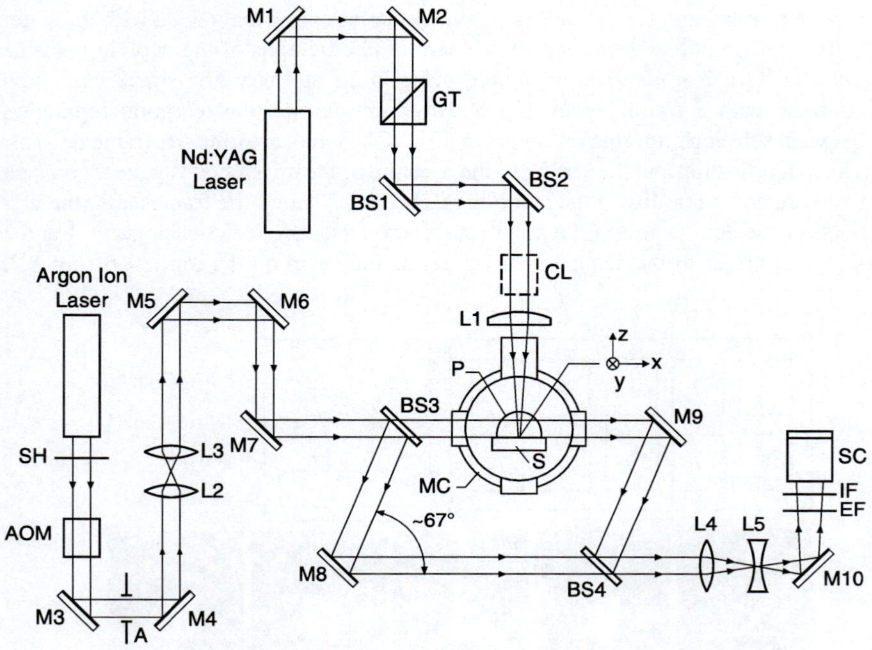

Fig. 8.2 Experimental setup for the interferometric measurement of the laser-induced plasma. SH = shutter, AOM = acousto-optic modulator, A = aperture, BS1 to BS4 = beam splitters, MC = measuring chamber, EF = edge filter, IF = interference filter. For further abbreviations see Fig. 8.1

is used for the interferometer, whereas the zeroth and higher orders are blocked by an aperture (see A in Fig. 8.2). The telescope (L2 and L3 in Fig. 8.2) serves to adjust the diameter of the illuminated area at the location of the plasma. To cover the total plasma geometry during the expansion phase, an illumination diameter of 20 mm was chosen. The second telescope (L4 and L5) and the internal optic of the streak camera form an image of the plasma on the photocathode. The beam deviation of the argon laser beam due to refractive index gradients in the interaction region of the Nd:YAG laser beam may cause uncertainties in the Abel inversion of the interferograms; however, accurate results are achieved by proper imaging of the plasma plane on the photocathode of the streak camera. The interference filter has a maximum transmission at the wavelength of the argon ion laser and a spectral width of 10 nm (FWHM). The edge filter (Schott, type OG 515) attenuates additionally the intense emission of the laser-induced plasma.

Nd:YAG laser, shutter, AOM and streak camera are synchronized using a personal computer as master trigger and three digital delay generators. Before the start of each measurement, a minimum of 50 flashlamp discharges of the laser is activated to stabilize the pulse energy to a standard deviation of 2%.

The framing and streak images shown in the following are gained with the setup of Fig. 8.1. Figures 8.3 and 8.4 show framing photographs of the evolving plasma induced by a 2 × 40 mJ laser double pulse in air at 1 bar. The streak camera is operated with a framing rate of 2 × 10^7 frames/s, i.e., the temporal separation between subsequent frames amounts to 50 ns. The exposure time per frame is 10 ns. The magnification and the setting of the aperture of the microscope are kept constant to assure comparability of the recordings. Figure 8.3 shows the frames after the first pulse of the double pulse. The first pulse reaches the sample between frame 1 and 2 (Numbers refer to the sequence of frames as indicated on the top left of Fig. 8.3.

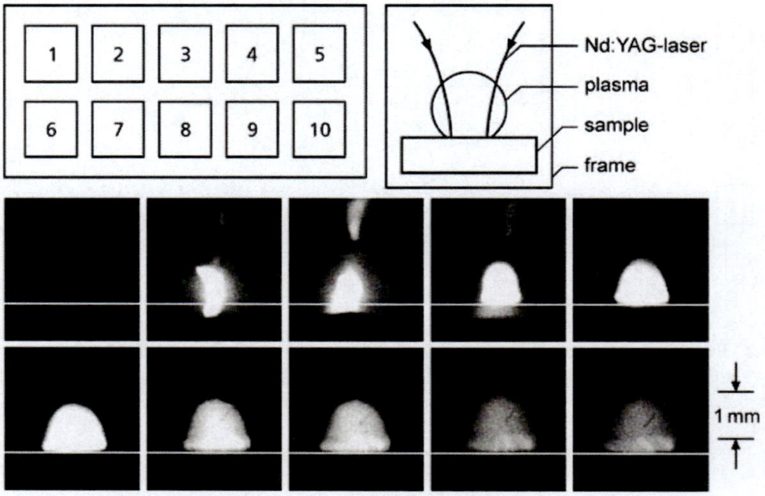

Fig. 8.3 Framing images of the laser-induced plasma after the first pulse of a double pulse. Exposure time 10 ns per frame, time between subsequent frames 50 ns. The time sequence of the frames is shown on the *top left* with the *numbers 1–10*. The view detected by the frames is illustrated schematically on the *top right*. The *white horizontal line* in each frame indicates approximately the position of the sample surface. Laser parameters: 2 × 40 mJ, interpulse separation 6 μs

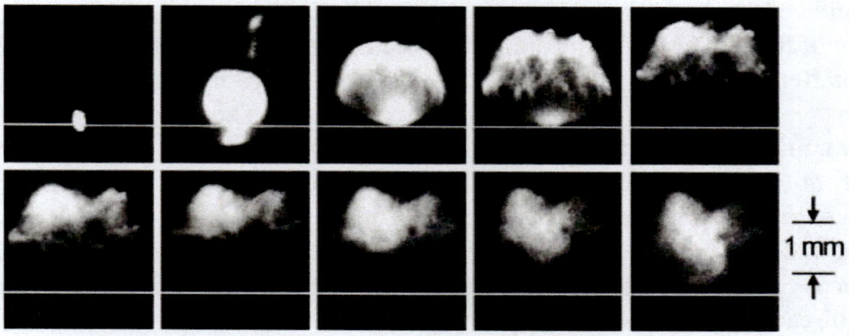

Fig. 8.4 Framing images of the laser-induced plasma after the second pulse of a double pulse. Experimental settings and arrangement of frames are the same as for Fig. 8.3

The arrangement of frames shown in Fig. 8.3 was chosen according to the usual orientation of the time axis from left to right. The original arrangement of the frames yielded by the streak camera is different.) Frames 2–4 are overexposed due to the high luminosity of the plasma at this phase. Figure 8.4 shows the frames after the second pulse of the double pulse. The second pulse of the double pulse reaches the sample at the exposure of frame 1 (frame numeration analog to Fig. 8.3).

The plasma geometry and dynamics observed in Figs. 8.3 and 8.4 differ significantly from each other. After the first pulse, a nearly hemispherical plasma geometry appears staying in contact with the sample surface for times up to 400 ns after arrival of the first laser pulse. The extension of the plasma in the direction of the laser beam axis amounts to ~1 mm (frame 5 corresponding to an elapsed time of 150 ns after the first visible plasma emission in frame 2). After the second pulse, the plasma emission starts at the sample surface (see frame 1 in Fig. 8.4). This is a clear indication that the second laser pulse primarily interacts with the sample surface and not with a residual plasma originating from the first pulse. The plasma then splits into two regions, where one region lasts at the sample surface and disappears after about 100 ns. The other region attains a layer-like shape, propagating toward the incoming laser beam. At ~150 ns after the irradiation of the second pulse of the double pulse, an inhomogeneous distribution of the luminosity in this layer is observed looking like a turbulent structure which tends to decrease at later times. After 200 ns, the expansion toward the incoming laser beam stops. After 350 ns, the luminous plasma starts to expand in the backward direction toward the sample surface. At 450 ns, the plasma shows a more homogeneous luminosity distribution and its shape differs significantly from the one observed in the time interval between 150 and 250 ns. The plasma has nearly reached the sample surface again.

The further temporal development can be observed by streak photographs as shown in Fig. 8.5. The observation geometry is the same as for the Figs. 8.3 and 8.4. The streak velocity amounts to 50 ns/mm on the screen of the streak camera.

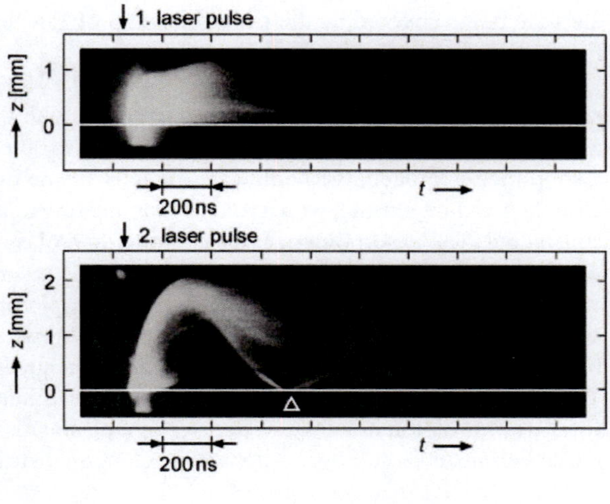

Fig. 8.5 Streak photographs after first and second laser pulse of a double pulse. Experimental parameters are the same as for Fig. 8.3

The time scale runs from left to right (In the original streak images delivered from the streak camera, the time scale runs from right to left. These images are presented here side-inverted according to the usual direction of the time axes). The white line illustrates the position of the surface of the sample. The luminous regions beneath the white line are due to an overexposure of the streak images for times close to the arrival of the laser pulse. The upper image shows the laser-induced plasma after the first pulse, and the lower image shows the plasma after the second pulse. After the first pulse, the plasma stays in contact with the sample surface, while the boundary of the luminous plasma expands to about 1 mm above the sample, as observed in the framing images of Fig. 8.3. The luminous layer evolving after the second pulse starts at the sample surface and expands up to a distance of about 2 mm from the sample surface. After 300 ns, the front of the luminous layer changes the direction and propagates back toward the sample. Approximately 600 ns after the irradiation of the second laser pulse, the luminous front reaches again the sample surface (marked by a white triangle in the lower image of Fig. 8.5). At the surface, a partial reflection of the plasma front is visible. The observed dynamics of the plasma expansion after the second pulse is a clear indication that the first laser pulse has led to a transient modification of the ambient particle density above the sample thus enabling a greater expansion of the sample material ablated by the second pulse. The observed reflection-like behavior of the luminous front at about 300 ns after the second laser pulse may be caused by a radially increasing particle density in the surrounding atmosphere as a result of the interaction of the first laser pulse. Streak photographs taken with a greater aperture of the microscope show that for times beyond 1 μs after the second laser pulse the luminosity of the plasma becomes more and more homogeneous and the structure and geometry of the plasma stays nearly constant for more than 5 μs.

In the time range shown in Figs. 8.3–8.5, the emission spectrum of the plasma changes from a continuum emission at early times to pronounced line emissions after about 800 ns (see Chap. 6, Fig. 6.3). The exponential decay of the continuum emission can be described by a decay time constant of about 90 ns corresponding to the time range covered by the frames 2 and 3 of Fig. 8.3 and frames 1, 2, and 3 of Fig. 8.4.

Framing and streak images of Figs. 8.4 and 8.5 show clearly that the dominant effect of the second pulse is again an interaction with the sample surface and not a re-heating of a residual plasma above the sample originating from the first pulse. In a simplified approach, the nonlinear character of the interaction of the laser beam with the residual plasma, which may lead to heating and ionization processes is not considered. The absorption of laser radiation in a plasma can then be described by the absorption coefficient using the formula (5.11) assuming $\nu \ll \omega$ and $\omega_p < \omega$.

For this estimate, the absorption by microscopic particulates originating from condensed iron vapor or solidified melt droplets thrown out of the interaction region by the first pulse is neglected. The collision frequency ν is given by relation (5.12). The electron–ion collision frequency ν_{ei} can be estimated using the formula (5.13) for the electron–ion momentum transfer in a plasma using the screened Rutherford cross section and assuming a Maxwellian electron distribution. The electron-neutral collision frequency is estimated taking the geometric cross section of neutral iron atoms. With the following plasma parameters determined spectroscopically at a time delay of 1 μs after a single pulse of 80 mJ: electron density $n_e = 7 \times 10^{16}\,\mathrm{cm}^{-3}$, neutral iron density $n_{\mathrm{FeI}} = 6 \times 10^{16}\,\mathrm{cm}^{-3}$, electron temperature $T_e = 8{,}300\,\mathrm{K}$, we obtain: $\nu_{ei} = 1.3 \times 10^{12}\,\mathrm{s}^{-1}$, and $\nu_{en} = 1.4 \times 10^{9}\,\mathrm{s}^{-1}$. Using relation (5.11), the estimated absorption coefficient due to ablated material originating from the first pulse amounts to $\alpha = 0.3\,\mathrm{m}^{-1}$. Taking as an upper limit a plasma dimension of 5 mm, this corresponds to a transmission of more than 99% of the irradiated laser intensity of the second pulse. In this case, a dominant part of the laser energy may propagate to the sample surface again where it interacts with a preablated sample surface. Hence, the plasma emission induced by the second pulse should start at the sample surface, which is consistent with the experimental observations of Figs. 8.4 and 8.5.

The expansion of the laser-induced plasma can be described by the Sedov-modelindexSEDOV-model for shock waves after a strong explosion in a homogeneous atmosphere [8.2, 8.3]. A dimensional analysis leads to the following relationship for the expansion z of the shock wave as a function of time t:

$$z = \lambda_0 (E_0/\rho_0)^{\frac{1}{2+\xi}}\, t^{\frac{2}{2+\xi}}, \tag{8.1}$$

where E_0 is the energy deposited in a small volume, ρ_0 is the density of the homogeneous undisturbed ambient gas, and λ_0 is a function of the specific heat ratio of the surrounding gas.

The parameter ξ describes the geometry of the energy release and the shock wave expansion with $\xi = 3$ for a point-like energy release and a spherical shock wave expansion and $\xi = 1$ for a planar shock wave expansion. Hence, for a spherical shock wave, the expansion should follow the relation $z \propto t^{0.4}$ and for a planar expansion $z \propto t^{0.66}$. The exponent of t is always expected to be twice that of E_0/ρ_0.

To determine the expansion velocity of the luminous plasma with a higher accuracy, streak images of the expanding plasma are taken with streak velocities of 1 and 5 ns/mm, while keeping the other experimental parameters constant. Figure 8.6 shows a double-logarithmic plot of the determined expansion after the first and the second laser pulse of a double pulse. The time zero was chosen at the start of the plasma emission detected in the streak images. The error bars in the z-direction indicate the reading error in determining the front position on the enlarged streak images. Uncertainties on the time axis are caused by deviations from the nominal streak velocity which are specified to be less than 5% and by reading errors. The corresponding error bars are not shown in Fig. 8.6, since they are significantly smaller than those in the z-direction. After the first pulse, two regions of different dynamics can be distinguished. For times up to 25 ns after the start of the interaction of the first laser pulse, the data points can be fitted by a linear regression described by the function $z \propto t^{(0.76 \pm 0.02)}$. A linear fit to the data points after 25 ns yields $z \propto t^{(0.44 \pm 0.04)}$. Under the assumption that the luminous plasma front coincides with the position of the shock front at the early expansion phase, we can compare the determined dynamics with the prediction of Sedov's model. For times ≤ 25 ns Sedov's model cannot be applied, since the laser pulse still persists and couples

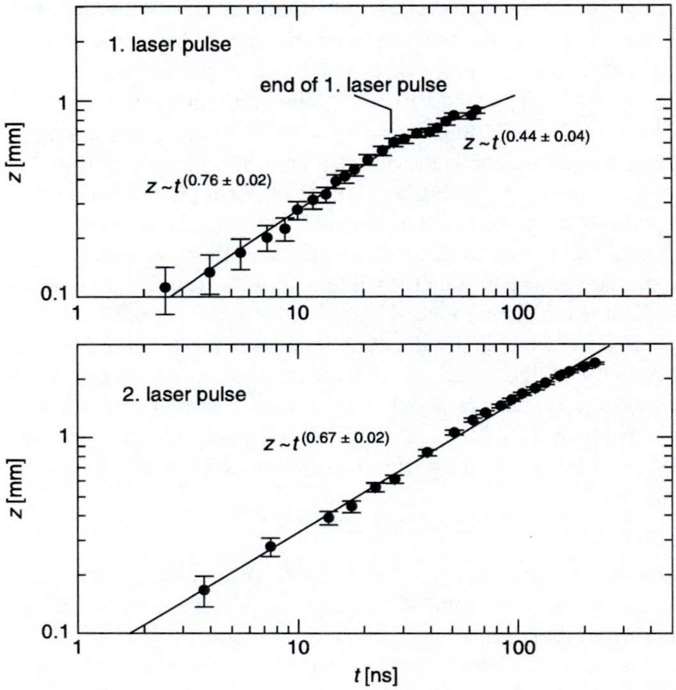

Fig. 8.6 Height z of the luminous plasma front above the sample surface as a function of time determined from side-on streak images. *Top*: after the first laser pulse, *bottom*: after the second laser pulse. Experimental parameters are the same as for Fig. 8.3

energy into the interaction region. Beyond 25 ns, the determined exponent is close to the theoretical value for a spherical expansion of a shock front.

For the second laser pulse, the determined data points are shown in the lower diagram of Fig. 8.6. Applying a single fit using the function $z \propto t^{\beta}$ yields an exponent of $\beta = 0.67 \pm 0.02$ which is consistent with a planar shock front expansion. (A close look to the data points in the lower diagram of Fig. 8.6 shows that there might be a small change in the dynamics after a time of about 60 ns. Applying the fit only to the data points with $t > 60$ ns yields $\beta = 0.53 \pm 0.02$. However, the change in the dynamics after the second pulse at a time of about 60 ns is significantly smaller than the one observed at about 25 ns after the first laser pulse.) These values indicate different expansion geometries for the plasmas induced by the first and the second laser pulse and are in qualitative agreement with the results of the framing images shown in Figs. 8.3 and 8.4.

The derivatives of the regression curves of Fig. 8.6 yield the expansion velocities of the luminous plasma front in the z-direction. In the time interval from $t = 30$ ns to $t = 60$ ns after the first laser pulse, this velocity decreases from 0.8×10^6 to 0.6×10^6 cm/s (corresponding to Mach numbers of 23 and 17), whereas in the same time interval after the second laser pulse the expansion velocities are 1.5×10^6 and 1.2×10^6 cm/s, respectively. Hence, after the second pulse the expansion is by about a factor of 2 faster than after the first pulse.

The experimental setup shown in Fig. 8.2 enables the observation of changes of the refractive index n caused by the expanding plasma in the measuring arm of the Mach–Zehnder interferometer. In a first step, without the laser-induced plasma, the mirrors of the interferometer are tilted to adjust the number and orientation of the interference fringes. A fringe orientation parallel to the sample surface is chosen. The slit of the streak camera crosses these fringes perpendicularly. Figure 8.7 shows streak images of the interferograms for a laser double pulse with 2×80 mJ, $\Delta t = 6\,\mu$s and $\Delta s = 5$ mm. The sample surface is located at $z = 0$ mm. The expanding plasma and shock front leads to a displacement of the interference fringes. After the first laser pulse, a disturbance of the fringe pattern propagates in positive z-direction. At the leading edge of this disturbance, a strong displacement of the interference fringes toward the sample surface is observed. Reference measurements with glass plates in the measuring arm of the interferometer show that this direction of fringe displacement corresponds to an increased index of refraction. Hence, the displacement of the fringes observed at the leading edge of the expanding fringe disturbance is attributed to an increased particle density caused by the shock wave. After the second laser pulse, an analog displacement of the interference fringes is observed at the leading edge of the evolving disturbance in the fringe pattern indicating the formation of a second shock front propagating behind the shock front induced by the first laser pulse. The reflection-like behavior of the luminous plasma observed in Fig. 8.5 at about 300 ns after the impact of the second pulse does not show up in a corresponding structure of the fringe patterns observed in Fig. 8.7.

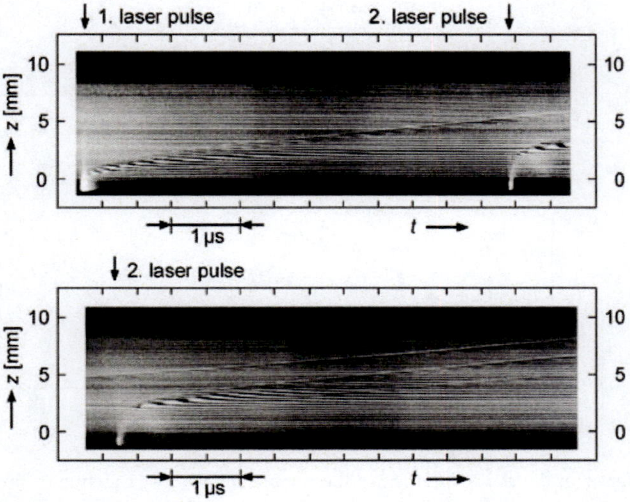

Fig. 8.7 Interferograms registered with the streak camera after the first and second laser pulse of a double pulse. Laser parameters: 2×80 mJ, interpulse separation $6\,\mu$s, $p_0 = 1{,}013$ mbar

A potential reason might be a too small change in the refractive index caused by the back-traveling plasma, which cannot be resolved by the setup used.

For a comparison of the experimental data with the Sedov model, the expansion z of the shock front as a function of time t is shown in a double-logarithmic plot in Fig. 8.8. The distortion of the interference fringes can be detected clearly for more than 12 μs after the laser pulse interaction (see Fig. 8.7 bottom, shock front induced by the first laser pulse at about $z = 7$ mm) in contrast to the luminosity of the plasma, which can be observed for a few microseconds only. The exponent describing the expansion of the shock front in the time range from 200 ns to 12 μs after the first pulse amounts to 0.45 ± 0.01. This result agrees with the exponent determined from the streak images of the luminous plasma at the early times between 30 and 65 ns after the first laser pulse, cf. Fig. 8.6. The time exponent of the shock front expansion after the second laser pulse is 0.35 ± 0.04 and hence is significantly smaller than the one determined from the luminous front at early times, cf. Fig. 8.6.

For the determination of the time exponent for the shock front propagation after the first laser pulse, functions of the following form are fitted to the measuring data:

$$z = At^a. \tag{8.2}$$

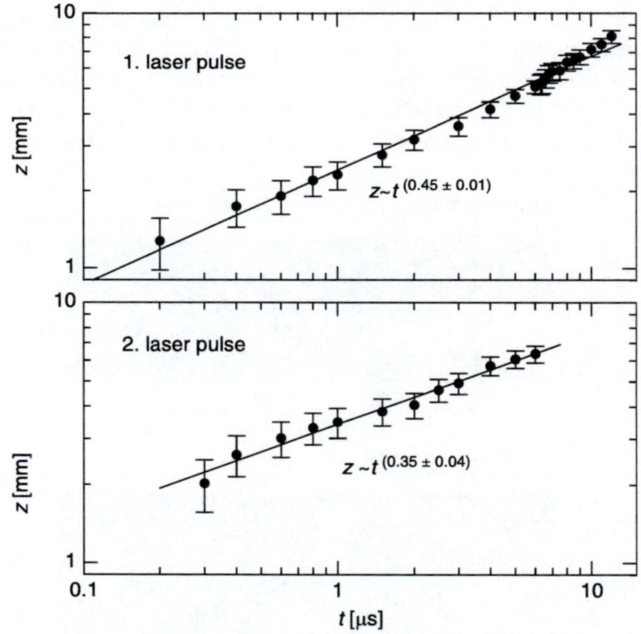

Fig. 8.8 Height z of the shock front above the sample surface as a function of time determined from interferograms observed in side-on direction with the streak camera, cf. Fig. 8.7. *Top:* after the first laser pulse, *bottom:* after the second laser pulse

Since the shock front propagates into an undisturbed ambient atmosphere, the preconditions of Sedov's model are fulfilled and a comparison of (8.2) with (8.1) yields the following relations: $a = 2/(2+\xi)$ and $A = \lambda_0(E_0/\rho_0)^{a/2}$. For the latter parameter holds also the relation: $A \propto (E_0/p_0)^{a/2}$, where p_0 is the ambient pressure, which can be changed in the experiment, cf. Fig. 8.2. In the following paragraphs, it is assumed that the energy E_0 deposited locally is proportional to the irradiated laser pulse energy E_1 (E_1 denotes the pulse energy of a single pulse or the pulse energy of the first pulse of a double pulse). The fit parameters A and a are determined for measurements with the following sets of parameters: (a) single laser pulses with $E_1 = 20$, 40, 80, 120, and 160 mJ at $p_0 = 1$ bar ambient pressure, (b) single laser pulses with $E_1 = 40$ mJ at ambient pressures of $p_0 =$ 100, 500, and 1,013 mbar. For these parameters, the determined exponent a varies between 0.36 and 0.50 with an average value of $\bar{a} = 0.43 \pm 0.05$. This value agrees with the expected exponent of 0.44 for a spherical shock front propagation according to Sedov's model. Relation (8.2) describes well the dependence of the shock wave position as a function of time for distances of up to 7 mm from the sample surface, which is still below the critical radius $r_c = (E/p_0)^{1/3}$, which amounts to $r_c = 9.3$ mm for $E = 80$ mJ and $p_0 = 1{,}013$ mbar [8.3]. At r_c, the counterpressure of the ambient gas can no longer be neglected and the self-similar solution of the problem of a strong explosion no longer holds.

Figure 8.9 shows the values of the fit parameter A determined for the parameter sets described above in a double-logarithmic plot. The linear regression shown in Fig. 8.9 yields: $A \propto (E_1/p_0)^{0.23\pm0.02}$. As expected by the Sedov model, the exponent is about half of $\bar{a}$. Hence, the dynamics of the shock front propagation after the first laser pulse can be described in a wide range of laser pulse energies and ambient pressures by Sedov's model for a spherical expansion.

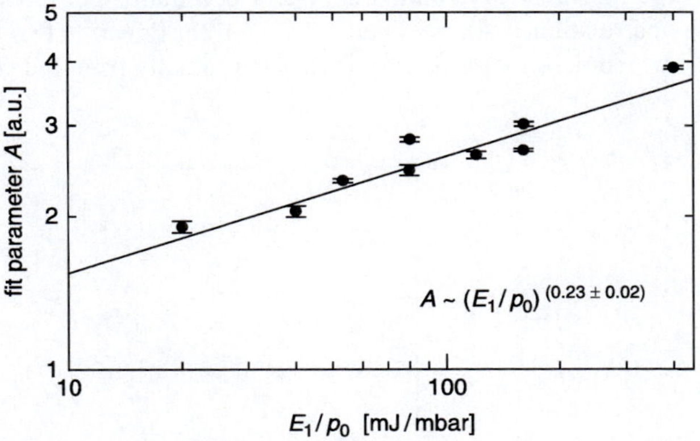

Fig. 8.9 Fit parameter A as a function of the ratio of the laser pulse energy E_1 and the ambient pressure p_0 for the first laser pulse

For the description of the expansion of the shock wave after the second laser pulse, in analogy to relation (8.2), the following function is used: $z = Bt^b$. Since the ambient atmosphere is not homogeneous for the second shock wave, the parameters B and b cannot be interpreted with respect to (8.1). The following measuring parameters were used to determine B and b: (a) double pulses with $E_1 = E_2$ (E_2 denotes the pulse energy of the second pulse of a double pulse) and $E_2 = 20$, 40, 60, 80, and 100 mJ at $p_0 = 1{,}013$ mbar ambient pressure, (b) double pulses with $E_1 = E_2 = 40$ mJ at ambient pressures of $p_0 = 100$, 500, and 1,013 mbar. The determined fit parameter b varies between 0.27 and 0.41, with an average value of $\bar{b} = 0.32 \pm 0.05$.

Figure 8.10 shows the determined parameter B as a function of the laser pulse energy of the second pulse E_2 at a pressure of $p_0 = 1{,}013$ mbar (pressure in the ambient atmosphere before the impact of the first pulse). A regression curve yields: $B \propto (E_2)^{0.30 \pm 0.02}$. The parameter B is not plotted as a function of (E_2/p_0) since there is no homogeneous pressure distribution in the interaction region at the time the second laser pulse is irradiated onto the sample. The determined exponent is approximately equal to $\bar{b}$. This behavior differs significantly from the shock wave propagation after the first laser pulse.

The propagation of shock waves in an inhomogeneous atmosphere was described by Sedov assuming the following relation for the density variation of the ambient atmosphere by a power law [8.2]:

$$\rho = \rho(r) = Dr^{-\varphi}, \tag{8.3}$$

where D is a positive constant, r is the radius, and φ is a dimensionless constant. The propagation of a spherical shock wave as a function of time is then given by the following power law:

$$r \propto t^{2/(5-\varphi)}. \tag{8.4}$$

Taking the experimentally determined exponent $\bar{b}$ for the shock wave after the second pulse and relation (8.4), we obtain $\varphi = -1.25$. Equation (8.3) then yields $\rho(r) \propto r^{1.25}$, which describes a density increasing radially from the center. While

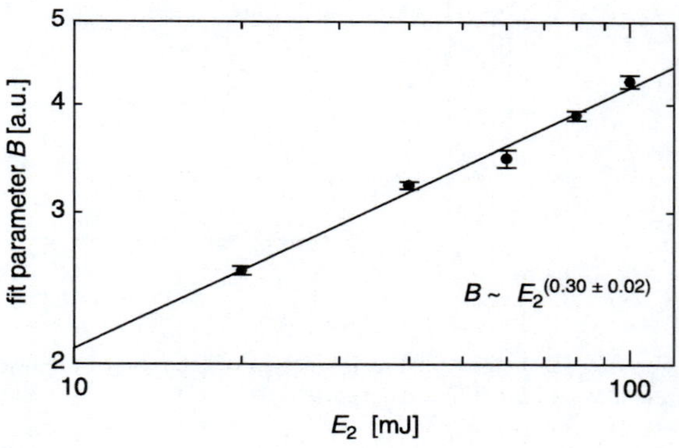

Fig. 8.10 Fit parameter B as a function of the energy E_2 of the second laser pulse at a pressure of $p_0 = 1{,}013$ mbar

the first shock wave propagates into a spatially homogeneous ambient atmosphere, the second shock wave encounters – as a result of the interaction of the first laser pulse – a locally rarefied ambient atmosphere, allowing for higher propagation velocities of the second shock wave.

So far the expansion of the plasma from a flat surface of the sample was studied. If a series of laser pulses is irradiated on the same spot on the surface of a sample, then a crater is formed, whose geometry influences the plasma expansion and dynamics. Instead of a free expansion into a half space the plasma is partially confined by the crater walls formed by the preceding laser pulses. Figure 8.11 illustrates this change in the plasma dynamics with side-on images of the luminous plasma. The exposure time of the images taken with an intensified high-speed CCD-camera was set to 50 ns. The position of the sample surface is marked by the white vertical line on the left of the images, and the laser beam is irradiated from the right. The experimental parameters were: single pulses, 160 mJ, $\Delta s = 5$ mm, $t_{\text{delay}} = 5\,\mu$s, ambient atmosphere: air. Figure 8.11a is taken for a flat sample surface and Fig. 8.11b in case of a crater formed by preceding laser pulses having a diameter of about 1 mm and a depth of 0.9 mm. The change in the plasma geometry is clearly visible. Obviously the crater effects a confinement of the plasma leading to faster expansion toward the incoming laser beam. The images were taken at the same delay time t_{delay} showing that the plasma volume, the center of the plasma, and the expansion velocity are changed depending on the two considered cases of a plasma induced on a flat surface or inside a crater.

The partition of the laser burst energy in a collinear double pulse studied so far was 1:1, i.e., each single pulse of the double pulses has the same energy. A change in this partition has an influence on the plasma dynamics as well. Figure 8.12 shows side-on images of the luminous plasma at different delay times, where the three images of column (a) are taken for a single pulse with 80 mJ, column (b) for a symmetric double pulse with 2 × 40 mJ and column (c) for an asymmetric double pulse with $E_1 = 10$ mJ and $E_2 = 70$ mJ. While the single pulse causes the largest plasma in the beginning (for $t_{\text{delay}} = 0\,\mu$s), it decays faster leaving a small emitting volume after 0.5 μs which is still in contact with the sample surface. In case of a

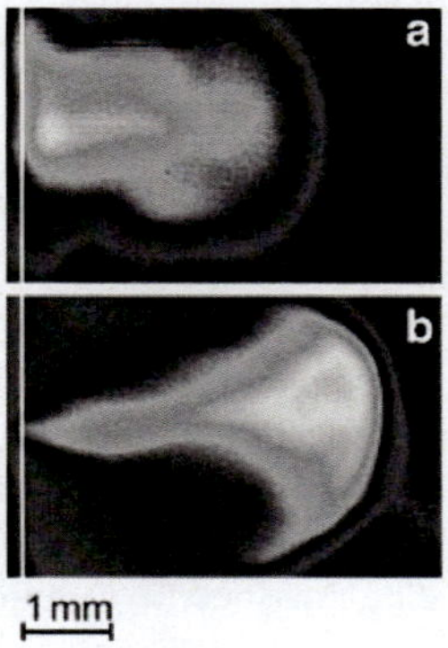

Fig. 8.11 Side-on images of the luminous plasma evolving from the sample surface (*white vertical line* on the *left*) for different local sample surfaces. The laser beam is irradiated from the *right*. (**a**) Flat sample surface, (**b**) surface with a crater formed by preceding laser pulses. For further details see text

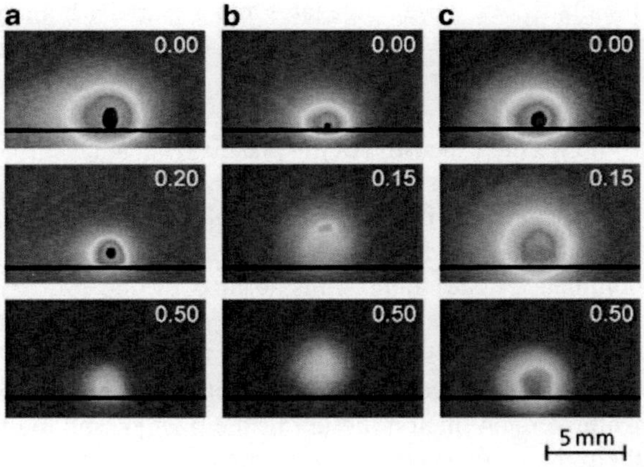

Fig. 8.12 Comparison of the plasma emission for (**a**) a single pulse of 80 mJ, (**b**) a symmetric double pulse with 2 × 40 mJ, and (**c**) an asymmetric double pulse with $E_1 = 10\,\text{mJ}$ and $E_2 = 70\,\text{mJ}$. The *numbers* in the *upper right corner* of each image show the delay time t_{delay} in microseconds of the start of the exposure with respect to the last pulse irradiated

symmetric double pulse shown in column (b) of Fig. 8.12, the plasma dimensions are smaller in the beginning and after 0.5 μs a spherical plasma is visible which is clearly detached from the sample surfaces and shows a volume being greater than the one in case (a) at the same delay time. For the asymmetric double pulse, the energy of the second pulse is significantly greater than in case (b); hence, the plasma is larger and emits stronger in the beginning. At later times, a spherical plasma is formed covering an even larger volume than in case (b) and still staying in contact with the sample surface at 0.5 μs. Furthermore, the plasma generated with asymmetric double pulses shows the strongest emission after 0.5 μs compared with the other cases.

The exposure time itself of each frame was kept constant and amounts to 50 ns. The further measuring parameters are: $\Delta s = 5\,\text{mm}$, $\Delta t = 6\,\mu\text{s}$. The laser is irradiated from the top, and the black horizontal line marks the position of the sample surface.

This behavior shows that a changed partition of the burst energy among the two pulses of a collinear double pulse allows to optimize the plasma dynamics with respect to reduced decay times, enlarged plasma volumes, and increased emission features.

8.2 Spatially Integrated Electron Density and Temperature

In order to investigate the effects of single- and double-pulse bursts on the laser-induced plasma, spatially and temporally integrated emission intensities, electron temperatures, and electron densities are measured. The time interval for integration, shown in Fig. 6.3, is chosen in accordance with the needs of quantitative multielement analysis with LIBS, namely high line-to-background ratio of the element-specific radiation. The delay of 0.8 μs with respect to the peak emission excludes the contribution of the continuum radiation. The width of the integration interval of 2 μs is in the range of the decay times of the line radiation, see Fig. 6.3. Hence, the temporal integration covers the major part of the line emission.

To determine the electron temperature and density, iron emission lines in the spectral region around 536 nm were selected (the other experimental parameters are described in Sect. 6.1). The plasma radiation is collected by an optical fiber bundle without additional optical elements. Due to the numerical aperture of the fiber bundle, the received radiation is spatially integrated over the whole emitting plasma volume. The fiber bundle is divided into four arms with about 40 fibers each. A fiber bundle output having a rectangular cross section is mounted at the entrance slit of a Czerny–Turner spectrometer (Jobin Yvon HR 640) with 640 mm focal length, 2,400 lines per millimeter grating and an intensified photodiode array detector (Princeton Instruments). The spectral resolution is 0.045 nm at 540 nm. The spectra of the 81st to the 100th pulse on a ground surface are accumulated to improve the signal-to-noise ratio. The electron temperature is deduced from Boltzmann plots, assuming that the plasma is in local thermal equilibrium as stated by Adrain et al. [8.4], and the electron density by measuring the line width of a Stark-broadened emission line [8.5]. Spectral line selection methods for Boltzmann plots are described in Sect. 19.6. The Fe lines used for the evaluations are listed in Table 8.1.

Figure 8.13 shows the whole spectrum detected simultaneously covering the spectral range of the lines listed in Table 8.1. Figure 8.14 shows a part of this spectrum around the iron line Fe I 538.34 nm with the data points and fit curves.

The intensities of the lines are determined by a fit of Lorentzian curves to the measured spectrum. The uncertainty in the electron temperature measurement is estimated by the evaluation of the temperature from the maximum and minimum slopes of straight lines through the error-affected measuring points in the Boltzmann plot. Figure 8.15 shows the Boltzmann plot for the spectra of Fig. 8.13. The quantity shown on the ordinate is:

$$y = \ln\left(\frac{I\lambda}{A_{nm} g_n} c_B\right) \tag{8.5}$$

with I measured intensity of a spectral line assumed to be proportional to the emission coefficient, cf. (9.3) in Chap. 9, unit: 1; λ wavelength, A_{nm} Einstein coefficient of the transition $n \rightarrow m$, g_n statistical weight of the upper level, c_B

Table 8.1 Wavelength λ, excitation energy E_n, Einstein coefficient A_{nm}, and statistical weight g_n of the upper level n of the selected iron lines for the determination of the electron temperatures from Boltzmann plots

λ (nm)	E_n (eV)	A_{nm} ($10^8\,s^{-1}$)	g_n
532.418	5.54	0.15	9
532.804	3.24	0.0115	7
537.150	3.27	0.0105	5
538.337	6.62	0.59	13
539.713	3.21	0.00259	9
540.578	3.28	0.0109	9

The width of the Fe line at 538.34 nm caused by Stark broadening is used to evaluate the electron density [8.6, 8.7]

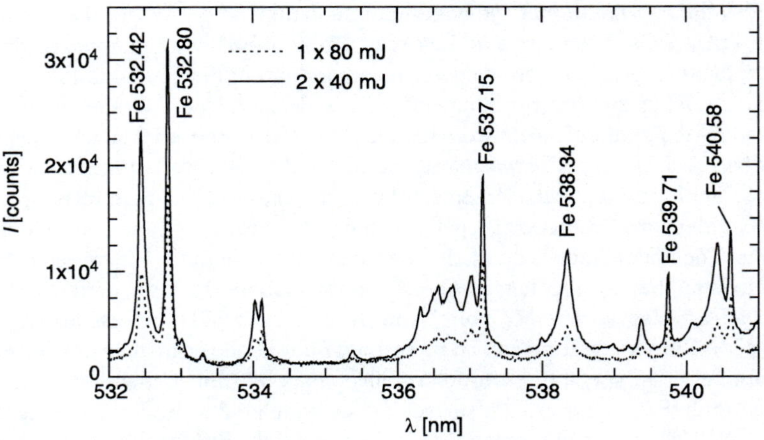

Fig. 8.13 Simultaneously detected spectral range, cf. Table 8.1. Measuring parameters: SP with $1 \times 80\,\text{mJ}$ and DP with $2 \times 40\,\text{mJ}$, $\Delta t = 6\,\mu\text{s}$, $t_{\text{delay}} = 0.8\,\mu\text{s}$, $t_{\text{int}} = 2\,\mu\text{s}$, air atmosphere at 1,013 mbar, $\Delta s = 5\,\text{mm}$

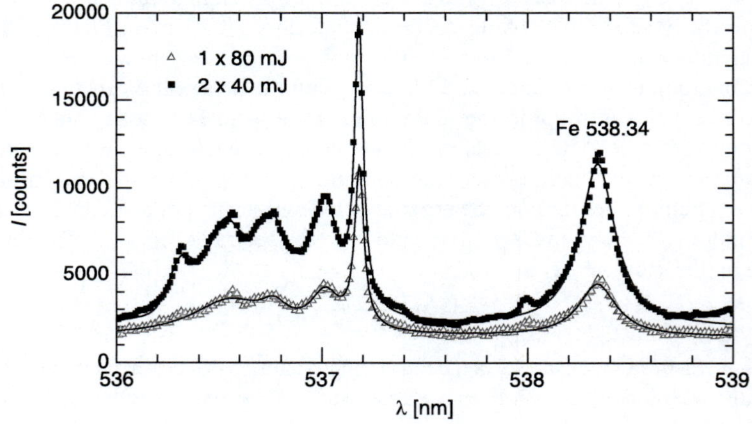

Fig. 8.14 Measured spectra (data points) and Lorentz fit curves (lines) in the spectral range of the line Fe I 538.34 nm, measuring parameters are same as for Fig. 8.13

constant factor for a given species in a given ionization state and fixed process and instrumental parameters, unit: $\text{m}^{-1}\text{s}^{-1}$.

The errors in each measuring point arise from the standard deviation of the line intensities of 5–10% and the stated errors of the Einstein coefficients in the range of 30%. The resultant measurement uncertainty in the temperature is about 10%. The repeatabiltiy in terms of standard deviation of the temperature values of subsequent measurements determined from least square fits of a straight line to the measured points in the Boltzmann plot amounts to less than 5%.

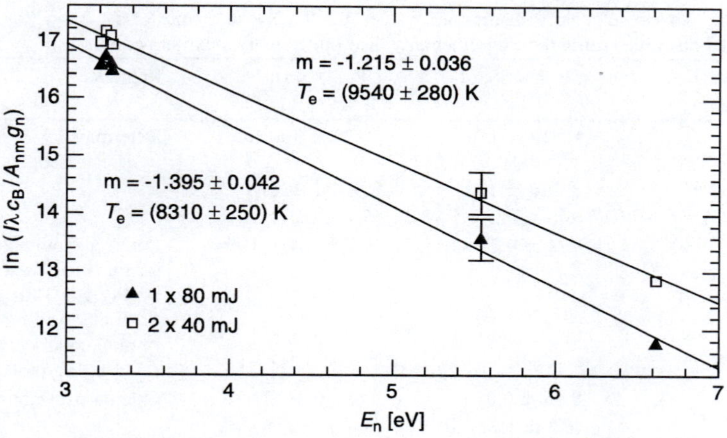

Fig. 8.15 Boltzmann plots of the spectra shown in Fig. 8.13 to determine the electron temperature

For the Boltzmann plots shown in Fig. 8.15, the determined slopes of the fit curves yield the temperatures of the plasmas as follows: $T_e = 9{,}540 \pm 280$ K for double pulses and $8{,}310 \pm 250$ K for single pulses (the errors given here refer to the uncertainty in fitting the data points to straight lines).

The error in the electron density measurement is influenced by the uncertainty in the spectral line fit and the errors in the Stark coefficients taken from Freudenstein and Cooper [8.7]. Additionally, it has to be considered that the Stark broadening constant stated in [8.7] depends on the temperature, cf. Sect. 10.3. The estimated total error amounts to about 20%. The repeatability in terms of standard deviation is 5% and less. The error bars in the following diagrams refer to the repeatability of the electron temperature and density measurements.

Table 8.2 summarizes the measuring results for single and double pulses with a burst energy of 80 mJ at a delay time $t_{delay} = 0.8\,\mu$s. The quantities listed in the first column are: the intensity of the iron emission at 532.42 nm determined from a Lorentz fit in arbitrary units, the electron temperature, the electron density, the total number of iron particles (atoms and ions), the degree of ionization ($r = n_e/n_{Fe}$), the total number of iron atoms in the neutral state (i.e., the number of Fe I atoms), the Boltzmann factor and the total number of iron atoms in the neutral state populating the excited state n.

The ratio of the measured intensities of Fe 532.42 nm for double pulses and single pulses amounts to $3{,}120/1{,}410 \approx 2.2 \pm 0.4$. In the following, this value will be compared with the expected ratio based on the values for T_e and n_e determined spectroscopically and N_{Fe}. The simplifying assumptions for this calculation are: complete atomization of the ablated material in the plasma, population of the excited levels and the ionization can be described by the Boltzmann distribution, Saha equation, respectively; the plasma is optically thin for the considered emission line. In a first step, the degree of ionization $r = n_e/n_{Fe}$ is calculated using the Saha

Table 8.2 Spectroscopically determined values for T_e, n_e at a time of $t_{delay} = 0.8\,\mu s$ and calculated quantities using the experimentally determined mass ablation per burst

Burst mode, burst energy	SP, 1 × 80 mJ	DP, 2 × 40 mJ	Remarks
$I^S_{Fe\ 532.42}$, $I^D_{Fe\ 532.42}$ (a.u.)	$1,410 \pm 150$	$3,120 \pm 300$	Determined spectroscopically
T_e (K)	$8,310 \pm 100$	$9,540 \pm 100$	
n_e (cm^{-3})	$(6.8 \pm 0.2) \times 10^{16}$	$(6.3 \pm 0.2) \times 10^{16}$	
N_{Fe} [1]	$(1.4 \pm 0.2) \times 10^{15}$	$(3.5 \pm 0.4) \times 10^{15}$	Deduced from ablated mass per burst (cf. Fig. 7.5, 130 ng for SP and 330 ng for DP and the molar weight of iron 56 g/mol)
r [1]	0.56 ± 0.05	0.857 ± 0.007	Calculated quantities
$N_{Fe\,I}$ [1]	$(6.2 \pm 1.2) \times 10^{14}$	$(5.0 \pm 0.6) \times 10^{14}$	
f_B [1]	$(4.4 \pm 0.4) \times 10^{-4}$	$(1.19 \pm 0.09) \times 10^{-3}$	
$N_{Fe\,In}$ [1]	$(2.7 \pm 0.8) \times 10^{11}$	$(6.0 \pm 1.2) \times 10^{11}$	
n_{Fe} (cm^{-3})	1.2×10^{17}	7.4×10^{16}	
$n_{Fe\,I}$ (cm^{-3})	5.3×10^{16}	1.1×10^{16}	
n_N (cm^{-3})	6.8×10^{17}	6.2×10^{17}	
n_N/n_{Fe}	5.6	8.4	

Experimental parameters: see Fig. 8.13. For the error of the temperature the repeatability in terms of the standard deviation of the temperature values determined under repeatability conditions is taken

$f_B = \exp(-E_n/kT_e)$ Boltzmann factor, $N_{Fe\,In}$ number of atoms in the upper state n; n_{Fe}, $n_{Fe\,I}$ iron and iron atom density; n_N nitrogen density

equation. With the determined parameters T_e and n_e, the partition functions and the lowering of the ionization energy follow $r = 0.56 \pm 0.05$ for single pulses and $r = 0.857 \pm 0.007$ for double pulses. Furthermore, the Saha equations shows that the second ionization stage is negligible, i.e., $n_{Fe\,III}/n_{Fe\,II} < 10^{-3}$; hence, the relation $n_{Fe\,II} \approx n_e$ holds. With $n_{Fe\,I} = n_{Fe} - n_{Fe\,II}$ follows:

$$n_{Fe\,I} = n_{Fe}(1 - r). \tag{8.6}$$

Assuming that this local relation holds for the whole plasma volume (which is a strong simplification, since the plasma is not homogeneous in densities and temperatures) allows to rewrite (8.6): $N_{Fe\,I} = N_{Fe}(1 - r)$. With r and N_{Fe}, we obtain $N_{FeI} = (6.2 \pm 1.2) \times 10^{14}$ for single pulses and $N_{FeI} = (5.0 \pm 0.6) \times 10^{14}$ for double pulses (see row no. 6 in Table 8.2).

The population of the upper level is given by the Boltzmann distribution for the measured values of T_e and the energy of the upper level E_n of the transition, i.e., 5.54 eV for the Fe line 532.42 nm (cf. Table 8.1). The expected intensity ratio is:

$$\frac{I^D_{Fe\,532.42}}{I^S_{Fe\,532.42}} = \frac{N^D_{Fe\,I}\,e^{-\frac{E_n}{kT^D_e}}\,u^S}{N^S_{Fe\,I}\,e^{-\frac{E_n}{kT^S_e}}\,u^D} = 1.8 \pm 1.0, \tag{8.7}$$

where u^S, u^D is the temperature dependent partition function for single and double pulses excitation, cf. Sect. 9.2, (9.3), (9.6). Within the error limits, this value coincides with the experimentally determined intensity ratio of 2.2 ± 0.4.

The total particle density can be calculated with the values for n_e and r: $n_t = n_e + n_{\mathrm{Fe\,I}} + n_{\mathrm{Fe\,II}} = (1 + 1/r)n_e$ yielding $\approx 1.9 \times 10^{17}\,\mathrm{cm}^{-3}$ for single pulses and $\approx 1.4 \times 10^{17}\,\mathrm{cm}^{-3}$ for double pulses. With the equation of state for ideal gases, a pressure inside the plasma of $\approx$ 220 hPa for single pulses and $\approx$ 185 hPa for double pulses can be estimated. These values are about a factor of 5 below the ambient pressure. If a pressure equilibrium with the ambient gas is assumed after a delay time of 0.8 μs, then there must be additional particles from the ambient atmosphere inside the plasma with a density of $6.8 \times 10^{17}\,\mathrm{cm}^{-3}$ for single pulses and $6.2 \times 10^{17}\,\mathrm{cm}^{-3}$ for double pulses. These values are by a factor of 5.6 and 8.4 higher than the respective iron particle density for single and double pulses. For the measured electron temperature T_e and electron density n_e, the degree of ionization of nitrogen is below 1%; hence, the electron density is mainly a consequence of the ionization of the iron particles. The approximation $n_{\mathrm{Fe\,II}} \approx n_e$ given above is still valid. The density of particles in the ambient atmosphere amounts to $2.5 \times 10^{19}\,\mathrm{cm}^{-3}$. Hence, more than 97% of the ambient atmosphere particles are displaced by the plasma (cf. Annex, Sect. A.1).

The size of the plasma can be estimated on the basis of the spectroscopically determined iron particle density n_{Fe} and the ablated mass assuming that the plasma parameters and the iron particle density are homogeneous inside the plasma and the ablated material is completely atomized in the plasma. For the iron particle density follows $n_{\mathrm{Fe}} = 1.2 \times 10^{17}\,\mathrm{cm}^{-3}$ for single pulses and $n_{\mathrm{Fe}} = 7.4 \times 10^{16}\,\mathrm{cm}^{-3}$ for double pulses. Assuming a hemispherical plasma yields a radius of 1.8 mm and 2.9 mm for single pulses and double pulses, respectively. Qualitatively this is consistent with the observations shown in Fig. 8.5, where a greater extension of the plasma after the interaction with a double pulse is detected compared with the plasma extension for a single pulse. The plasma radii estimated here are greater than the observed extension of the luminous plasma in the framing and streak images of Figs. 8.3–8.5. However, we have to consider that the ablated mass used for this estimation (cf. Fig.7.5) is an upper limit for the mass ablated for a semi-infinite body (instead of a thin foil) and furthermore that not all particles ablated are excited inside the plasma volume.

The temporal variation of the line intensities of Fe I 532.42 nm and Fe I 532.80 nm, the electron temperature, and electron density after the second pulse of a double pulse is shown in Fig. 8.16. Since the integration time of 2 μs is of the order of the times estimated in the following, the latter are upper limits for the real decay times, which have to be determined by a deconvolution procedure.

The line intensity of the Fe I 532.42 nm having an upper energy level of 5.54 eV decays monotonously, whereas the line Fe I 532.80 nm with a smaller upper energy level of 3.24 eV attains a maximum after $\approx$ 3 μs. The measuring points of the electron density can be fitted by an exponential curve with a time constant of 5.7 μs. The semi-logarithmic plot of the electron density shows that the temporal variation

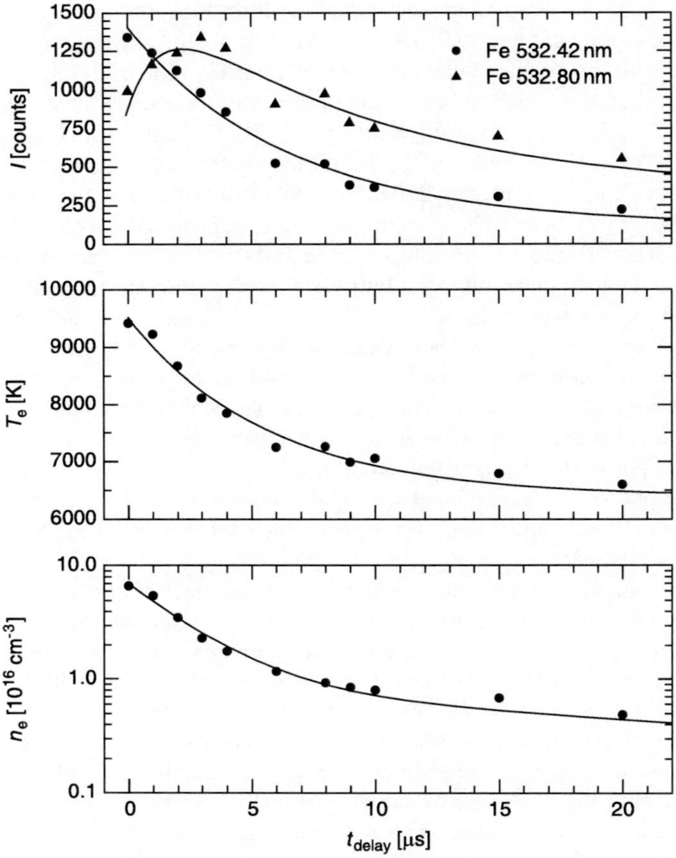

Fig. 8.16 Line intensity I of Fe 532.42 nm, Fe 532.80 nm, electron temperature T_e and density n_e as a function of the time delay t_{delay} after the second pulse of a double pulses with 2×40 mJ ($t_{int} = 2\,\mu$s, $\Delta t = 6\,\mu$s, $\Delta s = 0$ mm, ambient atmosphere: air at normal pressure)

can be described by two time constants. Up to about 6 μs, the time constant is $\approx 2.5\,\mu$s, whereas for later times the time constant is about 30 μs.

In the following, we compare the measured intensity of the Fe I 532.42 nm line with the intensity to be expected from the measured plasma parameters. Table 8.3 shows the measured values for the delay times 1 and 20 μs after the second pulse of a double pulse.

The experimentally determined intensity $I^{D}_{Fe\,532.42}$ decays after 20 μs to a value, which is $(18 \pm 5)\%$ of the initial value at 1 μs. Assuming an optically thin plasma, the line intensity is proportional to the total number of emitting particles. For the estimation, we consider the following two simplifying cases of a homogeneous plasma: (a) the total number of iron particles N_{Fe} is constant in the time period from 1 to 20 μs and (b) the plasma volume contributing to the line intensity is constant.

Table 8.3 Spectroscopically determined values of $I^{D}_{Fe\,532.42}$, T_e, n_e at the delay times of 1 and 20 μs and calculated quantities

t_{delay} (μs)	1	20	Remarks
$I^{D}_{Fe\,532.42}$ (a.u.)	1,240 ± 150	220 ± 30	Determined spectroscopically
T_e (K)	9,200 ± 100	6,600 ± 100	
n_e(cm^{-3})	(5.5 ± 0.2) × 10^{16}	(0.5 ± 0.02) × 10^{16}	
r [1]	0.86 ± 0.02	0.47 ± 0.05	Calculated quantities
f_B [1]	(9.1 ± 0.7) × 10^{-4}	(5.8 ± 1.6) × 10^{-5}	
$n_{Fe\,I}$(cm^{-3})	(9.0 ± 1.5) × 10^{15}	(5.5 ± 1.7) × 10^{15}	
n_{Fe}(cm^{-3})	(6.4 ± 0.4) × 10^{16}	(1.0 ± 0.2) × 10^{16}	

$n_{Fe\,I}$ = density of iron atoms, n_{Fe} = density of iron particles (atoms and ions), further abbreviations see Table 8.2

For case (a), we use $N_{Fe\,I} = N_{Fe}\,(1 - r)$, where r is calculated with the Saha equation, then the following relation holds:

$$\frac{I^{20\,\mu s}_{Fe\,532.42}}{I^{1\,\mu s}_{Fe\,532.42}} = \frac{(1 - r^{20\,\mu s})\mathrm{e}^{-\frac{E_m}{kT_e^{20\,\mu s}}}\;u^{1\,\mu s}}{(1 - r^{1\,\mu s})\,\mathrm{e}^{-\frac{E_m}{kT_e^{1\,\mu s}}}\;u^{20\,\mu s}} = 0.36 \pm 0.08. \tag{8.8}$$

The intensity of the line Fe 532.42 nm at 20 μs is then (36 ± 8)% of the value at 1 μs. Since the experimentally measured intensity decreases to 18%, the total number of iron particles must have been reduced by condensation, diffusion out of the radiating volume, or cluster formation to a value of about 50% of the initial total number of iron particles at 1 μs.

For case (b), a constant plasma volume, we take $n_{Fe\,I}$ instead of $N_{Fe\,I}$:

$$\frac{I^{20\,\mu s}_{Fe\,532.42}}{I^{1\,\mu s}_{Fe\,532.42}} = \frac{n^{20\,\mu s}_{Fe\,I}\,\mathrm{e}^{-\frac{E_m}{kT_e^{20\,\mu s}}}\;u^{1\mu s}}{n^{1\,\mu s}_{Fe\,I}\,\mathrm{e}^{-\frac{E_m}{kT_e^{1\,\mu s}}}\;u^{20\,\mu s}} = 0.06 \pm 0.03 \tag{8.9}$$

showing a decrease to (6 ± 3)% of the initial value. This value is smaller than the measured decrease to (18 ± 5)% which is an indication that the iron particles are dispersed over a greater volume in the time interval from 1 to 20 μs. Hence, from (8.9) follows a volume increase by a factor of 3 to get an intensity decrease consistent with the experimental findings.

Table 8.3 shows that the density of iron atoms decreases in the considered time interval; however, this decrease is much smaller than the decrease of the total iron particle density. This effect is attributed to the recombination of Fe ions.

The ratio of the line intensities $I_{Fe\,532.80}(t)/I_{Fe\,532.42}(t)$ (see Fig. 8.16 top) – assuming optically thin emission – allows also to determine the temperature $T(t)$. If we further assume that the plasma is homogeneous, then the intensity of an iron atom line is proportional to the total number of emitting iron atoms $N_{Fe\,I}$ [cf. (9.3)]: $I_{Fe\,532.42} \propto N_{Fe\,I}\,\left(\mathrm{e}^{-E_j/kT}/u^0_{Fe}(T)\right)$. Knowing $T(t)$, $I_{Fe\,532.42}(t)$ allows then

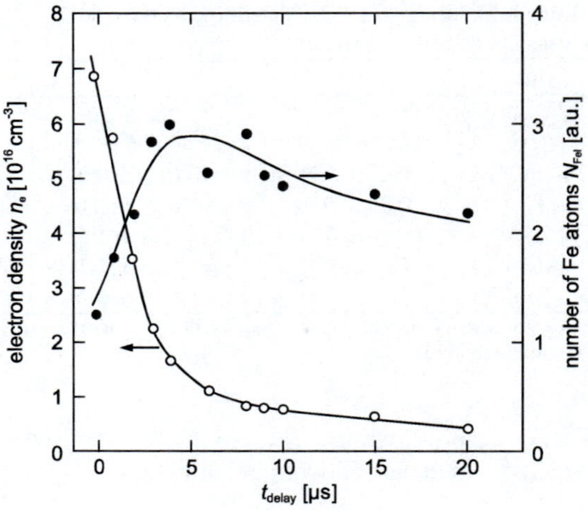

Fig. 8.17 Electron density and number of iron atoms as a function of time. The electron density data is the same as in Fig. 8.16, *bottom*. The number of iron atoms is deduced from the intensities of two iron atom lines (see Fig. 8.16, *top*)

to calculate $N_{\mathrm{Fe\,I}}(t)$. Figure 8.17 shows the electron density and the total number of iron atoms as a function of time. The strong monotonous decay of the electron density is accompanied by an increase in the number of iron atoms for delay times up to 5 μs. This increase is mainly fed by recombining iron ions.

The electron temperature decrease as a function of time shown in Fig. 8.16 is plotted in Fig. 8.18 again in a double logarithmic diagram showing that this temperature decrease can be described by a power law as $T_e \propto t^{-0.11}$. This decrease is much smaller than a cooling expected for a pure adiabatic expansion of the plasma volume, which is described by:

$$\frac{T_1}{T_2} = \left(\frac{V_2}{V_1}\right)^{\chi-1}, \tag{8.10}$$

where T_1, V_1 and T_2, V_2 are the temperature and volume of the plasma at time t_1, t_2 and χ is the adiabatic coefficient. Taking the volume increase of a factor of 3 estimated above and the adiabatic coefficient of a monoatomic gas of 1.67 relation (8.10) yields a temperature drop by a factor of 2.1 (see open square in Fig. 8.18), which is much stronger than the observed decrease by a factor of 1.4. Obviously the cooling of the plasma is reduced which is interpreted as a consequence of the decreasing degree of ionization releasing recombination energy contributing to a post-heating of the plasma.

Figure 8.19 shows the spatially integrated intensity of the Fe 532.42 nm line, the electron temperature, and density for single and double pulses (interpulse separation of double pulses $\Delta t = 6\,\mu\mathrm{s}$, $\Delta s = 5\,\mathrm{mm}$) as a function of burst energy under 1,013 mbar atmospheric pressure [6.14]. The intensity of the iron line Fe 532.42 nm is a factor of about two higher if a double pulse of 2×40 mJ is used instead of a single 80 mJ pulse (cf. also Fig. 6.3). For single pulses, the intensity increases up to pulse energies of about 70 mJ and remains nearly constant for higher energies. With double pulses, the intensity grows monotonically over the investigated energy range 50–350 mJ.

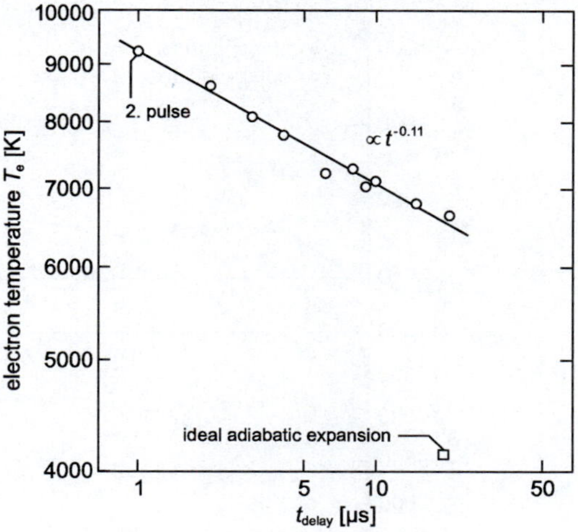

Fig. 8.18 Temperature of the laser-induced plasma after the second pulse of a double pulse as a function of time in a double logarithmic diagram (same data as Fig. 8.16, *middle*) compared with the temperature decrease estimated for an adiabatic plasma expansion with a volume increase by a factor of 3

The electron temperature determined for single pulses shows a saturation behavior for energies greater than 40 mJ at 8,300 K; for double pulses at 160 mJ at 10,200 K. The electron density is approximately equal for single and double pulses with energies up to 150 mJ. For greater burst energies, it amounts to about 9×10^{16} cm^{-3} for double pulses and (6–7) $\times 10^{16}$ cm^{-3} for single pulses.

The plasma parameters for double pulses of 2 × 40 mJ with variable interpulse separations are shown in Fig. 8.20 [6.14]. The experimental parameters are the same as for Fig. 8.19. Two limits were taken into account: (a) an interpulse separation of zero corresponds to a single pulse of 80 mJ and (b) an infinite interpulse separation corresponds to quasi-single pulses of 40 mJ (interpulse separation 33 ms, corresponding to the burst repetition rate of 30 Hz).

Again, the Fe 532.42 nm intensity and the electron temperature show significantly higher values for double pulses than for single pulses. For the double pulses, the electron temperature is approximately constant at about 9,300 K in the investigated range of interpulse separations. The line intensity attains a minimum for double pulses with interpulse separations of about 5–10 μs. The behavior of the electron density is similar to that of the line intensity, with the exception that the electron density generated by a single 80 mJ pulse is higher than those for double pulses with interpulse separations below about 15 μs.

The influence of the focal position relative to the sample surface is shown in Fig. 8.21 for single and double pulses of 80 mJ (interpulse separation of double pulses 6 μs, 1,013 mbar atmospheric pressure). The focusing lens is shifted for these measurements, so the distance between sample surface and fiber optics remains constant (see Sect. 6.1). For single pulses, the maximum of the Fe 532.42 nm intensity is observed close to $\Delta s \approx 9$ mm. For smaller or even negative Δs, the line intensity drops to a very small value. This corresponds to Fig. 6.2, which shows a plasma separated from the sample for negative Δs. In contrast to this,

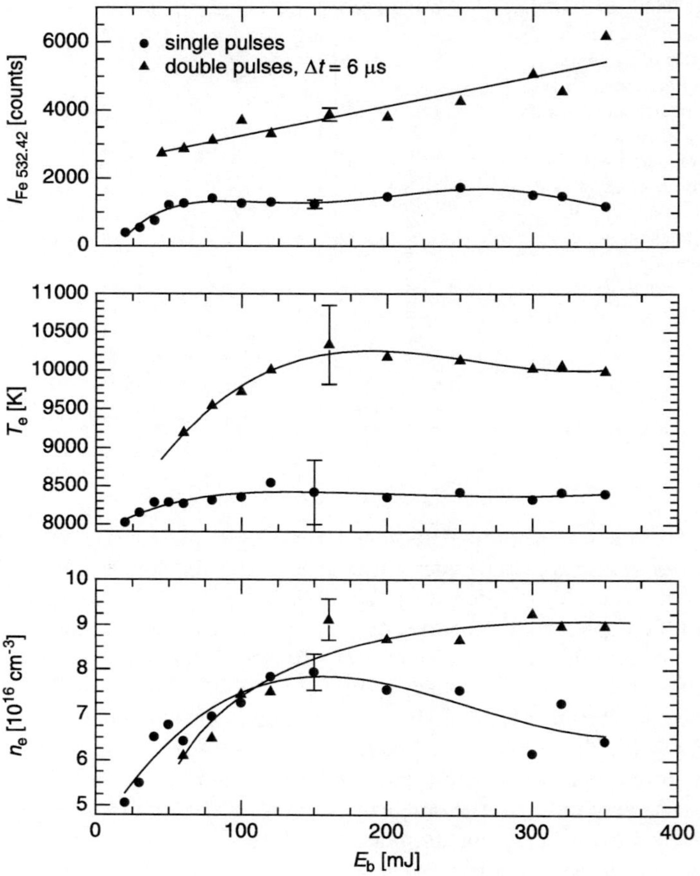

Fig. 8.19 The spatially integrated intensity of the Fe 532.42 nm line, electron temperature, and density for single and double pulses as a function of the burst energy in the time period shown in Fig. 6.3

the behavior for double pulses is nearly symmetric to $\Delta s \approx -1$ mm, revealing the symmetry of the intensity distribution of the laser beam with respect to the focal plane. Three local maxima are observed: at $\Delta s \approx -1$, -12, and 11 mm. The maxima at $\Delta s = -12$ and 11 mm have nearly the same width. Obviously at $\Delta s = -12$ and 11 mm, the intensity level at the target surface and the size of the irradiated area induce a plasma dynamic which offers optimum conditions for the emission of the iron line at 532.42 nm. The maximum at -12 mm is slightly smaller than the one at 11 mm, which may be the consequence of a small shielding effect of an atmospheric plasma in the case of $\Delta s < 0$ mm. The reason for the central peak at $\Delta s \approx -1$ mm is not yet clear. Its half width is about 2 mm, which

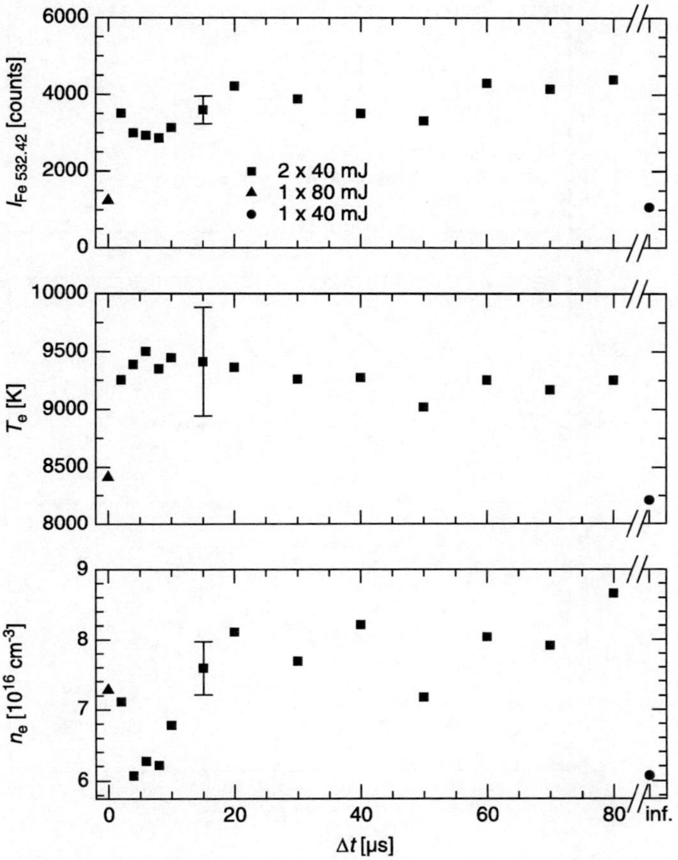

Fig. 8.20 The spatially integrated intensity of the Fe 532.42 nm line, electron temperature, and density for double pulses of 80 mJ burst energy as a function of the interpulse separation

is comparable with the Rayleigh length of the laser beam of about 1 mm, evaluated under the assumption of a Gaussian beam profile.

The electron temperature and density with single pulses attain absolute maxima for $\Delta s \approx 3$ and 7 mm, respectively, and local maxima at $\Delta s \approx 21$ mm. For double pulses, there are two maxima in T_e, of which the maximum at $\Delta s = 2$–3 mm is significantly greater than the one at $\Delta s = -6$ mm. Again this is an indication that, for negative Δs, an atmospheric plasma evolves, which attenuates the laser intensity at the sample surface and leads to reduced electron temperatures. Unlike the Fe intensity, the temperature for double pulses is not symmetric with respect to $\Delta s \approx -1$ mm, but at $\Delta s \approx -12$ and 11 mm, corresponding to the maxima of the Fe line intensity, its value is equal to about 8,800 K. The electron density also seems to attain maxima at $\Delta s \approx -12$ and 11 mm.

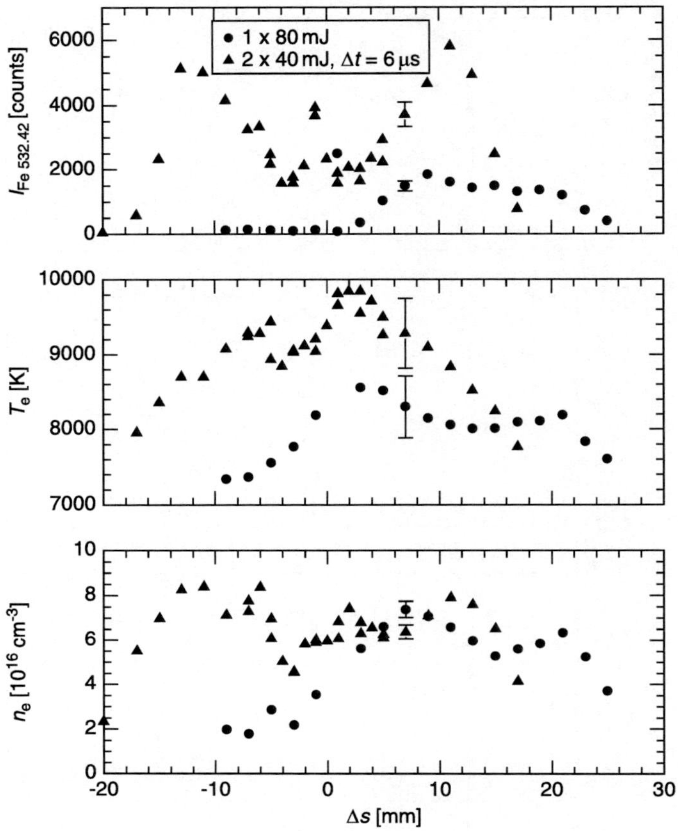

Fig. 8.21 The spatially integrated intensity of the Fe 532.42 nm line, electron temperature, and density for single and double pulses of 80 mJ burst energy as a function of the focus position Δs

With the setup shown in Fig. 7.11, the dependence of the plasma parameters on the ambient pressure was studied. The sample used is a pure iron sample. The following parameters were used: single pulses 80, 160 mJ, pulse width 20 ns; double pulses 2×40, 2×80 mJ, width 40 ns; interpulse separation $\Delta t = 6\,\mu$s, energy ratio 1:1; position of the beam waist relative to the sample surface $\Delta s = 10$ mm, detection after the single pulse or the second pulse of a double pulse with $t_{\text{delay}} = 1\,\mu$s, $t_{\text{int}} = 1\,\mu$s. Excitation temperatures are determined by Boltzmann plots using Fe I lines in the range between 427 and 542 nm with upper excitation energies between 3.21 and 6.76 eV. In the case of local temperature equilibrium (LTE, see Sect. 8.5), the excitation temperature is equal to the electron temperature. Evaluation of the line width of the Stark broadened atomic iron line at 538.337 nm yields electron densities. To assess the degree of absorption of this line in the plasma, we estimated the effective absorption coefficient, which is proportional to the value

k_t defined in [8.8], see also (9.31). For temperatures in the range of 7,500–9,700 K, we obtain $k_t < 0.8 \times 10^{-30}\,\text{m}^3$ and hence absorption plays only a minor role and no significant deviations from the Boltzmann plot are to be expected. For each burst energy and pressure value measurements at five different sample locations were conducted. At each sample location, five measurements are performed, where each measurement comprises a series of ten bursts. Hence, for each data point shown in the following diagrams, the spectral signals generated by $5 \times 5 \times 10 = 250$ bursts are evaluated. Figure 8.22 shows the intensity of the Fe I line at 538.34 nm and the continuum intensity measured at 534.6 nm – which lies within a spectral region where no line emission was observed – as a function of the pressure for a burst energy of 80 mJ.

The line intensity at 1 bar for double pulses is by a factor of 4.6 greater than the one for single pulses and decreases monotonously for reduced pressures. Single pulses show a different behavior. The line intensity increases for reduced pressures, attains a maximum at about 100 mbar, and then decreases for further pressure reductions. Error bars are shown enlarged for better visibility. For double pulses,

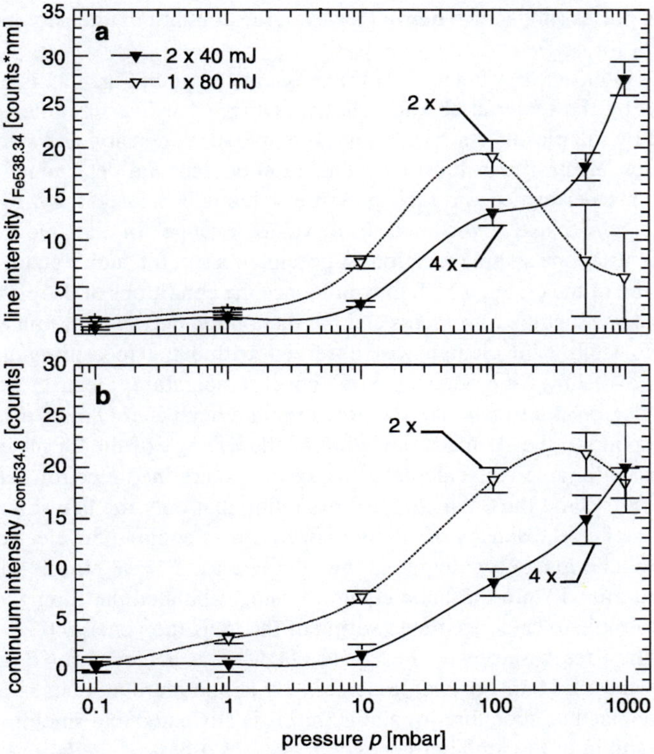

Fig. 8.22 Line intensity of the iron line Fe I 538.34 nm **(a)** and continuum intensity at 534.6 nm **(b)** as a function of pressure for a burst energy of 80 mJ

the error bars are significantly smaller than for single pulses. The physical reason of these reduced error bars is not clear.

The maximum in the line intensity of atomic iron at a reduced air pressure was also observed by Y. Iida, who studied the interaction of single pulses with a standard Al alloy containing 0.97% iron in the pressure range from 5 to 760 torr [8.9]. At an air pressure of 200 mbar, a maximum of the line emission Fe I 374.949 nm was found, which is by about a factor of 10 higher than the value measured at 1 bar, whereas here an enhancement by a factor of 3.2 was measured (cf. Fig. 8.22).

The continuum intensity measured in the line-free region at 534.6 nm decreases for reduced pressures with the only exception for single pulses while going from 1,000 to 500 mbar. For all data points taken between 0.1 mbar and 1 bar, the continuum intensity is higher for single pulses than for double pulses. At 0.1 mbar and at 1 bar, the continuum intensity does not differ significantly between single and double pulses. The ratio of line-to-continuum intensity is for all pressures greater than 0.1 mbar higher for double pulses than for single pulses. The greatest improvement of the line-to-continuum ratio amounts to a factor of 4.2 and is achieved for double pulses at 1 bar. An increase of the burst energy from 80 to 160 mJ does not significantly change the line intensities measured at 1 bar for single and double pulses (not shown here). However, the continuum intensity increases by a factor of 3 for single and double pulses.

The line intensity as a function of the pressure shown in Fig. 8.22 is not directly determined by the material ablation shown in Fig. 7.13. The line intensity is also influenced by the plasma state in terms of temperature, density, and size. Electron temperatures, electron densities, and ionization degrees are determined for all 25 measurements performed for each pressure value and burst energy. First, it was checked to which extent the measuring values change for the five consecutive measurements at one sample location. A possible reason for such a change could be the formation of the crater, which may influence the conditions of the plasma generation for subsequent measurements and the interaction between plasma and sample surface. No significant change was observed within the measuring uncertainty. Figure 8.23a–c shows the determined electron temperature, electron density, and ionization degree as a function of the pressure for a burst energy of 80 mJ. The error bars correspond to the standard deviation of the average of the 25 measurements. The ionization degree was calculated using the determined electron temperature, electron density, and the Saha equation assuming that only the first ionization step is significant, i.e., the density of singly ionized iron is equal to the electron density: $n_{\mathrm{Fe\,II}} = n_{\mathrm{e}}$. The ionization degree is then defined as: $\delta = n_{\mathrm{e}}/(n_{\mathrm{Fe\,I}} + n_{\mathrm{e}})$, where $n_{\mathrm{Fe\,I}}$ is determined from the Saha equation using tabulated partition function and Unsöld's formula to calculate the lowering of the ionization energy [8.10].

The highest temperatures of nearly 10,000 K were achieved for double pulses and a pressure of 1 bar. For single pulses, the temperature at 1 bar is about 15% less. The reduced temperature for single pulses is attributed to a shielding effect of a plasma ignited in the ambient gas which absorbs a part of the laser energy thus reducing the available energy for the ablation of the sample material and formation of the iron plasma. For double pulses, a reduced shielding occurs since the material

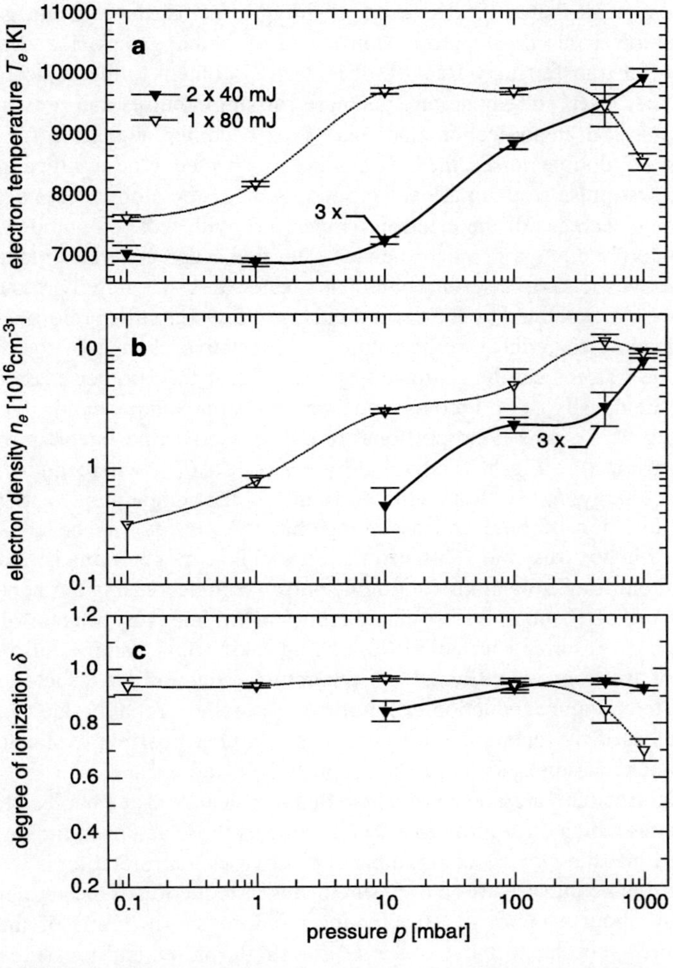

Fig. 8.23 (**a**) Electron temperature, (**b**) electron density, and (**c**) degree of ionization as a function of pressure for a burst energy of 80 mJ

ablated by the first pulse partly pushes the ambient gas out of the interaction region [8.1]. Hence, the energy of the second pulse can be transferred more effectively to the ablated iron vapor leading to higher temperatures of the iron plasma (although the energy of the second pulse is just half of that of the single pulses).

For double pulses, the electron temperature decreases monotonously while lowering the pressure and stabilizes at about 7,000 K for pressures less than 10 mbar. For single pulses, the temperature rises to 9,700 K being nearly constant for pressures between 500 and 10 mbar and then drops again for pressures $<$10 mbar. The rise of the electron temperatures observed for single pulses and pressures below 1 bar is attributed to a reduced shielding effect of the ambient gas plasma due to the lower initial gas particle density. The ambient gas plasma acts also as a energy buffer transferring a fraction of its energy content to the adjacent material plasma. The observed temperature decrease for single pulses and pressures below 10 mbar may be an indication that this effect becomes negligible for such low pressures. For double pulses, the buffer effect is expected to be significantly smaller since the first pulse leads to a localized reduction of the ambient gas density. The monotonous decrease of the electron temperature with reduced ambient pressures may be interpreted as a geometric effect. The first pulse generates a depleted gas volume above the sample surface, which is greater for lower initial pressures.

The plasma generated by the second pulse expands into this preformed enlarged volume leading to reduced temperatures. The relative change of the volume of emitting Fe I atoms can be estimated by the ratio of the number of emitting Fe I atoms to the density of Fe I atoms calculated from the Saha equation. The number of emitting Fe I atoms is proportional to the measured line intensity of the Fe I 538.34 nm line (see Fig. 8.22) divided by $e^{-E_j/kT}/u_o(T)$, where E_j is the upper excitation energy, k the Boltzmann constant, T the temperature, and $u_o(T)$ the partition function of Fe I and assuming that the plasma can be considered as optically thin for this line. This estimation yields a monotonous increase of the volume of emitting iron atoms for double pulses with decreasing ambient pressure. At a pressure of 10 mbar, the estimated emission volume is by a factor of 8 greater than at 1 bar. Assuming an adiabatic cooling taking this volume ratio and three degrees of freedom would cause a temperature reduction by a factor of 4. The observed temperature reduction is significantly smaller probably due to additional energy released by recombination processes. Another possible explanation is the lack of shock heating as the ambient gas pressure is reduced.

The electron temperature curves show that double pulses at 1 bar lead to electron temperatures which are approximately the same as those achieved for single pulses at reduced pressures of about 100 mbar. The obvious interpretation is that the first pulse in a double pulse causes a transient localized reduction of the ambient particle density by about a factor of 10, creating conditions – in terms of the electron temperature – similar to those observed for single pulses and pressures of about 100 mbar.

The electron temperatures and electron densities reported here for single pulse excited plasmas at 1 bar are similar to those found in an experiment, where single pulses of 100 mJ were focused on a low-alloy steel sample in an air atmosphere at 1 bar yielding temperatures in the range between 8,000 and 10,000 K and electron densities of $4–11 \times 10^{16}\,\mathrm{cm}^{-3}$ at a delay time of 3 μs [8.11].

The electron density at 1 bar is about the same for double and single pulses, see Fig. 8.23b. For double pulses, the electron density decreases for lower pressures. For pressures <10 mbar, the resolution of the spectrometer is not sufficient to determine the line width. For single pulses, the electron density attains a maximum at 500 mbar and then decreases monotonously. The electron density for double pulses at 1 bar is about the same as the one determined for single pulses at pressures around 100 mbar. The material ablation increases for reduced pressures (cf. Fig. 7.13), the degree of ionization is nearly constant or increases slightly (see Fig. 8.22c); hence, the reduction of electron densities is linked with an increase of the plasma volume. The decreasing electron density with reduced ambient pressure corresponds to the course of the intensity of the continuum emission shown in Fig. 8.22, bottom. The electron density decays by more than an order of magnitude while reducing the pressure from 1,000 to 1 mbar; hence, an even stronger decrease of the continuum radiation is expected, since the emission coefficient for free–free Bremsstrahlung scales as $\varepsilon_{\mathrm{cont}} \sim n_e^2$ [cf. (9.1)]. However, at the same time, the emitting volume increases, leading to the diminished reduction of the continuum emission with lower pressures (cf. Fig. 8.22b).

Figure 8.23c shows the degree of ionization as a function of the pressure. The degree of ionization is higher at 1 bar for double pulses than for single pulses. The curves cross at 100 mbar, indicating again that double pulses cause a plasma state comparable with the one generated with single pulses at a reduced pressure of 100 mbar.

Argon was used as ambient gas to study the influence of collinear double pulses on the excitation of the surrounding atmosphere as a function of the pressure. Figure 8.24a–c shows the investigated spectral range centered around 430 nm. The spectra shown in Fig. 8.23a, b are LIBS spectra gained with single pulses of 160 mJ and the pure iron sample in air and argon atmosphere, respectively. The spectrum shown at the bottom was detected with the same spectrometer setup observing the emission of a low-pressure argon lamp. The emission spectrum of the latter comprises only emission lines of atomic argon. The line detected at 434.81 nm in the LIBS spectrum shown in the middle is attributed to an Ar II emission substantiated as follows. The line at 434.81 nm is not observed with air as ambient atmosphere (cf. spectrum at the top in Fig. 8.24). Hydrogen is a potential contaminant for argon. The only line of hydrogen in this spectral range is the Balmer H_γ line at 434.05 nm. However, the deviation of the spectral position of 0.8 nm from the detected emission is too big to be attributed to a Stark shift. Furthermore, hydrogen lines show a stronger broadening at these plasma conditions than argon or iron lines. Estimating the Stark width (FWHM) of the H_γ line for a temperature of 10,000 K and an electron density of $10^{16}\,\mathrm{cm}^{-3}$ yields 1.06 nm, which is significantly greater than the width of the line observed at 434.81 nm in the spectrum shown in the middle of Fig. 8.24 [8.5]. The ratio of the intensities of the two Ar I lines (spectrum in Fig. 8.24c) at 434.52 and 430.01 nm amounts to 0.3. Such a relation would be expected for an excitation temperature of about 4,600 K in the argon discharge lamp (using line data from [8.12]). For higher temperatures to be expected in the LIBS plasma this ratio increases. However, in the LIBS spectrum taken in argon atmosphere a line pair with this intensity ratio at these spectral positions is not observed. The line observed in the LIBS spectrum at an argon pressure of 750 mbar at 434.81 nm lies – within the uncertainty of the wavelength calibration of the spectrometer – at the position of an Ar II line. This identification is further supported by the small line appearing on the right side of the Fe I 432.58 nm line. This line lies at 433.12 nm, which is also an Ar II line. Taking the parameters of these two Ar II lines, the expected intensity ratio $I_{433.12}/I_{434.81}$ is for temperatures between 10,000

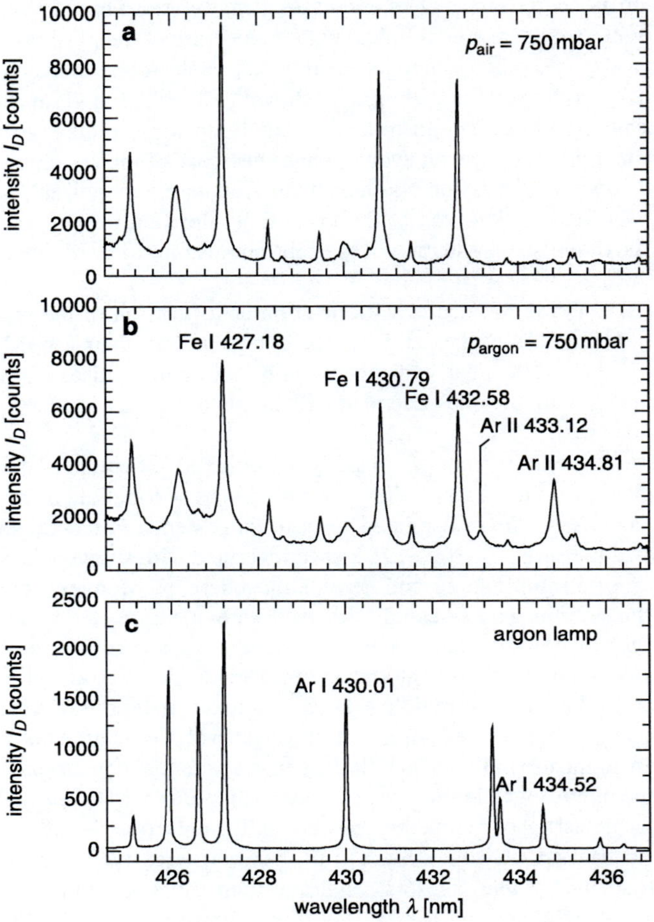

Fig. 8.24 Spectra of the pure iron sample gained with single pulses of 160 mJ in air atmosphere (*top*) and argon atmosphere (*middle*). *Bottom*: spectrum of a low-pressure argon lamp in the same spectral region

and 20,000 K about 0.15–0.16, which is in agreement with the observed ratio. For temperatures in the range of 7,000–20,000 K, the Ar II 434.81 line has a k_t value of $<5 \times 10^{-32}$ m^3, and hence this line is also not affected by self absorption (see above discussion for the Fe I 538.34 line).

Figure 8.25 shows the intensity of the Ar II 434.81 nm line as a function of the argon pressure for single and double pulses with a burst energy of 160 mJ. The line intensity attains a minimum for single pulses at 10 mbar. In the pressure range from 10 to 1,000 mbar, the intensity increases linearly with the pressure. For double pulses, the line was not observable at a pressure of 50 mbar. In the range from 100 to

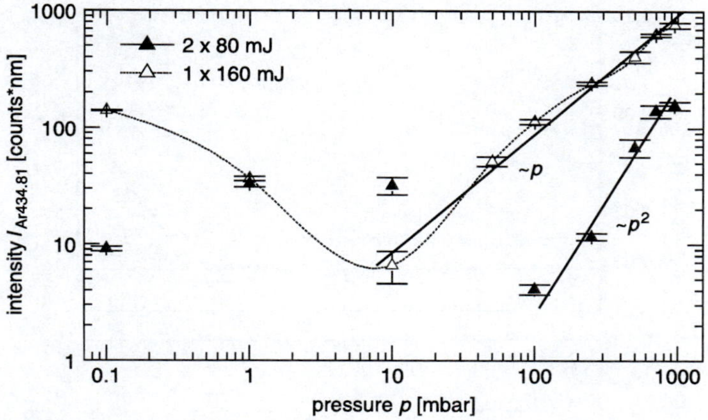

Fig. 8.25 Intensity of the Ar II 434.81 nm line as a function of the argon pressure for single and double pulses with a burst energy of 160 mJ

1,000 mbar, the line intensity increases with pressure following a power law $\sim p^2$. At 1 bar, the line intensity of the Ar II line is for double pulses about a factor of 6 smaller than for single pulses. The Ar II line intensity induced by single pulses reaches this level at a pressure between 100 and 200 mbar. Obviously double pulses lead to plasma conditions, which are similar to those generated with single pulses at reduced ambient particle density.

For pressures below 10 mbar, the intensity rises again for single and double pulses. The cause of this behavior is not known.

Figure 8.26a, b shows the line widths (HWHM) of the lines Ar II 434.81 and Fe I 432.58 nm as a function of the argon pressure. For pressures below 1 mbar, the line width converges to the instrumental width of 0.027 nm for that spectral range. The Ar II 434.81 nm line shows Stark broadening. Taking the broadening parameters given in the literature for this line, the electron density can be determined, see vertical scale on the right side of Fig. 8.26 [8.13]. Stark parameters for the Fe I 432.58 line are not available [This line was studied to compare its behavior with the Ar 434.81 line. The Fe 432.58 line lies in the same spectral window detected (cf. Fig. 8.24b) and shows no strong overlap with any other lines]. The electron density decreases at lower pressures. For double pulses, the electron density is always smaller than for single pulses. Comparing the values shown in Fig. 8.26 with those of Fig. 8.23 shows that the electron density determined from the Ar II line is significantly greater than the one determined with the Fe I 538.34 nm line. Determining electron densities with Ar II line implies that this electron density persists in regions, where Ar II dominates the plasma and these regions are expected to be spatially separated from the iron containing core of the plasma. The expanding plasma consisting of ablated iron species pushes the surrounding ambient gas particles outward, inducing a shock wave and collecting the ambient gas in a shell surrounding the core plasma. If for this argon shell, a minimum argon particle density corresponding to the initial gas pressure of 1 bar at room temperature

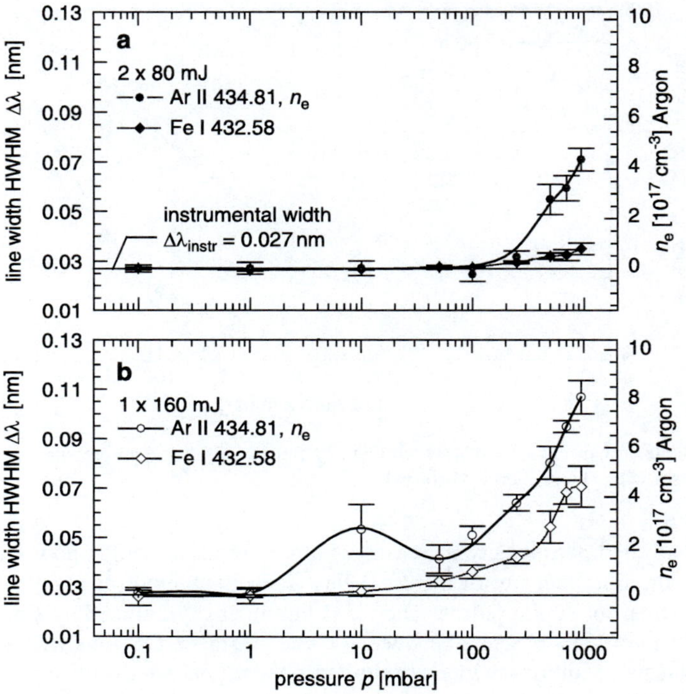

Fig. 8.26 Line widths of the Ar II line 434.81 nm and Fe I line 432.58 nm (see Fig. 8.24b) as a function of the argon pressure. *Top*: double pulses 2 × 80 mJ, *bottom*: single pulses 1 × 160 mJ. The *right scale* refers to the electron density determined from the Stark width of the Ar II line

($2.5 \times 10^{19}\,\text{cm}^{-3}$) is assumed, then the electron density for a given temperature using the Saha equation can be estimated. For 13,000 K (which is in the range to be expected from the spectrum in Fig. 8.24b, looking at the ratio of the two Ar II lines), an electron density of $8.7 \times 10^{17}\,\text{cm}^{-3}$ is obtained, which is approximately equal to the value determined for single pulses at 1 bar (see Fig. 8.26b). Since the second plasma induced by a double pulse expands into a locally rarefied argon particle density as an effect of the preceding laser pulse, the electron density level achieved in the argon shell is expected to be smaller than for single pulses in agreement with the measurements (cf. Fig. 8.26a).

8.3 Spatially Resolved Electron Density

The index of refraction caused by the laser-induced plasma is determined from the interference patterns detected in the framing mode of the streak camera. Figures 8.27 and 8.28 show framing images of the interference patterns after the first and the

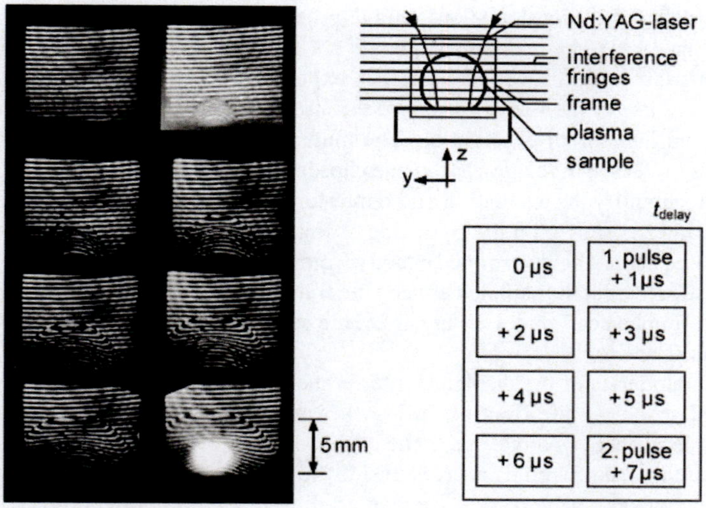

Fig. 8.27 Framing images of interferograms detected in a time interval comprising the first and the second laser pulse of a double pulse. In the *upper right*, a schematic view of the orientation of the fringes relative to the plasma and the sample is shown. The diagram in the lower right shows the temporal sequence of the frames with the delay times t_{delay} starting at 0 µs for the first frame. Experimental parameters are the same as for Fig. 8.7

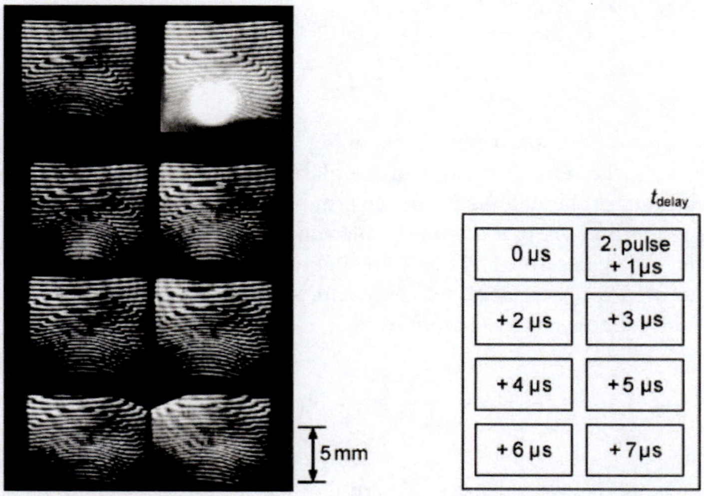

Fig. 8.28 Framing images of interferograms detected after the second laser pulse of a double pulse. All other experimental parameters are the same as for Fig. 8.27

second laser pulse of a 2×80 mJ double pulse and an ambient pressure of $p_0 =$ 1,013 mbar [8.1]. The framing rate is $10^6\,\mathrm{s}^{-1}$, corresponding to a time separation between subsequent frames of 1 μs. The exposure time amounts to 200 ns/frame. For a better use of the image format of the streak camera, the camera is rotated by 90° around the axis of observation. The mutual distance between the interference fringes is increased in relation to the one chosen for the streak images in the previous section to simplify the evaluation of the interferograms. The geometry and sequence of the frames is shown on the right side of Fig. 8.27. The time of irradiation of the laser pulses cannot be determined exactly from the framing images. In Fig. 8.27, the first pulse arrives at the sample between the frames no. 1 and 2, and the second pulse between frames no. 7 and 8. In Fig. 8.28, the second pulse hits the sample between frame no. 1 and 2.

The boundaries of the displaced interference fringes show a well-defined hemispherical shape. At the front of the shock wave, the fringes are displaced toward the sample surface, which is due to the increase in the refractive index caused by the compressed air and sample particles. Inside the sphere – defined by the leading edge of the shock wave – the fringes are displaced in the opposite direction, i.e., toward the positive z-direction, compared with their original position. This displacement corresponds to a refractive index <1 indicating the presence of free electrons and/or a depletion of the ambient atmosphere. The fringe displacements are symmetric with respect to the z-axis; hence, the method of Abel inversion can be applied to determine the refractive index distribution [8.14].

The fringe displacement is given by:

$$\delta(y,z,t) = \frac{1}{\lambda}\int_{V_\mathrm{p}} (n_\mathrm{p} - n_\mathrm{a})\mathrm{d}x, \tag{8.11}$$

where x, y, z are the coordinates shown in Fig. 8.2, λ is the wavelength of the argon ion laser, n_p is the refractive index of the plasma, and n_a is the refractive index of the ambient atmosphere in the reference arm of the interferometer. The integration runs along the line of sight across the plasma volume V_p. The refractive index of the plasma can be described by the following relation [8.14] (in this chapter, capital letters are used for the electron density and particle densities in order to avoid a mix-up with refractive index quantities):

$$n_\mathrm{p} - 1 = -\frac{e^2\lambda^2}{8\pi^2 m_\mathrm{e} c^2 \varepsilon_0} N_\mathrm{e} + \sum_j 2\pi N_j \alpha_j, \tag{8.12}$$

where N_e is the electron density, N_j are the densities of neutral atoms and ionic species in the plasma, α_j the polarizabilities.

The polarizability of atomic iron is $\alpha_{\mathrm{Fe\,I}} = 13 \times 10^{-30}\,\mathrm{m}^{-3}$ [8.15]. The contribution of the iron atoms to the refractive index can then be estimated with (8.12): $(n_\mathrm{p} - 1)_{\mathrm{Fe\,I}} = +8.2 \times 10^{-23}\, N_{\mathrm{FeI}}/\mathrm{cm}^{-3}$, where N_{FeI} is the particle density of atomic iron. The contribution of the free electrons amounts to $(n_\mathrm{p} - 1)_{N_\mathrm{e}} = -12 \times 10^{-23}\, N_\mathrm{e}/\mathrm{cm}^{-3}$.

With an average degree of ionization in the plasma determined spectroscopically of 90% the ratio $|(n_{\mathrm{p}}-1)_{N_{\mathrm{FeI}}}|/|(n_{\mathrm{p}}-1)_{N_{\mathrm{e}}}|$ amounts to 0.08; hence, the contribution of the atomic iron particles can be neglected. No data for the polarizability of singly ionized iron atoms have been found in the literature. However, calculations of polarizabilities for the transition element copper showed that the polarizability of a copper ion is less than one-tenths of the value calculated for copper atoms [8.16]. Assuming that similar relations hold for iron only a small contribution of the iron ions to the refractive index is expected, which is in the order of the one estimated for the iron atoms.

A reduction of the ambient gas particle density behind the shock wave will lead to a reduced refractive index. The order of magnitude of this effect is estimated assuming that the shock wave reduces the ambient gas particle density completely, i.e., the shock wave leaves a vacuum behind. Under this simplified assumption, the ratio of the fringe displacements due to vacuum δ_{v} and free electrons δ_{e} amounts to: $\delta_{\mathrm{v}}/\delta_{\mathrm{e}} = |1-n_{\mathrm{a}}|/|(n_{\mathrm{p}})_{N_{\mathrm{e}}}-n_{\mathrm{a}}| = 0.7$, for $n_{\mathrm{a}} = 1.0002779$, $\lambda = 514.5\,\mathrm{nm}$, and $N_{\mathrm{e}} = 10^{18}\,\mathrm{cm}^{-3}$. If a complete displacement of the ambient atmospheric particle density occurs, then the expected fringe shift is of the same order of magnitude as the one expected by the free electrons of the plasma. Since both effects go into the same direction, the experimentally determined fringe shifts can only be attributed to the free electrons of the plasma by the ratio $\delta_{\mathrm{e}}/(\delta_{\mathrm{e}}+\delta_{\mathrm{v}}) = 0.6$. Figure 8.29 shows the determined distribution of the refractive index 4 μs (left) and 5 μs (right) after the irradiation of one 80 mJ laser pulse. The formation of the shock front in a spherical geometry can be clearly seen by the values of the refractive index $n > 1$. Behind the shock front, a volume of reduced refractive index becomes visible.

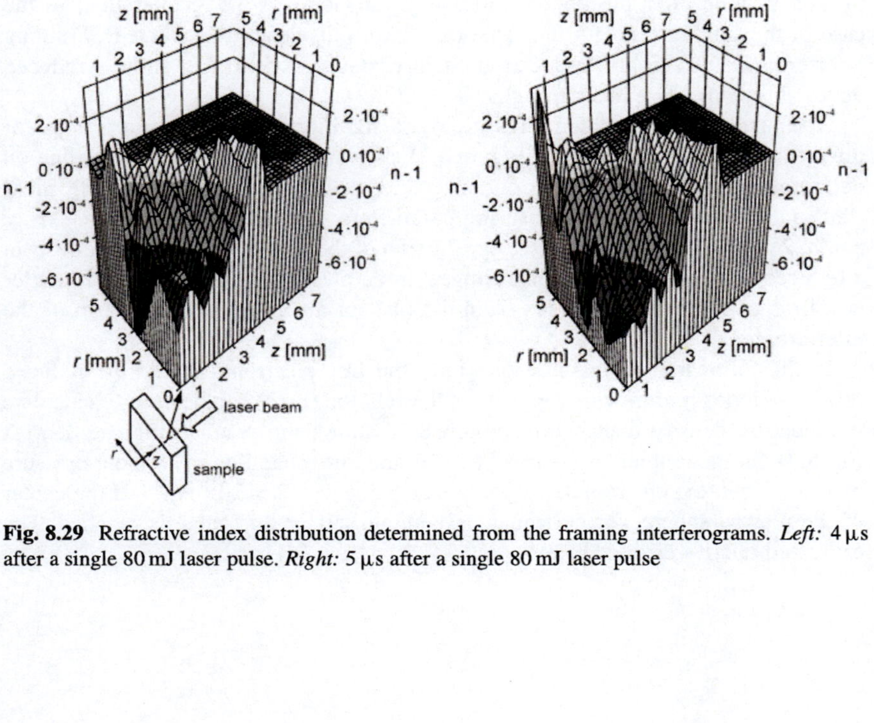

Fig. 8.29 Refractive index distribution determined from the framing interferograms. *Left:* 4 μs after a single 80 mJ laser pulse. *Right:* 5 μs after a single 80 mJ laser pulse

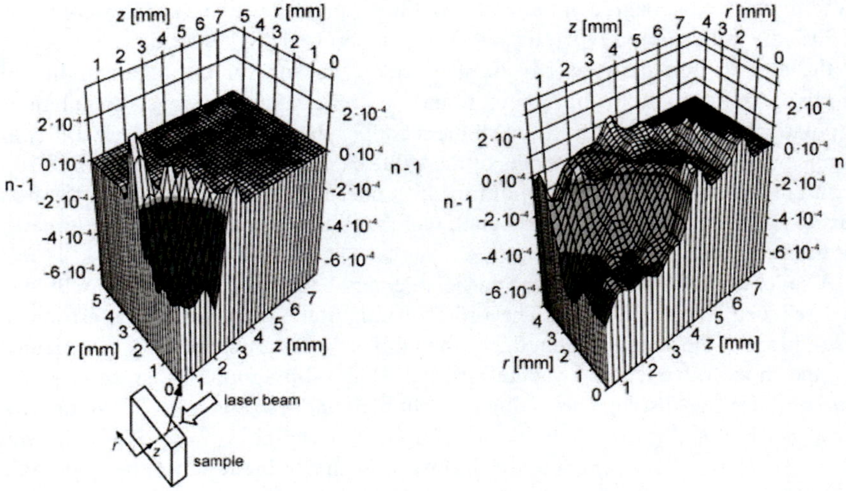

Fig. 8.30 Refractive index distribution determined from the framing interferograms. *Left:* 2 μs after a single 80 mJ laser pulse. *Right:* 2 μs after the second pulse of a 2 × 40 mJ double pulse

Figure 8.30 shows refractive index distributions 2 μs after a single 80 mJ laser pulse (left) and after the second pulse of a 2 × 40 mJ double pulse (right). The formation of two concentric shock waves can be seen in the right diagram, where the second shock wave is weaker than the first. Interpreting the central region with $n < 1$ as the plasma volume, the volume ratio amounts to 3.3, i.e., the plasma volume after the double pulse is by a factor of 3.3 greater than in the case of the single pulse (for the same total laser pulse energy irradiated). Framing interferograms taken at a reduced ambient pressure of 500 mbar show a reduced depth of the refractive index trough.

A reduction of the ambient pressure has a significant influence on the depth of the refractive index distribution. Figure 8.31 shows the refractive index distribution determined from framing interferograms gained at ambient pressures of 1,000 mbar (left) and 500 mbar (right), measuring parameters: 2 × 40 mJ, $\Delta t = 6\,\mu s$, $\Delta s = 5\,mm$, 2 μs after the second pulse. A reduction of the ambient pressure manifests in a reduced shift of the interference fringes in the plasma and therefore to a smaller deviation of the refractive index from the one for air in the reference arm of the interferometer.

In the following, we assume that only the free electrons lead to the reduced refractive index region with $n < 1$, then the relation (8.12) can be used to calculate the electron density distribution. Figure 8.32 shows the result using the data of Fig. 8.31 for the ambient pressures of 1,000 and 500 mbar. For an ambient pressure of 1,000 mbar, the electron density attains a value of $\sim 3 \times 10^{18}\,cm^{-3}$ at the center of the plasma sphere, for 500 mbar only about half of this value is reached. The estimated relative error is 20%.

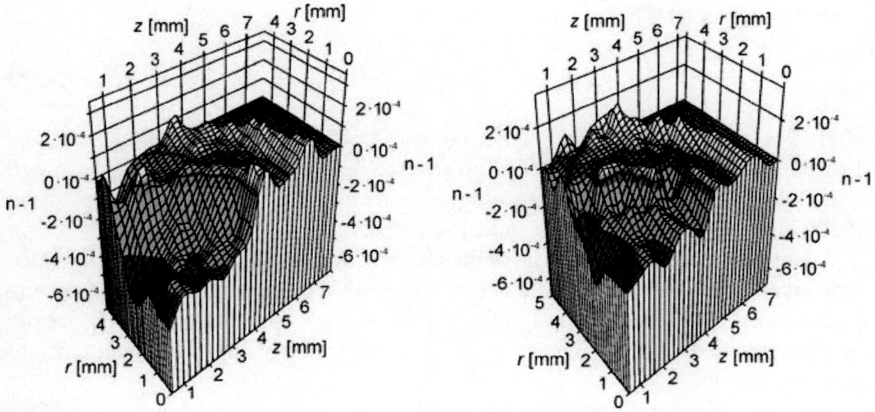

Fig. 8.31 Refractive index distribution determined from framing interferograms gained at an ambient pressure of 1,000 mbar (*left*) and 500 mbar (*right*), 2×40 mJ, 2 μs after the second pulse of a double pulse (the *left diagram* is identical to the *right diagram* of Fig. 8.30)

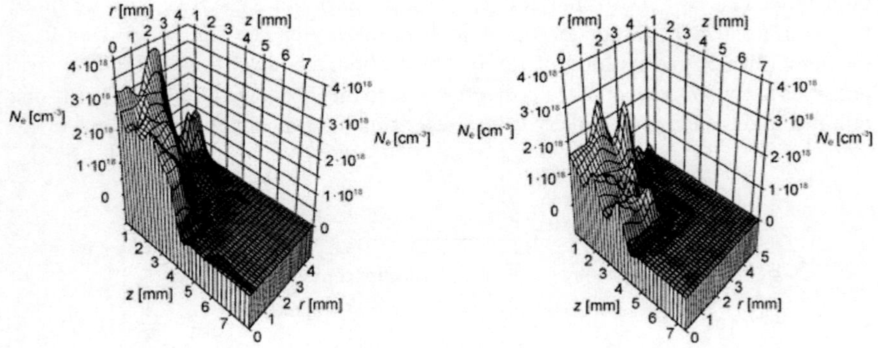

Fig. 8.32 *Left*: Electron density distribution calculated from Fig. 8.31, *left* for 1,000 mbar ambient pressure; *right*: from Fig. 8.31, *right* for 500 mbar, assuming that the refractive index inside the second shock wave $n < 1$ is caused by free electrons only. Measuring parameters: 2×40 mJ, 2 μs after second pulse of double pulse

Taking into account the above-discussed reduction of the ambient gas particle density behind the shock wave, the electron density is only ~60 % of $\sim 3 \times 10^{18}$ cm^{-3}, i.e., 2×10^{18} cm^{-3}. The electron density determined interferometrically after a single pulse of 80 mJ and 2 μs is about the same. For the comparison of the electron densities measured interferometrically with those determined by spatially integrated emission spectroscopy of a Stark broadened iron line, the electron density distribution – as shown in Fig. 8.32 – is approximated by parabolic profiles along the z-direction at $y = 0$ and along the y-direction at $z = \Delta$ – where Δ is the minimum distance between the boundary of the evaluable interferogram and the surface of the sample being about 0.5 mm – as follows:

$$N_e = N_{e,max}\left[1 - (\xi/r_e)^2\right], \tag{8.13}$$

where ξ is the spatial coordinate z or y (see Fig. 8.2), r_e is the radius of the parabolic distribution, and $N_{e,max}$ is the maximum electron density. The parameters $N_{e,max}$ and r_e of both parabolas for $y = 0$ mm and $z = \Delta$ are averaged. For an ambient pressure of 100 mbar, only one parabola can be evaluated.

An integration over the volume of the electron density distribution approximated in this way yields the total number of electrons within the plasma half-sphere having a radius of r_e. By division through the plasma volume an average electron density can be estimated. For the integrated average electron densities, an uncertainty of ± 20 % is assumed.

The effect of neutral iron atoms on the refractive index requires a correction of the determined electron densities to higher values. This correction amounts to about 7% and is therefore significantly smaller than the error of the single measuring values of the refractive index of 20 %. Hence, this correction is neglected.

Figure 8.33 shows the integrated average electron densities 2 μs after the second laser pulse ($p_0 = 1{,}000$ mbar and $p_0 = 500$ mbar) and 2 μs after a single pulse ($p_0 = 100$ mbar) as a function of the ambient pressure. The values deduced from the parabolic approximation and volume integration, the lower limits of the electron densities taking into account the correction due to the displaced air inside the plasma volume, and the electron density values determined by spectroscopy are shown.

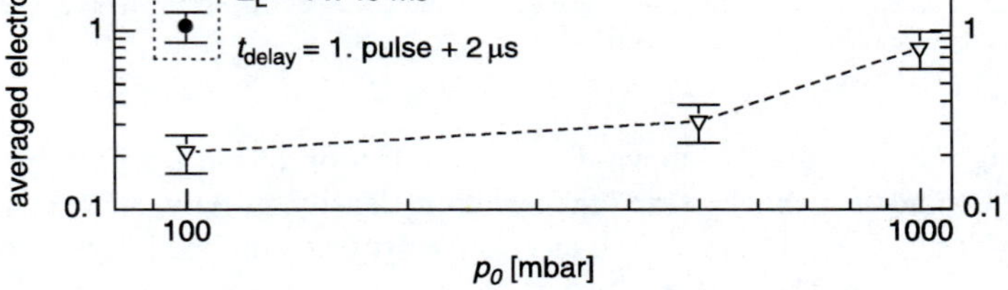

Fig. 8.33 Average electron density as a function of the ambient pressure determined by interferometry and spectroscopy. Measuring parameters for the interferometric data, 500 and 1,000 mbar: 2×40 mJ, $\Delta t = 6$ μs, $t_{delay} = 2$ μs, $\Delta s = 5$ mm; 100 mbar: 1×40 mJ, $t_{delay} = 2$ μs, $\Delta s = 5$ mm; for spectroscopy: Stark broadening of the iron line Fe I 538.34 nm, 2 × 40 mJ, $\Delta t = 6$ μs, $t_{delay} = 2$ μs, $\Delta s = 10$ mm (see also Fig. 8.23)

Interferometric measurements performed at $\Delta s = 10$ mm showed that within the measuring uncertainty there is no difference in the electron density compared with the values determined for $\Delta s = 5$ mm. The electron densities gained from interferometry are an order magnitude greater than those deduced spectroscopically. This discrepancy cannot be attributed to the different Δs values. On the other hand, the pressure dependence shows qualitatively the same behavior.

Figure 8.34 illustrates a model to describe qualitatively the observed difference in electron density values found. The curves show the density distribution of electrons, neutral iron atoms (Fe I), and singly ionized iron ions (Fe II) as a function of the plasma radius for single (left) and double pulses (right). The plotted numbers are values determined by the interferometric measurements. The electron density distribution can be approximated by a parabolic distribution. The maximum electron density at the plasma center is the same for single and double pulses, but the plasma is greater in case of the double-pulse excitation. If there is no difference in the electron density at the center of the plasma, then the density of singly ionized iron (Fe II) at the center is also the same due to the charge neutrality condition. Spectroscopic studies showed that the average degree of ionization is lower for single pulses than for double pulses. Hence, the density of iron atoms shown for double pulses is smaller than for single pulses.

The spectroscopically determined Stark broadening of the neutral iron atom line is the result of a superposition of all emission regions inside the plasma. However, the outer regions will contribute dominantly to this broadened line since there greater densities of Fe I prevail and the absolute number of Fe I atoms is also

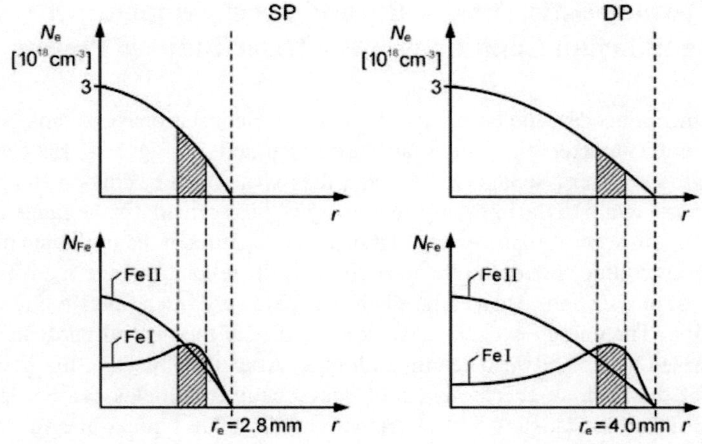

Fig. 8.34 Qualitative presentation of the density distribution of electrons, neutral iron atoms (Fe I), and singly ionized iron ions (Fe II) in laser-induced plasmas for single and double pulses of the same total pulse energy. The density distributions are shown as a function of the plasma radius. SP = single pulse, DP = double pulse. Parameters: SP 1×80 mJ, $t_{delay} = 2\,\mu$s, DP 2×40 mJ, $\Delta t = 6\,\mu$s, $t_{delay} = 2\,\mu$s; for both: $p_0 = 1$ bar, $\Delta s = 5$ mm

greater, as the volume of a spherical shell scales as $dV = 4\pi r^2 dr$ (see hatched region in Fig. 8.34 illustrating the interval where the neutral iron particle density attains a maximum). As a result, the electron density determined spectroscopically by a neutral iron line cannot be considered as a representative average electron density of the plasma, but as an electron density to be allocated to the outer plasma regions. Hence, the values determined spectroscopically for the electron density are underestimating the real average electron density. In this way, the spatial separation of the density distributions of Fe I and Fe II may explain qualitatively the found difference. The interferometric measurement is – taking into account the simplifying assumptions described above – independent of the distribution of neutral particles and ions inside the plasma.

The investigations presented in this chapter have shown that the effect of collinear double pulses is a local transient reduction of the particle density in the vicinity of the sample surface caused by the first pulse. The second pulse interacts predominantly with the sample surface. No re-heating of a residual plasma from the first pulse was observed. The material ablated by the second pulse expands faster into the cavity of reduced particle density originating from the first pulse attaining a significantly enlarged plasma volume. The electron densities are approximately the same for single and double pulses of the same total energy. Considering the larger plasma volume and the fact that the mass ablated using double pulses is also increased significantly (cf. Sect. 7.2) favors a more intense and longer lasting emission of element specific radiation to be used for LIBS analysis.

8.4 Characteristic Time Scales and Local Temperature Equilibrium Conditions of the Laser-Induced Plasma

The plasma induced by the laser pulse for LIBS is a nonstationary plasma governed by different characteristic time scales as illustrated in Fig. 8.35 ranging from picoseconds to several tenths of a second, thus spanning 11 orders of magnitude. The temporal width of the initiating laser pulse τ_L lies typically in the range of a few nanoseconds to several 10 ns. After ablation and ionization of the irradiated material, the electrons collide with each other described by the relaxation time τ_{ee}, which is of the order of picoseconds. Hence, the electrons attain very fast a Maxwellian velocity distribution. The energy exchange via collisions with the neutral particles or ions characterized by τ_k^0 and τ_k^1 takes much longer. After the time $\tau_{e,0}$, the population density of the discrete energy levels of heavy neutral particles can be described by the Boltzmann statistics with a temperature T_e. The typical plasma life time τ_{Plasma} amounts up to several $10\,\mu s$. At early times the spectral plasma emission is dominated by free–free (ff) and free–bound (fb) transitions of electrons being taken over by line emission at later times. The bar marked with τ_{mod} describes the time scale of double and multiple pulses (in the case of separated single pulses within a double-pulse burst this time corresponds to the interpulse separation Δt or for

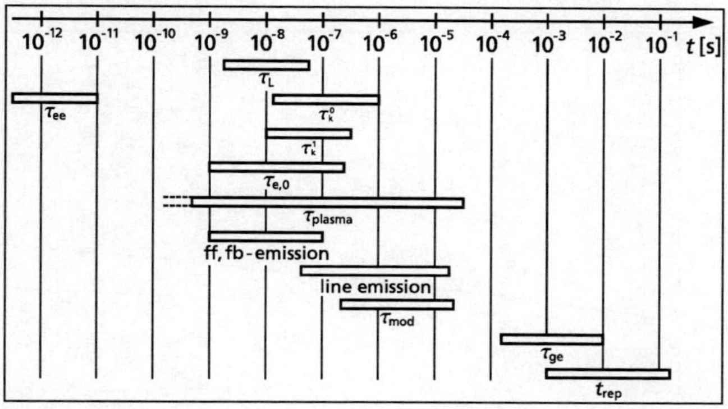

Fig. 8.35 Time scales of the laser-induced plasma for LIBS. τ_L = temporal width of laser pulse, τ_{ee} = electron–electron relaxation time, τ_k^0 = electron–neutral relaxation time, τ_k^1 = electron–ion relaxation time, $\tau_{e,0}$ = time for population of discrete levels according to Boltzmann statistics, τ_{Plasma} = life time of plasma, ff = free–free emission, fb = free–bound emission, τ_{mod} = time scale of multiple pulses, τ_{ge} = gas exchange time, t_{rep} = repetition time of laser pulses

multiple pulses to the sum of interpulse separations, cf. Fig. 3.9). This time domain overlaps with those of the plasma life time and the line emission. For repetitive plasma ignition, the gas exchange time τ_{ge} [cf. Chap. 3, relation (3.1)] has to be considered to assure an effective removal of the material ablated by the previous laser pulse. The bar t_{rep} describes the time between consecutive laser pulses or laser bursts corresponding to laser pulse repetition rates between a few hertz up to 1 kHz applied so far for LIBS.

In the following, characteristic spatial and temporal quantities describing the state of the laser-induced plasma will be discussed in more detail. A plasma can be described as classical if the thermal De Broglie wavelength of the electrons,

$$\lambda_B = \frac{h}{2\pi\sqrt{m_e k_B T_e}} \tag{8.14}$$

with h Planck constant, m_e electron mass, k_B Boltzmann constant, T_e electron temperature, is smaller than the mean distance λ_n between the free electrons $\lambda_n \approx n_e^{-1/3}$. In this case, quantum effects, e.g., degeneration of the electron gas, can be neglected. Furthermore, the average distance between particles is compared with the average collision length given by the Landau length

$$\lambda_L = \frac{e^2}{4\pi\varepsilon_0 k_B T_e} \tag{8.15}$$

with e elementary charge, ε_0 dielectric constant. Figure 8.36 shows these characteristic length scales in a $n_e - T_e$ diagram together with the plasma parameters determined experimentally (cf. Fig. 8.23a, b for double and single pulses).

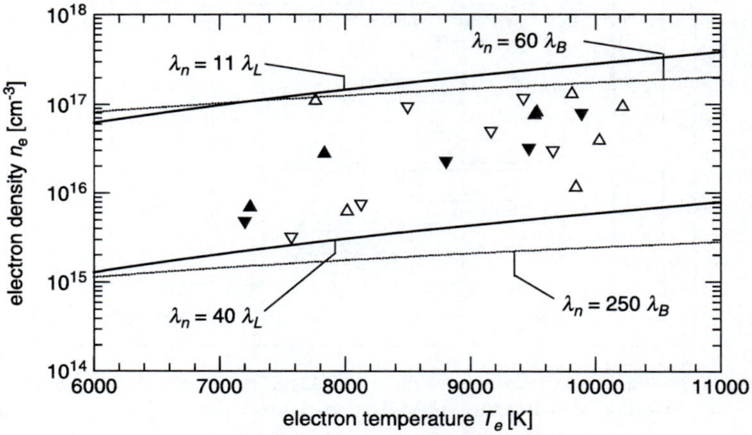

Fig. 8.36 Electron density and electron temperature measured for double and single pulses. The *curves* illustrate the characteristic plasma length scales De Broglie wavelength λ_B and Landau length λ_L in relation to the mean particle distance λ_n. *Filled triangles* = double pulses, *open triangles* = single pulses. *Triangles pointing upward*: burst energy 160 mJ, *triangles pointing downwards*: burst energy 80 mJ

The measured parameters lie in a parameter field where the De Broglie wavelength is by a factor of 60–250 and the Landau length by a factor of 11–40 smaller than the average particle distance. Hence, the laser-induced plasma can be treated as a classical and ideal plasma, where the interaction energy between charged particles is small compared with the kinetic energy.

Another description of ideality can be defined by the condition that the number of electrons in the Debye sphere is significantly greater than 1. The radius of the Debye sphere is given by the length along which the potential of an ion inside the plasma decays by a factor of $1/e$. For the number of electrons in the Debye sphere holds

$$N_D = \frac{4\pi}{3}\left(\frac{\varepsilon_0 k_B}{2e^2}\right)^{3/2}\sqrt{\frac{T_e^3}{n_e}}. \tag{8.16}$$

Figure 8.37 shows the $n_e - T_e$ diagram with curves for different N_D. For the laser-induced plasmas studied N_D varies between 1 and 8; hence, the condition $N_D \gg 1$ is not fulfilled for all plasmas. Hence, the assumption of a continuous charge density distribution inside the Debye sphere is not completely fulfilled. Such a continuous charge distribution is, e.g., assumed to calculate the reduction of the ionization energy or to calculate cross sections. For simplification, the fulfillment of the Debye approximation will be assumed in the following paragraphs.

For a local temperature equilibrium (LTE) in an optically thin plasma, the probability of a collision-induced transition is much greater than the probability of a radiating transition. This condition is fulfilled, if the following relation holds

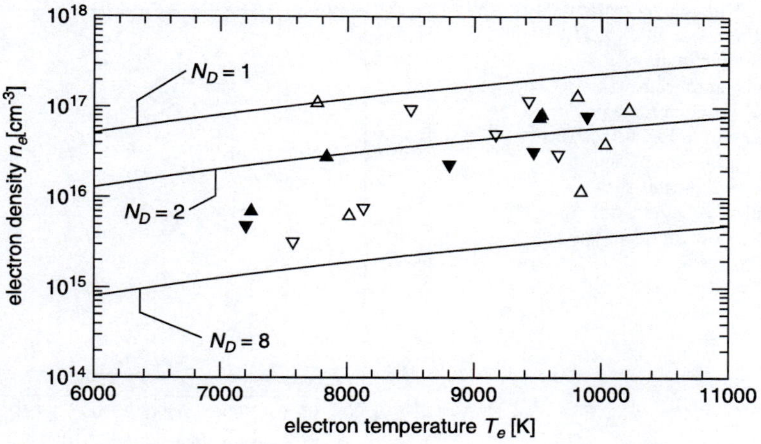

Fig. 8.37 Electron density and electron temperature measured for double and single pulses (same data points as in Fig. 8.36). The *curves* illustrate the Debye number N_D

for all energy levels of an atom with a principal quantum number greater or equal n [8.5]:

$$n_e \geq 7 \times 10^{18} \frac{(z+1)^7}{\mathrm{n}^{17/2}} \left(\frac{k_B T_e}{E_H} \right)^{1/2} \mathrm{cm}^{-3}, \tag{8.17}$$

where z is the ionization stage, $z = 0$ for neutral particles, $z = 1$ for singly ionized particles, etc., n is the principal quantum number, E_H is the ionization energy of hydrogen.

Relation (8.17) is valid exactly only for the hydrogen atom. For transitions of iron atoms studied, e.g., in Sect. 8.2, we have $\mathrm{n} \geq 3$ and $z = 0$. All experimentally found electron temperatures of the material plasma were less than $T_e = 11,000$ K (cf. Figs. 8.15, 8.16, 8.19–8.21, 8.23). Relation (8.17) yields then: $n_e \geq 2 \times 10^{14}$ cm^{-3}. This value is less than the determination limit of the electron density using the Stark broadened iron lines, which amounts to about 2×10^{15} cm^{-3}. Hence, in these cases, where the electron density can be determined using Stark broadened iron lines, the LTE condition is fulfilled.

Figure 8.38 shows the relation (8.17) for the considered iron atoms in a $n_e - T_e$ diagram together with the plasma trajectory taking the data of Fig. 8.16. Parameter of the data points is the time delay t_{delay} after the second pulse of the double pulse. The bar alongside the plasma trajectory indicates the typical integration time window used for quantitative analysis. Figure 8.38 illustrates clearly that the LTE condition is fulfilled for all relevant time delays.

For the Ar II line at 434.81 nm (cf. Fig. 8.26), we have $\mathrm{n} = 4$ and $z = 1$, so relation (8.17) results in: $n_e \geq 5 \times 10^{15}$ cm^{-3}. Again this condition is fulfilled for electron densities determined with the help of the Stark broadened Ar II line (cf. Fig. 8.26).

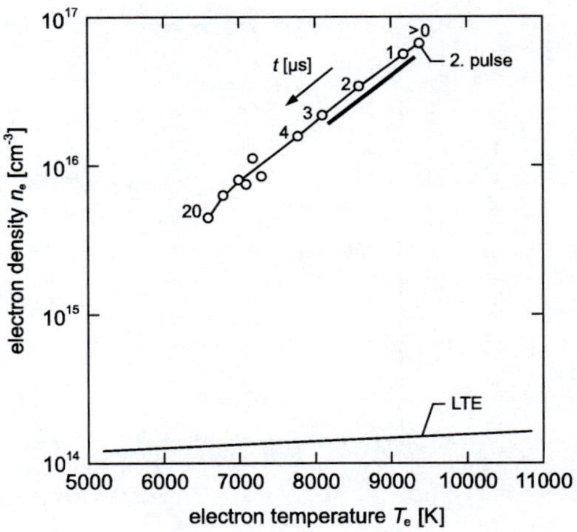

Fig. 8.38 Local temperature equilibrium condition (LTE) according to relation (8.17) and plasma trajectory in a $n_e - T_e$ diagram (data points are taken from Fig. 8.16). The *bar* alongside the plasma trajectory indicates exemplarily the integration time window for quantitative LIBS analysis

For nonstationary plasmas, the relaxation times of the various plasma species are considered to estimate whether a LTE persists. For electrons, the relaxation time to establish a Maxwell velocity distribution is given by [8.17]:

$$\tau_{ee} \approx 3.16 \cdot 10^4\, W_e^{3/2} n_e^{-1} \cdot \text{s}, \tag{8.18}$$

where W_e is the electron energy in eV and n_e is the electron density in cm^{-3}.

For $W_e = 1$ eV, $n_e = 10^{16}$ cm^{-3}, relation (8.18) yields $\tau_{ee} \approx 3 \times 10^{-12}$ s. Hence, for the electrons, an electron temperature is established very fast compared with the typical life time of a laser-induced LIBS plasma of several microseconds. The equilibration time to obtain LTE between electrons and atoms is given by [8.18]:

$$\tau_k^0 \approx \left[7.5 \cdot 10^{-7} \left(\frac{E_H}{k_B T} \right)^{3/2} n_e \right] \frac{n_t}{n_1} \frac{M}{m_e} \cdot \text{s}, \tag{8.19}$$

where n_t is the total density of heavy particles, n_1 is the density of ions, n_e is the electron density in cm^{-3}, and M is the mass of atoms or ions.

The corresponding time for singly ionized particles is:

$$\tau_k^1 \approx \left[7.5 \cdot 10^{-7} \left(\frac{E_H}{k_B T} \right)^{3/2} n_e \right] \frac{M}{m_e} \cdot \text{s} \tag{8.20}$$

Equations (8.19) and (8.20) are approximations, and they are exactly valid only in the case of hydrogen atoms.

So far we have discussed the time necessary to obtain a kinetic temperature of the particles. The equilibration time for a distribution of energy over the discrete terms of heavy particles according to Boltzmann's law is given by:

$$\tau_{e,z-1} \approx 1.1 \cdot 10^7 \frac{z^3}{f_{12} n_e} \frac{n_z}{n_z + n_{z-1}} \left(\frac{E_{z-1,2}}{z^2 E_H}\right) \left(\frac{k_B T_e}{z^2 E_H}\right)^{1/2} \exp\left(\frac{E_{z-1,2}}{k_B T_e}\right) \cdot \mathrm{s}, \quad (8.21)$$

where z is the stage of ionization, $z = 1$ for neutrals, $z = 2$ for singly ionized atoms, etc., f_{12} is the oscillator strength of the resonance line, and $E_{z-1,2}$ is the excitation energy of the resonance line.

Figure 8.39 shows again the $n_e - T_e$ diagram with curves for τ_k^0, τ_k^1, and $\tau_{e,0}$. For the calculation of τ_k^0, an ionization degree of 40 % is assumed. For the mass M in the relations (8.19) and (8.20), the value for iron is taken $M = 56\, m_p$, where m_p is the proton mass. The time $\tau_{e,0}$ is calculated for the Fe I line at 549.713 nm. Assuming $n_1 \gg n_0$ for the plasma, then $n_1/(n_1+n_0)$ is the degree of ionization. As upper limit for the degree of ionization 1 is taken. Figure 8.39 shows that the relaxation times for the kinetic energy between electrons and neutral particles τ_k^0 are in the range between 30 ns and 1 μs for an ionization degree of 40%. The corresponding times for ions τ_k^1 amount 12–400 ns. However, for those plasmas (data points in Fig. 8.39) with relaxation times τ_k^0 close to 1 μs the degree of ionization is between 80% and 100% and not 40% as assumed for the calculation (cf. Fig. 8.23c). According to relation (8.19), the relaxation time of the kinetic energy is by a factor of 2–2.5 shorter and amounts 0.4–0.5 μs. Hence, for the investigated plasmas, a kinetic equilibrium between electron and heavy particles is achieved 500 ns after the laser pulse irradiation.

The relaxation times for the excited states $\tau_{e,0}$ of the neutral particles lie in the range between 2 and 300 ns. For the spectroscopic results shown, e.g., in Figs. 8.22 and 8.23, the emission spectrum was detected with $t_{delay} = 1$ μs and $t_{int} = 1$ μs,

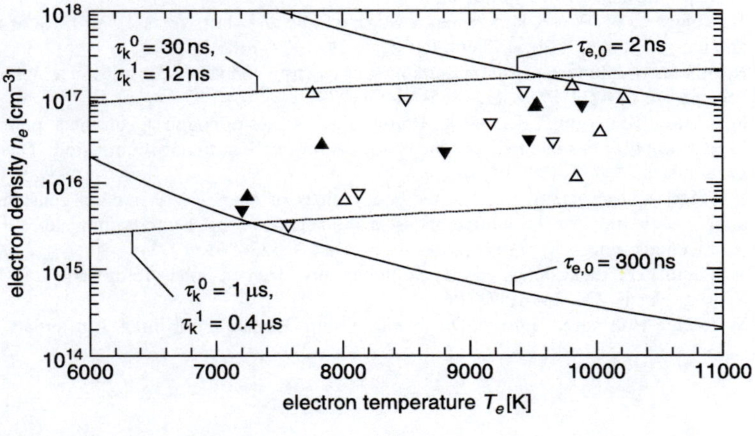

Fig. 8.39 Relaxation times in the n_e–T_e diagram, see text

so a LTE can be assumed. For those plasma states where the relaxation time is not significantly smaller than 1 μs, deviations of the LTE are to be expected.

References

8.1. R. Noll, R. Sattmann, V. Sturm, S. Winkelmann, Space- and time-resolved dynamics of plasmas generated by laser double pulses interacting with metallic samples. J. Anal. At. Spectrom. **19**, 419–428 (2004)
8.2. L. Sedov, *Similarity and Dimensional Methods in Mechanics*, 4th edn. (Gostekhizdat, Moscow, 1957), English transl.: ed. by M. Holdt (Academic, New York, 1959)
8.3. Y. Zel'dovich, Y. Raizer, in *Physics of Shock Waves and High-Temperature Hydrodynamic Phenomena*, vol. I, ed. by W. Hayes, R. Probstein (Academic, New York, 1966)
8.4. R. Adrain, J. Watson, P. Richards, A. Maitland, Laser microspectral analysis of steels. Opt. Laser Technol. **12**, 137–143 (1980)
8.5. R. Griem, *Plasma Spectroscopy* (McGraw-Hill, New York, 1964)
8.6. J. Fuhr, G. Martin, W. Wiese, S. Younger, Atomic transition probabilities for iron, cobalt, and nickel (a critical data compilation of allowed lines). J. Phys. Chem. Ref. Data **10**, 305–565 (1981)
8.7. S. Freudenstein, J. Cooper, Stark broadening of Fe I 5383 Å. Astron. Astrophys. **71**, 283–288 (1979)
8.8. J. Aguilera, J. Bengochea, C. Aragón, Curves of growth of spectral lines emitted by a laser-induced plasma: influence of the temporal evolution and spatial inhomogeneity of the plasma. Spectrochim. Acta B **58**, 221–237 (2003)
8.9. Y. Iida, Effects of atmosphere on laser vaporization and excitation processes of solid samples. Spectrochim. Acta **45B**, 1353–1367 (1990.)
8.10. H.-W. Drawin, P. Felenbok, *Data for Plasmas in Local Thermodynamic Equilibrium* (Gauthier-Villars, Paris, 1965)
8.11. K. Monge, C. Aragón, J. Aguilera, Space- and time-resolved measurements of temperatures and electron densities of plasmas formed during laser ablation of metallic samples. Appl. Phys. **A69**, S691–S694 (1999)
8.12. http://physics.nist.gov/PhysRefData
8.13. N. Konjevic, W. Wiese, Experimental Stark widths and shifts for spectral lines of neutral and ionized atoms. J. Phys. Chem. Ref. Data **19**, 1307–1385 (1990)
8.14. R. Boxmann, Interferometric measurement of electron and vapor densities in a high-current vacuum arc. J. Appl. Phys. **45**, 4835–4846 (1974)
8.15. S. Fraga, J. Karwowski, K. Saxena, Hartree-Fock values of coupling constants, polarizabilities, susceptibilities and radii for the neutral atoms, helium to nobelium. Atom. Data Nucl. Data Tables **12**, 467–477 (1973)
8.16. S. Fraga, K. Saxena, B. Lo, Hartree-Fock values of energies, interaction constants, and atomic properties for the ground states of the negative ions, neutral atoms, and first four positive ions from helium to krypton. Atom. Data **3**, 323–361 (1971)
8.17. A. Eberhagen, Elektromagnetische Strahlung aus Plasmen hoher Temperatur und Dichte. Z. Ang. Physik **XX**, 244–257 (1966)
8.18. W. Lochte-Holtgreven, *Plasma Diagnostics* (North-Holland Publishing, Amsterdam, 1968)

Chapter 9
Plasma Emission

This chapter describes the dominant radiation processes of the laser-generated plasmas such as Bremsstrahlung, recombination radiation, and line radiation. For the line emission being the relevant emission signature used for LIBS, the various line broadening mechanisms are presented and absorption processes are discussed.

9.1 Bremsstrahlung and Recombination Radiation

Transitions between free states of electrons moving in the Coulomb fields of the ions generate radiation. The emission coefficient of free–free transitions is given by [9.1]:

$$\varepsilon_{\mathrm{ff}}^{\omega,\mathrm{d}\Omega} = 16\frac{(\alpha a_0)^3 E_{\mathrm{H}}}{3(3\pi)^{1/2}}\sqrt{\frac{E_{\mathrm{H}}}{kT}}\mathrm{e}^{-\frac{\hbar\omega}{kT}} N_{\mathrm{e}} \sum_{z,a} z^2 N_a^z, \tag{9.1}$$

where α is the fine structure constant, a_0 is the Bohr radius, E_{H} is the ionization energy of hydrogen, T is the plasma temperature, $\hbar\omega$ is the photon energy, N_{e} is the electron density, z is the charge state of the ions, $z = 0$ corresponds to neutral atoms, $z = 1$ to singly ionized species, etc., and N_a^z is the density of ion species a with charge z, if Gaunt corrections, reduction of the ionization energy, and advance of the series limit are neglected. The unit of the emission coefficient (9.1) is $[\varepsilon_{\mathrm{ff}}^{\omega,\mathrm{d}\Omega}] = \mathrm{W/(m^3\,sr\,rad\,s^{-1})}$, i.e., power per unit volume per steradian per angular frequency interval. Relation (9.1) can be transformed to an emitted power per unit volume per wavelength interval $\varepsilon_{\mathrm{ff}}^{\lambda}$, $[\varepsilon_{\mathrm{ff}}^{\lambda}] = \mathrm{W/(m^3\,m)}$ by multiplying the right side with $(4\pi)2\pi c/\lambda^2$, where c is the velocity of light and λ the wavelength of the emitted radiation. Figure 9.1 illustrates the Bremsstrahlung emission described by the calculated emission coefficient $\varepsilon_{\mathrm{ff}}^{\lambda}$ using experimentally determined plasma parameters. The Bremsstrahlung emission shows a strong decay toward wavelengths below 400 nm, and it decreases by more than an order of magnitude within the first 4 μs. This type of emission generates a continuous background in the LIBS spectra. To achieve high ratios between line emission coefficients and continuous

R. Noll, *Laser-Induced Breakdown Spectroscopy*, DOI 10.1007/978-3-642-20668-9_9,

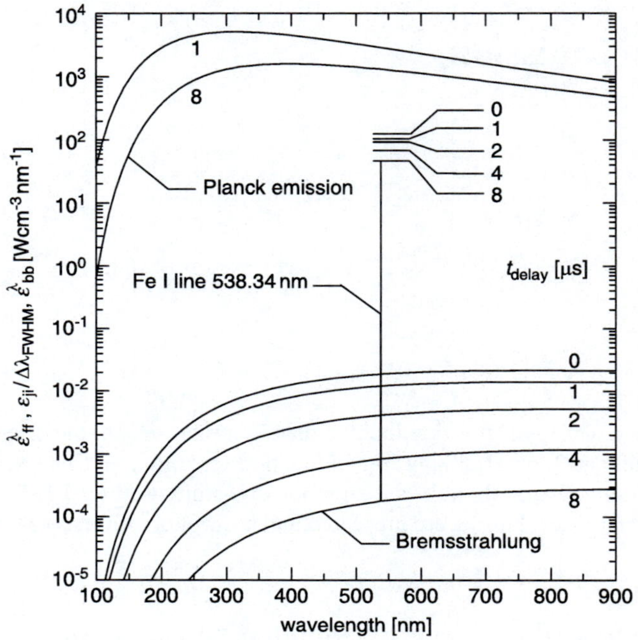

Fig. 9.1 Calculated Bremsstrahlung emission described by the emission coefficient for free–free transitions $\varepsilon^{\lambda}_{\rm ff}$, line emission coefficient $\varepsilon^{\lambda}_{ji}$, and equivalent Planck emission coefficient $\varepsilon^{\lambda}_{\rm bb}$ as a function of the wavelength using the plasma parameters from Fig. 8.16 for delay times $t_{\rm delay}$ from 0 to 8 µs after the second pulse of a double pulse. The calculated peak height of the emission line is plotted for $t_{\rm delay} = 8$ µs. The peak heights of this line for the different delay times are indicated by *horizontal dashes*

Bremsstrahlung emission, it is favorable to use short line emission wavelengths – as far as is by other restrictions possible – and a sufficient delay time $t_{\rm delay}$ after the plasma ignition.

Transitions between free states of electrons and bound states generate the recombination radiation described by the following emission coefficient [9.1]:

$$\varepsilon_{\rm fb}^{\omega,{\rm d}\Omega} = \frac{32}{3}\frac{(\alpha a_0)^3 E_{\rm H}}{3(3\pi)^{1/2}}\left(\frac{E_{\rm H}}{kT}\right)^{3/2} {\rm e}^{-\frac{\hbar\omega}{kT}} N_{\rm e} \sum_{z,a,{\rm n}} \frac{z^4}{{\rm n}^3} {\rm e}^{z^2 E_{\rm H}/{\rm n}^2 kT} {\rm N}_a^z, \tag{9.2}$$

where n is the main quantum number.

The overall contribution of the recombination radiation can be obtained by integrating relation (9.2) over all emitted frequencies and comparing this result with the overall contribution of the Bremsstrahlung radiation. It turns out that the recombination radiation is larger by a factor of $E_{\infty}^{z-1,a}/kT$, where $E_{\infty}^{z-1,a}$ is the ionization energy of the species a for the ionization stage $z-1$. Considering an iron plasma where only the first ionization stage is of relevance, we have $E_\infty^{0,\mathrm{Fe}} = 7.9\,\mathrm{eV}$. Comparing this with a typical plasma temperature of 0.7 eV (8,100 K, cf. Fig. 8.16) yields a factor of about 10. Hence, the total power radiated over all frequencies is by about an order of magnitude higher for recombination radiation than for the Bremsstrahlung emission.

9.2 Line Emission

In an optically thin plasma, the atoms and ions emit line radiation described by the line emission coefficient [9.2]:

$$\varepsilon_{ji}^{\nu,\mathrm{d}\Omega} = \frac{1}{4\pi}\Gamma(\nu)\,A_{ji}\,h\nu_{ji}\,N_a^z\frac{g_j}{u_a^z(T)}\mathrm{e}^{-\frac{E_j}{kT}}, \tag{9.3}$$

where $\Gamma(\nu)$ is the line profile as a function of the frequency, A_{ji} is the Einstein coefficient for the transition from upper level j to lower level i, h is the Planck constant, $h\nu_{ji}$ is the photon energy of the transition from j to i, N_a^z is the density of species a with charge z, g_j is the statistical weight of upper level j, $u_a^z(T)$ is the partition function of species a with charge z, and E_j is the energy of the upper level j.

The unit of $\varepsilon_{ji}^{\nu,\mathrm{d}\Omega}$ is $\mathrm{W}/(\mathrm{m}^3\,\mathrm{sr}\,\mathrm{s}^{-1})$. The line shape function $\Gamma(\nu)$ depends on the type of broadening being effective for the respective line. The function $\Gamma(\nu)$ is normalized as follows:

$$\int_{-\infty}^{+\infty}\Gamma(\nu)\mathrm{d}\nu = 1. \tag{9.4}$$

In case of natural line broadening or a pressure broadened line in the impact approximation, the function is given by the Lorentz expression [9.3]:

$$\Gamma_\mathrm{L}(\nu) = \frac{1}{2\pi}\frac{\Delta\nu_\mathrm{H}}{(\nu-\nu_0-\Delta\nu_\mathrm{V})^2+(\Delta\nu_\mathrm{H}/2)}, \tag{9.5}$$

where $\Delta\nu_\mathrm{H}$ is the full frequency width at half maximum, $\Delta\nu_\mathrm{V}$ is the frequency shift of the line maximum in relation to an unperturbed line, and ν_0 is the frequency of the line maximum for the unperturbed line.

The partition function $u_a^z(T)$ of species a in charge state z is given by:

$$u_a^z(T) = \sum_{i=1}^{i=n^*} g_{a,i}^z\mathrm{e}^{-\frac{E_{a,i}^z}{kT}}, \tag{9.6}$$

where $i = 1$ denotes the ground state and $i = n^*$ the highest level which is still bound. To estimate n^*, the lowering of the ionization energy is calculated using Unsöld's formula:

$$\Delta E_z^{\mathrm{ion}} = 3e^2(z+1)^{2/3}\left(\frac{4\pi}{3}n_\mathrm{e}\right)^{1/3}, \qquad (9.7)$$

where n_e is the electron density. The values $u_a^z(T, \Delta E_z)$ are tabulated [9.4]. Further data for iron and several ionization states of iron are given in [9.5]. Figure 9.2 shows the partition function of neutral (solid line) and singly ionized iron (dashed line); parameter is the lowering of the ionization energy. For electron densities between 0.5 and $7 \times 10^{16}\,\mathrm{cm}^{-3}$, ΔE_{Fe0} varies between 0.12 and 0.28 eV. For plasma temperatures below 10,000 K, this has only a minor influence on the partition function of neutral iron atoms. For singly ionized iron, the influence of ΔE_{Fe1} is negligible in the relevant temperature range.

Between the Einstein coefficient A_{ji} and the oscillator strength f_{ij}, the following relation holds [9.6]:

$$A_{ji} = \frac{2\pi e^2}{m_\mathrm{e} c \varepsilon_0} \frac{g_i}{g_j} \frac{f_{ij}}{\lambda_{ji}^2}, \qquad (9.8)$$

where e is the elementary charge, m_e is the electron mass, c is the velocity of light, ε_0 is the dielectric constant, g_i, g_j is the statistical weight of lower and upper level, f_{ij} is the oscillator strength, and λ_{ji} is the wavelength of the transition from j to i.

As an example the line emission coefficient for the Fe I line at 538.34 nm (cf. Fig. 8.14) is estimated with (9.3) by:

$$\varepsilon_{ji}^{\lambda} = \frac{\varepsilon_{ji}}{\Delta\lambda_{\mathrm{FWHM}}} = \frac{\iint \varepsilon_{ji}^{\nu,\mathrm{d}\Omega}\,\mathrm{d}\nu\,\mathrm{d}\Omega}{\Delta\lambda_{\mathrm{FWHM}}} = \frac{A_{ji}\frac{hc}{\lambda_{ji}} N_a^z \frac{g_j}{u_a^z(T)} \mathrm{e}^{-\frac{E_j}{kT}}}{\Delta\lambda_{\mathrm{FWHM}}},$$

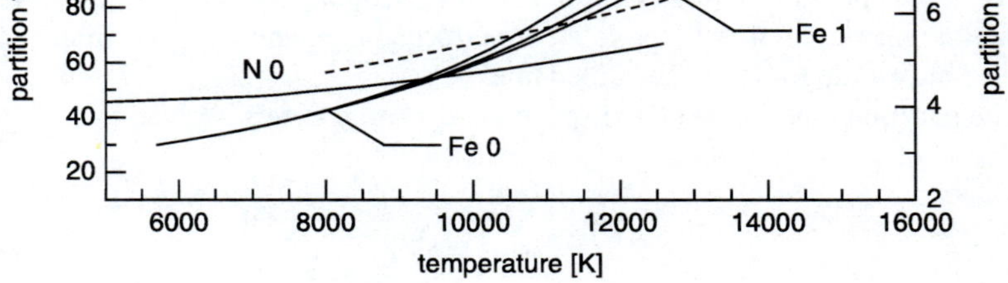

Fig. 9.2 Partition function for neutral (Fe 0) and singly ionized iron (Fe 1) as a function of the temperature, parameter: lowering of the ionization energy according to (9.7). Partition function of neutral (N 0) and singly ionized nitrogen (N I) for $\Delta E_\mathrm{N} = 0.1\,\mathrm{eV}$

where the integration runs over the line and the full solid angle. The line width (FWHM) is estimated by the Stark width (cf. Sect. 9.4) of that line. The line data is: $A_{ji} = 5.9 \times 10^7\,\mathrm{s}^{-1}$, $g_j = 13$, and $E_j = 6.63\,\mathrm{eV}$ [9.3]. The neutral iron density $N_a^z = n_{\mathrm{Fe}}^0$ is determined from n_e and the Saha equation. For the temperature and the electron density as a function of time, we take the data of Fig. 8.16. The partition function $u_a^z(T) = u_{\mathrm{Fe}}^0(T)$ is taken from [9.4]. The determined line emission coefficient for the 538.34 nm line is plotted in Fig. 9.1 for delay times from 0 to 8 μs. While the Bremsstrahlung decays over the first 8 μs by many orders of magnitude, the line emission shows only a slight decrease. The neutral iron line decay is less pronounced since the neutral iron number density is additionally fed by recombining iron ions.

The Bremsstrahlung and line emission is compared with the black body emission in case of an optically thick plasma. The spectral radiance of a black body is given by Planck's law [9.7]:

$$L^{\lambda,\mathrm{d}\Omega} = \frac{2hc^2}{\lambda^5}\frac{1}{\mathrm{e}^{h\nu/kT}-1}. \tag{9.9}$$

The unit of $L^{\lambda,\mathrm{d}\Omega}$ is $\mathrm{W}/(\mathrm{m}^2\,\mathrm{sr\,m})$. The spectral radiance of a black body is an emitted power per unit surface area per steradian per wavelength interval. For the comparison with the Bremsstrahlung and line emission coefficient, an equivalent emission coefficient is defined assuming that the plasma is described by a sphere with a radius r:

$$\varepsilon_{\mathrm{bb}}^{\lambda} = 4\pi\frac{A_\mathrm{p}}{V_\mathrm{p}}L^{\lambda,\mathrm{d}\Omega} = 4\pi\frac{3}{r}L^{\lambda,\mathrm{d}\Omega}, \tag{9.10}$$

where $A_\mathrm{p} = 4\pi r^2$ is the surface area of the plasma sphere with a radius r and $V_\mathrm{p} = (\frac{4\pi}{3})r^3$ is the volume of the plasma sphere.

The unit of the quantity $\varepsilon_{\mathrm{bb}}^{\lambda}$ is $\mathrm{W}/(\mathrm{m}^3\,\mathrm{m})$. Assuming $r = 2\,\mathrm{mm}$, $\varepsilon_{\mathrm{bb}}^{\lambda}$ is plotted in Fig. 9.1 taking the temperature for the delay times of 1 and 8 μs. The Planck curve describes the upper limit of the plasma emission. However, in this case, there is no line information available and the emission depends only on the temperature. If the real laser-induced plasma is not optically thick, its emission will always be below the Planck curve. Hence, for LIBS, the plasma state has to be optimized in such a way that high line intensities are achieved which are well above the continuum emission and still far enough away from the Planck limit, where self-absorption dominates and the spectral information is lost.

9.3 Absorption and Optical Thickness

For the absorption coefficient including the contribution of the induced emission holds the following relation [9.7]:

$$\alpha^\nu(\nu) = \frac{h\nu}{c}N_{a,i}^z\Gamma(\nu)B_{ij}\left(1-\mathrm{e}^{-\frac{h\nu}{kT}}\right), \tag{9.11}$$

where $N^z_{a,i}$ is the density of species a with charge z in the lower state i and B_{ij} is the Einstein's transition probability for absorption.

The unit of $\alpha^\nu(\nu)$ is m^{-1}. Inserting $B_{ij} = A_{ji}(g_j/g_i)(c^3/8\pi h\nu^3)$ and relation (9.8) in (9.11) yields:

$$\alpha^\nu(\nu) = \frac{e^2}{4m_e c\varepsilon_0} f_{ij} N^z_{a,i} \Gamma(\nu)\left(1 - \mathrm{e}^{-\frac{h\nu}{kT}}\right). \tag{9.12}$$

The population density of the lower state $N^z_{a,i}$ can be expressed by the Boltzmann formula:

$$N^z_{a,i} = N^z_a \frac{g_i}{u^z_a(T)} \mathrm{e}^{-\frac{E_i}{kT}}. \tag{9.13}$$

Furthermore, we convert (9.12) to $\alpha^\lambda(\lambda)$, i.e., the absorption coefficient in the wavelength interval $\mathrm{d}\lambda$ by use of the relation $\Gamma(\lambda) = \Gamma(\nu)(c/\lambda^2)$. Then, we get with (9.13):

$$\alpha^\lambda(\lambda) = \frac{e^2}{4m_e c^2\varepsilon_0} f_{ij}\Gamma(\lambda)\lambda^2 N^z_a \frac{g_i}{u^z_a(T)} \mathrm{e}^{-\frac{E_i}{kT}}\left(1 - \mathrm{e}^{-\frac{hc}{\lambda kT}}\right). \tag{9.14}$$

As an example this expression is evaluated for the Fe I 538.34 nm line by approximating the line profile function as follows: $\Gamma(\lambda) = 1/\Delta\lambda^{\mathrm{Stark}}_{\mathrm{FWHM}}$, where $\Delta\lambda^{\mathrm{Stark}}_{\mathrm{FWHM}}$ is the Stark width (cf. Sect. 9.4). Taking the data of Fig. 8.16, we obtain for $\alpha^\lambda(\lambda)$ in the first 8 μs values ranging from 51 to 59 m^{-1}.

For local temperature equilibrium (cf. Sect. 8.4), the ratio of the line emission coefficient (9.3) to the absorption coefficient (9.12) is given by Kirchhoff's law:

$$\frac{\varepsilon^{\nu,\mathrm{d}\Omega}_{ji}}{\alpha^\nu} = L^{\nu,\mathrm{d}\Omega}, \tag{9.15}$$

where $L^{\nu,\mathrm{d}\Omega}$ is the spectral radiance according to Planck's law [cf. eq. (9.9)], with respect to the frequency interval: $L^{\nu,\mathrm{d}\Omega} = L^{\lambda,\mathrm{d}\Omega}(\lambda^2/c)$, which can be verified easily by use of the relations (9.3), (9.12), and (9.13).

The optical thickness for a nonhomogeneous plasma along a line of sight in the x-direction is defined as:

$$\tau^\lambda_s = \int_0^s \alpha^\lambda(\lambda, x)\mathrm{d}x, \tag{9.16}$$

where s is the geometrical thickness of a plasma layer and $\alpha^\lambda(\lambda, x)$ is the absorption coefficient as a function of x.

For a geometric plasma extension of $s = 2$ mm, a homogeneous plasma, and the calculated absorption coefficients, an optical thickness of $\tau^\lambda_s \approx 0.11$ results for the chosen example of the Fe I 538.34 nm line. For a plasma layer of thickness s and a constant plasma temperature, the following spectral radiance is emitted along the line of sight running parallel to the plasma extension s:

$$I_\nu(s) = \int_0^{\tau^\nu} L^{\nu,\mathrm{d}\Omega}\mathrm{e}^{-t_\nu}\mathrm{d}t_\nu = L^{\nu,\mathrm{d}\Omega}\left(1 - \mathrm{e}^{-\tau^\nu}\right), \tag{9.17}$$

$$I_\nu(s) \approx \tau^\nu L^{\nu,\mathrm{d}\Omega} = \alpha^\nu s L^{\nu,\mathrm{d}\Omega} = \varepsilon_{ji}^{\nu,\mathrm{d}\Omega} s \quad \text{for} \quad \tau^\nu \ll 1,$$

$$I_\nu(s) \approx L^{\nu,\mathrm{d}\Omega} \quad \text{for} \quad \tau^\nu \gg 1,$$

where $\mathrm{d}t_\nu = \alpha^\nu \mathrm{d}x$ and $L^{\nu,\mathrm{d}\Omega}$ is Planck's law, see (9.9), here related to the frequency interval.

The first approximation holds for an optically thin plasma, and the latter for an optically thick plasma. Hence, the plasma can be considered as optically thin for the case of the Fe I 538.34 nm line.

9.4 Line Broadening

The relevant broadening mechanisms for emission lines of LIBS plasmas are the Doppler broadening and the Stark broadening. The Doppler broadening is an inhomogeneous broadening and the line profile can be described by a Gaussian function. The line width (FWHM) is given by [9.7]:

$$\Delta\lambda_\mathrm{D} = \left(\frac{8kT \ln 2}{M_\mathrm{a} c^2}\right)^{1/2} \lambda_{ji\,0}, \tag{9.18}$$

where M_a is the mass of the emitting atom species a and $\lambda_{ji\,0}$ the central wavelength of the transition from j to i. For an iron atom, a temperature of 8,500 K, and $\lambda =$ 400 nm, we have $\Delta\lambda_\mathrm{D} = 3.5$ pm.

Collisional broadening in a plasma by collisions with charged particles causes the so-called Stark broadening. The line profile can be described by a Lorentz function [cf. eq. (9.5)]. Multielectron species do not have a permanent dipole moment. The external electric field induces a dipole moment. The change of the emission frequency is in this case proportional to the square of the electrical field strength. This behavior is called quadratic Stark effect. The broadening of isolated transitions of neutral atoms and singly ionized ions is mainly caused by collisions with electrons. Hence, half-widths can be calculated by the electron collision approximation. The quasi-static broadening by ions is taken into account as a small correction. The line width (FWHM) for neutral atoms is approximately given by [9.8]:

$$\Delta\lambda_\mathrm{Stark} = \left[1 + 1.75A\left(1 - \frac{3}{4}N_\mathrm{D}^{-1/3}\right)\right] W_\mathrm{FWHM} \left(\frac{n_\mathrm{e}}{10^{16}\,\mathrm{cm}^{-3}}\right), \tag{9.19}$$

where $\Delta\lambda_\mathrm{Stark}$ is the full width at half maximum (FWHM) of the line, A is the dimensionless coefficient, and N_D is the number of particles in the Debye sphere, cf. (8.16), and W_FWHM is the Stark broadening parameter for the full width (FWHM).

In relation (9.19), we used the parameter W_FWHM instead of $W = W_\mathrm{FWHM}/2$, which is usually tabulated in theoretical calculations [9.9], since often experimentally determined Stark broadening parameters are given as the full width at half maximum, see, e.g., [9.10, 9.11]. The line shift is given by [9.7]:

$$\Delta\lambda_{\text{shift}} = D\left(\frac{n_\text{e}}{10^{16}\,\text{cm}^{-3}}\right) \pm A\left(\frac{n_\text{e}}{10^{16}\text{cm}^{-3}}\right)^{1/4} \\ \times \left(1 - \frac{3}{4}N_\text{D}^{-1/3}\right) W_{\text{FWHM}}\left(\frac{n_\text{e}}{10^{16}\,\text{cm}^{-3}}\right), \tag{9.20}$$

where D is the Stark shift parameter.

The coefficients W_{FWHM}, A, and D are independent of the electron density n_e, and are slowly varying functions of the electron temperature. A wavelength shift is defined to be positive, if the wavelength shifts to longer wavelength (red shift). The minus sign in the shift formula (9.20) applies to the high temperature range of those few lines that have a negative value of D/W_{FWHM} at low temperatures. For singly ionized ions, the factor (3/4) is to be substituted by 1.2.

The formulas (9.19) and (9.20) are good approximations (20–30%) in the following parameter range [9.8]:

$$N_\text{D} \geq 2 \quad \text{and} \quad 0.05 < A\left(\frac{n_\text{e}}{10^{16}\,\text{cm}^{-3}}\right)^{1/4} < 0.5. \tag{9.21}$$

The first condition is estimated for a temperature of 7,000 K and an electron density of $10^{17}\,\text{cm}^{-3}$ (the temperature is at the lower limit of the observed plasma temperatures – see Figs. 8.16 and 8.18 – and the electron density is at an upper limit) yielding $N_\text{D} = 2.5$. For the second condition, the range of observed electron densities between $3 \times 10^{15}\,\text{cm}^{-3}$ and $10^{17}\,\text{cm}^{-3}$– see, e.g., Fig. 8.23b – is taken into account, which results in $0.05 < A \cdot (0.74, \ldots, 1.8) < 0.5$. From this, we have $0.07 < A < 0.3$. The dimensionless parameter A is tabulated for isolated neutral atom lines in Appendix IV, pp. 320–362 of [9.9].

Neglecting the ionic part of the broadening relation, (9.19) reduces to:

$$\Delta\lambda_{\text{Stark}} = W_{\text{FWHM}}\left(\frac{n_\text{e}}{10^{16}\,\text{cm}^{-3}}\right). \tag{9.22}$$

This relation can be used to determine the electron density taking a measured line width – (FWHM) – after correction due to apparative and Doppler broadening – and a known Stark broadening coefficient W_{FWHM} [9.10–9.12].

The different normalized line profiles [cf. (9.4)] are plotted for Doppler broadening $\Gamma_\text{D}(\lambda)$ and Lorentz broadening $\Gamma_\text{L}(\lambda)$ in Fig. 9.3 taking as an example the Fe I, 538.34 nm line with the same widths (FWHM) for both profiles of 0.05 nm (this value is chosen for illustration only). The additionally shown Voigt profile $\Gamma_\text{V}(\lambda)$ has the same width (for the calculation of $\Gamma_\text{V}(\lambda)$ a Doppler width of 0.03 nm and a Lorentz width of 0.031 nm is taken), this function will be described in the following.

In general, the Doppler broadening and Stark broadening are superimposed. The resulting line shape is described by the normalized Voigt profile as follows:

$$\Gamma_\text{V}(\lambda) = \frac{2\sqrt{\ln 2/\pi}}{\Delta\lambda_\text{D}} K(u, a) \tag{9.23}$$

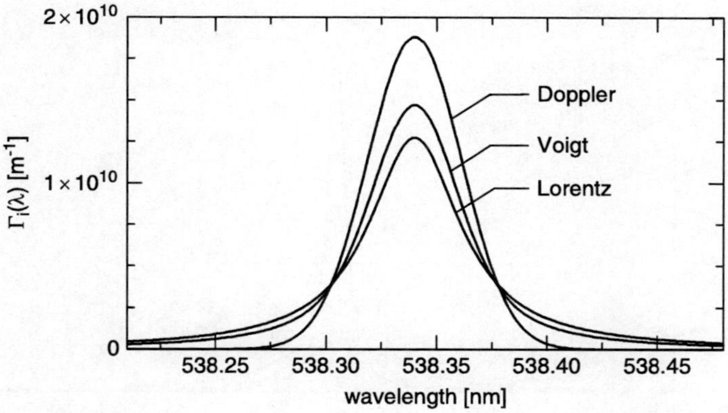

Fig. 9.3 Calculated normalized line profile functions of the same width (FWHM) for Doppler, Lorentz, and Voigt

with the Voigt profile function:

$$K(u,a) = \frac{a}{\pi} \int_{-\infty}^{+\infty} \frac{e^{-t^2}}{(u-t)^2 + a^2} dt \tag{9.24}$$

and the variables:

$$u = \frac{2\sqrt{\ln 2}}{\Delta\lambda_D} \left(\lambda - \lambda_{ji0}\right), \tag{9.25}$$

$$a = \sqrt{\ln 2}\frac{\Delta\lambda_N + \Delta\lambda_L}{\Delta\lambda_D} \approx \sqrt{\ln 2}\frac{\Delta\lambda_L}{\Delta\lambda_D}, \tag{9.26}$$

where $\Delta\lambda_N$ is the natural line width, $\Delta\lambda_L$ is the Lorentz line width, and $\Delta\lambda_D$ is the Doppler line width.

Figure 9.4 shows the Voigt profile function calculated for the experimentally determined temperature and electron density of Fig. 8.16 at the delay times 1, 4, and 8 μs. The line profile is narrowing for later delay times due to the strongly decreasing electron density which results in a reduced Stark broadening. On the other hand, the Doppler widths are nearly constant in the time interval from 1 to 8 μs; hence, the parameter a [cf. (9.26)] drops significantly.

Experimentally measured line shapes can be fitted by the Voigt function, see Fig. 9.5. The coefficient of determination in the shown example of the Fe 438.35 nm line is better than 0.999. The fit yields the widths of the Gaussian and the Lorentzian part of the Voigt profile [cf. (9.26)]: $\Delta\lambda_G^m = 0.021$ nm, $\Delta\lambda_L^m = 0.018$ nm. The index "m" indicates that these values are gained from the measurement.

The measured widths have to be corrected by the apparative broadening to obtain the correct line width of the Doppler and Lorentz part using the following relations:

$$\Delta\lambda_G^{m^2} = \Delta\lambda_D^2 + \Delta\lambda_G^{a^2}, \tag{9.27}$$

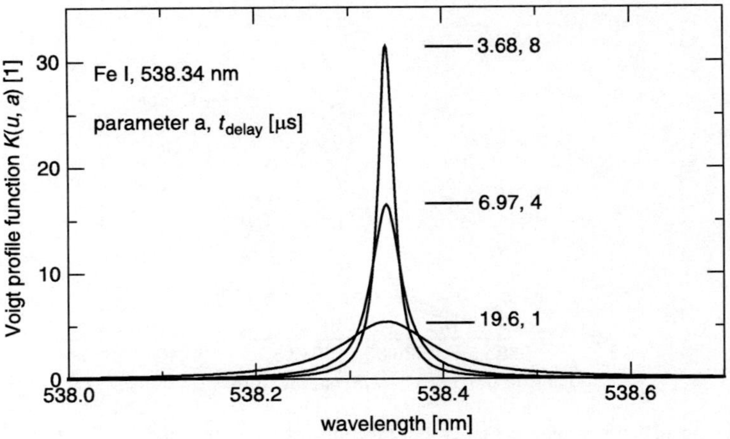

Fig. 9.4 Calculated Voigt profile function for the Fe I, 538.34 nm line and the parameters of Fig. 8.16

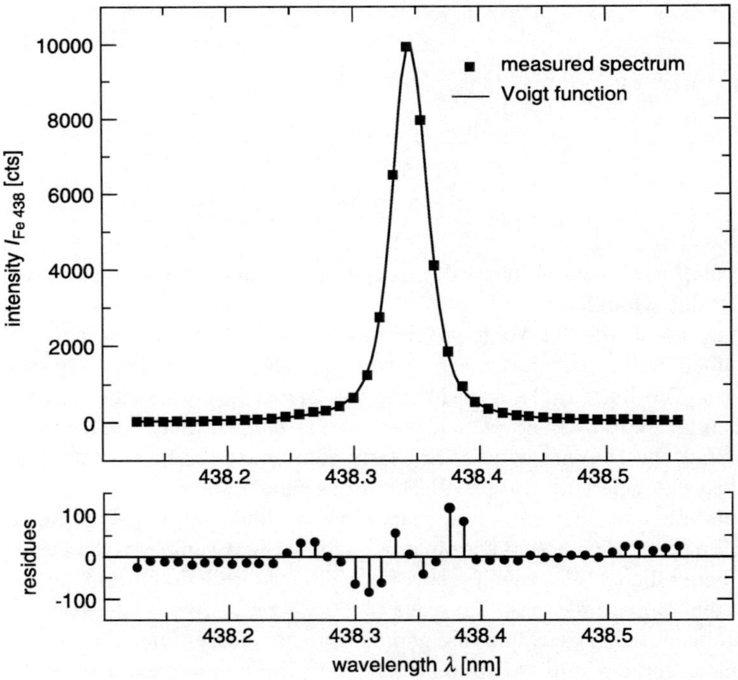

Fig. 9.5 Experimentally measured line shape of the Fe I line 438.34 nm and fit of the data by a Voigt function

$$\Delta\lambda_{\mathrm{L}}^{\mathrm{m}} = \Delta\lambda_{\mathrm{L}} + \Delta\lambda_{\mathrm{L}}^{\mathrm{a}}, \tag{9.28}$$

where $\Delta\lambda_{\mathrm{G}}^{\mathrm{m}}$, $\Delta\lambda_{\mathrm{L}}^{\mathrm{m}}$ are the measured Gaussian, Lorentzian line widths (FWHM); $\Delta\lambda_{\mathrm{D}}$, $\Delta\lambda_{\mathrm{L}}$ are the correct Doppler, Lorentz line widths (FWHM), and $\Delta\lambda_{\mathrm{G}}^{\mathrm{a}}$, $\Delta\lambda_{\mathrm{L}}^{\mathrm{a}}$ are the apparative broadening described by a Gauss, Lorentz function, respectively.

For the determination of the apparative broadening, the spectrometer-detector system should be operated in a configuration that is also used for the LIBS measurements. This refers, e.g., to the width of the entrance slit, the spectral range studied, and the settings of the detector (amplification, gating). In the interesting spectral range, the Lorentz and Gauss widths of a set of lines of the irradiated sample are determined as a function of the burst energy. Figures 9.6 and 9.7 show the results for an echelle MCP–CCD system (cf. Sect. 4.2). In Fig. 9.6, the measured Lorentz line width is plotted – determined after fitting the emission line by a Voigt profile – as a function of the burst energy for four iron lines from which the 381.30 nm line shows the smallest broadening. The Lorentz line widths decrease for smaller burst energies due to a decreasing electron density. The data of the narrowest line 381.30 nm is fitted by a quadratic polynomial yielding the Lorentz width for a burst energy of 0.5 mJ: $\Delta\lambda_{\mathrm{L}}^{\mathrm{m}} = (2.7 \pm 1.5)$ pm. This value is an upper limit of the Lorentz part of the apparative broadening, i.e., $\Delta\lambda_{\mathrm{L}}^{\mathrm{a}} \leq (2.7 \pm 1.5)$ pm.

Figure 9.7 shows the Gauss widths of the same lines. These widths do not depend on the burst energy but are approximately constant. The Doppler broadening being of the order of 3.5 pm and less does not contribute significantly to these widths [cf. (9.18) and (9.27)]. Hence, the average Gauss width is: $\Delta\lambda_{\mathrm{G}}^{\mathrm{m}} \approx \Delta\lambda_{\mathrm{G}}^{\mathrm{a}} = (20.1 \pm 0.8)$ pm. The total apparative broadening $\Delta\lambda_{\mathrm{V}}^{\mathrm{a}}$ can then be estimated using

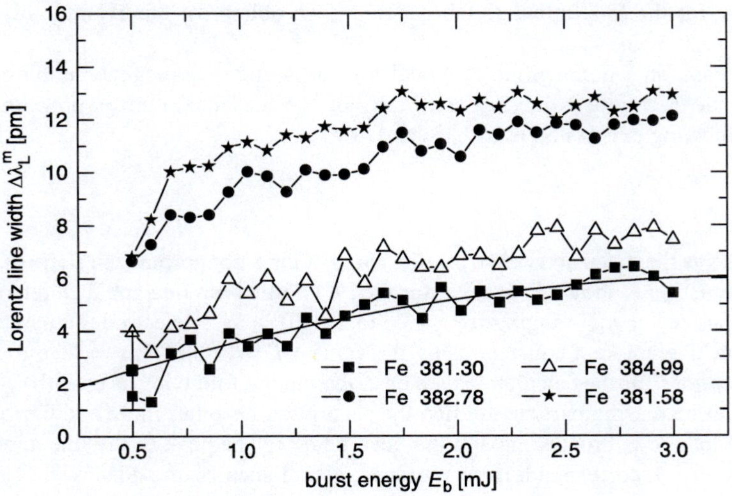

Fig. 9.6 Lorentz line widths $\Delta\lambda_{\mathrm{L}}^{\mathrm{m}}$ of a set of iron lines as a function of the burst energy for an echelle MCP–CCD system to determine the Lorentz part of the apparative broadening

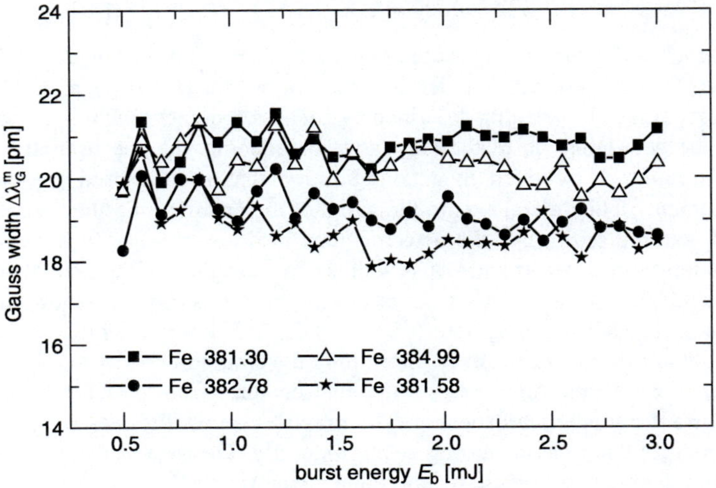

Fig. 9.7 Gauss line widths $\Delta\lambda_G^m$ of a set of iron lines as a function of the burst energy for an echelle MCP–CCD system to determine the Gaussian part of the apparative broadening

the following approximation of the width (FWHM) of a Voigt profile for given Gauss and Lorentz parts [9.13]:

$$\Delta\lambda_V^a = 0.5346\Delta\lambda_G^a + \sqrt{0.2166\Delta\lambda_G^{a^2} + \Delta\lambda_L^{a^2}}. \tag{9.29}$$

Hence, for the total apparative broadening, we obtain for the example discussed: $\Delta\lambda_V^a \leq 20.5\,\text{pm}$.

In case of a nonnegligible optical thickness and a homogeneous plasma, the shape of the emitted line in the vicinity of the line maximum is proportional to the following expression [cf. (9.16) and (9.17)]:

$$\gamma(\lambda) = 1 - e^{-\alpha^\lambda(\lambda)s}, \tag{9.30}$$

where s is the extension of the plasma and $\gamma(\lambda)$ is a nonnormalized shape function. This function is shown in Fig. 9.8 for the Fe I 538.34 nm line and different particle densities $N_a^z = n_{\text{FeI}}$ ranging from 10^{17} to $2 \times 10^{19}\,\text{cm}^{-3}$ (electron temperature and plasma size are kept constant using the values: $T = 7{,}300\,\text{K}$, $s = 2\,\text{mm}$. Further assumptions for the calculations are described in the following as case (b).)

With increasing particle density, the absorption becomes more and more important. The line is broadening and the peak flattens. The peak of the line approaches a value of 1 corresponding to the spectral radiance of the Planck function [cf. (9.17)]. The n_{FeI} values taken for Fig. 9.8 are much higher than the experimentally determined values, which are typically in the range of 5×10^{15} to $5 \times 10^{16}\,\text{cm}^{-3}$ (cf. Tables 8.2 and 8.3).

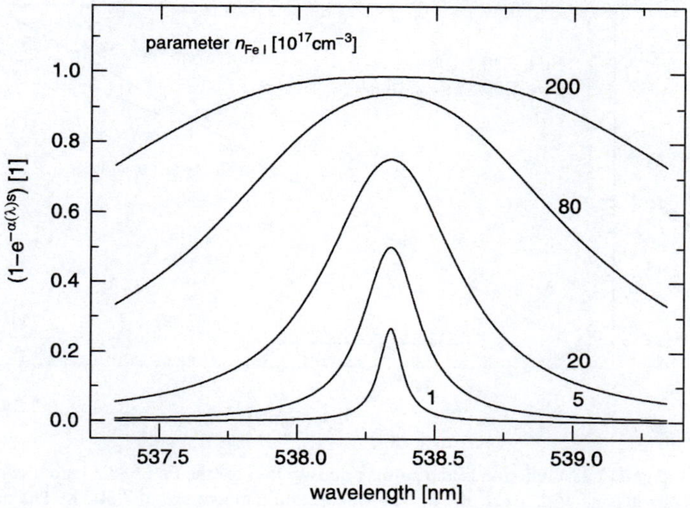

Fig. 9.8 Calculated line shape of the Fe I, 538.34 nm line according to (9.30), parameter: iron particle density

The radiance of the emission line can be calculated by integrating relation (9.17) over the line assuming that the variation of $L^{\lambda,\mathrm{d}\Omega}$ is small in the narrow wavelength range of the spectral line and that the plasma is homogeneous:

$$I(s) = L^{\lambda_{ji},\mathrm{d}\Omega} \int_0^{\infty} \left(1 - \mathrm{e}^{-\alpha^{\lambda}(\lambda)s}\right) \mathrm{d}\lambda = L^{\lambda_{ji},\mathrm{d}\Omega} \Lambda\left(N_a^z, T, s\right). \qquad (9.31)$$

The integral $\Lambda(N_a^z, T, s)$ is a function of the particle density N_a^z of the emitting species, the temperature, and the plasma extension. This integral is calculated as a function of the iron atom density $N_a^z = n_{\mathrm{FeI}}$ for the Fe I, 538.34 nm line assuming a constant temperature of 7,300 K and a constant plasma size of 2 mm. Two cases were considered: (a) the parameter a [see (9.26)] is kept constant at the value 3.68 corresponding to the experimentally observed Stark widths after a delay time of 8 μs (cf. Figs. 8.16 and 9.4); (b) while varying n_{FeI} in a constant plasma volume at a fixed temperature T, the electron density cannot be constant but varies according to the Saha equation. Hence, the Stark width changes as well and therefore the parameter a too. Figure 9.9 shows the results. When the particle density n_{FeI} is increased, the integral Λ and consequently the line radiance increases as well. In case (a), the slope changes for high particle densities, which is a consequence of self-absorption. However, in case (b) where the line broadening induced by the higher electron density is considered, this change is much less pronounced. In the range of particle densities n_{FeI} shown in Fig. 9.9, the parameter a varies in this case between 2.5 and 37.

For quantitative analysis with LIBS, the analyte emission lines have to be selected carefully to achieve a good response of the line intensity versus the analyte

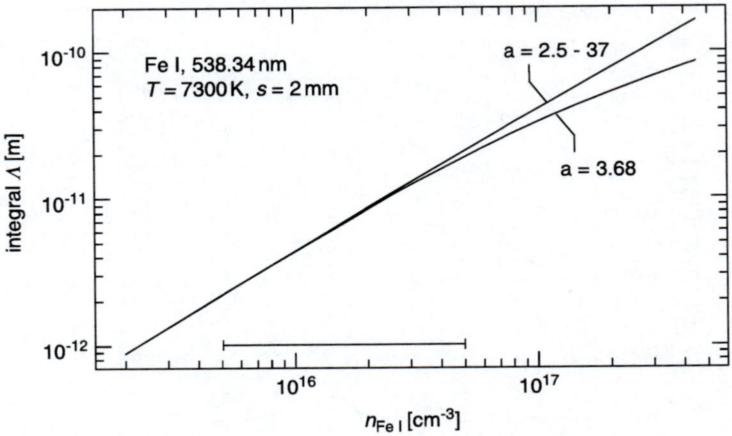

Fig. 9.9 Integral Λ as a function of the particle density n_{FeI} for the Fe I 538.34 nm line assuming a homogeneous plasma with a size of 2 mm and a constant temperature of 7,300 K. The parameter a is kept constant or varies, see text. The *horizontal bar* indicates the range of n_{FeI} values determined spectroscopically (cf. Tables 8.2 and 8.3)

concentration and to avoid self-absorption. Figure 9.10 shows a section of LIBS emission spectra of a set of aluminum samples with a Mg content varying over more than four orders of magnitude. Two Mg lines are visible where the line at 279.078 nm increases monotonously with increasing Mg concentration, whereas the line at 279.553 nm changes its shape showing a dip at high Mg contents. The inset diagram shows the response as a function of the Mg concentration. The 279.553 nm line shows a strong saturation behavior.

The dip is an indication that the LIBS plasma is not homogeneous but consists of regions having different particle densities and temperatures. Outer colder plasma regions absorb radiation from inner parts. The spectral width of this absorption is smaller than the emission from the core due to reduced line broadening at the outer plasma regions. At the center of the line, the optical thickness is greater; hence, we look only at the outer regions of the plasma, where the temperatures are lower and therefore the emission is weaker.

In order to assess the sensitiveness of a line for self-absorption, the absorption coefficient (9.14) is factorized as follows:

$$\alpha^{\lambda}(\lambda) = \Gamma(\lambda) N_a^z k_{\mathrm{t}}, \tag{9.32}$$

where the quantity k_{t} depends on the atomic data of the line and the temperature only:

$$k_{\mathrm{t}} = \frac{e^2}{4 m_{\mathrm{e}} c^2 \varepsilon_0} f_{ij} \lambda^2 \frac{g_i}{u_a^z(T)} \mathrm{e}^{-\frac{E_i}{kT}} \left(1 - \mathrm{e}^{-\frac{hc}{\lambda kT}}\right). \tag{9.33}$$

Table 9.1 shows the atomic data of the two Mg lines of Fig. 9.10 and the calculated k_{t} values for two temperatures [9.14, 9.15].

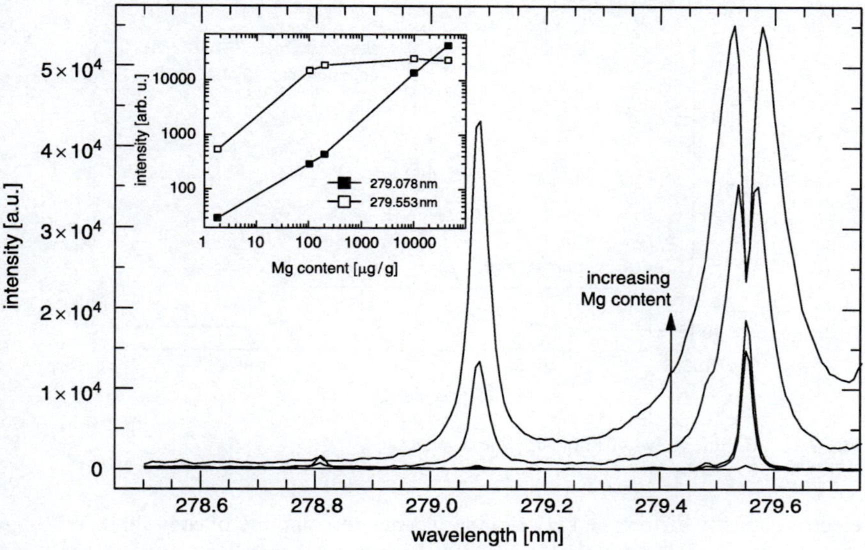

Fig. 9.10 LIBS spectra of a set of aluminum samples with varying Mg concentration

Table 9.1 Atomic data of the Mg II lines of Fig. 9.10 and k_t values for two temperatures $T_1 =$ 9,000 K and $T_2 = 6{,}000$ K

Mg line (nm)	Ionization stage	g_i	E_i (eV)	f_{ij}	$k_t(T_1)(m^3)$	$k_t(T_2)(m^3)$
279.078	II	2	4.423	0.953	2.16×10^{-30}	1.26×10^{-31}
279.553	II	2	0	0.629	4.30×10^{-28}	4.35×10^{-28}

g_i = statistical weight of the lower level, E_i = energy of the lower level, f_{ij} = oscillator strength

The Mg II 279.553 nm line is a resonance line and the k_t values are by orders of magnitude greater than for the line 279.078 nm indicating their strong sensitiveness to self-absorption. As a first estimate whether a line is affected by self-absorption in a LIBS plasma, its k_t value calculated for typical temperatures should be less than about 5×10^{-30} m^3.

The Mg II line at 279.078 nm is Stark broadened and theoretical as well as experimental Stark width data is available [9.8, 9.14]. Considering the apparative broadening, the determined electron density is in the range of 1.2–1.5 $\times 10^{17}$ cm^{-3}. Using relation (9.3), we can estimate the temperature from the line ratio of the 279.078 nm line to the 279.553 nm line for the sample having a Mg content of 100 μg/g. Due to the fact that the latter line is affected by absorption, this estimate yields an upper limit of the temperature. Taking into account the different Stark widths of these lines, we obtain $T < 13{,}000$ K.

Figure 9.11 shows another part of the LIBS spectrum with an Al I resonance line with line reversal demonstrating the shift between the absorption and emission maximum of 26 pm. The emission of the plasma core is shifted to longer wavelength. By use of relation (9.20), neglecting the ion correction term, we can estimate the

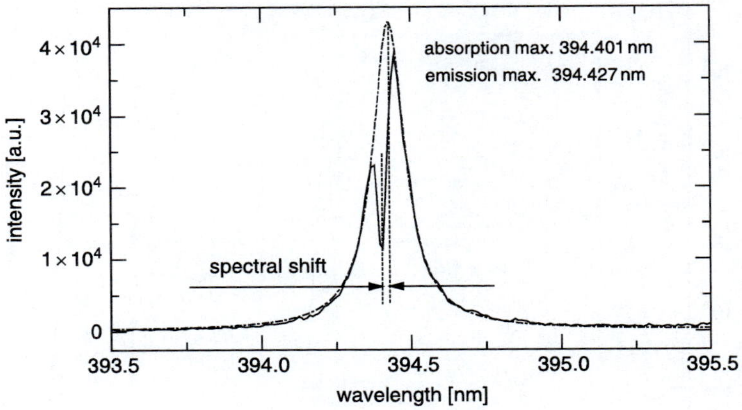

Fig. 9.11 Al I line at 394.401 nm with line reversal

electron density difference between the plasma core and the plasma shell. With the Stark shift parameter for the Al I 394.401 nm line taken from reference [9.9], we obtain for the electron density difference $\Delta n_e = 1.07 \times 10^{17}\,\mathrm{cm}^{-3}$, a value which is in the same order as the electron density determined from the Stark width of the Mg II line 279.078 nm (cf. Fig. 9.10).

References

9.1. H. Griem, *Plasma Spectroscopy* (McGraw-Hill, New York, 1964)
9.2. C. Breton, J. Schwob, Ch. 8, in *Vacuum Ultraviolet Radiation Physics*, ed. by N. Damany, J. Romand, B. Vodar (Pergamon Press, Oxford, 1974)
9.3. J. Fuhr, G. Martin, W. Wiese, S. Younger, Atomic transition probabilities for iron, cobalt, and nickel. J. Phys. Chem. Ref. Data **10**, 305–524 (1981)
9.4. H.-W. Drawin, P. Felenbok, *Data for Plasmas in Local Thermodynamic Equilibrium* (Gauthier-Villars, Paris, 1965)
9.5. J. Halenka, J. Madjev, Tables of partition functions of iron, Fe I–Fe X. Acta Astronom. **52**, 199–206 (2002)
9.6. Bergmann-Schäfer, *Lehrbuch der Experimentalphysik*, Band IV, Teil 2, *Aufbau der Materie*, Hrsg. H. Gobrecht, Walter de Gruyter, Berlin, 1981
9.7. A. Anders, *A Formulary for Plasma Physics* (Akademie-Verlag, Berlin, 1990)
9.8. G. Bekefi, *Principles of Laser Plasmas* (Wiley, New York, 1976)
9.9. H. Griem, *Spectral Line Broadening by Plasmas* (Academic, New York, 1974)
9.10. N. Konjevic, M. Dimitrijevic, W. Wiese, Experimental Stark widths and shifts for spectral lines of neutral atoms (a critical review of selected data for the period 1976 to 1982). J. Phys. Chem. Ref. Data **13**, 619–647 (1984)
9.11. N. Konjevic, W. Wiese, Experimental Stark widths and shifts for spectral lines of neutral atoms (a critical review of selected data for the period 1983 to 1988). J. Phys. Chem. Ref. Data **19**, 1307–1385 (1990)
9.12. S. Freudenstein, J. Cooper, Stark broadening of Fe I 5383 Å. Astron. Astrophys. **71**, 283–288 (1979)

9.13. J. Olivero, R. Longbothum, Empirical fits to the Voigt line width: a brief review. J. Quant. Spectrosc. Radiat. Transf. **17**, 233–236 (1977)
9.14. R. Kurucz, B. Bell, *Atomic Line Data* (Smithsonian Astrophysical Observatory, Cambridge, 1995), http://cfa-www.harvard.edu/amp/tools.html
9.15. S. Bukvic, A. Sreckovic, S. Djenize, Mg II h and k lines Stark parameters. New Astron. **9**, 629–633 (2004)

Chapter 10
Modeling of Plasma Emission

A modeling of the emission spectra of laser-induced plasmas for LIBS can contribute to a better understanding of the relevant parameters influencing the observed measuring radiation and potentially be an instrument for a quantitative analysis as well. In a rigorous treatment, such a modeling of the plasma emission should be part of a physical model of the LIBS method including plasma generation and plasma emission. Such a description is beyond the scope of this chapter.

Figure 10.1 illustrates schematically a physical model with the input quantities given by the experimental conditions and the desired output quantities.

The laser beam parameter $I_{\mathrm{L}}(R,t)$ is the irradiance of the incident laser beam as a function of coordinate R and time t. For simplicity, a rotationally symmetric laser beam profile is assumed and R is a radial coordinate giving the distance from the beam propagation axis. In practice, the quantity $I_{\mathrm{L}}(R,t)$ is often not known or only known approximately. The laser pulse or laser burst energy E_{L} can be measured quite easily, so at least the spatially and temporally integrated value of $I_{\mathrm{L}}(R,t)$ is known.

The wavelength of the laser radiation λ has an effect on the beam focusing and the absorption of the radiation by the specimen, the material vapor, and the plasma. Further laser parameters such as the polarization state and beam pointing stability may also have an effect on the interaction process with the specimen.

The ambient atmosphere can be characterized by the pressure $p_{\mathrm{g}}(r,t)$. Prior to the irradiation of the first laser pulse, this pressure can be assumed to be constant. However, in case of pulse bursts consisting of two or more pulses irradiated within a time interval of a few microseconds, the preceding pulses induce a local variation of the ambient particle density and pressure which influences the dynamics of the plasma expansion for the subsequent pulses (cf. Chap. 8) [10.1]. In this case, the pressure depends on time t. The ambient atmosphere consists of different species a, whose densities are described by $N_{a,g}(r,t)$. Besides the dependence on t, these densities are a function of the coordinate r. In an even more rigorous treatment, the mixing of particles ablated by the laser and particles from the initial ambient atmosphere has to be considered. Hence, in a pulse burst, the subsequent pulses

R. Noll, *Laser-Induced Breakdown Spectroscopy*, DOI 10.1007/978-3-642-20668-9_10,

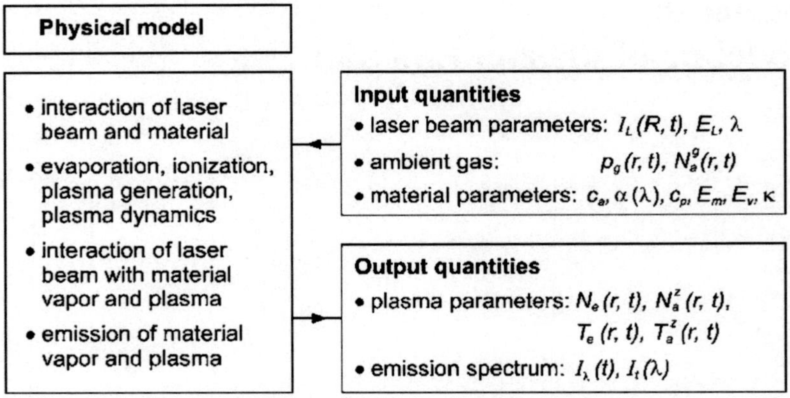

Fig. 10.1 A physical model for the LIBS method with the input and output quantities

encounter a pressure $p_g(r,t)$ to which not only the species of the ambient gas but also the ablated and vaporized particles will contribute. Further effects – not shown in Fig. 10.1 – arise from residual particulates originating from ablations by previous laser pulses which are still in the interaction volume and act as scattering and absorption centers for subsequent laser pulses. These scattering and absorption processes again depend on the wavelength of the irradiated laser beam. A gas flow in the interaction region will also influence the extent to which such particulates are moved out of the propagation path of the laser beam (cf. Sect. 3.1, Fig. 3.4).

Last but not least the material parameters are essential input quantities too. The concentrations of the species a in the specimen are described by the quantities c_a. The initial absorption of the laser beam depends on the natural absorption coefficient $\alpha(\lambda)$. The specific heat capacity c_p influences the temperature rise at the surface of the specimen. The specific melt and evaporation enthalpies E_m, E_v as well as the thermal diffusivity κ are necessary input quantities to describe the melting and evaporation of the specimen leading to the material ablation and crater formation (in case of, e.g., a metallic solid specimen). Not shown in Fig. 10.1 are further material properties such as the surface roughness, crater formation, or the inclination of the surface of the specimen versus the incident laser beam axis. Additionally most of these material parameters are a function of the temperature T and these dependencies are not known for many materials.

The desired output parameters are the electron density $N_e(r,t)$, the particle densities of species a with charge z: $N_a^z(r,t)$, the electron temperature $T_e(r,t)$, the particle temperatures $T_a^z(r,t)$, and last but not least the spectrum emitted by the plasma $I_\lambda(t)$ as a function of time and – for comparison with experimental data – the quantity deduced thereof $S_t(\lambda)$ describing the spectrum integrated over a time interval t_{int} starting after a delay time t_{delay} after the last irradiated laser pulse. The quantities $T_e(r,t)$ and $T_a^z(r,t)$ require the establishment of local equilibrium conditions among the ensemble of particles concerned.

The main interactions and processes to be considered in the physical model are listed on the left side of Fig. 10.1. Some aspects of these processes have already been studied, e.g., evaporation [10.2], plasma generation [10.3], ambient gas pressure [10.4, 10.5], plasma dynamics [10.1], and interaction of laser radiation and plasma [10.6]. Several papers were published treating approaches for a physical model of the LIBS plasma, e.g., [10.7, 10.8]. Considering the fact that many of the necessary material parameters for a physical model of the LIBS plasma formation are not known, the focus is put here on a heuristic modeling of the spectral plasma emission after the interaction of the laser pulse with the specimen.

Several papers were published in the last years presenting models to describe the postbreakdown phase of a LIBS plasma [10.9–10.12]. A spherical plasma is considered, which is assumed to be in local thermodynamic equilibrium (LTE). The idea is to describe the dynamic plasma evolution after the interaction of the laser pulse with the target.

In a simplified theoretical approach to model an inhomogeneous optically thick laser-induced plasma, it is assumed that the initial temperature distribution can be described by a parabolic function with the temperature maximum at the center of the plasma [10.9]. Furthermore, an expansion law for the plasma radius into an ambient vacuum is defined. The number of fitting parameters for the temperature profile and the plasma size amounts to 5. The plasma composition is assumed to be constant. An approximate solution of the radiation transport equation was used and spectra were calculated for a binary system of Si and C in the wavelength interval from 287 to 289 nm comprising a single strong Si line at 288.16 nm. Further model outputs are spatial and temporal distributions of atom, ion, and electron number densities. No direct experimental verifications were shown. The authors stated that the model inputs could easily be measured in an experiment. This may be possible for a fundamental study, for a practical application of the model the additional effort to determine experimentally the input parameters – such as total number of atoms in the plasma volume, the temperature function, the plasma expansion function, and the atomic stoichiometry of plasma species – is assessed to be prohibitive, since this would require an extensive additional apparative equipment.

In a further model expansion, the gas dynamic equations were coupled with the equation of radiative transfer for a spherical plasma evolving into an ambient vacuum [10.10]. Partition functions from the tables of Drawin and Felenbok were used [10.13]. The authors stated uncertainties of ratios of partition functions being less than 50%. This means that for a comparison of calculated line ratios to measured line ratios, deviations may occur of up to 50% even in the case of availability of a perfect postbreakdown plasma model. In this regard, expectations of the practical usefulness of a model have to be oriented to a more qualitative description as long as significantly more accurate partition function data are not available. The model output is a series of synthetic spectra in the spectral range of 280–290 nm which includes three spectral lines from the binary Si–C system taken. The intention of the authors is to vary the initial conditions until a close fit between the synthetic and the experimental spectra is obtained; however, no experimental verification was presented.

An experimental verification of a radiative model of laser-induced plasma expanding into vacuum was reported for a pure iron foil and a set of eight Al samples [10.11]. The goal was to predict the composition of the plasma from fits of synthetic spectra to measured spectra. Instrumental parameters such as slit function, diffraction grating efficiency, and detector response function were considered. The experiments were carried out at an ambient pressure of 0.5 torr, which results in rather large observed plasma diameters of about 3 mm achieved after 200 ns. At a delay time of 40 ns and an integration window of 55 ns, a spatially integrated temperature was determined using Fe II lines, yielding 16,500 K with an error of 13%. A qualitative comparison of a synthetic iron spectrum in the range from 220 to 300 nm calculated for an average temperature of 16,500 K with the experimental spectrum showed similarities; however, a detailed look reveals strong deviations in the range of 50% to several 100%. Using Fe II lines for the experimental temperature determination certainly has led to an overestimation of the temperature.

For the aluminum alloys, the authors could demonstrate fits with correlation coefficients greater than 95% for a spectral region between 278 and 288 nm comprising up to six lines [10.11]. They found a systematically higher intensity for the calculated Mg II 280.27 nm line than in the experiment and attributed this to an imprecise value of the line transition probability. However, taking the data from the Kurucz-files for the two Mg II lines 279.55 and 280.27 nm, assuming the same line profile, negligible self-absorption and calculating the Boltzmann factors a ratio of $I_{\mathrm{Mg\ II279.55}}/I_{\mathrm{Mg\ II280.27}} = 2.05$ is obtained for a temperature of 22,000 K (the influence of the temperature for this estimation is very low, since the two upper energy levels of the two Mg II lines differ only by about 0.01 eV) [10.14]. From the experimental spectra shown in Fig. 7 of [10.11], a line ratio of about 1.9 is deduced, which is quite close to the expected value.

The two Mg II lines considered are resonance lines, hence they may be affected by self-absorption. The ratio of the two lines in the synthetic spectrum of sample AA3 shown in Fig. 7 [10.11] is significantly less than 2 indicating the effect of self-absorption implemented in the applied model. However, the measured ratio for sample AA3 is consistent with the assumption of negligible self-absorption of these two Mg II lines at this level of Mg concentration (for sample, SM 10 having a fivefold higher Mg concentration the experimental line ratio shown in Fig. 7 [10.11] indeed decreases to about 1.75 showing that self-absorption is becoming important at these Mg concentration levels). Hence, the discrepancy found by the authors cannot be attributed to imprecision of the available line transition probabilities.

In a further paper of the same group, the radiative model of the postbreakdown laser-induced plasma expanding into ambient gas was studied [10.12]. A qualitative comparison between synthetic spectra and measured spectra at different delay times (0.1–4 μs) was shown for a SiC plasma in an argon atmosphere at atmospheric pressure. The spectral range covered two resolved emission lines of silicon and carbon (Si I 288.16 nm and C II 283 67 nm/283.76 nm). The authors concluded that the determination of intrinsic plasma characteristics from the experimentally determined parameters can in principle be solved. As a limitation they proposed to introduce at least two temperatures for electrons and heavy particles.

The profiles of single Ti ion lines generated by irradiating UV laser pulses at 308 nm on titanium targets in low-pressure nitrogen atmospheres ($\leq$1 mbar) were studied at the initial high-density plasma evolution ($\leq$200 ns) [10.15]. At this phase, the plasma expands fast into the low-density ambient atmosphere. Single observed asymmetric and self-reversed line profiles of Ti II lines could be described by a two zone one-dimensional plasma approach. (The oscillator strength of 0.0165 given in [10.15] for the Ti II 346.150 nm line is not correct, and the correct value is 0.0141, cf. e.g., [10.14].)

A one-dimensional two region plasma description was studied to fit a set of curves of growths (COG) of five iron atomic lines in the spectral region of 373–382 nm [10.16]. The number of parameters used to fit the observed COG is 7 for the two region plasma model. The authors could show that for intense lines, at high concentrations of the element in the sample and at initial times of the plasma evolution it is necessary to use a model of the laser-induced plasma that includes regions with different values of temperatures, electron densities, and atom densities. From the fitting of the set of COGs, a ratio of number densities of emitting neutral atoms of the inner to the outer region of 200 were determined, while the length of the central plasma region is stated as 1 mm and the length of the outer regions as 4 mm. Ambient gas effects were not studied.

Instead of modeling the postbreakdown dynamics of the LIBS plasma and calculating emission spectra linked to this dynamics in the following a simplified heuristic model is presented where no plasma expansion is considered and the inhomogeneity of the plasma is described by a shell model of the plasma surrounded by an ambient gas. This approach is led by experimental observations. For quantitative LIBS analysis, the spectrum emitted by the plasma expanding into the ambient atmosphere is typically detected at a delay time being in the range of 1 μs after the interaction of the laser pulses with the specimen. At this time region, the plasma dynamics has already slowed down significantly.

10.1 Heuristic Model

In the heuristic model, it is assumed that the composition of the specimen, the plasma, and the plasma state is known. The goal is to calculate the spectral plasma emission for an inhomogeneous plasma surrounded by an ambient atmosphere.

The plasma can be described by a simplified spherical shell model as illustrated in Fig. 10.2 [10.17]. A spherical description is assumed, since the experimentally observed plasma geometry 500 ns after the irradiation of the laser pulse (or the last pulse of a pulse burst) has nearly a spherical shape. Deviations from a spherical geometry certainly exist in the time interval up to 500 ns, see, e.g., Figs. 4 and 5 in [10.1]. However this time domain is usually not used for quantitative LIBS measurements. While for single pulses the plasma stays in contact with the sample surface, the plasma is detaching from the sample surface for collinear double pulses as shown in Fig. 8.12 by high-speed images of the laser-induced plasma observed in side-on direction. Especially for double pulses, the assumption of a spherical plasma

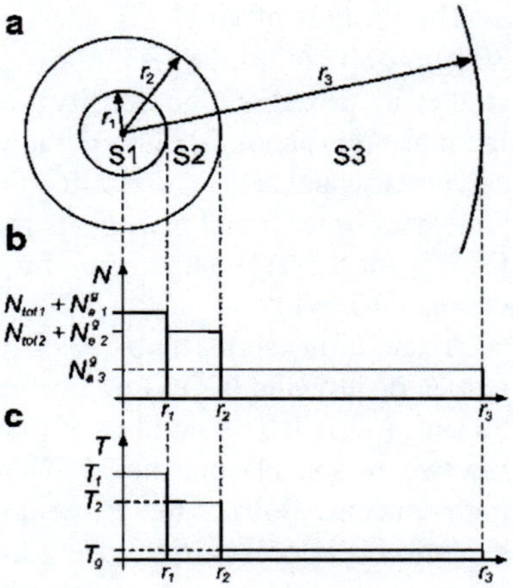

Fig. 10.2 Simplified plasma model for the heuristic approach. (**a**) The plasma is described by two shells: an inner sphere or plasma core S1 and an outer shell S2. The ambient atmosphere is described by shell S3; (**b**) densities, and (**c**) temperature distribution in the regions S1–S3. "Copyright 2009, American Institute of Physics [10.17]. Reprinted with permission"

geometry for delay times greater than 0.5 μs is well justified by these experimental observations.

The laser-induced plasma is described by two shells: the inner sphere or plasma core S1 and an outer shell S2. By this assumption the experimental fact of a nonhomogeneous plasma is described in the most simple approach analogous to [10.15]. Within these shells, spatially homogeneous values of the physical quantities are assumed. For example, sphere S1 contains mainly the ablated material, whereas in shell S2 particles of the ablated material and the ambient atmosphere can be taken into account. Furthermore, it is assumed that these shells contain only atoms, ions, and electrons. Molecules and particulates are not taken into account. Shell S3 is used to describe the stationary ambient atmosphere in the region between the plasma and the detector plane of a spectrometer operated under the same atmospheric conditions or the entrance slit of a vacuum spectrometer. This atmosphere absorbs a part of the emitted radiation of the shells S1 and S2 depending on the wavelength. The ambient gas may consist of mixtures of atoms and molecules.

Figure 10.3 illustrates schematically the heuristic model with the input quantities given by the experimental conditions and the desired output quantities. The input quantities and assumptions for the heuristic model are listed in detail in Table 10.1.

The assumption of a constant plasma geometry is a strong simplification. In a self-consistent physical model, the plasma dynamics in terms of the plasma expansion and the temporal evolution of the particle densities and temperatures are interrelated. In order to reduce the number of input parameters for the heuristic model, the plasma geometry is kept constant. From the experimental point of view, the spectral emission used for LIBS is observed after a delay time of several 100 ns after the laser pulse irradiation. At this time regime, the plasma has already attained its maximum extension and starts to decay with a moderate velocity. The characteristic time of this decay can be estimated as follows:

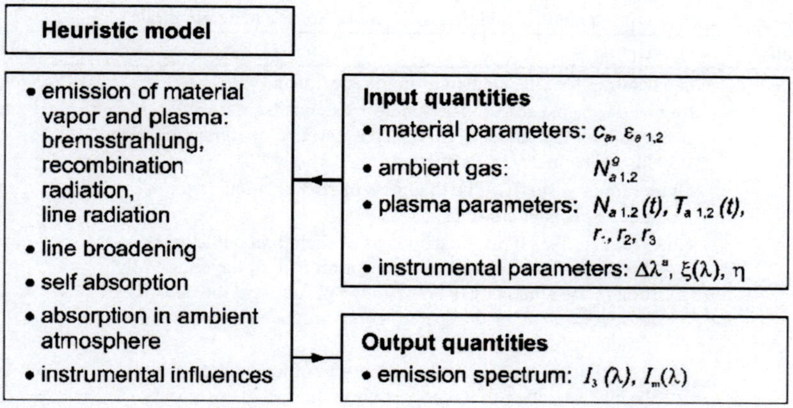

Fig. 10.3 The heuristic model to calculate the spectral plasma emission

$$\tau = d_{\text{Plasma}}/v_{\text{th}}, \tag{10.1}$$

where d_{Plasma} is the diameter of the spherical plasma and v_{th} is the thermal velocity of the atoms and ions in the plasma. Relation (10.1) yields the order of magnitude of the time after which the plasma disintegrates due to the nondirectional movement of the heavy particles. For a plasma of iron atoms and ions at a temperature of 9,000 K and a diameter of $d_{\text{Plasma}} = 2$ mm, the lifetime estimated by (10.1) is $\tau = 1.1\ \mu$s. For integration times of the plasma emission being smaller than this value, the change of the plasma geometry will not play an important role.

The particle densities are described as follows:

$$N_{a\,1,2} = \left[\frac{\dfrac{c_a\,\varepsilon_{a\,1,2}}{m_a}}{\sum\limits_b \dfrac{c_b\,\varepsilon_{b\,1,2}}{m_b}} \right] N_{\text{tot}1,2}, \tag{10.2}$$

$m_{\text{a,b}}$, mass of an analyte atom; and $N_{\text{tot}1,2}$, total particle density (all species) originating from the specimen in sphere 1, shell 2.

For the ambient gas density in sphere 1 and shell 2 holds:

$$N^{\text{g}}_{a\,1,2} = k^{\text{g}}_{a\,1,2} N^{\text{g}}_{\text{tot}\,1,2}, \tag{10.3}$$

$k^{\text{g}}_{a\,1,2}$, fraction of the density of species a of the ambient gas in sphere 1, shell 2; and $N^{\text{g}}_{\text{tot}1,2}$, total particle density of the ambient gas in sphere 1, shell 2.

For the density of species a of the ambient gas in shell 3, we set:

$$N^{\text{g}}_{a\,3} = k^{\text{g}}_{a\,3}\, N^{\text{g}}_{\text{tot}3}, \tag{10.4}$$

Table 10.1 List of input quantities and their description for the heuristic model

Quantity	Description	Units
c_a	Concentration of the element a in the specimen	kg/kg
r_1	Radius of plasma core 1. The sphere 1 extends from $r = 0$ to r_1. For simplicity the radius r_1 is kept constant, i.e., no dynamics of the plasma geometry is considered	m
r_2	Outer radius of shell 2. The thickness of shell 2 amounts to: r_2–r_1. The radius r_2 is kept constant	m
r_3	Outer radius of shell 3. The thickness of shell 3 amounts to: r_3–r_2. This distance describes the propagation path of the emitted light through the ambient gas between the plasma and the detector of the spectrometer or the spectrometer entrance slit in case of a vacuum spectrometer	m
$\varepsilon_{a\,1,2}$	Weighting factor describing the proportion of an element a in the plasma in relation to its proportion in the specimen. If the laser-induced evaporation process does not depend on the constituents, structure or chemical bonds, then the concentration relations inside the plasma are identical to those of the original specimen. In this case holds: $\varepsilon_{a,1,2} = 1$. The relation $\varepsilon_{a,1,2} < 1$ corresponds to a depletion of the respective element in the plasma and $\varepsilon_{a,1,2} > 1$ to an enrichment	1
$N_{a\,1,2}$	Particle density of species a originating from the specimen in sphere 1 or shell 2. Total density of particles originating from the specimen in sphere 1 or 2: $N_{\text{tot}\,1,2} = \sum_a N_{a\,1,2}$	$1/\text{m}^3$
$N^{g}_{a\,1,2}$	Particle density of species a originating from the ambient gas in sphere 1 or shell 2	$1/\text{m}^3$
$N^{g}_{a\,3}$	Particle density of species a (atoms or molecules) of the ambient gas in shell 3. For simplification, it is assumed that this particle density is stationary	$1/\text{m}^3$
$T_{1,2}$	Temperature in sphere 1 and shell 2. A local temperature equilibrium is assumed	K
T_g	Temperature of the ambient gas assumed to be stationary	K
$\Delta\lambda^a$	Apparative broadening given as full width at half maximum (FWHM). Typical values are in the range of 10–70 pm	pm
$\xi(\lambda)$	Spectral characteristics of the observation channel given, e.g., by the features of a light receiving and transmitting optics such as lenses, mirrors or fiber optics, reflectivity of the grating, and spectral sensitivity of the detector. The function $\xi(\lambda)$ is normalized: $0 < \xi(\lambda) \leq 1$. $\xi(\lambda)$ can be defined in an interval $(\lambda_{\min}, \lambda_{\max})$ by a series of data points $\xi(\lambda_i)$	1
η(order)	Weighting factor between 0 and 1 describing the fraction of diffracted power for a given order of the grating spectrometer. If $\eta(1.) = 1$, then only the first diffraction order is considered; if $\eta(1.) = 0.8$ and $\eta(2.) = 0.2$ then the first order contributes 80% and the second 20% to the observed total spectrum	1

$k^{g}_{a\,3}$, fraction of the density of species a of the ambient gas in shell 3; and $N^{g}_{\text{tot}3}$, total density of the ambient gas in shell 3.

10.2 Calculation of Emission Spectra

The key assumption in what follows is that the plasma is in LTE. The electron and ion densities can thus be determined using a system of Saha equations:

$$\frac{N_e N_a^{z+1}}{N_a^z} = \frac{2}{\Lambda_e^3} \frac{u_a^{z+1}\left(T, \Delta E_{a,z+1}^{\text{ion}}\right)}{u_a^z(T, \Delta E_{a,z}^{\text{ion}})} \exp\left(-\frac{E_{a,z}^{\text{ion}} - \Delta E_{a,z}^{\text{ion}}}{k_B T}\right), \tag{10.5}$$

subject to the constraint of constant total number of heavy particles:

$$N_a = \sum_{z=0}^{n_{z_a}-1} N_a^z, \quad z = 0, \ldots, n_{z_a} - 1 \tag{10.6}$$

and of charge neutrality:

$$N_e = \sum_{a=0}^{n_a-1} \sum_{z=0}^{n_{za}-1} z N_a^z, \tag{10.7}$$

where n_{z_a} is the number of considered ionization states of species a.

The $u_a^z(T, \Delta E_{a,z}^{\text{ion}})$ are the partition functions of the corresponding ions [10.13]. Λ_e is the thermal De Brogli wavelength of the electrons and $\Delta E_{a,z}^{\text{ion}}$ is the reduction of the ionization energy, depending on the plasma density. The population density of ionization stage z in the excited state n is given by:

$$N_{a,n}^z = N_a^z \frac{g_{a,z;n}}{u_a^z\left(T, \Delta E_{a,z}^{\text{ion}}\right)} \exp\left(-\frac{E_{a,n}^z}{k_B T}\right), \tag{10.8}$$

$g_{a,z;n}$ is the degeneracy and $E_{a,n}^z$, the energy of the excited state n. For a given particle density and temperature, the unknowns are the ion densities. The system of equations (10.5) is solved by the Newton method. For molecules, e.g., nitrogen N_2, the dissociation reactions and molecular ionization processes are not considered. In the case of nitrogen, only atomic nitrogen and its ionization stages are taken into account. Detailed calculations of atmospheric plasmas show that for temperatures beyond 8,000 K and pressures of 1 bar more than 90% of the nitrogen molecules are dissociated [10.30].

The variation of the spectral radiance in the plasma is described by the radiation transport equation:

$$\frac{dI(\lambda)}{ds} = \varepsilon(\lambda) - \alpha(\lambda) I(\lambda), \tag{10.9}$$

where $I(\lambda)$ is the spectral radiance with the unit $W/(m^2\,sr\,m)$, ds the incremental propagation path, $\alpha(\lambda)$ the absorption coefficient with the unit 1/m and $\varepsilon(\lambda)$ the emission coefficient. Relation (10.9) is a simplification of the general case of a radiation bundle propagating inside a plasma, which is attenuated not only by absorption but also by scattering of the radiation out of the direction of propagation. The total attenuation is then governed by the extinction coefficient given by the sum of the absorption coefficient and the scattering coefficient: $k(\lambda) = \alpha(\lambda) + \sigma(\lambda)$. In the following, the contribution of the scattering coefficient is neglected and the relation (10.9) is used. The emission coefficient and the absorption coefficient comprise contributions of line transitions (bound–bound), free–free transitions (Bremsstrahlung), as well as free–bound transitions. The emission coefficient is thus given by:

$$\varepsilon(\lambda) = \varepsilon_{\mathrm{l}}(\lambda) + \varepsilon_{\mathrm{ff}}(\lambda) + \varepsilon_{\mathrm{fb}}(\lambda), \tag{10.10}$$

with the line emission coefficient $\varepsilon_{\mathrm{l}}(\lambda)$, the emission coefficient of free–free transitions $\varepsilon_{\mathrm{ff}}(\lambda)$ and the emission coefficient of free–bound transitions $\varepsilon_{\mathrm{fb}}(\lambda)$. The index $\mathrm{d}\Omega$ is not used here for simplicity of notation, cf. Sect. 9.1, (9.1), (9.2) and Sect. 9.2, (9.3). The line emission coefficient itself is a sum of line emission coefficients of all species, ionization states, and all transitions inside the plasma:

$$\varepsilon_{\mathrm{l}}^{\lambda}(\lambda) = \sum_{a,\,z;\,nm} \varepsilon_{a,\,z;nm}^{\lambda}(\lambda), \tag{10.11}$$

where $\varepsilon_{a,\,z;\,nm}^{\lambda}$ is the line emission coefficient of the species a in the charge state z for the transition $n \rightarrow m$. The index λ explicitly shown in (10.11) indicates that this emission coefficient has the unit emitted power per unit volume per wavelength interval.

The line emission is broadened by Doppler and Stark effect. The superposition of these broadening contributions is described by Voigt profiles [cf. Sect. 9.4, (9.23)]. For the quadratic Stark broadening of multielectron systems, the change in the emission frequency is proportional to the square of the local electric field. Stark broadening is taken into account for those lines for which experimental or theoretical data is available [10.16, 10.19–10.26]. Stark broadening of hydrogen has to be considered separately. Due to the degeneration of the hydrogen levels, the existing dipole moment of the hydrogen atoms aligns in the external electric field and the shift of emission frequencies is proportional to the local electric field. The coefficients describing the relation between electron density and Stark width are tabulated in [10.27].

In case of a bound–bound transition for a single line of species a in charge state z (the indices a, z are omitted for simplicity on the right sides of the following equations) holds:

$$\begin{aligned} \varepsilon_{a,\,z;\,nm}(\lambda) &= \frac{1}{4\pi} A_{nm} N_n h \frac{c}{\lambda} \Gamma_{nm}(\lambda), \\ \alpha_{a,\,z;\,nm}(\lambda) &= (B_{mn} N_m - B_{nm} N_n) \frac{h\lambda}{c} \Gamma_{nm}(\lambda), \end{aligned} \tag{10.12}$$

or with:

$$\begin{aligned} B_{nm} &= A_{nm} \frac{\lambda^3}{8\pi h}, \\ g_n B_{nm} &= g_m B_{mn} \end{aligned}$$

$$\alpha_{a,z;nm}(\lambda) = \left(\frac{N_m}{g_m} - \frac{N_n}{g_n}\right) g_n A_{nm} \frac{\lambda^4}{8\pi c} \Gamma_{nm}(\lambda), \quad (10.13)$$

with h, the Planck's constant; c, the vacuum speed of light; $\Gamma_{nm}(\lambda)$, the line profile; A_{nm}, the Einstein's coefficient of spontaneous emission; B_{nm}, the Einstein's coefficient of stimulated emission; and B_{mn}, the Einstein's coefficient of absorption. N_n is the density of the upper level of the line and N_m, the population density of the lower level and g_n and g_m, the degeneracies.

Besides line radiation, the contributions of continuous free–free and free–bound radiation are included. The contribution of free–free transitions (Bremsstrahlung) is in a hydrogen approximation given by [10.27]:

$$\varepsilon_{\text{ff}}(\lambda) = C_{\text{ff}} \sum_a \sum_z z^2 N_{a,z}, \quad (10.14)$$

$$C_{\text{ff}} = \frac{128}{3} \pi^2 \frac{(\alpha a_0)^3 E_{\text{H}}}{\sqrt{3\pi}} \sqrt{\frac{E_{\text{H}}}{k_{\text{B}} T}} \frac{c}{\lambda^2} \exp\left[-\frac{hc}{\lambda k_{\text{B}} T}\right] N_{\text{e}}, \quad (10.15)$$

α is the fine structure constant; a_0, the Bohr radius; and E_{H} is the Rydberg constant. The bound–free (recombination) contribution to the total radiation is given by:

$$\varepsilon_{\text{fb}}(\lambda) = C_{\text{fb}} \sum_{a=0}^{n_a-1} \sum_{z=1}^{n_{za}-1} \sum_{n=n*}^{n*+\Delta n_{\max}} \frac{z^4}{n^3} \exp\left[\frac{z^2 E_{\text{H}}}{n^2 k_{\text{B}} T}\right] N_a^z, \quad (10.16)$$

$$C_{\text{fb}} = C_{\text{ff}} \frac{2}{3} \frac{E_{\text{H}}}{k_{\text{B}} T}. \quad (10.17)$$

The lower summation limit in relation (10.16) for the principal quantum number n is given by:

$$n^* = \sqrt{\frac{z^2 E_{\text{H}} \lambda}{hc}}. \quad (10.18)$$

Instead of summing up over all quantum numbers $n \geq n^*$, the summation is truncated at $\Delta n_{\max} = 30$.

In case of LTE, the absorption coefficients $\alpha\lambda$ can be deduced from Kirchhoff's law:

$$L^{\lambda,\text{d}\Omega} = \frac{\varepsilon(\lambda)}{\alpha(\lambda)}, \quad (10.19)$$

with the Planck function (cf. Sect. 9.2):

$$L^{\lambda,\text{d}\Omega} = \frac{2hc^2}{\lambda^5} \frac{1}{\exp\left[\frac{hc}{\lambda k_{\text{B}} T}\right] - 1}. \quad (10.20)$$

Assuming that the emission coefficient and the absorption coefficient are constant within core 1 or shell 2, the solution of the radiation transport equation (10.9) is:

$$I(\lambda, s) = I(\lambda, s = 0)\exp(-\alpha s) + \frac{\varepsilon(\lambda)}{\alpha(\lambda)}\left[1 - \exp(-\alpha\, s)\right]. \tag{10.21}$$

In the limiting case $\alpha s \ll 1$, the plasma is optically thin and the radiation increases linearly with the path length s (initial value $I(\lambda, s = 0)$ not considered in the following):

$$I(\lambda) = \varepsilon(\lambda)s, \tag{10.22}$$

whereas in case that $\alpha\, s \gg 1$ the plasma is optically thick and the spectral radiance becomes independent of the path length:

$$I(\lambda) = \frac{\varepsilon(\lambda)}{\alpha(\lambda)}, \tag{10.23}$$

which in case of thermodynamic equilibrium results in Kirchhoff's law, cf. (10.19).

Equation (10.21) is evaluated for every volume element of the plasma including reabsorption along the path through the plasma and the contributions of all volume elements are added to get the total irradiance. Besides line radiation, the contributions of continuous free–free and free–bound radiation are included. The necessary atomic line data such as Einstein coefficients, lower and upper energy of the transition levels, and lower and upper statistical weights for that transition are taken from the Kurucz data base [10.14].

Figure 10.4 shows the shell model with the integration paths applied. The radiation transport equation (10.9) is integrated from points 1 to 2 using the emission coefficient and the absorption coefficient of shell 2. The spectral radiance $I(\lambda, s = 0)$ is set to zero at point 1. From points 2 to 3, the radiation transport equation is integrated using the emission coefficient and absorption coefficient of the plasma core 1. The initial radiance is set to the result of the first integration path. Finally, the integration runs from points 3 to 4 using the values of shell 2 again and taking the result of the irradiance from the previous integration path. If the propagation path does not intersect the plasma core, then the integration runs from points 1 to 4. The obtained radiance is multiplied by the cross section $\Delta A = \pi(r_\mathrm{o}^2 - r_\mathrm{i}^2)$, see Fig. 10.4, yielding a spectral radiant intensity [unit: W/(sr m)] considering $r_3 >> r_2,\ r_1$. Between $r = 0$ and $r = r_2$ ten bundles of equal width $\Delta r = r_\mathrm{o} - r_\mathrm{i} = r_2/10$ are taken for the calculation.

Between point P_4 and the entrance of the spectrometer P_5 at r_3, the absorption of the calculated emission spectrum of the plasma in the ambient atmosphere is calculated. In case of argon as ambient gas, the absorption is calculated from the available atomic line data: Einstein coefficients, wavelengths, and statistical weights [10.14]. For the gases NO, N_2, O_2, O_3, and H_2O, the wavelength-dependent absorption coefficient is taken from experimental data [10.29].

The calculated spectral radiant intensity at the distance r_3 cannot directly be used to simulate a measured spectrum. The receiving solid angle (e.g., the slit aperture of

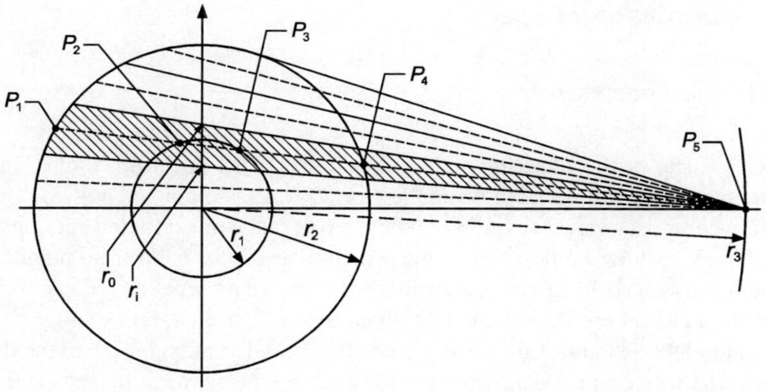

Fig. 10.4 Shell model of the plasma and integration paths of the radiation transport equation, see text. "Copyright 2009, American Institute of Physics [10.17]. Reprinted with permission"

the spectrometer at the distance r_3), the apparative broadening of the spectrometer, and the spectral characteristics have to be taken into account to calculate the simulated spectral radiant flux to be compared with a measured spectrum (assuming $r_3 \gg r_2, r_1$):

$$I_m(\lambda) = \frac{dA_s}{r_3^2}\xi(\lambda)\int_0^\infty f(\lambda, \lambda', \Delta\lambda^a) I_3(\lambda')d\lambda', \tag{10.24}$$

where $I_m(\lambda)$ is the calculated spectral radiant flux, unit: W/m; $I_3(\lambda)$, the calculated spectral radiant intensity at radius r_3, unit: W/(m sr); dA_s, the receiving entrance aperture of the spectrometer, unit: m^2; $\xi(\lambda)$, the spectral characteristics of the observation channel, unit: 1; $f(\lambda, \lambda', \Delta\lambda^a)$, the function describing the apparative broadening, unit: m^{-1}; and $\Delta\lambda^a$ is the full width at half maximum (FWHM) of the apparative broadening.

The number of free parameters of the presented heuristic model is eight (assuming only a single species of the ambient atmosphere and $\varepsilon_{a1,2} = 1$ corresponding to a stoichiometric laser ablation): $N_{tot1,2}$, $N_{a\,1,2g}$, $T_{1,2}$, r_1, and r_2. From the experimental point of view, there are further restrictions assuming that the following integral parameters can be determined experimentally with a moderate effort: the total mass of ablated particles (links r_1, r_2 with $N_{tot1,2}$), the size of the plasma (corresponds to r_2), the average plasma temperature gained with a Boltzmann plot using optically thin lines (links $N_{tot1,2}$, $N_{a\,1,2g}$, r_1, r_2, and $T_{1,2}$). If furthermore a pressure equilibrium with the ambient gas at the detection time of the spectrum (links $N_{tot1,2}$, $N_{a\,1,2g}$, $N_{e1,2}$, and $T_{1,2}$) is assumed in total five additional equations are available; hence, the number of free adjustable parameters can in principle be reduced to three.

The developed software for the heuristic model is called SPES which stands for *sp*ectra *e*valuation and *s*imulation.

10.3 Simulation of Spectra

10.3.1 Fe Spectrum

Figure 10.5 shows the experimental double pulse LIBS spectrum (solid line) of an iron sample in the wavelength range from 531 to 541 nm (cf. Fig. 8.13). This spectrum comprises various iron lines with a broad range of excitation energies from 3.2 to 6.6 eV and the Fe line 538.34 nm which is nearly isolated from neighboring lines and shows a strong Stark broadening with known parameters [10.26, 10.28].

For the calculation, the parameters given in the figure caption of Fig. 10.5 are used. In total 92 Fe I and II lines were taken for the calculation based on the data of [10.14]. The iron particle densities and the shell geometry result in a total Fe mass of 64 ng which is close to the measured ablated mass per burst of about 50–57 ng which can be estimated from published data [10.5]. The particle densities inside the plasma core and the plasma shell result in a total plasma pressure of 1 bar which is to be expected for the time domain $t_{\text{delay}} > 500$ ns after irradiation of the laser burst, where the plasma expansion has slowed down significantly. Nitrogen contributes significantly to this pressure balance. Furthermore, the portion of the electron density originating from nitrogen N II in relation to the portion of Fe II amounts to about 20% in the plasma core.

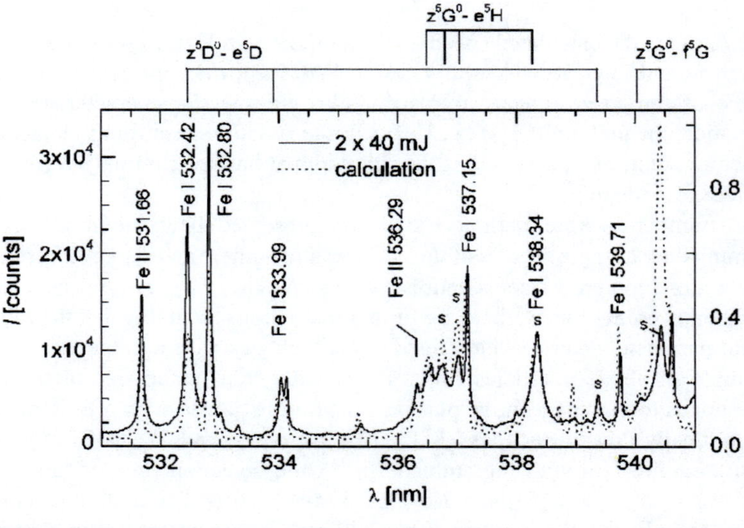

Fig. 10.5 Measured iron spectrum (*solid line*, measurement parameters: see text) and calculated spectrum (*dotted line*) using the following values: $T_1 = 12{,}000\,\text{K}$, $N_{\text{Fe}\,1} = 1.2 \times 10^{17}\,\text{cm}^{-3}$, $N_{\text{N}\,1\,\text{g}} = 3.5 \times 10^{17}\,\text{cm}^{-3}$, $N_{\text{e}\,1} = 1.4 \times 10^{17}\,\text{cm}^{-3}$, $T_2 = 9{,}500\,\text{K}$, $N_{\text{Fe}\,2} = 6 \times 10^{16}\,\text{cm}^{-3}$, $N_{\text{N}2\,\text{g}} = 6.5 \times 10^{17}\,\text{cm}^{-3}$, $N_{\text{e}\,2} = 5.5 \times 10^{16}\,\text{cm}^{-3}$, $r_1 = 1.0\,\text{mm}$, $r_2 = 1.2\,\text{mm}$, $\Delta\lambda^a = 72\,\text{pm}$. For the lines marked with an "S" Stark width broadening parameters are published (cf. Table 10.2, column 3).

For LTE the condition (8.17) should be fulfilled for all energy levels of an atom with a principal quantum number greater or equal n. For the electron densities and temperatures resulting from the simulation of Fig. 10.5 (see figure caption) condition (8.17) is fulfilled for iron atoms and singly ionized iron ions assuming as lowest principal quantum number 3.

The determined spatially integrated electron densities and temperatures gained in experiments, see Sect. 8.2, Table 8.2, are consistent with those of the simulation (see figure caption of Fig. 10.5). The experimental values lie between the core and shell values yielded by the simulation. They are closer to the shell values (temperature, total iron density, iron atom density, nitrogen atom density, and electron density), which is attributed to the fact that the experimentally determined temperature was based on atomic iron lines only. Furthermore, the total number of iron particles of core and shell from the simulation is 6.9×10^{14} which deviates less than 30% from the value 5×10^{14} stated in Table 8.2. For the ratio of the total nitrogen to total iron particle density in the shell, the simulation yields 10.8 to be compared with 8.4 in Table 8.2.

For the line 538.34 nm, the average Stark broadening constant (referring here to the FWHM) $C_s = 0.212 \times 10^{-17}$ nm cm^3 published by Freudenstein et al. was used, which was determined for an average temperature of 9,500 K [10.28]. Of course this is again a simplification since the quantity C_s itself depends on the temperature, see, e.g., Fig. 5 in [10.26], where measured line widths of the Fe I 538.34 nm line are shown for temperatures between 6,000 and 14,000 K with values of C_s spanning the range 0.06×10^{-17} to 0.28×10^{-17} nm cm^3. Table 10.2 shows an overview of Fe I multiplets and corresponding Stark broadening data – as far as available – in the spectral range considered for Fig. 10.5. For the multiplet z^5G^o–e^5H, it is obvious that the Stark widths differ significantly within the multiplet which is in contrast to statements saying that Stark widths should normally be nearly the same for all lines within multiplets (see, e.g., [10.25], p. 823). For the calculation, the Stark width broadening constants shown in column 4 of Table 10.2 are used.

For the first line of the multiplet z^5G^o–e^5H, the average of the values available for the lines 536.747 and 536.996 nm are taken for the calculation. For line 538.337 nm, the value of [10.28] is used, since this data is mostly deployed in LIBS studies and differs only by 5% from the average value calculated on the basis of the data given in [10.26]. For multiplet z^5D^o–e^5D where a Stark shift value is published for the third line, this value is used for the first two lines as well. For the line 540.415 nm, the average Stark width broadening constant is taken in column 3 of Table 10.2 using the values published in [10.24] for the multiplet 540.4 nm. Since there is only a negligible difference to the value for the 540.411 nm line, the same Stark broadening constant of 0.126×10^{-17} nm cm^3 is used for the calculation.

From this discussion, we have to notice that the available Stark data are fragmentary and have significant uncertainties. Hence, any comparison between calculated and experimentally observed spectra has to consider this starting position.

Also a single shell modeling was tested, but in this case it was not possible to achieve relative line intensities between the iron ion lines and atomic lines that are similar to the observed ones.

Table 10.2 List of Fe I lines in the spectral range shown in Fig. 10.5 belonging to multiplets where Stark broadening data is partially available

Multiplet	Fe I line (nm)	Stark width broadening constant (10^{-17} nm cm^3)	Stark width broadening constant used for calculation (10^{-17} nm cm^3)	STARK shift (nm)	Stark shift used for calculation (nm)	Reference; remark
z^5G^o–e^5H	536.486	–	0.132	–	–	[10.14], n.a.
	536.747	0.147	0.147	–	–	[10.26]
	536.996	0.116	0.116	–	–	[10.26]
	538.337	0.203	0.212	–	–	[10.26], ref. [10.28] states 0.212×10^{-17} nm cm^{-3}
z^5D^o–e^5D	532.418	–	0.090	–	0.050	[10.14], n.a.
	533.993	–	0.090	–	0.050	[10.14], n.a.
	539.317	0.090	0.090	0.050	0.050	[10.26]
z^5G^o–f^5G	540.050	–	0.126	–	–	[10.26]
	540.411	0.126	0.126	–	–	[10.26], this line overlaps with 540.415 nm belonging to another multiplet and having nearly the same upper energy level and oscillator strength
z^3G^o–e^3H	540.415	0.127	0.126	–	–	[10.24]

The Stark width broadening constant shown in the third column is the average of the ratios of the Stark width and the electron density measured for different temperatures for a given line taken from reference [10.26]. The Stark shift refers to an electron density of 10^{16} cm^{-3}. Only those data of [10.26] were taken, where an uncertainty classification of grades A, B$^+$ or B is stated corresponding to uncertainties being within 15%, 23% and 30%. *n.a.*, no Stark data available

Figure 10.5 shows that all experimentally observed lines can be reproduced in the calculation. Comparing the measured and calculated line shape of the Fe I 538.34 nm line shows a high degree of overlap. Also a small satellite on the left wing of this line shows up in the calculation being approximately in the right relation to the Fe I 538.34 nm line. The continuum base level of the calculated spectrum is slightly lower than the measured one. The origin of this discrepancy is not yet clear. A potential reason could be light scattering in the spectrometer which could raise the continuous background and is not considered in the calculation so far.

The peak positions of the calculated lines of the multiplet z^5D^o–e^5D coincide approximately with those of the experimental spectrum showing that the assumed

Stark shift is appropriate. If the Stark shift is not taken into account offsets between the calculated and measured peak positions were clearly visible.

The small double peak close to 533.99 nm is reproduced also in the calculation. The relative height of the peaks depends on the chosen radius r_2, a slight change of this parameter by 10% leads to a visible change in the relation of the peak heights. Discrepancies in the peak height and width of the line Fe I 532.42 nm of the multiplet z^5D^o–e^5D occur.

While for multiplet z^5G^o–e^5H the measured and calculated line shapes of the 538.34 nm line coincide well, discrepancies occur for the other three lines of this multiplet. Again this may partly be attributed to different uncertainties given for the respective Stark data, where for the 538.34 nm line an uncertainty of 15% is stated, compared with an uncertainty of 30% for the other three lines.

A significant difference between calculated and measured spectrum is observed for the line 540.411 nm of the z^5G^o–f^5G multiplet. This line overlaps with 540.415 nm which is a Fe I line from another multiplet, for which however no Stark data is available.

Simulations of the shown Fe spectrum were undertaken to reduce these discrepancies by variation of the Stark parameters and Einstein coefficients for the lines 532.4178, 532.8038, 540.4112, 540.4149, and 540.5774 nm. The discrepancies could be reduced significantly; however, this requires variations of Stark parameters and Einstein coefficients being partly greater than 50%. Hence, it is concluded that the found discrepancies may be attributed partly to the uncertainty of the available atomic parameters and partly to the strong simplification of the spherical stationary plasma model chosen.

While the model is able to simulate even detailed structures of the studied Fe spectrum, discrepancies between simulation and experiment remain. Comprehensive and accurate Stark data are a necessary precondition to further improve the simulation capability for such complex multiline spectra.

10.3.2 Al Resonance Line with Self-reversal

Figure 10.6, top, shows a LIBS spectrum of an Al I resonance line with line reversal taken with a laser burst irradiating an aluminum sample and detected with a high-resolution echelle spectrometer with the following experimental parameters: collinear double pulses 2×100 mJ, $\Delta t = 31\ \mu s$, $t_{delay} = 2\ \mu s$, $t_{int} = 5\ \mu s$, air atmosphere at 1,013 mbar, and $\Delta s = 25$ mm. (This spectrum is identical to the one shown in Fig. 9.11.). Fitting the wings of this line with a Lorentz function allows to estimate the shift between the absorption and emission maximum being of the order of 26 pm. This Al line has a Stark red shift. By use of the relation describing the Stark shift as a function of the electron density given in [10.18] [cf. (9.20)], neglecting the ion correction term, the electron density difference between the plasma core and the plasma shell can be estimated.

Table 10.3 shows the available Stark data for this line. With the Stark shift parameter for the Al I 394.401 nm line 0.0242×10^{-17} nm cm^3, we obtain for the

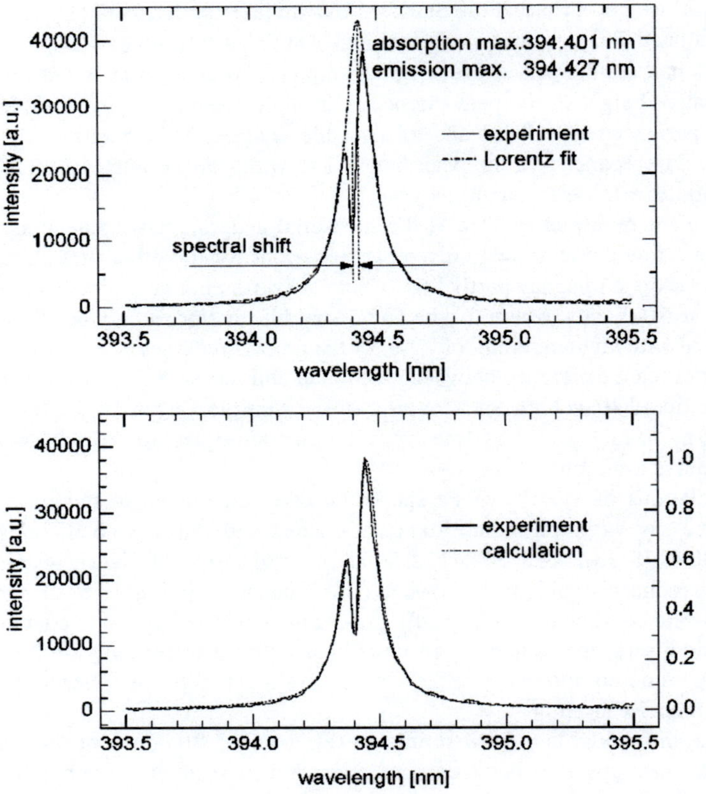

Fig. 10.6 Al I line at 394.4 nm with line reversal. *Top*: Experimental spectrum (*solid line*) and Lorentz fit curve to the wings (*dash-dot line*). *Bottom*: Experimental spectrum and calculated line profile (*dotted line*) using the following values: $T_1 = 10{,}500\,\mathrm{K}$, $N_{\mathrm{Al}1} = 1 \times 10^{17}\,\mathrm{cm}^{-3}$, $N_{\mathrm{N}1\mathrm{g}} = 5.90 \times 10^{17}\,\mathrm{cm}^{-3}$, $N_{\mathrm{e}1} = 1.4 \times 10^{17}\,\mathrm{cm}^{-3}$, $T_2 = 6{,}970\,\mathrm{K}$, $N_{\mathrm{Al}2} = 4.2 \times 10^{16}\,\mathrm{cm}^{-3}$, $N_{\mathrm{N}2\mathrm{g}} = 9.98\times10^{17}\,\mathrm{cm}^{-3}$, $N_{\mathrm{e}2} = 2.2\times10^{16}\,\mathrm{cm}^{-3}$, $r_1 = 1.0\,\mathrm{mm}$, $r_2 = 1.2\,\mathrm{mm}$, $\Delta\lambda^a = 10\,\mathrm{pm}$. "Copyright 2009, American Institute of Physics [10.17]. Reprinted with permission"

electron density difference between core and shell corresponding to a shift of 26 pm: $\Delta N_\mathrm{e} = 1.07 \times 10^{17}\,\mathrm{cm}^{-3}$.

Figure 10.6, bottom, shows a comparison between experimental and calculated line profile showing a good agreement. The simulation yields the electron densities in the core and shell: $N_{\mathrm{e}1} = 1.38 \times 10^{17}\,\mathrm{cm}^{-3}$ and $N_{\mathrm{e}2} = 2.22 \times 10^{16}\,\mathrm{cm}^{-3}$. This corresponds to a difference of $N_{\mathrm{e}1}{-}N_{\mathrm{e}2} = 1.16 \times 10^{17}\,\mathrm{cm}^{-3}$, which is within 10% equal to the result of the estimation given above.

In Sect. 3.5, Fig. 3.15 shows simulated emission spectra in the vacuum ultraviolet spectral range (VUV) calculated with the presented heuristic model demonstrating the strong effect of a small oxygen content on the absorption of emission lines in the VUV. For further details, see Sect. 3.5.

Table 10.3 Stark data of the Al I line 394.401 nm published for different electron densities and temperatures

Multiplet, Al I line 394.401 nm	T(K)	$n_e(10^{17} cm^{-3})$	Stark width broadening constant (10^{-17} nm cm^3)	Stark shift constant (10^{-17} nm cm^3)	Reference; accuracy
^{2}P^o–^{2}S	13,200	1.28	0.0422[a]	0.0242	[10.19]; 30–50%
	9,670	1.42	0.0327	0.0099	[10.19]; 50%
	11,000	0.1	0.0370[a]		[10.19]; 30%
	13,600	4.5	0.0311	0.0207	[10.19]; >50%
	11,700	2.5	0.0336	0.0168	[10.20]; 50%

Only those values marked with "a" having accuracies between 30% and 50% are taken to calculate an average Stark width broadening constant to be used for the simulation: 0.0396×10^{-17} nm cm^3. For the shift the value 0.0242×10^{-17} nm cm^3 is taken for the simulation

For significant improvements of the model with the aim to simulate precisely experimental LIBS spectra much more accurate and precise Stark data are needed.

10.4 Enthalpy of Plasma

The enthalpy of the plasma is calculated using the data gained from the simulation of the iron spectrum. The total enthalpy is the sum of the enthalpies of the plasma core (index 1) and the plasma shell (index 2):

$$H = H_1 + H_2.$$

Considering only the first ionization stage of the particles, the enthalpy of the plasma core is (and in an analogous way for the shell) given by:

$$\begin{aligned} H_1 = {} & N_{\mathrm{Fe0}} V_1 \left(\frac{5}{2} k T_1 + k T_1^2 \frac{\partial}{\partial T_1} \ln u_{\mathrm{Fe0}} \right) \\ & + N_{\mathrm{Fe1}} V_1 \left(\frac{5}{2} k T_1 + E_{\mathrm{Fe0}}^{\mathrm{ion}} + k T_1^2 \frac{\partial}{\partial T_1} \ln u_{\mathrm{Fe1}} \right) + \cdots \\ & \cdots + N_{\mathrm{N0}} V_1 \left(\frac{5}{2} k T_1 + \frac{E_{\mathrm{N_2}}^{\mathrm{D}}}{2} + k T_1^2 \frac{\partial}{\partial T_1} \ln u_{\mathrm{N0}} \right) \\ & + N_{\mathrm{N1}} V_1 \left(\frac{5}{2} k T_1 + \frac{E_{\mathrm{N_2}}^{\mathrm{D}}}{2} + E_{\mathrm{N0}}^{\mathrm{ion}} + k T_1^2 \frac{\partial}{\partial T_1} \ln u_{\mathrm{N1}} \right) + \frac{5}{2} N_{\mathrm{e1}} V_1 k T_1, \end{aligned} \tag{10.25}$$

where N_{Fe0}, N_{Fe1}, N_{N0}, N_{N1} are the densities for iron atoms, iron ions with $z = 1$, nitrogen atoms and nitrogen ions with $z = 1$, V_1 is the volume of the plasma core,

$E_{\mathrm{Fe0}}^{\mathrm{ion}}$, $E_{\mathrm{N0}}^{\mathrm{ion}}$ the ionization energies for iron and nitrogen atoms, $E_{\mathrm{N_2}}^{\mathrm{D}}$ the dissociation energy of nitrogen molecules and N_{e1} the electron density in the core. In relation (10.25), the enthalpy of molecular nitrogen is neglected. For a pressure of 1 bar and a temperature of 9,500 K as yielded in the simulation of the iron spectrum for the shell (cf. Fig. 10.5), calculations show that more than 99% of all nitrogen molecules are dissociated [10.30]; hence, the energy content of molecular nitrogen in the plasma can be neglected. The enthalpy of the plasma shell is calculated in an analogous way.

Taking the values of the simulation of the iron spectrum from Fig. 10.5 and evaluating the relation (10.25) by use of parabolic fit functions for the partition functions yields: $H_1 = 3.3\,\mathrm{mJ}$ and $H_2 = 2.7\,\mathrm{mJ}$. Referring the total enthalpy to the process energy irradiated initially, i.e., the burst energy of $2 \times 40\,\mathrm{mJ}$ used in the experiment to generate the iron spectrum, yields a percentage of 7.5%. Hence, $0.8\,\mu\mathrm{s}$ after the last laser pulse, the plasma enthalpy still comprises 7.5% of the total energy irradiated. In Sect. 19.1, the necessary energy to displace the ambient atmosphere is estimated.

References

10.1. R. Noll, R. Sattmann, V. Sturm, S. Winkelmann, Space- and time-resolved dynamics of plasmas generated by laser double pulses interacting with metallic samples. J. Anal. At. Spectrom. **19**, 419–428 (2004)

10.2. Z. Hwang, Y. Teng, K. Li, J. Sneddon, Interaction of a laser beam with metals Part I: quantitative studies of plasma emission. Appl. Spectrosc. **45**, 435–441 (1991)

10.3. S. Harilal, C. Bindhu, C. Riju, V. Nampoor, C. Vallabhan, Electron density and temperature measurements in a laser produced carbon plasma. J. Appl. Phys. **82**, 2140–2146 (1997)

10.4. Y. Iida, Effects of atmosphere on laser vaporization and excitation processes of solid samples. Spectrochim. Acta **45B**, 1353–1367 (1990)

10.5. L. Peter, R. Noll, Material ablation and plasma state for single and collinear double pulses interacting with iron samples at ambient gas pressures below 1 bar. Appl. Phys. B **86**, 59–167 (2007)

10.6. J. Aguilera, C. Aragon, F. Penalba, Plasma shielding effect in laser ablation of metallic samples and its influence on LIBS analysis. Appl. Surface Sci. **127–129**, 309–314 (1998)

10.7. M. Aden, E. Beyer, G. Herziger, H. Kunze, Laser-induced vaporization of a metal surface. J. Phys. D: Appl. Phys. **25**, 57–65 (1992)

10.8. A. Bogaerts, Z. Chen, D. Autrique, Double pulse laser ablation and laser-induced breakdown spectroscopy: a modeling investigation. Spectrochim. Acta B **63**, 746–754 (2008)

10.9. I. Gornushkin, C. Stevenson, B. Smith, N. Omenetto, J. Winefordner, Modeling an inhomogeneous optically thick laser induced plasma: a simplified theoretical approach. Spectrochim. Acta B **56**, 1769–1785 (2001)

10.10. I. Gornushkin, A. Kazakov, N. Omenetto, B. Smith, J. Winefordner, Radiation dynamics of post-breakdown laser induced plasma. Spectrochim. Acta B **59**, 401–418 (2004)

10.11. I. Gornushkin, A. Kazakov, N. Omenetto, B. Smith, J. Winefordner, Experimental verification of a radiative model of laser-induced plasma expanding into vacuum. Spectrochim. Acta B **60**, 215–230 (2005)

10.12. A. Kazakov, I. Gornushkin, N. Omenetto, B. Smith, J. Winefordner, Radiative model of post-breakdown laser-induced plasma expanding into ambient gas. Appl. Optics **45**, 2810–2820 (2006)

10.13. H.-W. Drawin, P. Felenbok, *Data for Plasmas in Local Thermodynamic Equilibrium* (Gauthier-Villars, Paris, 1965)
10.14. R. Kurucz, B. Bell, *Atomic Line Data, Kurucz CD-ROM No. 23* (Smithsonian Astrophysical Observatory, Cambridge, 1995). www.cfa.harvard.edu
10.15. J. Hermann, C. Boulmer-Leborgne, D. Hong, Diagnostics of the early phase of an ultraviolet laser induced plasma by spectral line analysis considering self-absorption. J. Appl. Phys. **83**, 691–696 (1998)
10.16. C. Colón, G. Hatem, E. Verdugo, P. Ruiz, J. Campos, Measurement of the Stark broadening and shift parameters for several ultraviolet lines of singly ionized aluminum. J. Appl. Phys. **73**, 4752–4758 (1993)
10.17. R. Wester, R. Noll, Heuristic modeling of spectral plasma emission for laser-induced breakdown spectroscopy. J. Appl. Phys. **106**, 123302-1–123302-10(2009)
10.18. G. Bekefi, *Principles of Laser Plasmas* (Wiley, New York, 1976)
10.19. N. Konjevic, M. Dimitrijevic, W. Wiese, Experimental Stark widths and shifts for spectral lines of neutral atoms. J. Phys. Chem. Ref. Data **13**, 619–647 (1984)
10.20. N. Konjevic, W. Wiese, Experimental Stark widths and shifts for spectral lines of neutral and ionized atoms. J. Phys. Chem. Ref. Data **19**, 1307–1385 (1990)
10.21. H. Griem, *Spectral Line Broadening by Plasmas* (Academic, New York, 1974), Appendix IV, p. 320
10.22. J. Aguilera, J. Bengoechea, C. Aragon, Curves of growth of spectral lines emitted by a laser-induced plasma: influence of the temporal evolution and spatial inhomogeneity of the plasma. Spectrochim. Acta B **58**, 221–237 (2003)
10.23. A. Bartecka, T. Wujec, J. Halenka, J. Musielok, Experimental Stark-broadening studies of the N I multiplet (1D) 3s2D – (1D) 3p2P0 at 7905.5 Å. Eur. Phys. J. D **29**, 265–271 (2004)
10.24. A. Lesage, J. Lebrun, J. Richou, Temperature dependence of Stark parameters for Fe I lines. Astrophys. J. **360**, 737–740 (1990)
10.25. V. Bakshi, R. Kearney, Measurements of Stark widths of some Ar I transitions in a Dc Argon plasma jet at atmospheric pressure. J. Quant. Spectrosc. Radiat. Transfer **42**, 405–413 (1989)
10.26. N. Konjevic, A. Lesage, J. Fuhr, W. Wiese, Experimental Stark widths and shifts for spectral lines of neutral and ionized atoms. J. Phys. Chem. Ref. Data **31**, 819–927 (2002)
10.27. H. Griem, *Plasma Spectroscopy* (McGraw-Hill, New York, 1964)
10.28. S. Freudenstein, J. Cooper, Stark broadening of Fe I 5383 Å. Astron. Astrophys. **71**, 283–288 (1979)
10.29. J. Pecker, *Space Observatories* (D. Reidel Publishing, Dordrecht, 1970)
10.30. J. Bacri, S. Raffanel, Calculation of some thermodynamic properties of air plasmas: internal partition functions, plasma composition, and thermodynamic functions. Plasma Chem. Plasma Process. **7**, 53–87 (1987)

Chapter 11
Quantitative Analysis

Quantitative analysis with LIBS is defined here as the determination of concentrations of chemical elements in a specimen. A more generic definition includes as well the application case of material identification where a decision is taken in terms of an allocation of a specimen to a material class by evaluating whether a set of determined concentrations falls into or is close to a set of predefined concentration intervals or is smaller or greater than predefined concentration thresholds or not.

This chapter is focused on quantitative analysis based on calibration. Calibration-free LIBS approaches were studied, but the achieved analytical performance in terms of accuracy levels and correctness of the determined concentrations - especially in the case of trace detection - are still significantly worse than those obtained with a quantitative LIBS analysis based on calibration [11.1].

The principle steps for a quantitative analysis with LIBS are shown in Fig. 11.1, which are described in more detail in the following sections.

11.1 Measuring Method

First of all, the measuring method has to be defined comprising the setup, process parameters, and evaluation procedure. The setup in turn is described by the instrumental components applied, cf. Chap. 4. Their respective settings belong to the process parameters of the measuring method, cf. Chap. 3. Table 11.1 shows a selection of the most important process parameters to be defined and documented.

Furthermore, the sample taking, sample preparation, and sample surface condition have to be defined.

The measured quantity is a line intensity of an analyte line i described by:

$$I_i = \iint\limits_{t_{\text{delay}},\, t_{\text{int}},\, \lambda_i} I^{\lambda}(\lambda, t)\,\mathrm{d}\lambda\,\mathrm{d}t, \tag{11.1}$$

where $I^{\lambda}(\lambda, t)$ is the spectral radiant flux detected by the spectrometer.

R. Noll, *Laser-Induced Breakdown Spectroscopy*, DOI 10.1007/978-3-642-20668-9_11,

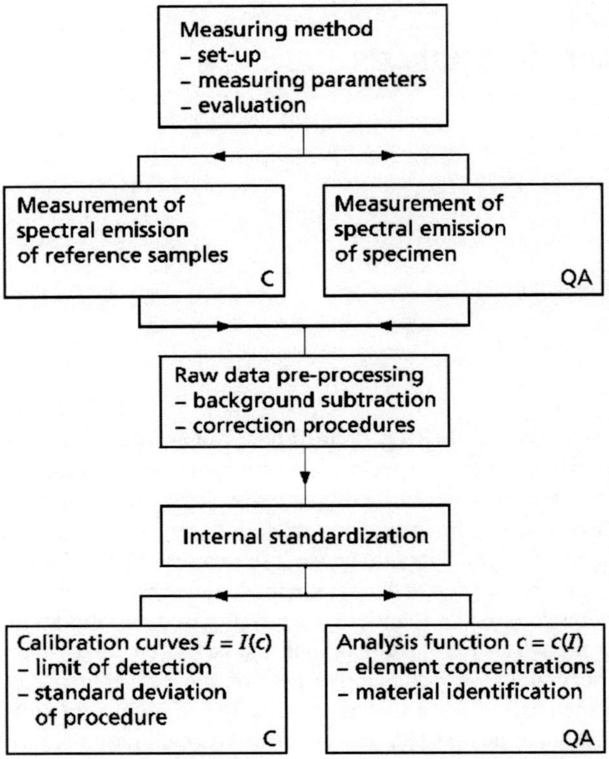

Fig. 11.1 Block diagram of the various steps for a quantitative LIBS analysis. The blocks marked with the letter "C" refer to the calibration procedure and those with "QA" to the quantitative analysis

Generally, the quantity $I^{\lambda}(\lambda, t)$ has to be corrected first to take into account, e.g., the spectral characteristics of the detector, continuum background radiation, and line interferences. The integral (11.1) runs over the analyte line i at the wavelength λ_i and the time interval $[t_{\text{delay}}, t_{\text{delay}} + t_{\text{int}}]$. The line emission depends on the line emission coefficient [cf. Sect. 9.2, relation (9.3)], which is in turn a function of the density of species and the temperature. The line emission coefficient is a local quantity and the radiation transport equation has to be solved [cf. Sect. 10.2, (10.9)] to determine the spectral radiant flux. The density of species in the laser-induced plasma may vary from laser pulse to laser pulse due to changing measuring conditions, e.g., the ambient atmosphere, the microstructure of the sample surface (roughness, crater formation), varying laser pulse energies, etc. Hence, for quantitative LIBS measurements, in general, line ratios are used for calibration, e.g.:

$$Q^{i,r} = \frac{I_{ia}}{I_{rb}}, \tag{11.2}$$

Table 11.1 Selection of process parameters to be defined for quantitative LIBS analysis

Process parameter	Symbol	Definition	Remarks
Laser wavelength	λ		Unit: m
Repetition frequency	ν_{rep}	The number of laser pulses or bursts per time interval	cf. Fig. 3.1 and Table 4.1. The gas exchange time has to be chosen according to relation (3.1). Unit: Hz
Beam propagation ratio	M^2	cf. Sect. 3.4	The beam propagation ratio is a measure of the beam quality. The minimum value is 1. ISO 111146–1 describes measurement arrangement and test equipment to determine M^2
Focusing optics	f	Focal length	
Aperture of focusing optics	D	Diameter of the illuminated aperture of the focusing lens	The quantities f, D, M^2, and τ_{L} determine the irradiance at the beam waist, see Sect. 3.4
Burst mode	SP, DP	SP = single pulse, DP = double pulse, cf. Chap. 6	For double pulses, the configuration has to be given: collinear, orthogonal preablation, orthogonal re-heating, dual pulse crossed beam, cf. Sect. 6.1
Burst energy	E_{b}	Total energy of a burst, i.e., the sum of the pulse energies within the burst: $E_{\text{b}} = \sum_j E_j$, see Fig. 3.9	Average and standard deviation of the burst energy should be measured. Unit: J
Pulse energy	E_j	Energy of the j-th pulse within the burst, see Fig. 3.9	Unit: J
Pulse duration	$\tau_{\text{L}}, \tau_{\text{L}_j}$	Pulse duration of a laser pulse given as FWHM. If bursts are applied τ_{L_j} denotes the pulse duration of the j-th pulse within the burst, see Fig. 3.9	Unit: s
Interpulse separation	Δt	Time between the two pulses of a double pulse (from maximum to maximum). If bursts with more than two pulses are used, an index is introduced so that Δt_j denotes the interpulse separation between the j and $j+1$ pulse, see Fig. 3.9	Unit: s

(continued)

Table 11.1 (continued)

Process parameter	Symbol	Definition	Remarks
Warming-up pulses	N_{wu}	Number of laser pulses generated to stabilize the thermal state of the laser. The laser pulses are blocked outside the laser by a shutter or beam dump and do not irradiate the sample	The warming-up pulses assure a sufficient degree of thermal stability of the laser operation with respect to beam profile, pulse energy, pointing stability
Prepulses	N_{pp}	Number of laser pulses irradiating the sample leading to a local ablation to remove surface contaminations or surface layers which are not representative for the bulk composition of the sample. The spectral emission of the plasmas generated with the prepulses is not used for the analysis	The spectral emission of the plasmas generated with the prepulses can be utilized to check whether a nonrepresentative surface layer is penetrated, cf. Sect. 13.1.2. If multiple pulses are deployed the analogous term is prebursts.
Measuring pulses	N_{mp}	Number of laser pulses irradiating the sample to generate plasmas, whose spectral emission is used to determine the composition of the sample	
Beam waist position	Δs	Position of the beam waist of the focused laser beam relative to the surface of the sample. $\Delta s > 0$ corresponds to a beam waist position inside the sample, see Fig. 2.2	In the description of the experimental setup, the procedure to determine the beam waist position shall be described. Unit: m
Delay time	t_{delay}	Time delay between the laser pulse and the start of the integration of the spectroscopic signal. This delay time denotes the true temporal delay considering pulse propagation times and electronic signal transit times. If bursts with two pulses and more pulses are used, an index is introduced so that $t_{delay\,j}$ denotes the delay time after the j-th pulse, see Fig. 2.4	Several delay times for a set of integration windows for one burst are used, e.g., for the experiments described in Sects. 13.1.3 and 16.2. Unit: s

(continued)

Table 11.1 (continued)

Process parameter	Symbol	Definition	Remarks
Integration time gate	t_{int}	Width of time gate for the integration of the spectral signal. The time of integration starts at the end of t_{delay}. If bursts with two and more pulses are used, an index is introduced so that $t_{\text{int}j}$ denotes the integration time after the j-th pulse, see Fig. 2.4	Unit: s
Gas pressure	p_g	The gas type and the ambient gas pressure or the gas pressure inside the measurement chamber should be given	Unit: bar
Flow rate	$\dot{V}_{\text{mc}}$	Gas flow rate in the measuring chamber	A gas flow in the interaction region assures that ablated or recondensated particulates are removed before the subsequent pulse irradiates the sample, cf. Sect. 3.1. Unit: m^3/s

where I_{ia} is the line intensity of analyte line i of species a and I_{rb} is the line intensity of reference line r of species b.

This ratio is called also internal standardization. The reference line r is usually an emission line of a dominant matrix element, e.g., iron for a steel sample. In the case of an optically thin and homogeneous plasma in a nonabsorbing ambient atmosphere, the measured line intensity is proportional to the line emission coefficient [cf. (9.3)]. Hence for the ratio $Q^{i,r}$ holds:

$$Q^{i,r} \propto \frac{N_a^z(T)}{N_b^z(T)} \frac{u_b^z(T)}{u_a^z(T)} e^{-\frac{E_i - E_r}{kT}}, \tag{11.3}$$

where $N_a^z(T)$, $N_b^z(T)$ is the density of species a, b in charge state z, $u_a^z(T)$, $u_b^z(T)$ are the partition functions of species a, b with charge z, and E_i, E_r is the energy of the upper level of line i, r.

Relation (11.3) shows that the ratio $Q^{i,r}$ depends on the temperature T of the plasma. In case of local temperature equilibrium, the temperature dependence of the densities is determined by a system of Saha equations, see Chap. 10, relation (10.5). The temperature influence on the last term of (11.3) can be reduced by choosing emission lines with a small difference in their upper energy levels. Such homologous line pairs are less sensitive to temperature variations of the laser-induced plasma.

In general, a series of laser pulses is irradiated onto one location of the specimen and averages are taken over these N_{mp} measuring pulses for a single location k on the sample after N_{wu} warming-up pulses and N_{pp} pre-pulses:

$$\left\langle Q^{i,r}_{m,j}(k) \right\rangle_{N_{wu},N_{pp},N_{mp}}, \tag{11.4}$$

where the index m denotes the m-th laser burst, and index j the pulse j within that laser burst. The average is taken from $m = 1$ to $m = N_{mp}$. If K locations are measured, then the following average is taken:

$$\left\langle \left\langle Q^{i,r}_{m,j}(k) \right\rangle_{N_{wu},N_{pp},N_{mp}} \right\rangle_{K}. \tag{11.5}$$

The average ratios (11.4) or (11.5) are used as measuring quantities for the calibration.

11.2 Calibration

Calibration is a standard topic for chemical analysis and well-established procedures exist which are described in the literature as well as in national and international standards [11.1, 11.2, 11.3]. These procedures can be deployed for LIBS as well. Hence, in the following paragraphs, the description is restricted to a short overview of the basic concepts and characteristic numbers used to characterize the performance of a calibration.

A set of reference samples with known composition is measured applying the defined LIBS measuring method (cf. Sect. 11.1). For a reference sample i having a known analyte concentration x_i the measurand – e.g., (11.4) or (11.5) – is determined, which is for simplicity denoted as y_i in the following [11.3]. Based on the obtained set of pairs of variates (x_i, y_i), a calibration curve is determined by regression analysis, examples for calibration functions describing the functional relationship between the response variable and the analyte amount are:

$$\hat{y} = a_0 + a_1\,x, \tag{11.6}$$

$$\hat{y} = a_0 + a_1\,x + a_2\,x^2, \tag{11.7}$$

$$\hat{y} = \frac{1}{2a_2}\left(\sqrt{4a_2(x-a_0)+a_1^2} - a_1\right), \tag{11.8}$$

where the coefficients a_0–a_2 are determined by the regression analysis. The type of calibration function to be chosen can be decided by statistical means, e.g., the Mandel test [11.4].

For a quantitative assessment of a calibration function, the working range has to be defined firstly, i.e., the lowest and highest concentration of the analyte considered. This concentration range has to be represented by a set of reference samples. If necessary, this concentration range can be further reduced by means of statistical methods or visual inspection of the calibration function to further improve the features of the calibration in sections.

Common assessment criteria for a calibration function are: (a) sensitivity, (b) standard deviation of procedure, (c) limit of detection, identification limit, and quantification limit.

The sensitivity is defined as the slope of the calibration function in the middle of the working range [11.5]:

$$\overline{y'} = f'(\bar{x}) = \left.\frac{\partial f}{\partial x}\right|_{x=\bar{x}}, \tag{11.9}$$

where

$$\bar{x} = \frac{1}{N}\sum_{k=1}^{N} x_k \tag{11.10}$$

is the average concentration of the N reference samples used to represent the working range. The average sensitivity $\overline{y'}$ can be easily calculated for the functions (11.6)–(11.8).

The standard deviation of procedure is defined as follows:

$$s_{x0} = \frac{s_y}{\overline{y'}}. \tag{11.11}$$

The quantity s_y is the residual standard deviation, in case of a linear calibration function, cf. (11.6) s_y is given by:

$$s_y = \sqrt{\frac{\sum_{k=1}^{N} [y_k - \hat{y}(x_k)]^2}{N-2}}. \tag{11.12}$$

For the calibration functions (11.7) and (11.8), the denominator in the radicand has to be set to $(N-3)$. The relative standard deviation of procedure is:

$$V_{x0} = \frac{s_{x0}}{\bar{x}} \times 100\%. \tag{11.13}$$

Figure 11.2 shows an example of a LIBS calibration to determine Si in an aluminum matrix where the calibration function is a root function. The coefficients of the calibration function, the standard deviation of procedure, and the calibration range are given in the diagram.

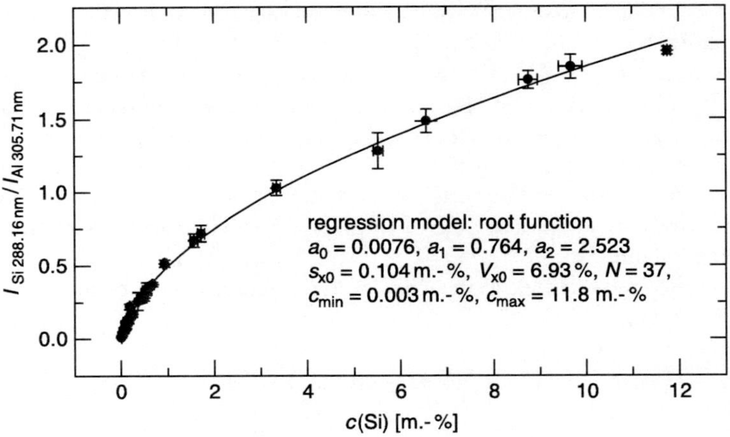

Fig. 11.2 Example of a calibration function for silicon in aluminum, for details see text

For the detection of traces, the limit of detection is a characteristic parameter often used in quantitative LIBS studies. The criterion for the detection of an analyte is that the measurand is greater than a critical value given by the sum of the average blank value of the measurand and the uncertainty of the blank value. The decision "detected" or "not detected" is taken on the y-axis only. The uncertainty of the blank measurement can be determined in two ways: (a) direct method, and (b) indirect method via the calibration straight line.

For method (a), the uncertainty is determined by a multiple measurement ($n \geq 10$) of the blank sample. The measured values have to be checked carefully to assess whether they fulfill the requirement of a normal distribution. If no blank sample is available, then method (b) has to be applied. In this case, the uncertainty of the blank value is determined in an indirect way by extrapolation of regression data of a calibration. This calibration is carried out in the proximity of the limit of detection, i.e., in the analyte concentration range from zero to about ten times the limit of detection.

The following simplifying assumptions are taken: there is no significant difference between the range of blank values and the range of measuring values of the sample containing the analyte; the measuring values of the blank sample follow approximately a normal distribution. The critical value is given by:

$$y_c = \bar{y}_b + s_b\, t_{f,\alpha} \sqrt{1 + \frac{1}{n}}, \tag{11.14}$$

where $\bar{y}_b$ is the arithmetic average of n measurements of the blank sample, s_b is the standard deviation of the measuring values of the blank sample, $t_{f,\alpha}$ is the t-value of a one-sided test, f is the degrees of freedom, and α is the level of significance corresponding to the probability of a type I error (the null hypothesis "the sample is a blank sample" is rejected when the null hypothesis is true, so called α-error or false positive).

The limit of detection (LOD) for a linear calibration function [cf. (11.6)] is then:

$$x_{\mathrm{LOD}} = \frac{y_{\mathrm{c}} - \bar{y}_{\mathrm{b}}}{a_1} = \frac{s_{\mathrm{b}}}{a_1} t_{f,\alpha} \sqrt{1 + \frac{1}{n}}. \tag{11.15}$$

For a level of significance of $\alpha = 0.01$ and $n = 10$ holds:

$$x_{\mathrm{LOD}} = 2.959 \times \frac{s_{\mathrm{b}}}{a_1} \approx \frac{3 s_{\mathrm{b}}}{a_1}. \tag{11.16}$$

The limit of detection determined in this way will be denoted with LOD_{3s}. The limit of detection should not be mixed-up with the smallest concentration that can be determined reliably. This limit of identification (sometimes also called detection limit) is the smallest concentration where the measurand y fulfills $y \geq y_{\mathrm{c}}$ with a probability $(1 - \beta)$, where β is the risk of error to interpret the signal as a blank value (type II error, false negative). For the limit of identification holds:

$$x_{\mathrm{ID}} = \left(1 + \frac{t_{f,\beta}}{t_{f,\alpha}}\right) x_{\mathrm{LOD}}. \tag{11.17}$$

For $\alpha = \beta$ relation (11.17) yields: $x_{\mathrm{ID}} = 2x_{\mathrm{LOD}}$. The limit of quantification (LOQ) describes an analyte concentration that can be determined with a given uncertainty. The LOQ can be estimated as follows:

$$x_{\mathrm{LOQ}} \approx k\, \phi_{n;\alpha} \frac{s_{\mathrm{b}}}{a_1}, \tag{11.18}$$

where $1/k$ is the relative measuring uncertainty, $k = 3$ corresponds to 33.33% uncertainty, $\phi_{n;\alpha}$ tabulated value, see, e.g., [11.3], for $n = 10$ and $\alpha = 0.01$: $\phi_{10;0.01} = 3.0$.

For the given example, we have: $x_{\mathrm{LOQ}} = 3x_{\mathrm{LOD}}$.

Figure 11.3a illustrates graphically the determination of the limit of detection x_{LOD}. A further quantity used to describe calibration curves is the background equivalent concentration (BEC) shown in Fig. 11.3b. For a linear calibration function [cf. (11.6)], the BEC is simply defined by:

$$\mathrm{BEC} = \frac{a_0}{a_1}. \tag{11.19}$$

With (11.16) and (11.19), the limit of detection can be expressed as follows:

$$x_{\mathrm{LOD}} = 3 \times \mathrm{BEC} \times \mathrm{RSD}_{\mathrm{b}}, \tag{11.20}$$

where $\mathrm{RSD}_{\mathrm{b}} = s_{\mathrm{b}}/a_0$ is relative standard deviation of the measurand for the blank sample.

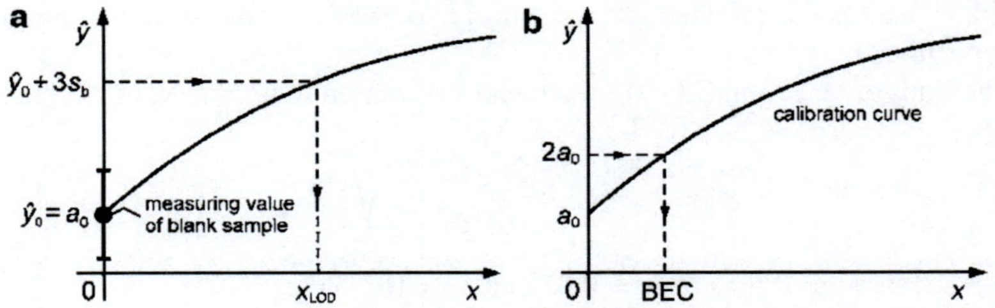

Fig. 11.3 Graphical illustration of (**a**) the detection limit x_{LOD} and (**b**) the background equivalent concentration BEC

11.3 Measurement and Determination of Analytes

The quantitative measurement is conducted according to the defined measuring method and the selected process parameters, cf. Sect. 11.1.

For a quantitative measurement, the inverse functions of the calibration functions [see (11.6)–(11.8)] are determined. The detected radiant flux $I^{\lambda}(\lambda, t)$ has to be corrected first to take into account effects such as continuum background radiation and line interferences. Then the line intensity is calculated by use of (11.1). With internal standardization and averaging procedures [cf. (11.2)–(11.5)], the input quantities for the analysis function are obtained. Finally, the analyte concentrations are determined.

The quality of an analysis can be assessed by the mean residual defined as follows:

$$R(i) = \sqrt{\frac{\sum\limits_{s=1}^{N} \left(c_{i,\mathrm{reference},s} - c_{i,\mathrm{measured},s}\right)^2}{N}}, \tag{11.21}$$

where i is the element, s is the sample, N is the total number of measured samples, $c_{i,\mathrm{measured},s}$ and $c_{i,\mathrm{reference},s}$ are the concentrations of element i and sample s determined with LIBS and the reference concentrations. The precision is defined as the mean relative standard deviations of the intensity ratios of, e.g., three replicate measurements:

$$\mathrm{RSD}(\langle Q^{i,r}\rangle) = \frac{\sum\limits_{s=1}^{N} \mathrm{RSD}\left\langle Q^{i,r}\right\rangle_s}{N}. \tag{11.22}$$

11.4 Recalibration

The term recalibration is used here for a correction procedure of potential drifts of a LIBS instrumentation used for quantitative measurements. Such an instrumental drift can, e.g., be caused by a progressing contamination of optical windows in the laser or measuring radiation beam path or temperature influences on instrumental components. The concept of recalibration is illustrated with the help of Fig. 11.4.

The aim of recalibration is to reduce the effort of a complete calibration procedure after a certain time of operation of a LIBS equipment which may be necessary due to instrumental drifts. Two samples are selected from the set of calibration samples representing analyte concentrations at the lower end of the working range and at the upper end to consider the sensitivity of the instrument for the recalibration. The intensity ratios measured during calibration for these two samples are documented and compared with those obtained with these two samples at a later time. For an effective conception of a recalibration procedure, the selection of recalibration samples should be taken in such a way to minimize the total number of samples for the correction of all analyte intensity ratios. Comparing the intensity ratios for an analyte line between calibration and a later recalibration action – cf. Fig. 11.4 right – yields a linear regression by use of which the new intensity ratios can be corrected to the intensity ratio scale of the original calibration. These corrected ratios are then the input quantities for the analysis functions.

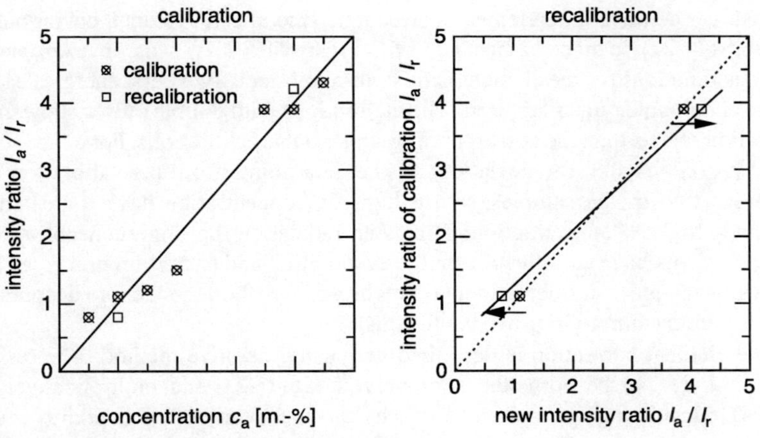

Fig. 11.4 *Left*: Line intensity ratio of an analyte line to a reference line as a function of the concentration. The *straight line* indicates the calibration function. *Right*: Line intensity ratio of the calibration as a function of the new line intensity ratio. The *arrows* indicate the shift of the ratios in relation to the original values obtained at the time of the calibration due to a drift of the instrument. The *dashed line* describes this relation if there is no drift; the *solid line* is used to correct the intensity ratios

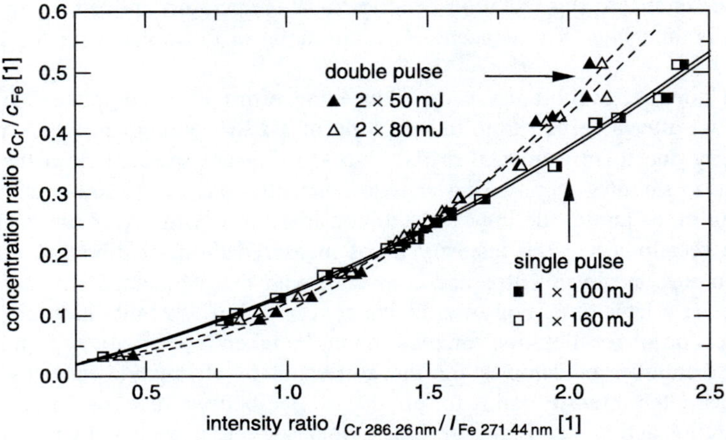

Fig. 11.5 Analysis functions for chromium for single (*squares*) and double pulses (*triangle*)

11.5 Interelement Correction

By use of iterative interelement corrections, the mean residual deviations [cf. relation (11.21)] can be reduced [11.6]. Figure 11.5 shows as an example four analysis functions for the element Cr in high-alloy steel. Two burst energies each for single and double pulses are used. The calibrations with double pulses show greater saturation effects than the corresponding single pulse calibrations. For concentration ratios $c_{\mathrm{Cr}}/c_{\mathrm{Fe}} > 0.3$, the deviations of the data points from the calibration curve increase. Two of the samples with a high Cr concentration have for example a relatively high Ni concentration. This is an indication that high concentrations of other elements have an influence on the evaporation and excitation processes in the laser-induced plasma. Interelement correction is a method to take into account such effects with the aim to improve the residuals.

Interelement correction is described here as an iterative method. The residuals (11.21) $R(i)$ can be corrected additively, see (11.23), and multiplicatively, see (11.24), considering spectral interferences and matrix effects. The additive factors correct the superposition of line intensities. The multiplicative factors take into account different excitation conditions in the plasma consisting of species ablated from the high-alloy steel samples. The following equations describe the analysis functions for concentrations ratios:

$$\hat{c}_i = b_{0,i} + b_{1,i}\, Q_i + b_{2,i}\, Q_i^2 + \sum_{j,j\neq i} k_{\mathrm{add},j}\, \hat{c}_{\mathrm{corr},j}\,, \tag{11.23}$$

$$\hat{c}_i = \left(b_{0,i} + b_{1,i}\, Q_i + b_{2,i}\, Q_i^2\right) \prod_{j,j\neq i} \left(1 + k_{\mathrm{mult},j}\, \hat{c}_{\mathrm{corr},j}\right), \tag{11.24}$$

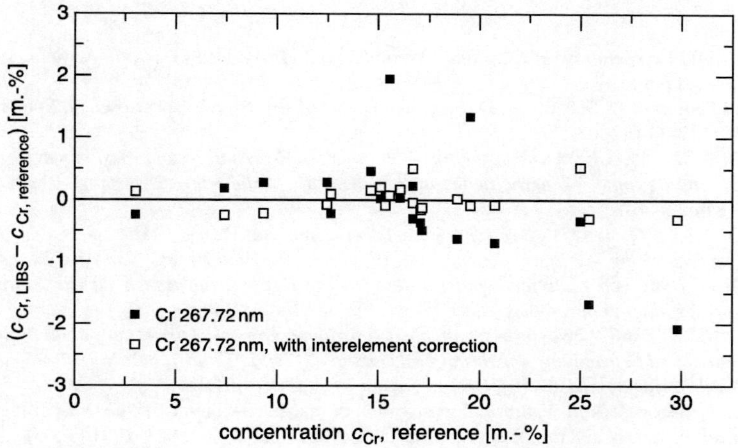

Fig. 11.6 Differences of Cr concentrations measured with LIBS to the reference concentrations as a function of the Cr concentrations with and without interelement correction. Single pulses, $E_b = 160\,\text{mJ}$

where $\hat{c}_{\text{corr},j}$ is the concentration ratio of element j to correct the residuals additively or multiplicatively, Q_i is the intensity ratio of I_i to I_r [cf. (11.2)], $k_{\text{add},j}$ is an additive correction factor for the element j, and $k_{\text{mult},j}$ is a multiplicative correction factor for the element j.

For example, the correction for Cr can be started multiplicatively with nickel. Thereby a regression of the Cr residuals of all calibrated samples is made with the corresponding Ni concentration ratios. The aim is to get a correction factor to calculate the residuals. Then the reduced residuals are taken for making a regression with the measured concentrations of another element considered as a disturbing factor. These corrections are continued with other elements until no further improvement is achieved. Then the absolute Cr concentrations can be calculated with the elemental concentration of the analysis functions of those elements which are contained in the samples.

Figure 11.6 shows the differences of the measured Cr concentration to the reference concentrations versus the reference concentration, where the filled squares represent uncorrected and open squares the corrected values.

The Cr 267.72 nm residual is reduced by a factor of 3 taking into account multiplicative corrections with the elements Ni and Ti for single pulses and 160 mJ. The coefficient of determination can be improved from $r^2 = 0.992$ to $r^2 = 0.999$. The effect of interelement correction for the quantitative analysis of other analytes of high-alloy steel will be presented in more detail in Sect. 13.2.

References

11.1. *IUPAC Compendium of Chemical Terminology*, 2nd edn. (1997). http://www.chemsoc.org/chbytes/goldbook/

11.2. V. Thomsen, D. Schatzlein, D. Mercuro, Limits of detection in spectroscopy. Spectroscopy **18**, 112–114 (2003)

11.3. DIN 32 645, *Chemische Analytik – Nachweis-, Erfassungs- und Bestimmungsgrenze – Ermittlung unter Wiederholbedingungen; Begriffe, Verfahren, Auswertung* (Beuth Verlag, Berlin, 2006)

11.4. DIN 38 402, *Teil 51, Deutsche Einheitsverfahren zur Wasser-, Abwasser- und Schlammuntersuchung – Allgemeine Angaben (Gruppe A) – Kalibrierung von Analyseverfahren, Auswertung von Analyseergebnissen und lineare Kalibrierfunktionen für die Bestimmung von Verfahrenskenngrößen (A51),* (Beuth Verlag, Berlin, 1986)

11.5. DIN ISO 8466–2, *Wasserbeschaffenheit – Kalibrierung und Auswertung analytischer Verfahren und Beurteilung von Verfahrenskenndaten – Teil 2: Kalibrierstrategie für nichtlineare Kalibrierfunktionen zweiten Grades* (Beuth Verlag, Berlin, 2004)

11.6. J. Vrenegor, R. Noll, V. Sturm, Investigation of matrix effects in LIBS plasmas of high-alloy steel for matrix and minor elements. Spectrochim. Acta B, **60**, 1083–1091 (2005)

Chapter 12
Combination of LIBS and LIF

An approach to further increase the sensitivity of LIBS for the determination of traces is the combination of LIBS and laser-induced fluorescence (LIF, also called laser-excited atomic fluorescence spectrometry, LEAFS) [12.1]. Limit of detections for, e.g., heavy metals in soil according to regulatory demands are required to be significantly below 1 μg/g, which is still difficult to achieve with LIBS only.

LIBS–LIF is a so-called hyphenated technique, which was first proposed by Measures and Kwong [12.2, 12.3]; the pulse of a second laser source, which is tuned to an optical transition wavelength of the analyte of interest, is directed into the plasma plume generated by laser-induced breakdown. The fluorescence radiation induced by this second laser pulse is then evaluated for chemical analysis. In addition to improved detection limits, a large linear calibration range and reduced matrix effects are reported for this technique with respect to LIBS [12.3, 12.4]. The hyphenated technique was used for basic investigations on the physical properties of laser-induced plasmas [12.5, 12.6] as well as for studies on the applicability for the inline detection of heavy metals in aerosols [12.7]. Yet, in most cases, the measurements were performed at ambient gas pressures reduced to values in the range of 100 mbar. Although this may be helpful to achieve a maximum signal-to-noise ratio and to provide matrix-independent measurements [12.4, 12.6], such experimental conditions are not acceptable as far as the applicability of the technique for in situ measurements is concerned. So far, only Gornushkin et al. used the LIBS–LIF technique for the determination of heavy metals in soils and demonstrated an LOD of 1 μg/g for cobalt measured in air at atmospheric pressure [12.8].

While LIF offers a high sensitivity, it is basically to be considered as a single species detection scheme. The combination of LIBS and LIF allows to link the multispecies capability of LIBS for a broad range of analytes with the high sensitivity of LIF for individual selected species.

A schematic drawing of the experimental arrangement used is shown in Fig. 12.1 [12.9]. Most of the measurements were performed in a vacuum-tight measurement chamber that was filled with air or with argon as a buffer gas. This chamber, shown in a perspective view in Fig. 12.2, was designed to allow investigations

R. Noll, *Laser-Induced Breakdown Spectroscopy*, DOI 10.1007/978-3-642-20668-9_12,

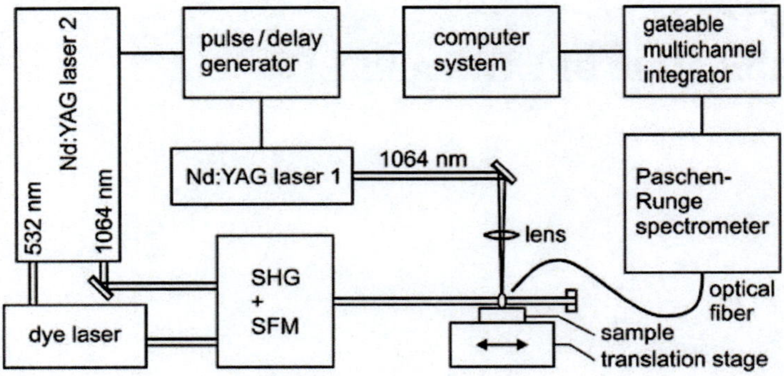

Fig. 12.1 Schematic setup for LIBS–LIF, for details see text. *SHG* = second harmonic generation, *SFM* = sum frequency mixing

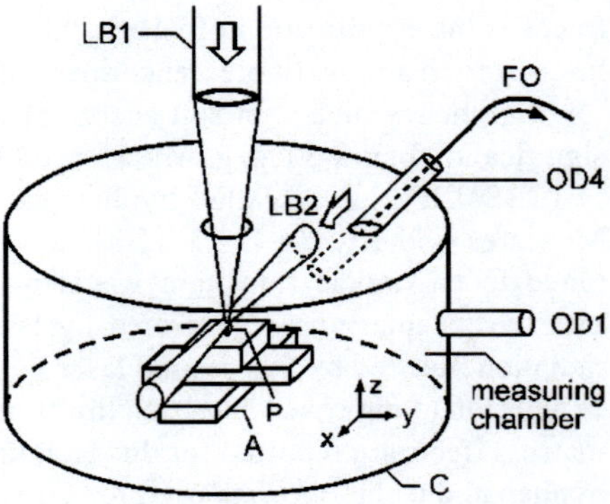

Fig. 12.2 Perspective view of the measurement chamber with the LIBS and LIF crossing laser beams

on the influence of the ambient gas pressure on the performance of the LIBS–LIF technique, because the importance of this experimental parameter has been demonstrated in other publications [12.4–12.6]. Although the gas pressure can be adjusted between 1 and 1,000 mbar within the chamber, the results presented were measured at atmospheric pressure. In the measurement chamber, the sample is positioned on a sample holder and the LIBS laser beam is focused onto the sample surface from the top. Depending on the mechanical properties of the samples to be studied, the LIBS measurements may lead to deep craters. In the case of, e.g., granular soil samples – cf. Sect. 14.3 – the craters that were generated in pressed soil pellets by 100 laser pulses with 80 mJ pulse energy were about 2 mm deep. Hence, the sample holder is mounted on a computer-controlled two-axis-translation stage within the chamber, which allows to perform studies on the effects of sample motion during the measurements.

A Q-switched Nd:YAG laser operating at 1,064 nm with a repetition rate of 10 Hz (Continuum, model Surelite I) delivered laser pulses with pulse energies of up to 450 mJ and a pulse duration of 5–7 ns, which were used for the ablation of sample material and the generation of the LIBS plasma (Nd:YAG laser 1 in Fig. 12.1).

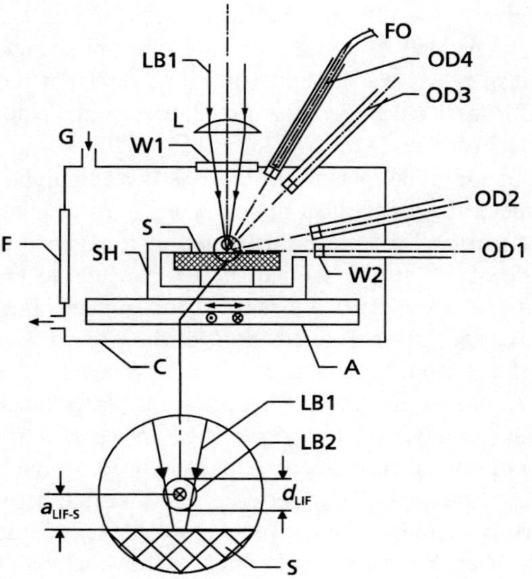

Fig. 12.3 Sectional view of the measurement chamber and the crossing laser beams. C = measurement chamber, A = translation stage, G = gas inlet and gas outlet, SH = sample holder, S = sample, F = flange to load the chamber, $OD1$–4 = observation directions for the fiber optics, $W1$ = window for the LIBS laser beam, $W2$ = detection window, FO = fiber optics to guide the radiation to the spectrometer, L = focusing lens, $LB1$ = LIBS laser beam, $LB2$ = LIF laser beam, d_{LIF} = diameter of LIF laser beam, $a_{\mathrm{LIF-S}}$ = distance between the axis of the LIF laser beam and the surface of the sample

The pulse energy was adjusted using a Glan-laser prism in the optical path and the LIBS pulses were focused onto the sample at normal incidence with a borosilicate lens ($f = 200$ mm).

A second, tunable laser source is needed for LIBS–LIF measurements to selectively excite analyte atoms or ions within the plasma volume. In this case, a frequency-doubled Q-switched Nd:YAG laser (Spectron, model SL 852 GSLM, 650 mJ at 532 nm, 10 Hz, 6–9 ns) was used to pump a dye laser (Lumonics, model HD-300) operating at 583 nm (a mixture of Rhodamin 6G and Rhodamin B was used as a dye) or 554 nm (Fluorescin 27 was used as a dye), respectively. The dye laser pulses were frequency-doubled by second harmonic generation (SHG) in a frequency-doubling and -mixing unit (Spectron, model SL 4000 EMX), thus generating laser radiation in the UV range at 291.5 or 276.8 nm.

For the resonant excitation of Cd at 228.80 nm, the laser pulses at 291.5 nm were converted by sum frequency mixing (SFM) with the 1,064-nm laser radiation from the Nd:YAG laser (laser 2 in Fig. 12.1) that is also used for pumping the dye laser. These LIF-laser pulses were guided into the measurement chamber via mirror optics and directed into the plasma volume parallel to the sample surface. Figure 12.3 shows a detailed sectional view of the measurement chamber and the crossing laser beams.

The distance between the LIF-laser beam axis (LB2 in Fig. 12.3) and the surface of the sample was adjusted to about half of the beam diameter, i.e., $a_{\mathrm{LIF-S}} \approx d_{\mathrm{LIF}}/2$. This distance $a_{\mathrm{LIF-S}}$ was around 4 mm for 228.80 nm and around 8 mm for 276.78 nm. The LIF-laser pulse energy at the plasma plume amounted to 16 μJ for 228.80 nm and 40 μJ for 276.78 nm.

The radiation of the LIBS-plasma as well as the LIF light were guided to a Paschen–Runge spectrometer via fiber optics. A single silica/silica fiber with a core diameter of 1,000 μm was therefore positioned at a distance of 10 cm from the plasma at an angle of 14° to the sample surface and 90° to the LIF-laser beam. No lenses were used between the plasma and the optical fiber, thus making the detection insensitive against small fluctuations of the plasma position and allowing the observation of the whole interaction volume probed by the LIF laser beam. This simple detection scheme was used, because it was intended to evaluate the applicability of the technique for fast in situ measurements. For this application, a minimum number of optical components should be used, which would increase the risk of misalignment otherwise. Furthermore, the distance between the detection optics and the plasma should not be too small to avoid the pollution of the optics by ablated sample material. A further improvement of the sensitivity of the technique is feasible, if, e.g., a collecting lens of 10 mm aperture is used, which would increase the solid angle of observation by two orders of magnitude.

A Paschen–Runge spectrometer equipped with exits slits and photomultipliers (PMTs) for the simultaneous detection of 29 spectral lines was used for the spectrally and temporally resolved detection of the plasma radiation (cf. Sect. 4.2). The diameter of the Rowland circle is 750 mm. The spectral bandwidth detected by each PMT can be estimated from the dispersion of the spectrometer and the width of the exit slits and amounts to less than 20 pm. The signals from the PMTs corresponding to the different spectral lines were processed with a fast gateable multichannel integrator (cf. Sect. 4.5) [12.10]. This multichannel integrator was controlled by a computer system that also triggered the two laser systems via a Stanford DG535 pulse/delay generator. A signal synchronized to the laser Q-switch was used for the triggering of the integrators to minimize timing jitter. The delay generator triggering the two laser systems allowed to adjust the time difference between the LIBS-pulse and the LIF-pulse ($\Delta t_{\mathrm{LIBS-LIF}}$) between 0 and 250 μs.

Figure 12.4 shows as an example simplified level schemes for (a) Cd [12.11] and (b) Tl illustrating the energy levels and optical transitions that are used for the detection of traces of these elements in soil, cf. Sect. 14.3 [12.9] (The term symbols indicate the total spin quantum number (first superscript), the total orbital quantum number, the total angular momentum quantum number (index), the second superscript "o" stands for odd parity of the level.)

Cadmium is directly excited from its ground level 5 $^1\mathrm{S}_0$ to the $5^1\mathrm{P}_1^{\mathrm{o}}$ level at 5.418 eV by UV laser pulses with the corresponding wavelength of 228.80 nm. The LIF was also observed at 228.80 nm. This excitation/detection scheme was already widely used for atomic fluorescence measurements with different atomizer units like graphite furnaces (GF) and excitation sources like electrodeless discharge

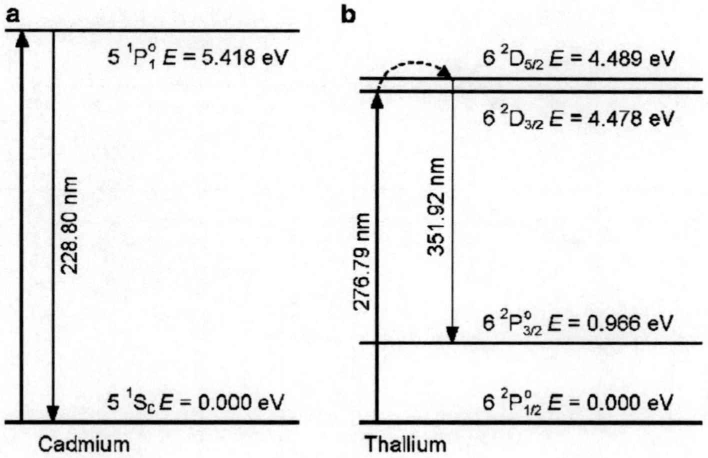

Fig. 12.4 Reduced level scheme for (**a**) cadmium and (**b**) thallium. Only those levels and transitions are shown, which are involved in the LIBS–LIF experiment. The *dashed arrow* indicates an energy transfer attributed to processes within the buffer gas plasma

lamps (EDL) [12.12] and also tunable lasers [12.13], showing very good limits of detection, especially for the latter case. A disadvantage of this excitation/detection scheme is the possibility of a poor signal-to-noise ratio due to laser light scattered by the ablated material. Assuming that a portion of the scattered laser light shows a dipole scattering behavior and that multiple scattering can be neglected, the scattered light intensity has a minimum in the direction parallel to the polarization of the laser radiation. In the experiments, the angle between the direction of observation and the horizontal polarization of the laser beam amounts to 14°, so that the intensity of this portion of scattered light is reduced to about $\sin^2(14°) = 6\%$ of its maximum. In this respect, the detection geometry was designed to reduce the negative effect of laser light scattering into the observation direction.

The excitation and detection scheme for Tl is given in Fig. 12.4b. The Paschen–Runge spectrometer used for the detection of the fluorescence light is equipped with a spectral line for Tl at 351.92 nm, because the strongest emission was observed in the UV range with spark excitation for this spectral line [12.14]. As there is no optical transition from the ground state $6^2P^o_{1/2}$ to the corresponding upper level $6^2D_{5/2}$ at 4.489 eV, the atoms were excited from the ground state to the $6\ ^2D_{3/2}$ at 4.478 eV level with laser pulses at a wavelength of 276.79 nm. Consequently, a transition from the $6^2D_{3/2}$ state to the $6^2D_{5/2}$ state is necessary to observe a fluorescence emission at 351.92 nm, either thermally assisted or by collisions with the buffer gas [12.5]. The spectral difference between the excitation wavelength and the detection wavelength of more than 75 nm makes sure that scattering of the laser light does not deteriorate the signal-to-noise ratio in this case. This excitation/detection scheme was already used for LIF measurements of Tl in a GF

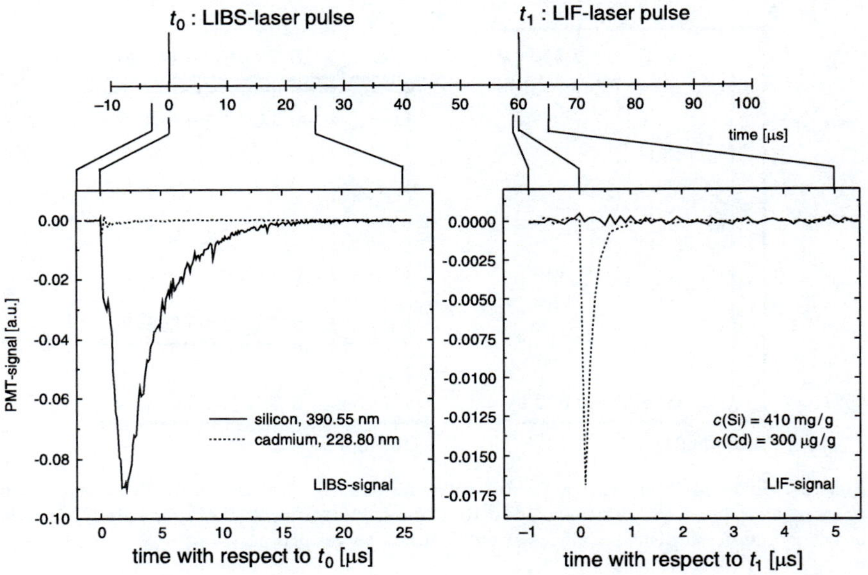

Fig. 12.5 Photomultiplier signals of a silicon (*solid curve*) and a cadmium (*dashed curve*) emission line as a function of time for a combined LIBS–LIF excitation. The species concentrations of Si and Cd in the soil sample are given in the *right* graph. Ambient atmosphere:argon at 950 mbar

[12.15]. Cadmium was also studied in the GF using a double-resonance excitation scheme to eliminate scattering problems, at the expense of increased complexity of the experimental apparatus [12.15].

Figure 12.5 shows the PMT signals of a silicon and a cadmium line as a function of time for a combined LIBS–LIF excitation of a Cd contaminated soil sample. At the time t_0, the LIBS laser beam having the fixed wavelength 1,064 nm is irradiated on the specimen and excites for both element channels a first peak attributed to plasma continuum emission (cf. Sect. 2.1, Fig. 2.4). After that a clear Si response is observed attaining a maximum at about 2 μs and decaying in the following ≈ 15 μs (LIBS-signal). For Cd, no significant element-specific signal was observed. At the time t_1, the LIF laser pulse is irradiated tuned to the Cd transition 228.80 nm (cf. Fig. 12.4, a). The result is a distinct Cd fluorescence response, whereas at the same time the Si signal rests around the base line.

Calibration curves were recorded by integration of the whole LIF signal ($t_{\text{LIF delay}} = -0.5\,\mu\text{s}$, $t_{\text{int}} = 5\,\mu\text{s}$) for different time distances $\Delta t_{\text{LIBS–LIF}}$ between the ignition of the LIBS plasma and LIF excitation [12.16]. The negative value $t_{\text{LIF delay}} = -0.5\,\mu\text{s}$ indicates that the integration of the PMT signal was started before the exciting LIF laser pulse was directed into the relaxing plasma. Thus, the whole fluorescence signal was recorded, including also the signal background due to residual laser radiation scattering. This integration scheme was chosen

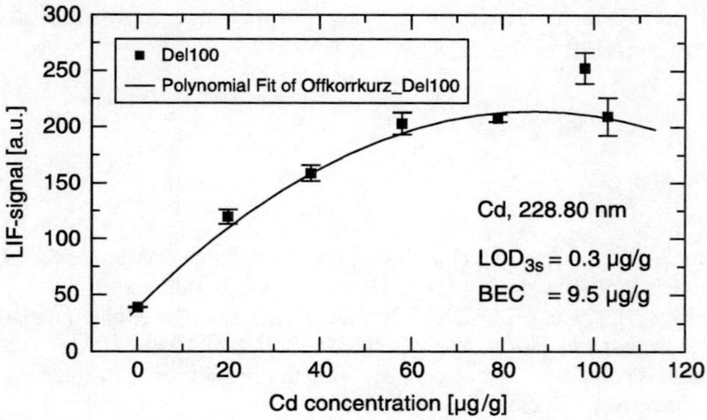

Fig. 12.6 LIBS–LIF calibration curve for cadmium. The analyte signal measured is plotted against the analyte concentration measured with ICP–OES. The calibration curve is fitted by a second order polynomial

taking into account the short fluorescence life time of the observed transition of $\tau_{\text{fl}} = 1.7\,\text{ns}$ [12.13]. The best detection limit was achieved for a large delay between the LIBS and the LIF laser pulse of $\Delta t_{\text{LIBS-LIF}} = 100\,\mu\text{s}$, which is attributed to the fact that a maximum of cadmium atoms must have relaxed to their ground state and that the plasma must have expanded to the volume probed by the LIF laser beam. Additionally, the spectral background due to the broadband emission of the plasma is negligible for these delay times, because the decay time of this continuum emission, which can be estimated from the time behavior of the PMT signal after the LIBS laser pulse, is only in the order of several 100 ns.

The corresponding calibration curve is shown in Fig. 12.6. A strong saturation behavior is observed for analyte concentrations higher than $50\,\mu\text{g/g}$. This is attributed to limitations of the signal processing electronics, because the gain of the electronics was tuned to a maximum in order to achieve a maximum slope of the calibration curve at low analyte concentrations. As a disadvantage, this led to a truncation of the LIF signals for high analyte concentrations. Thus, the saturation should not be confused with optical saturation in this case and an increased range of linearity is feasible if the amplification of the signal electronics is reduced. Nevertheless, the range of linearity might be restricted due to self-absorption, which may occur for the optical transition investigated for cadmium.

Notably, the background equivalent concentration (BEC), calculated as the ratio of the axis intercept of the calibration curve with the ordinate (cf. Sect. 11.2, Fig. 11.3 b) – on which the LIF-signal is plotted – divided by the slope of the calibration curve, amounted only to BEC $= 9.5\,\mu\text{g/g}$, showing that scattering of laser light gave only minor contributions to the LIF signal observed. The detection limit (cf. (11.16)) for Cd in soil determined under these experimental conditions resulted in $\text{LOD}_{3s} = 0.3\,\mu\text{g/g}$, which is the lowest value achieved with this technique so far.

LIBS–LIF analysis of thallium in soil samples was also studied and will be described in more detail in Sect. 14.3.

References

12.1. K. Niemax, W. Sdorra, Optical emission spectrometry and laser-induced fluorescence of laser produced plasma plumes. Appl. Optics **29**, 5000–5006 (1990)

12.2. R. Measures, H. Kwong, TABLASER: trace (element) analyzer based on laser ablation and selectively excited radiation. Appl. Optics **18**, 281–286 (1979)

12.3. H. Kwong, R. Measures, Trace element laser microanalyzer with freedom from chemical matrix effect. Anal. Chem. **51**, 428–432 (1979)

12.4. I. Gornushkin, S. Baker, B. Smith, J. Winefordner, Determination of lead in metallic reference materials by laser ablation combined with laser excited atomic fluorescence. Spectrochim. Acta **52B**, 1653–1662 (1997)

12.5. W. Sdorra, A. Quentmeier, K. Niemax, Basic investigations for laser microanalysis: II. Laser-induced fluorescence in laser-produced sample plumes. Microchim. Acta II **98**, 201–218 (1989)

12.6. W. Sdorra, K. Niemax, Temporal and spatial distribution of analyte atoms and ions in microplasmas produced by laser ablation of solid samples. Spectrochim. Acta **45B**, 917–926 (1990)

12.7. R. Neuhauser, U. Panne, R. Niessner, G. Petrucci, P. Cavalli, N. Omenetto, On-line and in situ detection of lead in ultrafine aerosols by laser-excited atomic fluorescence spectroscopy. Sensor Actuator B **39**, 344–348 (1997)

12.8. I. Gornushkin, J. Kim, B. Smith, S. Baker, J. Winefordner, Determination of cobalt in soil, steel and graphite using excited-state laser fluorescence induced in a laser spark. Appl. Spectrosc. **51**, 1055–1059 (1997)

12.9. F. Hilbk-Kortenbruck, R. Noll, P. Wintjens, H. Falk, C. Becker, Analysis of heavy metals in soils using laser-induced breakdown spectrometry combined with laser-induced fluorescence. Spectrochim. Acta B **56**, 933–945 (2001)

12.10. V. Sturm, L. Peter, R. Noll, Steel analysis with laser-induced breakdown spectrometry in the vacuum ultraviolet. Appl. Spectrosc. **54**, 1275–1278 (2000)

12.11. H. Didra, M. Baumann, Cadmium (5^1P1 $\rightarrow$ 5^3P_J) excitation transfer and 5^3P_J intramultiplett relaxation by collisions with molecular gases. Z. Phys. D **37**, 29–37 (1996)

12.12. B. Smith, M. Glick, K. Spears, J. Winefordner, A comprehensive table of atomic fluorescence detection limits and experimental conditions. Appl. Spectrosc. **43**, 376–414 (1989)

12.13. M. Bolshov, S. Rudnev, B. Hütsch, Determination of trace amounts of cadmium by laser excited atomic fluorescence spectrometry. J. Anal. At. Spectrom. **7**, 1–6 (1992)

12.14. A. Zaidel, V. Prokofev, S. Raiskii, V. Slavnyi, E. Shreider, *Tables of Spectral Lines* (IFI–Plenum, New York, 1970)

12.15. N. Omenetto, P. Cavalli, M. Broglia, P. Qi, G. Rossi, Laser-induced single-resonance and double-resonance atomic fluorescence spectrometry in a graphite tube atomizer. J. Anal. At. Spectrom. **3**, 231–235 (1988)

12.16. F. Hilbk-Kortenbruck, C. Rieck, R. Noll, Multi-elemental analysis of heavy metals in soil samples using laser-based analytical techniques, Abstracts of the lecture groups “Environmental Technology and Fundamentals of Laser-aided In-situ Soil Analysis”, Int. Meeting on Chem. Eng., Environmental Protection and Biotechnology, ACHEMA 2000, Conf. Proc. 2000, DECHEMA e.V., Frankfurt a. M. 386–388 (2000)

Chapter 13
Bulk Analysis of Metallic Alloys

This chapter describes LIBS investigations for a quantitative bulk analysis of metallic alloys such as steel in solid and liquid state, high-alloy steel, and aluminum. In case of aluminum, the focus is on LIBS analysis of moving Al specimens, which is of interest for inline identification of scrap pieces for material specific recycling (see also cf. Sect. 18.4).

13.1 Steel

The chemical analysis of steel is a demanding task for process control in steelmaking and for quality assessment of preproducts. The dominant analytical tools applied nowadays are based on physical methods such as spark discharge-optical emission spectrometry (SD-OES), X-ray fluorescence (XRF), or absorption spectroscopy of analytes in flames. These techniques are mostly used in off-site laboratories and they require preparational steps to present the steel sample in a physical state necessary for the respective method. The sample preparation includes steps such as cutting, milling, grinding, or drilling, which are time consuming and require high maintenance efforts even in automated container laboratories. Hence, a strong need is coming up to simplify or even avoid those steps of chemical analysis such as sample taking, sample transport, and sample preparation. Laser-based analytical methods will play a key role for the development of inline methods in metallurgical processing due to the following features: noncontact measurement at distances of centimeters to meters, high measuring speed, sample preparation, or conditioning by the laser beam itself.

13.1.1 Solid Steel Samples

For an inline analytical method the multielement capability is an important issue. Laser-induced breakdown spectrometry (LIBS) is a method enabling the

R. Noll, *Laser-Induced Breakdown Spectroscopy*, DOI 10.1007/978-3-642-20668-9_13,

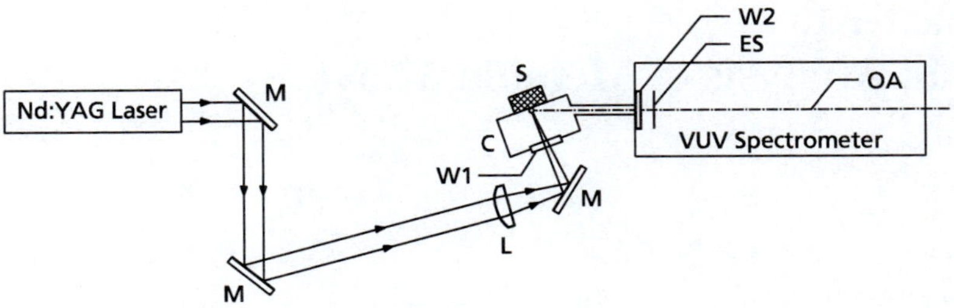

Fig. 13.1 Schematic setup for LIBS of steel samples in the VUV range. M = mirror, L = lens, W1 = entrance window of the measurement chamber, C = measurement chamber, S = sample, W2 = entrance window of the spectrometer made of MgF_2, ES = entrance slit of the spectrometer, width 15 μm

simultaneous analysis of various elements in a short time. However, the analytical performance in terms of limits of detection (LOD) reported for steel constituents so far does not achieve those of conventional methods such as SD-OES [13.1–13.3]. Hence, the main goal presented in the following was to improve the plasma excitation by multiple laser pulses and to make use of emission lines in the vacuum ultraviolet (VUV) spectral range [13.4]. The motivation of this work was to establish the basis for future inline applications of LIBS in the steel industry requiring LOD for light elements such as carbon, phosphorus, and sulfur in the range below 10 μg/g.

The experimental setup is shown in Fig. 13.1 (cf. also Sect. 4.6). A Q-switched Nd:YAG laser operating at 1,064 nm was used to excite the plasma. The Q-switch electronics of this laser was modified to generate up to three separated laser pulses within a single flashlamp pulse instead of a single laser pulse (SP; cf. Sect. 3.2, Fig. 3.9 and Sect. 6.2). The flashlamp repetition rate is 10 Hz. The influence of multiple laser pulses on the dynamics and physical state of the induced plasma was described in Chap. 6. The laser emits three equal-energy pulses of ~16 ns duration (FWHM) and with a time separation of 25 μs between the first and the second pulse, and 40 μs between the second and the third, respectively. The burst energy amounts to 300 mJ. The laser beam is guided to a plano-convex lens with a focal length of 80 mm. The converging laser radiation is then guided via a mirror and an optical window into a measurement chamber. This chamber has a cover plate on top with a circular opening having a diameter of ~12 mm. The sample to be analyzed is put onto this cover plate and located in such a way that the opening is closed. The focus of the laser beam was adjusted to position the beam waist inside the sample with a distance of $\Delta s = 6$ mm from the sample surface. The diameter of the craters formed on the sample surface is about 300 μm.

The optical axis of the VUV vacuum spectrometer describes the line between the center of the diffraction grating and the middle of the entrance slit. This optical axis intersects the propagation axis of the laser beam on the surface of the sample. The angle between the surface of the sample and the optical axis of the spectrometer amounts to $\alpha_D \sim 18°$ (cf. Fig. 4.18). With this orientation, it is assured that radiation

Table 13.1 List of elements, wavelengths, and energies E_n of the upper levels of the observed lines [13.30]

Element	Ionization stage	Wavelength (nm)	E_n (eV)
P	I	178.28	6.95
S	I	180.73	6.86
Fe	II	187.75	9.13
C	I	193.09	7.68
Ni	II	231.60	6.39
Cr	II	267.72	6.18
Fe	II	271.44	5.55
Si	I	288.16	5.08
Mn	II	293.31	5.40

from the whole laser-induced plasma can be received by the spectrometer aperture. The chamber has an internal volume of about 42 cm^3. This volume is flushed with Argon 5.0 at 1 atm and a flow rate of about 8 l/min. The entrance window of the spectrometer is made of VUV transparent magnesium fluoride.

The spectrometer has a Rowland circle diameter of 750 mm and is equipped with a holographic diffraction grating with 2,400 lines/mm. The optical system of the spectrometer is installed inside a vacuum vessel which is evacuated to $<10^{-3}$ mbar. The spectral resolution amounts to $\leq$25 pm in first order. Table 13.1 lists the emission lines where exit slits are positioned along the Rowland circle. The lines Fe 187.75 and Fe 271.44 nm are used as internal standards. The dispersed radiation passing the exit slits is detected by photomultiplier tubes (PMT).

The signals of 16 PMTs are processed by a multichannel integrator electronics (MCI; see Fig. 4.16, Sect. 4.5). The anode currents of the PMT are fed into current/voltage-converters that are positioned close to the sockets of the PMT inside the vacuum vessel to reduce as far as possible a capacitive load of the current source. The transformed signals are then transmitted via coaxial cables and BNC-feed through out of the vessel to the MCI. The bandwidth of the i/u-converters and the integrators is in the range of 10 MHz to enable an undistorted signal processing of the transient PMT currents, which are characterized by significant signal variations on a timescale down to 25 ns.

The gate for the integrators is generated by a delay/pulse generator which is triggered by the Pockels cell synchronization monitor of the laser. The outputs of the integrators are transmitted to a multiplexer and an A/D-converter. The digitized charges of the integrators for each laser pulse are stored and processed on a computer.

For the calibration of the LIBS measurements, a set of 14 reference samples with certified concentrations of the investigated elements was used. The samples are not prepared mechanically unless the surface is covered by craters from former measurements. In that latter case, the sample is simply ground manually with an Al_2O_3-grinding paper of grit size 60 for about 5 s. The laser beam itself is used to ablate additionally residual surface layers or contaminations. For the quantitative

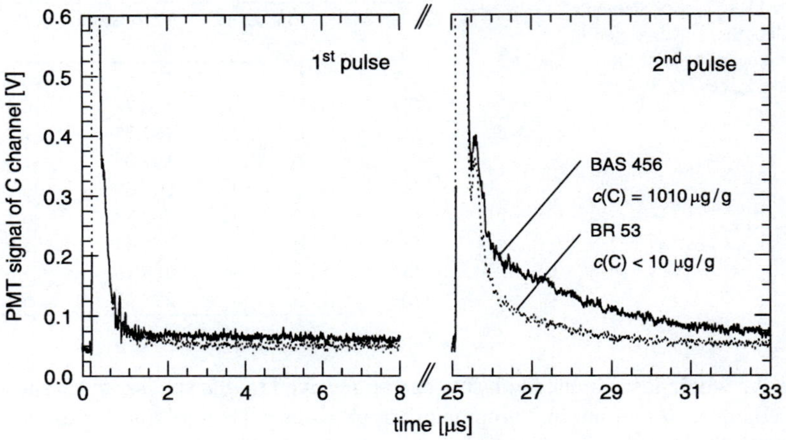

Fig. 13.2 Carbon emission as a function of time after the first and the second pulse of a laser burst. The *solid line* shows the signal obtained for the sample BAS 456 with a carbon content of 1,010 μg/g, and the *dotted line* belongs to the sample BR 53 with a carbon content below 10 μg/g

analysis, the number of prebursts (cf. Chap. 11, Table 11.1) applied to prepare the sample surface amounts to $N_{pp} = 100$.

Figure 13.2 shows time-resolved PMT signals of the carbon channel at 193.09 nm picked up at the current/voltage-converters with a Tektronix TDS 784D oscilloscope. The dotted line was gained using a sample with a carbon content <10 μg/g; the solid line belongs to a sample with a carbon content of 1,010 μg/g. With the beginning of the first laser pulse, the carbon signal shows a steep increase with a rise time of less than 10 ns. After achieving a maximum, the PMT signal decays within about 500 ns. In the decaying phase, only a small difference between the signals for the different samples is observed. After the second pulse of the burst, the situation is very different. Whereas the rise time is comparable the decay time differs significantly for the two samples. Moreover, the signal level is much higher compared to the signals after the first laser pulse. After the third laser pulse of the burst about the same signal levels as after the second pulse are achieved.

The signal evolution observed for different elements, element concentrations, and/or experimental parameters, e.g., laser burst energy, pulse separation, and focusing is qualitatively the same as described above. However, the decay times and thus the optimum position and width of the integration gate depend on these parameters. The best results with respect to the LOD were gained with a delay time of $t_{delay} = 650$ ns and an integration gate width of $t_{int} = 12$ μs.

The experimental setup was calibrated for seven elements for low-alloy steel. Figures 13.3–13.5 show calibration curves for carbon, phosphorus, and sulfur.

The LOD was calculated on the basis of the $3s$-criterion, where s is the empirical standard deviation of the intensity ratio for the sample with the lowest analyte content [cf. Chap. 11, relation (11.16)]. Each sample was measured five

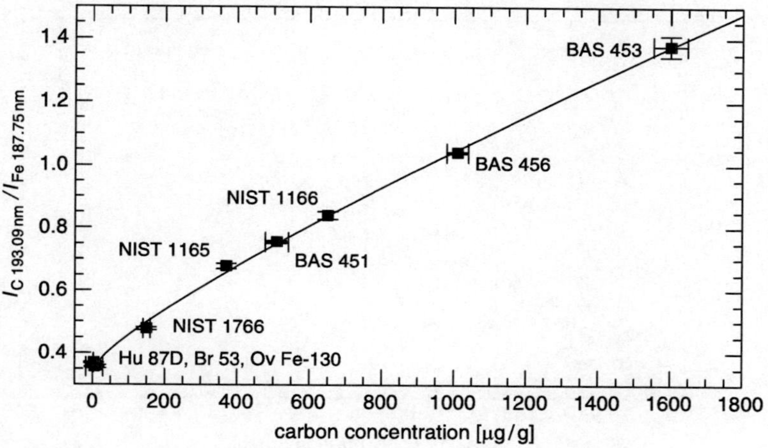

Fig. 13.3 Calibration curve of carbon for the emission line 193.09 nm. Sample denominations: *NIST*, National Institute of Standards & Technology; *BAS*, Bureau of Analysed Samples Ltd.; *Hu*, Hanau; *Br*, Breitländer; *Ov*, Ovako Steel

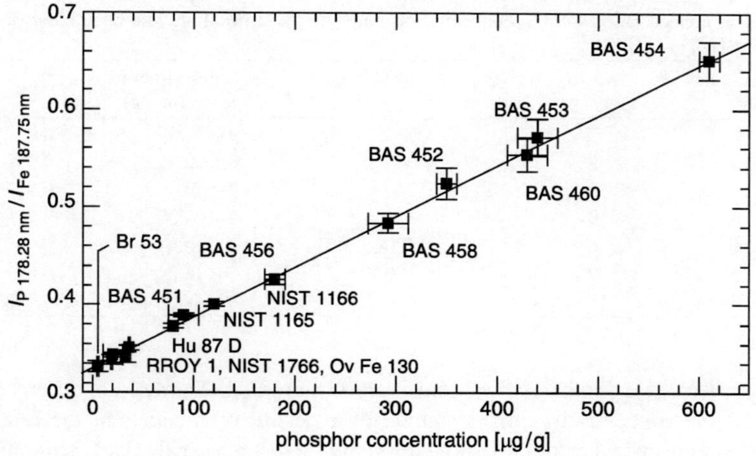

Fig. 13.4 Calibration curve of phosphorus for the emission line 178.28 nm. For abbreviations see Fig. 13.3. *RROY*, Rautaruukki Oy

times. Between the measurements, the sample was shifted on the top cover of the measurement chamber to expose another surface part to the laser beam. Each measurement comprises 600 laser bursts, where $N_{pp} = 100$ are applied as prebursts to locally condition the sample surface and $N_{mp} = 500$ measurement bursts to gain the spectral signals to be evaluated. The data points plotted in the Figs. 13.3–13.5 denote the mean values from five measurements. Numerical corrections such

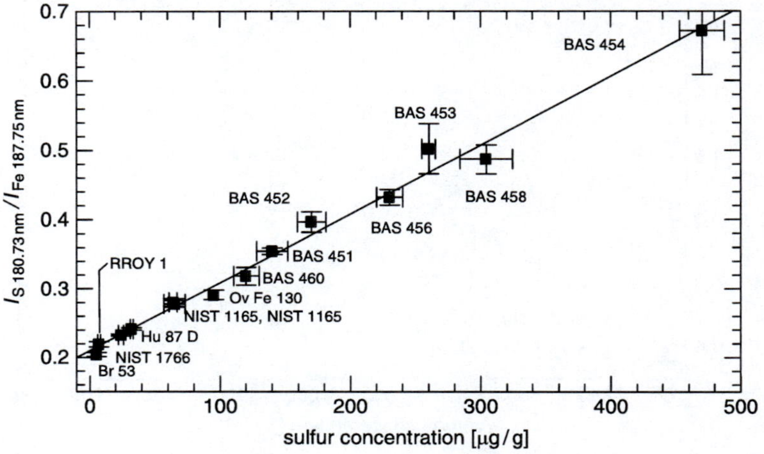

Fig. 13.5 Calibration curve of sulfur for the emission line 180.73 nm. For abbreviations, see Figs. 13.3 and 13.4

Table 13.2 Results of the calibration measurements: limit of detection (LOD_{3s}), background equivalent concentration (BEC), concentration range of the calibration, and coefficient of determination (r^2)

Element	LOD_{3s}(µg/g)	BEC (µg/g)	Concentration range (µg/g)	r^2
C	7	530	0–1,800	0.999
P	9	640	0–600	0.998
S	8	210	0–500	0.987
Mn	9	320	0–14,000	0.999
Ni	6	500	0–700	0.991
Cr	7	330	0–1,000	0.991
Si	11	680	0–6,000	0.990

as interelement or matrix corrections were not applied. The horizontal error bars indicate the concentration errors stated in the certification sheets of the reference samples; the vertical error bars show the standard deviation calculated from the five measurements. The curve plotted in Fig. 13.3 is a quadratic fit using a least squares algorithm describing a saturation behavior of the intensity ratio C/Fe with increasing carbon concentration.

Figures 13.4 and 13.5 show the calibration curves for phosphorus and sulfur. The calibration curves are linear functions fitted to the data points. In the setup of Fig. 13.1, the location of the laser irradiation is fixed; hence the only spatial averaging is given by the crater diameter of about 300 µm. Conventional methods such as SD-OES have an inherent spatial averaging over areas with a diameter of about 5 mm (cf. Fig. 3.17). For a comparison of results gained with these methods and LIBS, a spatial averaging, e.g., by a relative motion between sample and laser beam, has to be performed (cf. Sect. 3.6). Table 13.2 summarizes the results of the calibration for seven elements using as internal standard the Fe 187.75 nm line. The LODs are <10 μg/g for most of these elements. The coefficients of determination are always greater than 0.987. In general, using the line Fe187 instead of Fe271 as the internal standard resulted in a better analytical performance in terms of the LOD. For the elements C, P, S, Si, and Ni, the achieved LOD with Fe187 are approximately 20–50% smaller than with Fe271. In the case of Cr and Mn, the improvement is less significant. For a further improvement of the LOD, the background equivalent concentration (BEC) has to be reduced. The origin of the rather high BEC values observed is presumably due to the high signal level of the PMT immediately after the beginning of the laser pulse irradiation.

The applied gating of the integrators suppresses this signal part, but there is still the possibility that the electrical charge generated in the beginning of the PMT signal distorts the signal behavior at later times.

Multipulse excitation and observation of VUV emission lines allow to push down the LOD for light elements in low-alloy steel samples in the range below 10 μg/g. No or at least a reduced effort in sample preparation is required in comparison to SD-OES. The laser itself prepares the sample by a localized ablation of surface layers or contaminations. With these results, the capability of LIBS to be applied for inline analytical tasks in the steel industry is demonstrated.

13.1.2 Solid Steel Samples with Scale Layer

Production control samples are used in steel works to control steel composition. In certain phases of the steelmaking process, the chemical composition of the steel melt has to be determined with a certain accuracy in a short time of several minutes. The further treatment and alloying of the heat are regulated dependent on the determined composition. The interest in steel works analysis is a cost-effective analytical method, which reduces the required alloying material, shortens the process control time, and reduces maintenance costs of the analytical equipment.

For process control, the state of the art is to take samples with immersion sampling probes which are dipped into the melt in order to be filled with the liquid steel. During the solidification and cooling of the samples, a scale layer is formed at the surface having a thickness in the range of ~0.1–1 mm. In general, the composition of the scale layer deviates from the composition of the bulk material. In addition, the thickness of the scale layer depends on the type of the alloy steel. Especially, carbon shows an enhanced concentration in the inhomogeneous scale layer. Hence, the analysis of the bulk material requires the removal of the scale layer. The conventional approach is to remove the scale layers by automatic grinding or milling machines in the steel works. The subsequent analysis is carried out predominantly by OES using arc/spark excitation, XRF analysis, and/or combustion analysis (for the light elements). The analysis of the elements C, P, and S is of primary interest in steelmaking since these elements are important for the process

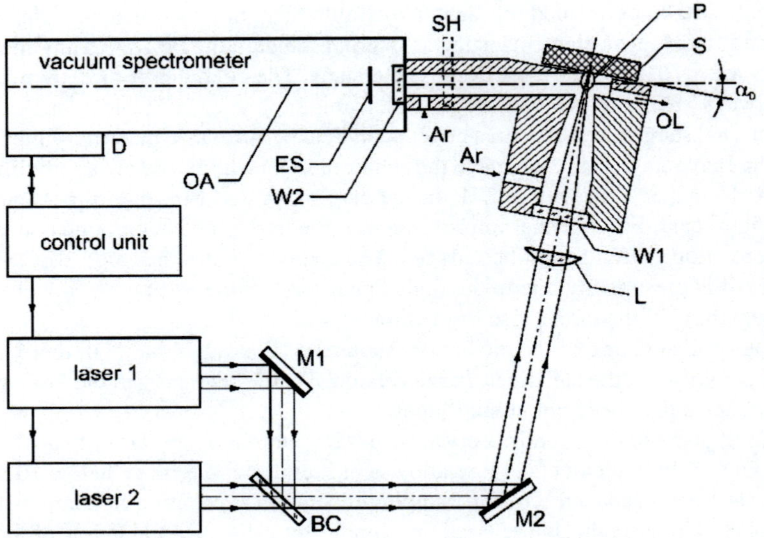

Fig. 13.6 Schematic view of the experimental setup. P = laser-induced plasma plume; S = sample; OL = outlet; Ar = argon gas; W1, W2 = optical windows; L = lens; M1, M2 = mirrors; BC = beam combiner; ES = entrance slit; OA = optical axis; D = detectors, signal electronics; SH = shutter

as well as are difficult to analyze spectrometrically. In the investigations presented in this section, further elements of interest are included such as Al, Cr, Cu, Mn, and Mo.

Elemental analysis of steel by LIBS is reported for liquid steel [13.5–13.8], as well as solid steel alloys, such as low-alloy steel [13.4, 13.9], or stainless steel [13.10, 13.11]. Some applications in steel analysis are summarized in [13.12]. LOD of less than 10 μg/g have been reported [13.4, 13.9, 13.13]. For bulk analysis of solid steel, samples with thin surface layers are measured or the samples are ground similar to sample preparation for arc/spark spectrometry. Nevertheless, before starting the LIBS signal acquisition, laser pulses are irradiated on the sample to ablate thin surface layers or contaminations (cf. Sect. 3.8). The ablated thicknesses using a typical LIBS setup with SP are in the range of 0.05 mm, which is not sufficient for the removal of thicker scale layers regarded here. In the following triple pulse laser bursts are applied for an enhanced local laser ablation of the scale layer and subsequent elemental analysis of the bulk material is carried out by LIBS using double pulse (DP) laser bursts (cf. Chap. 6). The sample preparation and the analysis are combined in a single setup avoiding a mechanical sample preparation and reducing sample transportation times [13.14].

To enhance the laser ablation of the scale layer during the preparation step, 10-Hz repetitive bursts consisting of three laser pulses (triple pulse (TP) bursts) are generated by the superposition of two Q-switched Nd:YAG laser beams (wavelength 1,064 nm); see Fig. 13.6. Laser 1 emits SP and laser 2 produces bursts of two laser

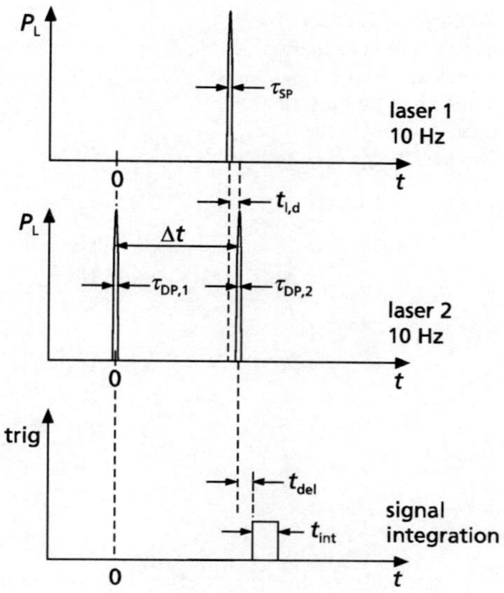

Fig. 13.7 Temporal sequence of laser pulses and signal integration. The single pulse output of laser 1 is superposed with the double pulse output of laser 2 for scale ablation. P_L = laser power; τ_{SP}, $\tau_{DP,1}$, $\tau_{DP,2}$ = pulse widths of single pulse or double pulses, respectively; Δt, $t_{l,d}$, t_{del} = time delays; t_{int} = integration time interval of PMT signal

pulses (DP bursts, interpulse separation $\Delta t = 50\,\mu s$); see Fig. 13.7. The widths of the laser pulses are ~6–10 ns depending on the operating parameters of the laser. The time delay of the second laser pulse to the third laser pulse within the TP burst of the superposed beams is set to $t_{l,d} = 1\,\mu s$. The subsequent LIBS analysis is carried out with DP bursts of laser 2.

The laser bursts are focused on the sample positioned at the sample stand, which is flushed by argon gas. The emitted light of the laser-induced plasma plume (P in Fig. 13.6) is detected directly without further optical imaging by a vacuum Paschen–Runge spectrometer (Model RS 1000 from OBLF, Witten, Germany, modified for this application) equipped with PMT for detection of the spectral line signals and the zeroth order signal. The distance of the sample surface to the entrance slit of the spectrometer is ~300 mm. The detection angle between the optical axis of the spectrometer to the sample surface is $\alpha_D \sim 10°$ (cf. Sect. 4.6, Fig. 4.18). The diameter of the Rowland circle of the spectrometer is 0.5 m and the spectral resolution amounts to ~0.02 nm (grating, 2,700 grooves per millimeter). The PMT signals are directly fed into an MCI electronics (cf. Sect. 4.5) for integration within a preset time interval, t_{int} (in the following $t_{int} = 10\,\mu s$) after the laser pulses. The time delay of the integration is set to $t_{del} = 4\,\mu s$ after the second pulse of the DP. The spectral lines installed are listed in Table 13.3.

The superposed laser beams are focused on the sample by a plano-convex lens with a focal length $f = 200$ mm. The diameters of the laser beams amount to ~6 mm. The sample surface is positioned perpendicular to the laser beam. The energy of an SP is $E_{SP} = 200$ mJ. The energies of the DP are $E_{DP,1} = E_{DP,2} = 200$ mJ; i.e., the energy of the DP bursts is $E_b = E_{DP,1} + E_{DP,2} = 400$ mJ. Analogously, the energy of a TP burst is 600 mJ. The energy E_{total} denotes the burst energy times the number of bursts irradiated on the sample.

Table 13.3 Element, wavelength λ, excitation energy E_j, $\log(g_i f_{ij})$, cf. (9.8), of the spectral lines installed in the Paschen-Runge spectrometer

Element	λ (nm)	E_j (eV)	$\log(g_i f_{ij})$
Al	396.152	3.143	−0.323
C	193.091	7.685	−0.211
Cr	267.715	6.180	0.36
Cr	286.257	5.856	−0.21
Cu	324.754	3.817	−0.062
Fe	271.441	5.553	−0.44
Fe	187.747	9.126	0.148
Mn	293.306	5.401	−0.12
Mo	281.615	6.071	0.51
Ni	225.385	6.822	−0.043
P	178.283	6.955	−0.389
S	180.731	6.861	−0.22
Si	251.611	4.954	−0.24

Some of the spectral lines are selected primarily for high-alloy steel analysis which is not regarded in this section

Each side of the production control sample is measured three times at different sample positions. If applied, the number of laser pulse bursts for ablation is set to $N_{abl} = 1{,}000$. The subsequent LIBS analysis at one sample position is performed using $N_{pp}(DP) = 100$ for additional preparation of the surface and $N_{mp}(DP) = 500$ for analysis.

Production control samples of low-alloy steel, so-called lollypop samples, were provided by IRSID including a spark OES analysis [13.15]. The samples were prepared by grinding one side, while the other side is left with the original scale layer. The samples are measured by spark OES on the ground side and with the laser method on both sides. The results are compared to evaluate the analytical performance of the laser method including preparation. The concentration ranges of the analyzed production samples and the reference samples for determination of the LOD are listed in Table 13.4. The reference samples used for calibration are standard certified reference materials of carbon steel from BAS (Bureau of Analysed Samples, Middlesbrough, UK) and low-alloy steel from NIST (National Institute of Standards and Technology, Gaithersburg, MD, USA). In addition, two high purity iron samples: BR53 (setting up sample, not certified) from Breitländer referensmaterial, Sweden, and Fe 130–35 from Ovako Steel, Sweden, with certificate of Ovako Steel, are included. The samples used for each specific calibration curve are listed in Table 13.6. It should be mentioned that the homogeneity of the BAS and NIST samples is not certified at a scale relevant to LIBS analysis, although the NIST samples are tested for homogeneity by atomic emission and X-ray spectrometry at NIST.

A high ablation rate is essential to remove the scale layer fast and efficiently. To illustrate the enhancement of material ablation by using multipulse bursts, the depths of the laser-induced craters on the ground side of production samples are measured for SP, DP, and TP laser irradiation. The total energy of the pulses irradiated on the sample is held constant at $E_{total} = 600$ J. The craters are generated

Table 13.4 Concentration ranges of the reference samples and the production control samples of low-alloy steel

Element	N_{LOD}	Concentration range of reference samples (m.-%)	Concentration range of production samples (m.-%)
C	9	0.001–1.03 (0.25)	0.004–0.80
S	18	0.0005–0.06	0.004–0.02
Mn	13	0.0005–1.58 (1.00)	0.040–1.60
P	18	0.0005–0.06	0.002–0.20
Si	14	0.0005–0.63 (0.04)	0.002–0.40
Cu	9	0.0005–0.51 (0.10)	0.005–0.03
Cr	10	0.0015–1.48 (0.26)	0.001–0.04
Al	12	0.0005–0.11	0.001–0.20
Ni	9	0.002–1.99 (0.20)	0.010–0.05
Mo	9	0.001–0.14 (0.5)	0.001–0.05

N_{LOD}, number of reference samples used to determine the LOD by a linear calibration curve. The maximum concentration of this subset of samples is given in brackets

Table 13.5 Crater depth h_c and crater diameter d_c for different laser pulse sequences applying the same total laser energy of $E_{total} = N_{abl} \times E_b = 600\,J$

	N_{abl}	E_b (J)	h_c (mm)	d_c (mm)
SP	3,000	0.2	0.10	1.20
DP	1,500	0.4	0.19	1.21
TP	1,000	0.6	0.27	1.19

with $N_{abl}(SP) = 3{,}000$, $N_{abl}(DP) = 1{,}500$, and $N_{abl}(TP) = 1{,}000$. The presented values are the average of three different sample positions for each pulse sequence (see Table 13.5). The crater depth of $h_c = 0.27$ mm achieved by TPs is greater than the depth achieved by SP by a factor of 2.7 and greater than the depth achieved by DP by a factor of 1.4. The crater diameters do not change significantly for these pulse sequences.

The calibration with reference samples has been evaluated by the LOD_{3s} (cf. Sect. 11.2). For the assessment of the combined preparation and analysis of the production control samples, the congruence R_i and the coefficients of determination r_i^2 are evaluated. R_i is defined as the mean quadratic difference of the determined concentrations on the ground and on the scale side of element i [in analogy to the mean residual, see Sect. 11.3, relation (11.19)]. The aim of the analysis is to improve the congruence, i.e., to reduce R_i:

$$R_i = \sqrt{\frac{1}{N}\sum_{a=1}^{N}(c_{\text{scale},a,i} - c_{\text{ground},a,i})^2}, \qquad (13.1)$$

Table 13.6 LOD_{3s} values without (1) and with (2) laser preparation

	Refer. Line	$LOD^{(1)}$ (μg/g)	$LOD^{(2)}$ (μg/g)	Reference samples
Al	Fe 187	7	7	BR53; Ovako; BAS: 456/1–460/1; NIST: 1761–1764, 1767
C	Fe 187	17	9	BR53; Ovako; BAS: 451/1, 453/1, 456/1, 458/1; NIST: 1763, 1765, 1767
Cr	Fe 187	7	27	BR53; Ovako; BAS: 451/1–455/1; NIST: 1761, 1765, 1767
Cu	Fe 187	6	10	BR53; Ovako; BAS: 453/1–455/1; NIST: 1762, 1763, 1765, 1767
Mn	Fe 271	4	10	BR53; Ovako; BAS: 451/1, 454/1–460/1; NIST: 1761, 1765, 1767
Mo	Fe 187	3	9	BR53; Ovako; BAS: 451/1–453/1, 455/1; NIST: 1761, 1765, 1767
Ni	Fe 187	31	35	BR53; Ovako; BAS: 451/1–454/1; NIST: 1764, 1765, 1767
P	0th order	4	22	BR53; Ovako; BAS: 451/1–460/1; NIST: 1761–1765, 1767
S	0th order	9	19	BR53; Ovako; BAS: 451/1 - 460/1; NIST: 1761 – 1765, 1767

$N_{pp} = 100$, $N_{mp} = 500$. The reference samples used for each calibration are listed in the last column. The notation, e.g., 451/1–453/1, means samples No. 451/1, 452/1, and 453/1 from BAS
(1) $N_{abl} = 0$; (2) $N_{abl} = 1{,}000$

where i is the element, N is the total number of production control samples, a is the index of the sample, and c_{scale}, c_{ground} are the concentrations (μg/g) determined with LIBS on the scale or the ground side, respectively.

To determine the element concentrations of the samples, a suitable calibration method has to be found. Therefore, it is necessary to investigate the influence of the crater geometry, e.g., crater depth and crater diameter, on the analysis step. For the analysis, the analytical performance is compared with and without the ablation step concerning the sensitivity and precision of the calibration. The calibration measurements have been carried out on 18 low-alloy reference samples (see Tables 13.4 and 13.6) with two different pulse sequences: (a) $N_{abl}(TP) = 0$, $N_{pp}(DP) = 100$, $N_{mp}(DP) = 500$, for normal calibration without preablation, and (b) $N_{abl}(TP) = 1{,}000$, $N_{pp}(DP) = 100$, $N_{mp}(DP) = 500$, including the laser preparation. After the ablation step, the analysis starts in a crater with a depth of ~0.27 mm. The crater depth was measured using a white-light interferometric microscope (NewView 200, Zygo Corporation). The calibration curves are evaluated concerning the LOD, sensitivity, and precision. For every analytical line and pulse sequence, calibration curves are taken in intensities each referenced to the reference lines Fe 271.441, Fe 187.747 nm, and the zeroth order signal. The best reference line for each element is listed in Table 13.6. Figure 13.8 shows the calibration curves for carbon and molybdenum with and without the ablation step. Comparing the precision and sensitivity of calibration for both pulse sequences,

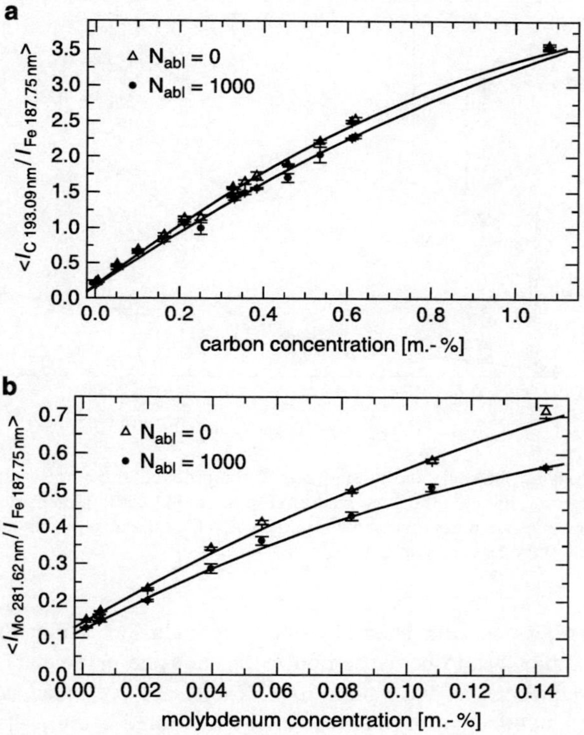

Fig. 13.8 Calibration curves without ($N_{abl} = 0$) and with ($N_{abl} = 1{,}000$) laser preparation for (**a**) carbon and (**b**) molybdenum

there is a general slight deterioration when the ablation step is included. All listed element lines show this behavior. The smallest difference is observed for the element carbon.

The LOD_{3s} for most of the observed elements are below 10 μg/g calibrating without laser preparation; see Table 13.6. For Ni and C, the LOD values amount 31 and 17 μg/g, respectively. Calibration with laser preparation, see above, resulted in higher LOD. For most of the considered elements, there is an increase of the LOD_{3s} by factors between 2.1 (sulfur) and 5.5 (phosphorus). There is no significant change for Al and Ni. The LOD_{3s} value for carbon decreases from 17 to 9 μg/g. The obtained accuracy and precision of analysis depend on the reference line. The best analytical performance for Mn is attained for the referencing to the iron line Fe 271. For phosphorus and sulfur, the zeroth order signal is found to be the better reference signal. For the other six listed elements, the Fe 187 line is the best reference line.

The calibrations described above showed that there is an influence of the crater depth and diameter on the sensitivity and precision of the analytical results. The calibration without the laser ablation step is more precise and more sensitive than

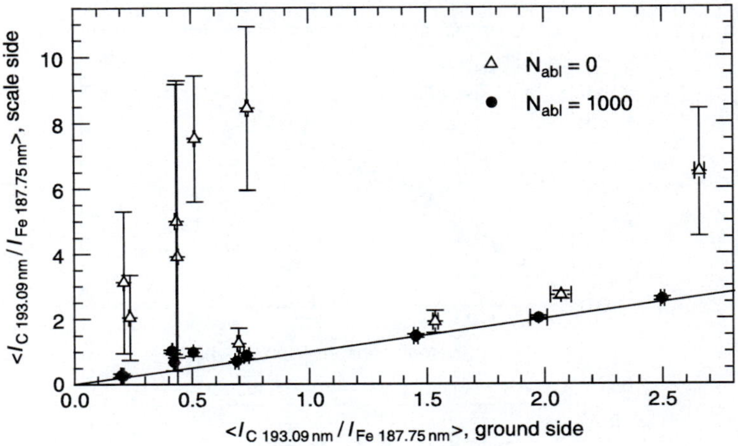

Fig. 13.9 Analysis on the scale and ground side of ten production control samples. *I* (element, wavelength of spectral line) are the integrated PMT signals. The carbon intensities divided by the reference Fe intensities are measured on the scale side and the ground side without ($N_{abl} = 0$) and with ($N_{abl} = 1{,}000$) the laser preparation

the corresponding one with laser ablation. A conclusion is that the analysis of production samples has to be carried out with the same parameters including the same pulse sequences as the calibration. Results are presented concerning the analysis of production control samples that are measured both on the scale and on the ground side. The basis for the determination of the element concentrations of the production control samples is the calibration with five reference samples that cover almost the concentration range of the production control samples. The analysis is carried out with ten production control samples for the elements C, Si, Mn, Cr, Mo, Ni, and Cu. Ten repetitions at different positions of each ground and each scale side of the samples are performed during the measurements. Calibrations with the pulse sequences $N_{abl}(TP) = 0$, $N_{pp}(DP) = 100$, $N_{mp}(DP) = 500$, and $N_{abl}(TP) = 1{,}000$, $N_{pp}(DP) = 100$, $N_{mp}(DP) = 500$ are conducted. The determined concentrations on the ground side are defined as the reference values.

Figure 13.9 illustrates the congruence of the analysis on the scale and the ground side by comparing the spectrometric intensity ratios of carbon and iron. The bisecting line represents the ideal location of the referenced intensities or the determined concentrations (cf. Figs. 13.10 and 13.11). Carbon is the element with the greatest deviation concerning the concentration in the scale layer and the bulk material. The congruence for carbon is decisively improved using the ablation step; see Fig. 13.9.

In Figs. 13.10 and 13.11a–c, only the measurements with the ablation step are shown. Figure 13.10 gives the determined concentrations of a LIBS measurement on the scale side versus the arc/spark measurement on the ground side of the same sample. In Fig. 13.11a–c, the determined concentrations of the LIBS measurements

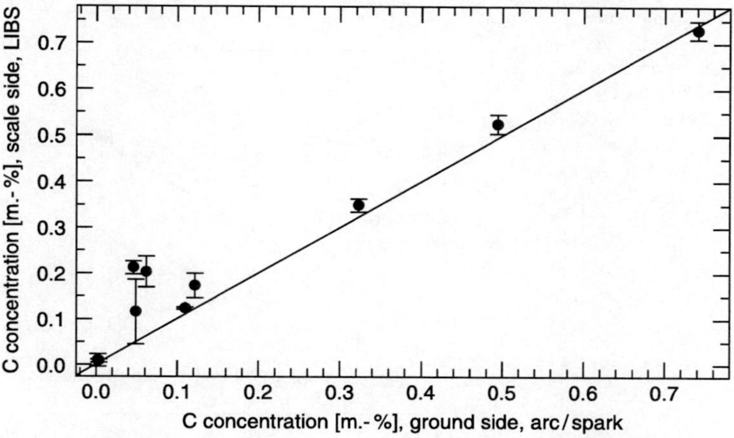

Fig. 13.10 Congruence between the determined concentrations measured by LIBS on the scale side and the determined concentrations measured by arc/spark spectrometry on the ground side for C. Coefficient of determination is $r^2 = 0.944$. Analysis with laser preparation of ten production samples. The calibration of the LIBS spectrometer was carried out with only five reference samples, while the arc/spark spectrometer was calibrated with a greater set of reference samples

of the scale side are compared with those of the ground side. The diagrams of Figs. 13.10 and 13.11a for C do not differ significantly and the coefficient of determination is approximately equal, which means that the comparison of the LIBS concentrations in Fig. 13.11 is suitable.

The coefficient of determination averaged over the observed elements amounts to $r^2 > 0.99$. The corresponding values for the congruence and precision are summarized in Table 13.7. For most elements such as C, Si, Cr, Cu, and Ni, the congruence R_i is reduced. The carbon analysis is strongly improved by a factor of nearly 25 to an absolute value of about $R = 730\,\mu g/g$ applying the ablation step. This factor rises to 50 when two samples assessed as outliers are neglected. This corresponds to a congruence of $R = 289\,\mu g/g$. For the elements Cr, Si, Ni, and Cu, the congruence values are decreased by factors between 1.2 for Ni and 4.1 for Cu. The absolute R_i values vary relatively strongly because of the different concentration ranges of the seven elements studied.

The results demonstrate the potential of the laser-based method for a fast preparation and analysis of production control samples. In an industrial application, the conventional steps of grinding the samples, their transport to a spark OES spectrometer, and analysis can be replaced and combined in a single setup for laser preparation and analysis by LIBS.

For a fast preparation of the sample, the scale removal was strongly improved by applying TP bursts. The laser-induced craters are deeper than the ones generated by SP using the same total laser energy by a factor of ~3. LOD below $10\,\mu g/g$ are achieved for most of the elements analyzed without the laser preparation step. The

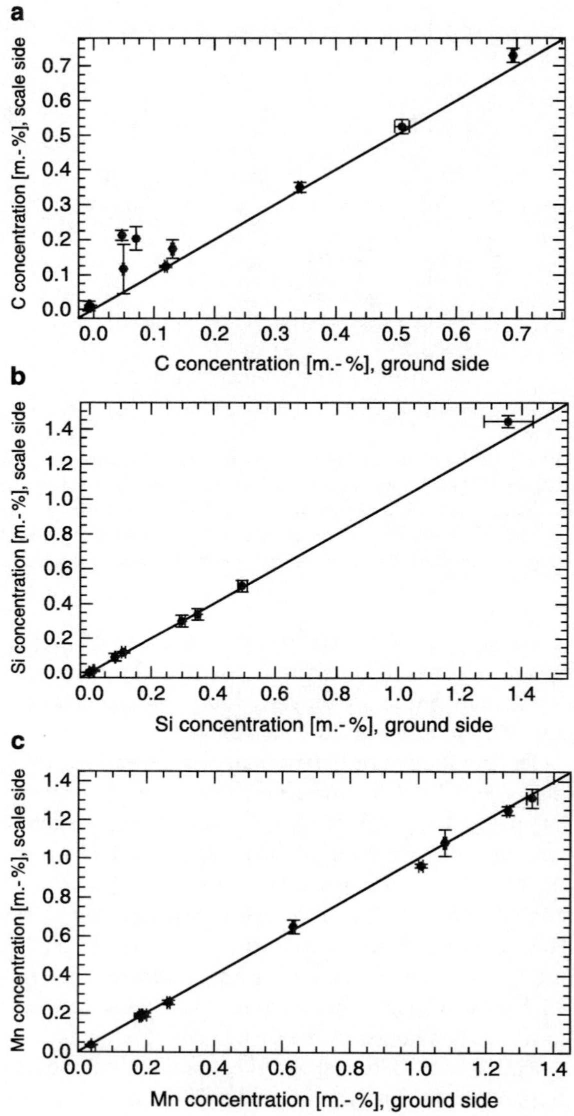

Fig. 13.11 Congruence between the determined concentrations on the scale and ground side for the elements: (**a**) C, $r^2 = 0.946$, (**b**) Si, $r^2 = 0.992$, and (**c**) Mn, $r^2 = 0.991$. Analysis with laser preparation of ten production samples

analytical performance deteriorates slightly when the ablation step is applied prior to the analysis step under the chosen experimental conditions.

The analysis on the scale side attains good congruences for all elements and high coefficients of determination concerning the analytical results on the conventionally ground side. Especially, the analysis of carbon is decisively improved.

Table 13.7 List of the congruence values R_i of seven elements including the analysis of ten production samples without (1) and with (2) the ablation step

Element	$R_i^{(1)}$ (μg/g)	$R_i^{(2)}$ (μg/g)	$R_i^{(1)}/R_i^{(2)}$
C	17,940	730	25
C[a]	14,511	289	50
Cr	110	50	2.1
Cu	160	38	4.1
Mn	190	190	1.0
Mo	19	19	1.0
Ni	48	39	1.2
Si	350	280	1.3

In the last column, the factor of improvement of analysis with laser preparation is listed

[a]Without two samples assessed as outliers. (1) $N_{abl} = 0$, (2) $N_{abl} = 1{,}000$

13.1.3 *Liquid Steel*

The increasing demands on high purity steel with lower impurity contents and a closely specified composition is a challenge to secondary metallurgy as well as to chemical analysis in steel industries. One major task is to adjust the multielement steel composition during ladle processing more precisely and to reduce the costs by minimizing the time effort, energy, and material consumption. It is essential for the competitiveness of steelmaking industry to develop new techniques, which enable the steel works to meet these requirements for ladle metallurgy and to produce more efficiently high purity steel within narrow tolerance limits.

Fast cycle times for primary melting units and the use of extended casting sequences result in net available residence time in a ladle, which can be as short as about 30 min. Conventionally, monitoring of the refining operations by a calibrated multielement analysis plays an important role to optimize the process control in ladle metallurgy [13.16, 13.17]. At present, monitoring of the melt composition takes place off-line and discontinuous by taking samples. The solidified samples are transported for subsequent analysis to the laboratory. Considerable progress has been made with regard to analysis duration and analytical performance of fast analysis of solid steel samples. With optimum local conditions, analysis durations of about 5 min including sample taking are achieved using conventional analyzing methods such as XRF spectrometry and SD-OES.

Nevertheless, the samples are only spot checks of the melt composition with regard to time and space, and reactions to irregularities of the process have a delay of at least 5 min. An instantaneous control of the process is not possible.

In order to overcome this limitation in future, it is necessary to get an increased data acquisition rate for the chemical analysis of the actual state of the steel melt, which can be provided to the steel worker for an efficient control of the refining processes. The homogenizing of the melt can be observed directly as well as irregularities and the end of the process are recognized earlier.

The fast multielement laser-based analysis performed directly by the immersion of a moveable lance into the melt is a methodical approach, which has the potential to carry out this task during ladle processing. Therefore, the goal of the investigations is to improve the sensitivity of LIBS and to demonstrate its capability for a fast quantitative multielement inline analysis for ladle metallurgy. The emphasis lies on the analysis of the light elements carbon, phosphorus, and sulfur in liquid steel.

Several papers were published presenting investigations of LIBS for the analysis of liquid steel. LIBS for inline analysis in a steel converter was studied using an access of the laser beam to the melt via a gas-flushed hole in the sidewall of the converter. In plant trials, an LOD for carbon of about 200 μg/g was demonstrated [13.18]. The analysis of molten steel in a small crucible heated with an induction furnace was investigated with a laser beam access irradiating a free melt surface from the top. LOD for carbon of 250 μg/g were reported [13.6]. First tests of an inline LIBS-based analyzer in secondary steelmaking were performed in the Scunthorpe Works in UK on an argon stir station where compositional trimming was carried out prior to continuous casting [13.19]. Changes of the carbon content in the steel melt of about 300 μg/g could be monitored inline. In a laboratory induction furnace, the changes of the melt composition by successive additions of the alloying elements Cr, Cu, Mn, and Ni were studied using LIBS [13.7]. The detection limits reported lie in the range between 530 and 2,070 μg/g for these elements. Light elements were not investigated. In conclusion, the analytical performance of LIBS fulfills for some elements the requirements at the converter, but is not sufficient for a quantitative analysis, especially of the crucial so-called light elements in ladle metallurgy. Typical concentrations of carbon, phosphorus, and sulfur to be measured at the ladle are within the range of 5–100 μg/g. Spectrometrical analysis of these elements is more challenging, because the most sensitive emission lines are in the deep UV, or even the vacuum UV wavelength range, e.g., C 193.1, P 178.3, and S 180.7 nm (cf. Sect. 13.1.1). An efficient transmission of this radiation from the laser-induced plasma on the melt surface to the spectrometer is a key issue to achieve high sensitivity.

Investigations of DP generated within a single flashlamp discharge of a Nd:YAG laser for plasma excitation showed a significant improvement of the detection limits [13.20]. With solid steel samples, an LOD for carbon of 90 μg/g and for boron of 6 μg/g was obtained. The plasma emission was observed with a Paschen–Runge spectrometer enabling the simultaneous detection of the line emissions of 15 elements covering a spectral range from 193.09 to 400.87 nm [13.2].

A further improvement of the LOD for light elements in solid steel samples was achieved by a TP excitation and a VUV Paschen–Runge spectrometer observing the plasma emission directly [13.4]. With this approach, detection limits for the elements C, P, S, Mn, Ni, and Cr in low-alloy solid steel samples of less than 10 μg/g were achieved for the first time using LIBS. An LOD for carbon in solid steel samples observing an emission line at 97.70 nm of 1.2 μg/g was reported [13.13]; however, in this investigation an ambient pressure in the target–laser

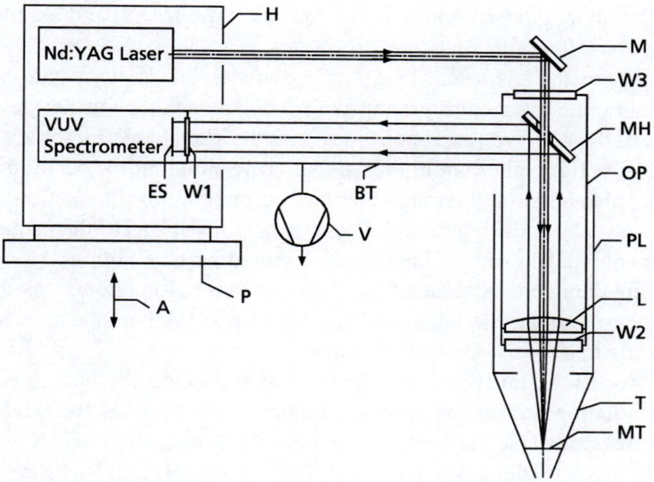

Fig. 13.12 Schematic setup for LIBS of liquid steel. M = mirror; L = lens; W1, W2, W3 = windows; MH = mirror with a hole; OP = optical probe; PL = permanent part of measuring lance; T = tip of lance; MT = steel melt; P = platform; H = housing; ES = entrance slit of spectrometer; A = vertical axis; V = vacuum pump; BT = beam tube

interaction region of 0.3 mbar was applied, which is not feasible for an inline analysis of liquid steel.

Figure 13.12 shows the experimental setup [13.8]. Laser source, spectrometer, and signal electronics are installed in a thermally stabilized housing. A Q-switched Nd:YAG laser operating at 1,064 nm with a repetition rate of 10 Hz was used to excite the plasma. The Q-switching electronics was modified to generate up to three separated laser pulses within a single flashlamp pulse. The laser emits three equal-energy pulses with a burst energy of 110–125 mJ [13.4]. The pulse width (FWHM) of each pulse amounts to about 16 ns. The interpulse separation between the first and the second pulse amounts to $\Delta t_1 = 25\,\mu$s, and between the second and the third pulse $\Delta t_2 = 42\,\mu$s. The laser beam is guided via a mirror to an optical probe (OP in Fig. 13.12) which contains the optical components to guide the laser beam and the plasma radiation. The OP is a gas-tight tube which is evacuated by a vacuum pump to a pressure of less than 0.1 mbar in order to achieve a high transmittance for the UV emission lines below 200 nm (cf. Fig. 3.15). The laser beam passes an entrance window (W3 in Fig. 13.12) of the OP and propagates from the rear side through a hole in an aluminum-coated mirror (MH) to the focusing lens. The focal length amounts to 216 mm (at 1,064 nm) and 165 mm (at 150 nm). The laser beam is focused on the surface of the steel melt inside the tip of the lance. The light emitted by the laser-induced plasma is collected by the focusing lens and propagates via the mirror MH and a concave aluminum-coated mirror (not shown in Fig. 13.12) to the entrance of the Paschen–Runge VUV spectrometer. The windows W1, W2, and the lens L in Fig. 13.12 were made of MgF_2 for UV transmittance (cf. Fig. 4.11).

The overall path length between the spectrometer entrance slit and the melt surface amounts to about 2 m.

The spectrometer has a Rowland circle diameter of 750 mm and is equipped with a holographic diffraction grating with 2,400 lines/mm. Further details of the spectrometer, the observed emission lines, and the signal electronics are described in Sect. 13.1.1. Each photomultiplier signal corresponding to the irradiation of a triple laser pulse burst is integrated using three integration time gates each with a width of $t_{int} = 12\,\mu s$. Each gate is delayed by $t_{delay} = 900$ ns to the respective laser pulse of the burst. The sum of the three integrated parts is digitized for each laser burst yielding the intensity value of the respective emission line for one laser burst. For each measurement, the integrated signals of 600 laser bursts are accumulated corresponding to a measuring time of 1 min.

The OP forms the inner part of a lance, which protects the optical components from heat radiation originating from the induction furnace and the steel melt. The lance was developed at the Institute of Ferrous Metallurgy (IEHK), University Aachen, Germany within the frame of the SELA project [13.21]. The lance consists of a permanent lance (PL in Fig. 13.12) manufactured of copper and a lance tip (T in Fig. 13.12) made of an Al_2O_3 refractory, which is immersed into the steel melt. The refractory material was optimized to reduce any disturbing outgassing during immersion of the tip in the melt especially with respect to the interesting analytes carbon, phosphorus, and sulfur. The permanent part is water cooled to keep the temperature inside the lance below 150°C to avoid damage of the optical components.

The whole measuring equipment is installed on a platform which can be lifted by a hydraulic system in the vertical direction (see P and A in Fig. 13.12). For a measurement, the lance tip is lowered step by step with the lifting apparatus toward the steel melt produced in an induction furnace. The liquid runs into the tip of the lance and forms a melt level. The vertical position of the lance is adjusted while observing the iron signal of the laser-induced plasma with the VUV spectrometer. At the optimum vertical position and thus the position of the melt level with respect to the focal position of the laser beam, the signal height of the iron emission attains a maximum. The estimated position of the beam waist is about 5–10 mm below the melt surface. The immersion depth of the tip in the steel melt amounts to about 20 cm.

Figure 13.13 shows a detailed view of the lower part of the lance and the tip with the flow directions of cooling water (WF in Fig. 13.13) and argon gas. Argon flushing of the volume inside the lance tip is necessary to protect the window of the OP from dust and melt splashes and to maintain a defined UV transmitting atmosphere in the interaction region. The selected argon purity was 4.8 (purity 99.998%); the flushing rate after immersion of the tip into the melt amounts to 18 l/min.

Based on the setup shown in Fig. 13.12, an analyzing device was designed and operated in a melting shop for evaluation. The laser analyzer with laser source, spectrometer, OP, and signal evaluation has been installed on the platform and the lifting device. Assessment of the analyzer was performed by calibrating the setup

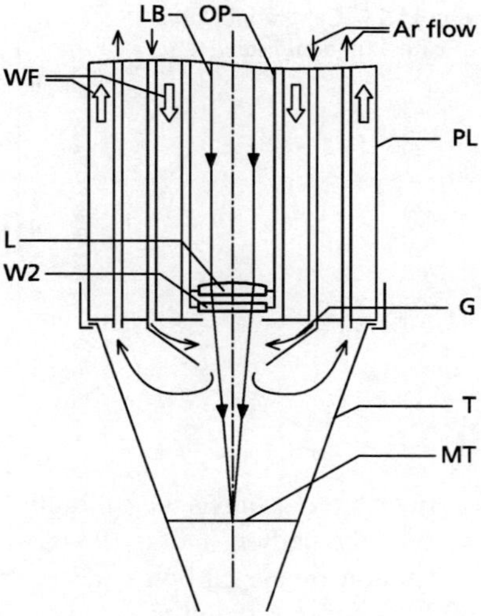

Fig. 13.13 Schematic view of the lower part of the lance. LB = laser beam; G = gas flow. Further abbreviations: see Fig. 13.12

for the elements carbon, phosphorus, sulfur, nickel, and chromium at a 100-kg induction furnace. The initial material Armco ingot iron was melted under an inert gas atmosphere. The Armco ingot iron which is practically the purest steel mill produced iron has a carbon content of about 80 μg/g.

Before starting the measurements in liquid steel, the adjustment of the laser analyzer was verified by the measurement of solid steel samples. The samples were positioned inside the lance probe in such a manner that the sample surface was approximately at the same position as the melt surface during melt analysis. A pure iron sample (e.g., Hu 85) and a low-alloy sample (e.g., BAS 456) were measured. Precipitations on the surface of the entrance window or misalignment of the optical setup due to vibrations during craning and lifting can be determined by this method. The entrance window (W2 in Figs. 13.12 and 13.13) was cleaned approximately every 5 h and replaced after approximately 20 h of accumulated immersion time. The optical setup had to be realigned after approximately 15 h of operation. Before the immersion of the lance the steel melt is deoxidized. Due to the enhanced formation of fumes, the gas-flow rate of Ar 4.8 is increased to 30 l/min during the immersion process. After reaching the required immersion depth the gas-flow rate is reduced to 18 l/min. After melting, alloying, and homogenization steps reference samples are taken from the melt and analyzed after solidification by conventional methods.

During melt analysis, one or more different materials were alloyed in steps of approximately 200 μg/g. After the alloying process, the steel melt is homogenized and the melt volume inside the lance tip is exchanged by lifting and lowering of the lance. The measuring time was 1 min at 10 Hz repetition rate for all elements. Altogether, measurements of 19 different steel melts were carried out at the 100-kg induction furnace. The average immersion time of a typical data collection experiment amounts to approximately 4 h.

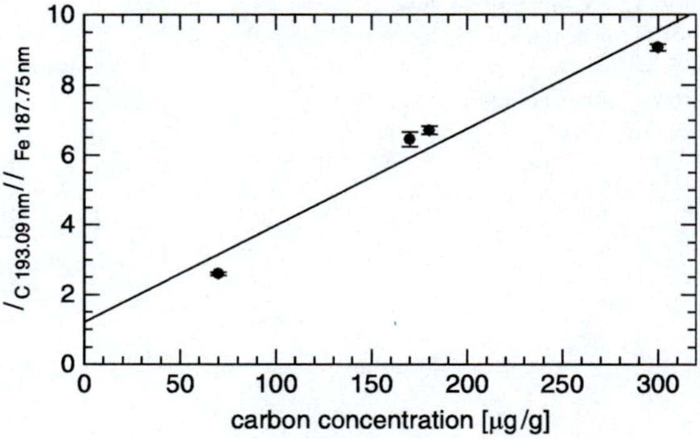

Fig. 13.14 Calibration curve of carbon in liquid steel

Figure 13.14 shows the calibration curve for carbon in liquid steel. The intensity ratio of the carbon line at 193.09 nm and the iron line at 187.75 nm is plotted as a function of the carbon concentration. Alloying of carbon to the melt volume of 100 kg was difficult to achieve since oxidization occurred and homogenization was difficult to establish. Hence, only four data points could be taken. The element concentration given on the abscissa is taken from the conventional analysis of the samples taken after the alloying steps. The error bars are given by the empirical standard deviation calculated from three to five measurements made after each alloying step. The average relative standard deviation of the intensity ratio amounts to approximately 2%. The LOD is estimated by fitting a linear function to the data points using the relation $3s/a_1$, where s is the standard deviation of the intensity ratio measured at the lowest carbon concentration and a_1 is the slope of the linear function [13.12, 13.22, 13.23] (cf. (11.16)). Since the preconditions for the determination of the LOD are not exactly fulfilled and this approach is based on an extrapolation well below the lowest measured carbon concentration, the calculated value for the LOD has to be considered as a rough estimate.

Calibrations were performed for the elements phosphorus, sulfur, nickel, and chromium in liquid steel all measured simultaneously within the same series of measurements. Figures 13.15 and 13.16 show exemplarily the calibration curves for phosphorus and nickel. The LOD are estimated by a fit of a linear function to the lower concentration range. These linear functions are intersecting the vertical axis at a positive value of the respective intensity ratio. This intensity ratio can be interpreted as the expected measuring value for a blank sample, where the intensity of the analyte line approaches a background value. However, the blank sample could not be realized experimentally in the given experimental setup for the liquid steel measurements. The results in terms of the estimated LOD, BEC, (cf. (11.19)), and the coefficient of determination r^2 are summarized in Table 13.8. For the first time, it was demonstrated that LIBS is capable to analyze light elements in liquid steel with estimated LODs in the range of 5–21 µg/g [13.8]. This is an improvement against the results reported before by more than one order of magnitude [13.18, 13.6, 13.19, 13.7]. The results obtained for liquid steel are similar to those gained at solid samples [13.4].

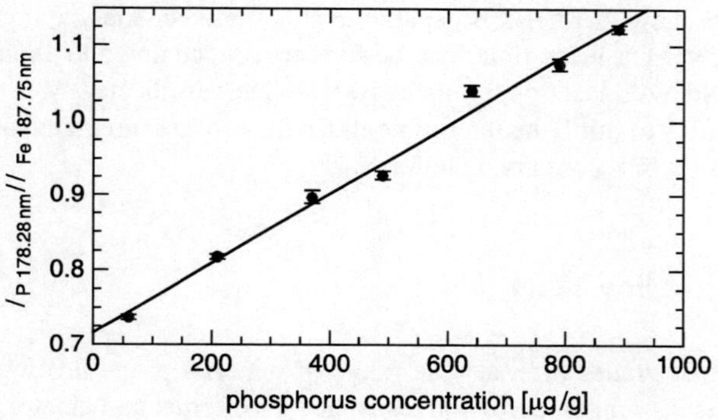

Fig. 13.15 Calibration curve of phosphorus in liquid steel

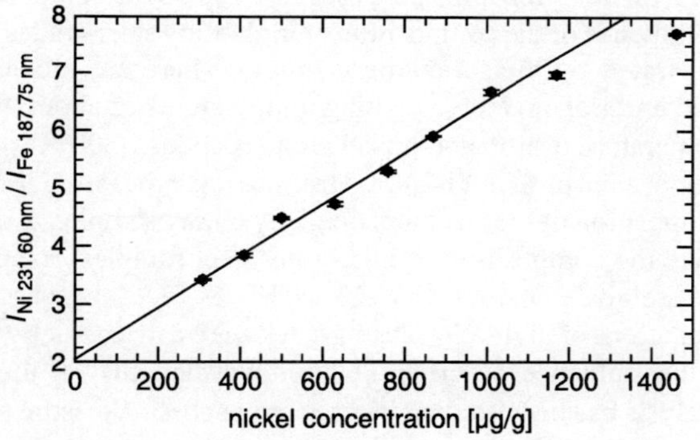

Fig. 13.16 Calibration curve of nickel in liquid steel

Table 13.8 Analyzed elements, estimated limits of detection, background equivalent concentration, concentration range of calibration, concentration range of linear fit, and coefficient of determination for the linear fitting curve for LIBS inline analysis of liquid steel

Element	Estimated LOD_{3s}(μg/g)	BEC (μg/g)	Concentration range of calibration (μg/g)	Concentration range of linear fit (μg/g)	r^2
C	5	45	70–300	70–300	0.950
P	21	1,550	60–890	60–890	0.990
S	11	870	50–1,100	250–1,100	0.993
Ni	9	460	310–1,500	310–1,010	0.989
Cr	9	310	90–900	90–610	0.984

The results show that LIBS is capable to quantitatively analyze liquid steel with high sensitivity. Measuring times can be further reduced down to about 6 s, e.g., by the use of a Nd:YAG laser operating at 100 Hz. The results clearly demonstrate the potential of LIBS to fulfill the requirements for the process integrated inline analysis of light elements in secondary metallurgy.

13.2 High-Alloy Steel

High-alloy steel grades offer a wide range of material properties with respect to, e.g., mechanical strength, corrosion resistance, or thermal and electric conductivity tailored to the needs of usage in industrial applications, home appliances, and medical technology. The constituents of high-alloy steel are mainly the elements Fe, Cr, and Ni with a total content of about 95 m.-%. For process control during steelmaking and for identification purposes, the chemical composition has to be analyzed. The influence of the composition of high-alloy steel grades on the material property is illustrated by the following example. There are nickel alloys, which consist of nickel and iron and have melting temperatures of about 900 °C, whereas the melting temperature of iron and nickel alone is about 1,500 °C.

For production control of high-alloy steelmaking processes, it is important to analyze the composition of the steel melt, e.g., by taking samples. LIBS is a suitable tool to determine the composition of high-alloy steel samples. Examples for LIBS to analyze steel are given in [13.4, 13.6–13.14, 13.18–13.21, 13.24, 13.26].

LIBS investigations of high-alloy steel grades were carried out by several groups. For example, the influence of different laser wavelengths on the accuracy and precision of analysis has been studied in air at atmospheric pressure for the elements Si, Ti, Nb, and Mo [13.11]. Palanco et al. have developed an instrument for a fast quality assessment in the steel industry with sample handling, surface preparation, and quantitative analysis [13.10]. The appearance of so-called matrix effects could be reduced by a multivariate calibration. A system identifying high-alloy steel grades of pipe fittings in routine industrial operation was developed [13.12]. LIBS is able to analyze stainless steel samples at high temperatures in air at atmospheric pressure [13.26]. In a remote LIBS approach, stainless steel has been analyzed at distances between the plasma and the detection system of more than 10 m [13.25]. Bassiotis et al. obtained calibration curves for the elements Cr, Ni, and Mn in stainless steel samples at atmospheric conditions using internal standardization [13.27]. Experimental parameters have been optimized to improve the linearities of the calibration curves over a broad range of concentrations.

In the following, a different experimental setup is described consisting of an Ar-flushed sample stand and a Paschen–Runge spectrometer having a high spectral resolution in a greater spectral range [13.28]. The influence of single and DP excitation and different laser burst energies on the mean residual deviation for the quantitative determination of concentrations is investigated.

The experimental setup is shown in Fig. 13.17. The laser beam generated by a Q-switched Nd:YAG laser working at the fundamental wavelength is guided via

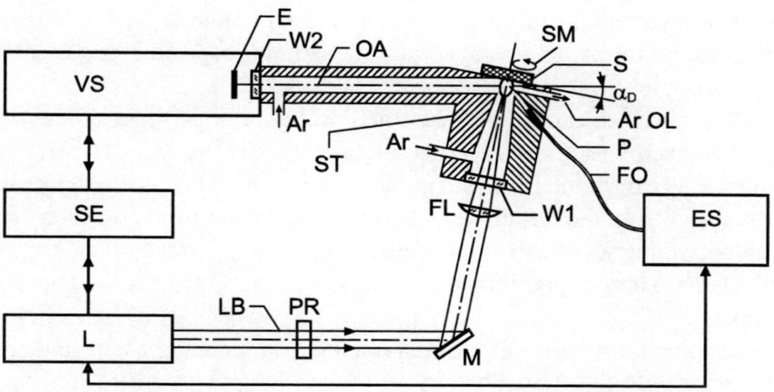

Fig. 13.17 Schematic view of the experimental setup. VS = vacuum spectrometer; SE = signal electronics; L = laser; LB = laser beam; PR = Glan-laser prism; M = mirror; FL = focusing lens; ST = argon-flushed sample stand; Ar = argon inlet; W1 = window of ST; P = laser-induced plasma; FO = fiber optics; ES = echelle spectrometer; Ar OL = Ar outlet; α_D = plasma detection angle; SM = sample movement; OA = optical axis of VS; W2 = entrance window of VS; E = entrance slit of VS

a Glan-laser prism and a mirror to a lens focusing the laser beam into an Ar-flushed sample stand onto high-alloy steel samples. The samples are moved during the measurement so that the laser beam propagates over the sample surface along a circle with a diameter of 3 mm and a frequency of about 1 Hz (cf. Sect. 3.6, Fig. 3.20). This leads to reproducible crater geometries over the whole measuring process. The stationary case – i.e., no relative movement between laser beam and sample surface – has been compared with the moving case yielding the result that the progression of the element intensities with increasing laser pulse number is more uniform.

The plasma light is detected via a direct light channel with a vacuum Paschen–Runge spectrometer (Model RS 1000 from OBLF) without further optical imaging and with a fiber optics and an echelle spectrometer (ESA 3000, LLA, Berlin, Germany). In the Paschen–Runge spectrometer, the plasma light is dispersed spectrally covering a spectral range from 178 to 406 nm (grating, 2,700 l/mm) with a spectral resolution of ~20 pm. The distance of the sample surface to the entrance slit of the spectrometer is ~300 mm. The plasma signals are detected under the detection angle of $\alpha_D \sim 10°$ between the optical axis of the spectrometer and the sample surface. Twenty photomultipliers are located on the Rowland circle (diameter 0.5 m) to detect 19 element lines and the zeroth order signal. The spectral lines installed for the elements Al, C, Cr, Cu, Fe, Mn, Mo, Ni, and Si are those listed already in Table 13.3. Additionally, the Ti line 337.280 nm with $E_j = 3.687$ eV, $\log(gf) = 0.18$ is detected. A signal electronics (MCI; cf. Sect. 4.5) integrates the signals in programmable time windows for each generated plasma.

The parameters kept constant for the calibrations and the determination of the plasma parameters are as follows: number of prepulses $N_{pp} = 50$, number of measuring pulses $N_{mp} = 300$, and flow rate of argon through the sample stand

20 l/min. The focal length of the plano-convex lens amounts to $f = 200$ mm and the focal position is $\Delta s = 5$ mm within the sample to avoid a gas breakdown in front of the sample surface.

The calibrations are carried out with the Paschen–Runge spectrometer in the SP and DP mode each with the laser burst energies $E_b = 100$ and 160 mJ; i.e., $E_b = 2 \times 50$ and 2×80 mJ for the DP (with pulse widths at FWHM of $\tau = 6$ ns for SP and $\tau = 10$ ns for DP). In the DP mode, the interpulse separation is set to $\Delta t = 30\,\mu$s. The signals are integrated with a time delay of $t_{\mathrm{delay}} = 2\,\mu$s and an integration width of $t_{\mathrm{int}} = 10\,\mu$s. Three repetitions on different sample locations are carried out on each sample.

For determination of the plasma parameters, the echelle spectrometer is used in combination with the Ar-flushed sample stand having an optical fiber port. The echelle spectrometer can detect quasi-continuous spectra in the range from 200 to 780 nm with a resolution of about 20 pm. The distance of the plasma to the optical fiber is about 60 mm. The following parameters are taken: SP with a burst energy of $E_b = 160$ mJ, $t_{\mathrm{delay}} = 6, 10, 20, 30\,\mu$s, and $t_{\mathrm{int}} = 400$ ns. Three high-alloy samples and a pure iron sample are measured covering iron concentrations in the range between 43 and 99 m.-%. Two repetitions on different sample locations are evaluated for each sample.

Twenty-one certified high-alloy reference steel samples were analyzed. Table 13.9 gives the reference concentrations of ten elements. The reference samples are certified reference materials of high-alloy steel from BAS, National Bureau of Standards (NBS, Washington D.C., USA), China National Analysis Center for Iron and Steel (NCS, Beijing, China), Carpenter Technology (Crawley, England), Metimex Reference Material (Pyskowice, Poland), MBH Analytical LTD (Barnet, England), Brammer Standard Company (BSC, Houston, USA), and Research Institute for Ferrous Metallurgy (Budapest, Hungary). Most of these samples represent Cr-/Ni-steel grades. For all samples, the iron concentration is greater than 42 m.-%, the chromium concentration is less than 30 m.-%, and the nickel concentration is less than 35 m.-%. The concentrations of the minor elements lie below 6 m.-% for Cu and for the other elements below 3 m.-%. The samples were prepared by grinding with Al_2O_3-grinding paper, grain size 60. The high matrix variation complicates the LIBS analysis as reported in [13.10]. Possible reasons can be spectral interferences from the elements nickel, iron, and chromium and the wide range of material properties, causing so-called matrix effects.

The calibrations are evaluated both for absolute and standardized intensities. For the latter case, the analyte intensities are divided by the intensity of the Fe 271.4 nm line, used as internal standard (cf. Sect. 11.1). The concentration c or concentration ratios $\hat{c} = c_{\mathrm{analyte}}/c_{\mathrm{reference}}$ are functions of the corresponding intensities I or intensity ratios $Q = I_{\mathrm{analyte}}/I_{\mathrm{reference}}$. For the analysis functions, the mean intensity values $\langle I \rangle, \langle Q \rangle$ averaged over three repetitions and second-order regressions are evaluated; see (13.2) and (13.3):

$$c_i = a_{0,i} + a_{1,i} \langle I_i \rangle + a_{2,i} \langle I_i \rangle^2, \tag{13.2}$$

$$\hat{c}_i = b_{0,i} + b_{1,i} \langle Q_i \rangle + b_{2,i} \langle Q_i \rangle^2, \tag{13.3}$$

Table 13.9 Composition of the reference samples of high-alloy steel in m.-%

Reference sample	Fe	Cr	Ni	C	Si	Mn	Mo	Cu	Al	Ti
CRM S 15	74.0	16.70	3.90	0.043	0.260	0.380	2.460	1.540	n.s.	n.s.
CRM S 19	76.1	7.00	12.80	0.260	2.320	0.320	0.110	0.190	n.s.	0.048
NBS SRM C 1288	42.8	19.55	29.30	0.056	0.410	0.830	2.830	3.720	0.003	0.012
NCS HS 23702–1	71.6	15.36	10.35	0.061	0.663	0.644	0.620	0.152	0.044	0.072
NCS HS 23702–2	72.7	12.46	11.52	0.164	0.413	0.981	0.350	0.363	0.074	0.210
NCS HS 23702–3	71.9	9.32	13.12	0.241	0.960	1.980	0.690	0.276	0.038	0.510
NCS HS 23702–4	70.7	7.40	16.41	0.184	0.920	1.570	0.500	0.303	0.240	1.030
NCS HS 23702–5	70.7	17.16	9.05	0.096	0.654	0.730	0.260	0.126	0.260	0.570
NCS HS 23702–6	71.4	20.76	6.93	0.017	0.171	0.156	0.079	0.049	0.034	0.310
NCS HS 23702–7	70.3	18.90	7.82	0.101	1.080	0.356	0.170	0.098	0.200	0.710
ECISS CRM 289–1	55.4	14.64	24.68	0.049	0.531	1.016	1.102	0.000	0.199	2.010
Metimex RM MW 27	45.2	15.57	35.23	0.230	1.950	1.750	n.s.	0.000	n.s.	n.s.
Metimex RM MW 37	58.9	25.00	12.26	0.250	1.100	1.550	0.340	0.050	n.s.	n.s.
BSC RM B.S. 184A	75.4	12.66	8.34	0.035	0.080	0.060	2.200	0.041	1.000	0.051
MBH RM PH 1	73.3	16.10	5.10	0.094	0.180	1.750	0.160	3.120	n.s.	n.s.
MBH RM PH 2	72.8	16.70	3.79	0.070	0.510	0.890	0.940	4.110	n.s.	n.s.
MBH RM PH 3	71.6	15.13	3.16	0.150	1.940	0.440	0.780	6.280	n.s.	n.s.
BSC RM B.S. 85D	67.7	17.09	10.03	0.049	0.550	1.690	0.590	0.450	0.130	0.480
BSC RM B.S. 179A	61.1	25.45	5.84	0.017	0.440	1.040	3.240	1.940	0.009	0.006
Carpenter RM CT 312	58.3	29.80	8.93	0.126	0.460	1.600	0.220	0.150	0.010	0.110
BAS CRM 407/2	93.6	3.03	0.53	0.490	0.660	0.195	0.830	0.397	0.040	n.s.
Mean	67.9	16.0	11.4	0.1	0.8	0.9	0.9	1.1	0.2	0.4

n.s., concentration value is not specified

where $a_{0,i}, a_{1,i}, a_{2,i}, b_{0,i}, b_{1,i}$, and $b_{2,i}$ are the regression coefficients of the second-order regressions of element i and $\langle I_i \rangle, \langle Q_i \rangle$ is the mean intensity value, mean intensity ratio of element i averaged over three measurements at one sample. A measurement on one sample location comprises $N_{mp} = 300$ laser bursts.

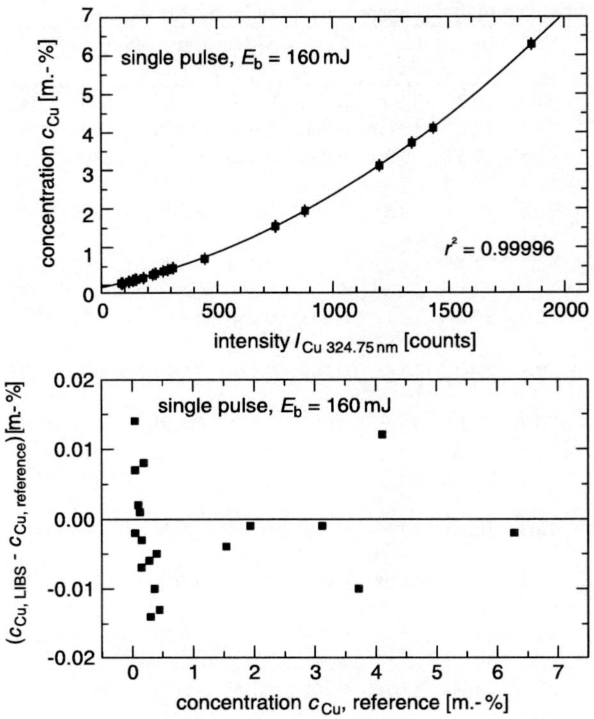

Fig. 13.18 *Top*: Analysis function for copper in the single pulse mode. RSD(I) $=$ 0.9%. R(Cu) $=$ 0.011 m.-%. *Bottom*: Absolute difference between the concentration measured with LIBS and the reference concentration versus the reference concentration

As a figure of merit for the accuracy, the mean residual deviations of ten elements $R(i)$ are determined; see relation (11.21) in Sect. 11.3. The precision is defined as the mean relative standard deviations of the intensity ratios of three replicate measurements; see relation (11.22).

Analysis functions for the element chromium were already shown in Fig. 11.5. Obviously, the differences between the DP and SP calibrations are greater than the differences concerning the different laser burst energies. Whereas the calibrations with DP show greater saturation effects than the corresponding SP calibrations, the sensitivity in the low concentration range is higher than the one for SP. A similar result was found for low-alloy steel samples, where detection limits can be improved by use of DP [13.12]. The ablation rates are higher for DP than for SP of the same laser energy [13.20]. The same behavior can be seen for the element lines of nickel, copper, manganese, silicon, and titanium. These elements are characterized by concentrations higher than 1 m.-%.

Figure 13.18 shows the copper analysis function using absolute intensities. Copper is an element with a small residual value of R(Cu) $=$ 0.011 m.-% in the SP

Table 13.10 Mean residual deviations $R(i)$ of ten element lines for two different laser burst modes and two different laser burst energies and ratio of residuals (DP to SP) using absolute intensities

Element	Samples[a]	Δc (m.-%)	$R(i)^{SP}$ (m.-%)		$R(i)^{DP}$ (m.-%)		$\langle R(i)^{DP}\rangle/\langle R(i)^{SP}\rangle$
			100 mJ	160 mJ	100 mJ	160 mJ	
Ni	21	34.7	0.39	0.37	0.87	1.59	3.2
Cr 267	21	29.7	0.63	0.62	1.45	2.09	2.8
Cr 286	21	29.7	0.50	0.57	1.12	1.63	2.6
Cu	19	6.2	0.038	0.011	0.143	0.127	6.0
Mo	20	3.2	0.055	0.050	0.086	0.127	2.0
Si	21	2.2	0.129	0.101	0.051	0.070	0.5
Ti	14	2.0	0.042	0.048	0.019	0.031	0.6
Mn	21	1.9	0.13	0.12	0.07	0.11	0.7
Al	14	1.0	0.019	0.029	0.018	0.023	0.8
C	21	0.5	0.020	0.023	0.014	0.018	0.7

Δc, concentration range; SP, single pulse; DP, double pulses
[a]For some elements not all of the 21 available reference samples are used, if the concentration is not specified for these elements

mode and for $E_b = 160$ mJ. Concerning the mean copper concentration of 1.1 m.-%, this corresponds to a relative accuracy of 1.0%.

Table 13.10 lists the mean residual deviations $R(i)$ of ten element lines. The elements are sorted by their concentration range Δc. For chromium two element lines, Cr 267.72 and Cr 286.25 nm are evaluated. Because not all of the reference samples have for all elements specified concentration values, the number of regarded samples varies. In the SP mode for most of the elements except for Cu, there is no significant difference between the calibrations for the two laser burst energies. The residuals in the DP mode are significantly greater especially for the matrix elements Cr and Ni but also for the minor elements Cu and Mo. The mean residual values for each element in the SP mode are compared with those in the DP mode. Considering the four different parameter sets SP lead to smaller residuals for the element lines Ni 225.38, Cr 267.72, Cr 286.26, Cu 324.75, and Mo 281.62 nm. The ratio $\langle R(i)^{DP}\rangle/\langle R(i)^{SP}\rangle$ is ≥ 2. The chosen DP parameters show better results for the elements Si, Ti, Mn, Al, and C by a factor of about 0.5–0.8. Comparing the four calibration parameter sets, there is no parameter set that is the best for all elements.

Besides the evaluation with absolute intensities, calibrations are carried out by dividing the absolute analyte intensities to the Fe 271.44 nm-line intensity as an internal standard (cf. Sect. 11.1). By measuring all the elements that can be contained in the steel samples, the real content of the elements can be calculated in m.-% [13.29]. Figures 13.19, top, and 13.20 show analysis functions for the elements Ni and Si. The coefficients of determination are close to one which results in good accuracies; see Table 13.11.

Figure 13.19, top, represents the analysis function for nickel in the SP mode for $E_b = 160$ mJ, which yields the smallest $R(\text{Ni})$-value and the smallest RSD($\langle Q_{\text{Ni}}\rangle$)-

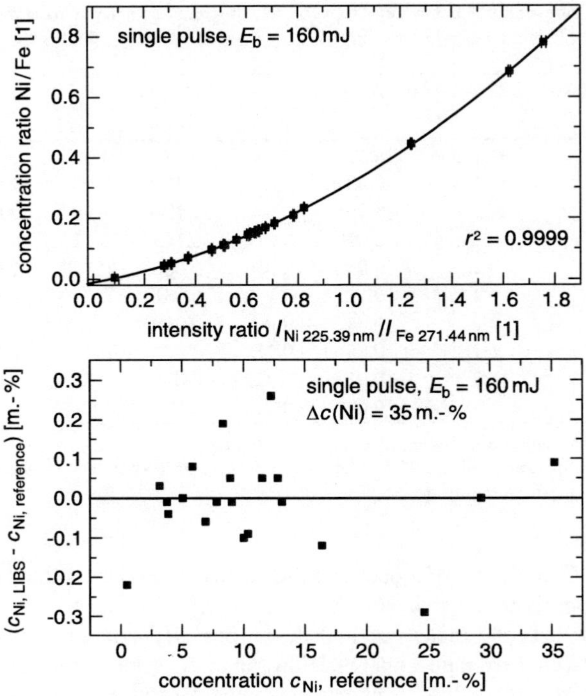

Fig. 13.19 *Top*: Analysis function for nickel in the single pulse mode *Bottom*: Absolute difference between the LIBS concentration and the reference concentration versus the reference concentration

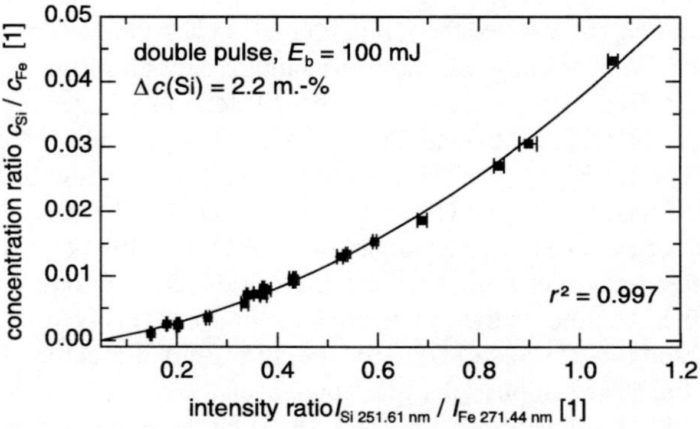

Fig. 13.20 Analysis function for silicon in the double pulse mode

value. The mean residual deviation amounts to 0.13 m.-%. Relating to the mean nickel concentration value of 11.1 m.-% this means a relative accuracy of 1.2%. The precision of three replicate measurements averaged over all samples is 0.3%. Figure 13.19, bottom, illustrates the R(Ni)-values in a different plot to visualize the

Table 13.11 Mean residual deviations $R(i)$ of ten element lines for two different laser burst modes and two different laser burst energies using internal standardization

Element	$R(i)^{SP}$ (m.-%)		$R(i)^{DP}$ (m.-%)		$\langle R(i)^{DP}\rangle/\langle R(i)^{SP}\rangle$
	100 mJ	160 mJ	100 mJ	160 mJ	
Ni	0.17	0.13	0.37	0.38	2.5
Cr 267	1.04	0.91	1.04	1.64	1.4
Cr 286	0.50	0.48	0.61	0.66	1.3
Cu	0.060	0.042	0.251	0.337	5.7
Mo	0.040	0.038	0.055	0.038	1.2
Si	0.150	0.112	0.041	0.058	0.4
Ti	0.063	0.065	0.018	0.020	0.3
Mn	0.144	0.134	0.085	0.050	0.5
Al	0.031	0.042	0.031	0.050	1.1
C	0.020	0.025	0.012	0.016	0.6

Same sample set as in Table 13.10

differences of the measured LIBS concentrations to the reference concentrations. Most of the residuals amount to less than 0.1 m.-%, but all are less than 0.3 m.-%.

In contrast to nickel silicon is an element for which DP of 100 mJ burst energy have the smallest R(Si)-values comparing the four parameter sets; see Fig. 13.20. The residual amounts to $R(\text{Si}) = 0.041$ m.-% with $r^2 = 0.997$. The precision is $\text{RSD}(\langle Q_{\text{Si}}\rangle) = 1.4\%$.

Table 13.11 summarizes the results of the calibrations using internal standardization. The ratios $\langle R(i)^{DP}\rangle/\langle R(i)^{SP}\rangle$ for all regarded elements except for Al exhibit the same tendencies (>1.0 or <1.0) as for absolute intensities. There is an improvement of $R(i)$ for Ni by a factor of about 3 achieving 0.13 m.-% for SP with $E_b = 160$ mJ. In the DP mode, the internal standardization leads to a significant reduction of the $R(i)$-values of Ni, Cr, and Mo for both laser burst energies by factors between 1.4 and 4.2. The $\langle R(i)^{DP}\rangle$-values of the elements Si, Ti, Mn, and C are in the same order of magnitude. In comparison to the calibrations with absolute intensities, the residuals are improved for six element lines (Ni, Cr 286, Mo, Si, Mn, and C) by about 50% comparing the smallest $R(i)$-values for the four parameter sets studied.

Table 13.12 lists the precision of the intensity ratios of three replicate measurements in terms of $\text{RSD}(\langle Q_i\rangle)$ for ten element lines. The precision for the elements Ni, Cr, Cu, and Mo is in the range from 0.3% to 1.2%. In contrast to that the RSD values of the other elements cover a range from 1.1% to 6.2%. Possible reasons for this behavior are the different properties of the analyte element lines compared to the internal standard line Fe 271.4 nm. The greatest difference between the upper energy levels of the analyte line and the reference line occurs for the element Al with $\Delta E_n = 2.4$ eV (cf. Sect. 11.1, relation (11.3) and discussion of homologous line pairs). Furthermore, the spectral distance is high with $\Delta\lambda = 396.152\text{ nm} - 271.441\text{ nm} = 124.711$ nm. These facts lead probably to the highest RSD values for Al.

Table 13.12 Relative standard deviations RSD($\langle Q_i \rangle$) in % for nine elements using ten element lines and internal standardization

	RSD($\langle Q_i \rangle$)^{SP} (%)		RSD($\langle Q_i \rangle$)^{DP} (%)	
	100 mJ	160 mJ	100 mJ	160 mJ
Ni	0.3	0.3	0.4	0.4
Cr 267	0.7	0.4	0.7	0.4
Cr 286	0.6	0.5	0.4	0.3
Cu	0.7	0.7	1.2	1.1
Mo	0.7	0.6	0.8	0.9
Si	3.8	2.5	1.4	1.1
Ti	3.5	3.0	2.6	2.1
Mn	2.3	1.6	0.8	0.6
Al	5.8	5.3	6.2	2.4
C	3.3	2.1	4.1	1.9

Bassiotis et al. determined the RSD values for Mn for eight high-alloy steel samples [13.27]. Five sample positions each irradiated with 150 laser pulses were measured. With the setup described here averaging over the RSD values of 20 samples, the mean values amount to 1.6% for SP and for $E_b = 160$ mJ and 0.6% for DP and for $E_b = 160$ mJ. Compared to the 3.1% reported in [13.27], this is a reduction by a factor of 2 for SP and by a factor of 5 for DP.

As already indicated in Figs. 11.5 and 11.6, the chromium analysis function shows increasing residual values for high Cr concentrations. Deviations observed at high concentrations are significant because of the high precision of the measured intensity ratios ($\leq$0.7% cf.; Table 13.12). In other studies, it has been shown that multivariate calibrations can reduce the residual values [13.10]. In contrast to multivariate calibrations interelement correction is an iterative method. The procedure of interelement correction is described in Sect. 11.5. Table 13.13 lists all multiplicatively and additively corrected residual values $R(i)$ for all four parameter sets and for ten element lines. In the column "disturbing elements" those elements are listed which have significant influence to reduce the residuals. The sign "*" means that this element is used for a multiplicative correction, whereas the sign "+" means an additive correction. In the last four columns, the factors of improvement are listed. By interelement corrections, the mean residual deviations $R(i)$ can be reduced on average for all ten element lines by a factor of 1.7–2.1. The greatest improvement is achieved for Cu by a factor of about 4 in comparison to the uncorrected residuals ($E_b = 160$ mJ, SP).

The question is whether the residuals can be explained with differences in the plasma parameters due to the wide range of material properties of the samples. Hence, the plasma parameters electron density and electron temperature are determined measuring samples with a high variation of the iron, chromium, and nickel content. The plasma parameters are determined for the plasmas of four samples with a high difference concerning the iron concentration. The plasma emission is detected with an echelle spectrometer. Four time delays $t_{delay} = 6, 10, 20,$ and $30\,\mu$s are studied. The spectra for the burst energy $E_b = 160$ mJ in the SP mode are evaluated. The electron densities are calculated by the Stark broadening of the Fe I-line 492.5 nm [cf. Sect. 9.4, (9.19)]. Three high-alloy steel samples and one pure

Table 13.13 Additive and multiplicative interelement corrections for ten elements

Element	Disturbing elements	$R(i)^{SP,corr}$(m.-%)		$R(i)^{DP,corr}$(m.-%)		Factor of improvement[a]			
						SP		DP	
		100 mJ	160 mJ	100 mJ	160 mJ	100 mJ	160 mJ	100 mJ	160 mJ
Ni	*Cr, *Mo, *Ti, *C	0.07	0.08	0.35	0.31	2.7	1.6	1.1	1.2
Cr 267	*Ni, *Ti	0.31	0.30	0.29	0.81	3.4	3.0	3.6	2.0
Cr 286	*Ni, *Ti	0.33	0.32	0.34	0.37	1.5	1.5	1.8	1.8
Cu	*Si, *Mo	0.017	0.010	0.086	0.142	3.5	4.2	2.9	2.4
Mo	+Cr, *Ni, *Cu, *Ti	0.021	0.020	0.041	0.026	1.9	1.9	1.3	1.5
Si	+Cr, *Ti	0.112	0.082	0.043	0.053	1.3	1.4	0.9	1.1
Ti	*Si	0.051	0.051	0.018	0.020	1.2	1.3	1.0	1.0
Mn	*Ti	0.085	0.085	0.084	0.042	1.7	1.6	1.0	1.2
Al	*Ni, +Mo, *Si	0.011	0.011	0.011	0.017	2.8	3.9	2.8	2.9
C	*Si	0.020	0.024	0.007	0.009	1.0	1.0	1.8	1.8
					Mean	2.1	2.1	1.8	1.7

Corrected $R(i)$ in m.-% and factors of improvement. Same sample set as in Table 13.9
[a]Factor of improvement = $R(i)^{uncorr}/R(i)^{corr}$, where $R(i)^{uncorr}$ are the values from Table 13.11

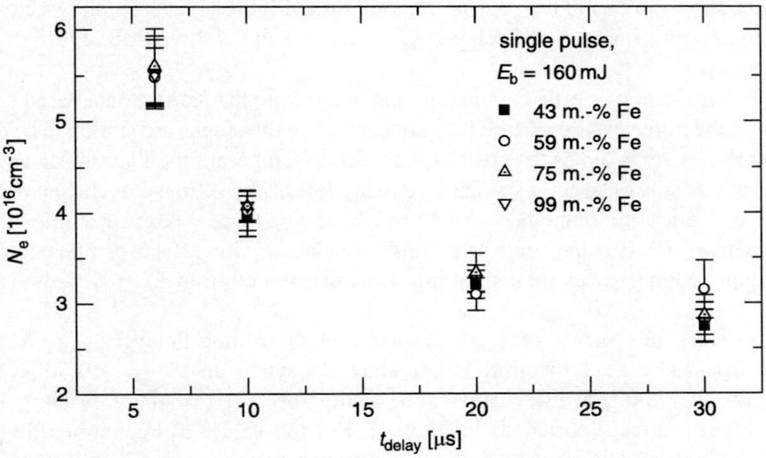

Fig. 13.21 Electron densities N_e for four time delays and four different samples showing a high matrix variation, Fe I-line 492.5 nm. Samples: C 1,288 (43 m.-% Fe), MW 37 (59 m.-% Fe), B.S. 184 (75 m.-%Fe), pure iron (99.99 m.-% Fe)

iron sample are measured covering iron concentrations in the range between 43 and 99 m.-%. Figure 13.21 shows the determined electron density N_e as a function of the delay time. The errors of electron densities N_e are between 4% and 10%. No significant difference of electron densities between the samples is observed.

Table 13.14 Line selection for the Boltzmann plot

No.	λ (nm)	$\log(g_m f_{mn})$	E_n (eV)	No.	λ (nm)	$\log(g_m f_{mn})$	E_n (eV)
1	273.36	−0.06	5.39	12	407.17	−0.022	4.65
2	278.81	−0.02	5.30	13	411.85	0.28	6.58
3	296.69	−0.404	4.18	14	413.47	−0.49	5.83
4	299.95	−0.47	4.99	15	419.91	0.25	6.00
5	356.54	−0.19	4.43	16	421.94	0.12	6.51
6	363.15	−0.036	4.37	17	427.18	−0.164	4.39
7	364.78	−0.194	4.31	18	430.79	−0.07	4.43
8	371.99	−0.431	3.33	19	432.58	−0.01	4.47
9	374.95	0.161	4.22	20	438.35	0.2	4.31
10	404.58	0.28	4.55	21	522.72	−0.969	3.93
11	406.36	0.07	4.61				

21 atomic iron lines from 273 to 523 nm

Besides the electron densities, the electron temperatures are determined for atomic lines of Fe. The line intensities are measured assuming that local thermodynamic equilibrium (LTE) conditions are satisfied in the plasma (cf. Sect. 8.5). The echelle spectrometer offers a wide spectral range for the line selection of the Boltzmann plot; 21 Fe I-lines are evaluated; see Table 13.14 (cf. also Sect. 19.6). As intensities for the Boltzmann plot, the line shapes are fitted by Gaussian and Lorentz profiles and from this fit the peak heights are used. Figure 13.22, top, shows a typical Boltzmann plot for the time delay $t_{\text{delay}} = 20\,\mu\text{s}$ and for three different high-alloy steel samples.

The transition properties of the lines are taken from the Kurucz data base [13.30]. Despite the high variation of the iron concentration the slopes are similar. The errors of the slopes are taken as the error of the electron temperatures. The significance for the evaluation is enlarged by regarding many lines. These lines are chosen because in the concentration range between 43 and 100 m.-% there is a linear relationship of the intensities versus the concentration. So self-absorption effects can be neglected. The upper energy levels are nearly uniformly distributed from $E_n = 3.4\,\text{eV}$ to about 6.5 eV.

The electron temperatures are evaluated at three time delays: $t_{\text{delay}} = 6, 20$, and $30\,\mu\text{s}$. The Fe I electron temperature decays from $T_\text{e} = 10{,}500\,\text{K}$ after $6\,\mu\text{s}$ to $T_\text{e} = 8{,}000\,\text{K}$ after $30\mu\text{s}$. The value for $t_{\text{delay}} = 6\,\mu\text{s}$ is comparable with temperatures reported so far [13.31]. But the values at later times lie about 1,000 K higher. The errors of the electron temperatures are between 3% and 5%. No significant difference between the samples is observed. So the origin for matrix effects – as, e.g., observed for Cr – does not show up in a significant way looking at the electron densities and electron temperatures. The reason for this might be that the effect of improvement by interelement corrections is in the same order of magnitude as the error of determination of the plasma parameters. For example, the Cr 267.72 nm residual can be reduced by 0.6 m.-% from 0.91 m.-% (see Table 3.11) down to 0.30 m.-% (see Table 13.11). Relating the value of the absolute improvement 0.6 m.-% to the mean Cr concentration of 16.0 m.-% of the measured sample set, this corresponds to an improvement by about 4%.

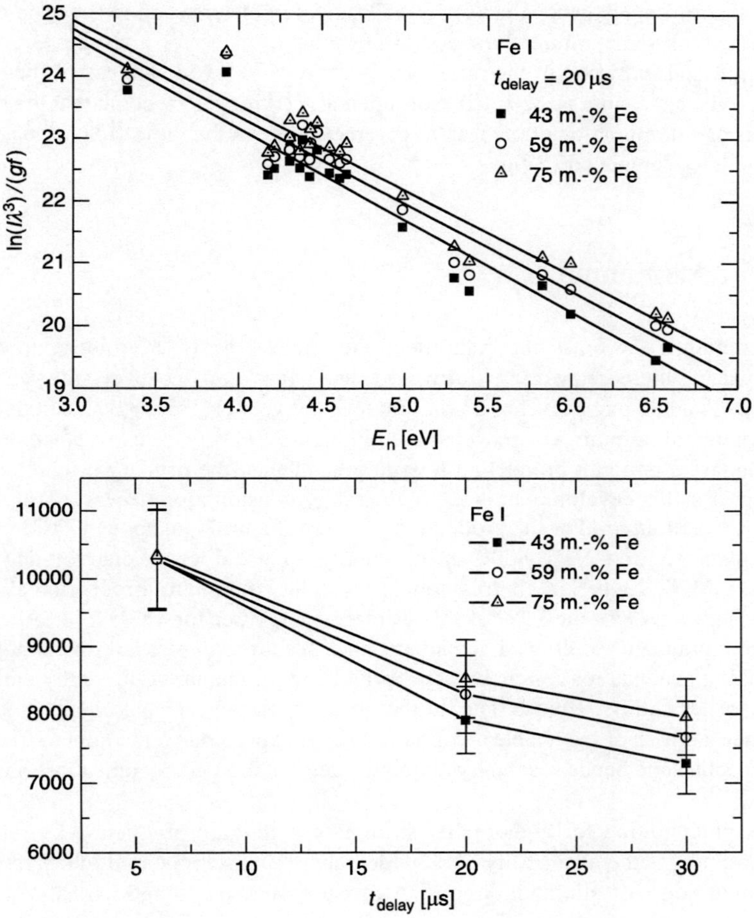

Fig. 13.22 *Top*: Boltzmann plot with 21 Fe I-lines. Relative error of the slopes about 3–5% [The argument of the ln-function is the same as in Fig. 8.15; however, instead of $A_{nm}g_n$ the term $g_m f_{mn}$ is taken, see (9.8)]. *Bottom*: Electron temperatures T_e for three time delays and three high-alloy steel samples

Calibrations were carried out for nine elements with single and double laser bursts and two different laser burst energies. The residuals are dependent on the laser burst mode. SP calibrations have the smallest residuals for the matrix elements Cr and Ni and for the minor elements Cu and Mo, and DP for the elements Si, Ti, Mn, and C. With the presented setup, precisions of intensity ratios are achieved in terms of RSD values lower than 1.0% for most analytes. Interelement corrections can improve the LIBS analysis on high-alloy steel samples on average by a factor of two for matrix and minor elements. As a trend interelement corrections reduce the residuals stronger for SP than for DP. Plasma parameters such as electron density and electron temperature show no significant difference for a strong variation of the iron concentration in the range of 40–75 m.-%. Maybe changes of the plasma state exist but cannot be detected experimentally taking into account that the relative errors of determination of the plasma parameters are of the same order of magnitude as the interelement corrections.

13.3 Aluminum

Aluminum is the most important metal after steel [13.32]. Aluminum production and use has increased worldwide by more than a factor of 10 between the years 1957 and 2003 achieving a level of 32 million tons per year [13.33–13.37]. About 25% is secondary aluminum, i.e., recycled aluminum [13.38]. In Germany, since 2003 the secondary aluminum production is even greater than the primary production. The reason for this development is the high energy consumption necessary to produce primary aluminum. For the production of 1 ton of aluminum about 170 GJ primary energy is required. Nearly 80% or this energy is needed for the electrolytic process [13.32, 13.39, 13.40]. In contrast to this secondary aluminum production allows to save between 88% and 95% of this primary energy per ton [13.32, 13.39]. This is an economic and ecological advantage since resources are saved; the charging of dumping grounds for residual materials and environmental pollution by emissions and wastes can be reduced. Due to the growing demand for aluminum products a further increase of recyclable aluminum scrap is expected in a medium to long-term perspective and hence secondary aluminum production will become more and more important.

A precondition for high-quality secondary aluminum production by remelters and refiners is the availability of shredded aluminum scrap sorted into fractions of wrought and cast aluminum alloys. A mixture of wrought and cast alloys would result in higher contents of alloying elements such as silicon and copper, which restricts the reuse to cast alloys only, since the separation of these interfering elements is barely possible or requires elaborate metallurgical processes. Hence for the production of wrought alloys from secondary aluminum, the aluminum scrap has to be melted with primary aluminum or has to be sorted [13.41].

Conventional analyzing and sorting techniques do not permit an economic separation of light metal alloys and their alloy groups. The task is to separate particularly aluminum alloys in fractions of cast and wrought alloys as well as in their different alloy subgroups under industrial conditions with a throughput of 2–4 t/h.

The methodical approach for this task is based on a combination of image processing, laser-based geometry detection, and LIBS. By means of the optical and geometrical features of the individual scrap particles, moving at a speed of 3 m/s on a conveyor belt, a pulsed laser beam is guided by a 3D-scan unit to the scrap particles and the chemical composition is determined by LIBS. Based on all measured quantities, the individual scrap particles are classified and automatically separated into two or more fractions. The pilot system developed for single particle aluminum scrap sorting based on this approach will be presented in Sect. 18.4. In the following, the combination of LIBS with a scanner and the analysis of static and moving aluminum samples is described.

The experimental setup is shown in Fig. 13.23 [13.42]. The laser beam can be directed onto selected surface positions by an *xy*-scanning unit with a field size of about $400 \times 400\,\text{mm}^2$. An autofocus unit consisting of a fast movable diverging and a fixed focusing lens with a focal length of $f = 1\,\text{m}$ in neutral position allows rapid changes of the focal point within a total range of 100 mm along the optical axis, so that the focal plane can be adjusted to different heights of the samples. In order to realize an economic process, the throughput of identified and sorted scrap particles has to be at least 2–4 t/h. Considering the speed of the conveyor belt (2–4 m/s) and the average mass of the particles, about 50 particles per second have to be identified. For this reason, a high-repetition Nd:YAG laser system with 50 Hz repetition rate and a Paschen-Runge spectrometer equipped with photomultipliers are used. The measuring rate allows for one to seven laser bursts per scrap particle for material identification. The plasma emission counterpropagating to the incoming optical axis is directed onto a dichroic mirror located on the laser beam axis between the *xy*-scanning unit and the autofocus telescope at an angle of 45 degrees. The emission is deflected by this mirror perpendicularly to the laser beam axis to a quartz lens (f–value 2) and then imaged onto the input aperture of an optical fiber coupled to the spectrometer. For the determination of spectral lines, spectroscopic investigations on different kinds of light metal alloys were carried out. A selection of 20 lines has been fixed to detect the elements Al, Si, Cu, Fe, Mg, Mn, Sn, Ni, Cr, Zn, Ti, Zr, Na, Ca, C, Cd, and Pb simultaneously. The signals are digitized and evaluated in real time. Reference samples of certified aluminum alloys with known compositions are analyzed statically and while moving on a belt conveyor at 3 m/s by LIBS.

With this setup, calibration measurements were carried out with static and moving certified reference samples. In the static case, the samples were measured at the center of the scan field (pos. A in Fig. 13.23). At each of four different sample positions, 25 measuring laser bursts were irradiated (cf. Sect. 11.1, $K = 4$, $N_{mp} = 25$). Each burst consisted of a DP with a total burst energy of $E_b = 170\,\text{mJ}$, a pulse energy ratio of $E_1 : E_2 = 1 : 2$, and a temporal interpulse separation of $\Delta t = 15\,\mu\text{s}$. In order to measure at moving particles, the samples are put centrally onto the conveyor belt moving along the y-direction in Fig. 13.23. A 3D-geometry detection unit acquires the temporal ($y = y(t)$) as well as the lateral sample position x and transmits the calculated target position for the laser focus to the scan head in real time. In contrast to the static case, the measurements at moving particles comprise $N_{mp} = 7$ measuring laser bursts at one sample position tracked by the scanning mirrors at a conveyor belt velocity of 3 m/s. The burst energy was set to $E_b = 190\,\text{mJ}$ with $E_1 : E_2 = 1 : 1$ and $\Delta t = 40\,\mu\text{s}$.

The measurements were carried out with 36 samples of certified aluminum alloys. This set of samples covers various concentration ranges of the major alloying

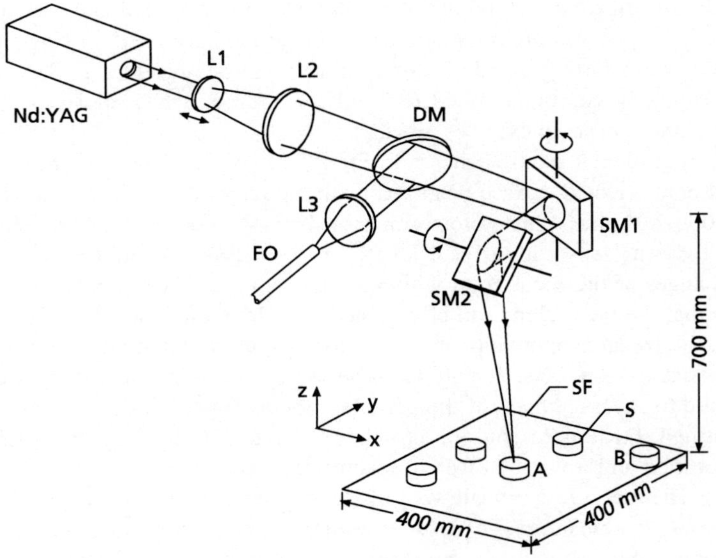

Fig. 13.23 Experimental setup for LIBS via an xy-scanner. L1–L3 = lenses; DM = dichroic mirror; SM1, SM2 = xy-scanning mirrors; FO = fiber optic to spectrometer; SF = scan field; S = sample. The conveyor belt moves along the negative y-direction

Table 13.15 List of analyte emission lines and concentration ranges covered by the reference samples

Element	λ (nm)	c_{min} (m.-%)	c_{max} (m.-%)
Si I	288.16/390.55	0.019	13.03
Fe II/Fe I	259.94/371.99	0.0167	4.93
Cu I	296.12/327.40	0.0006	12.2
Mn II	293.31	0.000004	10.8
Mg II	279.08	0.000062	5.1
Cr II	267.72	0.000002	0.37
Ni I	341.48	0.0007	2.03
Ti II	337.28	0.00034	0.26
Zn I	334.50	0.000008	11.6
Al I	305.71	80.0642[a]	99.99918[a]

[a]Concentration values for Al are not certified and calculated with: $c(\mathrm{Al}) = 100\,\mathrm{m.\text{-}\%} - \sum c_{\mathrm{certified\ elements}}$

elements Si, Fe, Cu, Mn, Mg, Cr, Ni, Ti, and Zn. Table 13.15 shows a list of these concentration ranges together with their corresponding analyte lines.

The goal of a calibration for sorting applications is to achieve the best possible accuracy and selectivity over a wide concentration range up to 15 or 20 m.-% with a few laser pulses. Hence as characteristic numbers of the analytical performance besides the LOD, the relative standard deviations of procedure (RSDP) will be considered [cf. Sect. 11.2, (11.13)].

The LOD is the threshold concentration at which the existence of a component can be confirmed with a defined level of significance α (e.g., $\alpha = 0.01$). According to the linear calibration method, LOD values can be estimated by [13.23]:

$$\mathrm{LOD}_{\mathrm{estim.}} = 1.2\Phi_{K,\alpha} s_{\mathrm{co}}, \tag{13.4}$$

where $\Phi_{K,\alpha}$ denotes a factor depending on the number of measurements K (sample positions) and s_{co} the standard deviation of procedure (in (11.16) the so-called blank value method was applied, whereas the relation given here refers to the linear calibration method). The linear calibration function $I(c) = a_0 + a_1 c$ is determined by pairs of mean values of the measured Intensity $\bar{I}_i = \langle\langle I_i \rangle_{N_{\mathrm{mp}}}\rangle_K$ and certified concentrations $c_i\,(i = 1, \ldots, N_{s,i};\ N_{s,i}$ = number of samples used for the calibration function). With the residue standard deviation $s_{I,c}$ the standard deviation of procedure s_{co} is given by

$$s_{\mathrm{co}} = \frac{s_{I,c}}{a_1} = \frac{1}{a_1}\sqrt{\frac{\sum\limits_{i=1}^{N_{s,i}} (I(c_i) - \bar{I}_i)^2}{N_{s,i} - 2}}. \tag{13.5}$$

In case of a quadratic regression function, the denominator of the root is replaced by $N_{s,i} - 3$. For $K = 4$ measurements per sample, the $\mathrm{LOD}_{\mathrm{estim.}}$ can be derived by use of $\Phi_{4,0.01} = 5.1$. Table 13.16 summarizes estimated values for LODs in comparison to other publications [13.43–13.46]. $\mathrm{LOD}_{\mathrm{min\,/\,max,ref}}$ denotes the lowest/highest detection limit given in those references. The $\mathrm{LOD}_{\mathrm{estim.}}$ were determined by the use of an individual subset $N_{s,i}$ for each analyte. In order to eliminate outliers, the intensity values used for the calibration functions were twice tested with a 3s-outlier criterion; i.e., $\bar{I}_i = \langle\langle I_{\mathrm{LIBS},i} \rangle_{N_{\mathrm{mp}},3s}\rangle_{K,3s}$. To each calibration function, reference samples were added in ascending order of concentration until the coefficient of determination exceeded $r^2 = 0.9$.

On the whole, the estimated $\mathrm{LOD}_{\mathrm{estim.}}$ are higher than the maximum given $\mathrm{LOD}_{\mathrm{max,ref}}$ values except for Cu. The reason for this difference lies primarily in the different experimental setup and measuring parameters used. In [13.43], the focal

Table 13.16 Estimated detection limits of major alloying elements for certified aluminum alloys in the static case in comparison to LOD values published elsewhere

Element	λ (nm)	$N_{s,i}$	$c_{i,\mathrm{min}}$ (m.-%)	$c_{i,\mathrm{max}}$ (m.-%)	r^2	s_{co} (m.-%)	$\mathrm{LOD}_{\mathrm{estim.}}$ (µg/g)	$\mathrm{LOD}_{\mathrm{min,ref}}$ (µg/g)	$\mathrm{LOD}_{\mathrm{max,ref}}$ (µg/g)
Si I	288.16	14	0.019	0.24	0.93	0.0056	341	14 [13.43]	284 [13.45]
Fe II	259.94	10	0.0167	0.1634	0.96	0.0039	241	50 [13.46]	–
Cu I	327.40	11	0.0006	0.042	0.91	0.0011	67	10 [13.43]	80 [13.46]
Mn II	293.31	17	0.0008	0.2	0.93	0.0022	135	2 [13.43]	90 [13.46]
Mg II	279.08	9	0.0001	0.028	0.90	0.0007	44	0.5 [13.43]	28 [13.46]
Cr II	267.72	26	0.00047	0.14	0.90	0.0039	242	10 [13.46]	–
Ni I	341.48	29	0.0007	1.23	0.96	0.0148	903	100 [13.46]	–
Ti II	337.28	10	0.00034	0.0285	0.91	0.0019	119	10 [13.46]	–

length for the laser beam focusing amounts to 25 cm, the laser repetition frequency is 1 Hz, and the plasma light was collected by a $f = 10\,\text{cm}$ quartz lens. In [13.45], the laser repetition frequency was 0.2 Hz, i.e., by two orders of magnitude smaller than the one used here. In [13.46], the repetition rate was 1 Hz, the lens to focus the laser beam had a focal length of 10 cm and the plasma light was observed by a fiber at a distance of 3 cm from the plasma.

Hence in [13.43–13.46], the influence of a breakdown on the particles ablated by previous pulses is greatly reduced by choosing a low laser repetition frequency and the light collection is maximized by keeping the distance of the collection optics close to the plasma ($\leq$10 cm). Thus, the reproducibility of the measurements is increased. However, such low laser repetition rates are not suited for the desired sorting task. Furthermore, the small focal length and observation distances used in these references are not adequate to cover a measuring field of 400 mm × 400 mm.

Furthermore, the factor $\Phi_{4,0.01}$ used in (13.4) differs from $\Phi_{10,0.01} = 3$ used in the stated references in Table 13.16. Finally, the $\text{LOD}_{\text{estim.}}$ are increased by the factor 1.2 in (13.4) in contrast to the reference values and are determined by raw intensities.

In order to characterize the experimental setup and its performance over large concentration ranges, calibrations for the analyte elements Si, Fe, Cu, Mn, Mg, Cr, Ni, and Ti were carried out. In the following, the relation between outlier corrected intensities ($3s$) and concentrations is approximated by the quadratic analysis function (cf. Chap. 11, Fig. 11.1) $c(I) = b_0 + b_1 I + b_2 I^2$, which is the inverse function of the calibration function. The ratio method is applied to improve the accuracy of the approximated function by internal standardization [cf. Chap. 11, (11.2)]. The intensity values of an analyte I_{analyte} are divided by the corresponding intensities of the reference element $I_{\text{reference}}$, which is in most cases the matrix element. For each sample, the concentration values are referenced as well. The analysis function is the approximation curve to $N_{s,i}$ pairs of $(c_{i,\text{ref}}, \bar{Q}_{i,\text{ref}})$ with

$$c_{i,\text{ref}} \equiv \frac{c_{i,\text{analyte}}}{c_{i,\text{reference}}}, \; \bar{Q}_{i,\text{ref}} \equiv \left\langle \langle Q_{i,\text{ref}} \rangle_{\text{N}_{\text{mp}},3s} \right\rangle_{K,3s}, \; Q_{i,\text{ref}} \equiv \frac{I_{i,\text{analyte}}}{I_{i,\text{reference}}}. \tag{13.6}$$

The diagrams in Fig. 13.24 show the analysis functions for the Si I 288.16 nm analyte line with and without internal standardization to the Al I 305.71 nm reference line. The improvement of the calibration by use of the ratio method can be determined comparing the residue standard deviations $s_{c,I}$ and $s_{c_{\text{ref}},Q_{\text{ref}}}$. To be correct in units, the nondimensional residue standard deviation $s_{c_{\text{ref}},Q_{\text{ref}}}$ has to be multiplied by the average Al-concentration of the used sample subset $\bar{c}(\text{Al}) \equiv \langle c_i(\text{Al}) \rangle_{N_{s,i}}$. In Table 13.17, the residue standard deviations of standardized and nonstandardized analysis functions for the different alloying elements are listed as well as their quotient as a factor of improvement. The column at the very right is an estimation of the RSDP relating to the average analyte concentration in the particular subsets of samples.

Table 13.17 Residue standard deviations without and with internal standardization and estimations of RSDP (last column)

Analyte element	λ (nm)	$N_{s,i}$	$\bar{c}(\mathrm{Al})$ (m.-%)	$s_{c,I}$ (m.-%)	$s_{c_{\mathrm{ref}},Q_{\mathrm{ref}}}\bar{c}(\mathrm{Al})$ (m.-%)	$\frac{s_{c,I}}{s_{c_{\mathrm{ref}},Q_{\mathrm{ref}}}\bar{c}(\mathrm{Al})}$	$\bar{c}_{\mathrm{analyte}}$ (m.-%)	$\frac{s_{c_{\mathrm{ref}},Q_{\mathrm{ref}}}\bar{c}(\mathrm{Al})}{\bar{c}_{\mathrm{analyte}}}$ (%)
Si I	288.16	31	91.1871	0.9652	0.7613	1.27	2.5358	30.0
Si I	390.55	31	91.1871	0.6991	0.2632	2.66	2.5358	10.4
Fe II	259.94	30	91.0898	0.1478	0.1189	1.24	0.4909	24.2
Fe I	371.99	30	91.0898	0.2129	0.1423	1.50	0.4909	29.0
Cu I	296.12	29	91.5514	1.2772	1.0337	1.24	0.8987	115.0
Cu I	327.40	29	91.5514	0.5385	0.2115	2.55	0.8987	23.5
Mn II	293.31	30	91.3330	0.0536	0.0483	1.11	0.2746	17.6
Mg II	279.08	29	91.1774	0.2793	0.2536	1.10	0.5284	48.0
Cr II	267.72	31	91.1871	0.0322	0.0328	0.98	0.0603	54.4
Ni I	341.48	31	91.1871	0.0519	0.0418	1.24	0.2357	17.7
Ti II	337.28	31	91.1871	0.0310	0.0174	1.78	0.0802	21.7

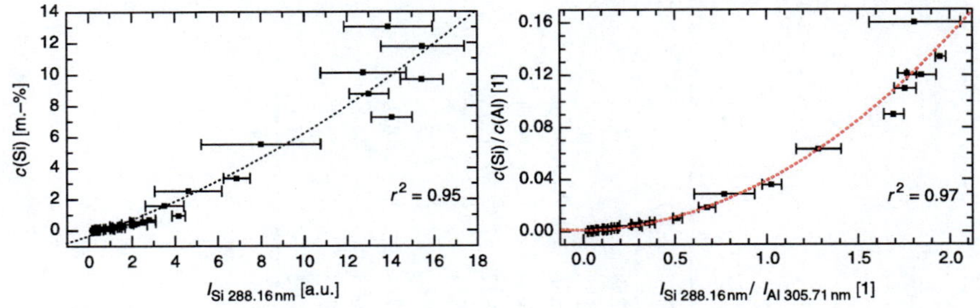

Fig. 13.24 Analysis functions of the Si I 288.16 nm emission line without (*left*) and with internal standardization to the Al I 305.71 nm emission line (*right*)

Table 13.18 Possible perturbing emission lines according to [13.30] with respect to the used analyte lines, interelement corrected residue standard deviations, and estimations of the RSDP (last column) for standardized analysis functions

Analyte element	λ (nm)	Possible perturbing emission line (nm)	log $(g_i f_{ij})$	A_{ji} $(10^6\ s^{-1})$	$s^{corr.}_{c_{ref},Q_{ref}}\bar{c}(Al)$ (m.-%)	$\frac{s_{c,I}}{s^{corr.}_{c_{ref},Q_{ref}}\bar{c}(Al)}$	$\frac{s^{corr.}_{c_{ref},Q_{ref}}\bar{c}(Al)}{\bar{c}_{analyte}}$ (%)
Si I	288.16	Ni II 288.15	−1.177	6.676	0.4495	2.15	17.7
Cu I	296.12	Mn II 296.12	−1.280	7.979	0.7029	1.82	78.2
Mn II	293.31	Cu I 293.31	−1.900	1.219	0.0442	1.21	16.1
Cr II	267.72	Mn II 267.73	−0.082	85.56	0.0074	4.34	12.3
Ti II	337.28	(Mg)	–	–	0.0125	2.48	15.6

For perturbing element in brackets no possible emission lines were found

For all analysis functions, the improvement factor by internal standardization is greater than 1 and attains more than 2.6 except for Cr (0.98). The best achieved RSDP ranges from 10.4% (Si) to 54.4% (Cr).

Besides the improvement of a calibration or analysis function by the ratio method, it is possible to find corrections of these functions due to matrix or interelement effects (cf. Sect. 11.6). The standardized analysis function of an analyte a for additive perturbing elements k is described by

$$c_{ref,a} = d_0 + d_1 Q_{ref,a} + d_2 Q^2_{ref,a} + \sum_{k,k \neq a} \delta_k c_{ref,k}. \tag{13.7}$$

This approach allows to improve the residue standard deviation of five emission lines listed in Table 13.18 by multilinear regression.

By combining internal standardization and interelement corrections for all analysis functions, improvement factors greater 1 and up to 4.34 were achieved.

To verify the general feasibility of LIBS measurements for samples moving on a conveyor belt velocity with 3 m/s, calibration measurements with slightly optimized excitation and detection parameters were carried out. In Figs. 3.25 (see

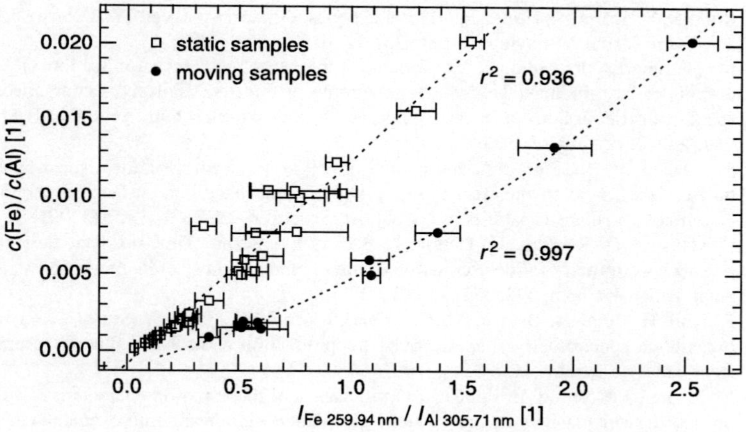

Fig. 13.25 Analysis functions of iron in aluminium for static and moving samples at 3 m/s with internal standardization for Fe II 259.94 nm

Sect. 3.9) and 13.25 standardized analysis functions for static and moving samples are shown for Cu I 327.40 nm and Fe II 259.94 nm, exemplarily. Although the moving samples were only analyzed once with seven laser bursts at a tracked position, the coefficient of determination of the quadratic regression is better than those for static samples. The improvement factor of the residue standard deviation is 3.2 for Cu and 1.5 for Fe.

References

13.1. J. Aguilera, C. Aragon, J. Campos, Determination of carbon content in steel using laser-induced breakdown spectrometry. Appl. Spectrosc. **46**, 1382–1387 (1992)

13.2. R. Noll, R. Sattmann, V. Sturm, S. Lüngen, H.-J. von Wachtendonk, Schnelle Multielementanalyse in der Stahlschmelze mit laserinduzierter Emissionsspektrometrie. Stahl u. Eisen **117**, 57–62 (1997)

13.3. T. Sugihara, Y. Funahashi, I. Fukui, T. Miyama, Rapid determination of ultra low carbon and nitrogen in steel with a modern simultaneous optical emission spectrometer, Progress in Anal. Chem. in the Steel and Metals Industry, Ed. Nauche, European Commission (1996), 229–235

13.4. V. Sturm, L. Peter, R. Noll, Steel analysis with laser-induced breakdown spectrometry in the vacuum ultraviolet. Appl. Spectrosc. **54**, 1275–1278 (2000)

13.5. E. Runge, S. Bonfiglio, F. Bryan, Spectrochemical analysis of molten metal using a pulsed laser source. Spectrochim. Acta **22**, 1678–1680 (1965)

13.6. C. Aragon, J. Aguilera, J. Campus, Improvements in quantitative analysis of steel composition by laser-induced breakdown spectroscopy at atmospheric pressure using an infrared Nd:YAG laser. Appl. Spectrosc. **47**, 606–608 (1993)

13.7. J. Gruber, J. Heitz, H. Strasser, D. Bäuerle, N. Ramaseder, Rapid in-situ analysis of liquid steel by laser-induced breakdown spectroscopy. Spectrochim. Acta Part B **56**, 685–693 (2001)

13.8. L. Peter, V. Sturm, R. Noll, Liquid steel analysis with laser-induced breakdown spectrometry in the vacuum ultraviolet. Appl. Opt. **42**, 6199–6204 (2003)
13.9. M. Hemmerlin, R. Meilland, H. Falk, P. Wintjens, L. Paulard, Application of vacuum ultraviolet laser-induced breakdown spectrometry for steel analysis – comparison with spark-optical emission spectrometry figures of merit. Spectrochim. Acta, Part B **56**, 661–669 (2001)
13.10. S. Palanco, J.J. Laserna, Full automation of a laser-induced breakdown spectrometer for quality assessment in the steel industry with sample handling, surface preparation and quantitative analysis capabilities. J. Anal. At. Spectrom. **15**, 1321–1327 (2000)
13.11. L. Cabalin, D. Romero, C. Garcia, J. Baena, J. Laserna, Time-resolved laser-induced plasma spectrometry for determination of minor elements in steelmaking process samples. Anal. Bioanal. Chem. **372**, 352–359 (2002)
13.12. R. Noll, H. Bette, A. Brysch, M. Kraushaar, I. Mönch, L. Peter, V. Sturm, Laser-induced breakdown spectrometry – applications for production control and quality assurance in steel industry. Spectrochim. Acta, Part B **56**, 637–649 (2001)
13.13. M. Khater, J. Costello, E. Kennedy, Optimization of the emission characteristics of laser-produced steel plasmas in the vacuum ultraviolet: significant improvements in carbon detection limits. Appl. Spectrosc. **56**, 970–984 (2002)
13.14. V. Sturm, J. Vrenegor, R. Noll, M. Hemmerlin, Bulk analysis of steel samples with surface scale layers by enhanced laser ablation and LIBS analysis of C, P, S, Al, Cr, Cu, Mn and Mo. J. Anal. At. Spectrom. **19**, 451–456 (2004)
13.15. IRSID, Voie Romaine, 57283 Maizières les Metz Cedex, France
13.16. H. Richter, H. Rzepczyk, D. Tembergen, Prozeßsteuerung sekundärmetallurgischer Verfahren. Stahl und Eisen **114**, 110–113 (1994)
13.17. H. Richter, H. Rzepczyk, D. Tembergen, Qualitätsbezogene Prozeßsteuerung in der Sekundärmetallurgie eines Oxygenstahlwerkes. Stahl und Eisen **115**, 83–87 (1995)
13.18. C. Carlhoff, Laserinduzierte Emissionsspektroskopie für die Direktanalyse von flüssigem Stahl im Konverter. Laser und Optoelektronik **23**, 50–52 (1991)
13.19. R. Noll, V. Sturm, L. Peter, I. Whiteside, Analysis using lasers, in *Proc. of the 49th Chemists' Conference*, 1997, ed. by R. Jowitt (Research and Development Department British Steel, Middlesbrough) pp. 22–27
13.20. R. Sattmann, V. Sturm, R. Noll, Laser-induced breakdown spectroscopy of steel samples using multiple Q-switch Nd:YAG laser pulses. J. Phys. D: Appl. Phys. **28**, 2181–2187 (1995)
13.21. R. Noll, V. Sturm, L. Peter, R. Hakala, J. Viirret, H.W. Gudenau, K. Mavrommatis L. Ernenputsch, Sensitivity-enhanced laser analysis (SELA) of steel melts for a fast multi-element online analysis during ladle processing in secondary metallurgy, EUR 19411 - Analytical techniques for processes, products and the environment, Technical steel research series; European Commission (Office for Official Publications of the European Communities), Luxembourg, 2001
13.22. V. Rai, F. Yueh, J. Singh, Study of laser-induced breakdown emission from liquid under double pulse excitation. Appl. Optics **42**, 2094–2101 (2003)
13.23. DIN 32 645, *Chemische Analytik – Nachweis-, Erfassungs- und Bestimmungsgrenze – Ermittlung unter Wiederholbedingungen; Begriffe, Verfahren, Auswertung* (Beuth Verlag, Berlin, 2006)
13.24. M. Corsi, G. Cristoforetti, M. Hidalgo, D. Iriarte, S. Legnaioli, V. Palleschi, A. Salvetti, E. Tognoni, Temporal and spatial evolution of a laser-induced plasma from a steel target. Appl. Spectrosc. **57**, 715–721 (2003)
13.25. C. Lopez-Moreno, S. Palanco, J. Laserna, Remote laser-induced plasma spectrometry for elemental analysis of samples of environmental interest. J. Anal. At. Spectrom. **19**, 1479–1484 (2004)
13.26. S. Palanco, L.M. Cabalin, D. Romero, J. Laserna, Infrared laser ablation and atomic emission spectrometry of stainless steel at high temperatures. J. Anal. At. Spectrom. **14**, 1883–1887 (1999)

13.27. I. Bassiotis, A. Diamantopoulou, A. Giannoudakos, F. Roubani-Kalantzopoulou, M. Kompitsas, Effects of experimental parameters in quantitative analysis of steel alloy by laser-induced breakdown spectrometry. Spectrochim. Acta Part B **56**, 671–683 (2001)
13.28. J. Vrenegor, R. Noll, V. Sturm, Investigation of matrix effects in LIBS plasmas of high-alloy steel for matrix and minor elements. Spectrochim. Acta B **60**, 1083–1091 (2005)
13.29. K. Slickers, *Automatic Atomic Emission Spectroscopy*, Brühlsche Universitätsdruckerei, Giessen, 2nd edn. (1992)
13.30. R. Kurucz, B. Bell, *Atomic Line Data*. Kurucz CD-ROM No. 23 (Smithsonian Astrophysical Observatory, Cambridge,, 1995)
13.31. J. Aguilera, C. Aragon, A comparison of the temperatures and electron densities of laser-produced plasmas obtained in air, argon, and helium at atmospheric pressure. Appl. Phys. A **69** [Suppl.], S475–S478 (1999)
13.32. K. Krone, *Aluminiumrecycling: Vom Vorstoff bis zur fertigen Legierung*, (Aluminium-Verlag, Düsseldorf, 2000)
13.33. *Metallstatistik 1957–1966*, Ed. Metallgesellschaft AG, vol. 54, Frankfurt am Main, 1967
13.34. *Metallstatistik 1967–1977*, Ed. Metallgesellschaft AG, vol. 65, Frankfurt am Main, 1978
13.35. *Metallstatistik 1978–1988*, Ed. Metallgesellschaft AG, vol. 76, Frankfurt am Main, 1989
13.36. *Metallstatistik, Metal Statistics 1989–1999*, ed. by World Bureau of Metal Statistics, vol. 87, Ware, England (2000)
13.37. *Metallstatistik, Metal Statistics 1993–2003*, ed. by World Bureau of Metal Statistics, vol. 91, Ware, England (2004)
13.38. *Aluminium-Taschenbuch*, ed. by Aluminium-Zentrale (Aluminium-Verlag, Düsseldorf, 1983)
13.39. C. Schmitz, *Handbook of Aluminium Recycling – Fundamentals, Mechanical Preparation, Metallurgical Processing, Plant Design* (Vulkan-Verlag, Essen, 2006)
13.40. K. Krone, J. Krüger, H. Orbon, H. Sommer, H. Vest, Ökologische Aspekte der Primär- und Sekundärerzeugung in der Bundesrepublik Deutschland. Metall **44**, 559–568 (1990)
13.41. G. Rombach, B. Friedrich, Aluminum recycling in Germany. Light Metal Age **59**, 66–75 (2001)
13.42. Ü. Aydin, R. Noll, J. Makowe, Automatic sorting of aluminium alloys by fast LIBS identification, in *7th Int. Workshop Progress in Analytica Chemistry in the Steel and Metal Industries*, ed. by J. Angeli (Glückauf GmbH, Essen, 2006) pp. 309–314
13.43. M. Sabsabi, P. Cielo, Quantitative analysis of aluminum alloys by laser-induced breakdown spectroscopy and plasma characterization. Appl. Spectrosc. **49**, 499–507 (1995)
13.44. G. Rieger, M. Taschuk, Y. Tsui, R. Fedosejevs, Laser-induced breakdown spectroscopy for microanalysis using submillijoule UV laser pulses. Appl. Spectrosc. **56**, 689–698 (2002)
13.45. M. Ismail, H. Imam, A. Elhassan, W. Youniss, M. Harith, LIBS limit of detection and plasma parameters of some elements in two different matrices. J. Anal. At. Spectrosc. **19**, 489–494 (2004)
13.46. M. Ismail, G. Cristoforetti, S. Legnaioli, L. Pardini, V. Palleschi, A. Salvetti, E. Tognoni, M. Harith, Comparison of detection limits, for two metallic matrices, of laser-induced breakdown spectroscopy in the single and double-pulse configurations. Anal. Bioanal. Chem. **385**, 316–325 (2006)

Chapter 14
Bulk Analysis of Nonconducting Materials

This chapter describes LIBS investigations for a bulk characterization or analysis of nonconducting materials, such as polymers, waste electric and electronic equipment, slag, soil, cement, droplets, gases, particulates, and aerosoles.

14.1 Polymers

In this section, chemical analysis of polymers with LIBS refers to the identification of plastic materials itself and to additives in technical polymers.

14.1.1 Identification of Plastic Materials

Polymers belong to the most wide spread packaging materials for consumer products as food and cosmetics. One approach to reduce the domestic waste streams of used polymers is a high-grade material recycling. Use of mixtures of polymers is restricted to the production of low-grade plastics, and the chemical recycling by, e.g., hydrogenation or energy recovery [14.1]. In the European Union, the share of recycled polymers (without energy recovery) is about 17% of the total annual polymer waste mass, so there is still potential to further increase this amount [14.2].

For a high-grade material recycling, the mixtures have to be separated into fractions of different polymers. For an economically interesting throughput of a sorting machine, a high-speed identification method with identification rates of several pieces per second is necessary.

The requirements for the purity of the separated fractions depend on the polymer type and the kind of contaminants. For the PE (polyethylene) fraction, e.g., a contamination of 5 m.-% of PP (polypropylene), PS (polystyrene), or PET (polyethylene terephthalate) may be tolerable, whereas PVC (polyvinyl chloride),

R. Noll, *Laser-Induced Breakdown Spectroscopy*, DOI 10.1007/978-3-642-20668-9_14,

should be close to 0% since corrosion problems will arise in the subsequent processing steps from decomposition of PVC forming hydrochloric acid.

It is well known that the molecular structure of the polymers influences the reflection and transmission spectra in the near infrared (NIR). Several approaches have been undertaken to use this fact for a high-speed identification of post-consumer plastics [14.3–14.7].

Surface contaminations of the polymers may deteriorate the measuring signals, thus reducing the identification accuracy. This is especially the case if contaminating surface layers are not transparent in the NIR and exceed a thickness of about 20 μm [14.7]. Furthermore, NIR is not applicable for black-colored samples. X-rays can be used to detect the chlorine content to identify PVC [14.8].

In the last years, NIR technologies became the state-of-the-art technology to identify bottles made of PVC, PET, or PP for automatic sorting lines [14.9, 14.10]. Medium infrared (MIR) techniques are capable to identify technical polymers, which are used as components of electric appliances [14.11]. Typical identification times for MIR are several seconds, which excludes such an approach for high-throughput applications.

The principal feasibility of LIBS to identify polymers and to detect additives in polymers will be presented in this and the following section. Compared with NIR, MIR, and X-ray methods, LIBS offers an extended broad spectral signature to identify polymers and additives in polymers at the same time for high-speed applications. In a first step, emission spectra of polymers in the near UV and visible spectral region using LIBS in combination with time-resolved spectroscopy were studied.

The experimental setup is shown in Fig. 14.1 [14.12]. A Q-switched Nd:YAG laser operating at 1,064 nm was used in the double pulse configuration mode, see, e.g., [14.13] (cf. Chap. 6). The laser source emits two 30 ns equal energy pulses with a time separation of 6 μs between them. The use of multiple pulses can significantly improve the analytical performance in terms of reproducibility of the signal and detection limits, cf. Chap. 13. The laser beam was focused by means of a 200 mm plano-convex lens onto the sample. The position of the focal plane was $\Delta s = 5$ mm below the surface of the sample to avoid an air breakdown, which can shield off a part of the laser energy. A bifurcated quartz fiber bundle collects the plasma radiation. One of the arms of the fiber bundle was mounted at the entrance slit of a spectrometer (Jobin Yvon HR 320 with a 600 l/mm grating) equipped with an intensified photodiode array (IPDA, 1,024 diodes) covering an effective spectral bandwidth of about 70 nm and gateable in the 10 ns–100 μs time range. The settings for the signal integration were chosen as follows: $t_{delay} = 1$ μs between the second laser pulse and the start of exposure of the photodiode array, $t_{int} = 10$ μs. The second arm of the fiber bundle was linked to a monochromator (Jobin Yvon HR 250 with a 2,400 l/mm grating) equipped with a photomultiplier tube (PMT, Hamamatsu R1547) and centered at the chlorine line Cl 725.66 nm. Due to the presence of strong continuum radiation in that spectral region and for normalization purposes, the continuum emission was also recorded by means of a positive-intrinsic-negative-photodiode (PIN) and an interference filter (IF, peak transmission 0.47 at 739.5 nm,

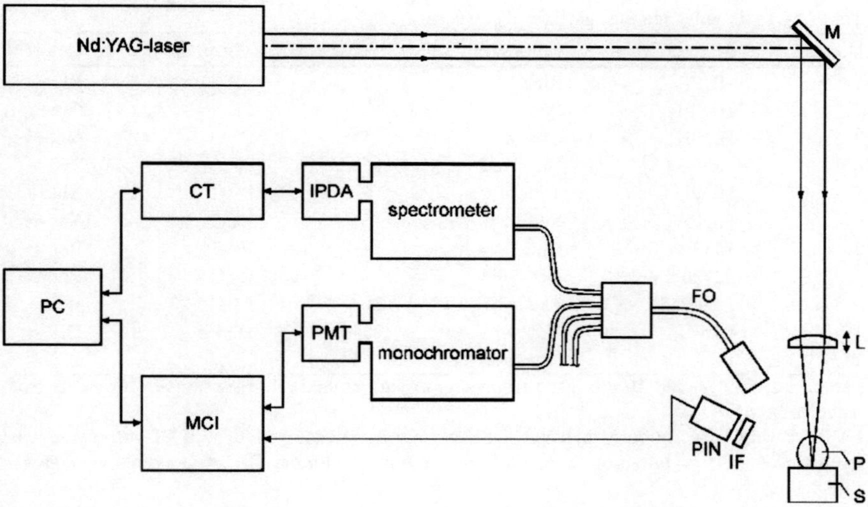

Fig. 14.1 Experimental setup. M = mirror, L = lens, P = plasma, S = sample, IF = interference filter, PIN = positive-intrinsic-negative photodiode, FO = fiber optics, IPDA = intensified photodiode array, CT = controller, PMT = photomultiplier tube, MCI = multichannel integrator electronics, PC = personal computer

bandwidth 4 nm). The outputs of the photomultiplier tube and the PIN-photodiode were fed to gated integrators (MCI) and their signals were integrated from 1 to 10 μs after the laser pulse and after the second laser pulse, respectively, when double pulses were applied.

Plastic bottles have great variations of colors and additives, which influence, e.g., the stiffness of the polymer. Within the frame of the investigations described in the following, it was not possible to measure all different bottles available. To restrict the number of samples for the experiments to 20 while keeping material variety as wide as possible, several bottles of a given material with similar color were mixed to produce for example a "medium green HDPE sample," which contains colorants and additives typical for green HDPE bottles. To generate these samples, the bottles have been washed and dryed for 3–4 h at 80–90°C to achieve a low moisture content. Then they were shredded to 3–5 mm flakes. These flakes were mixed to achieve a homogeneous melt and pressed into a mold. In this way, the samples listed in Table 14.1 were manufactured. Hence, most of these samples represent plastic mixtures of used bottles. Only the samples "HDPE blue 2" and "HDPE yellow 2" are processed by virgin material to obtain TiO_2-free samples. Since bottles made of LDPE with blue, red, or yellow color are rare, the corresponding samples are made from virgin material.

An additional pure reference sample of high-density polyethylene was used (delivered by Goodfellow GmbH, Germany, article no. ET323100/2). All samples had flat surfaces and were oriented perpendicularly to the laser beam. Finally, post-consumer plastic bottles of the polymers PE, PET, and PVC were used to

Table 14.1 Polymer samples used for the experiments

No.	Material	Color	No.	Material	Color
1	HDPE	Blue 1	12	PP	Blue
2	HDPE	Blue 2	13	PP	Green
3	HDPE	Brown	14	PP	Not dyed
4	HDPE	Not dyed	15	PP	Red
5	HDPE	Yellow 1	16	PVC	Blue
6	HDPE	Yellow 2	17	PVC	Not dyed
7	LDPE	Blue	18	PVC	Red 1
8	LDPE	Not dyed	19	PVC	Red 2
9	LDPE	Red	20	PVC	White
10	LDPE	Yellow	21	HDPE	GF
11	PET	Not dyed			

Samples 2, 6, 7, 9, and 10 are made from virgin material, and all other samples are made from recycled bottles
HDPE = high density polyethylene, LDPE = low density polyethylene, PET = polyethylene terephthalate, PP = polypropylene, PVC = polyvinyl chloride, GF = Goodfellow reference sample, not dyed

determine identification accuracies on unprepared samples. They were not washed. Measurements were performed on spots free of labels.

Artificial neural networks were applied for a classification of the spectra. Several networks like feedforward, dynamic learning vector quantization (DLVQ), or radial base function (RBF) networks were tested [14.14]. The best results were achieved with modularized feedforward networks trained by backpropagation. Modularization means that two or more subnetworks share the same input layer, but each subnetwork has its own hidden and output layer. Each output layer may have one or more output cells. The total number of output cells is the same as the number of different polymer classes. Besides the better results of the modularized networks compared with all other tested networks, the modularization has the advantage that the subnetworks can be retrained to adopt to new samples without influencing the other subnetworks.

Subnetworks with different numbers of hidden layers were compared, and it was found that one hidden layer is sufficient. All cells of the hidden layer of a subnetwork are connected with all input cells, and all output cells are connected with all cells in the hidden layer of its subnetwork. Details of the networks will be given below.

If input data of an unknown or test sample are supplied to the input cells, the sample is attributed to the class which has the highest activation at the output cell, regardless of the activation of the other output cells. The activation of an output cell corresponds to the similarity between actual input data and that input data for which this output cell was trained. Hence, the amount of incorrect classifications can be reduced, if a higher activation limit is defined which has to be exceeded. Otherwise the classification is assessed as ambiguous and the sample is not classified. It is obvious that there is a strong correlation between the activation limit, the amount of correct classifications, and undone classifications: the higher the activation limit is, the higher is the amount of not-classified samples and the amount of correct classifications. For practical reasons, the amount of not-classified samples has to be limited. The activation limit was fixed to that value, where 20% of the samples, averaged over all polymer classes, are not classified.

Molecular materials like polymers are partly atomized, when they are exposed to laser radiation, which is intense enough to create a plasma. This initially implies a limitation concerning LIBS application on polymers because of the loss of molecular information in the plasma. Table 14.2 shows the semi-structural formula of the investigated polymers. Some polymers differ in their stoichiometric C/H ratio. It is obvious to use this information to identify the polymers.

To simplify the experimental effort, it is advantageous to find a spectral region which permits a simultaneous detection of carbon and hydrogen lines. Figure 14.2 shows two spectra of a HDPE sample (sample 1 in Table 14.1) in the spectral region from 460 to 530 nm. In the upper spectrum, two intense clearly distinct lines appear at 486.1 and 495.7 nm. They are assigned to the H_β line of the Balmer series of hydrogen at 486.13 nm and to a carbon C I line at 247.86 nm, respectively. The carbon line arises in second order at 495.7 nm, as it is depicted in the lower spectrum of Fig. 14.2, which has been recorded adding a low pass filter in front of the fiber optic to avoid second-order contributions (Schott filter, type GG 375). The hydrogen line and the C/H ratio will be discussed below.

Table 14.2 Semi-structural formula and stoichiometric ratio of carbon and hydrogen atoms

Polymer	Formula	C/H ratio
PE	$(-CH_2-)_n$	0.5
PP	$(-CH(CH_3)-CH_2-)_n$	0.5
PET	$(-O-CH_2-CH_2-O-CO-C_6H_4-CO-)_n$	1.25
PVC	$(\text{-CHCl-CH}_2\text{-})_n$	0.66

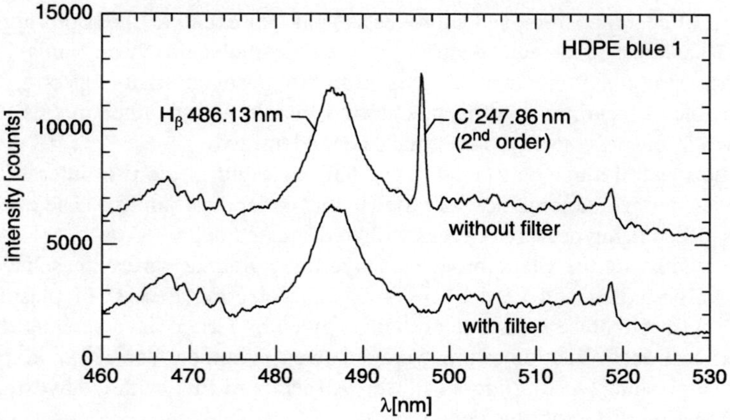

Fig. 14.2 Spectrum of HDPE blue 1 (sample 1 of Table 14.1) with and without the 247.86 nm carbon line in second order (double pulses 2 × 100 mJ). For better presentation the unfiltered spectrum is shifted in intensity by 4,000 counts

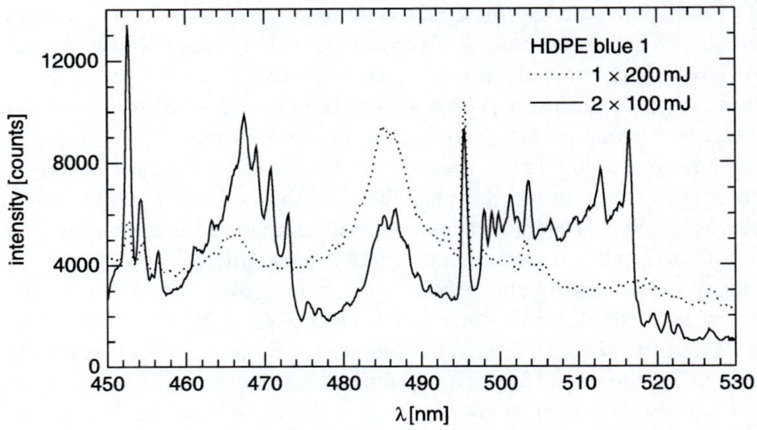

Fig. 14.3 Spectrum of HDPE blue 1 (sample 1 of Table 14.1) for single pulse and double pulse excitation

Figure 14.3 shows two spectra of a HDPE sample (sample 1 in Table 14.1) gained with single and double laser pulse excitation. In the spectrum generated with double pulse excitation, more spectral features appear and the line-to-continuum ratio is significantly enhanced. Hence, for the following measurements, mostly double pulse excitation is used.

Figure 14.4 presents spectra of different polymer samples in the spectral region between 440 and 530 nm. The experimental conditions for all these spectra are identical, i.e., 2 × 150 mJ double pulses, $t_{delay} = 1\,\mu s$ and $t_{int} = 10\,\mu s$. Within the spectra, distinct structures are observed. These structures arise from the excitation of the molecular species in the plasma plumes which constitute the polymer, their products of decomposition and additives as well. For example, the bands appearing at ≈ 470 and ≈ 516 nm can be attributed to the C_2-molecule (Swan bands) [14.15].

Concerning the atomic species present in the plasma emission spectra, various atomic lines of carbon, hydrogen, chlorine, titanium, and other metals used as additives in the polymeric products have been identified.

In the spectral range between 440 and 530 nm, eight "regions of interest" (ROI) have been defined, among these are the H_β line and the carbon line. The evaluation of these ROIs using neural networks will be discussed below.

The widths of the Stark broadened hydrogen Balmer series lines have been extensively studied and used to determine the electron density of plasmas (cf. Sect. 9.4). In the present case, the relations given by Griem have been used for an approximate determination of the averaged electron number density n_e in the laser produced plasma [14.16]. Figure 14.5 shows spectra of PET with the hydrogen and carbon line for different time delays.

With increasing time delay, the electron density decreases and hence the width of the H_β line. The electron density calculated from the line width of the H_β line decreases from ≈ $1.35\times10^{17}\,cm^{-3}$ at 1 μs time delay to ≈ $0.65\times10^{17}\,cm^{-3}$ at 3 μs.

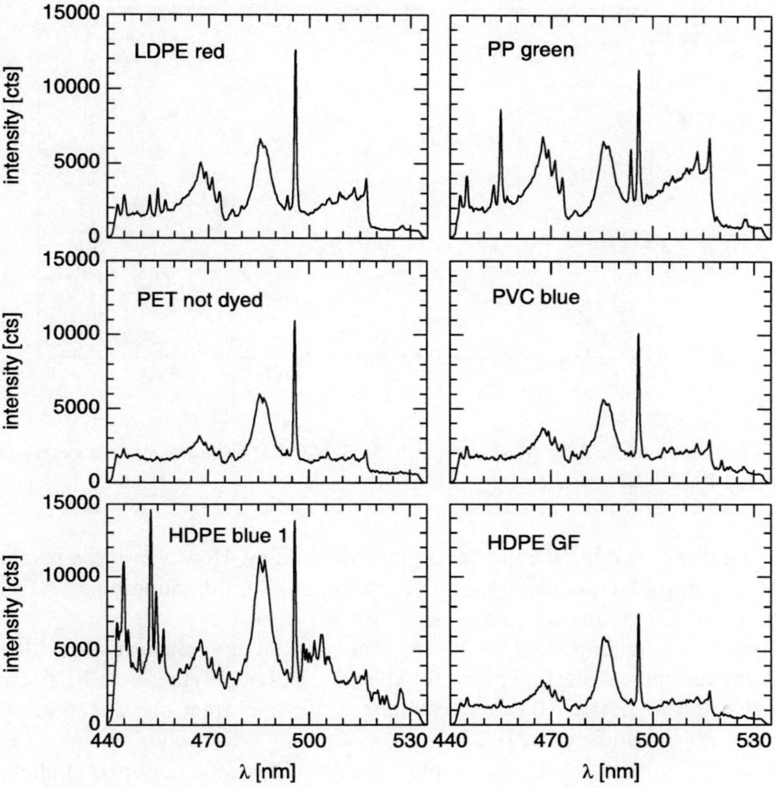

Fig. 14.4 Spectra of different polymers in the spectral region from 440 to 535 nm. HDPE GF denotes a pure high density polyethylene sample (Goodfellow GmbH), double pulses 2 × 150 mJ, $t_{del} = 1\,\mu s$, $t_{int} = 10\,\mu s$

At the beginning of the plasma formation, and because of the high electron densities, there is considerable Stark broadening which leads to FWHM of the H_β line in the order of 6 nm. At high electron densities, the red wing of the H_β line can interfere with the neighboring second-order carbon line. For a time delay longer than 3 μs, the width of the hydrogen line is so small that the carbon and the hydrogen lines are well separated.

The C/H intensity ratios were determined for different samples. The background contribution to the H_β and C I line intensities has been taken into account to determine the integrated line intensities. The carbon/hydrogen ratio depends on the time delay, probably due to the decreasing electron temperature and the different excitation energies, which amounts to 7.68 eV for the carbon line C 247.86 nm and 12.74 eV for the hydrogen line H_β 486.13 nm. With increasing time delay, the C/H ratio increases for the studied polymers LDPE, HDPE, PP, PVC, and PET. As expected from the stoichiometric C/H ratio, see Table 14.2, the PET sample yields

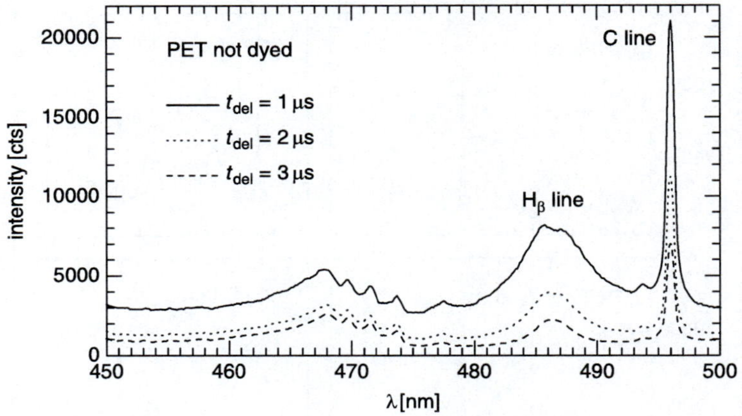

Fig. 14.5 Spectra of PET not dyed (sample 11 of Table 14.1) for different time delays (single pulses 1 × 100 mJ, $t_{int} = 5\,\mu s$)

the highest measured C/H ratio for a given time delay. However, these results are not sufficient for an unambiguous identification of all different polymers, since the values of the C/H ratio for some samples partly overlap.

However, it is important to notice that, despite the intrinsic difficulties in obtaining the appropriate C/H ratios for all categories of polymers, the PET samples exhibited the highest C/H values enabling a 100% correct classification of this category which will be described later.

Table 14.2 presents the basic chemical composition of the polymers studied. The only polymer which contains chlorine is PVC. Therefore, chlorine emission lines in the LIBS spectra were studied in order to identify PVC and to discriminate it from the other polymers.

Figure 14.6 shows two spectra of not dyed PVC (sample 17 in Table 14.1) in the spectral region from 710 to 760 nm. The distinct spectral features at 725.7 and 754.7 nm have been identified as chlorine lines. Both spectra have been obtained using the same laser energy of 300 mJ. The upper one corresponds to a single 300 mJ laser pulse, whereas the lower one was obtained with a double pulse having the same total energy distributed equally into two pulses separated by a time interval of 6 μs. Both spectra have been measured with a time delay of 0.5 μs after the laser pulse (respectively, the second pulse of the double pulse) and integrated over a time period of 10 μs. The 725.66 nm chlorine line in the double pulse spectrum is more intense by a factor of 4 in comparison with the single pulse excitation.

In Fig. 14.7, the ratio of the intensity of the Cl 725.66 nm line to the intensity of the nearby background is shown for various polymer samples. As a result, the PVC samples can be unambiguously identified by measuring that ratio.

Artificial neural networks were applied to classify the polymers to polymer categories by their LIBS spectra. Nine spectral ROI were evaluated to determine in arbitrary units: the intensity of the chlorine line Cl I 725.66 nm, the intensity

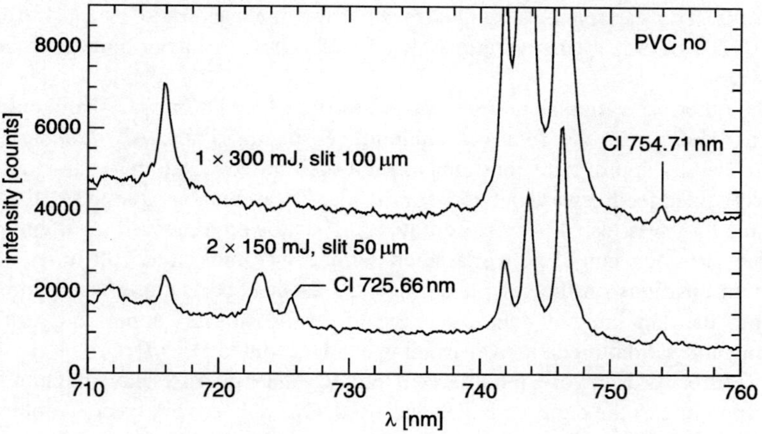

Fig. 14.6 Spectra of PVC not dyed (sample 17 of Table 14.1) with chlorine lines. For better presentation the single pulse spectrum is shifted in intensity by 2,000 counts

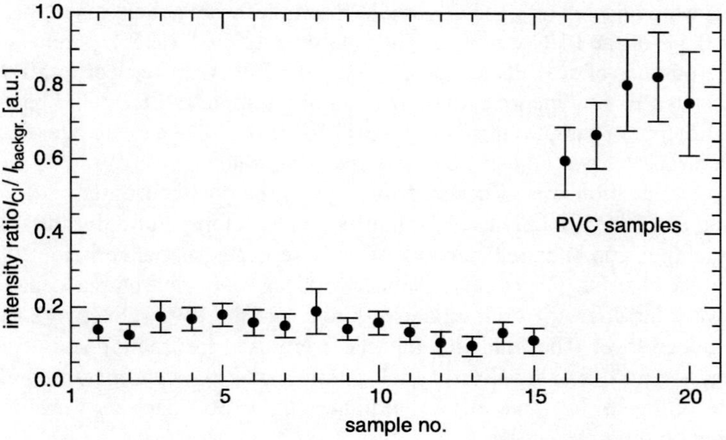

Fig. 14.7 Intensity ratio of the 725.66 nm chlorine line and the nearby background for polymer samples 1–20 listed in Table 14.1

of the background emission at 739.5 nm, the background corrected intensity of the hydrogen line H_β 486.13 nm, the background corrected intensity of the carbon line C I 247.86 nm, the background corrected intensity of the C_2-band emission at ≈516.5 nm, the background corrected intensities of four further lines in the spectral region between 440 and 535 nm. With the exception of Cl 725.66 nm and the background emission at 739.5 nm, all intensities were evaluated by integrating the values of pixels under the lines and applying a linear background correction. The nine intensities were further evaluated to achieve eight input variables for the neural networks. The variables are defined as follows: intensity ratio of H_β 486.13 nm to C I 247.86 nm, intensity ratio of Cl I 725.66 nm to background emission at 739.5 nm.

The other six variables are the intensities of C I 247.86 nm, C_2-band emission at ≈516.5 nm, and the four yet unidentified spectral lines. All variables were normalized by appropriate constants to limit their values to the range 0–1. At first, a modularized feedforward network for the identification of samples no. 1–20 listed in Table 14.1 was tested. Two different sets of data were recorded for training and validation of the neural network. Each sample was measured 100 times at 100 different positions on the sample surface. 50 datasets per sample were stored for training, the remaining 50 datasets are used for validation. By comparing results on training and validation data, over-training can be avoided [14.17].

A feedforward network, modularized into two subnetworks, with eight input cells corresponding to the eight variables was used. One subnetwork has one output cell for PVC and three cells in the hidden layer, and the other subnetwork has three output cells for PE, PP, and PET and eight cells in the hidden layer.

Table 14.3 summarizes the results. The column entitled "Not classified" gives the percentage of samples that could not be identified unambiguously by the network and thus were rejected (e.g., 28.6% of all PE samples were not classified, in contrast to only 0.4% of the PVC samples). The column entitled "Fraction pureness" gives, e.g., the number of real PE samples classified as PE with respect to all samples classified as PE. Any incorrect classification of a sample as PET could be avoided, and the amount of samples misclassified as PVC is 0.4%. The mean rate of 20% not classified samples was adjusted by the activation limit.

The classification was also tested for unprepared post-consumer bottles of PE, PET, and PVC. PP is barely used for bottles and therefore left out for the following experiment. A modularized network with a separate subnetwork for PE, PET, and PVC was used. Hence, each subnetwork has only one output cell. The PE subnetwork has four, the PET subnetwork five, and the PVC subnetwork two cells in the hidden layer. The following datasets were used for training and testing: for training, five bottles of PE, five of PET, and five of PVC were measured 20 times each, resulting in 300 datasets; for testing, similar 300 datasets were used, but measured on different bottles.

Table 14.3 Results of the automatic classification of the polymer samples using a feedforward neural network (values in %, the columns denoted with (1) refer to 100 samples, e.g., of PE; column (2) refers to 100 samples classified as, e.g., PE)

Sample material	Correctly classified, (1)	Incorrectly classified, (1)	Not classified, (1)	Fraction pureness, (2)
PE	65.0	6.4	28.6	93.6
PP	72.5	4.5	23.0	95.5
PET	72.0	0.0	28.0	100.0
PVC	99.2	0.4	0.4	99.6
Mean	77.2	2.8	20.0	97.2

Table 14.4 Results of the automatic classification of post-consumer polymer bottles using feedforward neural network (values in %, the columns denoted with (1) refer to 100 samples, e.g., of PE; column (2) refers to 100 samples classified as, e.g., PE)

Bottle material	Correctly classified, (1)	Incorrectly classified, (1)	Not classified, (1)	Fraction pureness, (2)
PE	76.0	6.0	18.0	100.0
PET	67.0	0.0	33.0	84.4
PVC	85.0	6.0	9.0	100.0
Mean	76.0	4.0	20.0	94.9

The results are summarized in Table 14.4. As for the previous results, the percentage of datasets not classified is adjusted to 20%. The mean fraction pureness is 94.9%. The PET fraction pureness of 84.4% is due to the fact that 6.0% of the PE bottles and 6.0% of the PVC bottles are classified as PET.

The applicability of LIBS for certain cases, like a high-speed separation of washed PET and PVC bottles, seems feasible with high accuracy. LIBS may also be used complementary to near-infrared spectroscopy, which can also be applied for the identification of polymers, but is not suited for dark colored samples.

14.1.2 Detection of Heavy Metals and Flame Retardants in Technical Polymers

During the recycling process of end-of-life waste electric and electronic equipment (EOL-WEEE), downgrading of valuable technical polymers has to be avoided by separating the collected material in fractions of high purity. Most EOL-WEEE pieces are doped with several additives to improve their mechanical, electrical, and chemical properties. Common additives are flame retardants, antioxidants, light stabilizers, fillers, dyes, and pigments. Their concentration varies typically from traces to several percent [14.18]. Waste pieces containing brominated flame retardants (BFR) and additives with heavy metals have to be automatically identified and removed from the waste stream to be recycled. This is due to the significant environmental problems during the waste management phase caused by these substances. To establish an economically feasible recycling process which meets these requirements, high-speed automatic sorting systems performing the identification of both the polymer and the detection of the critical additives at a rate of several parts per second are required. A prototype automatic identification and sorting line has been setup and evaluated for material-specific sorting of EOL-WEEE pieces [14.19]. The detection unit of this automatic sorting line is a multisensor system for the rapid identification of the polymer matrix and the detection of the contained additives. It comprises three spectroscopic modules based on LIBS, NIR (near infrared spectrometry) and MIR (mid-infrared spectrometry). This section presents investigations of LIBS for the analysis of heavy metals and brominated flame retardants in EOL-WEEE pieces before recycling, the setup of a LIBS analyzer for an automatic sorting line, and the evaluation of the LIBS analyzer for the inline detection of heavy metals and brominated flame retardants in moving EOL-WEEE pieces [14.20].

Frequently applied methods for the determination of metals and flame retardants are atomic absorption spectrometry (AAS), inductively coupled plasma atomic emission spectrometry and mass spectrometry (ICP-AES, ICP-MS), or electrochemical techniques [14.21]. However, these methods require the time-consuming step of sample dissolution prior to the analysis and are therefore not applicable for high-speed applications. Therefore, direct sampling analytical methods such as laser ablation mass spectrometry (LA-MS) [14.22], glow discharge spectrometry (GD-OES) [14.23, 14.24], sliding spark discharge optical emission spectrometry (sliding SD-OES) [14.25], and X-ray fluorescence spectrometry (XRF) have gained more interest. However, these methods are not feasible for inline applications with aspired detection rates of several parts per seconds. On the other hand, LIBS as a rapid analysis method without sample contact has the potential for inline high-speed applications. The identification of the polymer matrix using LIBS was described in Sect. 14.1.1 and studied by several groups [14.12, 14.26]. Bromine has been detected with LIBS in pure solid organic compounds and a helium atmosphere in the laser–target interaction zone has been found to significantly improve the detection [14.27]. The LIBS detection of heavy metals in polymers was reported for single pulse plasma excitation under laboratory conditions and applied to experiments at a double chain extruder [14.28, 14.29]. However, to sort EOL-WEEE waste according to the content of hazardous elements before recycling, waste pieces have to be analyzed while moving on belt conveyors at a speed of up to 1 m/s or higher (cf. Sects. 13.3 and 18.4). This requires a LIBS setup capable to compensate distance variations between the focusing optics and the surface of the waste piece.

The heavy metal detection with LIBS described in the following is focused on lead, chromium, cadmium, and mercury found in pigments, nucleating agents, and heat stabilizers. Brominated flame retardants are detected via Br and additionally via Sb as a component of the widely used synergist antimony trioxide Sb_2O_3. Due to the requirement to provide a LIBS analysis for inline measurement under industrial conditions, all measurements are constrained to be conducted in air under normal atmospheric conditions. Signal-to-noise enhancing detection in low-pressure or noble gas atmospheres is not applicable in this case.

A set of reference samples were used for the LIBS investigations and for calibration purposes [14.19]. The polymer matrix is acrylonitrile butadiene styrene (ABS) and the reference samples contain the elements Cd, Cr, Hg, Pb in concentrations of 50–10,000 μg/g, Sb in concentrations of 50–100,000 μg/g, and Br in concentrations of 50–300,000 μg/g. Furthermore, virgin samples with concentrations $c < 1\,\mu g/g$ for all these elements were provided for calibration.

The experimental setup to study the influence of detection parameters and excitation conditions is shown in Fig. 14.8 [14.20]. The pulses of a Q-switched Nd:YAG laser, which can be operated in single pulse and in double pulse mode, are focused onto the surface of the specimen by a single lens with a focal

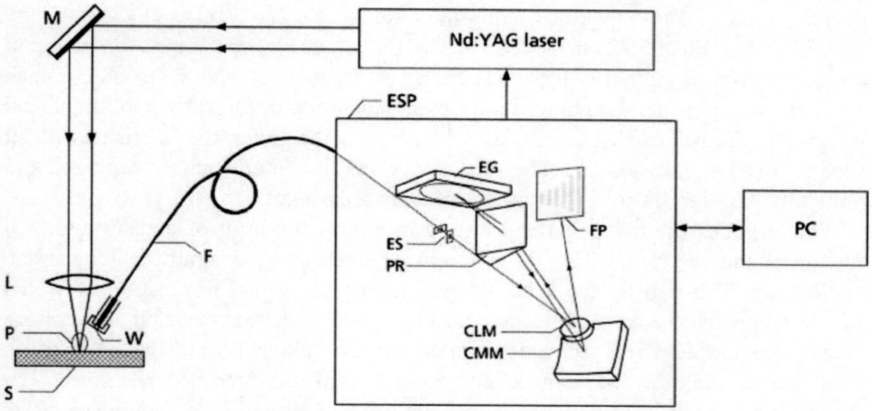

Fig. 14.8 Laboratory setup for LIBS experiments. M = mirror, L = lens, P = laser-induced plasma, S = sample, W = quartz window, F = fiber optic, ESP = echelle spectrometer, ES = entrance slits, CLM = collimator mirror, PR = prism, EG = echelle grating, CMM = camera mirror, FP = focal plane, PC = personal computer

length of $f = 220$ mm. The parameters of the laser are: wavelength $\lambda = 1{,}064$ nm, typical pulse energy $E_L = 350$ mJ, pulse duration (FWHM) $\tau_L = 7$ ns, repetition rate $\nu = 30$ Hz. For single pulse plasma excitation, the beam waist of the focus is positioned inside the sample volume at $\Delta s = 12$ mm underneath the surface of the specimen. The emission of the laser-induced plasma is received by a fiber optics with a core diameter of 600 μm and a length of 2 m. The receiving aperture of the fiber optics is protected by a quartz window and positioned at a distance of $d = 60$ mm to the laser-induced plasma. The received plasma radiation is guided to an echelle spectrometer equipped with a microchannel plate (MCP) electro-optic gateable image intensifier and a megapixel full-frame CCD camera (Kodak KAF 1001), cf. Sects. 4.2, 4.4. The plasma emission is detected during a defined gating interval at wavelengths from $\lambda = 200$ nm to $\lambda = 780$ nm with a spectral resolution of $\Delta\lambda = 5$ pm to $\Delta\lambda = 19$ pm. The integrated intensities are digitized and transferred to a personal computer for data processing.

The detection of bromium in air under atmospheric pressure is difficult since the most sensitive Br emission lines are in the vacuum ultraviolet (VUV) spectral region at wavelengths $\lambda < 170$ nm, e.g., 148.85, 153.17, or 163.34 nm [14.30]. Due to the strong absorption of these spectral emissions by the ambient atmosphere (cf. Sect. 3.5, Fig. 3.14), the application of these VUV lines is not feasible for an inline LIBS analyzer. Hence, bromium is detected using the $\lambda = 827.24$ nm emission line as the most sensitive Br emission line in the spectral range 200 nm $< \lambda <$ 900 nm. With the available echelle spectrometer, it was not possible to detect the Br emission at 827.24 nm.

To achieve high sensitivity and also to enable high-speed measurements, the echelle spectrometer is replaced by a Paschen–Runge polychromator. In this polychromator, the spectral emission lines Cd 228.80 nm, Cr 425.43 nm, Hg 253.65 nm, Pb 405.78 nm, Sb 259.80 nm, Br 827.24 nm, and the 0th-order signal are detected by photomultiplier tubes, cf. Fig. 4.7, Table 4.4, and Sect. 4.4. For each laser-induced plasma, the photomultiplier signals are processed by a multichannel integrator electronics (MCI, cf. Fig. 4.16), which integrates the intensities of all spectral lines simultaneously. The time-integrated line intensities are digitized and transmitted to the control PC for further data processing.

The target of the investigations was to minimize the limit of detection (LOD) for the elements Cd, Cr, Hg, Pb, Sb, and Br under the constraints of the inline application. This can be achieved by maximizing the signal-to-noise ratio (SNR) of the respective background-corrected integrated line intensities. If spectra are taken with the echelle spectrometer, both the line intensity and the SNR of an investigated emission line can be determined from one recorded spectrum. The spectral background and its noise within the wavelength range of the emission line are determined by the mean signal level and its standard deviation in wavelength regions adjacent to the emission line. The background corrected line intensity is then calculated by standard background correction procedures. The expectation value and its variation are gained by repeated measurements.

The features of the bromium line at 827.24 nm require the use of the Paschen–Runge polychromator which was also used for all Sb 259.80 nm measurements since Sb is contained in the same set of reference samples as Br. Using the Paschen–Runge polychromator, it is not possible to determine the spectral background and the background corrected line intensity within one measurement. Here, repeated measurements with one specimen of high- and with one specimen of low-analyte concentration are performed. Then, the analytical resolving power A of each analyte line is calculated, defined here as:

$$A = \frac{\bar{I}_{\text{high}} - \bar{I}_{\text{low}}}{\left[s^2\left(I_{\text{high}}\right) + s^2\left(I_{\text{low}}\right)\right]^{1/2}} = \frac{\bar{I}_{\text{analyte}}}{s(\bar{I}_{\text{analyte}})}, \qquad (14.1)$$

where $\bar{I}_{\text{high}}$ and $\bar{I}_{\text{low}}$ are the mean intensities of the analyte line measured with the specimens containing the analyte at high (c_{high}) and low concentration (c_{low}) and $s(I_{\text{high}})$ and $s(I_{\text{low}})$ are the respective empirical standard deviations. For a low analyte concentration, c_{low} close to or lower than the detection limit, the analytical resolving power A is similar to the expected mean SNR.

The influence of the laser pulse energy on the SNR was investigated for single pulse plasma excitation. For the elements Cd, Cr, Hg, and Pb, the echelle spectrometer and for Sb and Br the Paschen–Runge polychromator was used for signal detection. The plasma emission was detected in a fixed integration time gate $t_{\text{int}} = 10\,\mu\text{s}$ starting with a time delay $t_{\text{delay}} = 1\,\mu\text{s}$ after the laser-induced plasma ignition. The results for all investigated elements are summarized in Table 14.5. For Cd 228.80 nm, increasing the laser pulse energy from $E_{\text{L}} = 50$ mJ to $E_{\text{L}} = 300$ mJ by a factor of 6 improves the SNR by 94%. At the same time, the background-corrected line intensity increases by a factor of 20. The significantly smaller

Table 14.5 Relative change in the signal-to-noise ratio (SNR) or analytical resolving power (A) with increasing laser pulse energy

Element	λ (nm)	c_{low} ($\mu g/g$)	c_{high} ($\mu g/g$)	E_{L1} (mJ)	E_{L2} (mJ)	Relative change of *SNR* or $A(\%)$
Cd	228.80	–	10,000	50	300	94
Cr	425.43	–	10,000	50	300	−40
Hg	253.65	–	10,000	50	300	100
Pb	405.78	–	10,000	50	300	−40
Sb	259.80	50	100,000	100	350	0
Br	827.24	50	100,000	100	350	633

λ, emission wavelength of analyte; c_{low}, c_{high}, analyte concentrations in the ABS reference samples; E_{L1}, E_{L2}, laser pulse energy; A, analytical resolving power, see (14.1)

improvement of the SNR in comparison to the line intensity is due to the increasing electron density in the plasma yielding a higher spectral background level. Still, the expected detection limit is almost lowered by a factor of two. For mercury Hg 253.65 nm, the SNR could also be approximately doubled increasing the pulse energy from $E_L = 50$ mJ to $E_L = 300$ mJ. The most distinct improvement could be achieved for the Br 827.24 nm emission line, where detection was not possible with a pulse energy of $E_L = 50$ mJ. Increasing the pulse energy from $E_L = 100$ mJ to $E_L = 350$ mJ, the SNR could be improved by more than a factor of 6.

In contrast, for Sb 259.80 nm, no significant variation of the analytical resolving power A was found with increasing laser pulse energy. For the emission lines of Cr 425.43 nm and Pb 405.78 nm, even a decrease in SNR of about 40% was observed for higher pulse energies. Hence, the temporal evolution of the SNR has to be considered for these emission lines and suitable integration time gates for signal detection have to be determined.

Figure 14.9 shows the lead Pb 405.78 nm emission line for a width of the integration time gate of $t_{int} = 0.5\,\mu s$ at the time delays $t_{delay} = 1$, 3, and 20 μs.

Figure 14.10 shows the corresponding SNR as a function of the time delay ranging from 1 to 20 μs. While the overall signal intensity is decreasing as a function of time, the SNR is rising at early time delays of $t_{delay} < 1\,\mu s$ to $t_{delay} = 3\,\mu s$ and decreases slowly for time delays up to 20 μs.

Regarding signal integration, the temporal evolution of both the signal-to-noise ratio and the line intensity have to be considered. High line intensities with low SNR at early delay times contribute much more background to the integrated line intensity than the low line intensities at larger time delays which show a higher SNR. The noise contribution during the decay of the analyte signal with low SNR at very large delay times can be neglected due to the very low overall intensity of the line emission.

The target is to maximize the SNRs of the integrated background-corrected line intensities $I_{analyte\,i}(t_{delay}, t_{int})$ for each analyte line i by setting the optimum time delay t_{delay} and integration time gate t_{int}. Using a Paschen–Runge polychromator, the optimum parameters can be determined by sampling the photomultiplier signals directly with an oscilloscope and calculate the analytical resolving power A for any

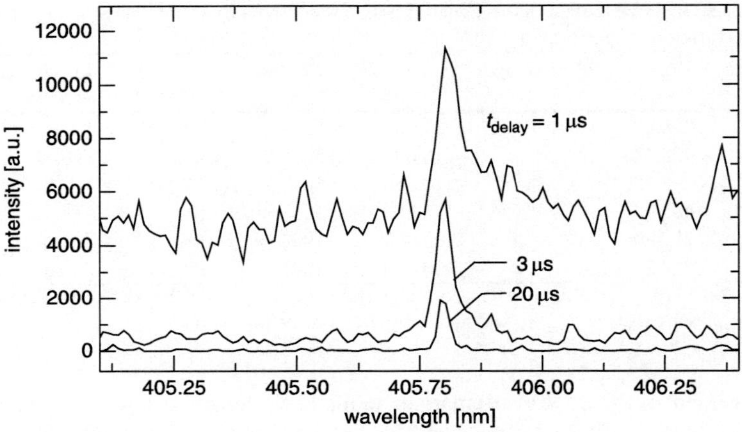

Fig. 14.9 Temporal evolution of the Pb 405.78 nm emission line for single pulse plasma excitation with laser pulse energy of $E_L = 250$ mJ. Detection with echelle spectrometer using an integration time gate of $t_{int} = 0.5\,\mu s$ at the stated time delays. ABS reference sample, $c_{Pb} = 10{,}000\,\mu g/g$

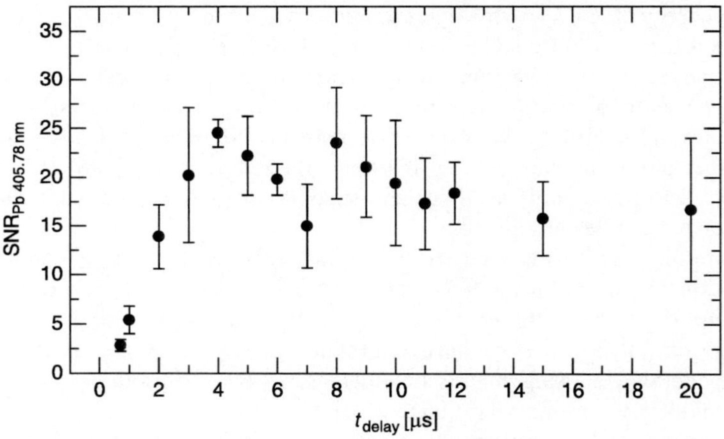

Fig. 14.10 Signal-to-noise ratio of the background-corrected Pb 405.78 nm emission line as a function of the time delay of the integration window for single pulse plasma excitation with a laser pulse energy of $E_L = 250$ mJ. Detection with echelle spectrometer using an integration time gate of $t_{int} = 0.5\,\mu s$. ABS reference sample, $c_{Pb} = 10{,}000\,\mu g/g$

possible time delay and integration time gate (t_{delay}, t_{int}) within the measured time range. This approach has been already demonstrated for high analyte concentrations [14.31]. However, for element concentrations lower than 1,000 μg/g, the photomultiplier signals are too weak to be recorded. Hence, the MCI electronics was used to sample the temporal evolution of the background-corrected analyte signal $I_{analyte}(t_{delay}, \delta t_{int})$ using a sampling interval δt_{int} for a series of time delays t_{delay}.

For each spectral emission line, repeated measurements were performed at different time delays t_{delay} varied in time delay steps of $\delta t_{\mathrm{delay}} = 1\,\mu\mathrm{s}$ with a fixed sampling interval $\delta t_{\mathrm{int}} = 1\,\mu\mathrm{s}$. The analyte signals $I_{\mathrm{analyte}}(t_{\mathrm{delay}}, t_{\mathrm{int}})$ and their empirical standard deviations $s_{\mathrm{analyte}}(t_{\mathrm{delay}}, t_{\mathrm{int}})$ are used to calculate the expected analytical resolving power A [see (14.1)] for those $(t_{\mathrm{delay}}, t_{\mathrm{int}})$–combinations, which can be formed by consecutive sampling intervals. For these $(t_{\mathrm{delay}}, t_{\mathrm{int}})$ combinations, the analyte intensity I_{analyte} is calculated by summing up the sampled intensities $I_{\mathrm{analyte}}(t_{\mathrm{delay}}, \delta t_{\mathrm{int}})$:

$$I^{\mathrm{acc}}_{\mathrm{analyte}}(t_{\mathrm{delay}}, t_{\mathrm{int}}) = \sum_{t_{\mathrm{d}}=t_{\mathrm{delay}}}^{t_{\mathrm{delay}}+t_{\mathrm{int}}-\delta t_{\mathrm{int}}} I_{\mathrm{analyte}}(t_{\mathrm{d}}, \delta t_{\mathrm{int}}), \tag{14.2}$$

where $I^{\mathrm{acc}}_{\mathrm{analyte}}(t_{\mathrm{delay}}, t_{\mathrm{int}})$ is the analyte signal calculated for the time delay t_{delay} and integration time gate t_{int} [this intensity is called accumulated intensity in the following to clarify the difference to the measured intensity in the sampling interval $I_{\mathrm{analyte}}(t_{\mathrm{delay}}, \delta t_{\mathrm{int}})$], $I_{\mathrm{analyte}}(t_{\mathrm{delay}}, \delta t_{\mathrm{int}})$ is the measured mean intensity in the fixed sampling interval δt_{int} at delay $t_{\mathrm{d}} = t_{\mathrm{delay}}$. The empirical standard deviation of the accumulated analyte signal is determined by

$$s^{\mathrm{acc}}_{\mathrm{analyte}}(t_{\mathrm{delay}}, t_{\mathrm{int}}) = \left\{ \sum_{t_{\mathrm{d}}=t_{\mathrm{delay}}}^{t_{\mathrm{delay}}+t_{\mathrm{int}}-\delta t_{\mathrm{int}}} s^2 \left[I_{\mathrm{analyte}}(t_{\mathrm{d}}, \delta t_{\mathrm{int}}) \right] \right\}^{1/2}. \tag{14.3}$$

In the calculation of the standard deviation, the covariance of measurements in consecutive sampling intervals is neglected. This assumption is based on the experimental observation that a temporally resolved photomultiplier signal of a single plasma emission shows a mean curve which is distorted by a strong white noise. This noise is due to statistical processes in the plasma emission, variations of the laser peak power, variations in the absorption of the laser pulses in the specimen and to photomultiplier noise. For single shot evaluation, the variations caused by this noise dominate the variations in the mean signal curve which is determined by averaging a high number of plasma emissions. For single shot evaluation, the covariance of the measured intensities in adjacent detection windows is therefore assumed to be very small. With $I^{\mathrm{acc}}_{\mathrm{analyte}}$ and $s^{\mathrm{acc}}_{\mathrm{analyte}}$, the analytical resolving power A for the parameters $(t_{\mathrm{delay}}, t_{\mathrm{int}})$ is then calculated by:

$$A(t_{\mathrm{delay}}, t_{\mathrm{int}}) = \frac{I^{\mathrm{acc}}_{\mathrm{analyte}}(t_{\mathrm{delay}}, t_{\mathrm{int}})}{s^{\mathrm{acc}}_{\mathrm{analyte}}(t_{\mathrm{delay}}, t_{\mathrm{int}})} = \frac{1}{\mathrm{RSD}^{\mathrm{acc}}_{\mathrm{analyte}}}. \tag{14.4}$$

It is equivalent to the inverse relative standard deviation $\mathrm{RSD}^{\mathrm{acc}}_{\mathrm{analyte}}$ of the accumulated analyte signal. The lowest LOD is expected at maximum analytical resolving power $A(t_{\mathrm{delay}}, t_{\mathrm{int}})$ or the lowest $\mathrm{RSD}^{\mathrm{acc}}_{\mathrm{analyte}}$.

The method described above allows to determine the optimum integration interval for raw LIBS signals. However, using internal standardization, i.e., evaluating the ratios of integrated line intensities of analyte and reference lines, yields higher stability for quantitative LIBS measurements (cf. Chap. 11). Hence, it is of interest, whether the optimum detection intervals for "raw" line intensities can also be applied to internal standardization. In this case, the parameters $(t_{\text{delay}}, t_{\text{int}})$ can be optimized individually for the analyte and the reference emission line. Internal standardization is described by

$$\hat{I}^{\text{acc}}_{\text{analyte}} = \frac{I^{\text{acc}}_{\text{analyte}}(t^{\text{a}}_{\text{delay}}, t^{\text{a}}_{\text{analyte}})}{I^{\text{acc}}_{\text{reference}}(t^{\text{r}}_{\text{delay}}, t^{\text{r}}_{\text{analyte}})}, \tag{14.5}$$

where the index "a" in $(t^{\text{a}}_{\text{delay}}, t^{\text{a}}_{\text{int}})$ denotes that these parameters belong to the analyte line and the index "r" refers to the reference line. For the relative variance holds:

$$\hat{\text{RSD}}^{\text{acc}\;2}_{\text{analyte}} = \left(\frac{\hat{s}^{\text{acc}}_{\text{analyte}}}{\hat{I}^{\text{acc}}_{\text{analyte}}}\right)^2 = \frac{(s^{\text{acc}}_{\text{analyte}})^2}{\left(I^{\text{acc}}_{\text{analyte}}\right)^2} + \frac{(s^{\text{acc}}_{\text{reference}})^2}{(I^{\text{acc}}_{\text{reference}})^2} - 2\frac{s^2(I^{\text{acc}}_{\text{analyte}}, I^{\text{acc}}_{\text{reference}})}{I^{\text{acc}}_{\text{analyte}}\, I^{\text{acc}}_{\text{reference}}} \tag{14.6}$$

The first two terms on the right side of (14.6) describe the relative variance of the integrated values for the analyte and the reference line, as they are obtained by individual optimization of the integration time gate. The last term is the normalized covariance of analyte and reference line, both integrated in individual integration time gates. In case of parameter fluctuations such as laser pulse energy, absorption of laser radiation or plasma emission, the relative standard deviation $\hat{\text{RSD}}^{\text{acc}}_{\text{analyte}}$ of the standardized signal will be significantly lower than the relative standard deviation $\text{RSD}^{\text{acc}}_{\text{analyte}}$ of the raw analyte signal. This is due to the covariance of analyte and reference signal when both reflect the influence of the parameter fluctuations in a similar way. The dependence of $\hat{\text{RSD}}^{\text{acc}}_{\text{analyte}}$ on the applied detection intervals $(t^{\text{a}}_{\text{delay}}, t^{\text{a}}_{\text{int}})$ and $(t^{\text{r}}_{\text{delay}}, t^{\text{r}}_{\text{int}})$ is not so obvious. The first two terms on the right side of (14.6) vary significantly with the setting of the time delay and integration time gate. This is evident from Figs. 14.9 and 14.10 and described in other publications [14.32]. On the other hand, the temporal evolution of the analyte and reference line intensity are comparable and differ mostly in parameters of the same principal, nonnegative intensity curve. Hence, the positive normalized covariance between analyte and reference signal is expected to vary much less with the choice of the individual integration time gates $(t^{\text{a}}_{\text{delay}}, t^{\text{a}}_{\text{int}})$ and $(t^{\text{r}}_{\text{delay}}, t^{\text{r}}_{\text{int}})$ than the sum of the individual RSDs of analyte and reference signal. Consequently, the optimized signal integration for internal standardization is expected to be achieved by the individual optimized detection intervals for the analyte and the reference line.

In the experiments, the temporal evolution of the line emission was sampled for time delays $t_{\text{delay}} = 0.5\,\mu\text{s}$ up to $80\,\mu\text{s}$ with a fixed sampling interval $\delta t_{\text{int}} = 1\,\mu\text{s}$. Five measurements with 50 repetitions each were conducted at different sample spots using laser pulses with a pulse energy of $E_{\text{L}} = 350\,\text{mJ}$ on high and low concentration samples. For the investigation of the emission lines of Cd 228.80 nm, Cr 425.43 nm, Hg 253.65 nm, and Pb 405.78 nm, the high concentration specimen contained $c_{\text{high}} = 1{,}000\,\mu\text{g/g}$ and the low concentration specimen contained

$c_{low} = 50\,\mu g/g$ of the respective analyte. For the emission lines of antimony Sb 259.80 nm and bromium Br 827.24 nm, the analyte concentration in the high concentration specimen was $c = 100{,}000\,\mu g/g$ and in the low concentration sample $c = 300\,\mu g/g$. Figure 14.11 shows the calculated, background corrected, accumulated intensity $I^{acc}_{analyte}(t_{delay}, t_{int})$ for Pb 405.78 nm [see relation (14.2)] as a function of the end of the integration time gate $(t_{delay} + t_{int})$ for different time delays t_{delay}. For simplification, this intensity is denoted in the graphs with $I_{Pb\,405.78\,nm}$. As expected, a higher accumulated, background corrected intensity is achievable setting earlier time delays t_{delay}. Increasing the integration time gate t_{int} for a fixed time delay t_{delay} yields a monotonous increase of the signal with a saturation behavior for large $(t_{delay} + t_{int})$.

Figure 14.12 shows the calculated relative standard deviation of the accumulated signal for different delays t_{delay} as a function of the end of the integration time gate $(t_{delay} + t_{int})$. Increasing $(t_{delay} + t_{int})$ reduces $RSD_{Pb\,405.78\,nm}$ for every measured time delay t_{delay} and hence the detection limit is improved. The lowest $RSD_{Pb\,405.78\,nm}$ for the Pb 405.78 nm emission line can be achieved for $t_{delay} = 2\,\mu s$. For earlier signal integration with $t_{delay} = 0.5$ or $1\,\mu s$, the lowest achievable $RSD_{Pb\,405.78\,nm}$ is significantly larger. This reflects the high amplitude of noise introduced by early signal integration, which cannot be compensated for by later integral contributions with lower amplitude. Figure 14.13 visualizes the minimum of $RSD_{Pb\,405.78\,nm}$ as a function of t_{delay} for a fixed end of the integration time gate of $(t_{delay} + t_{int}) = 80\,\mu s$. A minimum of $RSD_{Pb\,405.78\,nm}$ occurs at $t_{delay} = 2\,\mu s$.

The described method was used to determine the optimum parameters (t_{delay}, t_{int}) for all investigated emission lines and the results are listed in the column for single pulse excitation in Table 14.6.

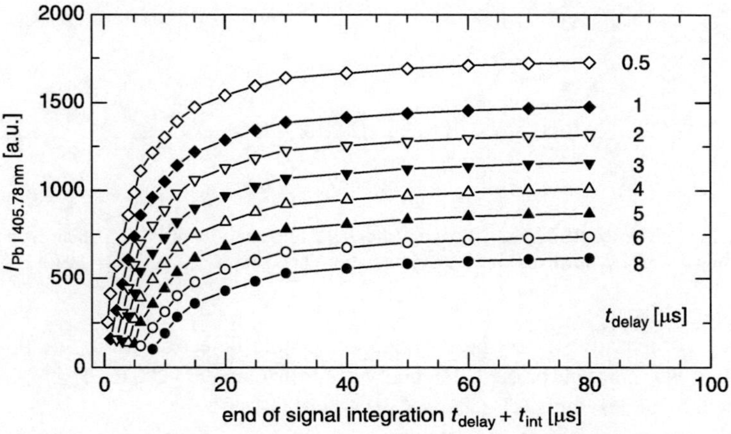

Fig. 14.11 Calculated, background corrected, accumulated line intensity of lead Pb 405.78 nm as a function of the end time of the integration time gate $(t_{delay} + t_{int})$ for different time delays t_{delay}

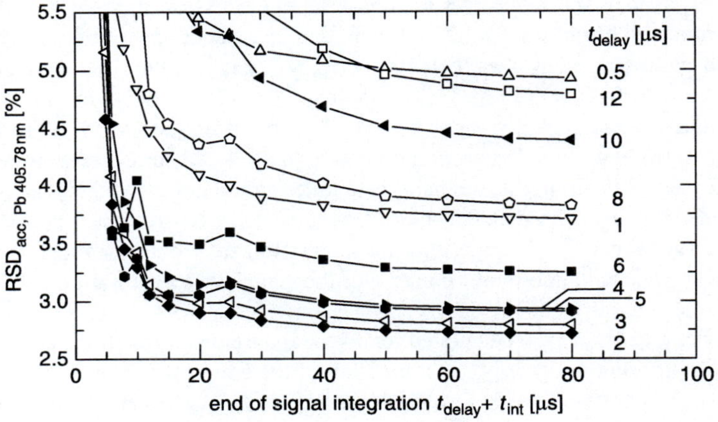

Fig. 14.12 Calculated relative standard deviation $RSD_{Pb\,405.78\,nm}$ of the Pb 405.78 nm integrated line intensity as a function of the end of the integration time gate ($t_{delay} + t_{int}$) for different time delays t_{delay}

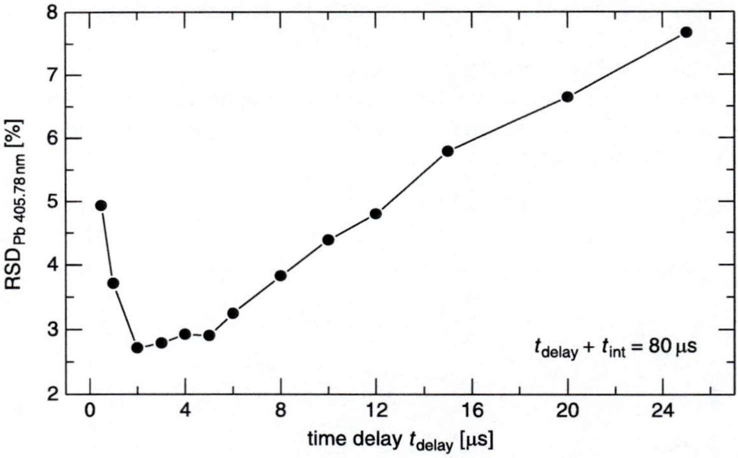

Fig. 14.13 Calculated $RSD_{Pb\,405.78\,nm}$ of the integrated line intensity of Pb 405.78 nm as a function of the time delay t_{delay} for fixed end of the integration time gate ($t_{delay} + t_{int}$) $= 80\,\mu s$

For a comparative assessment of the achievable detection limits, double pulse excitation was studied for the LIBS analysis of the elements Cd, Cr, Hg, Pb, Sb, and Br in technical polymers (cf. Chaps. 6, 7, and 8) [14.20].

In the experiments, two separate driver units for the Pockels cell in the laser head of the Nd:YAG laser generate the two laser pulses with an adjustable interpulse separation of $\Delta t = 18$–$30\,\mu s$. Some authors have demonstrated that for metal

Table 14.6 Detection limits according to the 3*s*-criterion for single and double pulse plasma excitation and the respective parameters (t_{delay}, t_{int}) under laboratory conditions

Analyte	Wavelength (nm)	Excitation energy (eV)	Single pulse			Double pulse		
			t_{delay} (µs)	t_{int} (µs)	LOD (µg/g)	t_{delay} (µs)	t_{int} (µs)	LOD (µg/g)
Cd	228.80	5.42	1	10	19	0.8	20	11
Cr	425.43	2.91	2	98	2	0.8	20	4
Hg	253.65	4.88	1.5	9.5	18	0.8	20	22
Pb	405.78	4.38	2	48	16	0.8	20	24
Sb	259.80	5.82	2	4	50	0.8	9.2	17
Br	827.24	9.36	0.8	5	15,000	0.8	20	11,000

matrices the line intensity and the detection sensitivity could be significantly improved for interpulse separations in the range of 0.5–80 µs depending on the matrix and the ambient gas conditions [14.13,14.33,14.34]. The signal enhancement of the line emission observed for double pulse excitation of steel samples is only a weak function of the interpulse separation in the interval between 2.5 and 80 µs [14.13]. If this is also valid for polymer samples is not known.

The typical pulse widths (FWHM) of the laser pulses amounts to $\tau_1 = 12$ ns and $\tau_2 = 15$ ns for the first and second pulse, respectively. The pulse energy ratio of the first to the second laser pulse is denoted with E_1/E_2. The burst energy is given by $E_b = E_1 + E_2$. Time delays t_{delay} given for double pulse excitation refer to the second pulse. Maximum plasma emission has been found for focusing the laser $\Delta s = 5$ mm below the sample surface.

The temporal evolution of the emission lines was found to differ significantly from the one in single pulse laser-induced plasmas. The spectral background of the Pb 405.78 nm emission line at time delays $t_{delay} < 3$ µs is reduced significantly using double pulse excitation, see Fig. 14.14 in comparison with Fig. 14.9. No significant emission could be detected for time delays $t_{delay} \geq 25$ µs, and it is evident that the background-corrected line intensity is decreasing faster for double pulse plasma excitation than for single pulse plasma excitation. The SNR of the Pb 405.78 nm line as a function of the time delay is shown in Fig. 14.15. As an important difference to single pulse plasma excitation (see Fig. 14.10), the SNR is not rising for increasing delay times between 1 and 4 µs. Both the intensity and the SNR of the Pb 405.78 nm emission are decreasing with the time delay. This was also observed for the emission lines Cd 228.80 nm, Cr 425.43 nm, and Hg 253.65 nm. According to the discussion above, it is clear that since both the line intensity and the signal-to-noise ratio SNR are decreasing with the time delay t_{delay}, the lowest $RSD_{Pb\,405.78\,nm}$ is achieved for the shortest time delay within the sampled time interval. Hence, the lowest detection limit is expected for the shortest time delay. As the integration time gate t_{int} was found to be of lower importance for the expected detection limit, it is adjusted to detect the significant parts of the line emission. The resulting settings for the time delay t_{delay} and the integration time gate t_{int} are listed in Table 14.6.

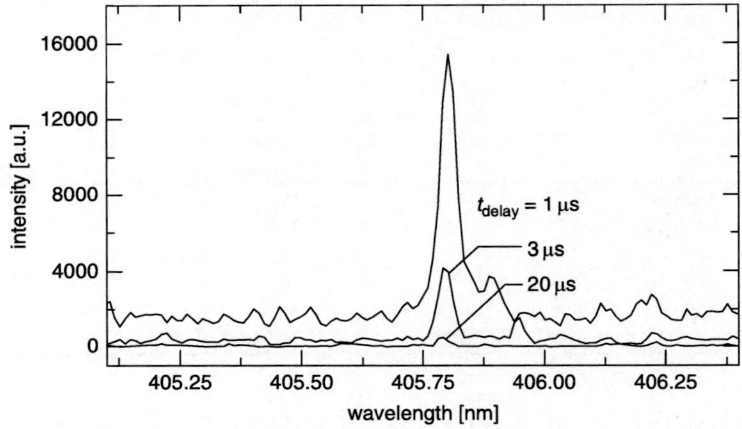

Fig. 14.14 Temporal evolution of the Pb 405.78 nm emission line for double pulse plasma excitation with the following parameters: $E_b = 250\,\text{mJ}$, $E_1/E_2 = 1$, $\Delta t = 30\,\mu\text{s}$. Detection with echelle spectrometer using an integration time gate of $t_{int} = 0.5\,\mu\text{s}$ at the stated time delays

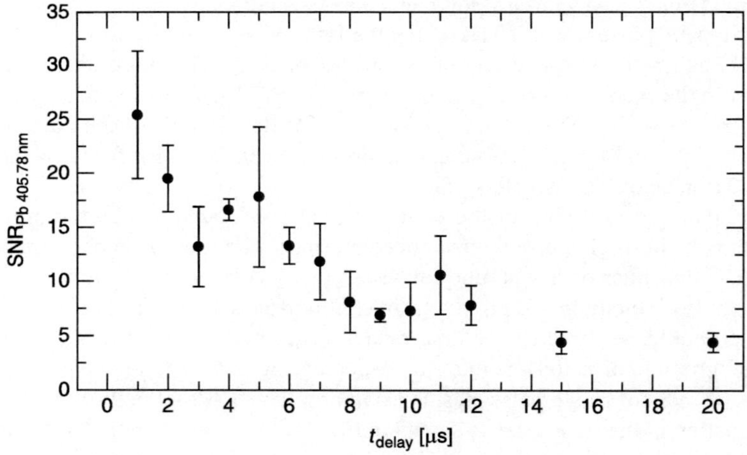

Fig. 14.15 Signal-to-noise ratio (SNR) of Pb 405.78 nm for double pulse plasma excitation as a function of the time delay t_{delay}. Laser burst parameters as for Fig. 14.14. Detection with echelle spectrometer using an integration time gate of $t_{int} = 0.5\,\mu\text{s}$ and time delays between $t_{delay} = 1$ and $20\,\mu\text{s}$

The detection limits for Cd, Cr, Hg, Pb, Sb, and Br in ABS polymer matrix with LIBS were determined using the Paschen–Runge polychromator. Fifty laser bursts onto one spot of the specimen with a frequency of $f = 6\,\text{Hz}$ were considered as one measurement. For each specimen, the measurement was repeated at five different spots. For calibration internally standardized line intensities were applied.

This means that for each single laser-induced plasma, the line intensities during the detection intervals discussed above were digitized individually and internally standardized on the 0th order which was integrated with a time delay $t_{delay} = 1\,\mu s$ and an integration time gate of $t_{int} = 10\,\mu s$. From the resulting 50 internally standardized line intensities of one measurement, outliers were removed if meeting the Grubbs outlier criteria [14.35]. Then, the average line intensities were calculated for this measurement. From the five mean values of the measurements taken at one specimen, the average and the standard deviation were used for calibration.

The spatial inhomogeneity of heavy metal concentrations in polymer matrix is a well-known phenomenon [14.36]. In order to lower the impact of spatial inhomogeneity on the calibration, a maximum of one mean value was removed before calibration in case it met the Nalimov outlier criteria, which is suitable for statistical populations with $N > 3$ [14.37].

For each emission line, the detection limit was determined according to the $3s$-criterion (cf. Chap. 11) [14.38]. Table 14.6 lists the determined detection limits (LODs) with the detection interval (t_{delay}, t_{int}) for each investigated emission line using a laser pulse energy of $E_L = 350$ mJ for single pulses and a burst energy of $E_b = 350$ mJ for double pulses. Example calibration curves for single pulse plasma excitation, the respective detection limits, background-equivalent concentrations (BEC), and mean coefficients of determination r^2 are shown in Fig. 14.16 for cadmium and in Fig. 14.17 for bromine. The corresponding calibration curves for double pulse plasma excitation are shown in Figs. 14.18 and 14.19.

The detection limits for Cd, Cr, Hg, and Pb are significantly lower than the maximum allowed concentration in plastic waste (see Table 14.7, last column) and therefore meet the requirements for EOL-WEEE recycling. The detection limits of

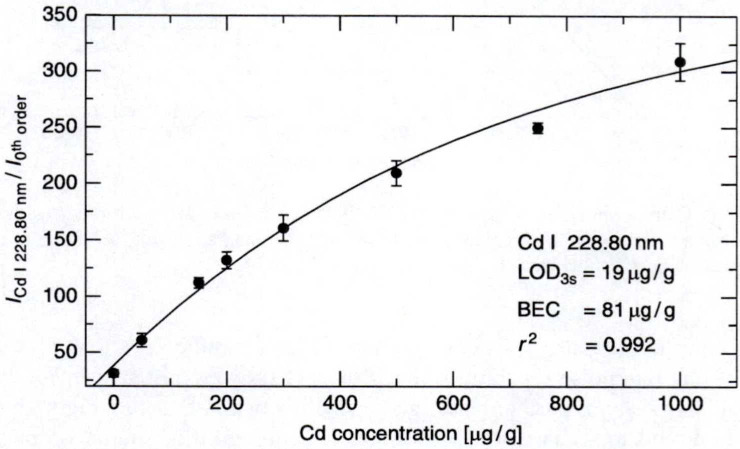

Fig. 14.16 Calibration curves for Cd 228.80 nm referenced to 0th order using single pulse plasma excitation with optimized time delay and time integration gate, see Table 14.6. Laser pulse energy: $E_L = 350$ mJ

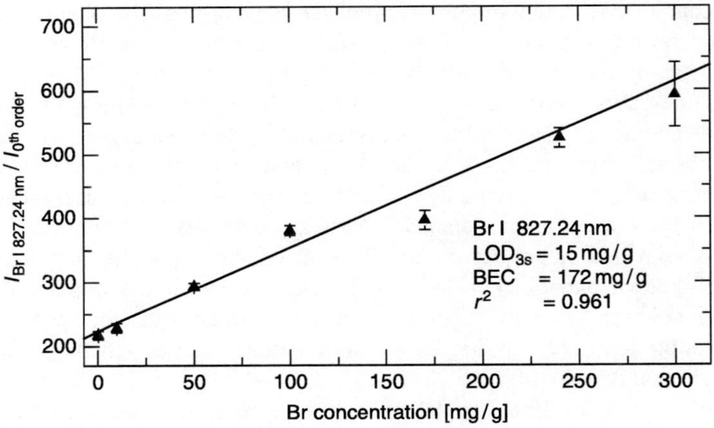

Fig. 14.17 Calibration curve for bromine Br 827.24 nm referenced to 0th order using single pulse plasma excitation with optimized time delay and time integration gate, see Table 14.6. Laser pulse energy: $E_{\mathrm{L}} = 350\,\mathrm{mJ}$

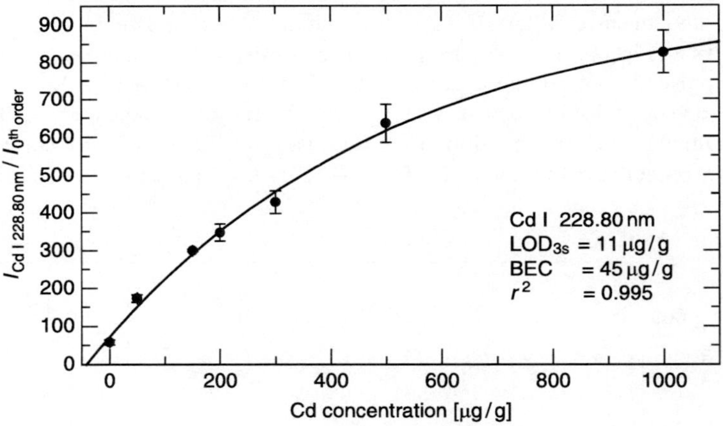

Fig. 14.18 Calibration curve for cadmium Cd 228.80 nm referenced to 0th order using double pulse plasma excitation with optimized time delay and time integration gate, see Table 14.6. Laser burst parameters: $E_{\mathrm{b}} = 350\,\mathrm{mJ}$, $E_1/E_2 = 1.2$, $\Delta t = 18\,\mu\mathrm{s}$

single and double pulse plasma excitation are comparable. This could be due to the fact that interpulse separations lower than 18 μs were not achievable with the applied laser system. Further investigations with a two laser setup might give more insight into this topic. For Br 827.24 nm, a detection limit of around 1.5 m.-% was achieved in ambient air. Although the detection limit is significantly higher than for other elements, it is sufficient to detect common concentrations of brominated flame retardants (BFR) in plastics which typically exceed 3–5 m.-%. Furthermore,

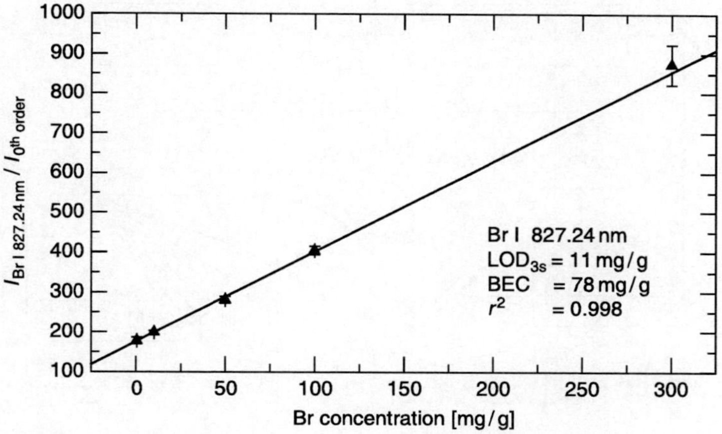

Fig. 14.19 Calibration curves for Br 827.24 nm referenced to 0th order using double pulse plasma excitation with optimized time delay and time integration gate, see Table 14.6. Laser burst parameters: $E_b = 350\,\text{mJ}$, $E_1/E_2 = 1.2$, $\Delta t = 18\,\mu\text{s}$

Table 14.7 Limits of detection (LOD) of the LIBS analyzer for the investigated elements in moving ABS samples

Element	λ (nm)	Dynamic focusing LOD (μg/g)	Required LOD (μg/g)
Cd	228.80	96	100
Cr	425.43	73	100
Hg	253.65	60	100
Pb	405.78	140	100
Sb	259.80	80	100
Br	827.24	n.d.	5,000

LOD calculated by the 3s-criterion
$n.d.$ = not determinable

the detection limit of 17 μg/g for Sb I 259.80 nm with double pulse laser excitation allows to detect very low concentrations of the BFR synergist Sb_2O_3 which is commonly added along in concentrations of several mass percent to enhance the flame retarding effect.

On the basis of the laboratory results, an inline LIBS analyzer was setup to provide the required sensitivity for heavy metal detection in polymers. The task of the LIBS analyzer is the rapid quantification of heavy metal and halogen concentrations in EOL-WEEE pieces moving at a speed of 0.5–1.0 m/s on a belt conveyor. The major problem for the application of LIBS to inline analysis of real waste EOL-WEEE pieces such as monitors, telephones, and keyboards is the large variation in size and shape. In contrast to laboratory experiments, no well-defined measuring distance to the surface of the specimen is guaranteed and large variations of the measuring distance for successive samples must be compensated by the LIBS analyzer. For this task, an autofocusing unit for LIBS was setup to keep the analyzing laser beam focused on the sample surface, see Fig. 14.20.

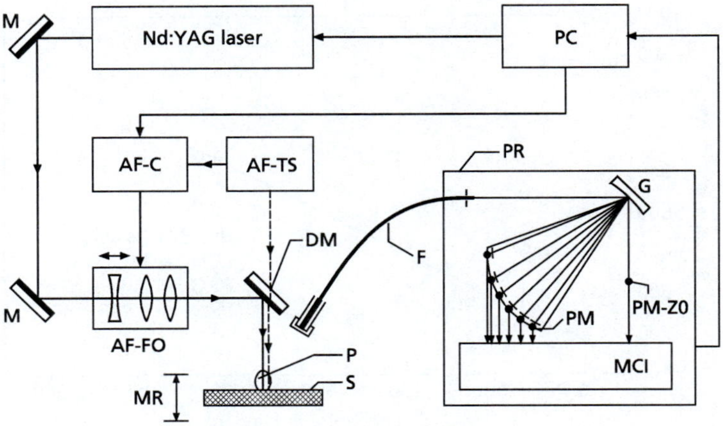

Fig. 14.20 Setup of the inline LIBS analyzer. PC = personal computer, PR = Paschen–Runge polychromator, G = grating, PM = photomultiplier, PM-ZO = photomultiplier at 0th-order, MCI = multichannel integrator electronics, F = fiber optics, P = plasma, S = sample, MR = measuring range, DM = dichroic mirror, M = mirror, AF = autofocusing unit, AF-TS = triangulation sensor, AF-C = controller, AF-FO = focusing optics

In the autofocusing unit, a laser triangulation sensor (Cyclop 220, nokra GmbH, Germany) measures the distance between the focusing optics and the sample surface with an accuracy of $\pm 70\,\mu m$ and a frequency of 50 Hz. The autofocus control processes the triangulation sensor output and generates the control signal for the variable focal length optical system (VarioScan, SCANLAB, Germany with an additional external $f = 200$ mm lens) with a mean focal length of about $f = 220$ mm.

An example for the signal stabilization achieved with the autofocusing unit is shown in Fig. 14.21. For rising distances of the sample surface to the detecting fiber, both the Pb 405.78 nm and the 0th-order signal show the expected intensity decrease. However, the Pb 405.78 nm signal internally standardized to the 0th-order signal is constant within 5% over a range of 50 mm. This result could not be achieved with static focusing. In Fig. 14.22, the internally standardized Pb 405.78 nm line intensity acquired with static focusing is compared with dynamic focusing for varying distances between the focusing optics and the sample.

Using static focusing, the LIBS signal shows a strong dependency on the deviation from the optimum distance to the sample surface. This dependency cannot be compensated by internal standardization which complies to earlier publications [14.39]. In contrast, dynamic focusing can provide standardized line intensities which are constant within 5% over a range of 50 mm.

To further reduce the measuring distance variations due to the shapes of the waste pieces, the waste pieces pass the LIBS analyzer on a tilted belt conveyor sliding on the side panel. The LIBS measurement is carried out through a gap in the conveyor side panel, see Fig. 14.23. However, while the waste pieces pass the LIBS analyzer on the belt conveyor, the focusing optics has to compensate for the distance variations due to the individual surface topology of the waste pieces.

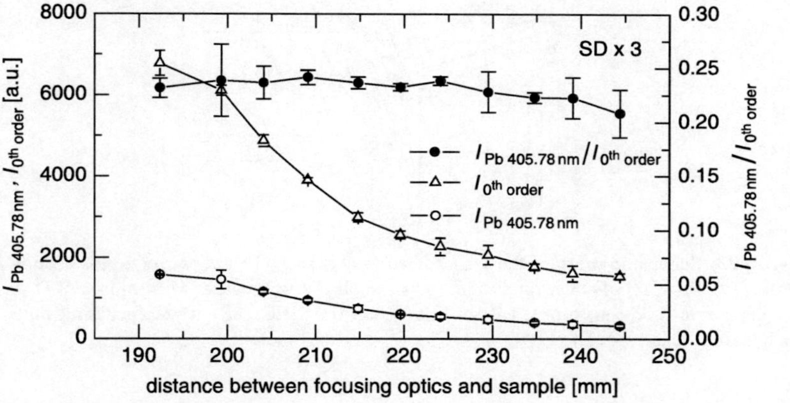

Fig. 14.21 Intensity of Pb 405.78 nm, 0th-order and ratio ⟨Pb I 405.78 nm/0th order⟩ as a function of the sample distance for a fixed fiber position using the autofocusing unit. Plasma excitation with double pulse laser in air on ABS polymer sample containing 2,000 μg/g lead. The *error bars* given by the standard deviation (SD) are shown enlarged by a factor of 3

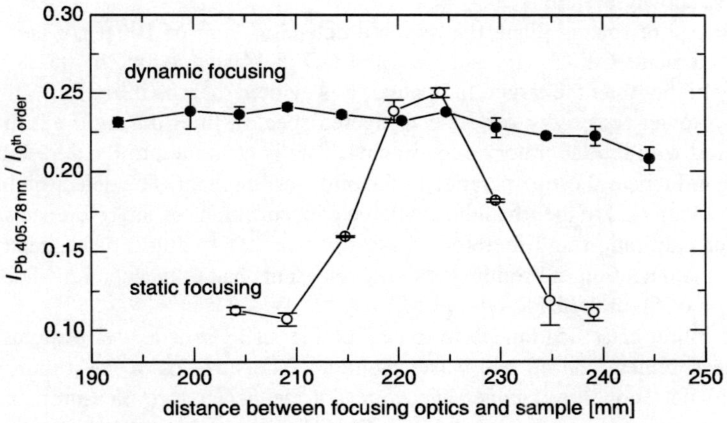

Fig. 14.22 Comparison of Pb 405.78 nm LIBS signal internally standardized to 0th-order for static and dynamic focusing

During inline measurements, double pulse laser bursts with a total burst energy of $E_b = 300\,\mathrm{mJ}$, a pulse energy ratio of $E_1/E_2 = 1.2$, a pulse separation of $\Delta t = 30\,\mu\mathrm{s}$, and a repetition rate of 30 Hz are irradiated onto the EOL-WEEE pieces. The emission of the laser-induced plasma is picked up by a fiber optics at a distance of 130 mm to the far end of the measuring range and is guided to a Paschen–Runge polychromator (cf. Sect. 4.2). The photomultiplier signals are integrated during an integration time gate of $t_{int} = 20\,\mu\mathrm{s}$ after a delay of $t_{delay} = 0.8\,\mu\mathrm{s}$ with respect to the second pulse of the laser burst.

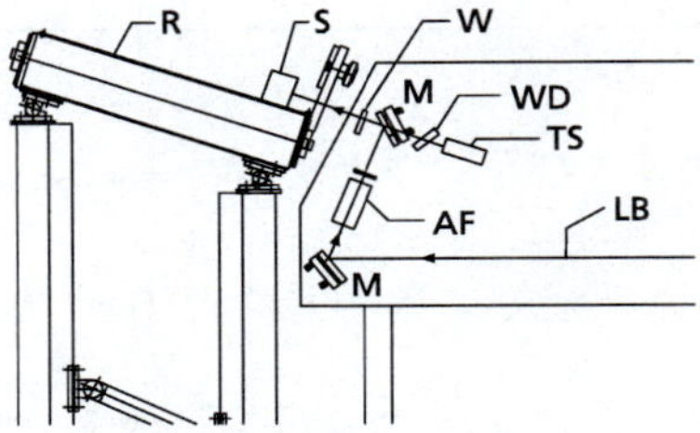

Fig. 14.23 Schematic setup of the sensor head (*right*) at the belt conveyor (cross section shown on the *left*). R = roll of conveyor system, S = sample, W = window, M = mirror, WD = wedge to compensate the beam offset, LB = laser beam for LIBS, AF = autofocussing unit, TS = triangulation sensor

The calibration of the inline LIBS analyzer was performed with reference samples moving on the belt conveyor at three different measuring distances. Three repeated measurements comprising 20 laser bursts were performed at each measuring distance. The limits of detection (LOD) for the investigated heavy metals and halogens determined with the inline LIBS analyzer operating at the pilot sorting plant are shown in Table 14.7.

At the pilot sorting plant, the required detection limit of 100 μg/g is achieved for the elements Cd, Cr, Hg, and Sb. The LOD of 140 μg/g for Pb 405.78 nm was slightly higher than the target. In general, it is evident that the inline LIBS analyzer shows a lower sensitivity for all investigated spectral lines during the inline tests compared with the laboratory experiments. This is attributed to the additional disturbing influences during operation at the pilot sorting plant. The detection limit for Br 827.24 nm was higher than the available concentrations of the reference samples used for calibration and therefore the respective LOD could not be determined. Hence, the detection of brominated flame retardants had to be accomplished by the detection of Sb in the BFR synergist Sb_2O_3.

The inline classification performance of the LIBS sensor was evaluated with 160 measurements on 40 real waste monitors. The surfaces of the monitor housings did not show any lacquer and were not cleaned before measurements. The reference analysis for the concentration of heavy metals and halogens in these monitor housings was provided by inductively coupled plasma atomic emission spectrometry (ICP-OES) [14.19]. In the real waste EOL-WEEE monitor housings, no mercury concentration >1 μg/g was detected with the ICP-OES reference analysis. Hence, the classification performance of the LIBS analyzer cannot be quantified for mercury. For the sorting of the monitor housings, the determined element concentrations were classified into four categories: (a) 0–100 μg/g, (b) 100–1,000 μg/g, (c) 1,000–2,000 μg/g, (d) >2,000 μg/g. For lead, the upper concentration limit of category (a) had to be increased to 140 μg/g due to the fact that the LOD of Pb 405.78 nm was 140 μg/g. The results of the inline classification obtained for monitor housings without surface preparation moving on the belt conveyor with a velocity of about 0.5 m/s are summarized in Table 14.8. For Cd,

Table 14.8 Results of the inline LIBS analyzer at the pilot sorting plant with real waste EOL-WEEE monitor housings moving at a velocity of 0.5 m/s

Classifications	Cd	Cr	Hg[a]	Pb	Sb
Number of correct classifications	135	152	(160)	152	152
Number of total classifications	160	160	160	160	160
Percentage of correct classifications (%)	84	95	(100)	95	95

[a]no mercury concentration > 1 μg/g was found in the real waste EOL-WEEE test samples with the ICP-OES reference analysis

Cr, Pb, and Sb, high classification ratios of up to 95% were achieved during the inline measurements at the pilot sorting plant.

14.2 Slag

Improvement of steel qualities is one object of the steel producing industry. For example, concentrations below 10 μg/g of sulfur and 100 μg/g of phosphorus are required to achieve the desired material properties of the pre-products. Hence, the goal is to continuously improve the metallurgical processing steps from the pre-product to the final product to achieve well-defined material qualities within narrow tolerances for the specified element concentrations. The composition of a steel melt is influenced by chemical reactions of the melt with slag components. Therefore, the chemical analysis of the slag provides essential information for a metallurgical process control. Thus, one important issue concerning steel production is the analysis of slag from the converter and the ladle. For the future, an increasing demand for slag analysis is expected to further optimize the metallurgical process guiding. For an efficient process control, the total time needed for slag analysis has to be reduced to less than 3 min.

XRF spectroscopy is a state-of-the-art measuring technique for slag analysis [14.40]. The slag taken in the plant is transported to a laboratory by, e.g., a pneumatic dispatch system. The time necessary for this procedure in an off-line laboratory is given in Table 14.9. Therein, it can be seen that nearly 40% of the total time is necessary for sample preparation and sample transportation. However, the analytical results during steel production are in most cases required within a short time scale of a few minutes, since the next processing step must be controlled on the basis of this analytical information. Thus, there is an increasingly strong need to simplify or even avoid those steps of sample transportation or sample preparation. Laser-based analytical methods will play a key role for the development of inline methods in metallurgical processing, since they offer noncontact measurements regardless of the material to be analyzed, high measuring speed, and sample preparation by the laser beam itself (cf. Sect. 13.1.2).

Several papers deal with basic LIBS investigations of oxidic materials such as inorganic minerals, concrete, soil, synthetic silicates [14.41–14.44]. First studies of slag analysis from blast furnace and liquid steel processes using LIBS demonstrated the principal suitability of this approach for process control purposes [14.45]. In

Table 14.9 Comparison of times needed for sampling, sample transportation, sample preparation, and analysis for the methods XRF and LIBS

	Time needed for XRF-analysis (min)	Time needed for LIBS analysis (min)
Sampling	2.0	2.0
Sample transportation	0.5	–
Sample preparation	2.0	–
Analysis time (including sample conditioning for LIBS analysis)	2.0	0.5–2.0
Total time	6.5	2.5–4.5

Since for LIBS a mechanical sample preparation is not necessary there is the option to position a laser analyzing system directly in the neighborhood of the metallurgical aggregate in the plant, thus avoiding a sample transportation to a laboratory

the following the analysis of converter slags with LIBS will be presented where a multivariate calibration model was applied for the first time [14.31].

14.2.1 Converter Slags

Before a measurement starts, the sample surface is irradiated by a number of prepulses to ablate a small amount of material from the surface, which may be not representative for the bulk composition (cf. Sect. 11.1, Table 11.1). The following laser bursts and the laser-induced plasma emission produced thereby are used for analysis. The necessary time for the surface preparation by the prepulses and the subsequent analysis is denoted with the term "analysis time" in Table 14.9. The comparison with the XRF standard analysis method shows that a significant improvement of the total time can be achieved.

The goal of the investigations described in the following was to provide a method based on LIBS for the quantitative analysis of slag samples from steel production. The tasks to be solved include the issue of sample inhomogeneity, as well as the issue of matrix effects due to the wide ranges of analyte concentrations in slag samples.

Figure 14.24 shows the experimental setup [14.31]. A Q-switched Nd:YAG laser operating at 1,064 nm was used to excite the plasma. The Q-switch electronics was equipped with a double-pulse option to generate within a single flashlamp pulse up to two separate laser pulses instead of a single laser pulse (cf. Sect. 3.2, Chaps. 7 and 8). The flashlamp repetition rate is 20 Hz. The laser emits within a single flashlamp discharge two laser pulses of approximately the same energy. The width of each pulse amounts to about 16 ns (FWHM). The time separation between the first and the second pulse is $\Delta t = 31\,\mu s$. The burst energy (cf. Sect. 3.2) amounts to $E_b =$ 210 mJ. The laser beam is guided via mirrors to a plano-convex lens with a focal length of 200 mm.

The converging laser radiation propagates via an optical window into a measurement chamber (cf. Sect. 4.6). This chamber has a cover plate on top with a circular

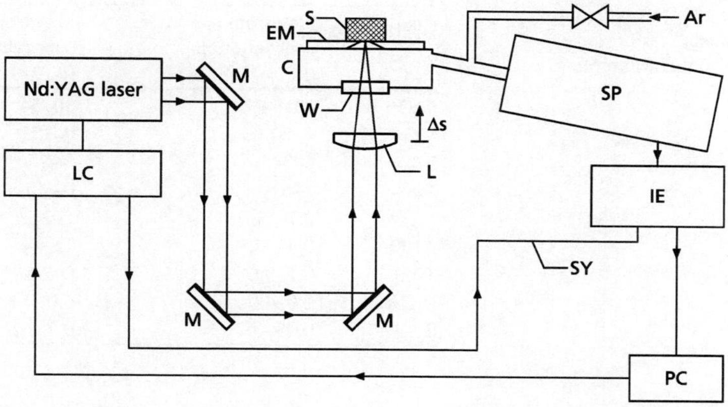

Fig. 14.24 Schematic setup for LIBS of slag samples. M = mirror, L = lens, W = window, C = measurement chamber, S = sample, EM = eccentric mechanism, LC = laser control, IE = integrator electronics, SY = signal cable for synchronization, SP = spectrometer, PC = personal computer

opening having a diameter of ~12 mm. The sample is put onto this cover plate to perform a measurement. During the measurement, the sample is fixed to the cover plate by a holder. The cover plate including the holder and the sample can be moved relative to the measurement chamber in a plane perpendicular to the optical axis of the laser excitation by an eccentric mechanism (cf. Sect. 3.6, Figs. 3.20 and 3.21). In this way, the sample is shifted relative to the laser beam axis on a circular trajectory with a diameter of typically 2 mm. Thus, a spatial averaging can be performed. The focus of the laser beam was adjusted in that way that the focal plane lies inside the sample with a distance of $\Delta s = 12$ mm from the sample surface.

The optical axis of the VUV vacuum spectrometer describes the line between the center of the diffraction grating and the middle of the entrance slit. This optical axis intersects the propagation axis of the laser beam on the surface of the sample. The angle between the surface of the sample and the optical axis of the spectrometer amounts to $\alpha_D \sim 11°$ (detection angle, cf. Fig. 4.18). With this orientation, the radiation from the laser-induced plasma can be coupled directly into the entrance aperture of the spectrometer. The internal volume of the chamber is flushed with argon 4.8 at 1.4 atm. and a flow rate [cf. Sect. 4.6, (4.7)] of about $\dot{V}_{in} = 520$ l/h.

The spectrometer has a Rowland circle diameter of 500 mm and is equipped with a holographic diffraction grating with 2,400 lines/mm (cf. Sect. 4.2, Fig. 4.7 and Table 4.4). The optical system of the spectrometer is installed inside a vacuum vessel which is evacuated to $<10^{-2}$ mbar. The dispersed radiation passing the exit slits aligned along the Rowland circle is detected by photomultiplier tubes (PMT). A signal electronics to simultaneously process up to 64 PMT signals in a pulse-by-pulse-mode (cf. Sect. 4.5) is triggered by the Pockels cell synchronization monitor of the laser, see connection SY in Fig. 14.24. The digitized charges of the integrators for each laser pulse are stored and processed in a personal computer.

Table 14.10 Concentration ranges of analytes of slag samples from converter and ladle

Analyte	Slag from converter concentration (m.-%)	Slag from ladle concentration (m.-%)
CaO	43–64	30–64
SiO_2	4–20	0–17
Fe_{tot}	6–34	0–5
Mn	2–6	0–0.5
MgO	0.7–11	1–10
Al_2O_3	0.5–2.0	15–42
TiO_2	0.1–0.6	0–2.5
P_2O_5	1.0–4.0	0.0–0.1
S	0.05–0.2	0.5–1.5

For the quantitative analysis, the number of pre-bursts applied to prepare the sample surface amounts to $N_{pp} = 200$ and the number of measuring bursts is $N_{mp} = 600$. Based on the setup shown in Fig. 14.24, an analyzing device was designed and operated in the laboratory of a steel plant for industrial evaluation.

Two different types of slag samples were investigated. The liquid slag is taken by a special sample probe where the liquid slag flows in a steel ring and solidifies in a glassy state. This sample can be directly analyzed with LIBS. This type of samples is called in the following "samples directly from converter." In order to suppress the influence of any inhomogeneities of these samples for the LIBS method development, a second type of samples were studied. In this case, solidified slag material is crushed and milled to obtain a powder, which is then pressed into a steel ring. The samples are manufactured by the use of an automatic press. Thereby, a constant pressure is applied for all samples. Thus, even though not experimentally verified, a similar sample density should be obtained for all samples. This type of sample is called "homogenized sample."

Table 14.10 shows typical concentrations of the analytes of slag samples from the converter and the ladle. All elements vary in a range of at least 10% and there is no single element whose concentration remains approximately constant. These variations of analyte concentrations result in matrix effects.

The influence of the varying composition of the slag samples on the properties of the laser-induced plasma was studied in terms of the radiation emitted by the plasma. The radiation reflected into the zeroth order of the grating was detected as a measure of the total radiation emitted from the plasma within the spectral sensitivity range of the spectrometer. Figure 14.25 shows the intensity of the zeroth order signal for different converter slag samples S1–S7 taken from a production line. Each data point corresponds to the average of ten measurements performed on the respective sample. If not otherwise stated, a single measurement comprises 200 pre-bursts and 600 measurement bursts. The error bars correspond to the standard deviation of the ten measurements. Obviously there are significant variations of the intensity of the zeroth order for the different samples. The standard deviation within the ten measurements on every sample is by a factor of 3.4 smaller than the standard deviation of the seven mean intensities from the seven samples.

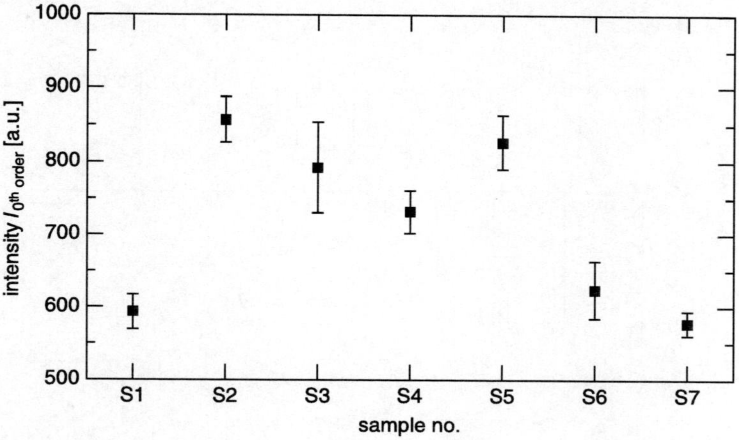

Fig. 14.25 Intensity of the zeroth order for seven different homogenized samples taken from the converter. Each data point represents the average of ten measurements under repeatability conditions. The error bars show the standard deviation of the ten measurements taken at one sample

Comparisons with experimental parameters, e.g., the laser burst energy for every measurement did not show any correlation to the observed variation of the plasma emission from sample to sample. Since these results are gained with homogenized samples – i.e., the slag material was chrushed, milled, and pressed – a variation of the density of the pressed samples, which may cause the observed variability in the intensity of the zeroth order, seems not to be probable. Therefore, it is assumed that the variations are caused mainly by matrix effects. In order to find out which element concentrations are favorable for an increasing intensity of the zeroth order, the data of Fig. 14.25 are analyzed with a partial least square algorithm (PLS) using the concentrations of the analytes in the samples determined with conventional methods.

Figure 14.26 shows the determined loadings for the different analytes of the measurement data of Fig. 14.25. Thereby, a positive loading p_1 indicates an increasing intensity of the zeroth order with an increasing concentration of the respective analyte. A negative loading p_1 corresponds to a decreasing intensity of the zeroth order with an increasing concentration of the respective analyte. No correlation was found between the behavior shown in Fig. 14.26 and physical properties of the analytes such as ionization energy, boiling temperature, and evaporation enthalpy. Obviously, the variations in the intensity of the plasma emission in terms of the zeroth order radiation are caused by matrix effects.

A series of experiments were performed with varying measuring parameters to minimize the influence of matrix effects and to optimize the sensitivity as well as the stability of the measurement for the quantitative determination of analyte concentrations.

The investigated measuring parameters are: the laser burst energy, the position of the focal plane with respect to the sample surface Δs, the time delay and gate width

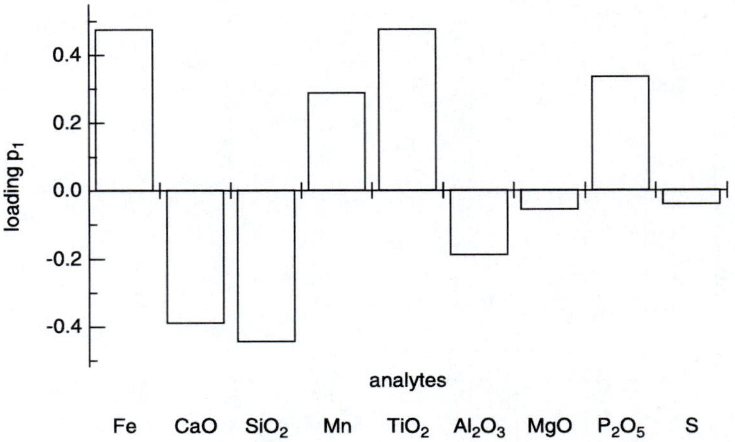

Fig. 14.26 Influence of the different analyte concentrations on the intensity of the zeroth order intensity described as loadings p_1 deduced from a data analysis by a partial least square algorithm taking the known analyte concentrations and the measured zeroth order intensity. Analytes showing positive p_1 values cause an increase of the zeroth order intensity with increasing analyte concentration, whereas negative loadings p_1 describe the opposite behavior

of the integration window for the individual element signals, the spatial averaging and the gas flushing rate of the measurement chamber. In the following, the influence of the time delay and gate width will be presented.

Figure 14.27 shows the intensity of the silicon line at 390.55 nm as a function of time recorded with a fast oscilloscope connected directly to the output of the corresponding photomultiplier tube. The time zero coincides with the irradiation of the second laser pulse of the double pulse burst. Three different time domains can be distinguished, denoted with the numbers 1, 2, and 3 in Fig. 14.27. By laterally shifting the entrance slit of the spectrometer while performing a series of measurements the spectral profile of this emission in the time domains 1, 2, and 3 was investigated. The intensity of the element emission was integrated in a time window of length t_{int} starting at t_{del} with respect to point zero. In the time domains 1 and 2 no distinct line profile is visible indicating that this emission is dominated by a superposition of continuum radiation and strongly broadened line emissions (cf. Figs. 2.4 and 15.6), whereas in the time domain 3 with $t_{del} \approx 5\,\mu s$ and $t_{int} = 3\,\mu s$ a clear line profile is detected with a spectral width (FWHM) of about 70 pm.

To find an optimum integration window an approach analogous to the one described in Sect. 14.1.2 was applied. The intensity of several emission lines as a function of time was recorded similar to Fig. 14.27 for two samples with high and low concentrations of the analytes to be determined. The samples were selected in such a way to cover the whole range of concentrations occurring in slag samples taken from a production line. The reference concentrations of each sample were determined with XRF analysis. Each sample was measured at three

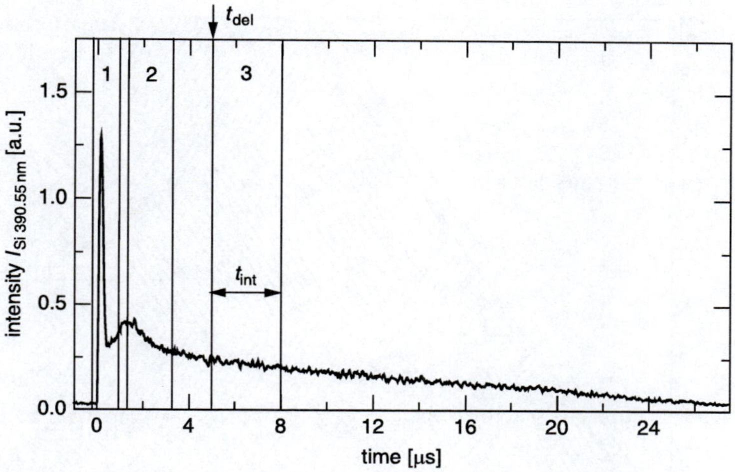

Fig. 14.27 Photomultiplier signal of the Si 390.55 nm channel as a function of time recorded with a fast oscilloscope (the negative signal voltage of the PMT was converted to a positive signal for this graph). Three different time domains are denoted with the numbers 1, 2, and 3, see text. t_{del} = delay time with respect to the start of the PMT signal, and t_{int} = width of the integration window

different locations. At each location, 50 pre-bursts and 100 measuring bursts were applied. The signals were digitized and numerically integrated within defined time windows. The position and width of the integration windows were varied and for each parameter combination the mean and the standard deviation of the three measurements per sample were calculated. The results were assessed with respect to the value of the following target function:

$$T_i = \frac{\Delta I_i / \Delta c_i}{\langle s(I_i) \rangle}, \tag{14.7}$$

where ΔI_i is the mean intensity of an emission line i for the sample containing the analyte at high concentration minus the mean intensity of the same emission line for the sample containing the analyte at low concentration, Δc_i is the difference of concentrations of the analyte i in the two samples, and $\langle s(I_i) \rangle$ is the average standard deviation of the intensity of analyte i for the two samples.

The target function (14.7) increases with higher sensitivity and lower standard deviation. This corresponds to a reduction of the uncertainty in the determination of a concentration using the analysis function. As an example, Fig. 14.28 shows the calculated target function $T_{Si\,390.55}$ of the silicon line at 390.55 nm for the numerically varied parameters: (a) start of integration t_{del} and (b) end of integration ($t_{del} + t_{int}$). At $t_{del} = 3.8\ \mu s$ and $t_{del} + t_{int} = 5.1\ \mu s$, the target function $T_{Si\,390.55}$ attains a maximum.

In order to further take into account the uncertainty of the analytical results, the target function was extended in the case of linear calibration functions, by

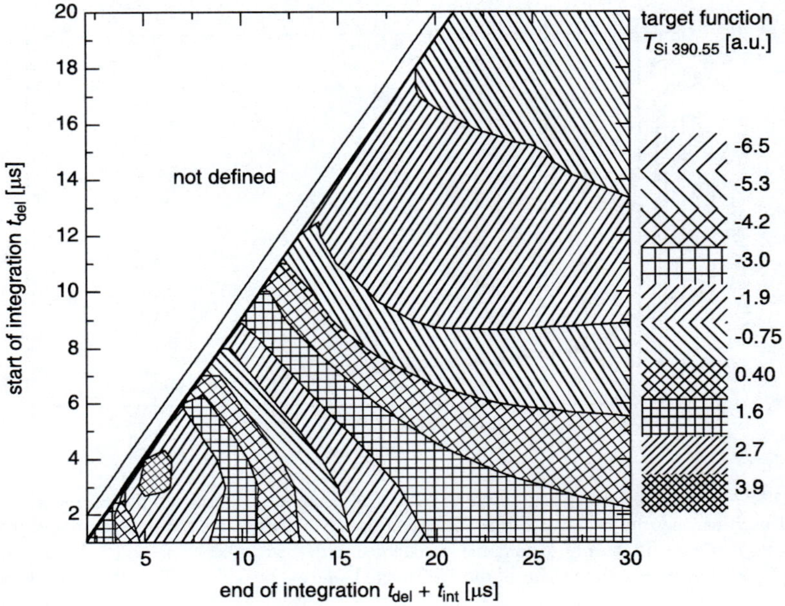

Fig. 14.28 Target function $T_{Si390.55}$ for the silicon line 390.55 nm in dependence on start of integration t_{del} and end of integration ($t_{del} + t_{int}$)

multiplying the numerator of the right side of equation (14.7) with the square of the correlation coefficient, i.e., the coefficient of determination. A series of parameter studies were performed and the results assessed using this target function yield the settings used in the following. Thereby, if possible, mutual interferences of the different parameters were taken into account.

With this set of parameters the experimental setup was calibrated for the analytes listed in Table 14.10 using different approaches: (a) univariate calibration with no internal standardization, i.e., only the measured mean line intensities of the analytes are used (denoted with "no internal standard" in Fig. 14.29); (b) univariate calibration with internal standardization: the measured line intensity is divided by the measured intensity of a single emission signal, e.g., the zeroth order signal, a background signal or another emission line (denoted with "internal standard." in Fig. 14.29, cf. Sect. 11.1); (c) univariate calibration where a sum of emission signals is used for internal standardization (denoted with "sum of internal standards" in Fig. 14.29); and (d) multivariate calibration (denoted with "multivariate calibration" in Fig. 14.29). The theory and the implementation of multivariate calibration models is described elsewhere [14.46]. In general, all integrated signals are chosen as input channels. The input channels span a vector space. The calibration model determines the direction in the vector space, which correlates best with the reference concentrations of the calibration samples. This direction is used for a linear

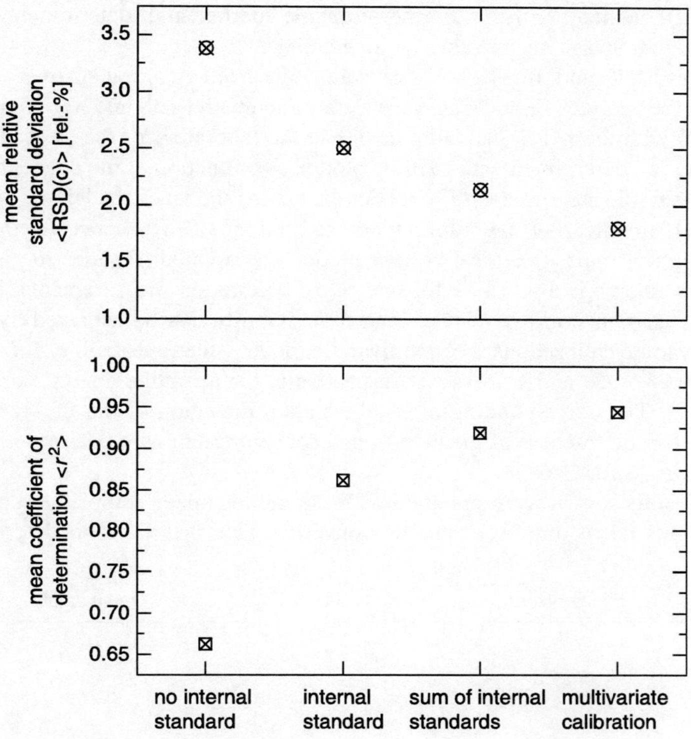

Fig. 14.29 Mean relative standard deviation <RSD(c)> (*top*) and mean coefficient of determination r^2 (*bottom*) for the analytes listed in Table 14.10 and homogenized samples from converter slag for different calibration procedures, see text

calibration function. The residuum, namely the variations in the dataset not included in this first linear calibration of the model, can be used as new input data for a further linear calibration. The further calibration constitutes a so-called further factor of the calibration model. Thereby, a model with several factors is implemented.

Figure 14.29 summarizes the results showing the mean relative standard deviation of the determined concentrations and the mean coefficient of determination for the above-described different calibration procedures for homogenized samples from converter slag. The mean values refer to the average for all analytes investigated. Figure 14.29 shows that for the univariate calibration without internal standardization the relative standard deviation amounts to 3.4% and the coefficient of determination is about 0.65. With internal standardization, the results could be improved significantly. However, the mean relative standard deviation could not be pushed down to less than 2%. Obviously, the matrix effects could not be suppressed completely by the selection of the measuring parameters using the optimization procedure described before. Figure 14.29 shows that a multivariate calibration is capable to reduce the relative standard deviation of the concentration as a mean value of all elements to 1.8%. Furthermore, the coefficient of determination can be increased to 0.95 as a mean value for all elements.

Figures 14.30 and 14.31 show measurements from a period of over 2 weeks, whereby the analytes Fe and CaO were determined with LIBS and XRF during shift operation with the LIBS analyzing device in the laboratory of the steel plant. The concentration determined with LIBS is plotted as a function of the concentration of the same sample measured with XRF. For the graphs shown on the left of Figs. 14.30 and 14.31, an univariate calibration model is used for LIBS, whereas for the graphs on the right a multivariate calibration model was applied. Besides the improved results, as shown in Fig. 14.29 for one set of calibration measurements, it can be seen that also an improvement in long-term stability can be achieved by the use of multivariate calibrations. For multivariate models, the repeatability for Fe_{tot} and CaO is 0.28 m.-% and 0.40 m.-%, respectively, for a routine analysis on a long-term basis. The corresponding residual standard deviation of the fit – which is a measure for the trueness of the determined concentration – are 0.85 m.-% for Fe_{tot} and 1.05 m.-% for CaO.

The results shown were obtained with the homogenized samples in contrast to the samples taken directly from the converter. This was done to determine the

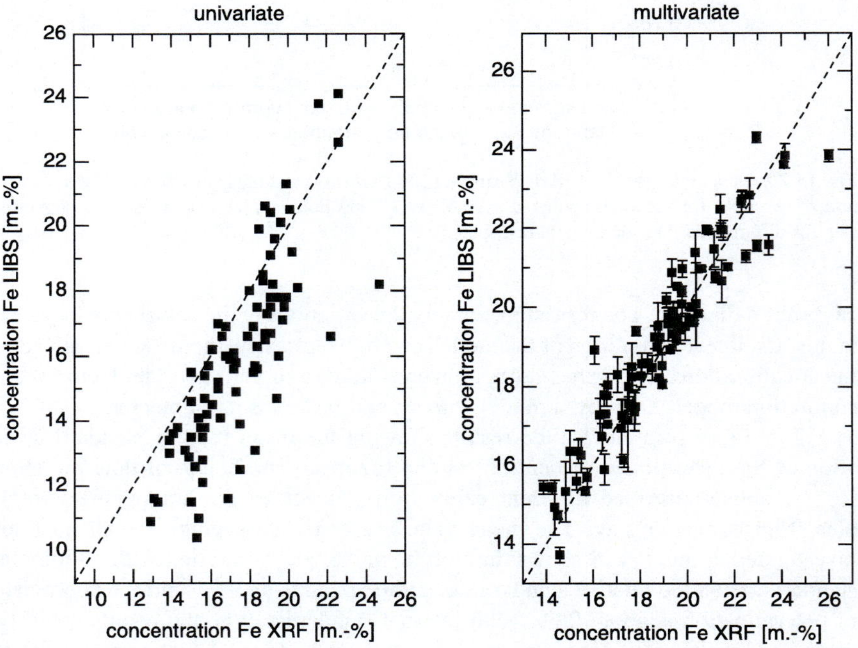

Fig. 14.30 Concentration of Fe determined with LIBS as a function of the concentration determined with XRF for homogenized slag samples from the converter. The measurements were performed in a period of over 2 weeks during shift operation. *Left*: univariate calibration model, *right*: multivariate calibration model

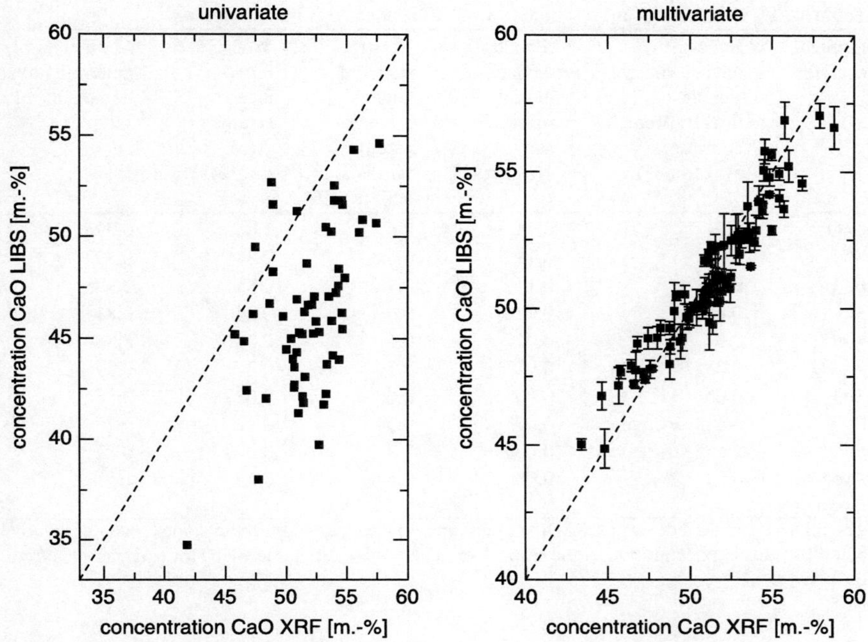

Fig. 14.31 Concentration of CaO determined with LIBS as a function of the concentration determined with XRF for homogenized slag samples from the converter. *Left*: univariate calibration model, *right*: multivariate calibration model

performance of the investigated LIBS procedure and setup without an additional influence originating from the inhomogeneity of the samples.

Table 14.11 summarizes the results obtained for samples taken directly from the converter and for the homogenized samples. In general, the standard deviation of the measured concentrations $s(c)$ is higher for the samples taken directly from the converter. With the exceptions of MgO and P_2O_5, the trueness is a factor 3.4 to 7.2 poorer than the repeatability for the homogenized samples. This is, if compared to standard XRF results, within the expectations of a routine analysis. The repeatabilities for the main elements of slag from the converter, namely CaO, SiO_2 and Fe_{tot}, on the homogenized samples in terms of the relative standard deviation amounts to 0.6%, 1.5%, and 0.7%, respectively.

In order to further improve these results, different measuring parameters could be used for the different analytes. In this case, a more specific optimization for the different elements is possible. With an extended spatial averaging of the sampled material, the repeatability of the results for the samples taken directly from the converter can be further improved (cf. Sect. 14.2.2).

By optimizing the measurement parameters as well as using multivariate calibration models, the relative standard deviation of the concentrations of all elements or analytes could be improved from 6.1% to 1.8%. The average RSD(c) for these main

Table 14.11 Summary of results gained with LIBS on converter slag samples

Element/ analyte	Standard deviation, samples directly from converter $s(c)$ (m.-%)	Standard deviation, homogenized samples $s(c)$ (m.-%)	Relative standard deviation, homogenized samples RSD(c) (%)	Trueness, homogenized samples $R(c)$ (m.-%)	Coefficient of determination, homogenized samples r^2
CaO	0.34	0.31	0.6	1.05	0.995
SiO_2	0.51	0.18	1.5	0.84	0.997
Fe_{tot}	0.8	0.14	0.7	0.85	0.998
Mn	0.08	0.036	0.9	0.24	0.81
MgO	0.07	0.027	0.5	0.48	0.88
Al_2O_3	0.03	0.029	2.3	0.21	0.89
TiO_2	0.05	0.007	2.0	0.029	0.88
P_2O_5	No measurements	0.013	0.5	0.12	0.75
S	No measurements	0.009	7.2	0.048	0.87
Average values	0.27	0.08	1.8	0.43	

The relative standard deviation RSD(c) in column 4 is calculated with the values given in column 3 and the mean concentration of the respective analyte given in Table 14.10 for slag samples from the converter

analytes CaO, SiO_2, and Fe_{tot} is less than 1% (see column 4 in Table 14.11). The average RSD(c) for all analytes amounts to 1.8%.

Additionally, further investigations have to be performed to find out the factors influencing the homogeneity of the slag samples and to improve the sample taking procedure to obtain more homogeneous slag samples from the converter and from the ladle.

The developed LIBS-based analyzing device was installed and set into operation in a steel plant for industrial evaluation. It allows to increase the number of samples to be analyzed within a given metallurgical processing step. This analyzing device is the first approach to bring LIBS into routine use for the analysis of slags from steel production.

14.2.2 Vacuum Slag

In steel works commonly, aluminum and/or silicon are used for an efficient deoxidation of steel melts where solid alumina and silica are formed. In order to obtain a liquid ladle slag, usually lime will be added as a slag forming agent. Furthermore, lime addition enhances the ability of ladle slag to absorb sulfur from the steel melt by the reaction [14.47]:

$$[S] + (CaO) + 2/3\,[Al] = (CaS) + 1/3(Al_2O_3). \qquad (14.8)$$

Long experience with steel desulfurization indicates that a maximum removal of sulfur from liquid steel can be obtained only by the use of lime saturated slags [14.48]. Therefore, a fast analysis of the major slag components calcium oxide CaO, silicon dioxide SiO_2 and aluminum oxide Al_2O_3 can improve the control of the steel melt and reduce the treatment times in the vacuum degasser. The mass fractions of the major components of the vacuum slags are in the range of 50–60% for CaO, 0.5–12% for SiO_2, and 20–40% for Al_2O_3. Samples are taken from the liquid slag layer at the surface of the steel melt by an automatic immersion slag sampler. Although great improvements are achieved in the reproducibility of the sampling from the moving liquid slag layer (thickness 5–12 cm, slag mass 2–4 t) in the vacuum degasser, the slag samples exhibit a significant heterogeneity in the species distribution as well as varying color, cracks, and ferrous oxide inclusions. After solidification, the slag samples are dismantled from the immersion sampler. Laboratory analysis requires at least 7 min including dismantling, pneumatic delivery of the sample from the steel works to the laboratory, crushing of the sample, and XRF analysis of a pressed pellet. A more precise analysis by a borate bead analysis takes additional time.

The goal of the investigations presented in the following is to demonstrate the feasibility of an onsite slag analysis by LIBS in the steel works without additional sample preparation and to reduce the analysis time significantly [14.49].

LIBS has been applied for slag analysis under laboratory conditions and/or for crushed and pressed samples [14.31, 14.45, 14.50]. The spatial averaging of the LIBS measurement by an optical microlens array is described in Sect. 3.6 (cf. Fig. 3.22) [14.51]. This is a key item due to the heterogeneity of the production slag samples without mechanical preparation. Another topic is the installation of a LIBS equipment in the harsh environment of a steel works (e.g., dust, temperature variation, stability of electricity, argon gas, and water supply) and its stable long-term operation (cf. Sect. 17.2).

The slag samples are taken by immersion samplers from the vacuum degasser during steel production. The color of the slag samples varies between dark gray, anthracite to pale gray and mixtures of brown-yellow-green-white. A majority of the sample surfaces are matt, but some samples are glossy. A variety of cracks and holes are present in the samples. Both sides of the samples can be used for analysis.

In order to find a representative average value for a sample, the standard analysis in the laboratory of the steel works is crushing and milling the slag material of the samples and pressing it to pellets for XRF analysis. In special cases, the samples are prepared as fused borate beads. For LIBS, a laser beam is focused to a diameter in the range of 20–500 μm. In the focal region, a laser-induced plasma is generated, which is used for spectrochemical analysis. Therefore, an efficient spatial averaging has to be applied to measure areas in the range of 1 cm^2. In this study, the spatial shaping of a Nd:YAG laser beam (wavelength 1,064 nm) by an optical microlens array (hexagonal packed microlenses of diameter of 0.9 mm, $f = 37$ mm) as described in Sect. 3.6 is used. Approximately 40 high-intensity spots are generated at the sample surface within the Nd:YAG laser beam diameter of 6 mm, see Fig. 14.32. In addition, the sample is moved eccentrically with the sample

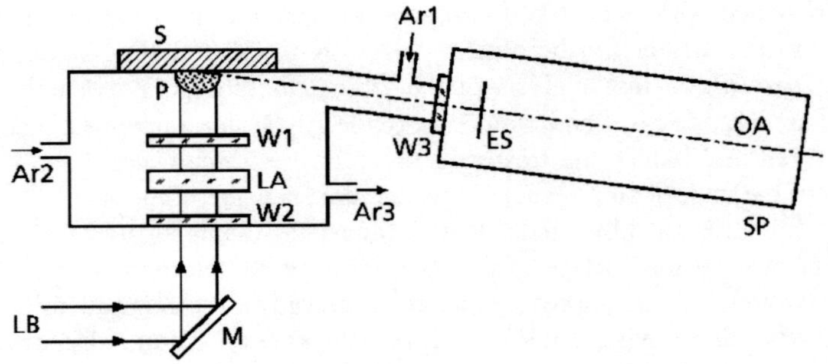

Fig. 14.32 Schematic setup of the laser beam optics, sample stand, and Paschen–Runge vacuum spectrometer (SP). ES, OA = entrance slit and optical axis of the spectrometer, LB = laser beam; M = mirror; LA = microlens array; W1 = window; W2 = window, AR coated at 1,064 nm; W3 = vacuum window of spectrometer, MgF_2; P = laser-induced plasma, S = slag sample; Ar 1, 2, 3 = inlets and outlet for argon gas flushing of the sample stand. The optical components are uncoated unless otherwise noted

plate of the argon flushed sample stand within the focal plane of the microlens array resulting in a surface area of ≈12 mm diameter covered by the LIBS measurement.

The LIBS instrument includes a vacuum Paschen–Runge spectrometer with photomultiplier tubes mounted at different positions for simultaneous spectral line detection and a Nd:YAG laser (pulse repetition rate $\nu_{rep} = 20$ Hz, pulse duration $\tau_L \approx 8$ ns, pulse energy 170 mJ). The LIBS plasma of the sample is generated in an argon flushed sample stand and the emitted light is directly coupled to the spectrometer observing the plasma light under an angle of about 12° to the sample surface.

The calibration is carried out by production samples of vacuum slag. For reference mass fractions w (in percent or equivalent 10^{-2} g/g = cg/g), the samples are analyzed directly by X-ray fluorescence as well as by borate bead analysis (BBA). Both analysis are carried out at the certified laboratory of the steel plant. For the borate bead analysis, the samples are crushed, and iron granules are removed by a magnetic separator or sieve. The powder material is then melted with borate at a furnace temperature of about 1,250°C. Finally, fused beads were obtained by casting on Ni-plates and distinct cooling.

In Figs. 14.33 and 14.34, analysis functions for silicon dioxide SiO_2 are shown. The radiant intensity of the spectral line Si I 251.4 nm, $I_{\text{Si 251.4 nm}}$, detected for each laser pulse by the corresponding PMT, is divided by the radiant intensity of the zeroth order, $I_{\text{0th order}}$, detected by a PMT mounted at the zeroth order position of the spectrometer grating (=position of reflected beam, cf. Fig. 4.7). Thus, small variations in the parameters (e.g., of laser, argon flush, sample and optical component positions) which are proportional to both signals are eliminated for the ratio $I_{\text{Si 251.4 nm}}/I_{\text{0th order}}$ of the signals. For each sample, the average $\langle I_{\text{Si 251.4 nm}}/I_{\text{0th order}}\rangle_{N_{mp}=800}$ is determined [cf. (11.4)].

The coefficient of determination is greater than 0.97 which is considered as good taking into account the sample variation and heterogeneity as well as the

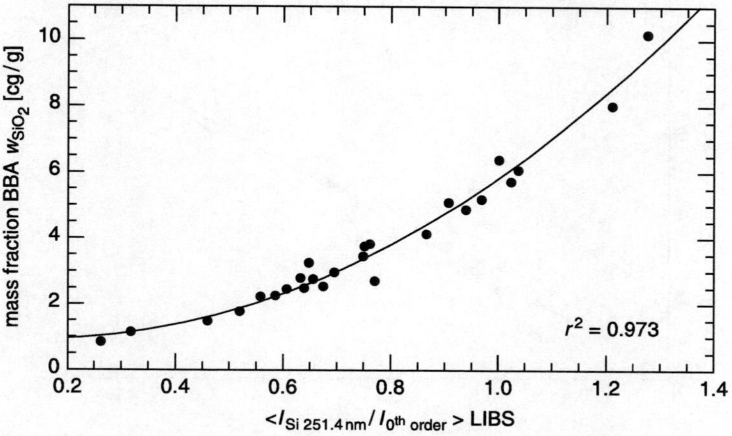

Fig. 14.33 Analysis function of SiO_2 with reference mass fractions by borate bead analysis (BBA) with a calibration set of 26 vacuum slag samples. The LIBS analysis is taken at one side of each sample. r^2 = coefficient of determination of the fitting curve (second-order polynomial)

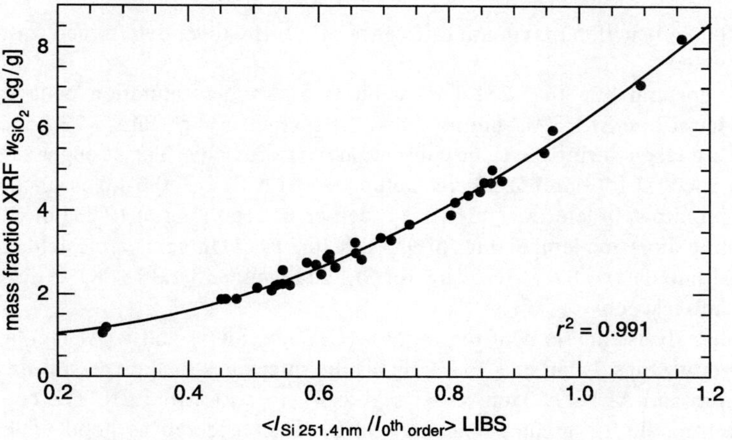

Fig. 14.34 Analysis function of SiO_2 with reference mass fractions by XRF with a calibration set of 37 vacuum slag samples. Both measurements are taken at the same sample side

measurement environment. The slightly better correlation to the XRF reference values in Fig. 14.34 compared with the borate bead analysis of Fig. 14.33 are possibly due to the fact that the BBA analysis gives a mean value of the total sample while the LIBS analysis is taken at one side of the sample. The relative difference of the mass fractions between side 1 (mass fraction w_1) and side 2 (mass fraction w_2) of the samples has been measured by XRF and it is in average $\langle(w_1 - w_2)/w_1\rangle = 1.9\%$

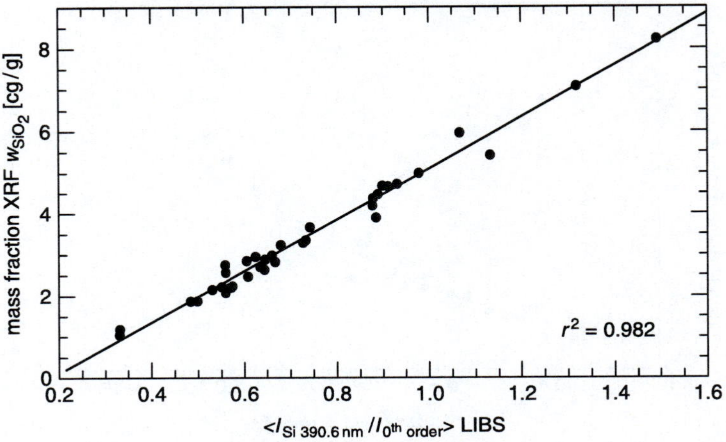

Fig. 14.35 Analysis function of SiO_2 with reference mass fractions by XRF with a calibration set of 37 vacuum slag samples. Same measurement as Fig. 14.34, except that the spectral line Si I 390.6 nm is evaluated. r^2 = coefficient of determination of the linear fitting curve

(rel.) for SiO_2 with a maximum difference of 9% (relative; determined for a set of 37 samples).

The spectral line Si I 251.4 nm exhibits a stronger saturation behavior than the spectral line Si I 390.6 nm as shown by comparison of Figs. 14.34 and 14.35 which are taken during the same calibration measurements. The stronger saturation of the spectral line Si I 251.4 nm compared with Si I 390.6 nm is expected by the approximately ten times higher k_t coefficient according to (9.33) of Sect. 9.4. Assuming a plasma temperature of 9,000 K, the k_t coefficients are calculated with spectral line data to $8.9 \times 10^{-30}\,m^3$ for Si I 251.4 nm and $0.85 \times 10^{-30}\,m^3$ for Si I 390.6 nm, respectively.

The analysis functions of the ratios $w(CaO)/w(SiO_2)$ and $w(Al_2O_3)/w(SiO_2)$ are given in Figs. 14.36 and 14.37 versus the intensities of the spectral lines Ca I 364.4 nm and Al I 309.3 nm referenced to $I_{Si251.4\,nm}$ [cf. (11.2)]. The coefficient of determination is greater than 0.97 which is considered as good under these conditions.

The analysis of unknown samples is carried out with a multivariate analysis [14.52]. This approach as well as the industrial application of the LIBS analysis of vacuum slags will be described in Sect. 18.2.

14.3 Soil

One of the main advantages of LIBS for chemical analysis purposes is the possibility to perform contactless measurements at solid substances without causing high demands on the sample preparation [14.53, 14.54]. This is especially promising

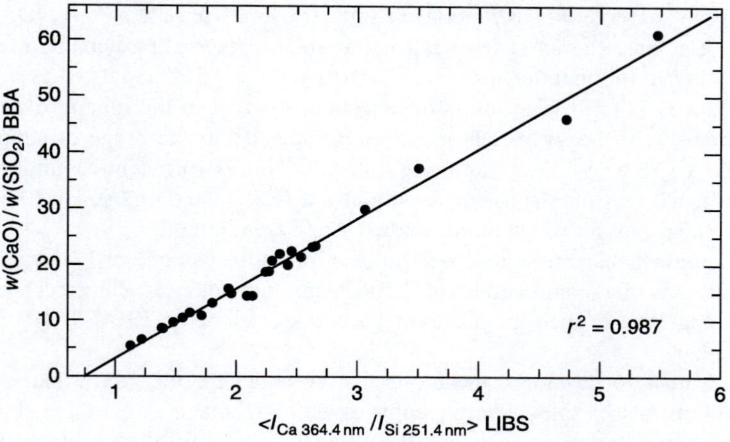

Fig. 14.36 Analysis function of CaO/SiO_2 with reference mass fractions by borate bead analysis (BBA) with a calibration set of 26 vacuum slag samples

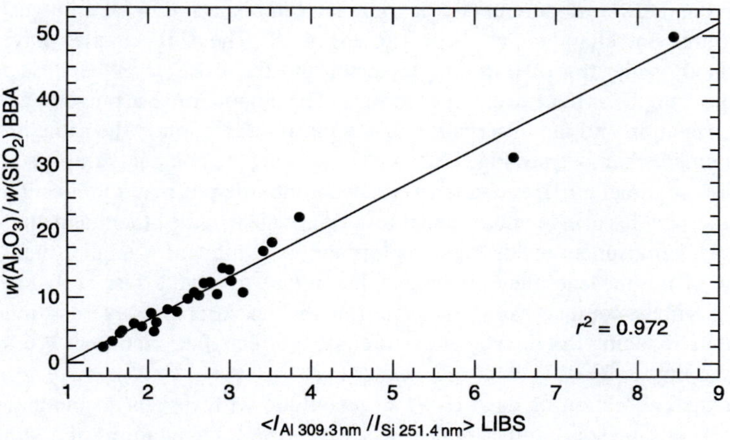

Fig. 14.37 Analysis function of Al_2O_3/SiO_2. Same parameters as Fig. 14.36

for the analysis of heavy metals in soils, because no chemical dissolution of sample material is required [14.55, 12.9], which in turn is the case for conventional techniques such as inductively coupled plasma optical emission spectrometry (ICP-OES) or atomic absorption spectrometry (AAS). Consequently, numerous investigations were performed in this field, ranging from studies on the influence of basic experimental parameters [14.56] to the evaluation of LIBS probes for the in situ subsurface detection of metals in soils using a cone penetrometer [14.57,14.58].

For most practical applications of analytical techniques, it is required that the analytical figures of merit are sufficient with respect to regulatory demands. In the case of the analysis of heavy metals in soils, the limits of detection (LOD) for the various elements should at least fall below the corresponding maximum content that is allowed for unpolluted soils. In Germany, these values are listed as so-called $Z0$-values in [14.59]. The limits of detection reported in the literature for LIBS measurements of heavy metals in soils are typically in the range of some μg/g to some 10 μg/g [14.53, 14.56, 14.60–14.62]. This is sufficient for most of the elements, but for some substances such as Cd ($Z0 = 0.6\,\mu g/g$), Hg ($0.3\,\mu g/g$) and Tl ($0.5\,\mu g/g$) detection limits below $1\,\mu g/g$ are required.

An approach to achieve improved limits of detection using laser-based analytical techniques is the combination of LIBS with laser-induced fluorescence (LIF, also called laser-excited atomic fluorescence spectrometry, LEAFS) [12.1]), see Chap. 12.

The aim is to develop a technique that is suited for the fast in situ analysis of heavy metals in soils. Consequently, as many elements as possible should be measured simultaneously, a requirement which can be fulfilled using LIBS. For two of those elements that cannot be determined with limits of detection below the regulatory limits using LIBS, namely Cd and Tl, the LIBS-LIF approach was studied. The experimental setup is described in Chap. 12.

For the LIBS measurements, a compact sample stand was used instead of a measurement chamber, cf. Sect. 4.6, Fig. 4.18. The soil samples have to be positioned on the top of this sample stand, and the LIBS laser beam is focused onto the sample surface from the bottom. The geometrical arrangement for the LIBS excitation and the observation of the plasma radiation is the same as for the measurement chamber, see Fig. 12.2. As a drawback, the compact sample stand does not allow to direct a LIF excitation laser beam into the plasma, nor does it allow to move the samples using a motorized translation stage. On the other hand, the volume of the gas atmosphere inside the sample stand is significantly smaller than the gas volume of the measurement chamber. This allows to have a fast exchange of the buffer gas in the compact sample stand during the measurements by a continuous gas flow, thus reducing the danger of accumulative effects [see also Sect. 4.6, relation (4.7) and Fig. 4.17].

For the calibration of the LIBS-LIF technique with respect to the analysis of heavy metals in solid soil samples, a set of samples containing the substances of interest over a broad range of concentrations was needed. As it is difficult to find certified reference material (CRM) with these characteristics, such a set of calibration samples was artificially produced by spiking standard soils from the Landwirtschaftliche Untersuchungs- und Forschungsanstalt (LUFA) in Speyer, Germany, with the heavy metals. The standard soil was dried, ground to a grain size < 0.1 mm, and suspended with metal–salt solutions. Afterwards, the sample was dried and homogenized again. The operating cycle for the preparation of the samples as well as the metal salts used for spiking are described in [14.63]. For reference, a portion of the samples was analyzed using the standard techniques ICP-OES or AAS after dissolution in aqua regia. The metal contents that were determined with this analysis were used as reference values for the calibration curves shown in the following. Finally, the sample material was mixed with a small amount of wax and pressed to pellets, so that an additional reference analysis using EDXRF spectroscopy (Spectro X-Lab 2000) was made possible with typical limits of detection in the low ppm-range for the heavy metals and proven accuracy by standard reference materials.

The applicability of LIBS in its own for the fast in situ analysis of heavy metals in soils was investigated. For this application, the Paschen–Runge spectrometer with photomultiplier (PMT) detection was deployed (cf. Sect. 4.2, Fig. 4.8).

The influence of experimental parameters such as the laser pulse energy E_L and the position of the laser focus in relation to the sample surface Δs was studied. These experiments were performed on a compact sample stand, on which the soil samples had to be positioned, instead of the measurement chamber shown in Figs. 12.2 and 12.3. The experiments were carried out in argon as buffer gas at atmospheric pressure.

The dependency of the analytical figures of merit of the LIBS measurements on the time delay t_{delay} between plasma formation and the start of the integration of the LIBS signals with the multichannel integrator electronics was investigated. The duration of the integration window t_{int} was adjusted to $t_{int} = 20\,\mu s$ according to the overall lifetime of the LIBS signal observed for the strong emission of the matrix element silicon at 288.2 nm. The gateable multichannel integrator electronics used for these experiments allows to integrate 16 channels with one integration window and 4 channels with another. As one of the integration windows was intended to be used for the integration of LIF signals in LIBS-LIF measurements, all elements that were to be analyzed using LIBS had to be integrated over the same time window. Thus, a compromise had to be found for the combination of the parameters: laser pulse energy E_L, focus position Δs, and time delay t_{delay}, which resulted in acceptable limits of detection for as many heavy metals as possible. For the chosen experimental setup, an optimum with respect to this demand was found for $E_L = 80$ mJ, $\Delta s = 5$ mm (i.e., the laser focus lies inside the sample), and $t_{delay} = 1\,\mu s$. The optimization of the conditions to a particular element allowed the improvement with respect to the limit of detection by about a factor of 2 or more. For example, an integration delay of $t_{delay} = 2\,\mu s$ resulted in an improved limit of detection of $LOD_{3s} = 3\,\mu g/g$ for cadmium instead of $6\,\mu g/g$ with $t_{delay} = 1\,\mu s$.

Calibration curves were recorded by measuring the LIBS signals for a set of soil samples prepared as described above. For each sample, 100 laser pulses were measured at three different positions on the sample surface. The number of laser pulses applied to one position on the sample surface was limited to 100 to make sure that the crater depth, which is approximately 2 mm for 100 laser pulses with a pulse energy of 80 mJ, does not reach the backside of the sample. As a result of the granular structure of the samples, soil particles can be found in the vicinity of the crater after a measurement, which are supposed to be ejected out of the interaction volume by the pressure of the laser-induced plasma. In consequence, only an upper limit for the amount of sample material which is atomized and therefore actually provided for the analysis can be estimated roughly. This estimate for the ablated mass per laser pulse – calculated from the crater volume generated by 100 laser pulses applied to the same sample position, the average mass density of the sample and the number of laser shots – amounts to 9 µg per laser pulse.

Two of the calibration curves obtained in this way are depicted in Fig. 14.38. As artificially spiked soil samples were used for the experiments, matrix effects were minimized. Thus, it was possible to directly plot the LIBS signal without any internal standardization against the analyte concentrations determined by conventional analytical techniques for calibration purposes. The error bars shown in the curves are given by the standard deviation of the three measurements performed under reproducible conditions.

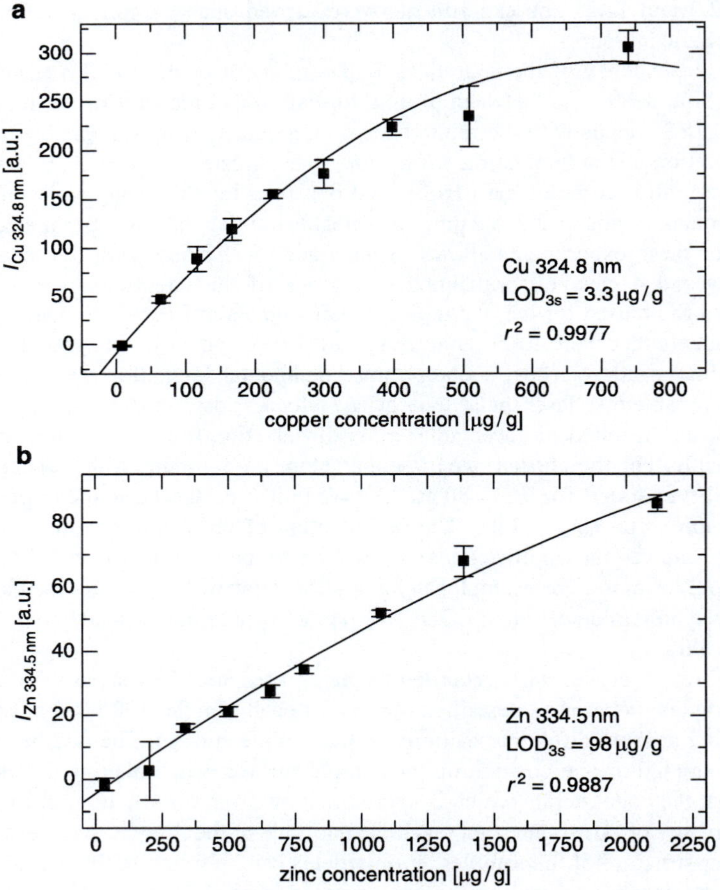

Fig. 14.38 LIBS calibration curves for (**a**) copper and (**b**) zinc. The analyte signal measured for the corresponding spectral line is plotted against the analyte concentration without using internal standardization. The ordinate is not corrected for a negative offset caused by the signal processing electronics. The calibration curves are second-order polynomials fitted to the experimental data

A slight saturation behavior was observed for the LIBS signals for many spectral lines with increasing analyte concentration, which was covered by the soil samples. This can be seen in Fig. 14.38a for Cu as well as for Zn in Fig. 14.38b. This behavior can be attributed to self-absorption of the element-specific line emission from the plasma at high analyte concentrations, which is supported by the fact that the nonlinearity is higher for resonance lines than for nonresonance lines, but additional effects such as limitations of the dynamic range by the detection and signal processing scheme cannot be excluded [cf. Sect. 9.4, relation (9.33)]. In consequence, one would have to select different spectral lines or to adjust the gain of the signal electronics, if an increased range of sensitivity is necessary to determine high analyte concentrations in certain applications. However, the goal of the investigations described is to achieve a high sensitivity for trace concentrations [14.44]. A second-order polynomial was fitted to the experimental data as a calibration curve. From this curve, the limit of detection was calculated by $LOD_{3s} = 3s_0/m$, where s_0 is the standard deviation of the three replicate measurements for the least concentrated sample and m is the slope of the calibration curve for low analyte concentrations (cf. Sect. 11.2). Calibration curves for all relevant heavy metals were evaluated in this way, resulting in the limits of detection listed in Table 14.12. Additionally, the spectral lines that were used for the calibration and the coefficients of determination of the calibration curves are given in this table. The results are illustrated in Fig. 14.39, where the regulatory limits for unpolluted soils are shown for comparison. The detection limits achieved with LIBS are sufficient for six elements, namely As, Cr, Cu, Ni, Pb and Zn, but have to be improved for Cd, Hg, and Tl. Experiments using air as ambient gas instead of argon have been carried out, indicating that similar limits of detection can be achieved.

For two elements, Cd and Tl, investigations on the applicability of the hyphenated LIBS-LIF technique were performed. A description of the setup and result for cadmium is given in Chap. 12. A LOD_{3s} of 0.3 μg/g was achieved for Cd which lies below the regulatory limit of 0.6 μg/g, cf. Fig. 14.39.

Table 14.12 Limits of detection (LOD_{3s}) for the analysis of heavy metals in soils using LIBS

Element	Wavelength of the spectral line λ (nm)	Limit of detection LOD_{3s} (μg/g)	Coefficient of determination r^2
As	235.0	3.3	0.9912
Cd	228.8	6	0.9993
Cr	425.4	2.5	0.9936
Cu	324.8	3.3	0.9977
Hg[a]	253.7	84	0.9858
Ni	231.6	6.8	0.9838
Pb	405.8	17	0.9863
Tl	351.9	48	0.9905
Zn	334.5	98	0.9887

The wavelength of the spectral line (λ) used for the calibration is given as well as the coefficient of determination of the calibration curve (r^2)

[a]Linear calibration based on three soil samples

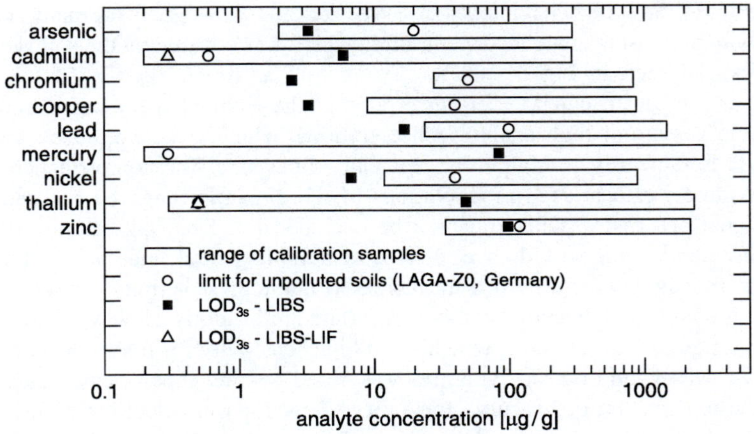

Fig. 14.39 Illustration of the detection limits (LOD_{3s}) achieved with LIBS and LIBS-LIF for the analysis of soils at atmospheric pressure in comparison with German regulatory limits for unpolluted soils (Z0). The *bars* depict the range of analyte concentrations covered by the calibration samples

LIBS-LIF measurements for Tl were performed in an air atmosphere at 950 mbar, for the practical reason that this experimental condition is even more promising for an implementation of the technique in an in situ measurement system than an analytical technique depending on argon as buffer gas. A variation of the LIBS-LIF delay $\Delta t_{\text{LIBS-LIF}}$ showed that distinct LIF signals can be observed for at least 250 μs. An example for the time behavior of the photomultiplier signal at 351.9 nm is given in Fig. 14.40. The signal was averaged over 30 laser shots. In the upper part (a), the signal for a soil sample containing 827 μg/g Tl is shown. The LIBS signal following the continuum emission after the plasma ignition can be observed, lasting for about 10 μs, as well as the strong LIF peak after 20 μs. The overshoot of the LIF signal to negative values after the excitation of laser-induced fluorescence is attributed to bandwidth limitations of the signal processing electronics. For comparison, the signal observed at the same spectral line is shown for a soil sample containing less than 0.3 μg/g of thallium Tl, which was the limit of detection of the analytical technique, which was used for the reference analysis. Neither an element-specific LIBS signal, nor any LIF signal can be observed in this case (note the different scaling of the y-axis in Fig. 14.40).

Figure 14.41 shows the calibration curve that was recorded for a LIBS-LIF delay of $\Delta t_{\text{LIBS-LIF}} = 20\,\mu\text{s}$. The detection parameters were the same as for Cd ($t_{\text{delay}}^{\text{LIF}} = -0.5\,\mu\text{s}$, $t_{\text{int}} = 5\,\mu\text{s}$). Similarly, a strong saturation behavior is observed. As the analyte concentration range was essentially larger than for Cd and thus the saturation behavior much more evident, an exponential saturation function was fitted to the experimental data instead of a second-order polynomial. The slope of this curve for

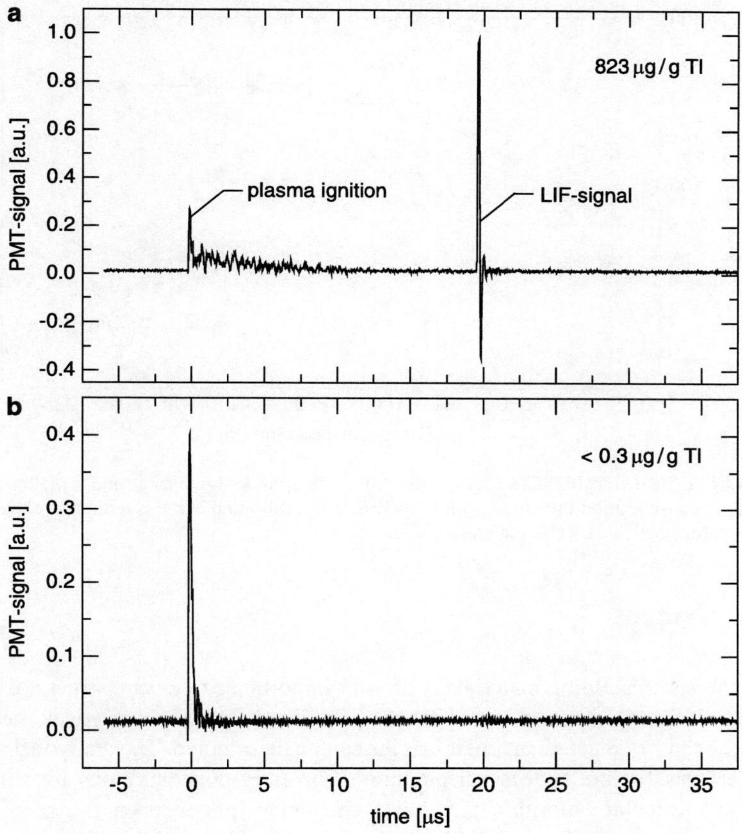

Fig. 14.40 Photomultiplier (PMT) signal of the thallium channel at 351.9 nm for LIBS-LIF measurements in air at atmospheric pressure, averaged over 30 laser pulses. The LIF excitation wavelength was tuned to the Tl transition at 276.8 nm. While distinct LIBS- and LIF-signals can be observed for a soil sample containing 823 μg/g of Tl (**a**), only the LIBS continuum peak is found for a sample containing less than 0.3 μg/g of Tl (**b**)

low analyte concentrations was calculated to determine the detection limit. In this case, a value of $LOD_{3s} = 0.5\,\mu g/g$ was achieved, matching the regulatory limit for unpolluted soils.

For further investigations on the influence of different experimental parameters such as the LIBS-LIF delay $\Delta t_{LIBS\text{-}LIF}$ on the analytical figures of merit of the technique with respect to the analysis of Tl in soils, additional calibration samples are required to expand the experimental data in the range of analyte concentrations below 100 ng/g.

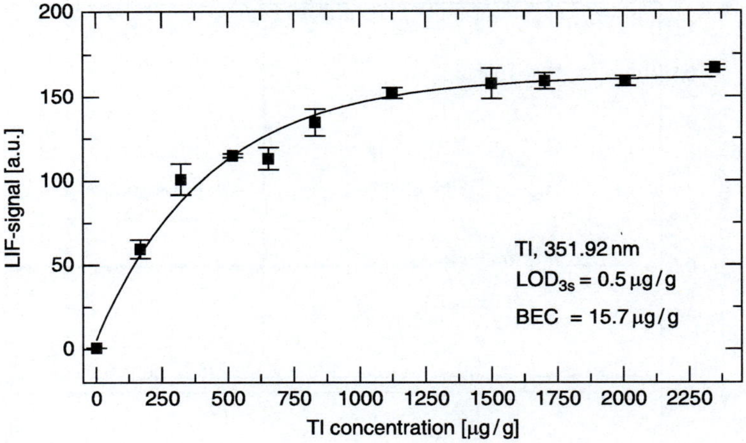

Fig. 14.41 LIBS-LIF calibration curve for thallium. The analyte signal measured is plotted against the analyte concentration measured with ICP-OES. The calibration curve is a fit of an exponential saturation function to the experimental data

14.4 Cement

The analysis of building materials is of vital importance to ensure structural safety and durability of existing constructions. By using reliable analytical methods, damages can be detected and their origins can be determined. The analytical results are the basis for the restoration planning. Currently concentrations for different salts, in particular chlorides, that have a negative influence on the stability of concrete constructions, are determined by wet chemical methods. These methods require an extensive sample preparation and are therefore time-consuming and expensive [14.64, 14.65]. Because of their complexity, this procedure is prone to faults. Additionally only one element within a pulverized drill core can be analyzed and the result is an average value representing the whole sample. Hence only a small number of wet chemical analysis are carried out in practice, resulting in higher risk of undetected stability and durability problems.

LIBS was studied for the determination of trace elements in building materials such as concrete and mortar by several groups [14.66, 14.67]. Weritz et al. studied the influence of the heterogeneity of concrete consisting of cement, fine and coarse aggregates and additives on the quantitative analysis of sulfur [14.66]. They could show that the uncertainty in the sulfur calibration can be improved by selecting data with a high Ca/O ratio indicating less heterogeneous constituents such as cement and cement mortar. Burakov et al. studied chlorine analysis in cement mortar (cement, fine-grained sand, gypsum, 1:2:1) using the IR chlorine line 837.59 nm [14.67]. They compared a LIBS setup with a combination of LIBS and an electrical discharge. The energy stored for the latter is in the order of 3 J; however, it seems that only a part of this energy is dissipated in the laser-induced plasma indicated by the shown voltage and current oscillations (cf. Fig. 1 in [14.67]). For the combined excitation, an increase in the chlorine signal by a factor of 11 on average was observed. A chlorine detection limit of 0.05% in cement mortar is stated for the combined excitation scheme using a 2*s* criterion. A calibration curve for laser excitation only required the use of internal standardization with the oxygen line O I 844.54 nm and averaging over 2,500 laser pulses per data point.

The aim of the investigations described in the following was to improve the limit of detection of chlorine in hydrated cement via the application of LIBS that is stated with 0.5 m.-% using He at a pressure of $p = 1{,}000$ mbar and the chlorine line at $\lambda = 837.59$ nm [14.68]. To increase the line emission of chlorine in the laser-induced plasma, several methods were studied by various authors. One of these methods is the use of an additional process gas like argon, air, nitrogen, oxygen, or helium [14.69, 14.70]. Tran et al. showed that the signal-to-noise ratio of the chlorine spectral line at $\lambda = 837.59$ nm can be improved in comparison with normal air at atmospheric pressure by a factor of 12 if a helium gas flow is used at atmospheric pressure to analyze solid organic compounds [14.71]. A further parameter studied to increase the chlorine emission is the ambient pressure. Asimellis et al. studied helium at pressures between $p = 0$ and 600 mbar inside a chamber to analyze NaCl-powder mixed up with CH_3COONa and Na_2SO_4 using a laser at a wavelength of $\lambda = 355$ nm [14.72]. At a pressure of about $p = 50$ mbar, they observed a maximum of the chlorine line emission at $\lambda = 837.59$ nm. In a further paper of Asimellis et al., the detection of bromine in pressed powder samples consisting of a mixture of KBr and K_2CO_3 was studied in a helium atmosphere with laser excitation at 355 nm [14.70]. A calibration curve for bromium ranging from 1 to 33.6 m.-% was determined at a helium pressure of 30 mbar. A limit of detection was not stated. The authors declare that an optimum He pressure appears to be in the range of 60–100 mbar.

Weritz et al. conducted measurements with different types of building materials under a 1 bar He atmosphere [14.68]. They estimated the limit of detection for chlorine in cement and mortar moving the sample continuously during the measurements to achieve an average value. The best limit of detection for chlorine in hydrated cement is stated with LOD = 0.5 m.-% using the 3*s* criterion.

LIBS-based detection of chlorine in artificial soil and geological samples was studied using the Cl line at 133.57 nm in a Mars atmosphere, i.e., in CO_2 gas at pressures below 10 torr [14.73]. The stated detection limit of Cl at 0.25 torr CO_2 is 7.5 m.-%, which is by far too great for the purpose stated in the introduction.

In the following investigations are described where the chlorine lines at $\lambda = 837.59$ nm – and for the first time – $\lambda = 134.72$ nm were studied to assess their analytical properties in a comparative way [14.74]. To detect the chlorine line in the UV, a direct light-channel is used to guide the plasma emission from the measurement chamber into the spectrometer. The total pressure inside the measurement chamber is varied between 10 and 1,000 mbar while the total gas inlet flow rate amounts 1.5–50 l/min. To reduce the LOD for chlorine in hydrated cement different settings for the integration time gate of the photomultiplier (PMT) signals t_{delay} and t_{int} are investigated and compared between the two chlorine lines.

Cement samples with different chlorine concentrations are used as calibration set for LIBS measurements. The set comprises seven cement samples with different contents of chlorine: 0.05, 0.2, 0.5, 0.8, 1, 2, and 4 m.-%.

The experiments are carried out with a Nd:YAG laser system operating at a wavelength of $\lambda_L = 1{,}064$ nm. The pulse duration amounts to $\tau_L = 8.5$ ns and the repetition rate is $\nu_L = 50$ Hz. The laser pulse energy in the experiments ranges between $E_b = 450$ mJ and $E_b = 480$ mJ. Figure 14.42 shows a schematic view of the experimental setup. The laser beam is guided into a measuring chamber passing an antireflecting coated window and is focused by a $f = 500$ mm lens. The chosen pulse energy level range and focal length are significantly greater than those values used in [14.67, 14.71, 14.72] (these authors used a focal length <150 mm and pulse energies <200 mJ) due to the following reason. The intended application is the analysis of concrete drill cores. These cores have to be split to obtain a fracture surface unaffected by the water flushing used during drilling. The water flushing may lead to wash out effects of the species to be analyzed. However, the fracture surface is uneven due to the included aggregates. Hence, the depth of focus in terms of the Rayleigh length (about 12.5 mm, cf. Sect. 3.4) of the focused laser beam has to be sufficiently large to assure a nearly constant irradiance. For the experiments described in the following plane samples were used.

The angle between the incident laser beam and the normal of the sample surface is $\alpha = 12°$ (cf. Sect. 3.9, Fig. 3.26). The measuring chamber has a net volume of

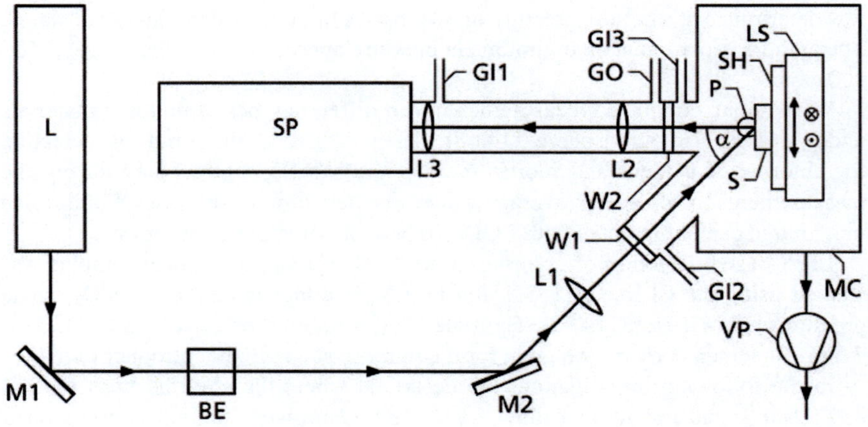

Fig. 14.42 A schematic view of the experimental setup. L = laser system, M1–M2 = mirrors, BE = beam expander, L1 = lens (f = 50 cm), W1–W2 = windows, GI1–GI3 = gas intake, GO = gas outlet, P = plasma, S = sample, SH = specimen holder, LS = linear stages, MC = measuring chamber, L2–L3 = MgF_2 lenses (each f = 20 cm), SP = spectrometer, VP = vacuum pump

50 l. Pressures between $p = 10$ and 1,000 mbar can be set. Inside the chamber two linear stages are mounted to move the samples without venting and opening the chamber. The emission of the laser-induced plasma propagates through a direct-light channel to a Paschen–Runge spectrometer (cf. Sect. 4.2). Inside this channel two MgF_2 lenses (L2 and L3 in Fig. 14.42) each with a focal length of $f = 201$ mm at $\lambda = 525$ nm collimate and couple the plasma light into the spectrometer. An additional magnesium fluoride MgF_2 window (W2 in Fig. 14.42) is used inside the channel between the chamber and the lenses to seal the chamber and to protect the MgF_2-optics of getting contaminated by ablated particles. The space inside the direct light channel between the lens L3 and the MgF_2-window (W2) is flushed with argon to avoid absorption in the UV due to residual oxygen gas or water vapor in the channel (cf. Sect. 3.5, Fig. 3.14). Between $\lambda = 121.5$ and 921.5 nm a total number of 55 spectral lines, including background channels can be detected by the Paschen–Runge spectrometer and a multichannel integrator electronics (MCI, cf. Sect. 4.5) being capable to integrate the channeltron signals in programmable time gates. Windows W1 and W2 are flushed during a measurement by a total helium gas flow rate of 1.5–50 l/min depending on the chosen pressure inside the chamber. The gas passes the surfaces of the optics and is guided onto the sample surface. By this setup, a contamination of the optical path by ablated particles between the plasma at the surface of the samples and the spectrometer is kept at a sufficient low level. A vacuum pump is connected to the measurement chamber to achieve a constant pressure inside the chamber while the optics are flushed. The pump evacuates the chamber and the pressure inside the chamber is set by the gas flow that can be adjusted by two valves. The gas flows for the entrance window and the MgF_2 window can be controlled separately. For all measurements, the ratio of gas flow rates at W1, W2 was set to 1:10 because of the greater aperture of the direct-light channel. The position of the beam waist was set underneath the sample surface $\Delta s > +1.5$ cm (cf. Chap. 2, Fig. 2.2 and Chap. 11, Table 11.1). The diameter of the laser beam at the sample surface amounts to $D_b > 800\,\mu$m. During all measurements, helium is used as process gas.

To be able to estimate the detection limit for chlorine in cement, samples with different chlorine concentrations are needed. Cement powder is mixed with water that contains a specific concentration of NaCl to achieve the wanted chlorine concentration in the sample. The cement water mixture is filled into plastic bottles. During the hardening process, the bottles are rotated constantly to ensure a homogenous distribution of chlorine in the sample. After some days, the plastic bottle is removed and the hydrated cement is divided into smaller pieces by a saw. The samples have a diameter of $D_s = 4.5$ cm and a thickness of $T_s = 1.5$ cm, see Fig. 14.43. The samples have many pores and by this they can take up to 8% of moisture due to the interaction with the ambient air, which was determined by weighing of the samples before and after a drying cycle in an oven heated to 60°C.

Single and collinear double pulse configurations are studied at different ambient pressures. The burst energy ranges from $E_b = 450$ mJ to $E_b = 480$ mJ and the pulse energy ratio is set to $E_1 : E_2 = 1 : 1$ in case of double pulses (cf. Sect. 3.2). The interpulse separation is varied between $\Delta t = 1$ and 8 μs. In a first step, the

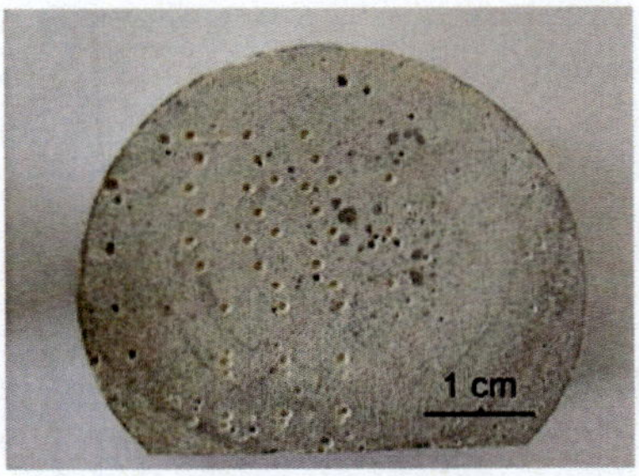

Fig. 14.43 Photograph of a hydrated cement sample that is used for the measurements. There are several craters visible on the sample surface that were generated by the laser bursts

delay time t_{delay} and integration time gate t_{int} have to be optimized for the detection of chlorine lines. Integration time gates reported so far vary between $t_{int} = 10$ and $20\,\mu s$ to detect the emission of the chlorine line $\lambda = 837.59$ nm using helium as inert gas to analyze organic compounds or pressed disks containing pure NaCl powder [14.71, 14.72]. Lancelin et al. used a delay time of $t_{delay} = 2\,\mu s$ and an integration time gate of $t_{int} = 20\,\mu s$ to study the signal-to-noise ratio of the chlorine 837.59 nm line of different gas mixtures containing chlorine compounds at ambient pressure [14.75]. No studies were published so far concerning the temporal evolution of the chlorine 134.72 nm line.

To determine the optimal delay time t_{delay} and the integration time gate t_{int}, the signals of the channeltrons inside the spectrometer are detected by an oscilloscope with an input impedance of $50\,\Omega$. Figure 14.44 shows the difference of the channeltron signals for the chlorine line 134.72 nm between a sample with a natural basic concentration of about 0.05 m.-% (U_l) and a sample with a high chlorine concentration (U_h) in cement as a function of time. About 100 ns after the interaction of the laser pulse with the sample surface, a distinct maximum of the difference is observed. The corresponding signal difference of the chlorine spectral line in the IR shows a similar behavior. In the following the starting point of the integration time gate of the MCI is set to $t_{delay} = 100$ ns. An integration time gate of $t_{int} = 10\,\mu s$ was chosen, where the signals of both channeltrons for the IR and UV chlorine line reach the base level. To reduce the influence of a possible inhomogeneous species distribution inside the calibration samples, the mean intensity of 20 subsequent spectra at 9 different positions are averaged to one value for each element detected by the spectrometer. The measuring positions are arranged in a grid of 10 mm × 10 mm, where each measuring point has a distance of $d = 5$mm to the adjacent point.

To investigate the influence of the gas pressure on the plasma emission, the pressure is varied between $p = 10$ and 320 mbar where the line intensity enhancement effect for the spectral line is expected [14.72]. Figures 14.45 and 14.46 show the signal-to-noise ratio of the chlorine lines in the IR and UV as a function of the helium pressure. Regarding the intensity for the Cl line at $\lambda = 837.59$ nm the ratio $I_{Cl\,837\,nm}/I_{Bg\,861\,nm}$ is increasing for both samples with growing p_{He} from 1.6 at $p_{He} = 10$ mbar to 1.9 for the 2 m.-% sample and to a value of 2.3 for the

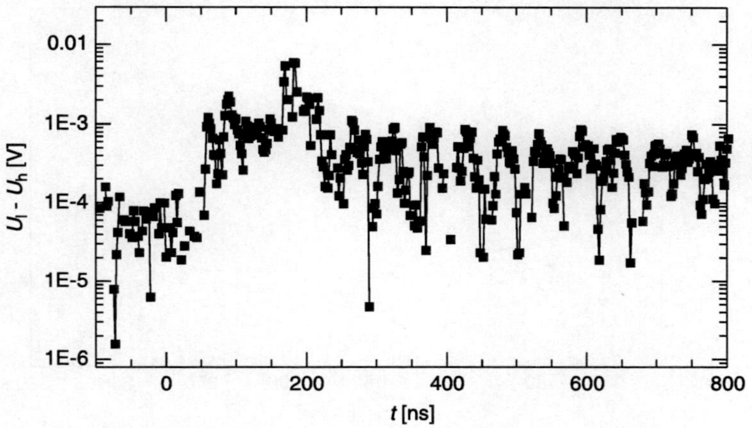

Fig. 14.44 The difference of the channeltron signals for a sample with a high chlorine concentration (4 m.-%) and a low concentration (0.05 m.-%) for the spectral line at $\lambda = 134.72$ nm as a function of time ($t = 0$ ns corresponds to the time of the laser pulse irradiation) at $p_{He} = 100$ mbar

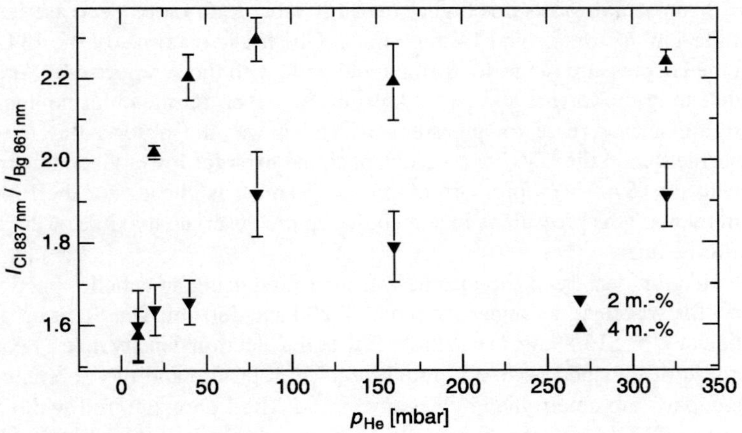

Fig. 14.45 Referenced spectral line intensity of the chlorine line $\lambda = 837.59$ nm as a function of the helium pressure for two cement samples with different chlorine concentrations

4 m.-% sample at a pressure of $p_{He} = 75$ mbar. A further increase in the pressure up to $p = 320$ mbar does not significantly change the signal ratio. Regarding the pressure regime between $p_{He} = 75$ and 320 mbar, the signal ratio seems to level off at a plateau. A different development is observed for the chlorine line at $\lambda = 134.72$ nm, see Fig. 14.46. At $p_{He} = 10$ mbar for both Cl concentrations different signal-to-noise ratios exist. By increasing the pressure both ratios increase. For the 4 m.-% sample, a maximum is attained at $p_{He} = 40$ mbar with a ratio of $I_{Cl\ 134\,nm}/I_{Bg\,861\,nm} = 2.5$. For the 2 m.-% sample, the maximum is located at about

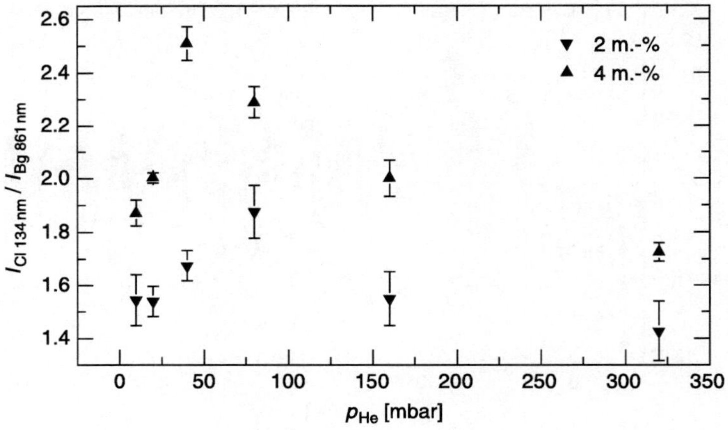

Fig. 14.46 Referenced spectral line intensity of the chlorine line $\lambda = 134.72$ nm as a function of the helium pressure for two cement samples with different chlorine concentrations

$p_{\mathrm{He}} = 80$ mbar. For pressures higher than 100 mbar both ratios decrease reaching 1.7 for the 4 m.-% sample and 1.4 for 2 m.-%. The found maxima for the 134.72 nm line in the He pressure range 40–80 mbar coincide with those reported by Asimellis et al. detecting the Cl-line at $\lambda = 837.59$ nm. However, for the chlorine line in the IR a significant decrease for pressures >75 mbar was not observed as found for the chlorine line in the UV. For a sufficient signal even for lower Cl concentrations in cement (<0.5 m.-%) a pressure of $p_{\mathrm{He}} = 60$ mbar is chosen for the following measurements. This pressure is located between the observed maxima of the IR and UV chlorine lines.

The electron density of the plasma was measured using an echelle spectrometer to detect the wavelength range between $\lambda = 200$ and 780 nm. The Stark broadened Ca II line at $\lambda = 315.89$ nm is used to estimate the electron density n_e(cf. Sect. 9.4). The profile of the spectral line is fitted by a Voigt profile and the Lorentzian and Gaussian parts are determined by deconvolution. The Lorentzian full width at half maximum (FWHM) of the line width is used to calculate the electron density [14.76]:

$$\Delta\lambda_{1/2} = 2w \frac{n_e}{10^{17}\,\mathrm{cm}^{-3}}, \tag{14.9}$$

where w is the half width at half maximum (HWHM) electron Stark parameter for the Ca II spectral line ($w = 29.2$ pm), $\Delta\lambda_{1/2}$ the FWHM of the line profile, and n_e is the spatially integrated electron density of the plasma [14.77].

Figure 14.47 shows the electron density as a function of the pressure p_{He}. At $p_{\mathrm{He}} = 5$ mbar, the electron density amounts to $n_e = 4.2\times10^{16}\mathrm{cm}^{-3}$. In the range of $p_{\mathrm{He}} = 5$–60 mbar, the increase in the electron density can be described by a linear regression curve with a coefficient of determination of $r^2 = 0.99$. In the regime between 120 and 1,000 mbar, the slope of the electron density increase is reduced

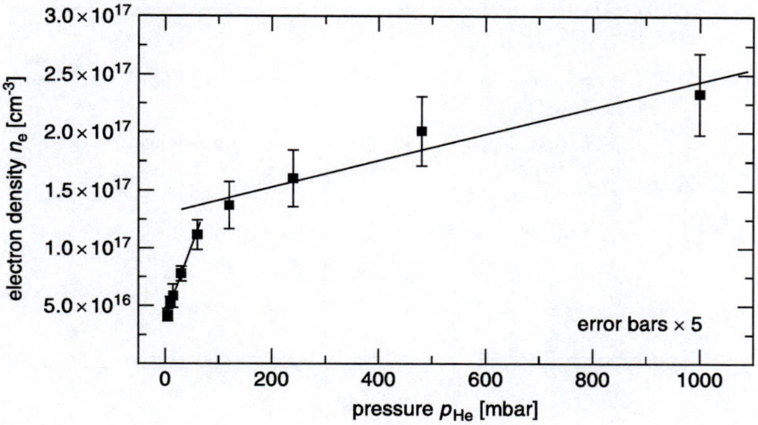

Fig. 14.47 Electron density for single pulses as a function of the helium pressure p_{He} inside the chamber

and at 1,000 mbar the electron density reaches a maximum of $n_e = 2.3 \times 10^{17}\,cm^{-3}$. At the observed maximum of the chlorine lines for an ambient pressure of about $p_{He} = 60$ mbar – see Figs. 14.45 and 14.46 – the electron density attains a value of $n_e > 1 \times 10^{17}\,cm^{-3}$.

After a suitable pressure regime for the chlorine signal was found, the influence of the integration time gate on the limit of detection is studied by varying the integration time gate t_{int} between 1.25 and 40 μs. Figure 14.48 shows the chlorine line intensity 837.59 nm for different integration time gates as a function of the chlorine concentrations. The slope is not as high as in the case of the chlorine line in the UV, cf. Fig. 14.49. For an integration time gate of $t_{int} = 1.25$ μs, the dynamic range of the counts between 0.05 and 4 m.-% amounts only ~900 counts. An increase in the integration time gate up to 10 μs results in an increased background that is indicated by a shift of the spectral line intensity for 0.05 m.-% from 8,200 to 10,000 counts. The dynamic range for the lowest and highest concentration varies between 9,500 and 17,000 counts. The increase in the detection time by a factor of 8 results in an increase in the dynamic range by a factor of 8. A similar case can be seen for the atomic chlorine line at $\lambda = 134.27$ nm in Fig. 14.49. For an integration time gate $t_{int} = 1.25$ μs, the signal varies between 4,600 and 6,600 counts, corresponding to a dynamic range of 2,000 counts for chlorine concentrations between 0.05 and 4 m.-% in cement. An increase in the integration time gate by a factor of 8 up to $t_{int} = 10$ μs results in a range of 5,500–15,000 counts. By this the dynamic range rises nearly by a factor of 5. A further increase in the detection time up to 20 and 40 μs does not significantly improve the maximum intensity neither the dynamic range.

The influence of the integration time gate on the limit of detection is studied applying the calibration curve method of the standard DIN 32 645 to estimate

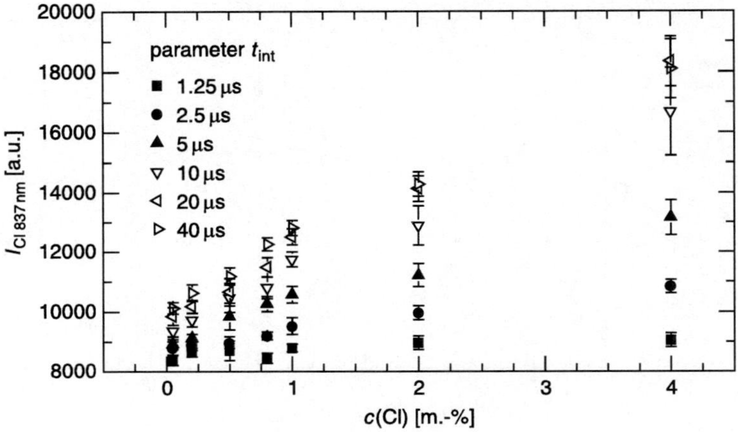

Fig. 14.48 Intensity of the Cl-line at $\lambda = 837.59$ nm as a function of the chlorine concentration for various integration times t_{int}, $p_{He} = 60$ mbar

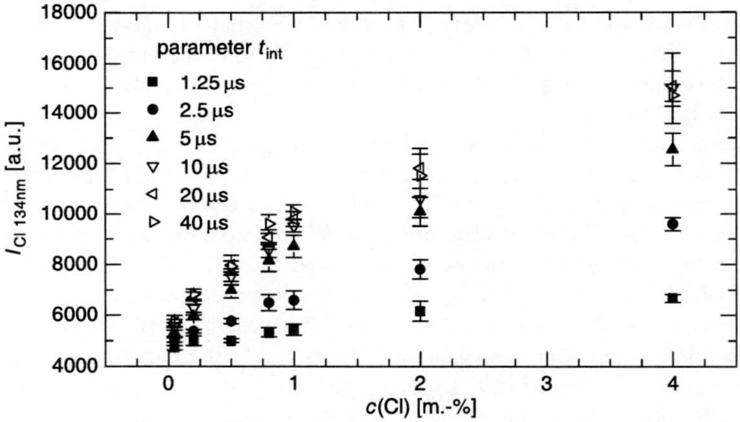

Fig. 14.49 Intensity of the Cl-line at $\lambda = 134.72$ nm as a function of the chlorine concentration for various integration times t_{int}, $p_{He} = 60$ mbar

the detection limit [11.3] (cf. Chap. 11). The $3s$-criterion to estimate the limit of detection for chlorine in cement is not used here because of the natural content of chlorine in cement generally. Figure 14.50 shows the detection limit for the Cl lines $\lambda = 134.72$ nm and $\lambda = 837.59$ nm as a function of the integration time gate t_{int} using the ratios $I_{Cl\,837\,nm}/I_{Bg\,861\,nm}$ and $I_{Cl\,134\,nm}/I_{Bg\,861\,nm}$. Starting with a detection limit of 0.9 m.-% for $t_{int} = 1.25\,\mu s$, a minimized detection limit of ≥ 0.1 m.-% is achieved for $t_{int} = 5$ and $10\,\mu s$ for the chlorine line in the UV. A further increase of t_{int} results in an increased detection limit of up to 0.28 m.-% for $t_{int} = 40\,\mu s$. The limit of detection for the chlorine line in the IR at an integration time $t_{int} = 1.25\,\mu s$

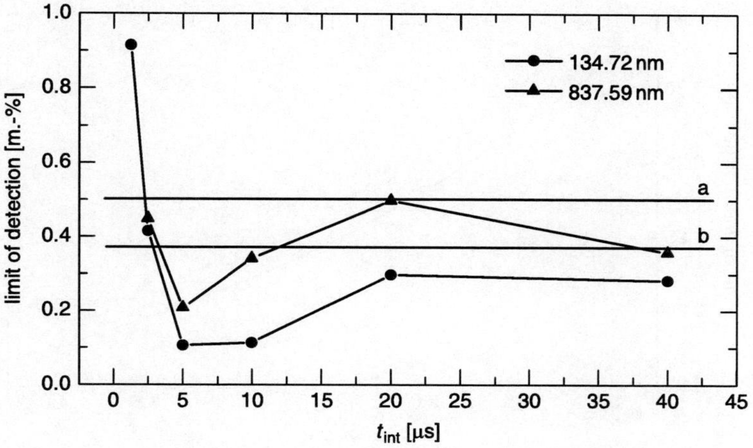

Fig. 14.50 The estimated limit of detection for single pulses and the Cl lines $\lambda = 134.72$ nm, $\lambda = 837.59$ nm as a function of the integration time gate t_{int}, $t_{delay} = 100$ ns. The *horizontal lines* indicate the achieved LODs for chlorine using the 837.59 nm line reported so far, a = [14.68], b = [14.72]

is not included into the diagram because the value is above 3 m.-%. Similar to the chlorine line in the UV, the limit of detection can be decreased by increasing the integration time t_{int}. For $t_{int} = 5$ μs, a minimum is achieved with 0.2 m.-%. A further increase in t_{int} results in a limit of detection of 0.5 m.-%. At a maximum integration time of $t_{int} = 40$ μs, the limit of detection for the chlorine line in the UV and the IR are nearly the same. Therefore, an integration time gate between $t_{int} = 5$ μs and $t_{int} = 10$ μs should be chosen for the further experiments. The 95% confidence interval of the limit of detection data shown in Fig. 14.50 (and also in Fig. 14.51) is given by multiplying the respective values by a factor of 0.57 and a factor of 3.5 (see Table 2 in [11.3] for $n - 2 = 3$ degrees of freedom).

To further improve the LOD, the influence of collinear double pulses was studied. The pulse ratio is set to $E_1 : E_2 = 1 : 1$ and interpulse separations between $\Delta t = 1$ μs and 8 μs are used in the measurements. Peter et al. reported that the use of collinear double pulses increases the spectral line intensity at ambient conditions [14.78] (cf. Sects. 6.1, 7.2, and 7.4). Delay time t_{delay} and integration time gate t_{int} are set to 100 ns and 10 μs, respectively. Because of the laser system, the energy is reduced to $E_b = 450$ mJ in the double pulse configuration. Process gas and pressure inside the chamber are the same like in the case of single pulses. Figure 14.51 shows the limits of detection achieved for the ratio $I_{Cl\,134\,nm}/I_{Bg\,861\,nm}$ using double pulses. All values are greater than those achieved with single pulses. The LOD ranges between 0.23 and 0.31 m.-%. The limit of detection for $\Delta t = 8$ μs is not shown in the diagram because the value is greater than 1 m.-%.

Figure 14.52 shows the detected spectral line intensities of the Cl 134.27 nm line as a function of the chlorine concentration for different interpulse separations

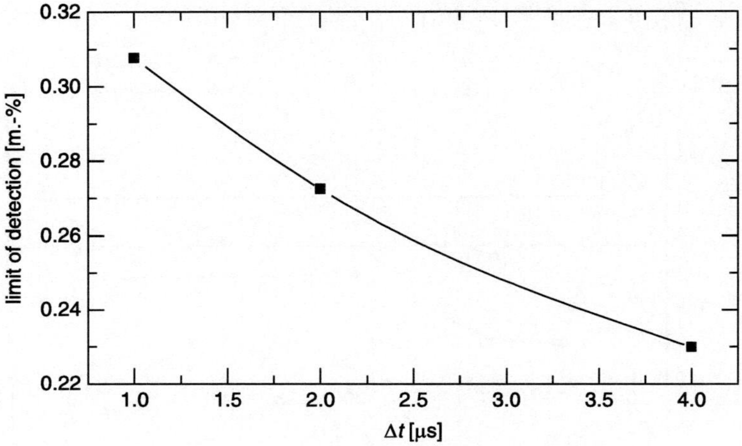

Fig. 14.51 The estimated limit of detection for double pulses and the Cl line $\lambda = 134.72\,\text{nm}$ as a function of the interpulse separation Δt

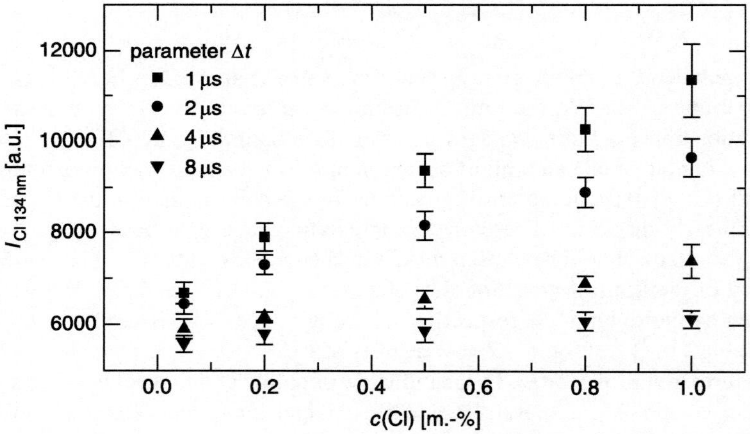

Fig. 14.52 The detected spectral line intensity of the chlorine spectral line a $\lambda = 134\,\text{nm}$ as a function of the chlorine concentration for different interpulse separations Δt, $p_{\text{He}} = 60\,\text{mbar}$, $t_{\text{delay}} = 100\,\text{ns}$, $t_{\text{int}} = 10\,\mu\text{s}$

Δt. Regarding the detected intensity for a chlorine concentration of 0.05 m.-% the line intensity drops from 6,500 counts for $\Delta t = 1\,\mu\text{s}$ down to 5,500 counts for $\Delta t = 8\,\mu\text{s}$. Additionally the dynamic range of the line intensity for concentrations between 0.05 and 4 m.-% is reduced for greater interpulse separations Δt. For $\Delta t = 1\,\mu\text{s}$, a range of $\Delta I_{\text{Cl}\,134\,\text{nm}} = 4{,}700$ counts exists whereas for an interpulse separation of $\Delta t = 8\,\mu\text{s}$ this value is reduced to $\Delta I_{\text{Cl}134\,\text{nm}} = 550$ counts. For $\Delta t \leq 2\,\mu\text{s}$, an increase of the spectral line intensity can be seen compared with single pulses. On the other hand, the dynamic range is reduced. At a He pressure of 60 mbar, the preconditioning effect of the first pulse of a double pulse burst does obviously not result in an increased spectral line intensity for interpulse separations $\Delta t \geq 4\,\mu s$.

This effect could be attributed to a further reduced electron density. In comparison with single pulses, the electron density reaches values that exist for single pulses at pressures below $p_{He} < 60$ mbar where a significant decrease in the spectral line intensities is seen for both lines.

The described investigations show that the estimated limit of detection for chlorine deploying the calibration curve criterion of DIN 32 645 can be reduced to LOD = 0.1 m.-% for the UV chlorine line, single pulse excitation, and an integration time gate of $t_{int} = 5\,\mu s$ corresponding to an improvement of a factor of >5 compared with the best values achieved so far.

14.5 Droplets

Under special conditions, LIBS allows to analyze the elemental compounds in liquids with high sensitivity [14.79]. This suggests the use of LIBS for a field of growing interest in the analytical sciences and speciation analysis [14.80, 14.81].

The chemical, biological, and toxicological properties of many elements are critically dependent on their chemical form, its oxidation number, and the compound the element is occurring in. This has a great impact on environmental and biological chemistry, occupational health, nutrition, and medicine. Typical targets of speciation analysis are metal ions in different oxidation states, metal complexes with inorganic and organic ligands, metalloid oxoanions, organometallic compounds, and ionic nonmetal species. An analytical technique yielding only the total amount of the analyzed element does not consider different possible forms. Coupled techniques combining chromatographic separation with element-specific detection have therefore become fundamental tools for speciation analysis. These so-called "hyphenated techniques" enable the elemental analysis of separated fractions of complex samples. Typical detection techniques are mass spectrometry or atomic absorption and atomic emission spectrometry and the most common separation techniques are gas chromatography (GC) for volatile compounds and high-pressure liquid chromatography (HPLC) for less volatile compounds. In recent years, capillary electrophoresis has also become more and more important in this field [14.82].

Table 14.13 gives an overview of the most important hyphenated techniques and their status of applicability.

The detection procedure should give maximum information with minimum sample consumption. The LIBS process as a detection technique offers several significant advantages: LIBS signals yield the elemental composition of complex samples, the plasma breaks up any molecular structure, and the individual elements lead to specific radiation. The information content is therefore higher than with the UV–VIS or fluorescence detectors that are frequently used in HPLC chromatography. LIBS is a true multielement technique. Depending on the spectrometer used,

Table 14.13 Different hyphenated techniques

	GC	HPLC	SFC	HPTLC	CE
MS	◇◇◇	◇◇◇	◇	◇	◇◇
IR	◇◇◇	◇◇	◇	◇◇	◇
OE	◇◇	◇	◇		
AAS	◇	◇◇			
NMR		◇◇			◇
Fluorescence	◇◇	◇◇◇		◇◇◇	◇◇◇
UV-VIS	◇	◇◇◇		◇◇◇	◇◇◇

Separation: GC (gas chromatography), HPLC (high-pressure liquid chromatography), SFC (supercritical fluid chromatography), HPTLC (high-performance thin layer chromatography), CE (capillary electrophoresis). Detection: MS (mass spectrometry), IR (infrared spectrometry), OE (optical emission), AAS (atomic absorption spectrometry), NMR (nuclear magnetic resonance), fluorescence spectrometry, UV–VIS spectrometry

◇◇◇ Standard technique

◇◇ Commercially available, less spread

◇ Experimental status, not commercially available

several elements can be monitored at the same time without loss of sensitivity. ICP-MS detectors (inductively coupled plasma mass spectrometry) are used to achieve highest sensitivity, making sample concentrations as low as pg/g (ppt) possible [14.83, 14.84]. Plasma generation is achieved with the aid of an inductively heated argon torch, and the sample is diluted in a large amount of gas needed for the plasma generation (typically several liters per minute). ICP-MS detectors achieve their highest sensitivity working with flow rates in the range of ml/min, which fit well to HPLC. Significantly lower flow rates lead to a loss in sensitivity.

With LIBS the sample is transferred directly into the plasma state, without the need for an additional gas flow. No dilution of the sample occurs and the method is capable to analyze sample amounts as low as a few nanoliters. This will turn out to be valuable for the coupling with low flow separation techniques such as capillary electrophoresis or micro-HPLC. In these separation techniques, only a few μ l/min are needed.

Water is one of the common solvents in HPLC chromatography and therefore the process of plasma generation using aqueous samples must be well understood. The mechanisms of plasma formation in water have been thoroughly examined and different theoretical models to describe the breakdown have been published (multiphoton model, cascade model) [14.85].

Different setups to record laser-induced breakdown spectra in a water matrix have been described in the literature: bulk analysis in cuvettes [14.86] or on the liquid surface [14.87], the analysis of liquid jets [14.88], aerosols [14.89] or the analysis of single, isolated droplets [14.90]. Plasma formation in bulk water generally suffers from short lifetimes such that the emission intensity falls steeply after approximately 1 μs, making sensitive measurements impossible [14.91]. Cremers et al. [14.92] compared single and double pulse LIBS spectra from metal samples placed under water and showed that only the double pulse arrangement provides sufficient excitation to identify the sample compounds. With a single pulse, the plasma formed is surrounded by an aqueous medium that leads to a fast decay of the plasma radiation. In the double pulse experiment, the first pulse produces a gaseous atmosphere in which the second pulse, reaching the target some 10 μs later, produces a plasma with a significantly longer lifetime. The excitation of the plasma in air or noble gas atmospheres yields larger signal intensities than plasma generation inside bulk water; this can also be seen in single pulse experiments [14.93]. Thin water jets, aerosols or single, isolated droplets are superior sample forms compared with bulk water, because the plasma can expand in a gaseous atmosphere.

Using a water film and a dual-beam dual-pulse setup, Kuwako et al. could demonstrate a limit of detection as low as 0.1 ppb for sodium in water [14.94]. An alternative approach to obtain a very high sensitivity was made by Cheung et al., who used an ArF-Laser at 193 nm for excitation instead of the commonly used YAG Laser (1,064 nm). Limits of detection in the low ppb region were achieved for sodium, calcium, and barium as a result of the efficient photoionization of hot water molecules by the UV photons [14.95, 14.96].

Figure 14.53 shows a diagram of the hyphenated technique HPLC–LIBS [14.97]. The HPLC output is fed into the droplet generator and transformed into a continuous stream of uniform droplets. In order to achieve the best possible transfer of sample material into the plasma, a single droplet generator based on a piezoelectric nozzle was applied. Unlike electro-spray or aerosol generators, a piezoelectric

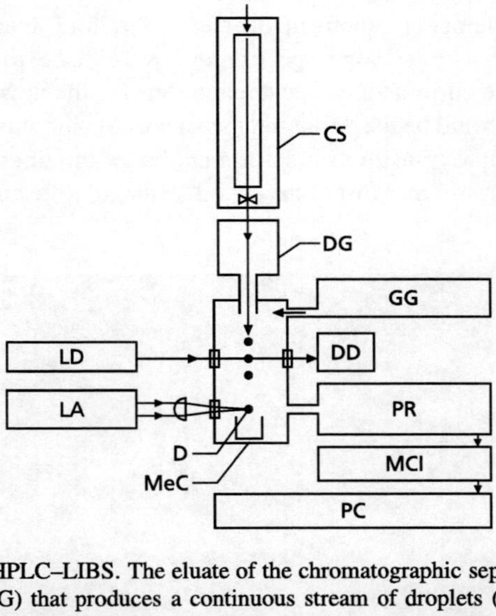

Fig. 14.53 Setup for HPLC–LIBS. The eluate of the chromatographic separation (CS) is fed into a droplet generator (DG) that produces a continuous stream of droplets (D). Individual droplets are detected by a unit consisting of a laser diode (LD), an achromatic lens, and a photodiode for droplet detection (DD). The detection of a droplet triggers a Q-switched Nd-YAG laser (LA), the laser pulse transforms a single droplet into the plasma state, the radiation is guided into a Paschen–Runge spectrometer (PR) and analyzed by a multichannel integrator electronics (MCI). Measurements are stored and processed in a computer (PC). A gas guiding system (GG) leads argon into the measurement chamber (MeC)

nozzle is a "drop on demand" device that produces single, isolated droplets with volumes typically between 100 pl and 1 nl and an initial velocity of 1–2 m/s. Figure 14.54 shows a stroboscopic picture of the generated sub-nanoliter droplets. By synchronizing laser pulses to single droplets plasmas are generated in a gaseous environment. The process of droplet formation is highly reproducible, leading to a uniform droplet size. Phase Doppler anemometry measurements performed with a piezoelectric droplet generator similar to the one used in this work revealed a relative standard deviation of the droplet volume of 3% for droplets of 86 μm diameter [14.98]. After detection of an individual droplet, the laser is triggered and emits a single Q-switched pulse that is focused onto the droplet and generates a plasma. The plasma radiation is collected and guided to the spectrometer.

Figure 14.55 shows the optical setup for the analysis of single picoliter droplets. For a better control of the experimental parameters, the HPLC was replaced by a stock of sample solution, having a constant analyte concentration. Prior to analysis all samples were filtered through 0.2 μm pore size membrane filters. The plasma radiation is resolved spectrally with a Paschen–Runge spectrometer equipped with 33 channel photomultipliers (type C 932, C 933, C 911 from Perkin Elmer), enabling the simultaneous detection of 31 different elements and the zeroth order signal. Among these elements are nonmetals such as Ar, H, C, O, N, P, S, alkali and alkaline earth metals (Na, Ca, Mg) and various metals that play a significant role in the fields of environmental chemistry, clinical biochemistry, industrial chemistry, nutrition and toxicology, e.g., As, Hg, Cr, Pb, and Pt.

Most of these elements can form different metalloid species. The choice of elements for the Paschen–Runge spectrometer was made to guarantee a broad spectrum of possible chromatographic applications for the new LIBS detector. The nonmetal elements could be used for the analysis of organic substances. Because all elements have multiple transitions a careful choice of the lines that are best suited for the analysis is important. Most of the stated elements were analyzed by recording

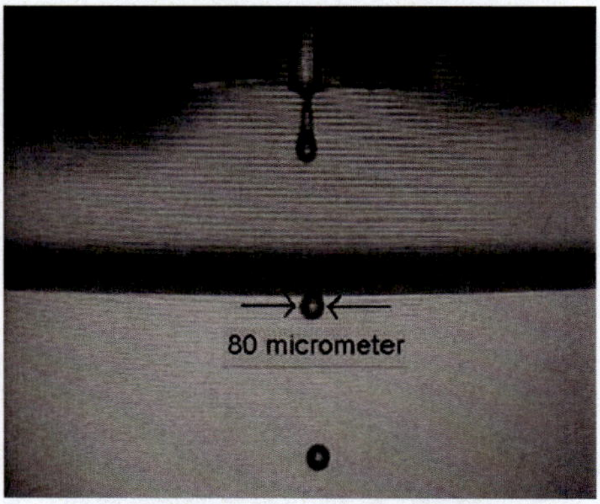

Fig. 14.54 Image of 270 pl droplets. The picture was taken with stroboscopic illumination, and the droplet formation frequency is approximately 800 Hz. A part of the glass capillary of the droplet generator can be seen in the upper part of the picture

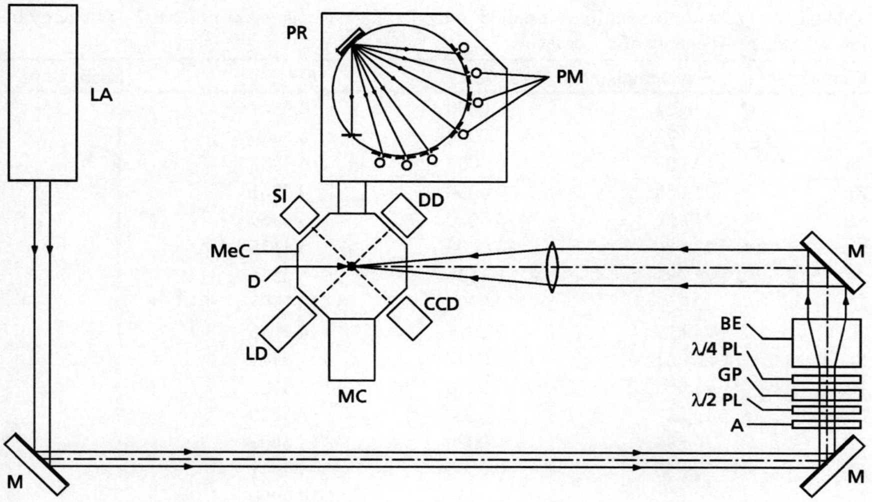

Fig. 14.55 Optical setup for LIBS experiments on droplets. M = mirror, LA = laser, A = aperture, $\lambda/2$ PL = $\lambda/2$ plate, $\lambda/4$ PL = $\lambda/4$ plate, GP = Glan-laser polarizer, BE = beam expander, D = droplet, MC = mirror chamber, MeC = measuring chamber, CCD = CCD-camera for droplet observation, DD = droplet detector (photodiode), SI = stroboscopic illumination, LD = laser diode (part of the droplet detector), PM = photomultiplier

laser-induced breakdown spectra with an echelle spectrometer. The matrix the element is surrounded by might affect the line intensities; therefore, all echelle spectra were recorded from small water droplets containing diluted samples. Usually the line showing the strongest intensity was chosen. A complete list of all elemental transitions detectable with the Paschen–Runge spectrometer is given in Table 14.14.

The plasma emission shows a strong continuum peak directly after the laser pulse irradiation that decays faster than the element specific line emission (cf. Chap. 2, Fig. 2.4). After a few microseconds, the element-specific line emissions dominate. For a better signal-to-noise-ratio, the continuum part of the plasma emission is excluded from the analytical signal with the aid of gated signal integration. Typical delay times between the laser pulse and the start of the signal integration are 0.5–1.5 μs, the integration gate width is between 20 and 50 μs. The photomultiplier signals are processed by a multichannel integrator electronics (MCI, cf. Sect. 4.5). The photocurrents from 33 photomultipliers are integrated simultaneously with individual time gates. For single pulse evaluation, the signal intensities of all channels are transmitted to a computer.

In order to select the elemental transitions with the best analytical performance, laser-induced breakdown spectra of droplets were studied between 200 and 780 nm with an echelle spectrometer (cf. Sect. 4.2, type ESA 3000 from Laserlabor Adlershof, Berlin, Germany) for each element. The experimental setup for the selection of the transitions was similar to the one described above except that the spectrometer was connected to the experiment with a fiber optics.

Table 14.14 List of element lines installed in the Paschen–Runge spectrometer. E_i = energy of lower level, E_j = energy of upper level

Element	Wavelength (nm)	E_i (eV)	E_j (eV)	Grating order
P	178.28	0.0000	6.9548	1
S	180.73	0.0000	6.8606	1
Se	196.09	0.0000	6.3232	1
Zn	213.86	0.0000	5.7961	1
Sb	217.59	0.0000	5.6969	1
Cd	228.80	0.0000	5.4175	1
As	234.98	1.3134	6.5884	1
H	121.57	0.0000	10.1995	2
B	249.77	0.0019	4.9646	1
Si	251.61	0.0277	4.9541	1
Hg	253.65	0.0000	4.8868	1
O	130.22	0.0000	9.5220	2
Pt	265.94	0.0000	4.6600	1
Mg	279.55	0.0000	4.4341	1
N	149.26	2.3837	10.6900	2
V	310.23	0.3680	4.3637	1
Sn	317.50	0.4250	4.3291	1
Cu	324.75	0.0000	3.8169	1
C	165.70	0.0054	7.4883	2
Ti	334.94	0.0488	3.7496	1
Ni	341.48	0.0254	3.6554	1
Co	345.35	0.4318	4.0212	1
Cr	357.87	0.0000	3.4638	1
Fe	371.99	0.0000	3.3322	1
Al	394.40	0.0000	3.1429	1
Ca	396.85	0.0000	3.1236	1
Ce	399.92	0.2954	3.3949	1
Pb	405.78	1.3206	4.3754	1
Ar	420.07	11.5491	14.5000	1
La	433.37	0.1729	3.0332	1
H	486.13	10.1995	12.7494	1
Na	589.00	0.0000	2.1024	1
0th order				0

The choice of the respective transitions was made on the basis of measurements with an echelle spectrometer

In order to achieve higher signal intensities, the plasma emission is collected with a spherical mirror and focused on the entrance slit of the Paschen–Runge spectrometer. The mirror is placed inside the measurement chamber and can be adjusted without opening the chamber. The droplet generator is also placed in the measurement chamber and produces a free-falling stream of droplets. It is made of a glass capillary with a small orifice (50 μm diameter) on one side that is enclosed by a tube-shaped piezoelectric element. The piezoelectric element is activated with a voltage pulse that induces a pressure wave in the liquid leading to the formation of a single isolated droplet with a volume between 100 pl and 1 nl and an initial velocity of 1–2 m/s. A CCD camera with a stroboscopic illumination is used to control and adjust the droplet generation (cf. Fig. 14.54). The piezoelectric nozzle is capable of delivering up to 2,000 droplets per second; however, droplet frequencies as low as 10–20 Hz are also possible. The droplet size can only be varied in a narrow range by adjusting the voltage pulse. A higher peak voltage leads to a higher droplet speed, and larger pulse length leads to larger droplets. Every individual piezoelectric nozzle has a set of parameters it is giving the best performance with (best stability and droplet uniformity) and the nozzles show little tolerance for deviations from these parameters. Therefore, systematic studies with droplets of different sizes were not accomplished.

A droplet detection unit consisting of a diode laser, an achromatic lens, and a photodiode is used to detect single droplets with a temporal precision of approximately 1 μs. The photodiode signal shows a characteristic "W" shape, when the droplet trajectory passes through the focus of the diode laser beam. This can be exploited for precise adjustment of the droplet in space and time. The photodiode signal is shown in Fig. 14.56. The shape of the signal can be explained with the droplet being a small sphere that does not refract light when the beam focus lies in the center of the sphere, so the loss of photodiode signal due to the droplet is minimized in this moment since all rays are propagating perpendicularly through the surface of the sphere.

After detection of a droplet, a synchronized laser pulse is emitted that transforms the droplet into a plasma. Temporal and spatial overlap between laser pulse and droplet are essential for a stable analytical signal. Temporal misalignment of a few microseconds or spatial misalignment of a few micrometers leads to a

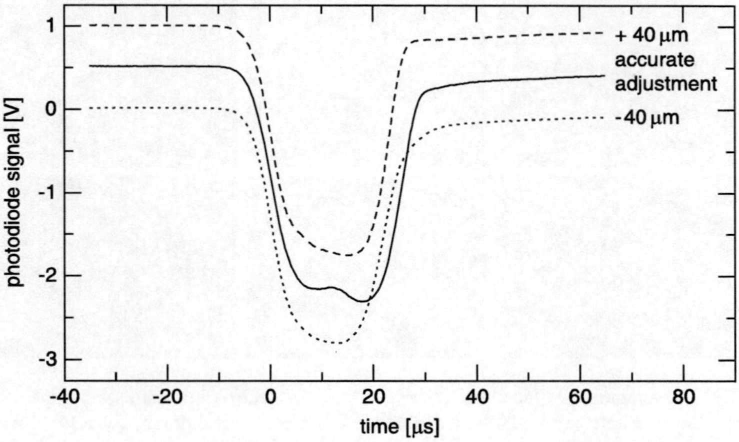

Fig. 14.56 Photodiode signal for different positions of the droplet. When the droplet passes through the focus of the diode laser beam, the characteristic "W" shape can be seen. A lateral displacement of ±40 μm leads to the two signals shown with dashed lines

significant decrease in signal intensity. A Q-switched Nd-YAG laser (10 ns pulse width) was used for the experiments. The repetition rate of the laser restricts the measurement frequency to 10 Hz, although the other components of the experiment (droplet generator, spectrometer, signal integration, and recording) could permit measurements at a rate of 1,000 Hz. The laser pulses were attenuated with the aid of a $\lambda/2$-plate and a Glan laser polarizer to 50–140 mJ. The laser beam is expanded to a diameter of approximately 3 cm and then focused with an achromatic lens having a focal length of 10 cm. For adjustment, the lens can be translated perpendicularly to the laser beam and in the direction of the laser beam. For the timing of the signal detection and the laser trigger relative to the droplet detection, a digital delay generator was used.

For a better understanding of the interaction between laser pulse and droplet, the plasma dynamics was studied with an electro-optical high-speed camera enabling exposure times down to 200 ps (Stanford Computer Optics, 4 Picos). A 500 ppm sodium solution was used to produce intense element-specific radiation. An interference filter (central wavelength 589 nm, FWHM 10 nm) placed in front of the camera transmits the radiation coming from the sodium transitions at 589.00 and 589.59 nm. A series of images recorded with a shutter time of 80 ns with varying

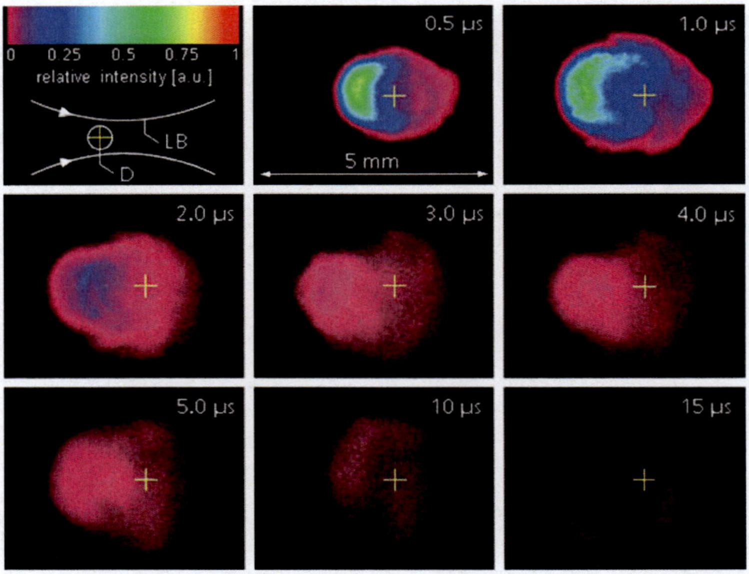

Fig. 14.57 Development of the plasma recorded with an electro-optic camera. Sample: 500 ppm sodium solution. Shutter time: 80 ns. Interference filter: center at 589 nm, FWHM 10 nm. Laser pulse: 50 mJ, 10 ns, focus 5 mm behind droplet. The *upper left* picture shows the relative positions of laser focus and droplet, and the initial droplet position is marked with a *reticle* in all pictures. Delay times relative to the start of the laser pulse are given for each picture. LB = laser beam, D = droplet

delay times is shown in Fig. 14.57; the hair cross indicates the initial position of the droplet. The laser beam (50 mJ pulses, 10 ns pulse width) enters from the left side and the focus position is a few millimeters behind the droplet. Every picture has a size of 5 mm × 3.75 mm.

The first breakdown can be detected at the rear side of the droplet because the beam is focused by the droplet itself leading to the highest laser irradiance at the rear side of the droplet. During the first 500 nanoseconds of expansion (not shown here), the breakdown moves in the direction of the laser beam, because evaporated material is further heated by the laser pulse. The intensity distribution of the plasma radiation is asymmetric with the side facing the laser emitting most of the radiation. The plasma grows within 500 ns to a diameter of approximately 2.5 mm. No fragmentation of the droplets was observed on the images. The diameter of the luminous plasma was measured as a function of time and the speed of propagation was calculated which decreases from 40 km/s during the first nanoseconds of the expansion to less than 1 km/s after 500 ns, see Fig. 14.58. Two regions with a different dynamic behavior can be distinguished. Between 15 and 30 ns the expansion can be fitted by a linear regression described by the function $d \sim t^{1.76}$(d = plasma diameter), between 100 and 500 ns the linear fit yields $d \sim t^{0.34}$. It must be pointed out that this is not the analysis of the expansion of the shock wave which can be described by the Sedov model [cf. Sect. 8.1, relation (8.1)] but the expansion of the luminous plasma front. Normally the plasma front is retarded relative to the shock wave. An example for the analysis of shock waves produced during a LIBS experiment is described in Sect. 8.1.

In consideration of the heat capacity of liquid water (4.18 J/gK) and the enthalpy of vaporization (2,258 J/g at 100°C) it takes approximately 0.7 mJ to vaporize 270 pl of liquid water. With the laser pulse having an energy of 50 mJ, sufficient energy is

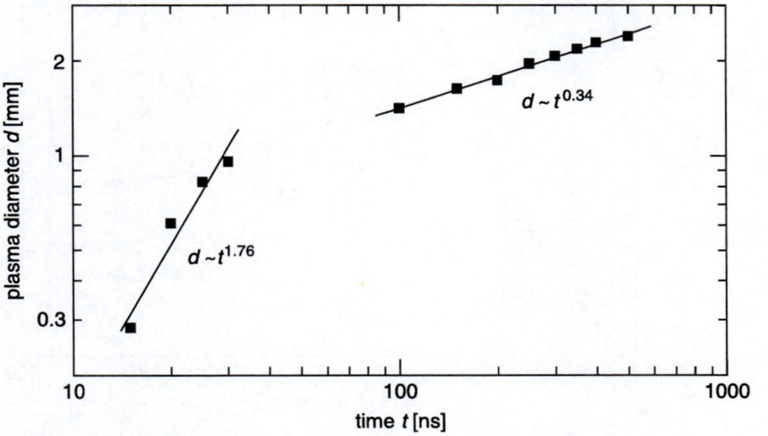

Fig. 14.58 Plot of the plasma expansion for the first 500 ns. The speed of the expansion drops from 40 km/s during the first nanoseconds to less than 1 km/s after 500 ns. Two different expansion regimes can be distinguished that follow different power laws

available for complete vaporization. From the fast expansion and the homogeneity of the plasma no hints are observed for any fragmentation into smaller subdroplets or parts of the droplet which are not transferred to the plasma state.

In a series of experiments the influence of the laser pulse energy and wavelength on the analytical sensitivity was studied. The fourth harmonic of a Nd-YAG Laser (266 nm) was applied to perform LIBS experiments with small droplets and the results were compared to the excitation with 1,064 nm. For both wavelengths calibration curves with different sodium concentrations between 0 and 10 $\mu g/g$ were measured and the limits of detection were determined (see Sect. 11.2). With 60 mJ pulses at 1,064 nm, a limit of detection of 0.75 $\mu g/g$ was calculated using the $3s$-criterion [11.3]. The same measurements were made with 25 mJ pulses at 266 nm and yielded a limit of detection of 2 $\mu g/g$. When using the same pulse energy (25 mJ) for both wavelengths, the signal intensities of the sodium 589.00 and 589.59 nm lines are approximately twice as high with the excitation at 266 nm compared with the 1,064 nm excitation. Since the laser could not deliver more than 25 mJ of the UV wavelength per pulse, the fundamental wavelength was used in all subsequent experiments.

The signal intensities show a strong correlation to the laser pulse energy. Figure 14.59 shows the rise of the line intensities of sodium lines for 500 ppm sodium droplets with the laser pulse energy. For more than 60 mJ/pulse, a saturation of the line emission is observed. Since very high pulse energies can lead to instabilities in the operation of the piezoelectric droplet generator, most measurements were performed in the range between 60 and 100 mJ/pulse.

In order to determine the analytical sensitivity, echelle spectra from different samples were recorded containing between 0 and 0.7 $\mu g/g$ calcium. Samples were diluted from a 100 $\mu g/g$ stock solution containing Ca stabilized with 5% nitric

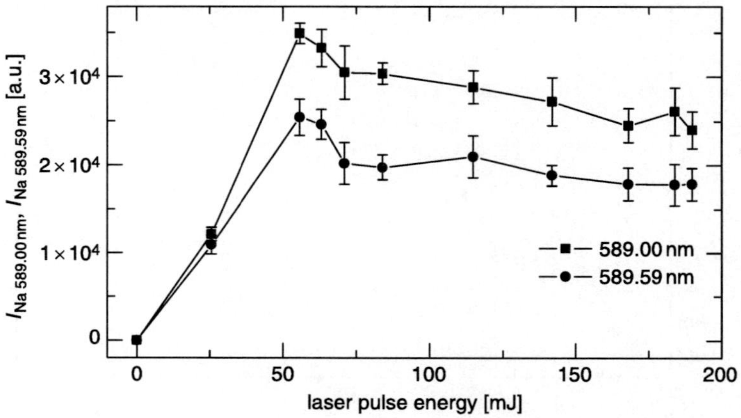

Fig. 14.59 Line intensities $I_{Na\,589.00\,nm}$ and $I_{Na\,589.59\,nm}$ versus laser pulse energy. Sample: droplets containing 500 $\mu g/g$ sodium. Fifty laser pulses were accumulated. A maximum is observed at approximately 60 mJ per pulse

acid. The average integral intensity of 20 measurements, with each measurement consisting of 100 droplets, is plotted as a function of the concentration in Fig. 14.60. Integral intensities were calculated by fitting a Gaussian function to the calcium peak at 393.37 nm and subtracting the background intensity (baseline subtraction). The concentration with the same signal intensity as three times the standard deviation of the blank sample, the limit of detection is calculated to be 20 ng/g [11.3] (cf. Sect. 11.2).

This detection limit is higher than those obtained with ICP-MS systems. Many elements can be detected in the range of low ppt values with modern ICP-MS systems [14.83, 14.84]. Nevertheless, the absolute sensitivity of the experimental setup described here is very high; for elements showing strong emission intensities it is comparable with ICP-MS systems, since 20 ng/g calcium contained in 100 droplets of 280 pl volume correspond to an absolute amount of only 560 fg calcium.

The analytical sensitivity is strongly dependent on the excitation and emission characteristics of the corresponding element. The excitation energy and the Einstein coefficient determine together with the plasma properties the emission intensity of the respective line of any element. Calcium belongs to the group of elements showing strong line emission yielding a low limit of detection. Similar elements are aluminum, boron, cadmium, chromium, copper, magnesium, sodium, and zinc. Here limits of detection below 1 μg/g can be expected. Weak emission intensities were detected for the elements arsenic, cobalt, mercury, lead, platinum, selenium, and tin. Here the corresponding limits of detection are higher, with the setup studied values between 20 and 100 μg/g are expected. Between these two groups, the elements iron, nickel, silicon, titanium, and vanadium can be found with limits of detection

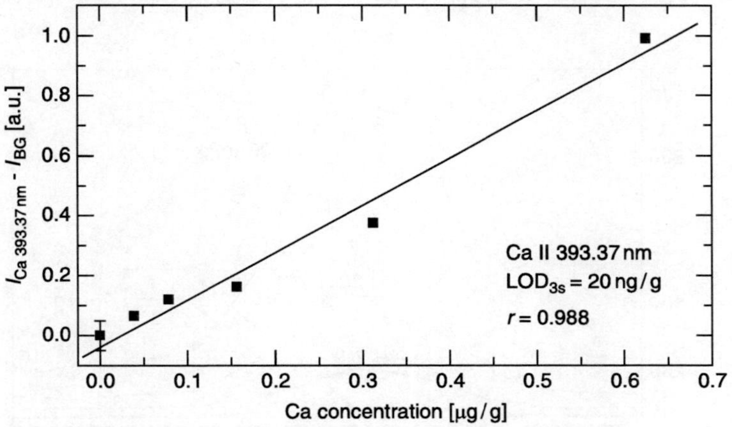

Fig. 14.60 Calibration curve for calcium. Experimental parameters: delay time $t_{delay} = 1.5\,\mu s$, integration gate width $t_{int.} = 50\,\mu s$, laser pulse energy 140 mJ, pulse width 10 ns, 100 droplets contribute to one measurement; 20 measurements were accumulated for each concentration. Limit of detection (LOD) = 20 ng/g, correlation coefficient $r = 0.988$. The *error bar* shown was multiplied by a factor of 5

in the low μg/g region. The elements were classified by comparing signal-to-noise values from echelle spectra recorded under identical conditions.

In the normal mode of operation the Paschen–Runge spectrometer can only detect the intensity of radiation in very narrow spectral windows defined by the resolution of the instrument and the positions of the photomultipliers respectively the exit slits. Scanning over a narrow spectral region (less than one nanometer) is possible by tilting a refractory plate positioned directly behind the entrance slit. The tilting moves the incoming beam and tunes all spectrometer channels to a shifted wavelength. The Paschen–Runge spectrometer is equipped with a stepper motor that allows the precise tilting of this plate. Plotting the photomultiplier intensity against the stepper motor position gives the spectral shape of the respective transition. Figure 14.61 shows the line shape for different concentrations of copper samples between 10 and 1,000 μg/g.

In order to simulate a chromatographic separation, continuous flow measurements were performed by connecting the droplet generator to a HPLC pump that forces a constant liquid flow (typically 5–15 μl/min) through the glass capillary. For the stability of the droplet production, it is important to adapt the volume flow of the droplet generator to the volume flow delivered by the pump. Otherwise the droplet generator may either run dry or droplet formation may fail because too much liquid is delivered to the glass capillary. In order to match the volume flows of the HPLC pump and the droplet generator, an active closed loop controlled the droplet frequency. A liquid flow of 13 μl/min is equivalent to a droplet frequency of 800 Hz (droplet volume 270 pl). Because the laser was operated at 10 Hz, a pulse divider was used to reduce the measurement frequency to 10 Hz. A six-way switching valve (Rheodyne) integrated into the flow system in front of the droplet generator allowed to inject small amounts of sample (20 μl) without interrupting the droplet stream.

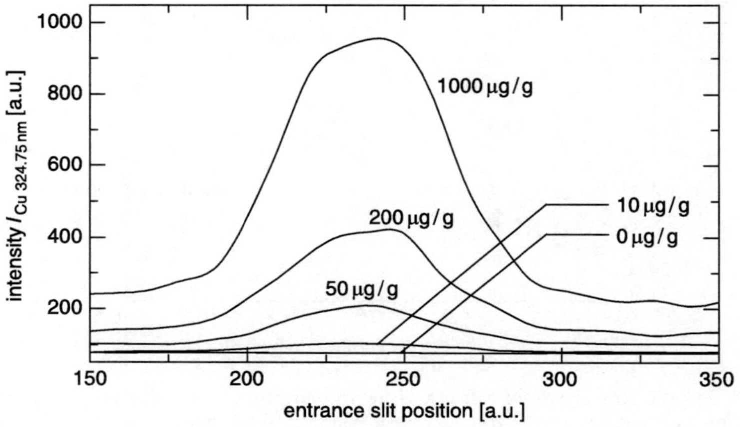

Fig. 14.61 Shape of the Cu I 324.75 nm line, recorded with the Paschen–Runge spectrometer by tilting the refractor plate behind the entrance slit. The concentrations of the different samples are given at the curves

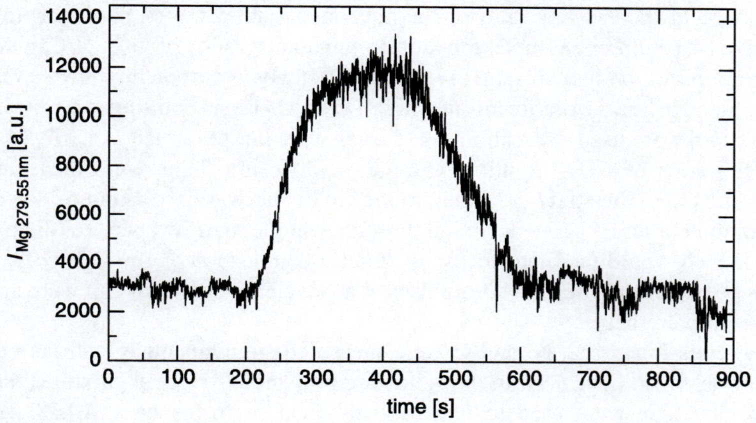

Fig. 14.62 Continuous flow measurement of 20 µl of 18 µg/g magnesium solution. The sample was injected in the continuous solvent flow coming from the HPLC-pump with the aid of a Rheodyne six-way valve. Temporal broadening of the peak can be seen as a result of dead volumes in the system. The droplet frequency was 850 Hz, and the measurement frequency was 10 Hz. The analyzed amount of sample is 4.2 ng. The data are smoothed with a ten point moving average function

Monitoring the signal intensity of the sample element as a function of time delivered a signal peak comparable with signals detected in chromatographic separations. Because of dead volumes in the droplet generator and in the adapter tubes used for connecting the capillary and the droplet generator a temporal peak broadening is detected. Figure 14.62 shows a continuous flow measurement of 20 µl of a sample containing 18 µg/g magnesium. The total amount of analyzed magnesium that gives rise to the signal shown is 4.2 ng because only one out of 85 droplets was analyzed due to the mismatch in droplet and laser frequency.

The results show the potential of the piezoelectric droplet principle for the development of new detectors for ultra low volume flow separation techniques such as micro-HPLC or capillary electrophoresis. Volume flows in the sub µl/min region will be accessible without dilution of the sample.

14.6 Gases

LIBS measurements of gas mixtures such as chlorine in air [14.99], fluorine in air [14.100], and NH_3 vapor in N_2 [14.101] have been described. Laser-induced breakdown in molecular gases initiates the fragmentation of molecules into elemental components, which are identified by detection of their atomic emission lines [14.102]. The detection of elements in gas mixtures is of interest for monitoring of gaseous components in process control, atmospheric pollution, and combustion systems [14.103]. LIBS detection of aerosols as solid particulates in gaseous components will be presented in Sects. 14.7 and 14.8. In combustion, hydrocarbon mixtures and the fuel–air ratio or the equivalence ratio are of particular interest [14.104, 14.105]. For example, the atomic abundance ratios of O/C, N/C, and H/O are measures of the fuel/air ratio [14.106, 14.107]. Hydrocarbon mixtures have been investigated in LIBS experiments in which TEA CO_2 lasers emitting at a wavelength of 10.6 μm were used, but calibration curves were not presented [14.102, 14.106]. Fuel/air ratios of CH_4–air mixtures of a jet diffusion flame were measured by LIBS [14.107]. The study described in the following was carried out to determine calibration curves of LIBS detection through a single, narrow aperture (diameter 7 mm), which would facilitate access to smaller combustion engines [14.108]. Five bursts of two closely spaced laser pulses at a wavelength of 1,064 nm were applied for the measurement.

The emissions of C, N, and O were measured simultaneously with an echelle spectrometer (cf. Sect. 4.2). In the following propane (C_3H_8) gas mixtures that are not combustible were used to find the analytical performance of LIBS without interference from ignition of combustion. However, the calibration curves cover atomic abundance ratios, which are given for the combustion of heptane (C_7H_{16}) and air near an equivalence ratio of $\Phi = 1$. Assuming an ideal homogeneous distribution of fine droplets, heptane was taken as an example with which to estimate the atomic abundance ratios for $\Phi = 1$ because heptane is a primary reference fuel and is often used as an engine model fuel [14.109]. Measurement of the equivalence ratio through a single narrow optical window is of interest, e.g., in automotive engineering for in-cylinder diagnostics before ignition at engine test stations [14.110] as well as in other combustion systems [14.105, 14.107].

Double pulse bursts with an interpulse separation of $\Delta t = 250$ ns are generated at a 10 Hz repetition rate by polarization coupling of orthogonally polarized output beams (beam diameter Ø6 mm) of two Nd:YAG lasers (Continuum Surelite I-10 and I-20). The energy of the double pulse bursts amounts to 230 mJ, with a first pulse of 130 mJ (pulse duration 8 ns) and a second pulse of 100 mJ (8.8 ns). The interpulse separation Δt of the two laser pulses is set by a trigger delay between the two lasers. The lasers are operated with reduced flashlamp power to permit the pulse energy to be adjusted.

The Q-switching Pockels cell of laser I-20 with a 20 Hz flashlamp rate is triggered at each second flashlamp pulse for a 10 Hz laser output. The laser beam is focused into the gas mixture in a sample chamber by a setup that comprises two 140-mm distant fused-silica lenses: lens 1 ($f_1 = 300$ mm; diameter 50 mm; plano–convex) and lens 2 ($f_2 = 50$ mm; diameter 25 mm; biconvex). Lens 2 is positioned in front of the fused-silica entrance window of the sample chamber. The sample chamber is an aluminum vacuum cross (material, AlCuMgPbF38; inner diameter 40 mm). The optical aperture at the entrance window is 7 mm. The spectra are recorded with an echelle spectrometer (Model ESA 3000EV/i, LLA GmbH, Berlin) covering a spectral range from ≈200 to 780 nm with some gaps without detection. A single UV-enhanced fiber of 0.8 mm core diameter and 0.8 m length is positioned beside the incoming laser beam at lens 1. With this configuration, only a small fraction of the emitted plasma light is captured and guided to the echelle spectrometer. This configuration is rather insensitive to misalignment. Nevertheless the optical signal can be increased further by imaging of the plasma light onto the fiber.

The CCD camera of the echelle spectrometer is equipped with a gated microchannel-plate (MCP) image intensifier (cf. Sect. 4.4, Table 4.8, Fig. 4.14). The exposure time is set to 500 ms, yielding an acquisition of five pulse bursts at a repetition rate of 10 Hz. The MCP gate is set to 10 μs per laser pulse at a delay of 0.5 μs to the second laser pulse of the burst. The Balmer H_α line ($\lambda = 656.2$ nm) is within a wavelength gap of 653–658 nm, which cannot be detected with the echelle spectrometer used. Therefore, the H_α line is measured with a monochromator (Jobin Yvon Model HR320; 600 grove/mm grating; slit, 50 μm; photodiode array detector) triggered simultaneously with the echelle spectrometer. An additional fiber bundle (cross section diameter 3 mm; length 3 m) is positioned beside the single fiber of the echelle spectrometer at lens 1 to guide the plasma light to the HR 320 monochromator. The MCP gate of the HR 320 monochromator is 2.8 μs at a delay of 0.3 μs.

Before filling, the sample chamber is evacuated to 3 Pa by a two-stage rotary vane vacuum pump via a fine vacuum adsorption trap with zeolite filling. The gases are carbon dioxide, nitrogen, and propane GA312. For gas mixtures with air, ambient air is used without any purification.

The gas pressure is measured by a test pressure gauge. The gas inlet is set by a metering valve. Dilution series were taken by starting with step $i = 0$ with a single gas component A in the sample chamber at a pressure $p_1 = 100$ kPa. Gas mixtures were set for the measuring points in successive steps numbered i by removing a portion of the gas or the gas mixture to a second pressure, $p_2 = 60$ kPa, and refilling the chamber with the second gas component B to the former gas pressure p_1. The dilution series results in partial pressure p_A for component A of $p_A/p_1 = (p_2/p_1)^i$ and for component B of $p_B/p_1 = 1 - (p_2/p_1)^i$. LIBS measurements are always taken at a total pressure of the gas mixtures of $p_1 = 100$ kPa and a room temperature of ≈293 K.

One spectrum of a laser-induced plasma in a carbon dioxide–air mixture near the spectral lines of C, N, and O included is shown in Fig. 14.63. For evaluation, the spectral peak area is determined by numerical integration at 247.54–248.22 nm for the C I line at 247.86 nm (denoted I_{C247}; excitation energy of the upper level, $E_u = 7.68$ eV), at 741.7–748.3 nm for the N I lines at 742.36, 744.23, and 746.83 nm (denoted I_{N742-8}; $E_u = 11.99$ eV), and 776.3–778.5 nm for the O I triplet at 777.19, 777.42, and 777.54 nm (denoted I_{O777}; $E_u = 10.74$ eV). The mean values and standard deviations of the spectral peak areas (or the spectral peak area ratios) from three spectra acquisitions are calculated and given in the calibration curves, whereas each spectrum results from the accumulation of five laser bursts at the CCD camera of the echelle spectrometer. Interference of other emission lines in a spectral region of at least 5 nm adjacent to these lines was not observed. This absence of interference permits the use of low-resolution spectrometers or spectral filters for detection. The composition of air is assumed to be the same as the standard atmosphere, i.e., 0.78084 N_2, 0.209476 O_2, 0.00934 Ar, and 0.000314 CO_2 as fractional volumes of the major gas species [14.111].

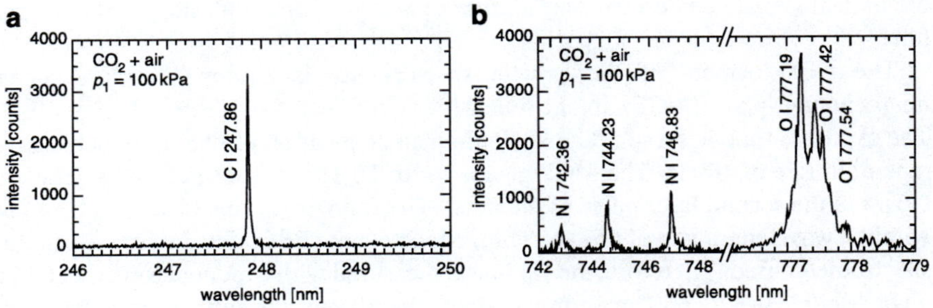

Fig. 14.63 Spectral emission of (**a**) carbon and (**b**) nitrogen and oxygen for partial pressures of 0.2177 kPa CO_2 and 99.782 kPa air, respectively. This mixture results from step $i = 12$ of the dilution series that starts with pure CO_2 to which air is added

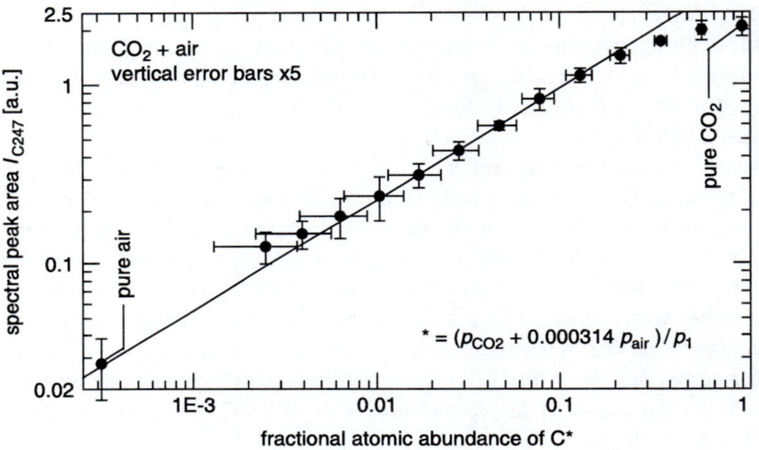

Fig. 14.64 Spectral peak area of the carbon C I 247.9 nm emission line versus the fractional atomic abundance of C. The *horizontal error bars* are estimated from the added relative errors (times step number i) of the measured pressures p_1 and p_2, with an uncertainty of the manometer of 1% of the final value (200 kPa) which means 2% at 100 kPa for each pressure value, taken into account

Figure 14.64 gives the spectral peak area versus the fractional atomic abundances of C that result from a dilution series starting with pure $CO_2 (i = 0)$ and with air added until step $i = 12$. The fractional atomic abundance of C is calculated as noted in Fig. 14.64 by the partial pressures of the CO_2 gas as well as the CO_2 fraction of the air filled in during the dilution series. It is assumed that nearly 100% of the CO_2 molecules are fragmented at the laser pulse energies irradiated. Additionally, pure air is measured after evacuation of the sample chamber.

A function $I \propto x^b$, where x denotes the fractional atomic abundance of C, is fitted to the data points of pure air and in the middle of the curve of Fig. 14.64. For low C abundances of the dilution series, probably systematic errors caused by the stepwise gas pumping and filling actions during a dilution series are the reasons for the deviations. For high C abundances, an overflow of the CCD camera occurs because the amplification of the MCP has been adjusted to produce sufficient sensitivity for low C abundances. Nevertheless, it is shown that the LIBS signal varies over more than three decades, approximately according to $I_{C247} \propto x^{0.62\pm0.01}$. In addition, absorption and plasma conditions may change when the mixture approaches a pure CO_2 mixture. A saturation behavior similar to that of C is observed for the O I 777 nm line, where no saturation of the camera occurs. This saturation leads to an O/C calibration curve versus the atomic abundance ratio of O and C – see Fig. 14.65 – that varies approximately according to $I_{O777}/I_{C247} \propto x^{0.63\pm0.03}$ (x is the atomic abundance ratio of O/C), with smaller deviations at high CO_2 fractions.

From this observation, the presumption is that the absolute atomic abundances or the number densities might also affect the calibration curves in such a way that the spectral peak area will saturate at higher number densities. Such an influence of the absolute atomic abundances would also result in an influence of gas temperature on the calibration curves, which should be considered when the gas temperatures vary.

Dilution series with propane gas mixtures were conducted by starting with pure propane and adding CO_2 until step number $i = 8$ was reached; see Figs. 14.66–14.68. The spectral peak area ratio of O/C in Fig. 14.66 approaches a constant value at high CO_2 fractions that is consistent with the ratio of 0.94 ± 0.03 measured for pure CO_2, cf. Fig. 14.65. The measured values are fitted by $I_{O777}/I_{C247} = ax_{O/C}/(d + x_{O/C})$, where $x_{O/C}$ denotes the partial pressure ratio of O and C, $a = 0.97 \pm 0.02$, and $d = 0.85 \pm 0.07$.

Figures 14.67 and 14.68 show the calibration curves for the hydrogen to carbon (H/C) and the oxygen to hydrogen (O/H) ratios, respectively. For the H_α line, the peak value $I_{pk,H656}$ was taken instead of the peak area.

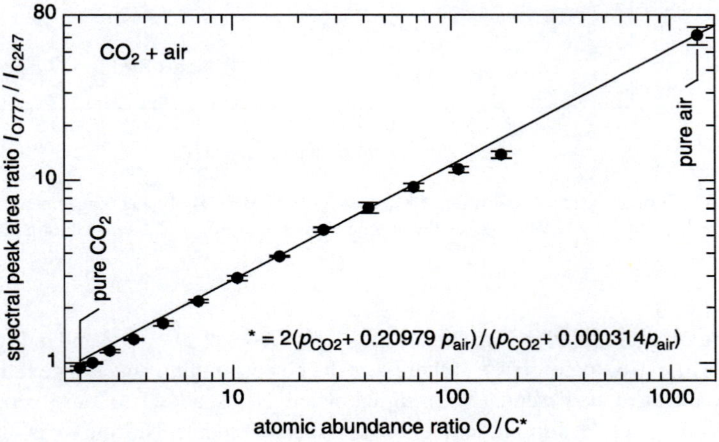

Fig. 14.65 Ratio of the spectral peak area of the O I 777 nm and the C I 247.9 nm emission lines versus the atomic abundance ratio O/C from the same dilution series as in Fig. 14.64

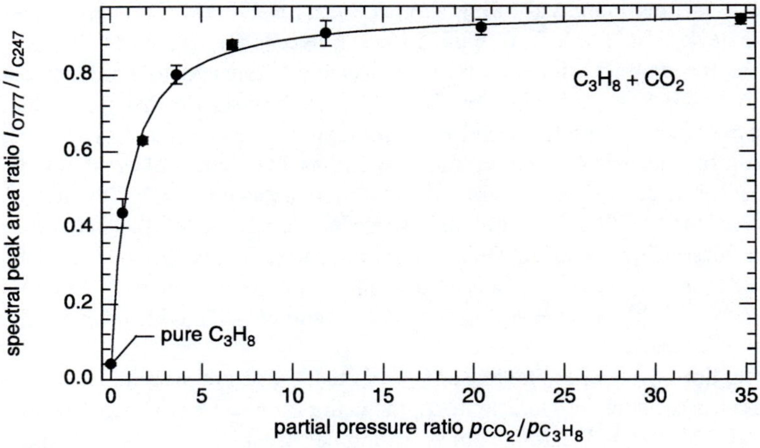

Fig. 14.66 Ratio of the spectral peak area of the O I 777 nm and the C I 247.9 nm emission lines versus the partial pressure ratio of CO_2 to C_3H_8 for a dilution series that starts with pure propane to which CO_2 is added

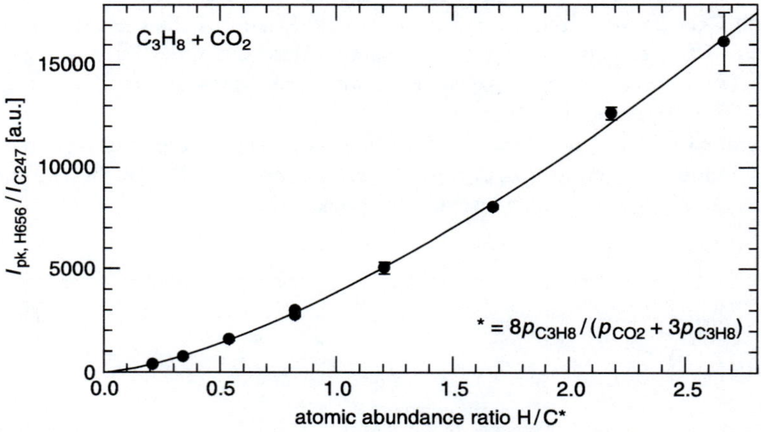

Fig. 14.67 Ratio of spectral peak value $I_{pk,H656}$ of the BALMER H_α line at 656.2 nm and spectral peak area I_{C247} of C I 247.9 nm versus the atomic abundance ratio H/C. Same dilution series as in Fig. 14.66

The curves are fitted in Fig. 14.67 by $I_{pk,\ H656}/I_{C247} \propto x^{1.47\pm0.03}$ and in Fig. 14.68 by $I_{O777}/I_{pk,\ H656} \propto x^{1.29\pm0.03}$, where x is the corresponding abundance ratio.

The nitrogen to carbon (N/C) signal when N_2 is added to pure propane in a dilution series is similar to the former measurement. The curve is fitted by $I_{N742\text{-}8}/I_{C247} \propto (x_{N/C} - x_o)^{1.13\pm0.03}$, where $x_{N/C}$ is the atomic abundance ratio of

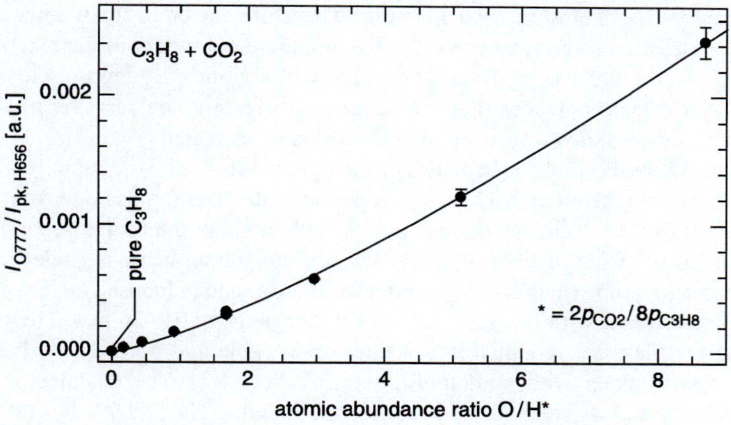

Fig. 14.68 Ratio of the spectral peak area of the O I 777 nm line and the spectral peak value $I_{pk,H656}$ of the Balmer H_α line at 656.2 nm versus the atomic abundance ratio O/H. Same dilution series as in Fig. 14.66

N/C and $x_o = 0.92 \pm 0.4$ is an offset value. The addition of N_2 to hydrocarbon mixtures results in the occurrence of nitrile (CN) emission bands in the LIBS spectra. The CN signal varies with the atomic abundance ratio of N/C and attains a peak for an N/C atomic abundance ratio of about 2.5.

The combustion of 1 kg of heptane (C_7H_{16}, 100.2 g/mol) with 15.2 kg of air (28.96 g/mol) according to the chemical reaction $C_7H_{16} + 11O_2 = 7CO_2 + 8H_2O$ results in an equivalence ratio of $\Phi = 1$, which is the desired value in the ignition region of a combustion engine. Therefore, the atomic abundance of C can be estimated to be $7 \times N_A \times 1$ kg/(0.1002 kg/mol) from propane and $0.000314 \times N_A \times 15.2$ kg/(0.02896 kg/mol) from CO_2 of air ($N_A = 6.02 \times 10^{23}$ mol^{-1}, the Avogadro constant). Similar calculations for the other elements result in atomic abundance ratios of 2.3 for H/C, 1.38 for O/H, 11.7 for N/C, and 3.1 for O/C, which correspond to $\Phi = 1$. The measurements of Figs. 14.67 and 14.68 and the calibration curve for N/C demonstrate that LIBS through a single narrow aperture allows to monitor the abundance ratios in regions covering these values.

14.7 Particulates

Alkalis, zinc and lead, are well known for their negative impact on the blast furnace process. As accompanying elements of the charged burden and coke, they affect its degradation behavior and can lead to higher coke consumption, accretions, and scaffoldings. These elements are partly removed from the blast furnace by the slag and the top gas, while another part forms inner circuits and remains in the blast furnace. The control and removal of alkalis, zinc and lead, can be improved by certain measures in blast furnace operation. Therefore, the build-up of inner circuits has to be detected as early as possible. The inline measurement of the alkalis, zinc and lead, in the top gas of the blast furnace will significantly help in closing the metallurgical mass balances. For the inline multielement analysis, the method of laser-induced breakdown spectroscopy (LIBS) is investigated [14.112].

Figure 14.69 shows the schematics of the optical setup. The output pulses of two Nd:YAG lasers (wavelength 1,064 nm, repetition rate 10 Hz, pulse duration ~8 ns) are superposed to collinear double pulses with a short interpulse separation of $\Delta t = 64$ ns (cf. Chap. 6 and [14.108]). The combined laser beam is guided through the measuring probe (length 2.7 m, diameter 76 mm) and is focused by lens L1 into the top gas stream with top gas velocities in the range of 10–20 m/s. The emitted light of the plasma is coupled into a fiber optic cable and guided to a Paschen–Runge spectrometer with photomultipliers (cf. Sect. 4.2). The channels installed are (element and wavelength of spectral line in nm): Na 588.99, K 766.49, Pb 405.78, Zn 334.50, Zn 213.86, C 247.86, N 746.83, O 777.19, H 656.28, Ca 393.37, Fe 271.44, and zeroth order. The signals of the photomultipliers are integrated by a multichannel integrator electronics (MCI, cf. Sect. 4.5). The closely spaced double pulses of the laser beam are used to increase the spectral signal for top gas pressures of about 3–4 bar [14.113]. The sensitivity of the measuring system was proofed with a reference gas cell for the gaseous components and with solid dust samples for other channels.

The access to the top gas is achieved via an optical measuring probe (MP in Fig. 14.69) inserted through a flange of the top gas tube before the dust separator of the blast furnace, see Fig. 14.70. The top gas tube diameter is approximately 3 m.

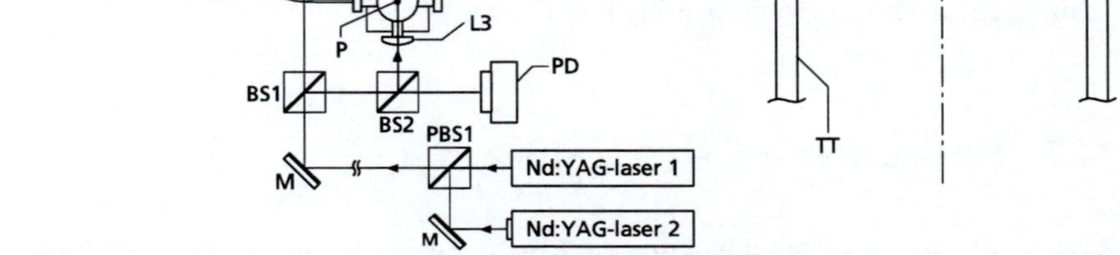

Fig. 14.69 Schematics of the optical setup. M = mirror; PBS1 = beam combiner; BS = beam splitter; PD = photodiode; FO = fiber optics; W1, W2 = optical windows; L1, L2, L3 = lenses; TT = top gas tube; P = laser-induced plasma; MP = measuring probe; LH = lens holder; RC = reference chamber; PR = Paschen–Runge spectrometer; MCI = multichannel integrating electronics; PC = computer

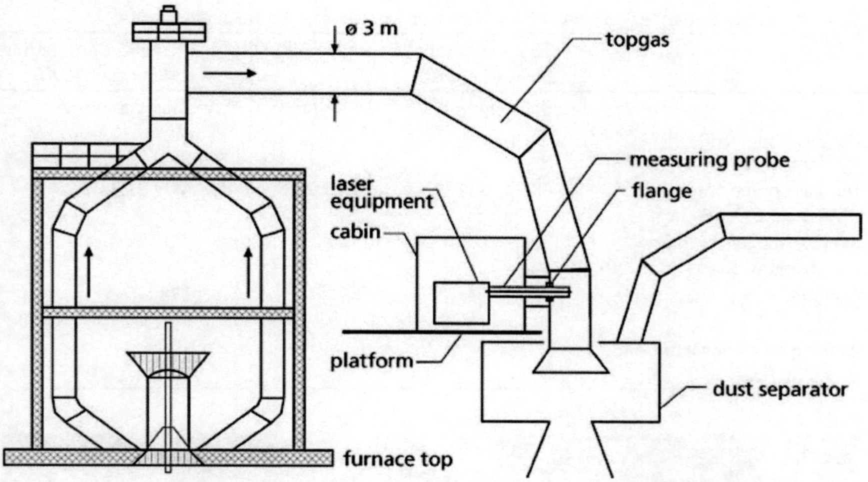

Fig. 14.70 Schematics of the top of the blast furnace with the measurement location. The cabin ($\sim 3 \times 2.4 \times 2.9\,m^3$) with the laser measuring equipment is installed at the top gas tube before the first dust separator

The measurement cabin includes the measuring probe handling, the optical setup with two lasers, the Paschen–Runge spectrometer, a computer and further measurement equipment for gas flow, pressure, and temperature. The CO concentration is monitored in the cabin in order to ensure that it is below the safety limit. The gas tightness of the measuring equipment was verified before the first top gas access for pressures up to 4 bar.

Table 14.15 summarizes key data of the blast furnaces involved in the onsite trials. The measuring cabin with the LIBS instrumentation is fixed at a platform at about 42–46 m height above ground.

The front side of the measuring probe and the measurement cabin installed at the top gas tube are shown in Fig. 14.71.

The measuring equipment was tested in onsite measuring campaigns at each blast furnace. The measuring system was able to monitor simultaneously and in situ the relative concentrations of sodium, potassium, zinc, lead (Na, K, Zn, Pb), as well as carbon, nitrogen, oxygen, hydrogen, calcium (C, N, O, H, Ca), and iron (Fe) in the top gas stream. The average concentration of the circulating materials is in the range of about 1–20 mg/m^3 top gas. For every 10 s a measuring value was taken being the mean value of 100 laser bursts. A quantitative measurement would require calibration by reference values of the top gas composition or a reference top gas stream at 3–4 bar with defined element concentrations which was not available at the time of the study.

The precision of the laser measurement is estimated by the relative standard deviation of the mean line intensity observed in a time interval not influenced by the charging process. The element-specific relative standard deviation was determined

Table 14.15 Summary of key data of the blast furnaces chosen for the onsite trials

	ThyssenKrupp Steel AG (TKS)	Rautaruukki Oyj (RR)	Aceralia Corporation Siderurgia (ACER)
Blast furnace	Schwelgern 2	Masuuni 2	Blast furnace A
Height of platform above ground (m)	46	42	42
Top gas composition N_2:CO:CO_2:H_2 in %	48:24:24:4	49:23:23:5	44.9:22.4:22.2:4.1
Average top gas pressure (absolute) (bar)	3.4	2.4	3.0
Average top gas temperature (°C)	120	180	178
Average dust concentration (g/m³)	8	5	10

Fig. 14.71 *Left*: measuring probe before insertion into the top gas tube at TKS. *Right*: measurement cabin installed at the blast furnace of RR

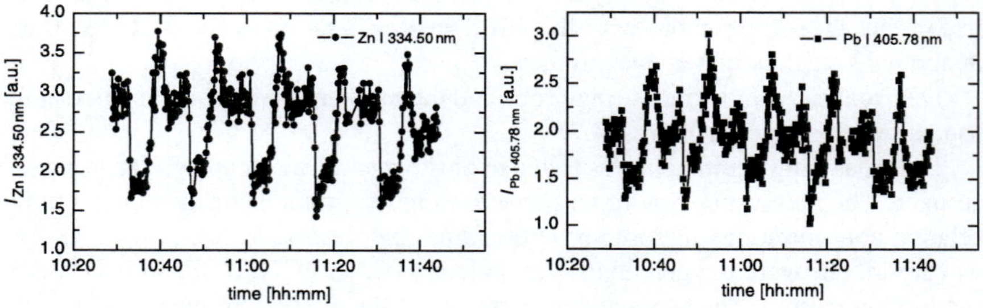

Fig. 14.72 Monitoring of Zn (*left*) and Pb (*right*) at the blast furnace of TKS

for ten consecutive measuring values, i.e., 1,000 laser bursts, and ranges between 2.3% and 7% for the circulating materials. Figure 14.72 gives an example of the time behavior of the spectral signals for Zn and Pb during smooth operation of the blast furnace. The peaks coincide with sinter charging having a period of 14 min.

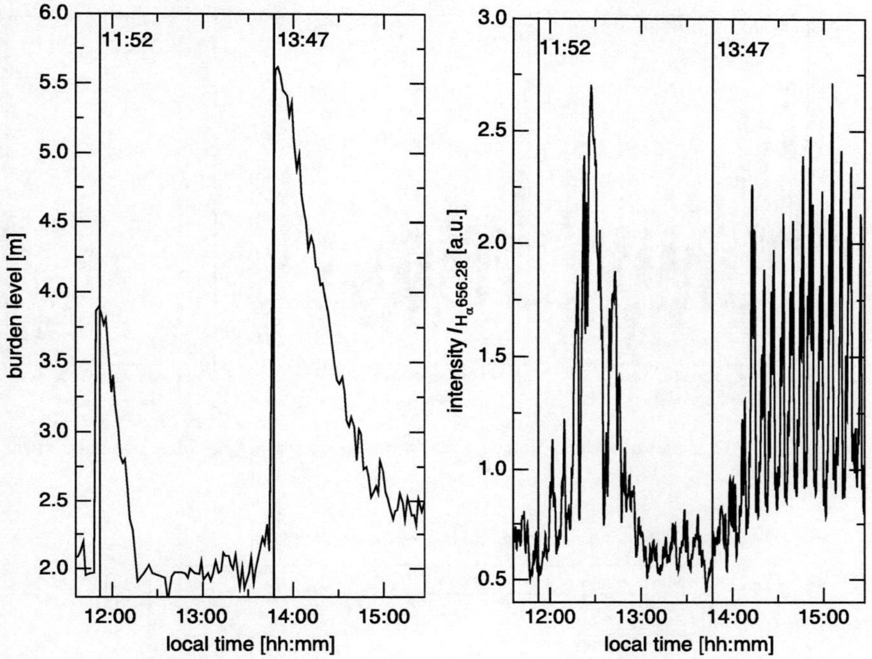

Fig. 14.73 The *left diagram* shows the burden level during several hours. Two slippings of the blast furnace occur. After the slipping, a high output of the dust and the gas amount was monitored as can be seen in the hydrogen signal (*right diagram*) measured with LIBS at the top gas tube of RR

All gained laser data were correlated to data of the blast furnace operation. The monitoring of the dust revealed information about the charging process and the correlation of the element concentrations to operating parameters like hot blast or differential pressure (pressure difference between the top and the hot blast). Moreover, it enabled the observance of differences in the dust discharge evoked by effects like the shutdown or the slipping of a blast furnace.

Figure 14.73 shows such a slipping behavior monitored at the blast furnace of RR: a sudden change in the burden level is followed by a high output of dust and gas which can be monitored by the LIBS signal of hydrogen (right graph of Fig. 14.73).

Figure 14.74 shows the relative concentrations of Na and Zn during operation of the blast furnace with decreasing hot blast which results in a smaller output of dust and a corresponding decrease in the Na and Zn signals.

The periodical changes in the circulating materials result from the charging process of the blast furnace with coke and burden. Sodium and potassium are coming mainly from the coke, while zinc is carried into the blast furnace by the ore and sinter. It was observed that the signals of Na and K as well as the signals of Zn and Pb are running parallel.

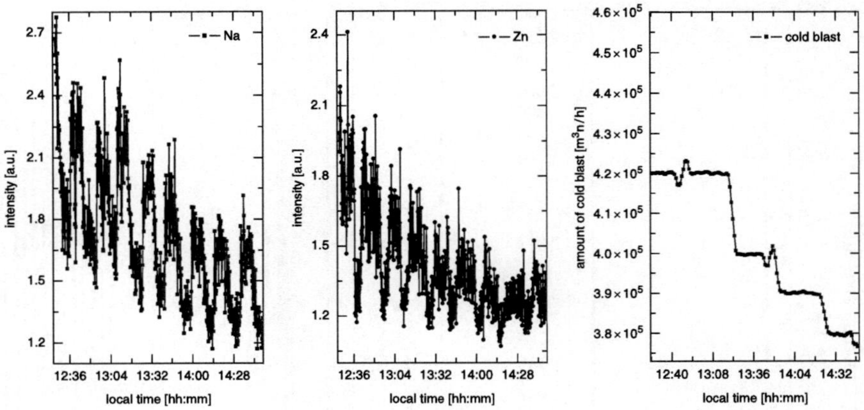

Fig. 14.74 LIBS signal of sodium Na and zinc Zn during decreasing hot blast operation of the blast furnace at TKS

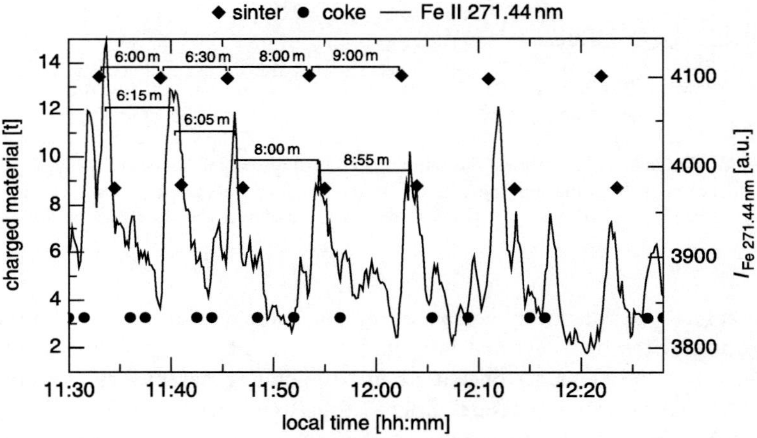

Fig. 14.75 Monitoring of iron at the top gas tube of RR. The time of putting sinter or coke into the blast furnace is marked by *symbols*. The periodical fluctuation of the LIBS intensity of Fe is mainly caused by charging of sinter. The time between two maxima is consistent with the time between two charging maxima of sinter

Comparing the monitored data of the blast furnace at RR with the data taken at the blast furnace of TKS, similarities and differences can be observed. Similar to TKS, the progression of the circulating materials is influenced by the charging process. But different from TKS, the blast furnace at RR is working with open charging with a charging process which runs irregular and much faster than at TKS. Figure 14.75 gives an example of a measurement of Fe. The periodical changes result from charging of sinter which is indicated by the maxima of the iron intensity.

The in situ monitoring of relative concentrations of the elements Na, K, Zn, Pb, as well as C, N, O, H, Ca, and Fe in the top gas has been demonstrated successfully at three blast furnaces. Continuous measuring periods of up to 10 h a day were carried out in the top gas stream of a blast furnace showing the capability of the LIBS instrumentation to measure in a harsh industrial environment. The optical window of the measuring probe is kept free from dust for more than 10 h continuous operation in the top gas stream due to an improved nitrogen flow. The LIBS instrumentation delivers for all ten elements every 10 s a measuring value which is the mean of 100 laser bursts. The standard deviation is between 2.3% and 7% for the circulating materials. Various safety tests of the measuring equipment had been carried out to ensure that the safety concept (valves, stuffing boxes, etc.) was suitable.

14.8 Aerosoles

The risk of fine (0.1–2.5 μm) and ultrafine (≤0.1 μm) particulates to affect the organism and human health depends on particle size [14.114, 14.115]. Ultrafine particulates can enter alveoli and, through them, penetrate the blood vessels and cells of the human body. Since the health risks are governed by the composition of the particles – especially when heavy metals are involved [14.116, 14.117] – there is a need to measure the chemical composition as a function of particulate size. In addition, the size-dependent composition may serve as a signature for the particle generating process and source apportionment. This information allows an identification of particle emitting processes and a focus upon reducing emissions from the most relevant processes of, e.g., large-scale industrial facilities.

14.8.1 Aerosoles Sampled on Substrates

A study is presented where particle sampling methods are combined with laser-induced breakdown spectroscopy [14.118]. While standard analytical methods applied to emission samples are costly and time consuming, the use of LIBS offers the possibility to gain composition information after sample collection much faster. Furthermore, LIBS can be applied on-site.

Panne et al. and Yamamoto et al. investigated particulate matter collected on substrates [14.119, 14.120]. Yamamoto used an electrical low-pressure impactor (ELPI) with Al foils for size classification, while Panne took quartz fiber as a substrate to collect the aerosol without any size classification. In the study described in the following, the laser pulse energies used to excite the plasma are an order of magnitude smaller than those used in the cited works. While Panne measured absolute emission intensities related to the mass of the collected material, the ratio of two intensities related to size classes was investigated here. As a reference for the line ratios Yamamoto added yttrium to the samples, which is not a constituent of the collected particulates. Line emissions from the elements iron and calcium were used as internal reference for those are the dominant constituents of the particulates studied. Furthermore, the possibility of source apportionment was not discussed by Panne and Yamamoto.

Arnold and Cremers analyzed thallium particles smaller than 20 μm on filters [14.121]. They did not classify the particles by size and related the intensity to the mass of the particles collected. To investigate the capabilities of LIBS for this application, the influence of various measurement parameters were studied in a laboratory setup. The elements Ag, Ba, Cd, Co, Cr, Cu, Hg, V, Mg, Mn, Na, Ni, Pb, and Zn were analyzed. Calibration curves were determined for 28 emission line ratios, where each analyte line was referenced to an iron and a calcium line. Finally, samples of particulate originating from steelmaking processes were analyzed based on this set of calibration curves.

For LIBS analysis, the airborne particles are collected using an electrical low-pressure impactor (ELPI, Dekati Ltd.). This system combines a cascade impactor with a charger and an electrometer to measure the particle load during the sampling process. The impactor collects the aerosol particles separately in 13 size fractions according to their different aerodynamic diameters ranging from 40 nm to 13 μm. As shown in Fig. 14.76, the impactor consists of a series of jet and collection plates. Large particles are deposited first, while smaller ones can follow the air stream to the lower stages with the cut-off size depending on the aerodynamic design of each stage. The particulate matter is collected on Al foils with a thin grease coating applied as exchangeable substrates on the collection plates. The conductivity of the substrates is necessary for the electrometer operation, while the grease increases the adhesion of the surface for an efficient collection of the particles. Since the grease layer is very thin, the electrical capacity of the ELPI is not influenced.

Figure 14.77 shows an allocation of mass and particle number, plotted as a function of the aerodynamic diameter of the particles, typical for calibration samples as obtained from electrometer measurements. The aerodynamic diameter is defined as the diameter of a sphere with a standardized density of 1 g/cm^3, which has the same descent rate as the particle itself. The data were taken during the collection of a calibration sample lasting 30 min. The width of the bars indicates the limits at which 50% of the particles with the respective diameter are deposited on a particular stage. On the substrates, the material is deposited not homogeneously, but only at the locations of the jets, as shown in Fig. 14.78. Depending on the design of the jet plates, there are up to several dozen depositions with diameters of $\geq$0.5 mm on a substrate area having a diameter of 20 mm.

The area covered by particles depends on the stage and can be of the order of some percent of the total substrate surface. Furthermore, the diameter of the spots is independent of the composition of the particles used for sampling. For the larger size fractions, a lower number of larger jets depositions is found.

Table 14.16 shows the number of jets as well as the respective jet diameter for each stage of the impactor [14.122]. In order to calibrate the LIBS measurements, a set of calibration samples were generated. Salt solutions were nebulized by an aerosol generator, guided through a diffusion dryer and sampled by the ELPI as described above. The produced aerosol is assumed to represent the salt composition in the solution, which is supported by the calibration measurements (see below).

The salt solutions were prepared from chlorides of the metals stated above. The concentration of these metals was varied in relation to a constant iron and calcium concentration which was included as reference material in all samples. Since a large range of concentrations is expected in the industrial samples, the concentrations for the calibration samples are varied according to a logarithmic scale covering three to

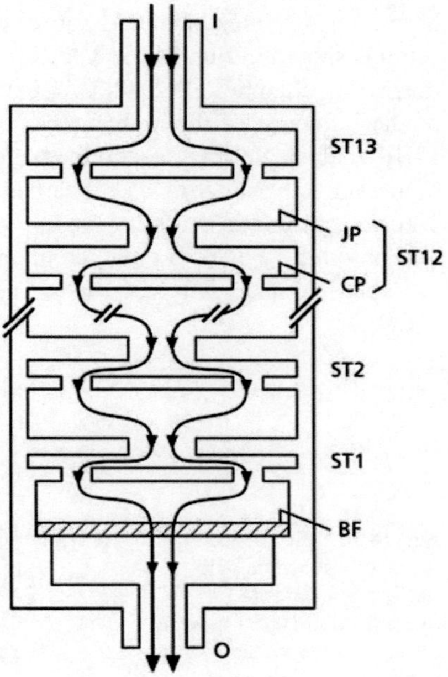

Fig. 14.76 Cascade impactor. ST1–ST13 = stages, JP = jet plate, CP = collection plate, BF = backup filter, I = inlet, O = outlet

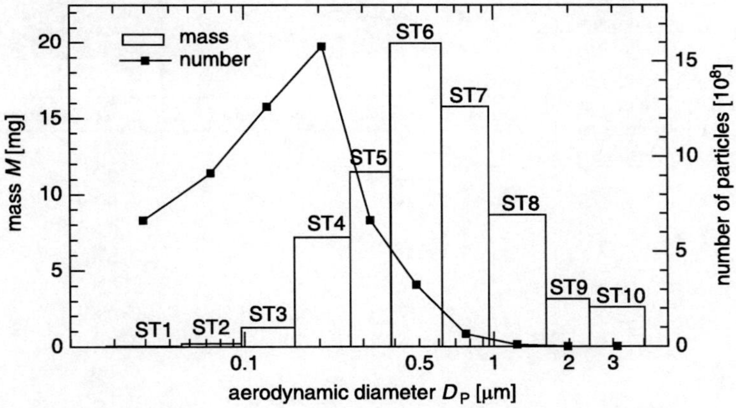

Fig. 14.77 Typical allocation of mass and particle number for the used impactor plotted versus the aerodynamic diameter, ST1–ST10 = stages of the impactor, cf. Fig. 14.76

five orders of magnitude. Thus, the molar mixing ratio for calibration ranges from 10^{-5} to 0.1.

An experimental LIBS setup was developed to analyze particulate matter on substrates. Due to the characteristics of the samples, laser pulse energy as well as the pressure of the ambient gas has to be adapted to achieve an efficient excitation of a thin layer of deposited particulates into the plasma state. An ablation of the substrate material has to be kept as small as possible to reduce interference by high intensity lines of the Al substrate. A schematic drawing of the experimental setup is shown in Fig. 14.79. The measurements were performed in a vacuum-tight measuring chamber filled with air, helium, or argon. This chamber allows the study of the influence of the ambient gas type and pressure on the performance of the LIBS analysis. In the measurement chamber, the sample is positioned on a sample holder and the excitation laser beam is focused onto the sample surface from the top with an achromatic lens system ($f = 60$ mm) as shown in more detail in Fig. 14.80.

The sample holder is mounted on xy-translation stages within the chamber. A Q-switched Nd:YLF slab laser operating at 1,047 μm with repetition rates of up to

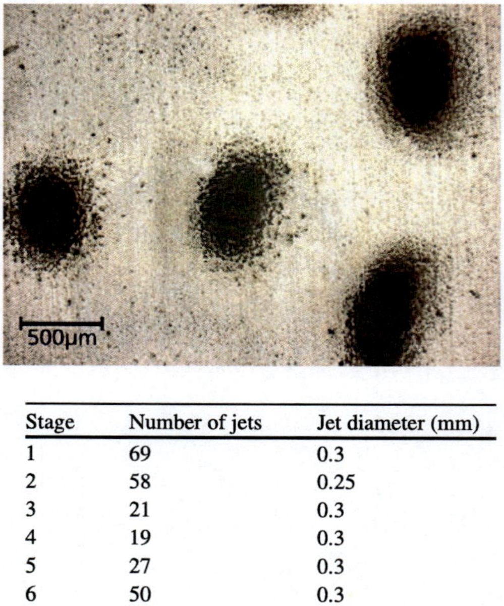

Fig. 14.78 Microscope image of a substrate with $FeCl_3$ particulates (ST7, magnification 100×) showing the areas of deposited particulates (*dark spots*) corresponding to the geometry of the respective jet plate

Table 14.16 The number of jets and the respective jet diameter for each stage

Stage	Number of jets	Jet diameter (mm)
1	69	0.3
2	58	0.25
3	21	0.3
4	19	0.3
5	27	0.3
6	50	0.3
7	48	0.4
8	20	0.7
9	17	1.0
10	14	1.4
11	3	3.2
12	1	6.3
13	1	8.3

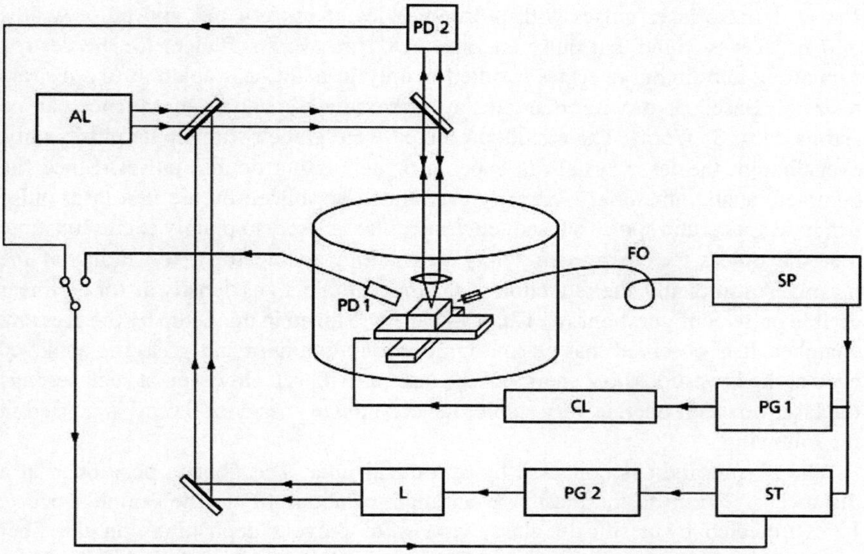

Fig. 14.79 Experimental setup. AL = alignment laser; CL = stage controller; L = laser; FO = fiber optic; SP = spectrometer; PG 1, 2 = pulse generator; PD 1, 2 = photodiodes; ST = threshold switch

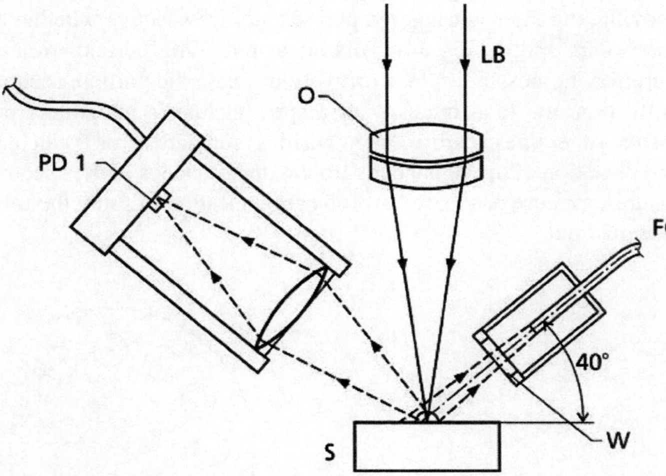

Fig. 14.80 Optical arrangement. FO = fiber optic, LB = laser beam, O = focusing objective, PD = photodiode, S = sample, W = protection window

1 kHz delivers laser pulses with pulse energies of up to 4 mJ, and pulse widths of 7 ns. It was found that pulse energies of 0.6 mJ were sufficient for the desired excitation, and higher energies resulted mainly in additional ablation of substrate material. Based on a crater diameter of approximately 30 μm, the fluence can be estimated to 85 J/cm^2. The results presented were gained with single pulses only, even though the laser system is capable of generating double pulses. Since the complete analyte material is already evaporated or ablated by the first laser pulse irradiating a certain spot, a subsequent laser pulse is likely to mainly excite substrate material unless its energy can solely be made available for a re-heating of the plasma. Additionally, the reduction of the local ambient gas density as for collinear double pulse configurations (cf. Chap. 8) can be realized in this setup by the pressure chamber. It is observed that plasma ignition on aluminum and grease is achieved even at the lowest possible energy of the laser at 0.03 mJ. However, at such settings the laser does not operate very stable; hence, a pulse energy of 0.6 mJ was used in the following.

The plasma light is collected by an optical fiber. The fiber is positioned at a distance of 20 mm to the plasma at an angle of about 40° to the sample surface. It is protected by a silicate glass window to prevent depositions on the fiber entrance face by ablated material. The received light is guided to an echelle spectrometer (ESA3000, LLA Instruments) covering a wavelength range from 200 to 895 nm with a spectral resolution of 5 to 20 pm. For this study 100 spectra, each taken at a different location, were accumulated following a laser track as illustrated in Fig. 14.81. Thereby, the tracks were chosen to be half circles covering the highest possible percentage of areas with particulate depositions. While the axes are moving, the laser is triggered periodically irrespective whether a particle accumulation (dark spot in Fig. 14.78) is hit or not. With the experimental setup chosen, there was no possibility to selectively irradiate the particle accumulations. For the evaluation, the total intensity of a spectral line is integrated, preventing influences from other lines nearby. The intensities' standard error is calculated from the standard deviation of the signal background on both sides of the spectral line. In a first step, uniform pure or greased Al foils are measured to study the influence of the substrate material.

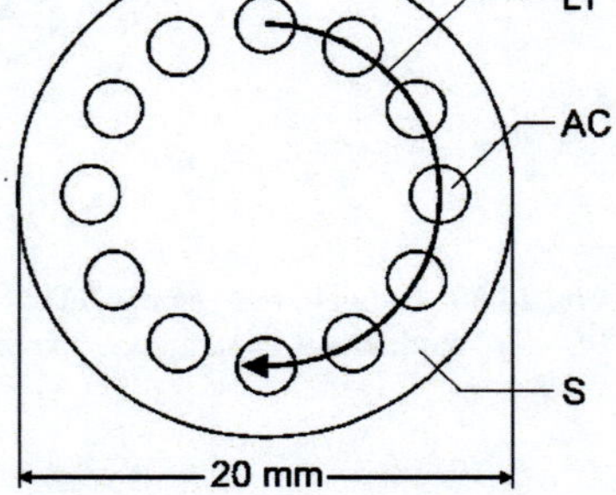

Fig. 14.81 Schematic view of a measurement track on a substrate, LT = laser track, AC = areas with particles deposited (*dark spots* in Fig. 14.78), S = sample foil

The influence of various ambient gases under different pressures on the line intensity was investigated. The importance of the ambient gas has been demonstrated in previous publications (cf. Sect. 7.3) [14.123]. The measurement chamber was evacuated and then refilled with laboratory air under defined pressures from 1 to 1,000 mbar. The intensity of the aluminum line at 257.509 nm as a function of the pressure is shown in Fig. 14.82. At a pressure of 5 mbar, the line intensity reaches a maximum, which is approximately a factor of five greater than at 1,000 mbar. For helium, the maximum line intensity is achieved at 10 mbar with an intensity three times higher than for air, whereas argon shows a maximum at 600 mbar with an intensity two times higher than the maximum for air. Tests performed with particulates on the substrate showed that the use of air or helium at low pressures had a negative effect on the line intensities of elements comprised in the particulate matter. Obviously, the low pressure facilitates a detaching of the particles in the vaporizing phase, thus dispersing the particles in a larger volume, so that fewer particles are enclosed in the plasma itself. Therefore, argon at 600 mbar was used for the further measurements.

Figure 14.83 shows the line intensities of Al 257.509 nm and C 247.856 nm as a function of the delay between the laser pulse and the start of the signal integration t_{delay} measured on greased Al foils. The integration time t_{int} was kept constant to 50 ns. With rising t_{delay}, the signal from aluminum decreases monotonically. On the other hand, the carbon signal is constant for delays up to about 300 ns. To keep the substrate signal (aluminum) low while simultaneously maintaining the particle signal (simulated by the grease coating) as high as possible, the delay time of $t_{delay} = 180$ ns was chosen for calibration and sample measurements. Additionally, it was observed that the signal-to-noise ratio shows a maximum for a delay of a

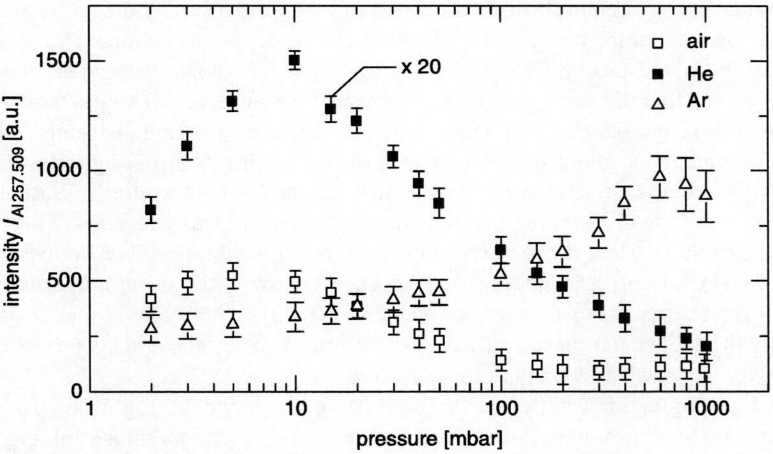

Fig. 14.82 Intensity of the aluminum Al 257.509 nm line as a function of the gas pressure for the gas types air, helium, argon for a pure, clean aluminum foil. For better visibility, all *error bars* are magnified by a factor of 20

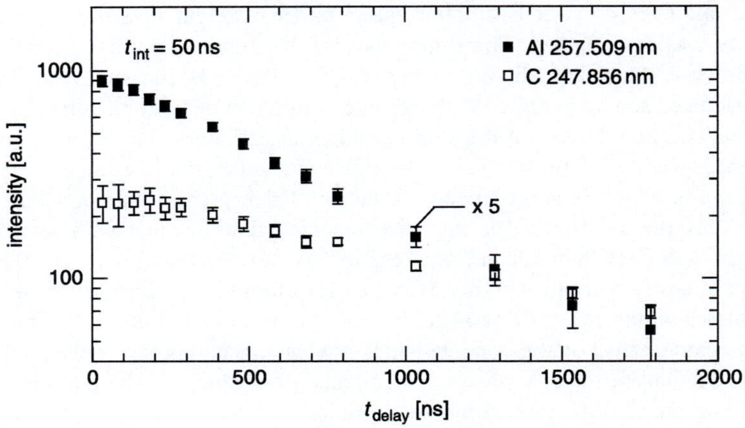

Fig. 14.83 Line intensity of Al 257.509 nm and C 247.856 nm as a function of the delay time t_{delay}. For better visibility, all error bars are magnified by a factor of 5

few hundred nanoseconds depending on the chosen line. This is due to the decrease of the continuum background shortly after plasma ignition. For the calibration and sample measurements, the integration time t_{int} of the spectrometer was set to 3 μs. No significant improvement of signal intensity was observed for longer integration times.

With the chosen parameters the prepared calibration samples were measured using concentric circles as shown in Fig. 14.81. One hundred pulses were accumulated to enhance the signal-to-noise ratio. The intensity ratio of an analyte line to a reference line was plotted versus the molar mixing ratio given by the preparation of the calibration samples. Figure 14.84 shows as an example the calibration curves for the line ratio of Zn 213.857 nm to Ca 422.673 nm for the stages three to eight. It can be observed that the curve progression is nearly the same for all stages. Obviously there is no significant dependence between particle size and the slope of the calibration curves, although the relative error for small molar mixing ratios can be up to 25%. This supports the assumption that the particle composition represents the salt composition of the solutions from which the aerosol was generated. Therefore, it is possible to use a mean curve for all stages for calibration (see dotted line in Fig. 14.84). Due to its independence on stage numbers, the mean curve was also taken for calibration of the stages other than ST3 to ST8. Potential matrix effects within the calibration curves caused by the substrate material could not be observed, since only one substrate material was studied.

An example for a mean calibration curve is shown in Fig. 14.85 for lead referenced to an iron line. The intensity ratio increases with the molar mixing ratio following a power law with an exponent of 0.7. Calibration curves were gained for all elements of interest – Ag, Ba, Cd, Co, Cr, Cu, Hg, V, Mg, Mn, Na, Ni, Pb, and Zn – referenced to iron or calcium lines.

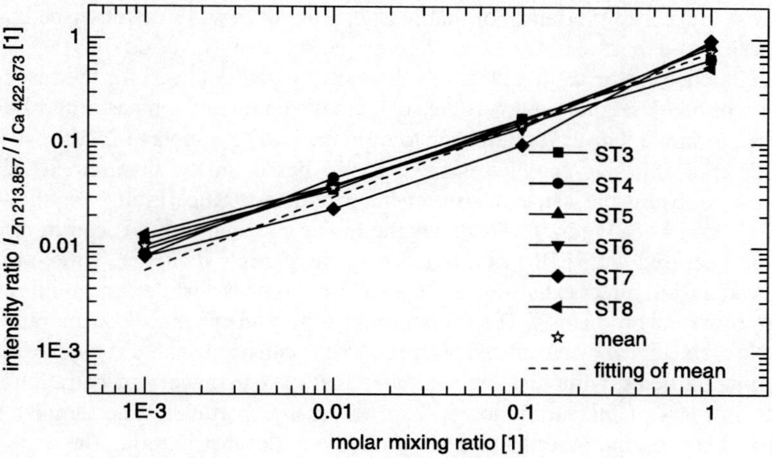

Fig. 14.84 Calibration curves of zinc and stages 3–8 showing monotonously increasing relations

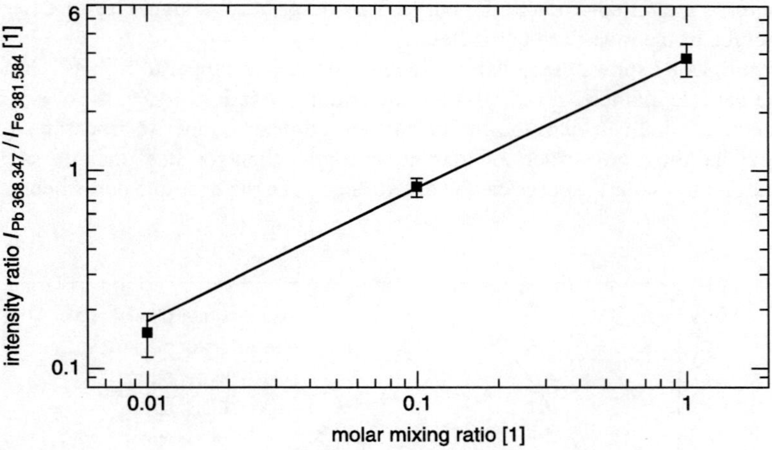

Fig. 14.85 Calibration curve of lead referenced to iron as a function of the molar mixing ratio

The limit of detection (LOD) is the lowest quantity of an element that can be distinguished from the absence of that substance. It is estimated from the mean of the blank samples (samples without the respective element, cf. Sect. 11.2), the standard deviation of the blank samples, and a confidence factor depending on the amount of blank samples. The LOD is usually estimated by the $3s$−criterion [11.3]. A linear relation between intensity ratio and molar mixing ratio is a precondition for this approach. In the study presented here, only a small set of molar mixing ratios could be realized which was not sufficient to fit a linear curve to the data points at low concentrations. For that reason, a fit curve was used described by the expression $y = \alpha\, x^d$ and the LOD is determined with $x_{\mathrm{LOD}} = \Phi(s_b/\alpha)^{1/d}$, where the confidence factor $\Phi = 2.9$ equals the factor given in [11.3] for the respective amount of blank samples and s_b is the standard deviation of the measuring values of the blank sample. For lead referenced to iron, the LOD amounts to 2.7%.

Industrial emission samples were analyzed based on the determined calibration curves using the same measurement parameters. The results are shown in Figs. 14.86–14.88. Figure 14.86 shows the molar concentration of lead in relation to iron measured with LIBS as a function of the particle diameter. Three samples were taken from direct exhausted air of a sintering process, while two samples were taken from steel production. The sinter samples show an enhanced concentration for particle diameters between 50 and 500 nm. A peak can be observed at approximately 200 nm. Particles with diameters greater than 1 μm have concentrations that are two orders of magnitude lower than for smaller particles. The samples from steel making, on the other hand, have a relatively flat distribution. The maximum concentration is attained for particles at about 500 nm, but the concentration level is three orders of magnitude lower than the concentrations determined for the sintering process. The size-dependent curves are characteristic for the process where the particles are emitted. Hence, the curves can be used as a signature to determine the source of the measured particulates.

Figure 14.87 shows the molar concentration ratio of copper to iron as a function of the particle diameter revealing a similar result like in Fig. 14.86, as for a particle diameter of about 0.2 μm the molar concentration of copper to iron for sintered samples is about two orders of magnitude higher than for steel-making samples. Therefore, the sample source can be determined as in the example described before.

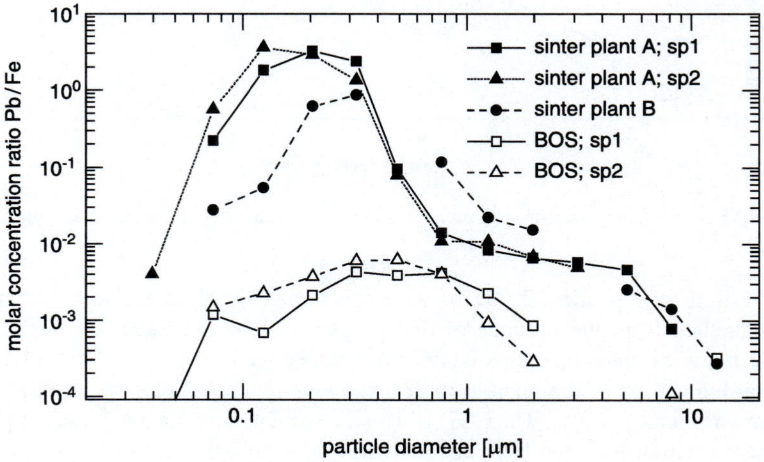

Fig. 14.86 Results of the LIBS analysis of collected samples showing the molar concentration ratio of lead to iron as a function of the particle diameter. sp = sample

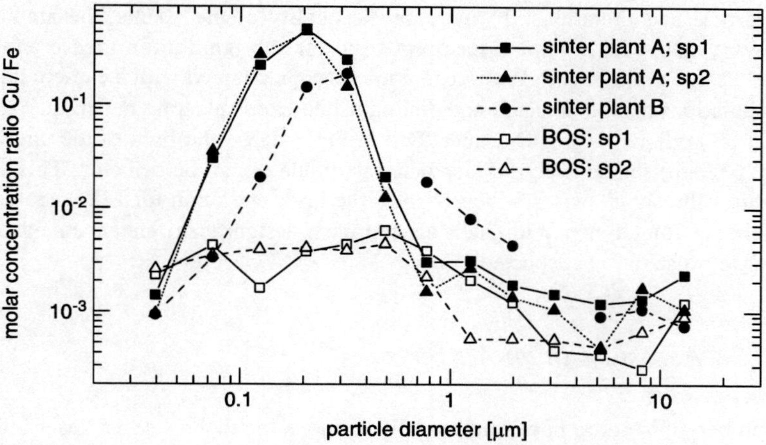

Fig. 14.87 Results of the LIBS analysis of collected samples showing the molar concentration ratio of copper to iron as a function of the particle diameter

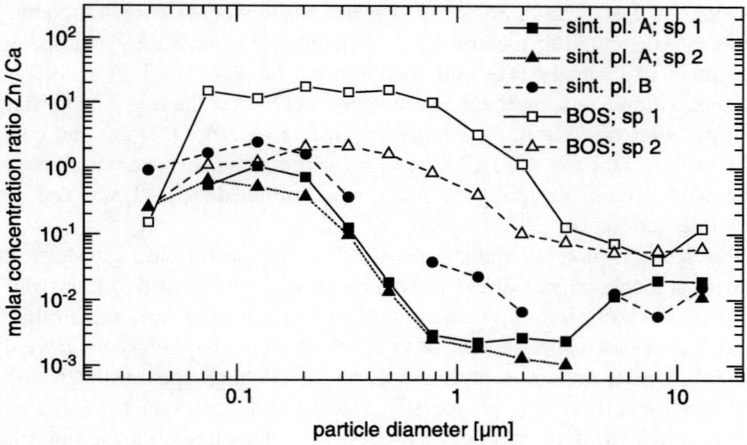

Fig. 14.88 Results of the LIBS analysis of collected samples showing the molar concentration ratio of zinc to calcium as a function of the particle diameter

Figure 14.88 shows the molar concentration ratio of zinc to calcium as a function of the particle diameter. For particle diameters between 0.3 and 5 μm, the molar concentration ratio for samples from steel making is on average one to two orders of magnitude higher than for sintered samples. Thus, the sample source can also be distinguished with a high level of significance.

In the experiments described, the majority of laser-induced plasmas were generated on the substrate only, because it was not possible to selectively irradiate the particle accumulations. To increase the quality of the results, the amount of irrelevant data – i.e., the data resulting from plasma ignition on grease and substrate – can be minimized. With, e.g., a microscope coupled with a camera particle accumulations can be detected and distinguished from substrate material. With the use of photodiodes – cf. PD 1 and PD 2 in Fig. 14.79 – the light of the alignment laser (632 nm) scattered by the deposited particulates can be detected. This signal triggers a threshold switch which enables the laser operation for LIBS analysis of particle accumulations. With such an approach a significant enhancement of the signal-to-noise ratio is expected.

14.8.2 Aerosoles in an Air Stream

A number of studies have successfully demonstrated the use of laser-induced breakdown spectroscopy (LIBS) for the analysis of aerosol particles [14.118, 14.119, 14.124–14.132].

While standard analytical methods like mass-spectrometry, e.g., ICP-MS, are time consuming, LIBS is a fast method that can be automated and applied inline and onsite. As described in Sect. 14.8.1 some of the works concern the analysis of particles sampled on filter substrates fractionated in size classes by an impactor, and a subsequent LIBS analysis is undertaken in the lab [14.118, 14.119, 14.128]. For inline applications, this approach is not adequate; hence, there is a need to analyze aerosol particles directly in an air stream. In this regard, D. Hahn and coworkers have investigated the use of LIBS for aerosol particle analysis in a number of studies dealing with aerosol sampling statistics, conditional data analysis, and plasma–particle interaction [14.129–14.132].

In this section, fundamental studies are described to develop a system for size-dependent aerosol analysis directly in an air stream with regard to industrial applications [14.133]. For inline process control, there is a need for a fast multielement analysis with a fast response-time of the system. In the following the development of a measurement system comprising a Paschen–Runge spectrometer with rapid multichannel electronics is presented enabling fast reactions to changes in the chemical composition of particulate matter. The focus lies on basic investigations of the system behavior, which are necessary for its calibration. Compared with the analysis of solid samples, the calibration is more complex in this case since it can be influenced by both the element concentrations in the particles and the total mass of particulate matter in the aerosol stream.

Additionally, the measurement process is influenced by the size of the individual particles, the carrier gas, and the particle location within the plasma [14.134–14.136]. In the following investigations focused on the effect of the particle mass concentration and the relative elemental concentration are presented. The temporal plasma evolution of the analyte response of particle matter and gas phase species is compared. Electron density and temperature are derived as a function of time. The system behavior is investigated for the analysis of monodisperse and polydisperse aerosols under different measurement conditions.

Artificial aerosol is nebulized by an aerosol generator (AGK-2000, Palas, Germany) into droplets of up to 2 μm diameter using N_2. The aerosol is then guided via a diffusion dryer by using of N_2 as a carrier gas into a measurement chamber to provide a stream of solid particles ranging between 20 and 800 nm in particle diameter. The generator can produce aerosol of different elements by nebulizing different salt solutions. In the following the analysis is restricted to the elements calcium and sodium. It can be assumed that the composition of the dried particles equals the composition of the dissolved material.

To provide size-resolved measurements, a commercially available SMPS (Scanning Mobility Particle Sizer, TSI, USA) is used to fractionate particles by size. The SMPS system comprises a DMA (differential mobility analyzer) model 3080L and a CPC (condensation particle counter) model 3785. The DMA separates particles in a gas stream in a size range from 10 to 1,000 nm aerodynamic diameter. Due to their electrical mobility, which is inversely proportional to their size, particles are fractionated by size, depending on the electrostatic field strength of the DMA and the sheath gas flow rate, and counted using the CPC. By scanning the electrostatic field strength of the DMA and simultaneously counting the particles, an aerosol size distribution is obtained. An example of a particle size distribution given in number and mass concentration is shown in Fig. 14.89. The particle size was scanned from 19 to 800 nm in 27 logarithmic steps. The line intensity for calcium measured with LIBS is also given.

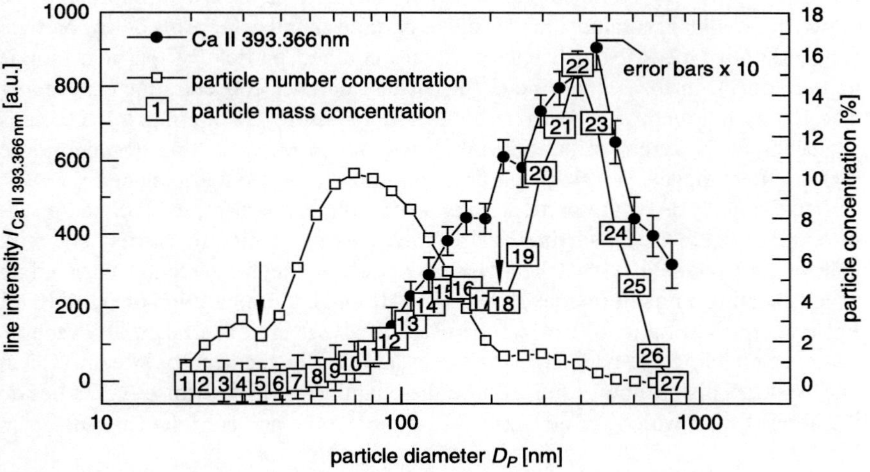

Fig. 14.89 Particle size distribution given in number- and mass distribution. The size distribution is scaled in percentage of the total particle number and the total mass, respectively (*right ordinate*). Additionally the obtained size-dependent LIBS-signal for Ca is plotted and scaled to the *left ordinate*. The measurement points of the mass concentration are marked with *numbers in ascending order*

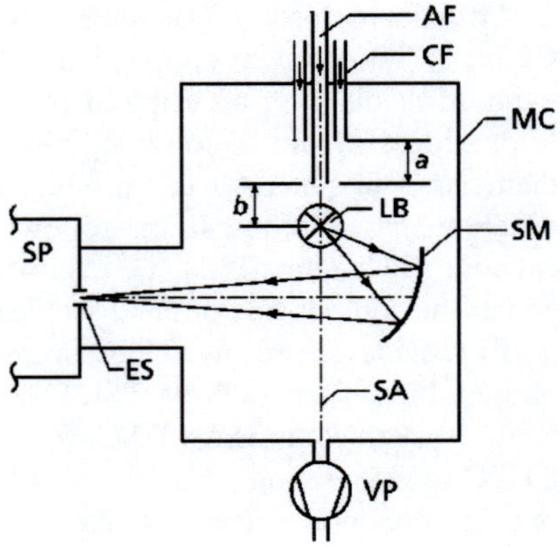

Fig. 14.90 Setup for the LIBS experiments. SP = spectrometer, ES = entrance slit, AF = aerosol flow, CF = co-flow, MC = measuring chamber, LB = laser beam propagating perpendicularly to the plane of projection, SM = spherical mirror, SA = symmetry axis, VP = vacuum pump, a = 50 mm, and b = 10 mm

The outlet of the DMA is connected via a conductive silicon tube of 2 m length and 5 mm diameter to the measurement chamber of the LIBS setup, cf. Fig. 14.90. The gas pressure in the measuring chamber is set to 980 mbar to generate a continuous aerosol flow of 1 l pm through the DMA for all experiments.

The number distribution shows a maximum at 70 nm particle diameter and the mass distribution at about 400 nm. The total number and mass of particles as a sum of all size classes were 4.5×10^6 particles/cm^3 and 10 mg/m^3, respectively, in this measurement. The dips occurring at 34 and 200 nm particle diameter (marked with arrows in Fig. 14.89) in the number distribution and in the mass distribution are due to two different measurement modes of the particle counter. Aerosols of low particle concentration ($<2 \times 10^4$ particles/cm^3) are counted particle by particle (single particle only). In the second mode for particle number concentrations exceeding 2×10^4 particles/cm^3, the particle number is measured by photometry. The dips appearing in the size distributions are indications of the switching between these measurement modes, i.e., single particle counting and photometric mode.

For the analysis of the particulate matter in the gas stream, a LIBS setup was developed which is shown in Fig. 14.90. All experiments are carried out with a flashlamp pumped Q-switched Nd:YAG laser at the fundamental wavelength λ = 1,064 nm, a repetition rate of ν_{rep} = 10 Hz and a pulse duration of τ_{L} = 10 ns. Pulse energies of up to 300 mJ can be used. Analyses are undertaken in a vacuum chamber enabling the use of a co-flow in addition to the aerosol stream (CF in Fig. 14.90). This co-flow is guided into the chamber via two diffuser inlets beside the aerosol inlet which are arranged 50 mm above the aerosol inlet (distance a in Fig. 14.90).

The expanded laser beam with a diameter of 18 mm is focused into the flow stream perpendicular to the aerosol flow at a distance b = 10 mm downstream of the aerosol exit aperture (2 mm diameter) using an achromatic lens having a focal length of f = 110 mm. Under these conditions, breakdown is initiated by each laser pulse. The generated plasma light emission is collected by a spherical mirror and focused directly onto the entrance slit of a spectrometer in Paschen–Runge configuration (cf. Sect. 4.2) in which channel electron multipliers are installed for each spectral line of interest. For multichannel measurements, the fast multichannel integrator electronics (MCI) is used (cf. Sect. 4.5). To detect spectral line profiles and the whole emission spectrum, an echelle spectrometer (cf. Sect. 4.2) is used in addition. In this case, the plasma light emission is collected by a quartz lens into an optical fiber.

To investigate the temporal resolution of the LIBS setup, aerosol generated from sodium and calcium chloride solutions was injected into the measuring chamber. Therefore, the pressure in the measuring chamber was held constant to 980 mbar to provide a continuous gas flow of 1 l pm as described above. NaCl and $CaCl_2$ particles were injected into the gas flow for nearly 10 s in succession. This measurement was carried out without and with the use of an additional argon co-flow. While using the argon co-flow with a flow rate of 5 l pm the pressure in the measuring chamber was still 980 mbar. The time in which the complete gas volume in the chamber is exchanged is estimated to be 8 min without the co-flow. Through the use of the co-flow with a flow rate of 5 l pm, the gas volume is exchanged within approximately 1.3 min.

The plasma emission was observed using the echelle spectrometer with a delay time of 800 ns and an integration time gate of 30 μs. Two pulses were averaged which indicates a temporal resolution of 0.2 s. Figure 14.91 shows the time-dependent line intensity for the two elements Na and Ca.

Without co-flow the time in which the signal rises between 10% and 90% of the maximum signal amounts to 3.5 s (Fig. 14.91, top). In the case of an argon co-flow this value can be reduced to 2.5 s. The $1/e$ decay time of the signal is in the order of 20 s without the co-flow, and the duration until the signal can no longer be distinguished from the background is longer than 90 s. In contrast to that, the use of a co-flow reduces this time to approximately 20 s: see Fig. 14.91, bottom. The time when the signal has reached $1/e$ of its maximum amounts to 10 s.

Using no additional co-flow, the nonaveraged line intensity of the calcium line is very noisy whereas with the Ar co-flow this noise level is reduced.

The fast response time is very important for industrial applications to be able to react to changes in the elemental composition. A fast decay time of the system is also important for the temporal resolution and the size resolution concerning SMPS measurements. Assuming a SMPS scanning time of 135 s, which is used here, approximately seven size classes can be swept consecutively.

The emission response to varying concentrations of Na within the salt solution was investigated. Figure 14.92 shows the obtained calibration curve. The laser energy was 110 mJ in this case, and the plasma emission was observed using the echelle spectrometer. Spectra were recorded using a delay time of 800 ns and an integration time gate of 20 μs. A series of 100 (10 × 10 pulses) spectra were recorded and averaged. Error bars are equal to the standard deviation of the ten-pulse averages. The fit curve of the double-logarithmic plot shows a linear behavior with a slope of 0.9 up to a concentration of Na of 0.2 m.-%. Beyond this concentration, the slope of the curve decreases to 0.5.

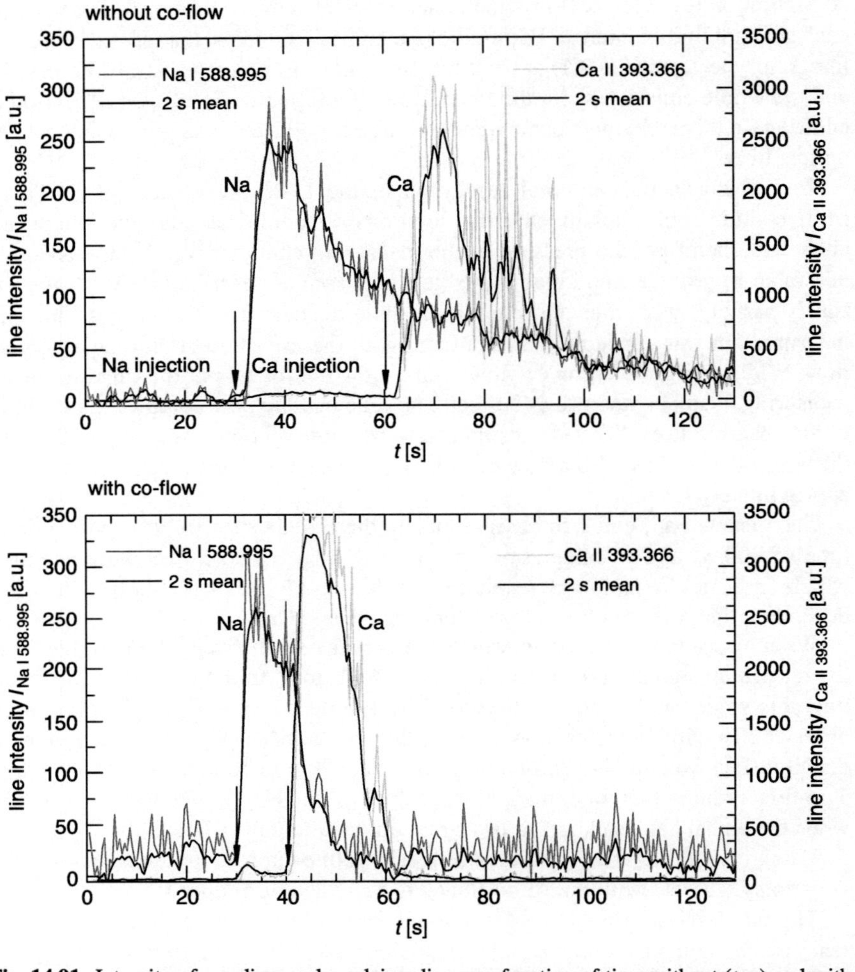

Fig. 14.91 Intensity of a sodium and a calcium line as a function of time without (*top*) and with (*bottom*) the use of an additional co-flow into the measurement chamber. Additionally the moving intensity average over 2 s is plotted. The *arrows* mark the time at which NaCl and $CaCl_2$ particles were injected into the measurement chamber via a tube of 2 m using N_2 as a carrier gas

There are two possible explanations for this behavior. On the one hand, it can be explained by the concept of the curve of growth (COG). According to Gornushkin et al., who applied the COG method to laser-induced breakdown spectroscopy, nonlinearity in the calibration function of solid samples for concentrations in the range between 0.1 and 1 m.-% is observed depending on the delay time and the laser energy used [14.137]. The double logarithmic plot of the theoretical curve of growth has an asymptote with slope 1 for concentrations up to 0.1 m.-%, and

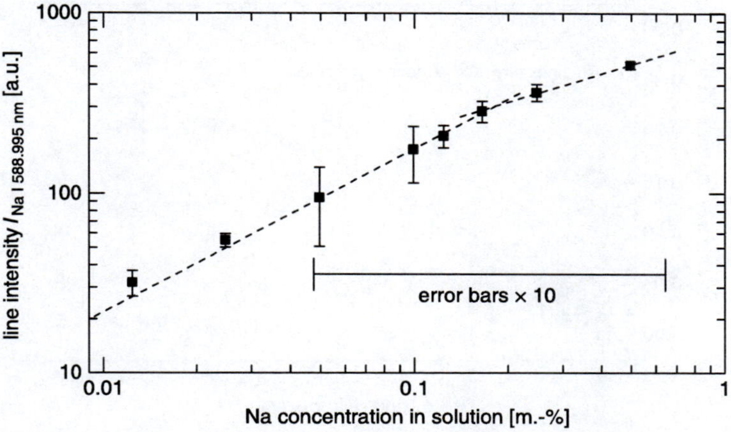

Fig. 14.92 Double-logarithmic calibration curve for particles of all size classes obtained from sodium solution using N_2 as carrier gas without co-flow. *Error bars* are multiplied by a factor of 10 for concentrations ≥0.05 m.-%

another with slope 0.5 for higher values. The intersection between these asymptotes marks the point of transition between low and high optical density of the plasma. Because the conditions differ from those used here no quantitative comparison can be made. Gornushkin et al. studied the curves of growth of Cr in a steel matrix for solid samples.

The emission of the two lines Na lines, i.e., Na I 589.592 nm and Na I 588.995 nm was tested for self-absorption. Saturation effects can be observed above a concentration of 0.1 m.-% supported by the line intensity ratio of these two lines, but the accuracy of the data is not high enough for quantification. For the Na I 589.592 line, which is expected to show a weaker self-absorption, a slope of 0.7 was estimated for concentrations ≥0.2 m.-%.

On the other hand, the observed behavior in Fig. 14.92 could be due to growing particle size with increasing concentration of the solution. The increase of the concentration results in the growth of the size of solid particles remaining when the liquid droplets dry out, which leads to a rise in the total mass concentration of the aerosol. Thus, on the one hand, the reduced increase in the signal above a concentration of 0.2 m.-% can be due to incomplete dissociation of larger particles or, on the other hand, due to stronger losses of the larger particles during the transport to the measuring chamber.

However, both effects, i.e., plasma properties and particle size effects, may play a role here. To investigate the influence of the argon co-flow on the LIBS signal, the signal response to a variation of sodium and calcium with and without the use of the co-flow was measured. In this experiment, solutions of sodium and calcium chloride with an elemental mass concentration of 1 m.-% each were generated. For the LIBS measurements, the solutions were mixed such that the content of

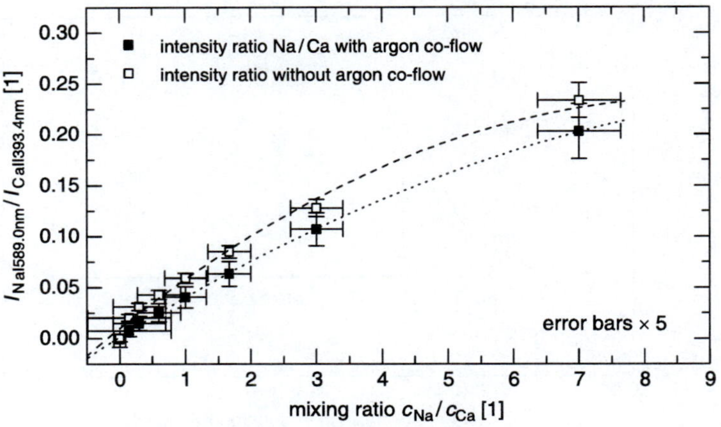

Fig. 14.93 Line intensity ratio as a function of the mass concentration ratio of sodium and calcium within the solution. *Error bars* are multiplied by a factor of 5

Na was successively increased and an aerosol stream was generated as described before. Spectra were recorded with the echelle spectrometer integrated with a delay time of 800 ns after the laser pulse and a gate width of about 30 μs. The laser energy was 200 mJ and a series of 200 (10 × 20 pulses) spectra were averaged. The measurements were undertaken with and without an additional argon co-flow. In Fig. 14.93, the obtained line intensity ratio of Na and Ca for both measurements is plotted versus the mixing ratio of Na and Ca cations of the solution. Figure 14.93 shows that the variation of the mixing ratio is well represented by the LIBS signal. The elemental mixing ratio can be clearly derived from the ratio of the measured line intensities. The measurement with an argon co-flow reduces the ratio of the line intensities by an offset. This can be explained by different plasma conditions and excitation conditions for the lines. Because of the enhanced plasma confinement for the argon co-flow, the degree of ionization is expected to increase, which causes a reduced Na I to Ca II signal ratio. By approximating a polynomial function to the data points, the standard deviation of procedure was estimated to be 0.16 for the measurement without argon according to [11.4]. In case of the argon co-flow, this value is reduced by a factor of 5 down to 0.032.

Size-resolved LIBS analysis of calcium particles were undertaken for different measurement conditions and compared with the aerosol size distribution and the particle mass concentration for each size class.

To provide a size-resolved analysis of particle composition, particles generated from $CaCl_2$solution as described above were fractionated by size using the DMA. The $CaCl_2$ solution had a calcium elemental mass concentration of 0.23 m.-%. The particle size was scanned from 19 up to 800 nm in 27 logarithmic steps, and the fractionated particles were subsequently analyzed by LIBS for each size class. The characteristic of the obtained emission line intensity for Ca II at 393.366 nm

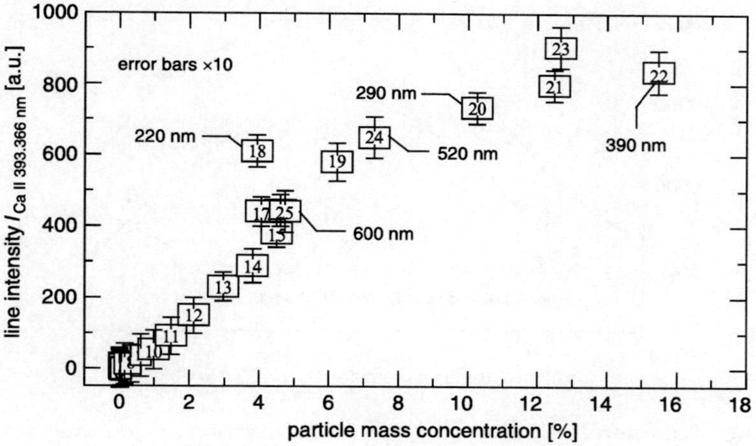

Fig. 14.94 Line intensity of Ca II 393.366 nm as a function of size-dependent particle mass concentration. Particle diameters are represented by *numbers in ascending order* as in Fig. 14.89. All *error bars* are multiplied by a factor of 10

with increasing particle size is shown in Fig. 14.89 together with the particle size distribution. The plasma emission was observed using the echelle spectrometer as described above and the laser pulse energy was set to 200 mJ.

Figure 14.89 allows a comparison of the line intensity of Ca with the particle size distribution. The Ca line emission intensity follows the particle mass distribution and attains a maximum in the particle diameter range between 350 and 450 nm as the mass distribution does.

In Fig. 14.94, the calcium line intensity is plotted as a function of the particle mass-concentration. Here, a correlation between the particle mass concentration and the obtained LIBS signal can clearly be seen. For small particles, the line intensity increases nearly linearly with mass up to a particle diameter of approximately 140 nm (no. 15) and a corresponding mass concentration of 4.5%.

From a particle mass concentration of 6% upwards, corresponding to a particle diameter of 220 nm (no. 18) the signal rises more slowly. The offset of measurement point no.18 is due to the two different measurement modes of the CPC (cf. discussion of Fig. 14.89). A possible reason for the slower increase may be the influence of the particle size in the laser-induced plasma. Laser-particle interaction of larger particles is different from that of smaller particles. In this regard, one aspect could be a different location within the plasma [14.134]. Furthermore, the interpretation given for the Na calibration curve (Fig. 14.92) may also be applied here, i.e., the particle size.

Figure 14.95 shows the line intensity of calcium Ca II 396.847 nm as a function of particle diameter with and without the use of a co-flow. The LIBS signal is analyzed by the Paschen–Runge spectrometer with a delay time of 2 μs and an integration

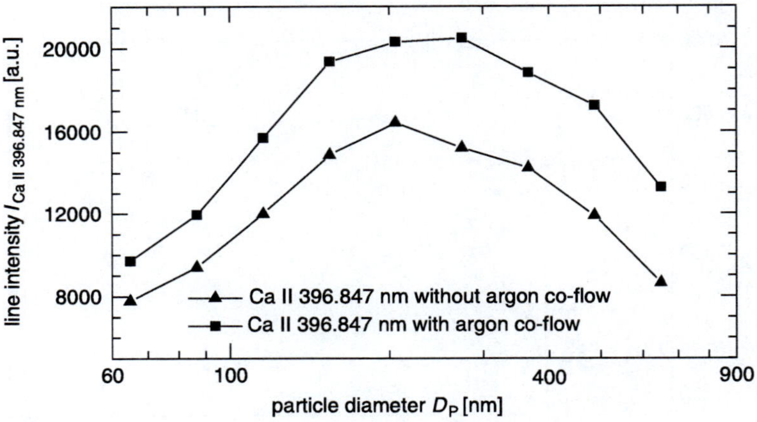

Fig. 14.95 Size-resolved LIBS signal of Ca particles with and without the use of an argon co-flow, measured with the Paschen–Runge spectrometer

time gate of 25 μs. The laser energy was 265 mJ in this case and 1,000 consecutive single spectra were accumulated. By the use of the Ar co-flow, the LIBS signal is enhanced by a factor of about 1.5. The use of the Ar co-flow does not influence the characteristic of the size-dependent curve significantly.

References

14.1. J. Brandrup, M. Bittner, W. Michaeli, G. Menges (eds.), *Die Wiederverwertung von Kunststoffen* (Carl Hanser Verlag, München, 1995)
14.2. C. Delgado, L. Barruetabeña, O. Salas, in *Assessment of the Environmental Advantages and Drawbacks of Existing and Emerging Polymers Recovery Processes*, ed. by O. Wolf, EC Joint Research Center, 22939 EN, 2007
14.3. H. Lucht, U. Plauschin, H. Dürr, Kunststoffe mit Infrarot-Messung sortenrein trennen. Umwelt **23**, 443–444 (1993)
14.4. D. Scott, A two-colour near-infrared sensor for sorting recycled plastic waste. Meas. Sci. Technol. **6**, 156–159 (1995)
14.5. M. Alam, S. Stanton, Sorting of waste plastics using near-infrared spectroscopy and neural networks. Process Contr. Qual. **4**, 245–252 (1993)
14.6. R. Feldhoff, D. Wienke, K. Cammann, H. Fuchs, On-line post consumer package identification by NIR spectroscopy combined with a fuzzy ARTMAP classifier in an industrial environment. Appl. Spectrosc. **51**, 362–368 (1997)
14.7. N. Eisenreich, T. Rhode, Identifizieren von Kunststoffen. Kunststoffe **86**, 222–224 (1996)
14.8. N. Basta, K. Fouhy, K. Gilges, A. Shanley, S. Hshio, Chem. Eng. **11**, 37–43 (1990)
14.9. see e.g., products of the following companies: LLA Instruments GmbH, Justus-von-Liebig-Str. 9, 12489 Berlin, www.lla.de; National Recovery Technologies Inc., 566 Mainstream Dr., Suite 300, Nashville, TN, http://www.nrtsorters.com, 2010
14.10. http://www.dradio.de/dlf/sendungen/firment/1020957, 2009/8/21

14.11. J. Jansen, C. van Hastenberg, Identifikation von Kunststoffteilen. Kunststoffe **84**, (1994) 51–54

14.12. R. Sattmann, I. Mönch, H. Krause, R. Noll, S. Couris, A. Hatziapostolou, A. Mavromanolakis, C. Fotakis, E. Larrauri, R. Miguel, Laser-induced breakdown spectroscopy for polymer identification. Appl. Spectr. **52**, 456–461 (1998)

14.13. R. Sattmann, V. Sturm, R. Noll, Laser-induced breakdown spectroscopy of steel samples using multiple Q-switch Nd:YAG laser pulses. J. Phys. D Appl. Phys. **28**, 2181–2187 (1995)

14.14. A. Zell, *Simulation Neuronaler Netze* (Addison Wesley, Berlin, 1994)

14.15. R. Pearse, A. Gaydon, *The Identification of Molecular Spectra* (Chapman and Hall, London, 1976), p. 83

14.16. H. Griem, *Plasma Spectroscopy* (McGraw-Hill, New York, 1964), p. 305, 538

14.17. R. Hecht-Nielsen, *Neurocomputing* (Addison Wesley, Berlin, 1990)

14.18. *Ullmann's Encyclopedia of Industrial Chemistry*, vol. A 20 (VCH Publishers, Weinheim, 1992)

14.19. European Project, Development of multipurpose industrial units for recycling of plastic wastes by on-line pattern recognition of polymer features (Sure-Plast), contract no. BRPR-CT98–0783, project no. BE97–4890, 2001, samples prepared by Gaiker, Bilbao, Spain

14.20. M. Stepputat, R. Noll, On-line detection of heavy metals and brominated flame retardants in technical polymers with laser-induced breakdown spectrometry. Appl. Optics **42**, 6210–6220 (2003)

14.21. J. Marshall, J. Carrol, J.S. Crighton, C.L.R. Barnard, Industrial analysis: metals, chemicals and advanced materials. J. Anal. At. Spectrom. **9**, 319–356 (1994)

14.22. R. Zenobi, Modern laser mass spectrometry, Fresen. J. Anal. Chem. **348**, 506–509 (1994)

14.23. C. Lazik, R.K. Marcus, Electrical and optical characteristics of a radio frequency glow discharge atomic emission source with dielectric sample atomization. Spectrochim. Acta B **48**, 1673–1689 (1993)

14.24. Z. Weiss, New method of calibration for glow discharge optical emission spectrometry. J. Anal. At. Spectrom. **9**, 351–354 (1994)

14.25. A. Golloch, D. Siegmund, Sliding spark spectroscopy – rapid survey analysis of flame retardants and other additives in polymers. Fresen. J. Anal. Chem. **358**, 804–811 (1997)

14.26. J. Anzano, I. Gornushkin, B. Smith, J. Winefordner, Laser-induced plasma spectroscopy for plastic identification. Polym. Eng. Sci. **40**, 2423–2429 (2000)

14.27. M. Tran, Q. Sun, B. Smith, J. Winefordner, Determination of F, Cl and Br in solid organic compounds by laser-induced plasma spectroscopy. Appl. Spectrosc. **55**, 739–744 (2001)

14.28. H. Fink, U. Panne, R. Niessner, Analysis of recycled thermoplasts from consumer electronics by laser-induced plasma spectroscopy. Anal. Chim. Acta **440**, 17–25 (2001)

14.29. H. Fink, U. Panne, R. Niessner, Process analysis of recycled thermoplasts from consumer electronics by laser-induced plasma spectroscopy. Anal. Chem. **74**, 4334–4342 (2002)

14.30. NIST Atomic spectra data base, http://physics.nist.gov/cgi-bin/AtData/main_asd

14.31. M. Kraushaar, R. Noll, H.-U. Schmitz, Slag analysis with laser-induced breakdown spectrometry. Appl. Spectrosc. **57**, 1282–1287 (2003)

14.32. R. Wisbrun, I. Schechter, R. Niessner, H. Schroder, K. Kompa, Detector for trace elemental analysis of solid environmental samples by laser plasma spectroscopy. Anal. Chem. **66**, 2964–2975 (1994)

14.33. L. St-Onge, M. Sabsabi, P. Cielo, Analysis of solids using laser-induced plasma spectroscopy in double-pulse mode. Spectrochim. Acta B **53**, 407–415 (1998)

14.34. F. Colao, V. Lazic, R. Fantoni, S. Pershin, A comparison of single and double pulse laser-induced breakdown spectroscopy of aluminium samples. Spectrochim. Acta B **57**, 1167–1179 (2002)

14.35. National Standard of Germany, *DIN 53 804–1, Statistische Auswertung, Teil 1: Kontinuierliche Merkmale* (Beuth, Berlin, April 2002)

14.36. A. Dobney, A. Mank, K. Grobecker, P. Conneely, C. de Koster, Laser ablation inductively coupled plasma mass spectrometry as a tool for studying heterogeneity within polymers. Anal. Chim. Acta **423**, 9–19 (2000)
14.37. R. Kaiser, W. Gottschalk, *Elementare Tests zur Beurteilung von Messdaten* (Bibliographisches Institut, Mannheim, 1984)
14.38. National Standard of Germany, *DIN 32 645, Chemische Analytik – Nachweis-, Erfassungs- und Bestimmungsgrenze* (Beuth Verlag, Berlin, September 2006)
14.39. A. Cremers, D. Romero, An evaluation of factors affecting the analysis of metals using laser-induced breakdown spectroscopy (LIBS), in *Remote Sensing*, ed. by R.T. Menzies, Proc. SPIE **644**, 7–12 (1986)
14.40. H. Bennet, G. Oliver, *XRF Analysis of Ceramics, Minerals and Allied Materials* (Wiley, New York, 1992)
14.41. J. Vadillo, J. Laserna, Laser-induced breakdown spectroscopy of silicate, vanadate and sulfide rocks. Talanta **43**, 1149–1154 (1996)
14.42. H. Wiggenhauser, D. Schaurich, G. Wilsch, LIBS for non-destructive testing of element distributions on surfaces. NDT&E Int. **31**, 307–313 (1998)
14.43. A. Knight, N. Scherbarth, D. Cremers, M. Ferris, Characterisation of laser-induced breakdown spectroscopy (LIBS) for application to space exploration. Appl. Spectrosc. **54**, 331–340 (2000)
14.44. F. Hilbk-Kortenbruck, R. Noll, P. Wintjens, H. Falk, C. Becker, Analysis of heavy metals in soils using laser-induced breakdown spectrometry combined with laser-induced fluorescence. Spectrochim. Acta B **56**, 933–945 (2001)
14.45. G. Doujak, R. Mertens, W. Ramb, J. Flock, J. Geyer, S. Lüngen, Slag analysis by laser-induced breakdown spectroscopy. Stahl Eisen **121**, 53–58 (2001)
14.46. H. Martens, T. Næs, *Multivariate Calibration* (Wiley, New York, 1998)
14.47. B. Bergmann, N. Bannenberg, Schlackenführung und Schlackenoptimierung in der Sekundärmetallurgie. Stahl Eisen **111**, 125–131 (1991)
14.48. H. Lachmund, Y. Xie, High purity steels: a challenge to improved steelmaking processes. Ironmak. Steelmak. **30**, 125–129 (2003)
14.49. V. Sturm, H.-U. Schmitz, T. Reuter, R. Fleige, R. Noll, Fast vacuum slag analysis in a steel works by laser-induced breakdown spectroscopy. Spectrochim. Acta B **63**, 1167–1170 (2008)
14.50. C. López-Moreno, S. Palanco, J.J. Laserna, Quantitative analysis of samples at high temperature with remote laser-induced breakdown spectrometry using a room-temperature calibration plot. Spectrochim. Acta B **60**, 1034–1039 (2005)
14.51. V. Sturm, Optical micro-lens array for laser plasma generation in spectrochemical analysis. J. Anal. At. Spectrom. **22**, 1495–1500 (2007)
14.52. K. Esbensen, *Multivariate Data Analysis*, 5th edn. (Camo Process, Oslo, 2002), p. 159
14.53. R. Wisbrun, I. Schechter, R. Niessner, H. Schröder, K.L. Kompa, Detector for trace elemental analysis of solid environmental samples by laser plasma spectroscopy. Anal. Chem. **66**, 2964–2975 (1994)
14.54. D. Cremers, J. Barefield II, A. Koskelo, Remote elemental analysis by laser- induced breakdown spectroscopy using a fiber-optic cable. Appl. Spectrosc. **49**, 857–860 (1995)
14.55. F. Hilbk-Kortenbruck, R. Noll, P. Wintjens, H. Falk, C. Becker, Spectrochemical analysis of heavy metals in soils with laser-based techniques, Abstracts of the lecture groups "Environmental Technology and Fundamentals of Laser-aided In-situ Soil Analysis", *Int. Meeting on Chem. Eng., Environmental Protection and Biotechnology, ACHEMA 2000, Conf. Proc. 2000*, DECHEMA e.V., Frankfurt a. M., 449–451 (2000)
14.56. R. Multari, L. Foster, D. Cremers, M. Ferris, Effect of sampling geometry on elemental emissions in laser-induced breakdown spectroscopy. Appl. Spectrosc. **50**, 1483–1499 (1996)
14.57. G. Theriault, S. Bodensteiner, S. Lieberman, A real-time fiber-optic LIBS probe for the in situ delineation of metals in soils. Field Anal. Chem. Technol. **2**, 117–125 (1998)

14.58. B. Miles, J. Cortes, Subsurface heavy-metal detection with the use of a laser-induced breakdown spectroscopy (LIBS) penetrometer system. Field Anal. Chem. Technol. **2**, 75–87 (1998)
14.59. Mitteilungen der Länderarbeitsgemeinschaft Abfall (LAGA) Nr. 20: *Anforderungen an die stoffliche Verwertung von mineralischen Reststoffen/Abfällen – Technische Regeln* (Erich Schmidt Verlag, Berlin, 1995)
14.60. K. Yamamoto, D. Cremers, M. Ferris, L. Foster, Detection of metals in the environment using a portable laser-induced breakdown spectroscopy instrument. Appl. Spectrosc. **50**, 222–232 (1996)
14.61. A. Eppler, D. Cremers, D. Hickmott, M. Ferris, A.C. Koskelo, Matrix-effects in the detection of Pb and Ba in soils using laser-induced breakdown spectroscopy. Appl. Spectrosc. **50**, 1175–1181 (1996)
14.62. A. Ciucci, V. Palleschi, S. Rastelli, R. Barbini, F. Colao, R. Fantoni, A. Palucci, S. Ribezzo, H. van der Steen, Trace pollutants analysis in soil by a time-resolved laser-induced breakdown spectroscopy technique. Appl. Phys. B **63**, 185–190 (1996)
14.63. C. Becker, F. Hilbk-Kortenbruck, R. Noll, Preparation of soil samples containing heavy metals for the calibration of laser-based analytical techniques, Abstracts of the lecture groups "Environmental Technology and Fundamentals of Laser-aided In-situ Soil Analysis", *Int. Meeting on Chem. Eng., Environmental Protection and Biotechnology, ACHEMA 2000, Conf. Proc. 2000*, DECHEMA e.V., Frankfurt a.M., 394–396 (2000)
14.64. R. Springenschmidt, Arbeitskreis "Prüfverfahren Chlorideindringtiefe" des Deutschen Ausschusses für Stahlbeton: Anleitung zur Bestimmung des Chloridgehaltes von Beton, Beuth Verlag GmbH, Berlin; Research group "Test methods for chloride penetration depth" of Deutscher Ausschuss für Stahlbeton: instruction for the determination of chloride content in concrete (Beuth Verlag GmbH, Berlin, 1989)
14.65. American Society for Testing and Materials (ASTM). *Standard Test Method for Electrical Indication of Concrete's Ability to Resist Chloride Ion Penetration, ASTM C 1202–97* (ASTM, West Conshohocken, 1997)
14.66. F. Weritz, D. Schaurich, A. Taffe, G. Wilsch, Effect of heterogeneity on the quantitative determination of trace elements in concrete. Anal. Bioanal. Chem. **385**, 248–255 (2006)
14.67. V. Burakov, V. Kiris, S. Raikov, Optimization of conditions for spectral determination of chlorine content in cement-based materials. J. Appl. Spectrom. **74**, 321–327 (2007)
14.68. F. Weritz, D. Schaurich, G. Wilsch, Detector comparison for sulfur and chlorine detection with laser induced breakdown spectroscopy in the near infrared. Spectrochim. Acta B **62**, 1504–1511 (2007)
14.69. J. Aguilera, C. Aragon, A comparison of the temperatures and electron densities of laser-produced plasmas obtained in air, argon, and helium at atmospheric pressure. Appl. Phys. A **69**, S475-S478 (1999)
14.70. G. Asimellis, A. Giannoudakos, M. Kompitsas, Near-IR bromine LIBS detection and ambient gas effects on emission line asymmetric Stark broadening and shift. Spectrochim. Acta B **61**, 1270–1278 (2006)
14.71. M. Tran, Q Sun, B.W. Smith, J. Wineforder, Determination of F, Cl and Br in solid organic compounds by laser-induced plasma spectroscopy. Appl. Spectrosc. **52**, 739–744 (1999)
14.72. G. Asimellis, S. Hamilton, A. Giannoudakos, M. Kompitsas, Controlled inert gas environment for enhanced chlorine and fluorine detection in the visible and near-infrared by laser-induced breakdown spectroscopy. Spectrochim. Acta B **60**, 1132–1139 (2005)
14.73. L. Radziemski, D. Cremers, K. Benelli, C. Khoo, R. Harris, LIBS-based Detection of As, Br, C, Cl, P and S in the VUV spectral region in a Mars atmosphere. Lunar and Planetary Science **XXXVI**, 1747 (2005)
14.74. C. Gehlen, E. Wiens, R. Noll, G. Wilsch, K. Reichling, Chlorine detection in cement with laser-induced breakdown spectroscopy in the infrared and ultraviolet spectral range. Spectrochim. Acta B **64**, 1135–1140 (2009)

14.75. H. Lancelin, L. Dudragne, P. Adam, J. Amouroux, Time resolved laser induced breakdown spectroscopy for fluorine; chlorine and sulfur detection using an optical fibre probe. High Temp. Mater. Proc. **4**, 109–126 (2000)
14.76. N. Konjevic, Plasma broadening and shifting of non-hydrogenic spectral lines: present status and applications. Phys. Rep. **316**, 339–401 (1999)
14.77. M. Dimitrijevic, S. Sahal-Brechot, Stark broadening of Ca II spectral lines. J. Quant. Spectrosc. Radiat. Transfer **49**, 157–164 (1993)
14.78. L. Peter, R. Noll, Material ablation and plasma state for single and collinear double pulses interacting with iron samples at ambient gas pressures below 1 bar. Appl. Phys. B **86**, 159–167 (2007)
14.79. R. Knopp, F. Scherbaum, J. Kim, Laser-induced breakdown spectroscopy (LIBS) as an analytical tool for the detection of metal ions in aqueous solutions. Fresen. J. Anal. Chem. **355**, 16–20 (1996)
14.80. B. Welz, Speciation analysis: where is it going? An attempt at a forecast. Spectrochim. Acta B **53**, 169–175 (1998)
14.81. R. Lobinski, J. Szpunar, Biochemical speciation analysis by hyphenated techniques. Anal. Chim. Acta **400**, 321–332 (1999)
14.82. A. Timerbraev, Element speciation analysis by capillary electrophoresis. Talanta **52**, 573–606 (2000)
14.83. http://www.agilent.com
14.84. D. Beauchemin, Inductively coupled plasma mass spectrometry. Anal. Chem.**74**, 2873–2894 (2002)
14.85. P. Kenedy, D. Hammer, B. Rockwell, Laser-induced breakdown in aqueous media. Prog. Quant. Electr. **21**, 155–248 (1997)
14.86. R. Knopp, F. Scherbaum, J. Kim, Laser-induced breakdown spectroscopy (LIBS) as an analytical tool for the detection of metal ions in aqueous solutions. Fresen. J. Anal. Chem. **355**, 16–20 (1996)
14.87. P. Fichet, P. Mauchien, J. Wagner, C. Moulin, Quantitative elemental determination in water and oil by laser-induced breakdown spectrometry. Anal. Chim. Acta **429**, 269–278 (2001)
14.88. W. Ho, C. Ng, N. Cheung, Spectrochemical analysis of liquids using laser induced plasma emission: effects of laser wavelength. Appl. Spectrosc. **51**, 87–91 (1997)
14.89. J. Carranza, B. Fisher, G. Yoder, D. Hahn, On-line analysis of ambient air aerosols using laser-induced breakdown spectroscopy. Spectrochim. Acta B **56**, 851–864 (2001)
14.90. H. Archontaki, S. Crouch, Evaluation of an isolated droplet sample introduction system for laser-induced breakdown spectroscopy. Appl. Spectrosc. **42**, 741–746 (1988)
14.91. G. Arca, A. Ciucci, V. Palleschi, S. Rastelli, E. Tognoni, Trace element analysis in water by laser-induced breakdown spectroscopy technique. Appl. Spectrosc. **51**, 1102–1105 (1997)
14.92. A. Pichahchy, D. Cremers, M. Ferris, Elemental analysis of metals under water using laser-induced breakdown spectroscopy. Spectrochim. Acta B **52**, 25–39 (1997)
14.93. R. Wal, T. Ticich, J. West, P. Householder, Trace metal detection by laser-induced breakdown spectroscopy. Appl. Spectrosc. **53**, 1226–1236 (1999)
14.94. A. Kuwako, Y. Uchida, K. Maeda, Supersensitive detection of sodium in water with use of dual-pulse laser-induced breakdown spectroscopy. Appl. Optics **42**, 6052–6056 (2003)
14.95. K. Lo, N. Cheung, ArF laser-induced plasma spectroscopy for part-per-billion analysis of metal ions in aqueous solution. Appl. Spectrosc. **56**, 682–688 (2002)
14.96. X. Pu, W. Ma, N. Cheung, Sensitive elemental analysis of aqueous colloids by laser-induced plasma spectroscopy. Appl. Phys. Lett. **83**, 3416–3418 (2003)
14.97. C. Janzen, R. Fleige, R. Noll, H. Schwenke, W. Lahmann, J. Knoth, P. Beaven, E. Jantzen, A. Oest, P. Koke, Analysis of small droplets with a new detector for liquid chromatography based on laser-induced breakdown spectroscopy. Spectrochim. Acta B **60**, 993–1001 (2005)

14.98. H. Ulmke, T. Wriedt, H. Lohner, K. Bauckhage, The piezoelectric droplet generator – a versatile tool for dispensing applications and calibration of particle sizing instruments, Precis. Eng. – Nanotechnology, *Proc. of the 1st Int. Euspen Conf.*, vol. **2** (Shaker Verlag, Aachen, 1999), pp. 290–293

14.99. C. Haisch, R. Niessner, O. Matveev, U. Panne, N. Omenetto, Element-specific determination of chlorine in gases by laser-induced-breakdown-spectroscopy LIBS. Fresen. J. Anal. Chem. **356**, 21–26 (1996)

14.100. M. Tran, B.W. Smith, D. Hahn, J. Winefordner, Detection of gaseous and particulate fluorides by laser-induced breakdown spectroscopy. Appl. Spectrosc. **55**, 1455–1461 (2001)

14.101. D. Plemmons, C. Parigger, J. Lewis, J. Hornkohl, Analysis of combined spectra of NH and N_2. Appl. Optics **37**, 2493–2498 (1998)

14.102. A. Schwebel, A. Ronn, Spectroscopy of laser-induced dielectric breakdown in gas mixtures. Chem. Phys. Lett. **100**, 178–182 (1983)

14.103. M. Casini, M. Harith, V. Palleschi, A. Salvetti, D. Singh, M. Vaselli, Time-resolved LIBS experiment for quantitative determination of pollutant concentrations in air. Laser Part. Beams **9**, 633–639 (1991)

14.104. I. Glassman, Combustion (Academic, San Diego, 1996), pp. 20–29

14.105. A. Sandrowitz, J. Cooke, N. Glumac, Flame emission spectroscopy for equivalence ratio monitoring. Appl. Spectrosc. **52**, 658–662 (1998)

14.106. R. Schmieder, Combustion applications of laser-induced breakdown spectroscopy, in *Proceedings of the Electro-optics Laser Conference*, Cahners, Chicago, 1981, pp. 17–27

14.107. X. Phuoc, F. White, Laser-induced spark for measurements of the fuel-to-air ratio of a combustible mixture. Fuel **81**, 1761–1765 (2002)

14.108. V. Sturm, R. Noll, Laser-induced breakdown spectroscopy of gas mixtures of air, CO_2, N_2, and C_3H_8 for simultaneous C, H, O, and N measurement. Appl. Optics **42**, 6221–6225 (2003)

14.109. K. Fieweger, R. Blumenthal, G. Adomeit, Self-ignition of S.I. engine model fuels: shock tube investigation at high pressure. Combust. Flame **109**, 599–619 (1997)

14.110. K. Kuwahara, H. Ando, Diagnostics of in-cylinder flow, mixing and combustion in gasoline engines. Meas. Sci. Technol. **11**, R95–R111 (2000)

14.111. D. Lide (ed.), *CRC Handbook of Chemistry and Physics*, 73rd edn. (CRC Press, Boca Raton, 1992), pp. 14–11

14.112. V. Sturm, A. Brysch, R. Noll, H. Brinkmann, R. Schwalbe, K. Mülheims, P. Luoto, P. Mannila, K. Heinänen, D. Carrascal, L. Sancho, A. Opfermann, K. Mavrommatis, H.W. Gudenau, A. Hatziapostolou, S. Couris, Online multi-element analysis of the top gas of a blast furnace, in *7th Int. Workshop Progress in Analytical Chemistry in the Steel and Metal Industries*, ed. by J. Angeli (Glückauf GmbH, Essen, 2006), pp. 183–188

14.113. A. Brysch, V. Sturm, R. Noll, H. Denecke-Arnold, H. Brinkmann, K. Mülheims, A. Opfermann, K. Mavrommatis, H.-W. Gudenau, K. Heinänen, Laser-based elemental analysis of the top gas of a blast furnace, *Int. Symp. on Photonics in Measurement*, 11–12 June 2002, Aachen, Germany, VDI-Berichte Nr. 1694, 2002, 117–123, ISBN 3-18-091694-X

14.114. European commission, Directorate C: information note: airborne particles and their health effects in Europe, ENV.C1/SAZr, March 2003

14.115. World Health Organisation, *World Health Report 2002* (WHO, Geneva, 2002)

14.116. D. Adriano, *Biogeochemistry of Trace Metals* (Lewis, Boca Raton, 1992)

14.117. J. Fergusson, *The Heavy Elements: Chemistry, Environmental Impact and Health Effects* (Pergamon, Oxford, 1991)

14.118. T. Kuhlen, C. Fricke-Begemann, N. Strauss, R. Noll, Analysis of size-classified fine and ultrafine particulate matter on substrates with laser-induced breakdown spectroscopy. Spectrochim. Acta B **63**, 1171–1176 (2008)

14.119. U. Panne, R. Neuhauser, M. Theisen, H. Fink, R. Niessner, Analysis of heavy metal aerosols on filters by laser-induced plasma spectroscopy. Spectrochim. Acta B **56**, 839–850 (2001)
14.120. Y. Yamamoto, R. Yoshiie, S. Uemiya, Simple and rapid analysis of heavy metals in sub-micron particulates in flue gas, *6th Int. Symp. and Exhibition on Gas Cleaning at High Temperature*, Osaka, Japan, 2005
14.121. S. Arnold, D. Cremers, Rapid determination of metal particles on air sampling filters using laser-induced breakdown spectroscopy. Am. Ind. Hyg. Assoc. J. **56**, 1180–1186 (1995)
14.122. M. Marjamäki, J. Keskinen, D. Chen, D. Pui, Performance evaluation of the electrical low-pressure impactor (ELPI). J. Aerosol Sci. **31**, 249–261 (2000)
14.123. G. Cristoforetti, S. Legnaioli, V. Palleschi, A. Salvetti, E. Tognoni, Influence of ambient gas pressure on laser-induced breakdown spectroscopy technique in the parallel double-pulse configuration. Spectrochim. Acta B **59**, 1907–1917 (2004)
14.124. L. Radziemski, T. Loree, D. Cremers, N. Hoffman, Time-resolved laser-induced breakdown spectrometry of aerosols. Anal. Chem. **55**, 1246–1252 (1983)
14.125. R. Neuhauser, U. Panne, R. Niessner, G. Petrucci, P. Cavalli, N. Omenetto, On-line and in-situ detection of lead aerosols by plasma-spectroscopy and laser-excited atomic fluorescence spectroscopy. Anal. Chim. Acta **346**, 37–48 (1997)
14.126. S. Yalcin, D. Crosley, G. Smith, G. Faris, Spectroscopic characterization of laser-produced plasma for in situ toxic metal monitoring. Hazard. Waste Hazard. Mater. **13**, 51–61 (1996)
14.127. M. Nunez, P. Cavalli, G. Petrucci, N. Omenetto, Analysis of sulfuric acid aerosols by laser-induced breakdown spectroscopy and laser-induced fragmentation. Appl. Spectrosc. **54**, 1805–1816 (2000)
14.128. R. Yoshiie, Y. Yamamoto, S. Uemiya, S. Kambara, H. Moritoni, Simple and rapid analysis of heavy metals in sub-micron particulates in flue gas. Powder Technol. **180**, 135–139 (2008)
14.129. D. Hahn, W. Flower, K. Hencken, Discrete particle detection and metal emission monitoring using laser-induced breakdown spectroscopy. Appl. Spectrosc. **51**, 1836–1844 (1997)
14.130. D. Hahn, M. Lunden, Detection and analysis of aerosol particles by laser-induced breakdown spectroscopy. Aerosol Sci. Tech. **33**, 30–48 (2000)
14.131. J. Carranza, D. Hahn, Sampling statistics and considerations for single-shot analysis using laser-induced breakdown spectroscopy. Spectrochim. Acta B **57**, 779–790 (2002)
14.132. J. Carranza, D. Hahn, Plasma volume considerations for analysis of gaseous and aerosol samples using laser-induced breakdown spectroscopy. J. Anal. At. Spectrom. **17**, 1534–1539 (2002)
14.133. N. Strauss, C. Fricke-Begemann, R. Noll, Size-resolved analysis of fine and ultrafine particulate matter by laser-induced breakdown spectroscopy. J. Anal. At. Spectrosc. **25**, 867–874 (2010)
14.134. V. Hohreiter, D. Hahn, Calibration effects for laser-induced breakdown spectroscopy of gaseous sample streams: analyte response of gas-phase species versus solid-phase species. Anal. Chem. **77**, 1118–1124 (2005)
14.135. G. Lithgow, S. Buckley, Influence of particle location within plasma and focal volume on precision of single-particle laser-induced breakdown spectroscopy measurements. Spectrochim. Acta B **60**, 1060–1069 (2005)
14.136. V. Hohreiter, D. Hahn, Plasma-particle interactions in a laser-induced plasma: implications for laser-induced breakdown spectroscopy. Anal. Chem. **78**, 1509–1514 (2006)
14.137. I. Gornushkin, J. Anzano, L. King, B. Smith, N. Omenetto, J. Winefordner, Curve of growth methodology applied to laser-induced plasma emission spectroscopy. Spectrochim. Acta B **54**, 491–503 (1999)

Chapter 15
Spatially Resolved Analysis

This chapter describes LIBS investigations for a spatially resolved microanalysis of samples using the capability to focus laser radiation to spot sizes below 10 μm. Applications of the developed method will be presented in Sect. 18.3.

15.1 High-Speed LIBS for Microanalysis

The parameter range of laser pulse energies used for bulk analysis of solid, liquid, or gaseous substances typically lies between 10 mJ and 1 J, the repetition frequency at 10–50 Hz [15.1–15.6], see Fig. 3.1. Several authors have studied spatially resolved measurements using LIBS [15.7-15.12]. In this case, the gained spectral information is linked to the position of irradiation of the laser pulse onto the sample surface. A relative motion between the focused laser beam and the sample enables a scanning across the surface. The result is for example, presented as an element line intensity of the laser-induced plasma versus one or two spatial coordinates. The latter situation yields maps of element-specific intensities of a sample surface. Two kinds of spatial resolution have to be distinguished: the lateral resolution and the longitudinal or depth resolution, see Sect. 3.6. For scanning LIBS, the lateral resolution is generally influenced by: (a) the beam diameter on the sample surface, (b) the laser pulse energy, (c) the laser pulse duration, (d) the material properties of the sample, and (e) the type, pressure and flow conditions of the ambient gas atmosphere.

A lateral spatial resolution of 6 μm was reported for a sharp aluminum–copper transition using a frequency quadrupled Nd:YAG laser at 266 nm and pulse energies below 100 μJ [15.8]. The repetition rate of the laser was 2 Hz, and the spectral range recorded simultaneously was 50 nm. Surface analysis of photovoltaic cells using a pulsed N_2 laser at 337.1 nm and 1 mJ pulse energy was performed with a lateral spatial resolution of 30 μm and a repetition rate of 10 Hz [15.9]. Single shot spectra were detected in the spectral range from 542 to 657 nm. Mapping of platinum group metals in automotive exhaust catalysts has been performed with a 20 Hz laser at a wavelength of 532 nm [15.10]. The spectral range detected simultaneously was

R. Noll, *Laser-Induced Breakdown Spectroscopy*, DOI 10.1007/978-3-642-20668-9_15,

15 nm. With a line-shaped laser focus and an imaging spectrometer an inclusion in a stainless steel sample was studied by scanning the laser line across a sample surface of 1.6 mm × 4.4 mm. The estimated spatial resolution achieved along the laser line is 17 and 15 μm in the direction perpendicular to the microline focus [15.11]. The acquisition rate was 2 Hz and the spectral range detected simultaneously was approximately 15 nm. LIBS-based microanalysis of tool steel and glass was studied using an echelle spectrometer allowing to observe a spectral range from 200 to 800 nm simultaneously [15.12]. However, in this case, the measuring frequency is limited by the read-out time of the applied megapixel CCD-sensor array to well below 1 Hz.

The repetition rates used so far for scanning LIBS lie in the range of 2–20 Hz, with the consequence that the scanning of macroscopic samples with small step sizes takes a long time. The use of a line-focus offers an opportunity to speed up scanning; however, in this approach an imaging spectrometer has to be used, limiting the spectral range detected simultaneously to typically less than 20 nm. This is a significant restriction for the selection of emission lines of analytes and reference elements most suitable for quantitative spectral analysis and for the number of elements to be determined simultaneously. Figure 15.1 shows an overview of the LIBS parameters measuring frequency and spectral resolution. The measuring frequency is strongly linked to the type of detector used for the acquisition of the spectral information. Common configurations of spectrometers and detector types are as follows (cf. Sect. 4.2): (a) Czerny–Turner spectrometer and intensified PDA or CCD, (b) echelle spectrometer and intensified CCD-camera, and (c) Paschen–Runge spectrometer and PMT. The requirement concerning the spectral resolution $\lambda/\Delta\lambda$ depends on the expected density of emission lines in the spectrum of the laser-induced plasma to clearly separate analyte emissions from interfering lines. A quantitative analysis of iron-based samples generally requires a spectral resolution of about 5×10^3 and more. Most of the existing LIBS setups are presented by the data points 1–10, where the measuring frequency is mostly limited by the detector read-out time or the laser repetition frequency to less than 60 Hz. The data points 11–14 illustrate the potential to push the measuring frequency into the range of 100 Hz to 1 kHz. The combination of a Paschen–Runge spectrometer with PMT detectors achieves the highest ever so far achieved measuring frequencies for LIBS, see data point 14 in Fig. 15.1.

In this chapter, R&D activities will be presented enabling to push the measuring speed of scanning LIBS to the 1 kHz range [15.24]. The spectral range detected simultaneously was extended to cover the vacuum ultraviolet, visible, and near-infrared spectral region.

A diode-pumped solid-state laser (DPSSL) was used for the excitation of the plasma, see Sect. 4.1, Figs. 4.2 and 4.3, and Table 4.3. Oscillator and amplifier are pumped longitudinally using a fiber optic link to two diode pumping modules. The oscillator is equipped with an electro-optic Q-switch. The beam is expanded using a Galilean telescope and has a beam diameter at the exit of about 10 mm. The laser is operated at a wavelength of 1,064 nm. The other laser parameters are: repetition rate up to 1,000 Hz, pulse energy 2 mJ, pulse width (FWHM) 5–6 ns, beam propagation ratio $M^2 < 1.3$ [cf. Sect. 3.4, relation (3.5)].

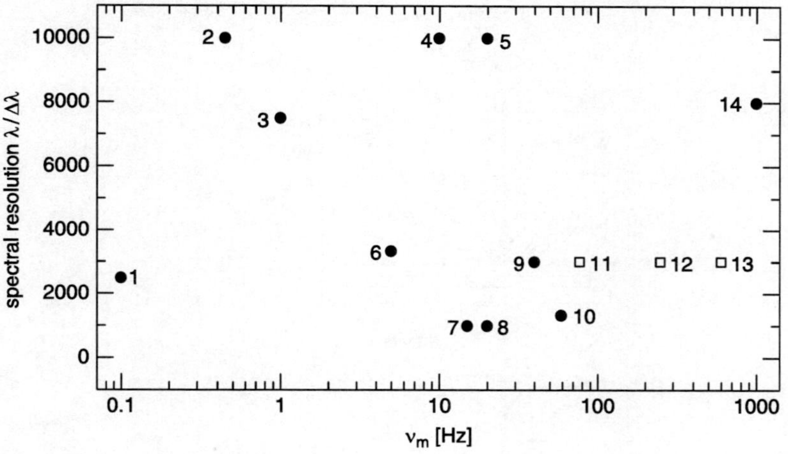

Fig. 15.1 Measuring frequency ν_m versus spectral resolution $\lambda/\Delta\lambda$ for different detector, spectrometer, and laser combinations used for LIBS are plotted with *filled bullets*. The numbers close to the data points in the graph refer to the following references: 1 = [15.13], 2 = [15.14], 3 = [15.15], 4 = [15.16], 5 = [15.17], 6 = [15.18], 7 = [15.19], 8 = [15.20], 9 = [15.21], 10 = [15.22]. Data point 14 is the high-speed LIBS system described in this chapter. The *open squares* for data points 11–13 represent commercially available detectors which can be used for LIBS [15.23], the assumed resolution of 3,000 can be achieved with standard spectrometers and monochromators

The laser beam is coupled to a measuring chamber shown schematically in Fig. 15.2. The measuring chamber is a gas-tight vessel made of high-alloyed steel. Inside the measuring chamber, an argon gas atmosphere is maintained at a pressure of about 900 mbar. The use of the inert gas allows to transmit the ultraviolet (UV) emission of the laser-induced plasma to the spectrometer, cf. Sect. 3.5. The sample to be analyzed is brought into the chamber through a transfer port (V). The sample is clamped in a sample cassette which is fixed to the x-, y-translation stages (A) installed inside the chamber (C). These translation stages allow to shift the sample with respect to the position of the laser focus in two perpendicular directions. The positioning accuracy of the axes is 1 μm, and the minimum step size amounts to 5 μm.

The laser beam is focused by an objective consisting of three lenses to correct the spherical aberration. The focal length is 60 mm, and the aperture 48 mm. The beam diameter on the focusing lens amounts to $\sim$ 26 mm. The calculated $1/e^2$-focus diameter taking into account the beam quality of the DPSSL of $M^2 = 1.3$ is $\sim$ 4 μm. The waist of the beam focus is positioned onto the surface of the sample ($\Delta s = 0$ mm).

The emission of the laser-induced plasma is collimated by a spherical mirror to the entrance slit of a Paschen–Runge vacuum spectrometer having a Rowland circle diameter of 1 m and covering a spectral range of 130–777 nm. Along the Rowland circle, exit slits and photomultipliers are positioned at predefined spectral positions

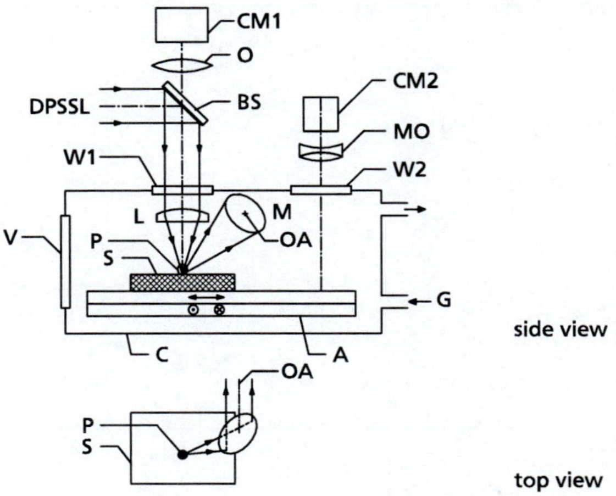

Fig. 15.2 Schematic setup for the high-speed scanning microanalysis. DPSSL = beam of the diode-pumped solid-state laser; BS = beam splitter; W1 = window for the laser beam; L = focusing lens; S = sample; A, x-, y = translation stages; P = laser-induced plasma; M = spherical mirror to collect the plasma emission; OA = optical axis of the spectrometer; W2 = window to observe the sample surface; MO = microscope zoom objective (10×); CM 1, 2 = CCD camera to observe the craters; G = gas flushing of the measurement chamber; C = measuring chamber; and V = transfer port

to detect the radiation of up to 48 emission lines, cf. Sect. 4.2, Fig. 4.7. The signals of the photomultiplier tubes detecting the transient radiation are transmitted to a multichannel integrator electronics (MCI, cf. Sect. 4.5, Fig. 4.16).

The sample surface can be observed by an optical system consisting of a microscope zoom objective and a CCD-camera. The optical axis of this microscope is oriented parallel to the optical axis of the analyzing laser beam. For an inspection of the produced craters, the sample can be moved by the translation stages toward the field of view of the microscope and the CCD-camera, see MO and CM2 in Fig. 15.2.

DPSSL, translation stages, and MCI electronics are synchronized to allow for single-shot data evaluation and a defined allocation of the spectral signals to the respective measuring position, i.e., the x- and y-position. At 1 kHz measuring frequency and a scanned area of 1×1 cm^2 with a step size of 20 μm, a data volume of about 48 MB is collected which is then processed to generate the element mappings (cf. Sect. 15.2).

Sample translation stages, laser source, pumping unit, Pockels cell, and integrator electronics have to be synchronized to achieve a correct measuring point position of better than 1 μm and a high signal stability. Figure 15.3 shows the timing diagram for high-speed scanning LIBS. The counter signal of the linear stages is counted by a synchronization electronics to determine the true position of the stages, see graph a in Fig. 15.3 (the nonequidistant pulse sequence of the counter signal is a consequence of the interpolation procedure applied in connection with the measures

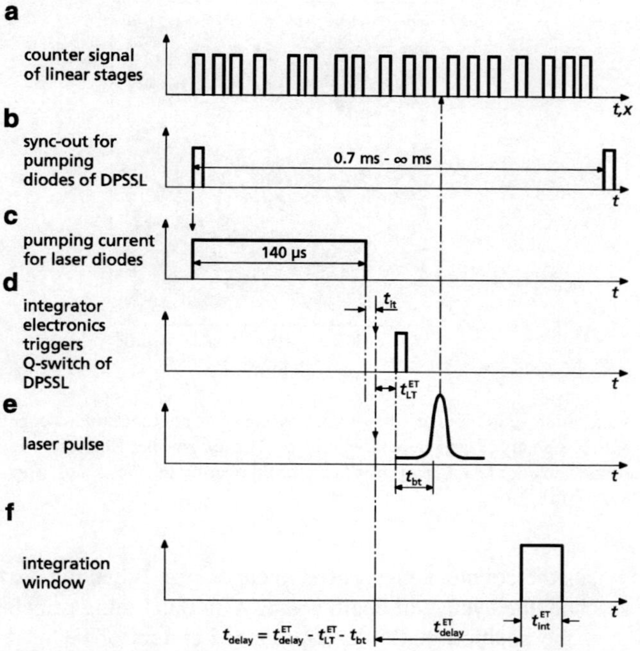

Fig. 15.3 Timing diagram for high-speed scanning LIBS. a–f: see captions on the left side. t_{it}, internal transit time; $t_{\mathrm{LT}}^{\mathrm{ET}}$, time set in the event table (ET) for the laser trigger (LT); t_{bt}, built-up time (bt) for the laser pulse after activation of the Q-switch including transit times; $t_{\mathrm{delay}}^{\mathrm{ET}}$, delay time set in the event table for the start of the integration window; $t_{\mathrm{int}}^{\mathrm{ET}}$, time set in the event table for the width of the integration window; and t_{delay}, true delay time between the start of the laser pulse and the start of the integration window

integrated in both translation axes). After a programmable number of counts, the pumping diodes of the DPSSL are triggered and pump the laser crystal for 140 µs (graphs *b* and *c* in Fig. 15.3). After the falling edge of the pumping current and a fixed internal transit time t_{it}, the integrator electronics is triggered and an event table – containing the temporal sequence for various trigger signals and the start and widths of the integration windows for the different signal channels – is started (graphs *c* and *d*). The Pockels cell is triggered by a signal initiated by the event table (graph *d*). This procedure ensures that the temporal jitter between the Pockels cell switching and the start of the integration windows is kept to less than 1 ns.

After the laser pulse built-up time $t_{\mathrm{bt}} = 400\,\mathrm{ns}$ (including transit times), the laser pulse is emitted and the plasma is generated on the sample surface (graph *e*). The vertical dashed arrow in Fig. 15.3 running from the laser pulse peak in graphs *e*–*a* illustrates that the laser ablation takes place at a defined position of the axes. The event table is processed step by step and at the end the measured values are digitized and transferred to a computer. During the whole time, the synchronization

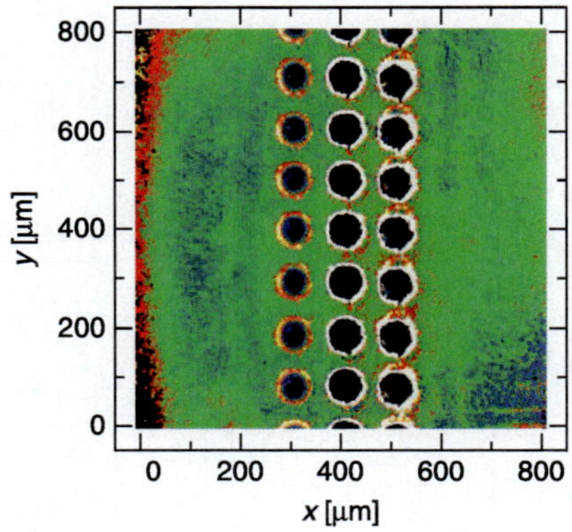

Fig. 15.4 Optical microscope image taken with a white light confocal microscope of a polished steel surface with line scans of laser generated craters. The *left row* is a line scan with a single laser pulse per position. The *middle* and *right row* is a tenfold overlay of lines scans. *Step size*: 100 µm, laser pulse energy 2 mJ

electronics tracks the counter signal of the linear stages. When the next measuring position is reached, the cycle will begin again. With this timing principle, a spatial repeatability of the position of the laser-produced craters of better than 1 µm is achieved within a scanning area of up to 45 mm × 110 mm.

This capability was proofed by comparing the optical microscopic image of a line scan, where only a single laser pulse is irradiated per position with a neighboring line scan where ten single pulse scans are overlaid, see Fig. 15.4 [15.25]. As can be seen from the microscope image, the ablation craters lie exactly on top of each other. By successive two dimensional scans, a three dimensional analysis can be performed. For each further laser pulse irradiation at the same position, the crater depth increases successively thus enabling to yield chemical information underneath the sample surface.

For a more detailed determination of the geometry of the craters produced by the pulses of the DPSSL, a ground steel surface was prepared using a grinding paper with a grain size of 80. Figure 15.5 shows images of craters generated by 1, 2, 4, and 8 laser pulses taken with an optical microscope. A central tiny crater is surrounded by a ring-shaped structure originating from residues of the melt pushed out of the interaction region by the pressure of the laser-induced plasma and the accompanying shock wave (cf. Sect. 7.1). The central crater diameter for a single shot can be estimated to be less than 15 µm. The crater depth is measured using an optical microscope determining the difference in the focusing positions of the microscope objective while sequentially focusing on the sample surface in the neighborhood of the crater and the bottom of the crater. However, this yields only a rough estimate of the crater depth, since the resolution of the microscopic images is limited by diffraction of the illuminating light at the crater structure itself. The crater depth amounts to about 2–8 µm for a single laser pulse depending on the pulse energy and

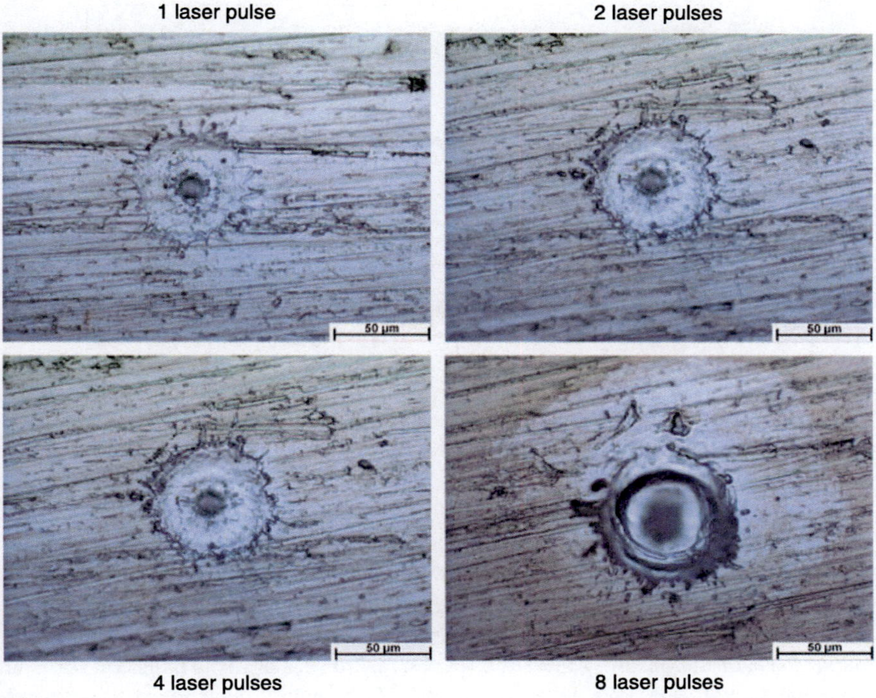

Fig. 15.5 Images of the impact area of 1, 2, 4, and 8 laser pulses of the DPSSL with a pulse energy of 2 mJ on a ground steel surface taken with an optical microscope

the number of pulses irradiated at one position. The diameter of the region enclosing the debris around the crater is about 50 µm.

The diameter of the central crater will mainly determine the lateral analytical resolution. The debris area will affect consecutive measurements in the sense of a crosstalk. However, an analysis of the element maps gained with this set-up and a step size for scanning of 20 µm has not revealed any significant crosstalk, which may be attributed to the fact that the ratio of the debris mass – contributing to the plasma signal of the subsequent laser pulse – to the mass originally present in the central crater volume is very small.

Stereo SEM images and white light confocal microscopy images of a single pulse crater generated with a 2 mJ laser pulse on a steel target show crater depths of about 2.3 µm [15.25]. With a scanning measurement comprising 3.75×10^6 laser pulses having each 2 mJ at a step size of 20 µm an average mass loss of 0.77 ng per pulse was determined. Taking the mass density of steel this corresponds to an effective crater volume of 98 μm^3. Assuming that the central crater geometry can be described by a right circular cone, this volume implies a diameter of the base of the cone of 13 µm, which is consistent with the observations of Fig. 15.5.

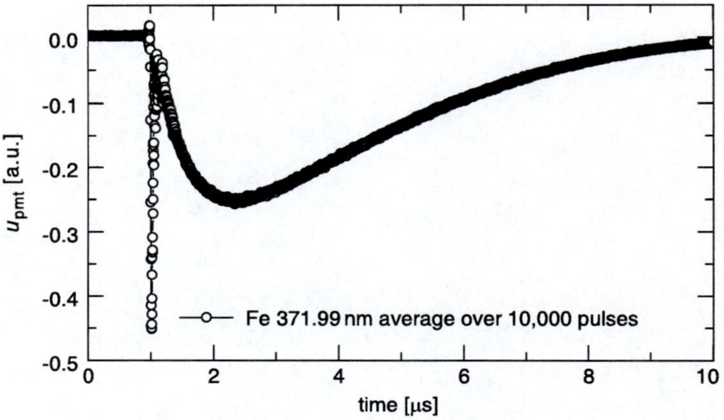

Fig. 15.6 Photomultiplier signal of the element channel for Fe at 371.99 nm as a function of time

Figure 15.6 shows the photomultiplier signal of the iron line at 371.99 nm as a function of time. The signal was transmitted from the socket of the multiplier by a coaxial cable RG 174 to an oscilloscope terminated with 50 Ω. At the beginning a narrow emission peak is observed corresponding to the continuum emission in the early phase of the plasma generation (cf. Chap. 2, Fig. 2.4). This emission starts in coincidence with the laser pulse irradiation. The temporal behavior of this peak is similar for all element channels. The width of this peak amounts to about 25 ns. The subsequent second emission maximum occurring at about 600 ns after the laser pulse irradiation belongs to the element-specific radiation component. For the element maps shown in the following, the element signals are integrated within defined time gates using the following parameters, if not stated otherwise: delay time with respect to the laser pulse 2 μs, gate width 8 μs.

The principal limitation of the plasma generation frequency for high-speed LIBS with single-pulse evaluation is among others defined by the temporal behavior of the line emission of the laser-induced plasma. The element-specific emission extends over a period of about 20 μs at maximum. Hence, the principal limit is estimated to be at about 50 kHz, which is nearly two orders of magnitude beyond the achieved maximum frequency for LIBS of 1 kHz described in this chapter. However, further limits are imposed by plasma residuals and gas exchange processes [cf. Sect. 3.1, Fig. 3.1, relation (3.1)]. The gas exchange is limited by turbulences, which may cause an enlarged laser beam focus due to local gradients of the refractive index of the neighboring ambient atmosphere, thus reducing the laser irradiance in the focus and decreasing the pulse-to-pulse stability of the intensity distribution irradiating the target. Therefore, a gas filtering system was integrated in the measuring chamber to remove ablated and recondensated particles from the argon atmosphere. The gas is exhausted via a nozzle positioned close to the interaction region. To study the efficacy of this approach, a steel sample was scanned with 100 and 1,000 Hz

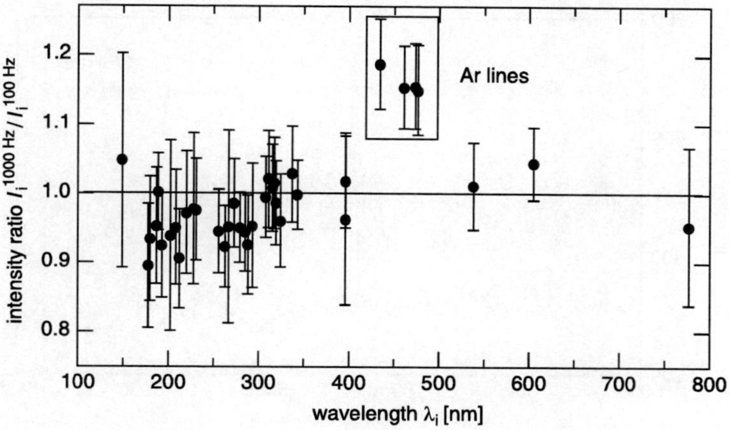

Fig. 15.7 Ratio of element intensities acquired at a measuring frequency of 1,000 Hz to those gained at 100 Hz at the same steel sample as a function of the wavelength λ_i of 37 atomic and ionic lines. The *error bars* are the standard deviation of the 250,000 measuring data for each plotted data point. The four data points between 430 and 480 nm are argon lines representing the ambient gas atmosphere in the measuring chamber

laser repetition rate. The overall number of measuring points in each case amounts to 250,000. The ratio of the average intensities of the 1,000 and the 100 Hz measurement was calculated for each element channel i and plotted in a diagram versus the wavelength λ_i of the respective emission line i, see Fig. 15.7. From all detected element lines, only the four argon lines – representing the ambient atmosphere – showed a significant change in the average signal intensity ratio. The upper energy level of the transitions of the atomic and ionic lines of the species originating from the steel sample lies in the range between 3.12 eV (Ca II at 396.85 nm) and 10.74 eV (O I at 777.19 nm). Figure 15.7 shows that for these lines no significant difference in the intensity is observed within the relative standard deviation of the intensity ratio. This is an indication that the conditions of the laser-induced plasma with respect to the excitation of the sample species do not significantly change while increasing the repetition rate from 100 to 1,000 Hz. As the excitation of the sample species does not change, it is assumed that the exhaust gas is equally effective at both repetition rates.

In contrast to this, the argon lines show an enhancement by about 15–20%. The physical reason is not yet clear. All argon lines plotted in Fig. 15.7 are argon ion lines with upper energy levels between 19.26 and 21.14 eV. A slight temperature rise in the argon atmosphere adjacent to the sample surface caused by the increased energy dissipation while irradiating a ten times higher average laser power (from $100\,\text{Hz} \times 2\,\text{mJ} = 200\,\text{mW}$ to $1,000\,\text{Hz} \times 2\,\text{mJ} = 2\,\text{W}$) may cause a layer of reduced argon particle density close to the surface. This density reduction favors a faster shock wave expansion after the irradiation of the laser pulse which in turn excites

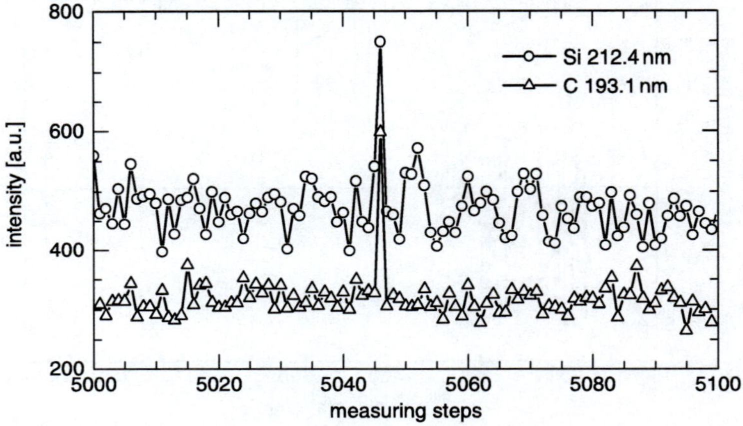

Fig. 15.8 Silicon and carbon signal for a line scan across the surface of a steel sample as a function of the measuring step. The step size amounts to 20 μm, so the total length of the line scan shown is 2 mm. The peaks at the measuring step 5,046 indicate a SiC inclusion

the ambient argon stronger leading to an increase in the argon line intensities (cf. Sect. 8.2, Figs. 8.24 and 8.26).

15.2 Element Mappings

With the SML system (cf. Sect. 18.3) so-called element mappings can be generated via one- or two-dimensional scanning. Element mappings are presentations of intensities of element channels as a function of one- or two-dimensional spatial coordinates. Since all element channels are measured simultaneously, a set of element mappings is generated for all channels. An example for a scanning along a single line – one-dimensional scanning – is shown in Fig. 15.8. The intensities of the elements, silicon and carbon, are plotted as a function of the measuring steps for a scan across a steel sample. In this figure, a step size of 20 μm was chosen. Hence, the length of the scan shown in Fig. 15.8 amounts to 2 mm. Both signal value are scattering around a base level. However, at measuring step 5,046 both signals attain a significantly higher level which returns to the base level again for the next step. Within a localized length of 20 μm, a clear spatial correlation of enhanced silicon and carbon signal is observed indicating that the laser beam has irradiated a silicon carbide (SiC) inclusion.

An example of a two-dimensional scan is shown in Fig. 15.9. The scanning area is $10 \times 10\,\mathrm{mm}^2$ of a low-alloyed steel sample. At a step size of the translation axis of 20 μm, the shown element maps for sulfur (top) and manganese (middle) comprise 250,000 measuring points each generated by a single laser pulse of 2 mJ

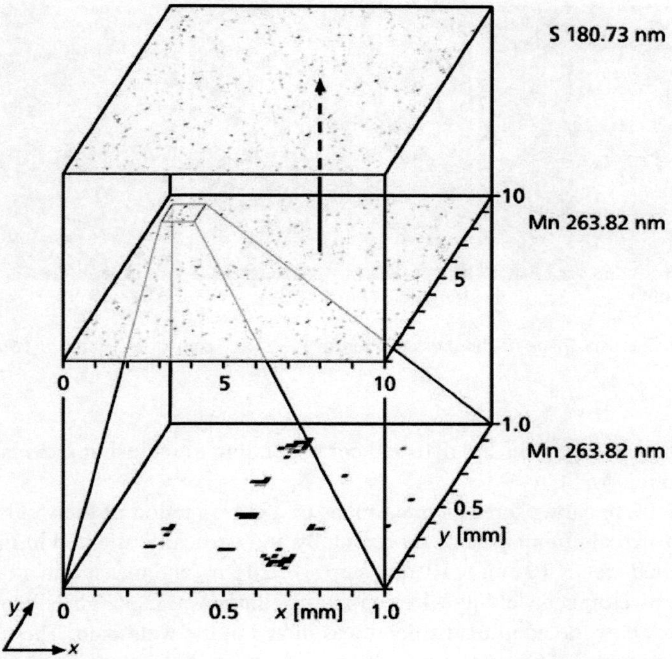

Fig. 15.9 Sulfur and manganese map of a steel sample in a scan field of 10 × 10 mm^2 (*upper and middle plane*). The *lower plane* shows a magnification of a detail of the manganese map with the dimensions 1 × 1 mm^2. The wavelengths given adjacent to the element denomination state the used emission line of the spectrometer

and a measuring frequency of 1 kHz. The maps show intensities of the sulfur line at 180.73 nm and of the manganese line at 263.82 nm. Only those intensities are plotted, which are greater than the average intensity value plus five times the standard deviation of the intensity distribution of all measuring points. By this those positions show up where a significantly higher element signal persists. Such signals are caused by, e.g., inclusions in the steel matrix, cf. Sect. 18.3.

A close look on the sulfur and manganese map shows that the dot patterns show a clear spatial correlation. This is exemplified for a pair of dots by the arrow shown in Fig. 15.9. This means that at positions of enhanced manganese concentration also an enhanced sulfur concentration exists. At these measuring positions, manganese sulfide (MnS) inclusions are identified. The element map shown in the lower plane of Fig. 15.9 shows a magnified part of the manganese map for a subarea of 1×1 mm^2. Single as well as groups of inclusions can be seen with lateral dimensions in the range of 50–100 μm. Within a cluster of manganese inclusions, different manganese concentrations exist, which is discernible by the different gray scales. The detection sensitivity for analytes of inclusions reaches down to the sub-femtogram range. Depending on the type of inclusion, smallest inclusions with volumes in the range

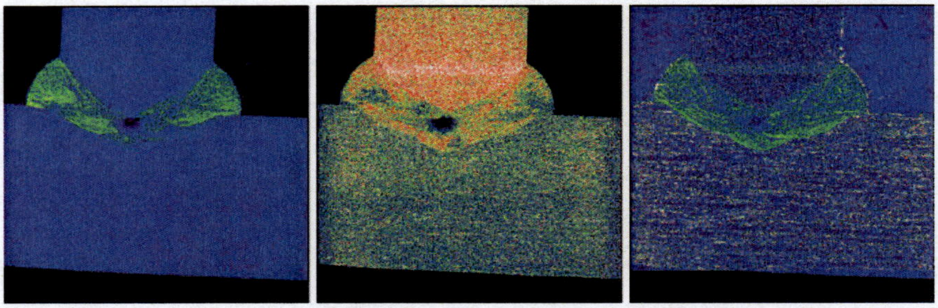

Fig. 15.10 Element maps of the cross section of the weld seam of a T-joint of two aluminum sheets

of 10^{-5} to $4 \times 10^{-2}\,\mu m^3$ are detected corresponding to inclusion sizes between 30 and 430 nm [15.25].

Figure 15.10 shows three element maps of a cross section of the weld seam of a T-joint of two aluminum sheets (horizontally and vertically oriented in Fig. 15.10). The scanned area is 10 mm × 10 mm, step size 20 μm, and measurement frequency 1 kHz. For welding, a welding wire containing silicon was used. The silicon element map shows how the additional silicon was mixed in the weld bath. The manganese and iron maps reveal that the two sheets consisted of two different aluminum grades.

The application of element mappings for segregation and inclusion analysis, as well as quantification methods, will be described in Sect. 18.3.

References

15.1. D. Cremer, Overview of applications of laser-induced breakdown spectroscopy (LIBS), in *Proceedings of the 6th International Congress on Applications of Lasers and Electro-optics, ICALEO 87*, San Diego, 1987, ed. by G. Kychakoff, pp. 45–50
15.2. J. Winefordner, I. Gornushkin, D. Pappas, O. Matveev, B. Smith, Novel uses of lasers in atomic spectroscopy – plenary lecture. J. Anal. At. Spectrom. **15**, 1161–1189 (2000)
15.3. M. Milan, P. Lucena, L. Cabalan, J. Laserna, Depth profiling of phosphorus in photonic-grade silicon using laser-induced breakdown spectrosmetry. Appl. Spectrsc. **52**, 444–448 (1998)
15.4. L. Berman, P. Wolf, Laser-induced breakdown spectroscopy of liquids: aqueous solutions of nickel and chlorinated hydrocarbons. Appl. Spectrosc. **52**, 438–443 (1998)
15.5. H. Zhang, F. Yueh, J. Singh, Laser-induced breakdown spectrometry as a multimetal continuous-emission monitor. Appl. Optics **38**, 1459–1466 (1999)
15.6. R. Adrain, J. Watson, Laser microspectral analysis: a review of principles and applications. J. Phys. D Appl. Phys. **17**, 1915–1940 (1984)
15.7. L. Moenke-Blankenburg, *Laser Microanalysis* (Wiley, New York, 1989), p. 93
15.8. C. Geertsen, J. Lacour, P. Mauchien, L. Pierrard, Evaluation of laser ablation optical emission spectrometry for microanalysis in aluminium samples. Spectrochim. Acta B **51**, 1403–1416 (1996)

15.9. J. Vadillo, S. Palanco, M. Romero, J. Laserna, Applications of laser-induced breakdown spectrometry (LIBS) in surface analysis. Fresenius J. Anal. Chem. **355**, 909–912 (1996)
15.10. P. Lucena, J. Vadillo, J. Laserna, Mapping of platinum group metals in automotive exhaust three-way catalysts using laser-induced breakdown spectrometry. Anal. Chem. **71**, 4385–4391 (1999)
15.11. M. Mateo, L. Cabalin, J. Baena, J. Laserna, Surface interaction and chemical imaging in plasma spectrometry induced with a line-focused laser beam. Spectrochim. Acta B **57**, 601–608 (2002)
15.12. K. Loebe, A. Uhl, H. Lucht, Microanalysis of tool steel and glass with laser-induced breakdown spectroscopy. Appl. Optics **42**, 6166–6173 (2003)
15.13. D. Cremers, R. Wiens, M. Ferris, R. Brennetot, S. Maurice, Capabilities of LIBS for analysis of geological samples at stand-off distances in a Mars atmosphere, in *Technical Digest, Conference: Laser-Induced Plasma Spectroscopy and Applications*, Orlando (September 25–28, 2002), pp. 5–7
15.14. U. Panne, R. Neuhauser, M. Theisen, H. Fink, R. Niessner, Analysis of heavy metal aerosols on filters by laser-induced plasma spectroscopy. Spectrochim. Acta B **56**, 839–850 (2001)
15.15. B. Charfi, M.A. Harith, Panoramic laser-induced breakdown spectrometry of water. Spectrochim. Acta B **57**, 1141–1153 (2002)
15.16. O. Samek, D. Beddows, H. Telle, J. Kaiser, M. Liska, J. Caceres, A. Gonzales Urena, Quantitative laser-induced breakdown spectroscopy analysis of calcified tissue samples. Spectrochim. Acta B **56**, 865–875 (2001)
15.17. Thermo-ARL 4460 with SparkDat. http://www.thermoarl.com
15.18. J. Carranza, B. Fisher, G. Yoder, D. Hahn, On-line analysis of ambient air aerosols using laser-induced breakdown spectroscopy. Spectrochim. Acta B **56**, 851–864 (2001)
15.19. D. Body, B. Chadwick, Optimization of the spectral data processing in a LIBS simultaneous elemental analysis system. Spectrochim. Acta B **56**, 725–736 (2001)
15.20. A. Whitehouse, J. Young, I. Botheroyd, S. Lawson, C. Evans, J. Wright, Remote material analysis of nuclear power station steam generator tubes by laser-induced breakdown spectroscopy. Spectrochim. Acta B **56**, 821–830 (2001)
15.21. K. Lo, N. Cheung, ArF laser-induced plasma spectroscopy for part-per-billion analysis of metal ions in aqueous solutions. Appl. Spectrosc. **56**, 682–688 (2002)
15.22. S. Palanco, M. Klassen, J. Skupin, K. Hansen, E. Schubert, G. Sepold, J. Laserna, Spectroscopic diagnostics on CW-laser welding plasmas of aluminium alloys. Spectrochim. Acta B **56**, 651–659 (2001)
15.23. HR 2000 from http://www.oceanoptics.com, DU 401 from http://www.andor-tech.com, Spec-10:100B from http://www.prinst.com
15.24. H. Bette, R. Noll, High-speed laser-induced breakdown spectrometry for scanning microanalysis. J. Phys. D Appl. Phys. **37**, 1281–1288 (2004)
15.25. G. Müller, *Lokale Elementverteilungsanalyse mittels rasternder laser-induzierter optischer Emissionsspektrometrie zur Charakterisierung von Einschlüssen und Seigerungen in Stählen*, Dissertation, Shaker Verlag, Aachen (2004)

Chapter 16
Depth Profiling

This chapter describes laser-induced breakdown spectroscopy (LIBS) investigations for the determination of the thickness of coatings or the depth profile of an element in a surface layer.

16.1 Measurement of the Thickness of Coatings

In the production process of galvanized and galvannealed sheet steel, inline monitoring of the coating thickness and the chemical depth profile of the zinc coating is of great interest. To assure reliable corrosion protection, the coating thickness and composition, e.g., the Zn, Fe, and Al content in hot-dip galvanized coatings, have to be kept constant over complete coils with a sheet length of up to 3 km. A small amount of Al (~ 0.15–0.25 m.-%) leads to improved adhesion of the coating and prevents iron diffusion from the substrate due to the formation of an Fe_2Al_5 inhibition layer at the interface between substrate and zinc coating [16.1].

A state-of-the-art method for coating thickness analysis in galvanizing plants is X-ray fluorescence (XRF) [16.2]. The X-ray line emission of iron from the substrate is measured, which is partially absorbed by the zinc coating. Depending on the coating thickness, the iron line is more or less absorbed. Online monitoring of the Al content is currently not realized as XRF is less sensitive to light elements such as Al. Depth profiles of the coatings are measured offline only, e.g., by glow discharge optical emission spectroscopy (GD-OES) [16.3].

Depth profiling of coated samples by LIBS was studied by different research groups mostly with the aim to achieve high depth resolution [16.4–16.12]. Therefore, the experiments were performed on stationary samples by irradiating a single spot or a microline on the sample with a series of laser pulses (see also the description in Sect. 3.6, Fig. 3.19) [16.10].

On a moving sheet steel in a production line, the irradiation of a series of laser pulses on the same surface position is not applicable because of the following reasons. First, the diameter of LIBS craters caused by a series of single pulses on

R. Noll, *Laser-Induced Breakdown Spectroscopy*, DOI 10.1007/978-3-642-20668-9_16,

one position with a beam profile optimized for high depth resolution is rather large (~1 mm) and would reduce the corrosion protection. The technical effort to irradiate one position on a moving sheet steel with several laser pulses would be tremendous. The laser focus would have to move with the sheet steel with velocities in the range of 1 m/s while keeping a spatial precision in the micrometer range. Therefore, a new method was developed to characterize the coating with a stationary laser beam and a series of single laser bursts each irradiated on a different laterally displaced sheet steel position (for the term "burst" see Sect. 3.2, Fig. 3.9). Depth information is obtained by tuning the ablation depth by variation of the burst energy. As the coating thickness is in the range of about 10 μm, the ablation depth per burst has to be of the same order, see Fig. 3.19 for $n = 1$ [16.13]. This increased ablation depth is achieved using bursts consisting of collinear double pulses (cf. Sect. 3.2 and Chap. 6). The burst parameters (i.e., interpulse separation, energy distribution within the burst) of the analyzing laser were optimized to minimize the impact of the LIBS craters on the sheet steel quality. These parameters led to craters with diameters smaller than 100 μm, which did not influence the corrosion behavior and were invisible for the naked eye on the typical surface texture of hot-dip galvanized sheet steel. The thickness resolution was determined by experiments performed on a set of reference samples of well-characterized electrolytic-coated sheet steel with different coating thicknesses in the range of 3.2–11.2 μm. The performance to gain depth information was evaluated on industrial hot-dip galvanized sheet steel.

Figure 16.1 shows schematically the experimental setup [16.13]. A flashlamp pumped Q-switched Nd:YAG-laser operating at 1,064 nm and modified for the emission of up to six pulses at 10 Hz repetition rate (>2 μs interpulse separation) is used. The widths of the individual laser pulses in the burst are ~ 15–20 ns depending on the number of pulses in the burst. In the experiments described here, the laser is operated in double pulse mode (interpulse separation of $\Delta t_1 = 4\,\mu s$, pulse energy ratio $E_1 : E_2 = 1 : 1$, burst energy $E_b = E_1 + E_2 = 0.2$–3 mJ, see Fig. 3.9) to achieve an ablation depth with a single burst in the range of the coating thickness (ca. 10 μm).

The laser light is guided via a dichroic mirror (high reflectivity for 1,064 nm, transparent for the plasma light) and focused by an achromatic lens L1 ($f = 60$ mm) onto the surface of the sheet steel, where a small amount of the sample is evaporated and the plasma is generated. The ambient atmosphere was air. The plasma light is collected and collimated with the same lens, passes the dichroic mirror, and is reimaged by a quartz lens ($f = 100$ mm) onto the core of the optical fiber of the spectrometer. An echelle spectrometer (cf. Sect. 4.2, Table 4.4) offering a broad spectral detection range (200–780 nm) with high spectral resolution (5–20 pm) detects the plasma emission. The time delays used for the signal integration were 0.5 and 2.0 μs, respectively, and the integration window was set to 10 μs. For energy control, the laser setup has been modified by inserting an additional Pockels cell as electro-optical attenuator between the laser resonator and the amplifier stages (not shown in Fig. 16.1). The Pockels cell attenuates the transmitted laser pulse dependent on the applied high voltage (0 to ~ 8 kV) by up to a factor of 10. The typical burst energy was in the range of 0.2–3 mJ. The chosen

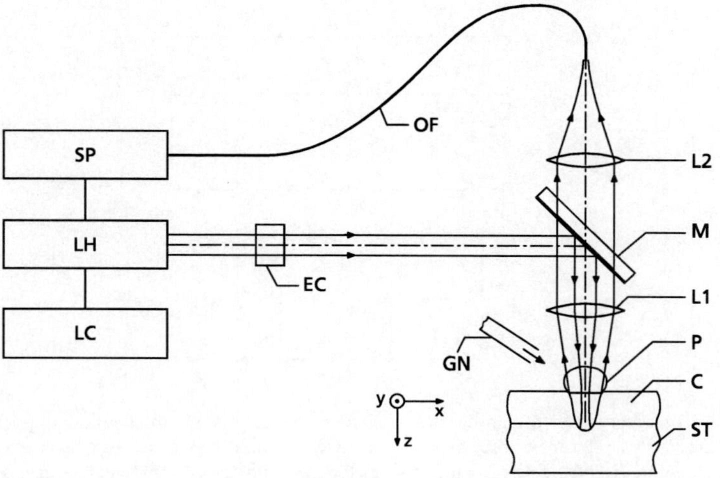

Fig. 16.1 Schematic view of the experimental setup. P = laser-induced plasma plume, C = coating, ST = steel substrate, L1 = focusing lens, L2 = imaging lens, M = mirror, OF = fiber optics, EC = energy control, LH = laser head, LC = laser control, SP = echelle spectrometer, GN = gas nozzle

laser parameters allow to penetrate the zinc coating with a single laser burst, which is a necessary precondition for the approach studied here. The crater diameters were $<100\,\mu m$, which is small enough not to reduce the corrosion protection of the coating.

The new method is developed for coating thickness measurement and depth profiling of moving sheet steel in a production line, i.e., under conditions where only single laser bursts can be applied to one distinct sheet steel position. The approach to obtain depth information is to control the crater depth by changing the burst energy in small increments and to measure the LIBS signal for every single laser burst, see Fig. 16.2, top. The signal gained for one distinct burst energy is a weighted mean of the chemical composition over the whole ablation volume and is evaluated further as described in the following.

The measuring conditions were simulated by moving the sheet steel samples (dimensions: 10 cm × 15 cm) with motorized *xy*-translation stages. For different burst energies, rows with equally spaced LIBS craters are generated as the sheet steel moves under the static optical setup. In each row in the x-direction, the burst energy was kept constant and in the y-direction the burst energy was attenuated. For each burst energy E_{bi}, a number of k ($k = 20 - 30$) replicates is applied to the sheet steel samples. Figure 16.3 shows a schematic diagram of a part of the resulting crater pattern. The distance s between adjacent craters represented by circles was set to about 0.5 mm. The number of craters per row was in the range of $20 - 30$ depending on the experiment. Each crater was generated with a single laser burst, and the complete spectrum was measured for each laser burst. From row

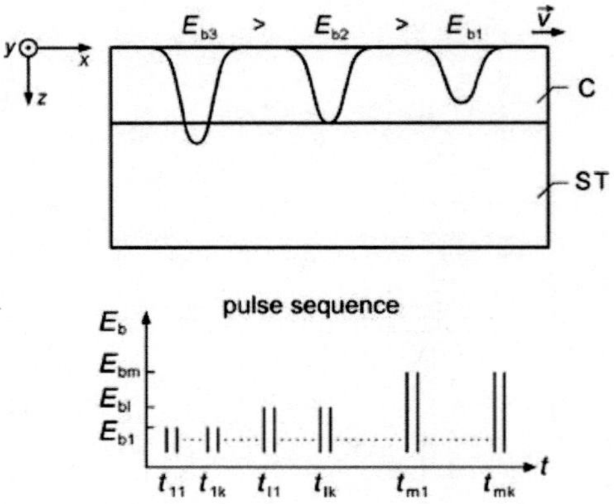

Fig. 16.2 *Top*: Procedure to gain depth information of the coating (thickness, composition) on moving sheet steel. The measuring depth is adjusted by variation of the laser burst energy E_{bi}, $i = 1, 2, \ldots$. For E_{b3}, the ablated depth reaches the steel substrate. Abbreviations as in Fig. 16.1. *Bottom*: Sequence of laser bursts. For each burst energy E_{bi} a number of $k (k = 20-30)$ replicates is applied to the sheet steel samples

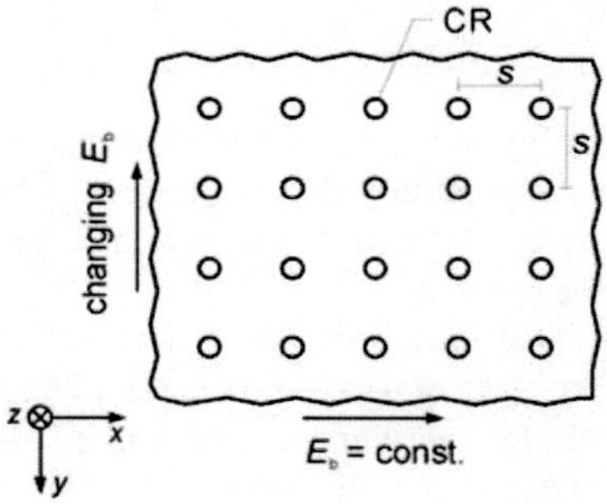

Fig. 16.3 Schematic view of the resulting crater pattern on the sheet steel. The burst energy is kept constant in x-direction and attenuated in y-direction. The separation s between adjacent craters (CR) represented by *circles* amounts to about 0.5 mm

to row the burst energy is varied in small steps (step size $\geq$ 0.05 mJ, depending on the experiment) within the range of 0.2–2.6 mJ using the laser energy control. The whole setup is automated to synchronize the sheet steel motion, the laser operation, and the spectrometer.

A set of reference samples of electrolytic galvanized sheet steel with well-defined coating thicknesses provided by ThyssenKrupp Steel AG (TKS, Dortmund, Germany) was used for the experiments on coating thickness determination. The set consists of sheet steel plates with different coating thicknesses ranging from 3.1 to 11.2 μm. Industrial hot-dip galvanized samples with an average coating thickness in the range of 10 μm were used for the experiments to gain depth information of the Al content of the coating. These samples were also provided by TKS.

The method for coating thickness determination was studied for the electrolytic galvanized sheet steel samples taking the same set of laser parameters and spectrometer settings to assure comparability of the results. The laser burst energy was

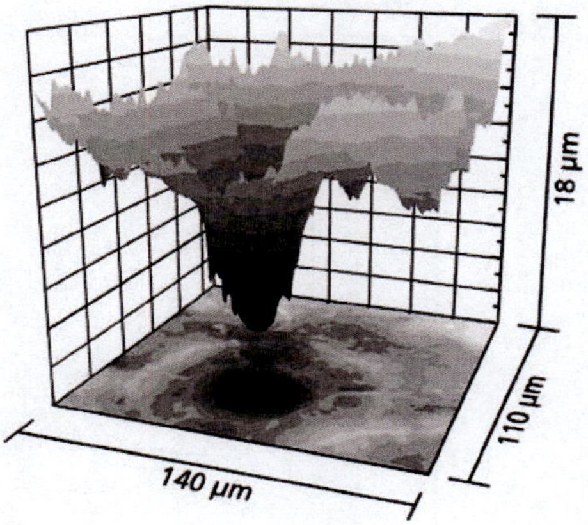

Fig. 16.4 Contour plot of a LIBS crater measured by white light interferometry. The crater was generated with a 2-mJ laser burst on an electrolytic galvanized sheet steel

Table 16.1 Crater depths of LIBS carters generated with a burst energy of 2 mJ on electrolytic galvanized sheet steel with different thicknesses of the zinc coating

Coating thickness (μm)	Average crater depth $h_c(E_b = 2\,\text{mJ})$ (μm)	Standard deviation of $h_c(E_b = 2\,\text{mJ})$ (μm)
11.2	12.6	2.0
9.2	11.7	2.0
6	11.7	0.7
4.7	9.8	1.6
3.2	9.7	1.9

The average crater depth is the mean value of five crater depths determined by white light interferometry

tuned in 18 steps from 0.22 to 2 mJ, and for each energy step 20 measurements were carried out. Figure 16.4 shows exemplarily a contour plot of a LIBS crater generated with 2 mJ on the sheet steel with 9.2 μm coating thickness. The topography was determined by white light interferometry. Obviously, the surface roughness complicates the determination of the crater depth and leads to large uncertainties.

In Table 16.1, the average crater depths of LIBS craters generated with a burst energy of 2 mJ on the sheet steel with different coating thicknesses are compared. For 5 of the 20 craters of each coating thickness, the crater depth was determined and the mean values were calculated. Due to the rough surfaces, the relative errors are up to 20%. There seems to be the tendency that the crater depth is lower for thinner coatings, which may be due to a lower ablation rate for the hard steel substrate compared to the soft zinc coating.

In the LIBS experiment, the measurement window starts at a delay time of 2.0 μs after the laser burst and has a width of 10 μs. For each laser burst, a complete spectrum was acquired, allowing to study various analyte lines for data evaluation. As an elemental line of the coating, the zinc line at 472.22 nm ($^3S_1 \rightarrow {}^3P_1$) and for the substrate the iron line at 438.35 nm ($^4F_5 \rightarrow {}^4F_4$) were selected, as these lines have strong line intensities and are not interfered by neighboring lines. The

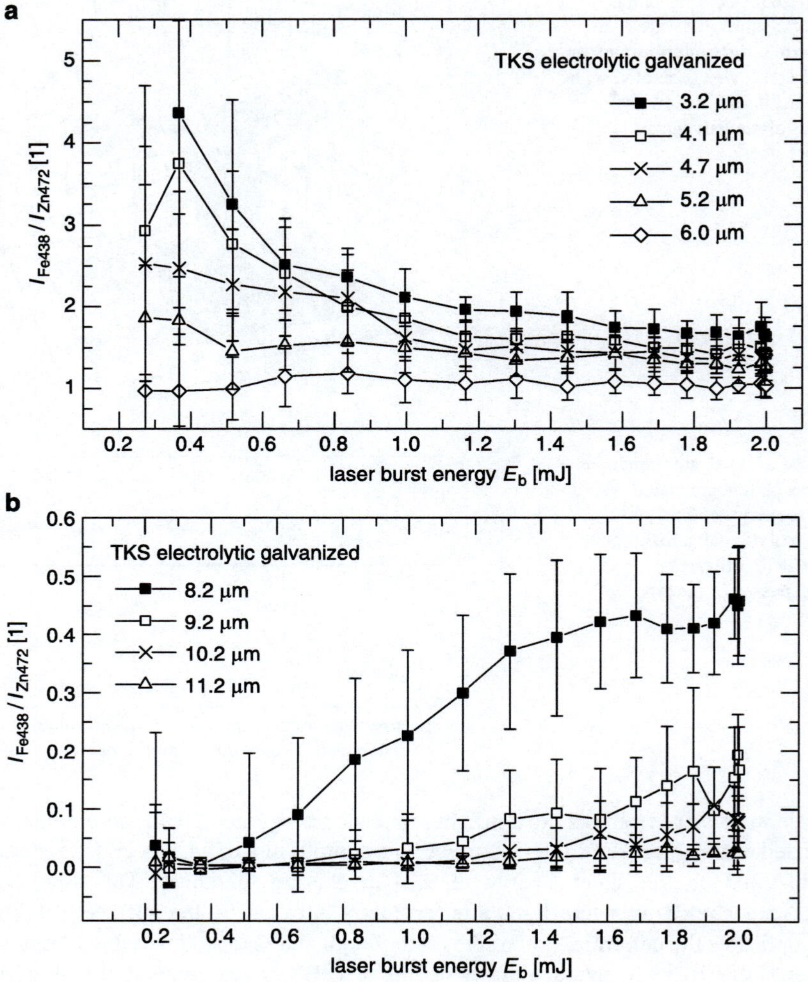

Fig. 16.5 Intensity ratio I_{Fe438}/I_{Zn472} as a function of the burst energy E_b for different coating thicknesses. Samples: electrolytic galvanized sheet steel; (**a**) 3.2–6.0 µm; (**b**) 8.2–11.2 µm. $t_{delay} = 0.5\,\mu s$, $t_{int} = 10\,\mu s$

line intensities I_{Fe438} and I_{Zn472} were computed by integrating the line profile and subtracting the background. For data evaluation, the ratio I_{Fe438}/I_{Zn472} was calculated for each laser burst and averaged over the 20 measurements of the same burst energy. In Fig. 16.5a, b, this ratio is shown with the standard deviations as a function of the burst energy.

The I_{Fe438}/I_{Zn472} value of the thinner coatings (thickness $d \leq 6.0\,\mu m$, Fig. 16.5a) is even at the lowest burst energy greater than zero, i.e., the coating is always

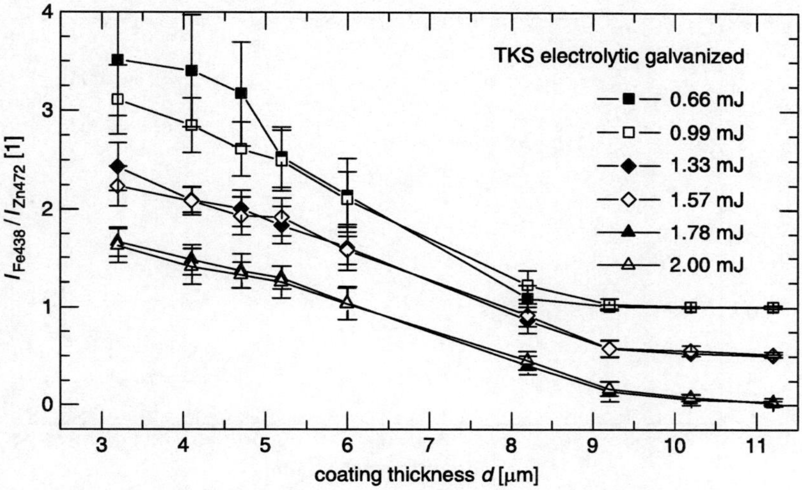

Fig. 16.6 Intensity ratio $I_{\text{Fe438}}/I_{\text{Zn472}}$ as a function of the coating thickness d for different burst energies E_b. To prevent overlapping of the graphs, the graphs for 0.66 and 0.99 mJ are shifted by 1.0 and the graphs for 1.33 and 1.57 mJ are shifted by 0.5 in the positive direction of the ordinate. Samples: electrolytic galvanized sheet steel (TKS); coating thickness: 3.2–11.2 μm

penetrated. The decline of the $I_{\text{Fe438}}/I_{\text{Zn472}}$ value with increasing burst energy seems to be in conflict with the expected higher relative number of Fe atoms but can be explained by the different temperature behavior of the population density of the excited states corresponding to the I_{Fe438} and $I_{\text{Zn472,22}}$ lines, according to the Boltzmann distribution. At low burst energies, the amount of ablated material and hence the plasma size is smaller and cools down more rapidly. This means that at a given time delay t_{delay}, the plasma temperature is lower for lower burst energies. As the excitation energy of the I_{Fe438} line ($E_j = 4.31$ eV) is much lower than that for the $I_{\text{Zn472.22}}$ line ($E_j = 6.65$ eV), the relative number of Fe atoms excited in the upper state of the Fe_{438} transition is higher in cooler plasmas than in hotter plasmas expected for higher burst energies. At small coating thicknesses, this effect seems to have a stronger impact than the increase due to the higher number of ablated Fe atoms. For the thicker coatings ($d \geq 8.2$ μm; Fig. 16.5b), the intensity ratio starts at zero and rises for higher burst energies. For both groups it is obvious that the intensity ratio at a fixed burst energy is larger for thinner than for thicker coatings. Hence, the intensity ratio at a fixed burst energy is a measure for the coating thickness, and the crater depths for different thick coatings vary only slightly, see Table 16.1.

To determine a proper burst energy, the intensity ratio is plotted as a function of the coating thickness for different burst energies, see Fig. 16.6. Coatings that are thinner than 6 μm can be distinguished for all burst energies, and coatings thicker than 8.2 μm can only be distinguished for burst energies >1 mJ as at a lower burst energy the coating is not penetrated. The standard deviation becomes smaller with

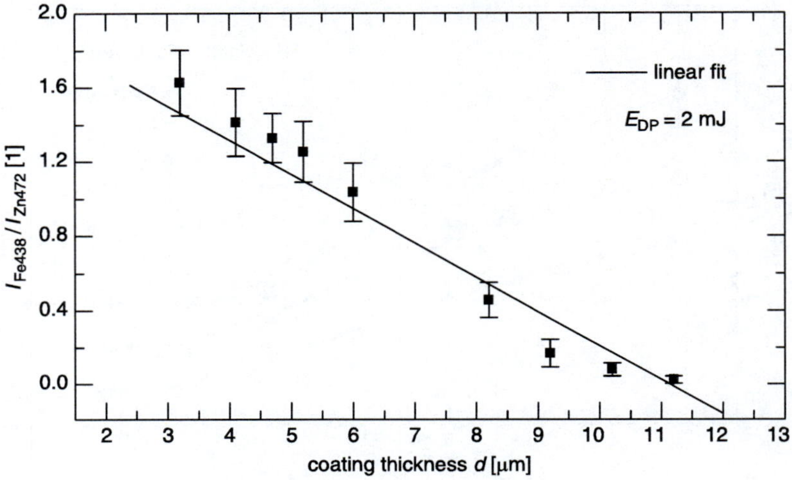

Fig. 16.7 Intensity ratio I_{Fe438}/I_{Zn472} as a function of the coating thickness d for a fixed burst energy of 2 mJ and linear fit. Samples: electrolytic galvanized sheet steel (TKS); coating thicknesses: 3.2–11.2 μm

increasing burst energy. Regarding the laser burst energy range studied, a burst energy of 2 mJ should be used to determine thickness variations.

For an estimation of the achieved depth-resolving power, the measured data for a burst energy of $E_b = 2$ mJ are fitted, and an approximately linear relation is found between the intensity ratio I_{Fe438}/I_{Zn472} and the coating thickness d, see Fig. 16.7, which can be described as follows:

$$\frac{I_{Fe438}}{I_{Zn472}}(d) = A + Bd, \tag{16.1}$$

where B is the slope and A the axis intercept. Calculating the inverse function and using standard error calculations with the known errors ΔA, ΔB, $\Delta(I_{Fe438}/I_{Zn472})$, the error of d can be estimated to about 400 nm.

The depth resolution can be further improved based on a systematic line selection [16.14] using multivariate data analysis (MVDA) [16.15–16.18], correction of plasma temperature [16.19, 16.20], and selection of ambient atmosphere [16.21–16.23].

To find optimum experimental conditions and the most sensitive I_{Fe}/I_{Zn} ratio, 21 iron lines and 6 zinc lines were selected in the spectral range from 300 to 500 nm, and all possible I_{Fe}/I_{Zn} ratios were evaluated in terms of thickness resolution using an MVDA software and applying a temperature correction.

The experimental parameters are those described above for Fig. 16.1. In the experimental setup a gas nozzle was added, mounted at a distance to the sample of 12 mm directing the argon (Ar) and nitrogen gas flows under an angle of 45°

onto the sample surface. The estimated gas flows are approximately 170 l/min for Ar and 215 l/min for N_2. The experiments in air were performed at normal pressure without a gas flow.

The time delays used for the signal integration were $t_{delay} = 1.0\,\mu s$ in air and N_2 and $t_{delay} = 1.0$ and $2.0\,\mu s$ in Ar, and the integration window was in all measurements set to $t_{int} = 1\,\mu s$. The times t_{delay} and t_{int} were not optimized. The burst energy was varied in the range of 0.5–2.6 mJ.

In the experiments, the different sheet steels were mounted on an *xy*-translation stage and were moved under the static optical setup. For each detector setting and ambient atmosphere, the burst energy is varied from 0.5 to 2.6 mJ in 22 equidistant steps, and for each energy step 40 measurements are performed. This was repeated for each of the different coating thicknesses, so for each parameter set 40 repetition measurements are available for each of the eight coating thickness giving in total a set of 3 (air, N_2, Ar) $\times$ 22 $\times$ 40 $\times$ 8 $=$ 21,120 spectra.

In the first step, the integrated line intensities of 21 iron lines and 6 zinc lines were calculated by fitting Voigt profiles to each of the lines of each echelle spectrum and subsequent numerical integration. The lines with their atomic data are listed in Table 16.2.

Then the average values of the 40 repetitions for all possible I_{Fe}/I_{Zn} ratios ($6 \times 21 = 126$) for the 22 burst energies and the eight coating thicknesses are calculated. For coating thickness determination, the I_{Fe}/I_{Zn} ratio at a fixed burst energy is plotted as a function of the coating thickness and a fit curve is determined.

The root mean square error (RMSE) value of the fits is defined as follows:

$$\text{RMSE} = 2\sqrt{\frac{\sum\limits_{i=1}^{N}\left(d^{i}_{\text{Zn},m} - d^{i}_{\text{Zn,c.t.v.}}\right)^2}{N}}, \tag{16.2}$$

where $d^{i}_{\text{Zn},m}$ is the thickness determined with the measured I_{Fe}/I_{Zn} ratio and the calibration function, $d^{i}_{\text{Zn,c.t.v.}}$ is the conventional true value of thickness determined by GD-OES, and N is the number of different coating thicknesses (eight in this case). The RMSE value is a measure for the accuracy of the thickness determination related to a specific I_{Fe}/I_{Zn} ratio for a given parameter set, i.e., burst energy, delay time, and ambient atmosphere.

The parameter combinations studied are: 22 burst energies, 3 ambient atmospheres with 2 delay times for Ar, and 1 delay time for N_2 and air, and 126 I_{Fe}/I_{Zn} ratios with and without temperature correction. From this follows a total number of 22,176 calibration curves, which has to be handled. This large number of calibration curves was split in eight groups, for each ambient atmosphere and the two delay times for Ar, with and without temperature correction. Each group comprises 2,772 different I_{Fe}/I_{Zn} ratios (=126 line ratios $\times$22 burst energies) and was evaluated by a commercial software for MVDA (Unscrambler 9.2 from CAMO Technologies), see Table 16.3.

Unscrambler calculates a linear model for the prediction of the coating thickness including all I_{Fe}/I_{Zn} ratios using the partial least squares (PLS) method and returns

Table 16.2 List of studied iron and zinc lines with atomic parameters in the spectral range from 300 to 500 nm

Element	λ (nm)	$\log(g_i f_{ij})$	E_j(eV)	$k_t(10^{-30}\,m^3)$
Group A: iron lines for temperature determination				
Fe I	385.26	−1.24	5.39	0.08
Fe I	392.29	−1.65	3.21	0.61
Fe I	410.75	−0.73	5.85	0.12
Fe I	410.98	−0.91	5.86	0.08
Fe I	411.85	0.28	6.58	0.46
Fe I	428.24	−0.81	5.07	0.27
Fe I	436.98	−0.73	5.88	0.1
Fe I	441.51	−0.62	4.42	0.99
Fe I	449.46	−1.14	4.96	0.14
Fe I	452.86	−0.82	4.91	0.3
Group B: additional iron lines for coating thickness measurement				
Fe I	355.85	−1.12	4.47	0.48
Fe I	382.78	0.06	4.80	3.81
Fe I	384.33	−0.14	6.27	5.3
Fe I	404.58	0.28	4.55	7.74
Fe I	407.17	−0.02	4.65	3.31
Fe I	420.20	−0.71	4.43	0.86
Fe I	430.79	−0.07	4.43	3.54
Fe I	432.58	−0.01	4.47	3.82
Fe I	438.35	0.20	4.31	7.52
Fe I	440.48	−0.14	4.37	3.13
Fe I	495.76	0.13	5.31	1.32
Group C: zinc lines				
Zn I	328.23	−0.38	7.78	1.61
Zn I	330.26	−0.06	7.78	3.3
Zn I	334.50	0.25	7.78	6.36
Zn I	468.01	−0.82	6.65	1.17
Zn I	472.22	−0.34	6.65	3.45
Zn I	481.05	−0.14	6.65	5.32

The lines in group A with low self-absorption ($k_t < 1 \times 10^{-30} m^3$ at $T_e = 8{,}500$ K) were used to determine the plasma temperatures. In the multivariate data analysis, all combinations of iron lines from group A and group B with the zinc lines from group C were used. The quantity k_t is a measure for the optical thickness calculated for 8,500 K [cf. Sect. 9.4, relation (9.33)]. $\log(g_i f_{ij})$ = logarithm of statistical weight of lower level and oscillator strength (cf. relation (9.8)), E_j = energy of upper level

the regression coefficient for each of the I_{Fe}/I_{Zn} ratios for the 22 different burst energies. In the next step all the 2,772 regression coefficient are sorted in ascending order, and an algorithm is applied to find the I_{Fe}/I_{Zn} ratio with the highest regression coefficients within five neighboring energy steps. Hence, I_{Fe}/I_{Zn} ratios with higher regression coefficients at only one burst energy are neglected. In the final step, the model is recalculated with the I_{Fe}/I_{Zn} ratio with the five neighboring energies and the RMSEs are calculated.

Table 16.3 Comparison of the accuracies obtained with the standard I_{Fe438}/I_{Zn472} ratio and the best I_{Fe}/I_{Zn} ratio determined by MVDA

Ambient atmosphere	Delay time t_{delay} (μs)	Temperature correction	Intensity ratio I_{Fe}/I_{Zn}	RMSE (μm)	E_b range (mJ)
I_{Fe438}/I_{Zn472}					
Air	1	No	438.35/472.22	0.51	0.7–1.1
Air	1	Yes	438.35/472.22	0.50	0.6–1.0
Ar	1	No	438.35/472.22	0.41	0.5–0.9
Ar	1	Yes	438.35/472.22	0.41	0.5–0.9
Ar	2	No	438.35/472.22	0.30	0.6–1.0
Ar	2	Yes	438.35/472.22	0.36	0.6–1.0
N_2	1	No	438.35/472.22	0.61	1.0–1.4
N_2	1	Yes	438.35/472.22	0.80	1.2–1.6
I_{Fe}/I_{Zn} *ratios with the highest accuracy*					
Air	1	No	440.48/468.01	0.37	0.5–0.9
Air	1	Yes	410.75/468.01	0.37	1.5–1.9
Ar	1	No	428.24/468.01	0.19	2.2–2.6
Ar	1	Yes	411.85/472.22	0.32	0.5–0.9
Ar	2	No	410.98/328.23[a]	0.14[a]	1.5–1.9
Ar	2	Yes	411.85/328.23	0.20	1.2–1.6
N_2	1	No	495.76/334.50	0.27	2.2–2.6
N_2	1	Yes	382.78/481.05	0.68	1.3–1.7

[a]The highest accuracy is obtained in an Ar atmosphere at $t_{delay} = 2\,\mu s$ without temperature correction using the ratio $I_{Fe410.98}/I_{Zn328.23}$

An issue to be considered arises because the upper levels of an iron line and a zinc line are different. Then the corresponding I_{Fe}/I_{Zn} ratio is a function of the ratio of the number of iron atoms N_{Fe} and zinc atoms N_{Zn} in the ablated crater volume and the temperature, since the line intensity of species i depends on the temperature assuming negligible optical thickness (cf. Sect. 9.2, relations (9.3) and (9.8)):

$$I_i \propto N_i \left(\frac{g_m f_{mn}}{\lambda^3 u(T)}\right) \exp\left(-\frac{E_n}{k_B T_e}\right), \tag{16.3}$$

where N_i is the number of atoms of species i, g_m the degeneracy of the lower level, f_{mn} the oscillator strength, λ the transition wavelength, $u(T)$ the partition function, E_n the energy of the upper energy level, k_B the Boltzmann constant, and T_e the electron temperature.

To take care of this effect and also to compensate temperature fluctuations due to fluctuations of the laser power or different absorbtivity of the sample surface, a temperature correction of the I_{Fe}/I_{Zn} ratios was applied as follows:

$$\frac{I_{Fe}(T_e)}{I_{Zn}(T_e)} \frac{u_{Fe}(T_e)}{u_{Zn}(T_e)} \frac{\exp\left(\dfrac{E_{n,Fe}}{k_B T_e}\right)}{\exp\left(\dfrac{E_{n,Zn}}{k_B T_e}\right)} \propto \frac{N_{Fe}}{N_{Zn}}. \tag{16.4}$$

For each spectrum T_e was determined by a Boltzmann plot using iron lines with low self-absorption, i.e., $k_t < 1 \times 10^{-30}\,m^3$ (at $T_e = 8{,}500\,K$), see Table 16.2, group A. T_e is determined separately for each spectrum yielding values in the range of 6,000–9,000 K for the various parameter settings.

A set of reference samples of electrolytic galvanized sheet steel with well-defined coating thicknesses provided by ThyssenKrupp Steel AG (TKS) was used for the experiments. The set consists of sheet steel plates with different coating thicknesses ranging from 3.1 to 11.2 μm. In the following experiments on samples with coating thicknesses from 4.1 to 11.2 μm were carried out.

Figure 16.8 shows the line intensity ratio I_{Fe438}/I_{Zn472} as a function of the burst energy for the coating thicknesses 4.1, 8.2, and 11.2 μm.

The Zn line at 472.22 nm ($^3S_1 \rightarrow {}^3P_1$, $E_j = 6.65\,eV$) and the Fe line at 438.35 nm ($^4F_5 \rightarrow {}^4F_4$, $E_j = 4.31\,eV$) are both strong lines, and the difference in the upper energy levels ΔE is 2.34 eV. Figure 16.8a shows the ratio I_{Fe438}/I_{Zn472} as a function of the burst energy without a temperature correction (for the interpretation of the shown curve progression see discussion given above in connection with Fig. 16.5). Figure 16.8b shows the same data applying a temperature correction as described above [cf. relation (16.4)]. Now the I_{Fe438}/I_{Zn472} values of all three coating thicknesses increase with the burst energy as expected. But it can be seen in both figures that at any burst energy the I_{Fe438}/I_{Zn472} ratio is higher for thinner coatings. Hence, the I_{Fe438}/I_{Zn472} ratio at a fixed burst energy can be used as a measure for the coating thickness, cf. Fig. 16.7. Table 16.3 lists the I_{Fe}/I_{Zn} ratios with the lowest RMSE values determined at a specific parameter set by MVDA together with the RMSE values of the standard I_{Fe438}/I_{Zn472} ratio used, e.g., in Fig. 16.7.

In each case, the RMSE value is given for the burst energy range leading to the lowest RMSE values. Table 16.3 shows the effects of the ambient atmosphere, the delay time, and the temperature correction on the RMSE values of the standard I_{Fe438}/I_{Zn472} ratio. For a delay time of 1 μs, and without temperature correction, the thickness resolution is best in Ar (0.41 μm) and worst in N_2 (0.61 μm). Physically, reasonable temperature correction does not enhance thickness resolution for an N_2 atmosphere – it makes it even worse. In Ar atmosphere, a delay time of 2 μs enhances the resolution from 0.41 to 0.3 μm.

The smallest RMSE values for a specific parameter set are found for the I_{Fe}/I_{Zn} ratios listed in the second part of Table 16.3. For each of the parameter sets, another I_{Fe}/I_{Zn} ratio was actually optimal, but most of the lines are lines with low self-absorption ($k_t < 1.5 \times 10^{-30}\,m^3$, $T_e = 8{,}500\,K$), cf. Table 16.2. In Ar, these values are more than a factor of 2 smaller than the RMSE values of the I_{Fe438}/I_{Zn472} ratios. Again, temperature correction does not improve the resolution. The best thickness resolution of 140 nm was achieved in Ar atmosphere at a delay time of 2 μs with the I_{Fe410}/I_{Zn328} ratio. Figure 16.9 visualizes the increase in thickness resolution in comparison to the I_{Fe438}/I_{Zn472} ratio. The plot shows the difference between the thicknesses $d_{Zn,m}$ determined with the calibration model and the measured I_{Fe}/I_{Zn} ratio and the conventional true coating thickness $d_{Zn,c.t.v.}$ as a function of $d_{Zn,c.t.v.}$

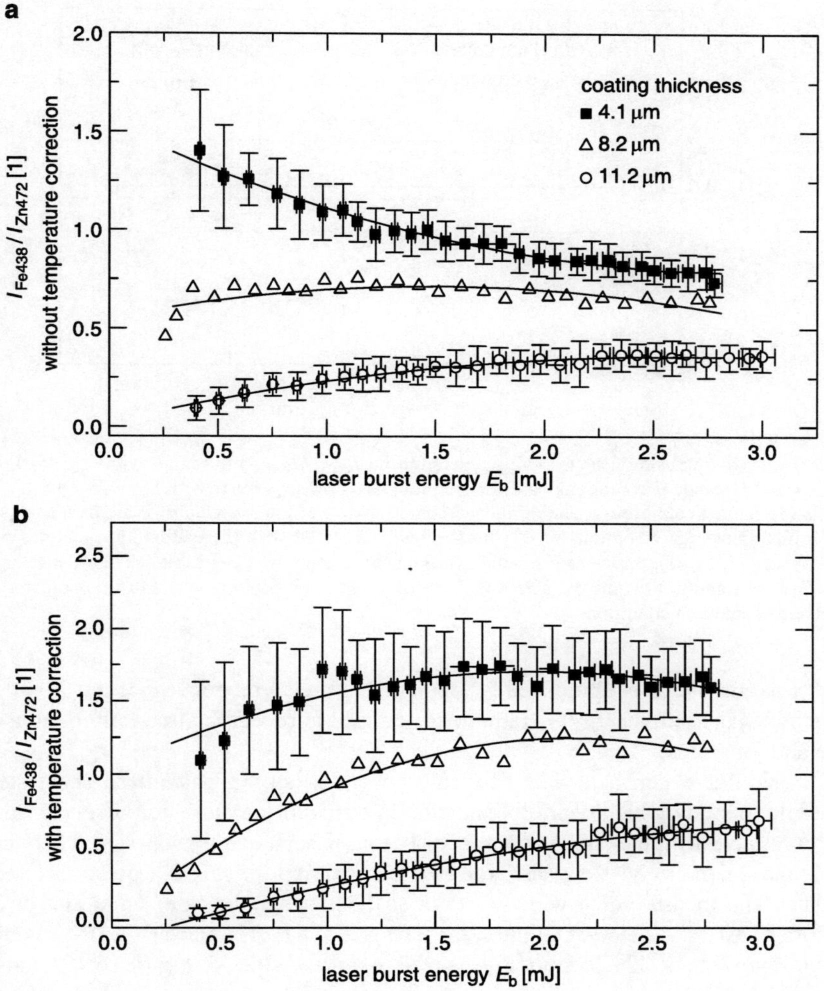

Fig. 16.8 I_{Fe438}/I_{Zn472} ratio as a function of the burst energy for different coating thicknesses: (**a**) without temperature correction, (**b**) with temperature correction. In both cases, the ratio is higher for thinner coatings and can be used as a measure for the coating thickness

The thickness resolution is further enhanced from 510 nm down to 140 nm, i.e., by more than a factor of 3.

Applying the temperature correction did not appear to lead to any improvement in the thickness resolution. An explanation for this is that the plasma temperature also depends on the coating thickness. For a particular burst energy, the plasma temperature was found to be higher for thinner coatings ($\Delta T_e \sim$ 500–1,000°K higher for $d_{Zn} = 4.1\,\mu$m than for $d_{Zn} = 11.2\,\mu$m). Hence, T_e contains also depth

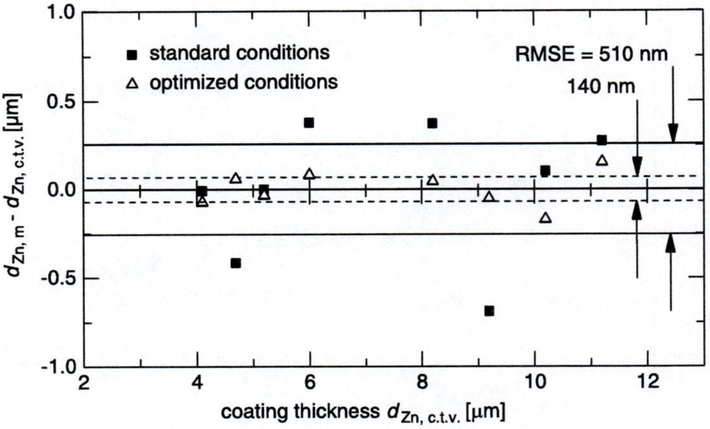

Fig. 16.9 Comparison of the resolving power for the I_{Fe438}/I_{Zn472} ratio in air with $t_{delay} = 1\,\mu s$, $t_{int} = 1\,\mu s$ (standard conditions) and for the ratio $I_{Fe410.9}/I_{Zn328}$ in Ar with $t_{delay} = 2\,\mu s$, $t_{int} = 1\,\mu s$ (optimized conditions). The optimized conditions were found by multivariate data analysis. The plot shows the deviations of the thicknesses determined with the calibration functions and the measured I_{Fe}/I_{Zn} ratios as a function of the conventional true coating thickness determined by GD-OES and wet-chemistry. The RMSE value at optimum conditions is 140 nm compared to 510 nm at standard conditions

information which is eliminated by the temperature correction. Since a calibration method was used, the I_{Fe}/I_{Zn} ratio does not need to reflect the true ratio of ablated Fe and Zn atoms.

A similar experiment was performed using a hot-dip galvanized sheet steel sample to study which depth information in terms of the aluminum concentration is accessible with the studied approach. The laser burst energy was varied between 0.2 and 2.6 mJ in 54 steps, and for each energy step 30 single acquisitions were taken. The measurement window starts at 0.5 µs after the laser burst and has a width of 10 µs. As elemental lines, the zinc line at 472.22 nm ($^3S_1 \rightarrow {}^3P_1$) and the aluminum line at 396.15 nm ($^2S_{1/2} \rightarrow {}^2P_{3/2}$) were selected. Figure 16.10a shows the line intensities I_{Al396} and I_{Zn472} as a function of the burst energy. Both element lines can be seen from the beginning as there is an Al enrichment mainly in the form of Al_2O_3 at the surface typical for hot-dip galvanized coatings [16.24]. The error bars indicate the standard deviation of 30 LIBS signals. In the referenced signal I_{Al396}/I_{Zn472} (Fig. 16.10b), a finger print of the Al depth profile can be seen. The decline in the ratio at low burst energies may be attributed to the penetration of an Al_2O_3 surface layer with a typical thickness of less than 10 nm [16.24]. A high Al concentration at the surface was also found in the LIBS depth profiles of galvannealed (annealed hot-dip galvanized) sheet steel by St-Onge et al. [16.5]. The I_{Al396}/I_{Zn472} ratio attains a minimum right before the ablation depth reaches the depth of the inhibition layer. Then the ratio increases as more and more Al from the inhibition layer and the steel substrate (Al is alloyed in the steel substrate, 2 m.-%)

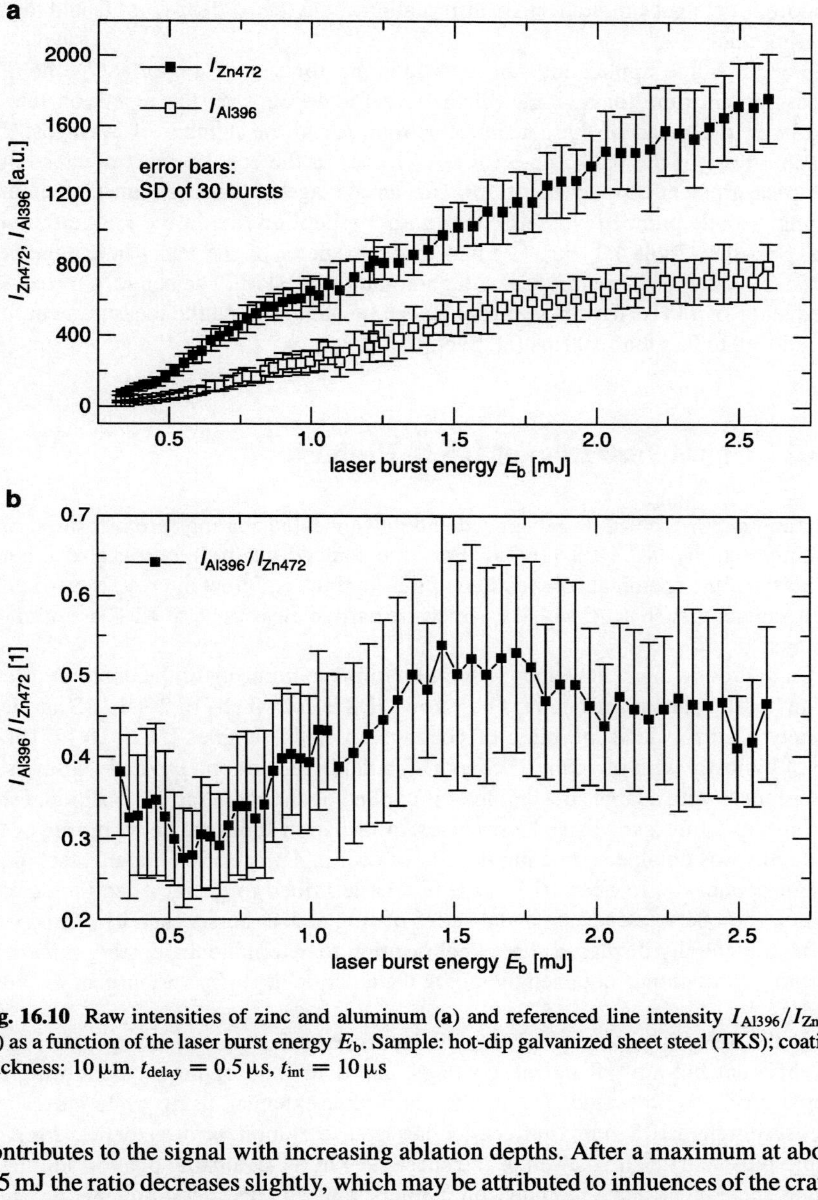

Fig. 16.10 Raw intensities of zinc and aluminum (**a**) and referenced line intensity I_{Al396}/I_{Zn472} (**b**) as a function of the laser burst energy E_b. Sample: hot-dip galvanized sheet steel (TKS); coating thickness: 10 μm. $t_{delay} = 0.5\,\mu s$, $t_{int} = 10\,\mu s$

contributes to the signal with increasing ablation depths. After a maximum at about 1.5 mJ the ratio decreases slightly, which may be attributed to influences of the crater geometry on the plasma dynamics and the plasma state. More details concerning the characterization of elemental depth profiles will be described in Sect. 16.2.

The results demonstrate the potential of LIBS for the inline analysis of galvanized sheet steel. The estimated thickness resolution of less than 200 nm is comparable to the resolution limit achievable with inline XRF gauges. In addition, LIBS offers the opportunity to get simultaneously information about the Al depth profile and the Zn coating thickness.

For an inline application, the measurement time is an important point. The measurement time for a single thickness value depends on the repetition rate of the laser, the speed of the detector electronics, and the number of averages. The readout time of the ICCD detector (~2 s) used in the echelle spectrometer limits the measurement time to about 40 s (for an average of 20 single measurements). Using a diode-pumped Nd:YAG laser with a repetition rate in the kilohertz range (cf. Sect. 4.1, Table 4.3, Fig. 4.3) and a spectrometer of the Paschen–Runge type (cf. Sect. 4.2, Fig. 4.7) equipped with photomultiplier (PMT) detectors, a measuring frequency of 1 kHz (for single readouts) can be obtained and the measurement time is reduced to less than 100 ms (cf. Sect. 15.1).

16.2 Characterization of Depth Profiles

Coating depth profiles of galvanized and galvannealed coatings are measured only offline, e.g., by GD-OES [16.25]. For an enhanced quality control, also a depth analysis of the chemical composition of the coating is of great interest. In particular, light elements such as Al and Mg, which are hard to measure with XRF, are of major interest.

In the following, a new approach for an inline monitoring of the depth profile of Al in hot-dip galvanized coatings based on LIBS is described [16.26]. LIBS has been widely used for depth profiling of coatings on static samples [16.5–16.12, 16.27, 16.28], mostly with the aim to achieve high depth resolution. Typically, a series of laser pulses with a large spot diameter is applied on the same sample position. From the spectra of the successive laser pulses, information about the depth profile of the elements was obtained. On a moving sheet steel in a production line this technique is not applicable. In Sect. 16.1, a method is described to characterize the coating with a stationary laser beam and a series of single laser bursts each irradiated on a different laterally displaced sheet steel position to determine the coating thickness. Depth information is obtained by tuning the ablation depth by variation of the burst energy.

In the following, the method is applied to characterize the aluminum depth profile of different hot-dip galvanized coatings, and a method of inline monitoring the depth profile is presented. The method is further extended using triple pulses and measuring the LIBS signals after each laser pulse to monitor three depths for each triple-pulse burst. The motion of the sheet steel in a galvanizing plant is simulated using sheet steel disks mounted on a rotary table. LIBS measurements on sheet steel samples moving with up to ~1 m/s were performed and compared to results obtained on slowly moving sheet steel.

Figure 16.11 shows the experimental setup [16.26]. The Q-switch pulses of a flashlamp pumped Nd:YAG-laser operating at 1,064 nm and modified for the emission of up to six pulses at 10 Hz repetition rate (interpulse separation $\Delta t \geq 2\,\mu s$,

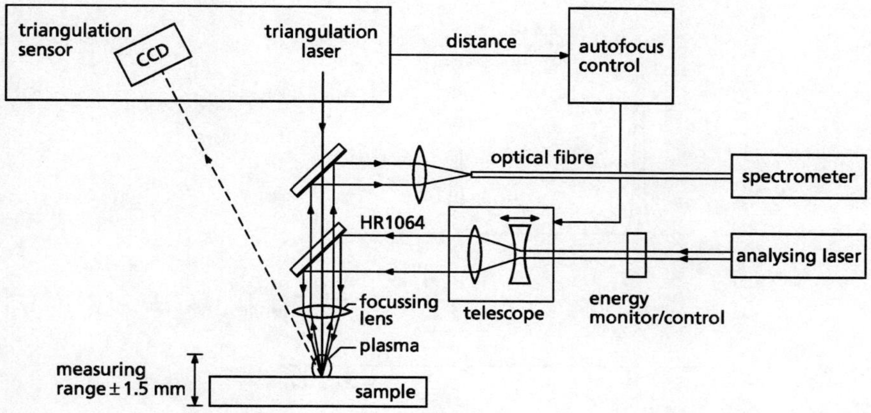

Fig. 16.11 Optical setup with autofocus system

cf. Sects. 13.1.1 and 13.1.3) are guided via a dichroic mirror (high reflectivity for 1,064 nm, transparent for the plasma light) and focused by an achromatic lens ($f = 60$ mm) onto the surface of the sheet steel, where a small amount of the sample is evaporated and the microplasma is generated. The laser is operated in double pulse mode (interpulse separation of $\Delta t_1 = 4\,\mu$s, energy ratio $E_1 : E_2 = 1 : 1$, burst energy $E_b = E_1 + E_2 = 0.1$–2.1 mJ, see Fig. 16.12a) or triple pulse mode (interpulse separations of $\Delta t_1 = 4\,\mu$s, $\Delta t_2 = 4\,\mu$s, energy ratio $E_1 : E_2 : E_3 = 1 : 1 : 0.7$, burst energy $E_b = E_1 + E_2 + E_3 = 0.165$–$1.65$ mJ, see Fig. 16.12b) to achieve an ablation depth with a single burst exceeding the range of the coating thickness (ca. 10 μm) to be measured. The attenuation of the burst energy is achieved by an external Pockels cell. Attenuation of up to a factor of 10 is possible. With the focus diameter of approximately 23 μm and the widths of the individual laser pulses in the burst of ~15–20 ns (depending on the number of pulses in the burst) power densities in the range of 1–20 GW/cm^2 are reached.

The plasma light is collected and collimated with the same lens, passes the dichroic mirror, is reflected by a second dichroic mirror, and is reimaged by a quartz lens ($f = 100$ mm) onto the core of an optical fiber which guides the light to a Paschen–Runge spectrometer (diameter of Rowland circle 750 mm, resolution ~30 pm) equipped with 16 PMTs placed on the Rowland circle at the positions of specific elemental lines. The elemental lines used are the Al line at 396.15 nm ($^2S_{1/2} \rightarrow {}^2P_{3/2}$), the Zn line at 472.22 nm ($^3S_1 \rightarrow {}^3P_1$), and the Fe line at 438.35 nm ($^4F_5 \rightarrow {}^4F_4$). The photocurrent of the PMTs is integrated and digitized by a multichannel integrator electronics (cf. Sect. 4.5), which allows for sampling rates of up to 1 kHz and programmable time delays t_{delay} and width of the integration window t_{int} with a resolution of 50 ns for each pair of channels. In addition, it is possible to set the integration windows after each pulse of one laser burst. The time delays and integration windows used are different for the various elemental lines and for the pulses within the laser burst. The specific settings are given below, but

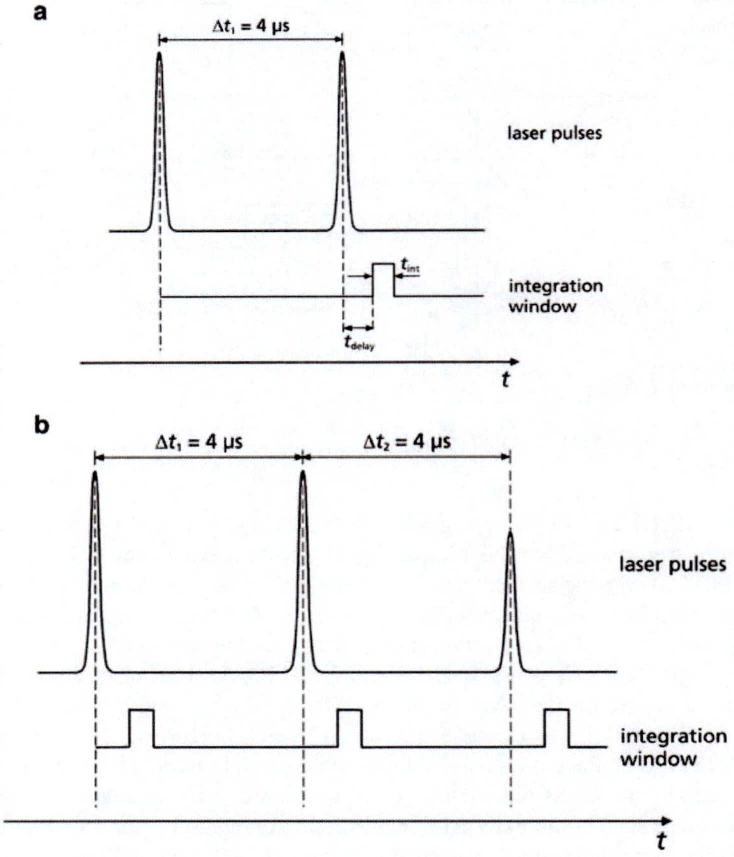

Fig. 16.12 Temporal structure of the laser bursts and the settings of the integration windows used in this study: (**a**) Double pulses with interpulse separation $\Delta t_1 = 4\,\mu\text{s}$ and burst energy distributed equally, signal integration after pulse 2; (**b**) Triple pulses with interpulse separation $\Delta t_1 = 4\,\mu\text{s}$, $\Delta t_2 = 4\,\mu\text{s}$, and $E_1 : E_2 : E_3 = 1 : 1 : 0.7$, signal integration after pulse 1, 2, and 3

generally the delay times t_{delay} were in the range $0.3 - 0.5\,\mu\text{s}$ and the widths of the integration windows were in the range $t_{\text{int}} = 0.5 - 3\,\mu\text{s}$. Short delay times are recommended to maximize the signals observed, as only small plasmas are generated. According to investigations on microplasmas performed by Detalle et al. with comparable focusing conditions, also LTE can be assumed at such delay times [16.29]. The optical setup contains also an autofocus system for the compensation of sheet steel vibrations of the rotating sheet steel disks. It is designed to compensate vibrations of the measuring object with amplitudes of up to 1.5 mm which is in the range of the vibrations occurring typically at S-rolls in galvanizing plants. The autofocus system is based on a laser triangulation sensor and a telescope with a motorized dispersing lens, see Fig. 16.11. The laser beam of the triangulation sensor passes the two dichroitic mirrors and is aligned orthogonal and central through the focusing lens. The distance of the sensor to the sample surface is measured and transferred to a control PC. The second component is a telescope with a motorized dispersing lens, whereby the beam divergence can be adjusted to shift the focus position of the analyzing laser beam. A control software calculates the shift of the dispersing lens to keep the focus position of the analyzing laser beam constant relative to the sample surface.

The new method is developed for coating thickness measurement and depth profiling of moving sheet steel in a production line, i.e., under conditions where only single laser bursts can be applied to one distinct sheet steel position. The approach to obtain depth information is to control the crater depth by changing the burst energy in small increments and to measure the LIBS signal for every single laser burst. Figure 16.2 (bottom) shows schematically the applied sequence of laser bursts. The number of craters per energy step is 80–200 and the energy is tuned in more than 50 steps. The signal gained for one distinct burst energy is a weighted mean of the chemical composition over the whole ablation volume. A set of measurements is performed with the same burst energy to determine an average value of the line intensities for the corresponding ablation depth. For the next set, the burst energy is tuned to the next value. The averaging area is the length of the sheet steel traversing during one measurement cycle and is determined by the speed of the sheet steel, the laser repetition rate, and the number of measuring bursts. By referencing the Al line intensity and the Fe line intensity to the Zn line intensity of the main element zinc, the influence of fluctuations of the measuring process on the spectroscopic information is reduced.

The measuring conditions were simulated by moving the sheet steel samples under the focusing objective. Two different translation stages were used. For quasi-static experiments and small sheet steel pieces (dimensions: 10 cm × 15 cm), a motorized *xy*-translation stage was used as described in Sect. 16.1. For the experiments on moving sheet steel disks (diameter 490 mm), a rotating table was used. The table rotates with a frequency of up to 1.66 Hz leading to velocities of more than 2 m/s in the outer part of the disk. In order to move to different measuring radii, the rotating table is mounted on a linear translation stage.

The resulting crater patterns are rectangular arrays (*xy*-translation stage) or spirals (rotating table). The whole setup is automated to synchronize sheet steel motion, laser operation, and PMT readout.

Different industrial hot-dip galvanized samples provided by ThyssenKrupp Steel AG (TKS) and by OCAS N.V. (OCAS, Zelzate, Belgium) were used as samples [16.30]. The samples from OCAS (OCAS II) had a coating thickness of about 10 μm. Rectangular pieces (10 cm × 15 cm) and sheet steel disks were available. A GD-OES analysis shows that 50% of the maximum Al concentration is attained at a depth of about 6.2 μm. The TKS samples had coating thicknesses of 11 μm (TKS I), 10 μm (TKS IV), and 7 μm (TKS III). The TKS samples were cut in rectangular pieces with the dimensions 10 cm × 15 cm (TKS I) and 5 cm × 5 cm (TKS III, TKS IV). The GD-OES depth profile of TKS I shows that also Al is present in the steel substrate, whereas the samples TKS III, TKS IV, and OCAS II

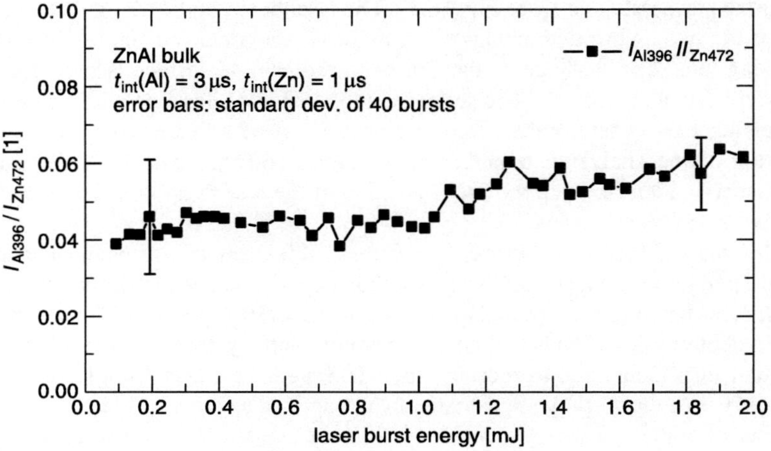

Fig. 16.13 I_{Al396}/I_{Zn472} versus burst energy measured on a ZnAl binary bulk sample (c_{Al} = 0.14 m.-%), t_{int} = 3 μs for the Al396-channel and t_{int} = 1 μs for the Zn472-channel, delay time t_{delay} = 0.5 μs. The I_{Al396}/I_{Zn472} ratio is nearly constant in the investigated burst energy range up to 1 mJ

had no Al in the substrate. For the TKS I sample, the Al concentration reaches 50% of its maximum value at a depth of 10 μm.

For comparative measurements, a certified binary reference ZnAl-bulk sample (Breitländer, MBH 41XZ5) with an Al concentration of c_{Al} = 0.14 m.-% comparable to hot-dip galvanized coatings was used.

In the measurement procedure, the delay time t_{delay} and the width of the measurement window t_{int} are kept constant while changing the burst energy over more than one order of magnitude. As the plasma parameters – like temperature or size – are changing with the burst energy, it is important to know how this influences the measured I_{Al396}/I_{Zn472} ratio. Also fractionation can have an influence, as found for zinc in brass for power densities lower than 0.3 GW/cm^2 [16.31]. Therefore, the procedure was initially applied to the ZnAl binary reference sample using double pulse bursts with burst energies of 0.1–2.1 mJ (burst energy equally distributed on double pulses with interpulse separation Δt_1 = 4 μs), see Fig. 16.12a. The delay time t_{delay} was set to 0.5 μs and the integration windows t_{int} were 3 μs for the Al channel and 1 μs for the Zn channel. The resulting referenced line intensity I_{Al396}/I_{Zn472} shown in Fig. 16.13 is nearly independent of the burst energy for energies up to 1 mJ, i.e., the ratio of the number of Al and Zn atoms in the crater volume is constant, and so it is not necessary to correct the measurements made on the sheet steel samples. The slight increase at E_b > 1 mJ is due to saturation of the Zn472 channel, whereas the signal of the Al396 channel continues to increase.

In the experiments on the hot-dip galvanized sheets, it was checked what depth information for the aluminum concentration is made available with the studied approach. The procedure described above was applied to the hot-dip galvanized sheet steel samples OCAS II and TKS I (coating thicknesses: 10 and 11 μm, respectively). For both samples, the laser burst energy was varied between 0.1 and 2.1 mJ as for the ZnAl-bulk sample. For each energy step, 200 single acquisitions were performed. The delay time t_{delay} was also set to 0.5 μs, the width of the integration window was 3.0 μs for the Al line at 396.15 nm and 0.5 μs for the zinc line at 472.22 nm, which is shorter as for the ZnAl-bulk sample to avoid a saturation of this channel. For the different sheet steels, the intensity ratio I_{Al396}/I_{Zn472} as a function of the burst energy is shown in Fig. 16.14. The data points represent the mean values measured at each energy step, and the black lines are interpolated curves calculated using a polynomial fit. In this representation, the samples OCAS II and TKS I can be distinguished clearly. Both curves start at a ratio of about 0.05, but the OCAS II curve increases much more quickly and reaches a slightly higher end value at 2.1 mJ. This behavior can be explained by the different Al depth profiles of the sheet steels. The Al concentration starts to increase at about 6 μm for OCAS II and at about 10 μm for TKS I. Assuming comparable ablation behavior at each particular burst energy for both sheet steels, the region of high Al concentration is reached at a lower burst energy for OCAS II. Therefore, the relative number of Al atoms in the ablated crater volume is also higher for OCAS II, which leads to a higher intensity ratio I_{Al396}/I_{Zn472}. Thus, the depth profile obviously influences the ratio I_{Al396}/I_{Zn472} measured as a function of the burst energy.

Based on these results, a method for a continuous inline detection of variations in the aluminum depth profile can be deduced. The target Al profile is represented by the ratio

$$R_t(E_b) = \frac{I_{Al396}}{I_{Zn472}}(E_b)_t. \tag{16.5}$$

The actual Al profiles are represented by the continuously measured ratio

$$R_m(E_b) = \frac{I_{Al396}}{I_{Zn472}}(E_b)_m. \tag{16.6}$$

Deviations from the target profile can then be determined by calculating the ratio $R_m(E_b)/R_t(E_b)$. To avoid artifacts arising from scatter in the data, it is best to determine polynomial fits to the data points $R_m(E_{bi})$ and $R_t(E_{bi})$ in a first step, and then to divide the two polynomial functions. In the ideal case of an unchanged profile, the ratio of the ratios would be constant and equals 1, see Fig. 16.15 (dotted line).

To simulate the case of changing profiles, Al depth profile of the OCAS II sample is defined as the target profile and the Al depth profile of TKS I as the profile to be monitored. The ratio of the two fitting functions (black lines in Fig. 16.14) is shown in Fig. 16.15. This curve shows the variation in the Al depth profile for the sample TKS I relative to that for sample OCAS II even better than the graphs in Fig. 16.14. At low burst energy, the curve falls due to the higher Al concentration of TKS I (~0.15 m.-%) compared to OCAS II (~0.1 m.-%) up to a depth of about 5–6 μm. Then it increases as the region of high Al concentration is reached for the OCAS

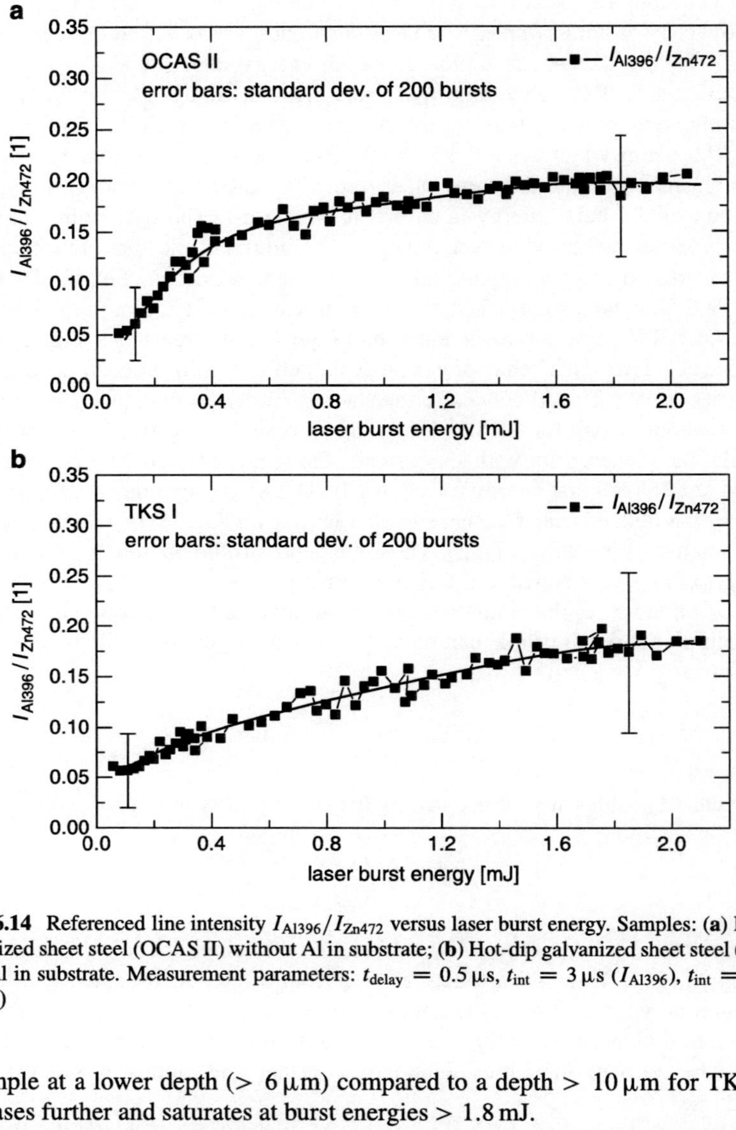

Fig. 16.14 Referenced line intensity I_{Al396}/I_{Zn472} versus laser burst energy. Samples: (**a**) Hot-dip galvanized sheet steel (OCAS II) without Al in substrate; (**b**) Hot-dip galvanized sheet steel (TKS I) with Al in substrate. Measurement parameters: $t_{delay} = 0.5\,\mu s$, $t_{int} = 3\,\mu s$ (I_{Al396}), $t_{int} = 0.5\,\mu s$ (I_{Zn472})

II sample at a lower depth ($> 6\,\mu m$) compared to a depth $> 10\,\mu m$ for TKS I. It increases further and saturates at burst energies > 1.8 mJ.

Since a laser burst consists of several pulses, one approach to enhancing the depth information is to measure also between the pulses of one laser burst as well. The first pulse ablates to an initial depth 1 and the LIBS signal contains the integrated depth information up to this depth. The following pulses of the burst ablate deeper and the LIBS signals correspond to larger depths. This approach was tested with hot-dip galvanized samples of different coating thickness (TKS III, $d_{Zn} = 7\,\mu m$; TKS IV, $d_{Zn} = 10\,\mu m$) using triple pulses. The burst parameters are shown in Fig. 16.12b. The delay times after a pulse was set to $0.3\,\mu s$. The widths of the

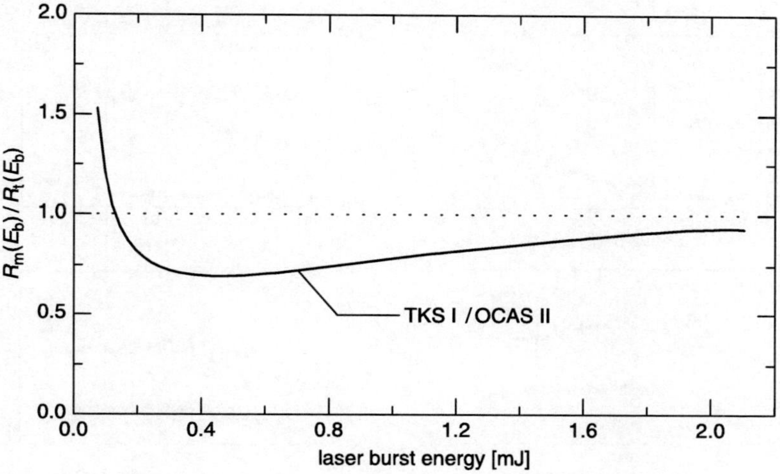

Fig. 16.15 Ratio of the intensity ratios $I_{Al396.15}/I_{Zn472.22}$ of the samples TKS I and OCAS II of Fig. 16.14 as a function of the burst energy. The curves visualize the deviation of the Al depth profiles of the TKS I sample from that of sample OCAS II

integration windows were varied; they were shorter for Zn472 and Fe438 to avoid saturation at higher burst energies, and longer for Al396 to get a sufficient signal at low burst energies. The integration windows were 0.5 μs for Zn472 and Fe438 and 2.0 μs for Al396 after pulse 1, 1.0 μs for Zn472, Fe438, and Al396 after pulse 2, and 0.7 μs for Zn472, and Fe438 and 2.0 μs for Al396 after pulse 3. The burst energy was tuned in 51 steps from 0.165 to 1.65 mJ, and 80 measurements were made at each energy step.

Figure 16.16 shows the corresponding I_{Al396}/I_{Zn472} versus E_b graphs and I_{Fe438}/I_{Zn472} versus E_b graphs of the 7–μm (Fig. 16.16, left) and the 10–μm (Fig. 16.16, right) samples measured after each pulse of the triple burst. The burst energy axis displayed for pulse 1 (depth 1) is the energy of the first pulse, for pulse 2 (depth 2) it is the total energy of the first pulse and second pulse, and for pulse 3 (depth 3) it is the complete burst energy. Figure 16.16a shows that both coatings are already penetrated with the first pulse, as the I_{Fe438}/I_{Zn472} ratio starts to increase at about 0.25 mJ for the 7–μm coating and at about 0.5 mJ for the 10–μm coating. As the Fe_2Al_5-inhibition layer is located at the interface between coating and substrate, it is approximately penetrated at the same energies. This is reflected in the trends seen in the I_{Al396}/I_{Zn472} ratios. For both coatings, the ratio starts at about 0.04 and increases with increasing burst energy, but the increase is faster for the thinner coating because the Al-rich inhibition layer is reached at 0.25 mJ. The corresponding ratio I_{Al396}/I_{Zn472} is ∼0.06, whereas for the thicker coating the value of 0.06 is reached at a significantly higher burst energy of approximately 0.35 mJ. The I_{Al396}/I_{Zn472} ratios and I_{Fe438}/I_{Zn472} ratios after pulse 2 are shown in Fig. 16.16b. Both ratios increase with increasing burst energy, but the increase

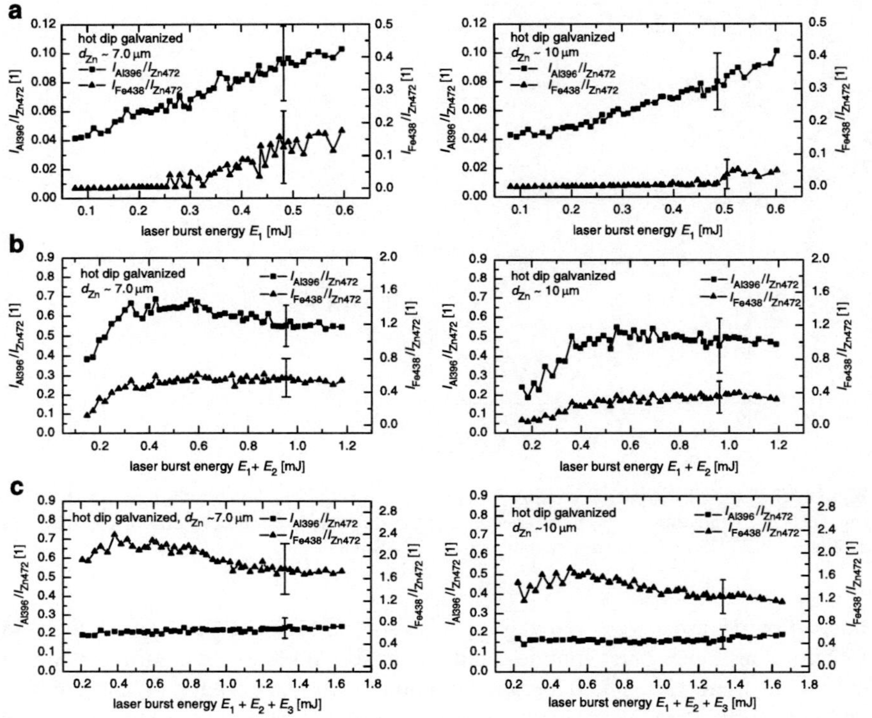

Fig. 16.16 $I_{\text{Al396}}/I_{\text{Zn472}}$ ratios and $I_{\text{Fe438}}/I_{\text{Zn472}}$ ratios as a function of the burst energy for coating thicknesses of $d_{\text{Zn}} = 7.0\,\mu\text{m}$ (TKS III) and $d_{\text{Zn}} = 10.0\,\mu\text{m}$ (TKS IV) for different ablation depths. (**a**) Pulse 1 (depth 1), $t_{\text{delay}} = 0.3\,\mu\text{s}$, $t_{\text{int}} = 2.0\,\mu\text{s}$ (I_{Al396}), $t_{\text{int}} = 0.5\,\mu\text{s}$ (I_{Zn472}, I_{Fe438}). (**b**) Pulse 2 (depth 2), $t_{\text{delay}} = 0.3\,\mu\text{s}$, $t_{\text{int}} = 1.0\,\mu\text{s}$ (I_{Al396}), $t_{\text{int}} = 1.0\,\mu\text{s}$ (I_{Zn472}, I_{Fe438}). (**c**) Pulse 3 (depth 3), $t_{\text{delay}} = 0.3\,\mu\text{s}$, $t_{\text{int}} = 2.0\,\mu\text{s}$ (I_{Al396}), $t_{\text{int}} = 0.7\,\mu\text{s}$ (I_{Zn472}, I_{Fe438})

is faster and the absolute values are higher for the thinner coating. This is due to the higher number of ablated Zn atoms in the case of the thicker coating. The coating is penetrated even at the lowest energy. After the third pulse (depth 3), the $I_{\text{Al396}}/I_{\text{Zn472}}$ ratios are approximately constant, i.e., the coating is completely penetrated. Again the $I_{\text{Al396}}/I_{\text{Zn472}}$ ratio and $I_{\text{Fe438}}/I_{\text{Zn472}}$ ratio are higher for the thinner coating.

The results discussed so far were all measured on a slow moving *xy*-translation stage. In order to check the applicability of the technique for an inline monitoring on moving sheet steel in a galvanizing plant and to study the influence of the sheet steel speed on the LIBS signals, experiments were performed on rotating sheet steels of the type OCAS II using the setup with the rotary stage. As the rotary axis is not perfectly orthogonal to the rotary plane, a slight variation in the distance of $300\,\mu\text{m}$ between the focusing lens and the sheet steel surface occurs within a full rotation. Four different sheet steel velocities were tested. In the first step, a measurement

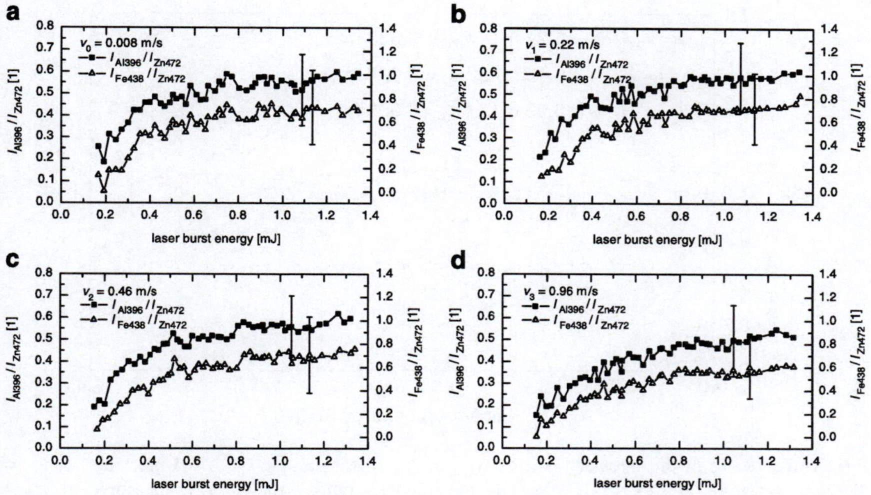

Fig. 16.17 Referenced line intensity I_{Al396}/I_{Zn472} and I_{Fe438}/I_{Zn472} versus laser burst energy for different sheet steel velocities. Sample: OCAS II. Measurement starts after pulse 2 of double pulse. Parameters: $t_{delay} = 0.25\,\mu s$, $t_{int} = 3\,\mu s$ (I_{Al396}), $t_{int} = 1.0\,\mu s$ (I_{Zn472}, I_{Fe438}). **(a)** $v_0 = 0.008\,m/s$, **(b)** $v_1 = 0.22\,m/s$, **(c)** $v_2 = 0.46\,m/s$, **(d)** $v_3 = 0.96\,m/s$

was taken on a slowly moving sheet steel ($v_0 \sim 0.008$ m/s), which is the reference measurement. Double pulses were used, and the burst energy was varied between 0.13 and 1.3 mJ in 51 steps; the intensities of the lines Al396, Zn472, and Fe438 were measured after the second pulse of the burst. In Fig. 16.17a, the I_{Al396}/I_{Zn472} ratio and I_{Fe438}/I_{Zn472} ratio are shown as a function of the burst energy. The result for the I_{Al396}/I_{Zn472} ratio is different to that obtained on the *xy*-translation stage, see Fig. 16.14a, due to the different settings of the integration delay and window. Figure 16.17b–d shows the results for sheet steel velocities of $v_1 = 0.22$, $v_2 = 0.46$, and $v_3 = 0.96\,m/s$. All the graphs look very similar, although both ratios are slightly lower at 0.96 m/s. To see the influence of the sheet steel speed in more detail, the I_{Al396}/I_{Zn472} ratios measured at $v_1 = 0.22$, $v_2 = 0.46$, and $v_3 = 0.96$ m/s are referenced to the I_{Al396}/I_{Zn472} ratio measured at $v_0 = 0.008$ m/s, see Fig. 16.18. The graphs were generated similar to Fig. 16.15, i.e., first a polynomial fit is applied to the curves in Fig. 16.17 and after that the fit curves are divided. For 0.22 and 0.46 m/s, and burst energies $\geq$0.2 mJ the ratio is $\sim$1.

The variations from this – in particular at low burst energies for 0.46 m/s – are due to the uncertainty associated with the fits. However, clear deviations can be seen for 0.96 m/s. On average the I_{Al396}/I_{Zn472} ratio measured at 0.96 m/s is approximately 15% lower than that seen at the other sheet steel speeds. This effect may be due to the limited speed of the autofocus system, which means that the sample surface is on average slightly out of focus, meaning that the penetration depth at a given burst energy and therefore the relative number of Fe and Al atoms is reduced.

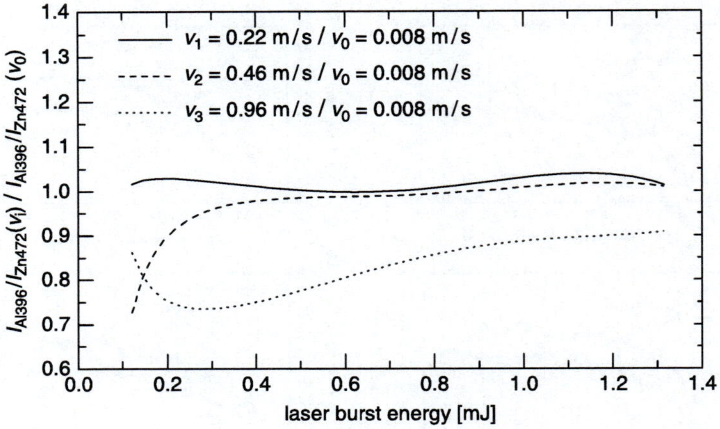

Fig. 16.18 Ratio of the intensity ratios I_{Al396}/I_{Zn472} of the sample OCAS II measured at $v_1 = 0.22$, $v_2 = 0.46$, and $v_3 = 0.96\,m/s$ to the intensity ratio $I_{Al396}/I_{Zn4,722}$ measured at $v_0 = 0.008\,m/s$ as a function of the burst energy. The graphs were generated similar to Fig. 16.15 and visualize the effect of the sheet steel speed on the intensity ratio I_{Al396}/I_{Zn472}, see text

The results demonstrate the potential of LIBS for inline monitoring of the aluminum depth profile of galvanized coatings and offers the possibility to further improve the quality control in galvanizing plants.

For an inline monitoring, the measurement time is an important issue. For the described experiments, this time is limited by the repetition rate of the applied flashlamp pumped Nd:YAG laser of 10 Hz. By replacing this laser by a diode-pumped solid state laser (DPSSL, cf. Sect. 4.1), the acquisition rate can be increased up to 1 kHz. Due to the better beam quality of the DPSSL of $M^2 \sim 1$ also the Rayleigh length of the focus is increased, and the method is expected to become less sensitive to sheet steel vibrations. The measurement time for an Al depth profile will be in the range of 5 s (50 burst energy steps and 100 averages). In combination with the method for inline measurement of the Zn coating thickness, the Al depth profile can be monitored simultaneously [16.13]. As LIBS is sensitive to other light elements such as Mg, this method is as well of interest for the inline analysis of magnesium zinc coatings and may become an alternative method to XRF for inline monitoring of metallic coatings [16.32].

References

16.1. The role of aluminum in continuous hot-dip galvanizing. http://www.galvinfo.com/GINotes/G_Note10.PDF (2004)

16.2. Metallic coatings – measurement of coating thickness – X-ray spectrometric methods, ISO-Norm 3497: (E) (2000)

16.3. Ed. R. Payling, D. Jones, A. Bengtson, *Glow Discharge Optical Emission Spectrometry* (Wiley, New York, 1997)
16.4. D. Anderson, C. McLeod, T. English, A. Smith, Depth profile studies using laser-induced plasma emission spectrometry. Appl. Spectrosc. **49**, 691–701 (1995)
16.5. L. St-Onge, M. Sabsabi, Towards quantitative depth-profile analysis using laser-induced plasma spectroscopy: investigation of galvannealed coatings on steel. Spectrochim. Acta B **55**, 299–308 (2000)
16.6. V. Margetic, M. Bolshov, A. Stockhaus, K. Niemax, R. Hergenröder, Depth profiling of multi-layer samples using femtosecond laser ablation. J. Anal. At. Spectrom. **16**, 616–621 (2001)
16.7. M. Mateo, J. Vadillo, J. Laserna, Irradiance-dependent depth profiling of layered materials using laser-induced plasma spectrometry. J. Anal. At. Spectrom. **16**, 1317–1321 (2001)
16.8. J. Vadillo, C. Garcia, S. Palanco, J. Laserna, Nanometric range depth-resolved analysis of coated-steels using laser-induced breakdown spectrometry with a 308 nm collimated beam. J. Anal. At. Spectrom. **13**, 793–797 (1998)
16.9. C. Garcia, M. Corral, J. Vadillo, J. Laserna, Angle-resolved laser-induced breakdown spectrometry for depth profiling of coated materials. Appl. Spectrosc. **54**, 1027–1031 (2000)
16.10. M. Mateo, L. Cabalín, J. Laserna, Line-focused laser ablation for depth-profiling analysis of coated and layered materials. Appl. Optics **42**, 6057–6062 (2003)
16.11. D. Papazoglou, V. Papadakis, D. Anglos, In situ depth and topography monitoring in LIBS elemental profiling of multi-layer structures. J. Anal. At. Spectrom. **19**, 483–488 (2004)
16.12. M. Mowery, R. Sing, J. Kirsch, A. Razaghi, S. Bechard, R. Reed, Rapid at-line analysis of coating thickness and uniformity on tablets using laser induced breakdown spectroscopy. J. Pharm. Biomed. Anal. **28**, 935–943 (2002)
16.13. H. Balzer, M. Hoehne, V. Sturm, R. Noll, Online coating thickness measurement and depth profiling of zinc coated sheet steel by laser-induced breakdown spectroscopy. Spectrochim. Acta B **60**, 1172–1178 (2005)
16.14. H. Balzer, S. Hölters, V. Sturm, R. Noll, Systematic line selection for online coating thickness measurements of galvanised sheet steel using LIBS. Anal. Bioanal. Chem. **385**, 234–239 (2006)
16.15. M. Martin, N. Labbe, T. Rials, S. Wullschleger, Developing laser-induced breakdown spectroscopy as a high throughput technique for quantifying the elemental composition of preservatives-treated wood. Spectrochim. Acta B **60**, 1179–1185 (2005)
16.16. M. Kraushaar, R. Noll, H.U. Schmitz, Slag analysis with laser-induced breakdown spectrometry. Appl. Spectrosc. **57**, 1282–1287 (2003)
16.17. H. Fink, U. Panne, R. Niessner, Process analysis of recycled thermoplasts from consumer electronics by laser-induced breakdown plasma spectroscopy. Anal. Chem. **74**, 4334–4342 (2002)
16.18. S. Palanco, J. Laserna, Full automation of a laser-induced breakdown spectrometer for quality assessment in the steel industry with sample handling, surface preparation and quantitative analysis capabilities. J. Anal. At. Spectrom. **15**, 1321–1327 (2000)
16.19. H.S. Huang, K.C. Lin, Laser-induced breakdown spectroscopy of liquid droplets: correlation analysis with plasma-induced current versus continuum background. J. Anal. At. Spectrom. **20**, 53–59 (2005)
16.20. V. Lazic, R. Fantoni, F. Colao, A. Santagata, A. Morone, V. Spizzichino, Quantitative laser induced breakdown spectroscopy analysis of ancient marbles and corrections for the variability of plasma parameters and of ablation rate. J. Anal. At. Spectrom. **19**, 429–436 (2004)
16.21. S. Lui, N. Cheung, Resonance-enhanced laser-induced plasma spectroscopy: ambient gas effects. Spectrochim. Acta B **58**, 1613–1623 (2003)
16.22. V. Detalle, M. Sabsabi, L. St-Onge, A. Hamel, R. Heon, Influence of Er:YAG and Nd:YAG wavelengths on laser-induced breakdown spectroscopy measurements under air or helium atmosphere. Appl. Optics **42**, 5971–5977 (2003)

16.23. L. Peter, V. Sturm, R. Noll, Liquid steel analysis with laser-induced breakdown spectrometry in the vacuum ultraviolet. Appl. Optics **42**, 6199–6204 (2003)

16.24. S. Feliu Jr., V. Barranco, XPS study of the surface chemistry of conventional hot-dip galvanised pure Zn, galvanneal, and Zn-Al alloy coatings on steel. Acta Materialia **51**, 5413–5424 (2003)

16.25. Ed.R. Payling, D. Jones, A. Bengtson, *Glow Discharge Optical Emission Spectrometry*, (Wiley, New York, 1997)

16.26. H. Balzer, M. Höhne, R. Noll, V. Sturm, New approach to monitoring the Al depth profile of hot-dip galvanised sheet steel online using laser-induced breakdown spectroscopy. Anal. Bioanal. Chem. **385**, 225–233 (2006)

16.27. D. Anderson, C. McLeod, W. English, A. Smith, Depth profile studies using laser-induced plasma emission spectrometry. Appl. Spectrosc. **49**, 691–701 (1995)

16.28. P. Pouli, K. Melessanaki, A. Giakoumaki, V. Argyropoulos, D. Anglos, Measuring the thickness of protective coatings on historic metal objects using nanosecond and femtosecond laser induced breakdown spectroscopy depth profiling, Spectrochim. Acta B **60**, 1163–1171 (2005)

16.29. V. Detalle, J. Lacour, P. Mauchien, A. Semerok, Investigation of laser plasma for solid element composition microanalysis. Appl. Surf. Sci. **138–139**, 299–301 (1999)

16.30. H. Balzer, M. Hoehne, S. Hoelters, V. Sturm, R. Noll, E. Leunis, S. Janssen, M. Raulf, P. Sanchez, M. Hemmerlin, Online depth profiling of zinc coated sheet steel by laser-induced breakdown spectroscopy, in 7th International Workshop Progress in Anal. Chem. in the Steel and Metal Industries, ed. by J. Angeli, (Glückauf GmbH, Essen, 2006), pp. 237–242

16.31. X. Mao, A. Ciocan, R. Russo, Preferential vaporization during laser ablation inductively coupled plasma atomic emission spectroscopy. Appl. Spectrosc. **52**, 913–918 (1998)

16.32. ZE-Mg Zink-Magnesium-Oberflächen, Die neue Generation von Zinklegierungsüberzügen: korrosionsbeständiger, verarbeitungsfreundlicher. http://www.thyssenkrupp-steel.com/upload/binarydata_tkscsauto/3695/ze_mg_de.pdf (2003)

Chapter 17
LIBS Instruments

This chapter describes LIBS instruments designed for different application fields, the requirements, setup, and performance. The topics chosen for the following sections refer mainly to industrial applications in R&D, inline process control, and quality inspection.

A comprehensive presentation of LIBS instruments for applications such as, e.g., security, space exploration, or in pharmaceutical industry is outside the scope of this chapter. The different types of LIBS instruments for industrial applications are described in the following with the help of typical examples.

17.1 Laboratory Systems

Laboratory systems are defined here as LIBS instruments designed for use in a laboratory of an industrial company. As an example, a LIBS system is presented for the analysis of steel samples having a scale layer. The industrial application case and the methodology to ablate the scale layer and to subsequently analyze the bulk steel sample with LIBS is described in detail in Sect. 13.1.2. In the following, the requirements, concept, and design of the LIBS system called ATLAS is presented [17.1].

In a first step, the requirements are defined with the end user of the ATLAS system, in this case a steel plant. Key points are the times for combined sample preparation and analysis to be attained, correctness and precision of the analysis, the procedure of production control in the steel plant (interval operation), the number of samples to be analyzed per day, sample types, thicknesses of scale layers, elements to be analyzed, and their concentration ranges.

Table 17.1 lists the times of the different process steps for a conventional sample analysis and those aspired for the ATLAS system. The laser method has the potential to reduce the total time by up to 60 s, which allows to shorten the feedback time for process guiding. The limit of detection to be achieved especially for the elements

R. Noll, *Laser-Induced Breakdown Spectroscopy*, DOI 10.1007/978-3-642-20668-9_17,

Table 17.1 Comparison of the times of the process steps needed for conventional sample analysis and those aimed for the LIBS-based ATLAS system

Process step	Conventional $t_{\min} - t_{\max}(\mathrm{s})$	ATLAS $t_{\min} - t_{\max}(\mathrm{s})$
Pneumatic dispatch from converter to laboratory of steel plant	20–35	20–35
Receipt and preparation by grinding	70	n/a
Transport from grinding machine to SD-OES system	12–15	n/a
SD-OES analysis (two to four sample positions)	40–80	–
Laser ablation and LIBS (two to four sample positions)	–	60–120
Total time (s)	142–200	80–140

n/a, not applicable; SD-OES, spark discharge optical emission spectroscopy

C, P, and S is $< 10\,\mu g/g$ and the relative standard deviation of procedure should be about 2 rel.% [cf. Sect. 11.2, relation (11.13)].

The LIBS system has to be able to analyze on average 420 samples per day. This number may vary between 320 and 550. On average, the time between two consecutive samples arriving at the laboratory of the steel plant amounts to 3.5 min. After taking of the production control sample in the plant, the sample is transported via pneumatic dispatch to the steel works laboratory, which takes 20–35 s. Within this time, the LIBS system has to be stabilized and set into the state "ready for measurement." A sample is measured at two to four sample positions [$K = 2$–4, cf. Sect. 11.1, relation (11.5)] depending on the mutual deviations of the measuring results. A measurement is divided into three steps:

1. Ablation of the scale layer at $K = 2$–4 sample positions, $t_{\mathrm{abl},K} = \sum_{k=1}^{2-4} t_{\mathrm{abl},k}$,
2. Analysis of the bulk material at $K = 2$–4 sample positions, $t_{\mathrm{LIBS},K} = \sum_{k=1}^{2-4} t_{\mathrm{LIBS},k}$,
3. Change of sample position for $K = 2$–4 sample positions, $T_{\mathrm{SP},K} = \sum_{k=2}^{2-4} T_{\mathrm{SP},k-1}$.

The total response time of the ATLAS system is then: $t_r = t_{\mathrm{abl},K} + t_{\mathrm{LIBS},K} + T_{\mathrm{SP},K}$, which should be in the range between 60 and 120 s, cf. Table 17.1.

The samples to be analyzed are production control samples, see Fig. 17.1, left, with a scale layer on the surface formed during the cool down phase of the sample. For comparison Fig. 17.1, right shows a production control sample after grinding. Conventional production control requires the removal of the scale layer at one side of the sample by grinding or milling prior to SD-OES analysis.

The scale layer covers the sample surface completely. It has a thickness of several tenths of millimeters. The chemical composition of the scale layer differs from the composition of the bulk material. The carbon content in the scale layer, e.g., is not representative for the carbon content of the bulk material. A standard procedure is to irradiate a series of laser pulses onto the same position on the sample surface

Fig. 17.1 Production control sample, *left*: original appearance with scale layer, *right*: after grinding; width of the sample: about 35 mm

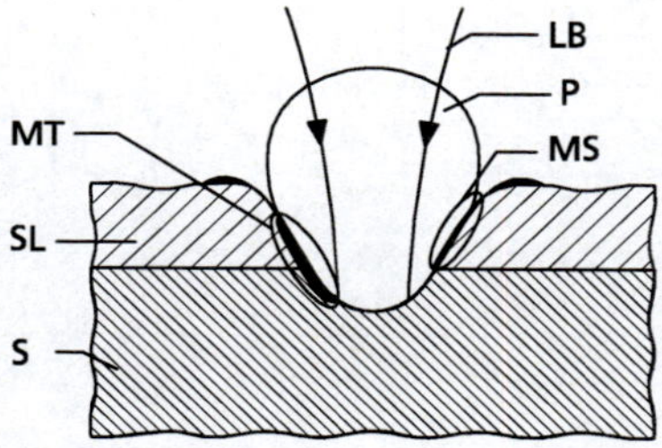

Fig. 17.2 Crater and plasma geometry of a sample covered with a surface layer of different composition. LB = laser beam, S = bulk sample, SL = surface layer, MT = melt phase of transition region, MS = melt phase of surface layer, P = plasma

(cf. Sect. 11.1, prepulses and measuring pulses). The laser beam ablates material and a crater is formed penetrating the surface layer and extending into the bulk material to be analyzed. In general, the result of the interaction with the prepulses is residues at the crater walls, such as, e.g., solidified melt phases of a metallic bulk sample. This situation is illustrated schematically in Fig. 17.2 [17.2]. These residues (MS in Fig. 17.2) contain constituents from the surface layer (SL) or from both the surface layer and the bulk material (MT). Similar issues have been discussed in various papers [17.3, 17.4, 17.5]. Since the plasma expands very rapidly and strikes the crater walls, there is a principal danger of crosstalk, i.e., the plasma emission originates from material from both the bulk material and the surface layer. This is caused by the mixing of melt phases in the transition region and subsequent excitation by the plasma. Apart from that, the plasma conditions are changing when penetrating pulse-by-pulse into a material forming a deep crater (depth > width), as shown schematically in Fig. 17.3 and studied, e.g., by [17.6].

Penetration of hard surface layers – such as scale layers on steel samples discussed here – involves focusing of a series of laser pulses on one spot for a stepwise ablation of the surface layer to access the bulk material to be analyzed. For the first few laser pulses, the sample surface can be considered as being flat

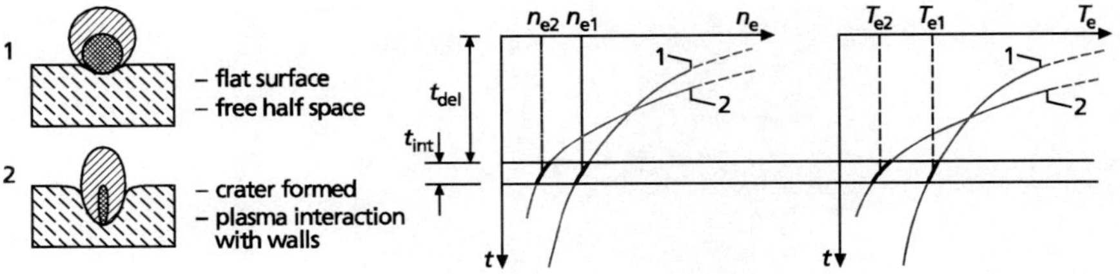

Fig. 17.3 Change of plasma parameters depending on the crater aspect ratio, case 1: crater depth < crater width, case 2: crater depth > crater width. n_e = electron density, T_e = electron temperature

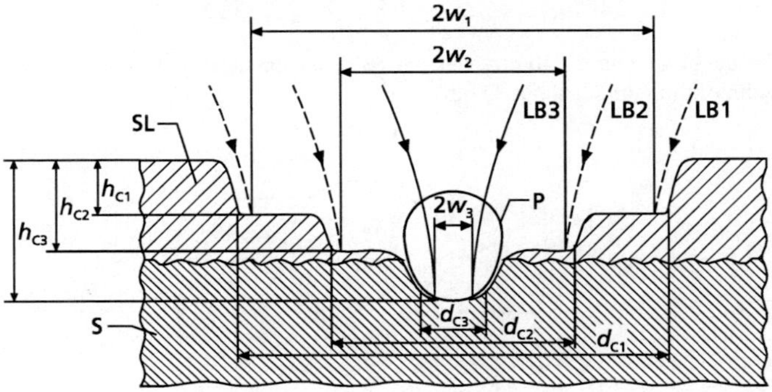

Fig. 17.4 Crater geometry generated by a step-by-step reduction of the laser beam width to minimize cross talk and to stabilize plasma conditions. The crater width is reduced successively with increasing depth (no. 1–3). LB, P, SL, S = see Fig. 17.2. $2w_i$, diameter of the laser beam; d_{Ci}, diameter of crater; h_{Ci}, depth of crater

and the laser-induced plasma expands into the free half space (case 1 in Fig. 17.3) [17.2]. After a series of laser pulses, a deep crater is formed and the interaction of the plasma with the crater walls becomes significant (case 2). The plasma state described by the temporal evolution of electron temperature and electron density is shown qualitatively for the two cases considered in Fig. 17.3 on the right.

In the crater, the induced plasma initially has both higher electron density and temperature due to the confinement by the crater. On the other hand, it is quenched by the contact with the crater wall and thus relaxes more rapidly. If the integration window parameters t_{delay} and t_{int} have been optimized for the flat surface, this corresponds to a plasma state (n_{e1}, T_{e1}). As the crater forms, the plasma state during the fixed integration window changes from (n_{e1}, T_{e1}) to (n_{e2}, T_{e2}). The integration window parameters are not optimal for the new plasma state; hence, this effect limits the precision of LIBS measurements if not taken into account adequately. The approach chosen here to overcome the limitations of cross talk and change of plasma conditions is the step by step generation of a defined crater geometry as illustrated in Fig. 17.4 [17.2, 17.7]. The deposition of the laser energy is controlled spatially by adjusting the beam width and focal position. This is performed by an x-, y-, z-scanner steering the laser beam onto the sample.

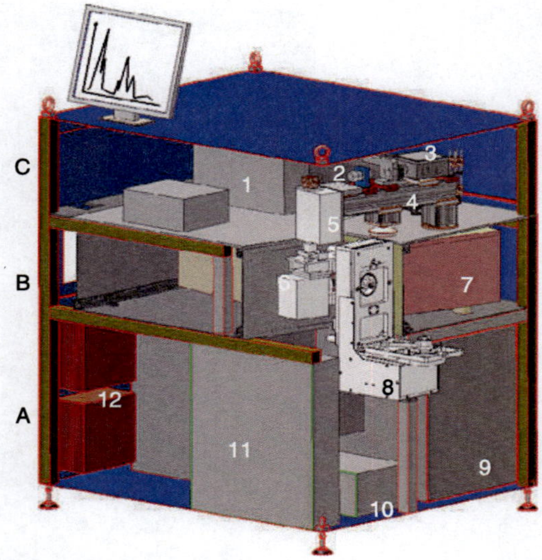

Fig. 17.5 Components of the ATLAS system for the analysis of steel samples with scale layers. The components are mounted in three levels A–C. 1, multichannel integrator electronics (MCI; cf. Sect. 4.5); 2, laser source; 3, shutter box; 4, Glan laser prism to attenuate the laser pulse energy; 5, beam shaping and beam diagnostic module; 6, x-, y-, z-scanner; 7, spectrometer; 8, sample stand (cf. Sect. 4.6); 9, control computer; 10, vacuum pump; 11, electric wiring board; 12, fan

Figure 17.5 illustrates the components of the ATLAS system comprising an x-, y-, z-scanner. The setup is structured in three levels, A–C. The laser source (number 2 in Fig. 17.5) is mounted in level C to assure good accessibility for, e.g., the exchange of the laser flashlamp. The laser pulse energy is attenuated externally by a Glan laser prism (4) to the pulse energy necessary for the ablation and analysis step. The attenuated beam enters a beam shaping and beam diagnostic module (5), which allows to adjust the spatial beam profile and an inline monitoring of the beam parameters. The laser beam propagates through the scanner (6), which deflects the laser beam for the ablation step prior to the analysis in the center of the ablation crater as illustrated in Fig. 17.4.

After the scanner, the beam passes the laser channel in the sample stand (8) and is focused onto the production control sample with the scale layer. This sample is positioned by a sample handling system that is an integral part of the sample stand. The measuring radiation is guided to a Paschen–Runge spectrometer (7), spectrally dispersed and detected by photomultiplier (PMT) tubes. The signals are processed by a multichannel integrator electronics (1), which also synchronizes the whole measuring process.

The dust-proof housing is equipped with an air conditioning to reduce the influence of changing ambient conditions on the stability of the measuring system

Fig. 17.6 Photograph of the ATLAS system

such as temperature, humidity, and dust load. Figure 17.6 shows a photograph of the LIBS system ATLAS for the analysis of scaled steel samples.

Figure 17.7 shows a photograph of the sample stand and its components, see figure caption. Figure 17.8 illustrates the operation in four steps. The first photo shows the sample stand with the pivoting plate (cf. Fig. 17.7) in horizontal position ready to take over the production-control sample. In the second photograph, the sample is put manually or by a robot onto the sample support and fixed by activation of an underpressure in the sample support. In the third photo, the pivoting plate is moving into an upright position driven by a pneumatic cylinder. Finally, the fourth photo shows the pivoting plate in vertical position. The sample is now fixed onto the rotating disk and ready to start the ablation and measurement phases.

After a first measurement, the rotating disk allows to rotate the sample by a defined angle to a further measurement position as illustrated in Fig. 17.9. This procedure is repeated several times depending on the number of measuring positions required. The radius of the white circle shown in Fig. 17.9 amounts to about 12 mm. The angular positions differ by 20°.

After the measurement, the pivoting plate moves again to the horizontal position, the underpressure is deactivated and the production control sample can be taken away.

The composition range of the production control samples to be analyzed by ATLAS is given in Table 17.2.

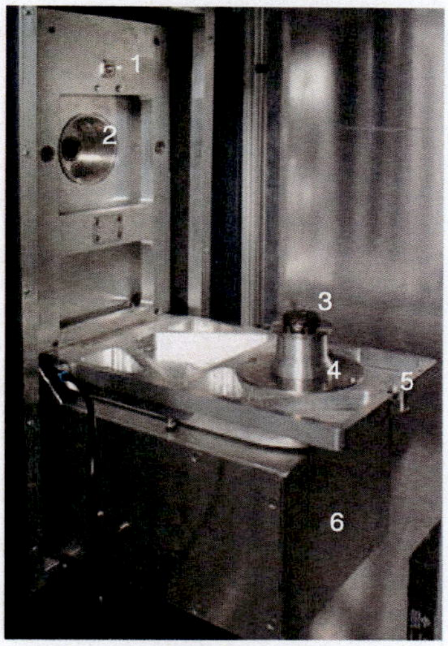

Fig. 17.7 Sample stand with sample handling system. The sample handling system is shown in the acceptance position. 1, connecting plate; 2, rotating disk; 3, production control sample; 4, sample support; 5, pivoting plate; 6, housing for pneumatic cylinder

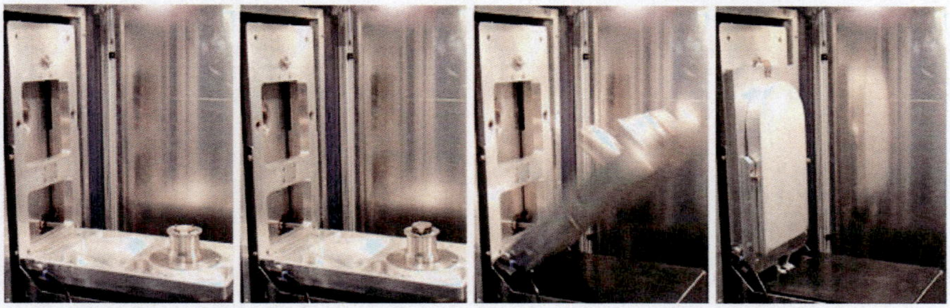

Fig. 17.8 Operation of sample stand shown in a sequence of four steps. From *left* to *right*: (**a**) The pivoting plate is in horizontal position ready to take over the production control sample. (**b**) The sample is put onto the sample support. (**c**) The pivoting plate moves into an upright position. (**d**) The pivot plate has reached the vertical position and is in contact with the connecting plate. The sample is fixed onto the rotating disk and the measurement starts

To detect the elements listed in Table 17.2, a Paschen–Runge spectrometer (cf. Sect. 4.2) was designed equipped with 27 detectors in the spectral range from 178 to 438 nm. Figure 17.10 shows a photograph of the spectrometer.

The signals of the PMTs are transferred to the multichannel integrator electronics (MCI; cf. Sect. 4.5), see Fig. 17.11.

The MCI has a modular structure comprising a set of A/D-converter electronic boards where each board has 16 inputs for PMT signals. Up to four A/D-converter boards are linked to a signal electronics having the following features: ethernet interface for communication with a PC, input for an incremental encoder, three trigger inputs, and eight trigger outputs with a resolution of 10 ns to control process sequences, digital input and outputs likewise to control process sequences.

Fig. 17.9 Production control sample with scale layer. By rotation of the sample in the rotating disk (no. 2 in Fig. 17.7), a series of measuring positions can be defined lying on a circle

Table 17.2 Ranges of chemical composition of production control samples to be analyzed with ATLAS

Element	c_{min} (m.-%)	c_{max} (m.-%)	Element	c_{min} (m.-%)	c_{max} (m.-%)
Fe	85	100	Mo	0	1.7
C	0	1.5	Ni	0	12
Si	0	3.5	Cr	0	4
Mn	0	2.5	V	0	0.6
P	0	0.1	Ti	0	0.45
S	0	0.1	B	0	0.13
Cu	0	1.5	Co	0	0.35
Al	0	0.5	Ca	0	0.015
N	0	0.03			

For each PMT signal, two independent integration channels are available with different sensitivity. Each of the 32 integrator channels of a single A/D-converter board is digitized by a 24-bit A/D converter. The position and width of the integration windows are freely programmable for each PMT signal.

The ATLAS system was tested with 40 reference samples to determine limits of detection (LOD) in different crater depths, see Table 17.3 (cf. Sect. 11.2). The number of prepulses applied is kept constant: $N_{pp} = 200$ (cf. Sect. 11.1). The number of measuring pulses is $N_{mp} = 3,000$. Table 17.3 shows the achieved LODs for different pulse ranges within this series of 3,000 pulses corresponding to different crater depths. The crater depths are in the range of 0–50 μm for 1–500 pulses, 50–100 μm for 501–1,000 pulses, and so on up to 250–300 μm for 2,501–3,000 pulses. No systematic change of LODs with increasing crater depths up to 300 μm is observed. For nine elements, LODs less than 10 μg/g are achieved for a crater depths range of 250–300 μm.

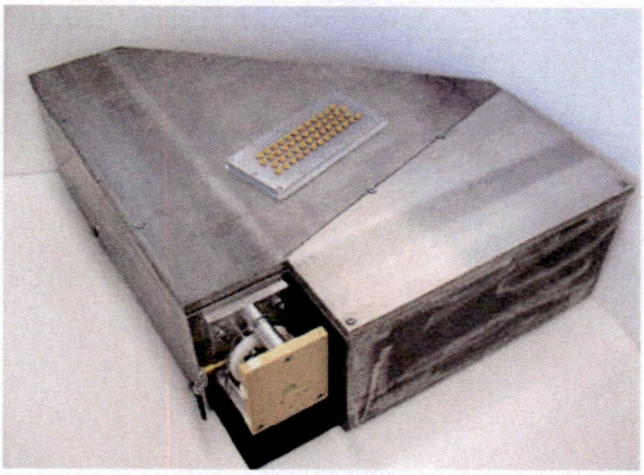

Fig. 17.10 Paschen–Runge spectrometer (no. 7 in Fig. 17.5) of the ATLAS system. *Above the middle*: feed-throughs of the photomultiplier signals, *bottom*: interface to the sample stand

Fig. 17.11 Multichannel integrator electronics to process the signals of the photomultipliers

The limit of detection depends on the number of measuring pulses used to average the measuring signal, see Table 17.4. For most elements, the best LODs were achieved for an averaging over 3,000 measuring pulses. For 14 elements, the LODs are better than 10 $\mu g/g$.

The ATLAS system was tested extensively on-site in a steel plant [17.2]. Figure 17.12 shows the concentration of manganese of scaled process control samples determined with ATLAS compared to the reference value from conventional analysis. For the conventional analysis, the samples have to be ground to remove the scale layer and are then analyzed by a spark optical emission spectrometer (OES). This graph shows the results of LIBS measurements of 276 scaled samples. The residual deviation as a measure for the accuracy is 2.3 rel.-% [cf. Sect. 11.3, relation (11.21)].

17.2 Inspection Systems for Inline Process Control

Inspection systems for inline process control are defined here as LIBS instruments designed for use in an industrial plant. They are integrated in a production line. The measuring results gained inline are evaluated to control preceding or subsequent

Table 17.3 Limits of detection for different crater depths up to 300 μm

Element	LOD_{3s} (μg/g), pulse range					
	1–500	501–1000	1,001–1,500	1,501–2,000	2,001–2,500	2501–3,000
Al	4	6	6	5	8	6
As	7	5	7	9	13	14
B	1	4	2	2	1	3
C*	7	9	11	9	7	12
C	6	17	18	7	19	12
Ca	9	7	3	2	1	6
Co	16	19	29	31	30	13
Cr	26	12	19	25	31	23
Cu	3	1	4	6	5	3
Mn	5	12	13	9	9	6
Mo	16	23	21	15	24	21
Nb	23	10	18	23	19	15
Ni	9	20	27	15	16	14
P	9	20	18	10	10	8
S	9	8	13	7	9	7
Si	45	14	19	11	15	24
Ti	2	2	2	3	3	1
V	11	7	12	14	18	8

The blank sample is measured ten times. Data are evaluated for a linear calibration curve in the lower concentration range. For carbon two lines are installed: C 193.1 nm and C* 165.8 nm. $\Delta s =$ 8 mm

process steps. The requirements on such machines can be summarized as follows: automated operation, capability to work under harsh industrial conditions, long-term stability and reliability, user-friendly operation and maintenance.

As an example for an inline inspection machine based on LIBS, a system for the identification of material mix-ups in a production line of pipe fittings and tubes made of high-alloy steel grades will be presented [17.8]. Especially in the oil and gas industry, pipe lines and process piping are increasingly exposed to corrosion as the fields become more sour, wells are deeper with higher pressures and hotter products. In order to meet the requirements with regard to corrosion resistance and mechanical stability, more than 30 different steel grades are used for the production of pipe fittings, in particular alloy steels, stainless steels, duplex, superduplex, 6Mo grades, high nickel alloys, titanium, and clad steels. The concentration range of Fe and Ni extends to 100 m.-%, the range of Cr, Mo, and Cu to 30 m.-%. The use of wrong steel grades can lead to corrosion with severe consequential damages. Growing quality requirements, particularly in the nuclear industry and environmental responsibilities demand a material identification of each produced pipe fitting. The developed inspection system is called LIFT, which stands for *L*aser *I*dentification of *F*ittings and *T*ubes.

Before LIFT was installed, the fittings were inspected by Spark-OES. The fittings had to be cleaned and a considerable percentage of the fittings had to be measured two or three times before they had been correctly identified. The duration of one

Table 17.4 Limits of detection for averages of 1,000, 2,000, and 3,000 measuring pulses for each sample position

Element	LOD_{3s} (μg/g), number of pulses		
	1,000	2,000	3,000
Al	4	2	3
As	5	5	6
B	2	2	1
C*	4	5	4
C	10	9	7
Ca	7	3	2
Co	16	20	19
Cr	6	12	15
Cu	2	2	2
Mn	7	7	5
Mo	13	13	9
Nb	14	14	12
Ni	10	11	8
P	14	10	7
S	7	7	5
Si	26	17	10
Ti	1	1	1
V	6	7	7

The blank sample is measured ten times

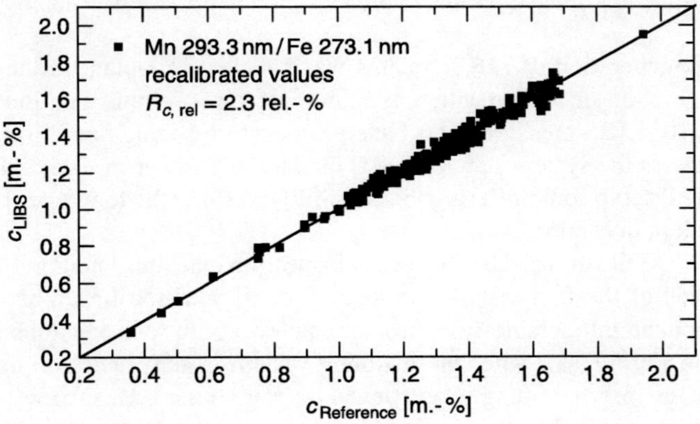

Fig. 17.12 Concentration of manganese in scaled steel samples measured with LIBS without any mechanical preparation compared to the reference concentration values gained with conventional methods requiring grinding of the sample to remove the scale layer

inspection was 4 s and the electrode of the Spark-OES had to be cleaned every three measurements. The throughput was 60 fittings per hour. The surface of the fittings had to be abraded after the spark discharge. In a third step, the fittings had to be marked with an ink jet printer or by electrolytic etching. In view of a

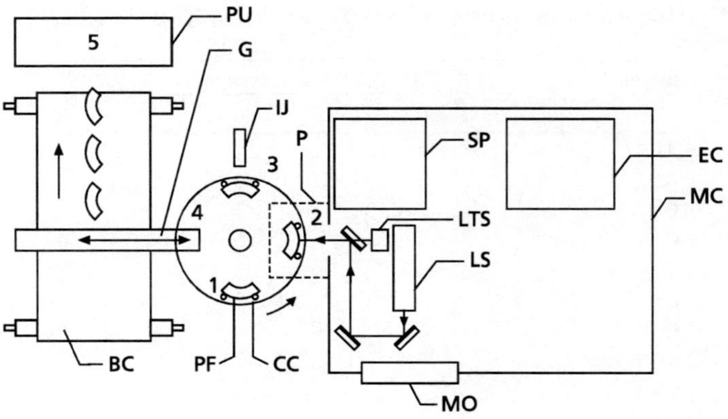

Fig. 17.13 Schematic setup of the inspection machine LIFT for the automated identification of steel grades of pipe fittings. PF = pipe fitting, CC = table type circular conveyor, P = protection cover, MC = machine cabinet, EC = electric cabinet, MO = monitor, operation panel, LS = laser, LTS = laser triangulation sensor, SP = spectrometer, IJ = ink jet marker, G = gripper, BC = belt conveyor, PU = packaging unit. The numbers one to five indicate the sequence of actions. 1 = input of pipe fitting to be inspected, 2 = inspection with LIBS, 3 = marking of the specimen, if result of inspection is okay, 4 = the gripper takes the specimen and puts it on the belt conveyor, 5 = packaging of the specimens

material inspection of each produced pipe fitting, the economic efficiency had to be improved.

The first generation of LIFT systems was installed in a plant in the year 1998 and is in operation since then with only minor interrupts. Within a period of 4 years, more than 10^8 LIBS measurements have been performed and monitoring data was stored to assess the system status. This is the largest number of data collected with a single LIBS setup so far offering the possibility to study the long-term behavior of a LIBS system in routine industrial use.

Figure 17.13 illustrates the setup of the inspection machine, and Fig. 17.14 shows a photograph of the first installed system in the production line. The fittings are inserted by hand into a table-type circular conveyor with four positions (numbered with 1–4 in Fig. 17.13). After the handling system is activated with a switch for two hands, the inserted fitting is positioned in front of the laser inspection machine (position 2). A protection cover encloses the fitting during the inspection and assures the laser safety according to IEC 825–1. The duration of the inspection is 2 s. If the result of the inspection is "OK," the fitting will be marked at the third position of the handling system with an ink jet printer. At the fourth position, the fitting is transferred by a gripper to a belt conveyor and transported to the packaging unit. When a material mixing is identified by LIFT, the fitting will not be marked. Instead, it will be sorted out reversing the direction of movement of the belt conveyor. The handling system can be used for sizes of the pipe fittings ranging from 0.5 to 8″ and the maximum throughput amounts to 450 fittings per hour.

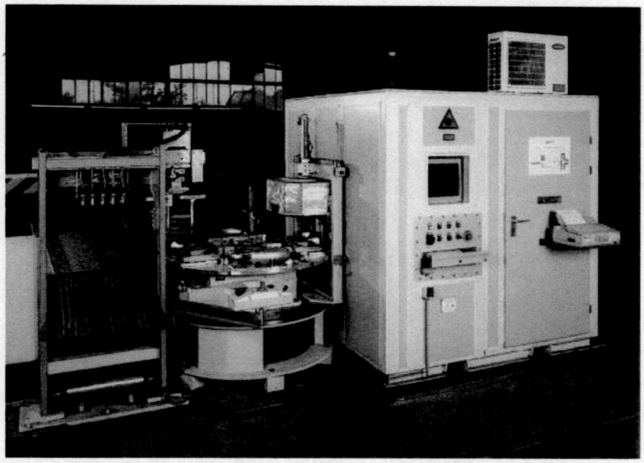

Fig. 17.14 First LIFT inspection system installed in the production plant. *Right*: inspection system, *middle*: table type circular conveyor to feed the pipe fittings, *left*: belt conveyor to transport the inspected parts to the packaging line, cf. Fig. 17.13

In order to operate LIFT in dusty surroundings at a humidity of up to 90% and at an ambient temperature of up to 40°C, all components of the laser inspection machine are installed in a closed air-conditioned cabinet. For maintenance purposes, the cabinet can be entered through a door that is usually locked.

A TCP/IP interface enables to backup measurement data and records regularly by the host computer. An interface for the remote maintenance by modem allows for an operational support, e.g., to set parameters of LIFT or for diagnosis and recovery of errors during operation.

Figure 17.15 shows the optical setup of LIFT. The emitted light of the Nd:YAG laser (50 Hz repetition rate, 100 mJ) passes a beam splitter where a small part is reflected to a photo diode to monitor the laser pulse energy. The laser is guided via two mirrors, shutter, quarter wave plate, focusing lens, beam combiner, and window to the specimen. The quarter wave plate protects the laser from back reflections originating from the target, which may cause damage to the laser system. A mechanical shutter blocks the laser beam when no measurement is activated. The position of the workpiece in relation to the beam waist of the Nd:YAG laser beam determines the irradiance in the interaction region. A triangulation sensor measures the sample position d between a reference plane and the surface of the specimen using a wavelength of 670 nm generated by a diode laser. The beam combiner transmits the laser beam of the triangulation sensor propagating collinearly to the Nd:YAG laser beam to the interaction region. Window W1 in Fig. 17.15 is part of a wall of the machine cabinet that separates the outer region from the internal volume of the machine. This window is heated in order to avoid condensation of water from the outer humid atmosphere.

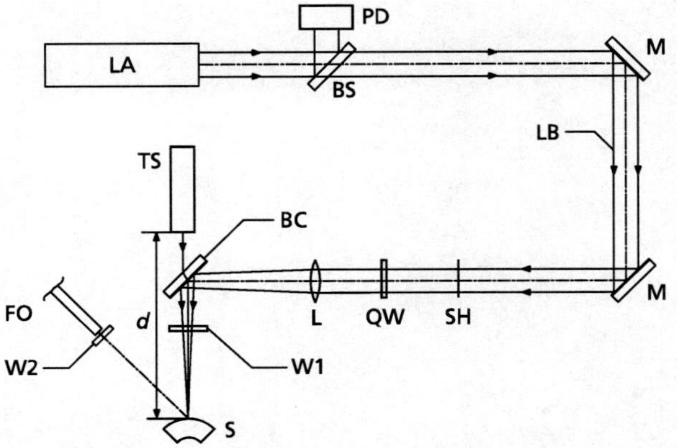

Fig. 17.15 Optical setup of LIFT. LA = Nd:YAG laser, BS = beam splitter, PD = photo diode, M = mirror, LB = laser beam, SH = shutter, QW, quarter wave plate, L = focusing lens, BC = beam combiner; W1, W2 = window; S = specimen, TS = triangulation sensor, FO = fiber optics, d = working distance

The emitted light of the plasma is guided with fiber optics (length 2 m, core diameter 1 mm, numerical aperture 0.22) – positioned at a distance of about 65 mm from the plasma – to a Paschen–Runge spectrometer equipped with 12 PMTs to detect the element-specific radiation of the elements Fe, Ni, Cr, Mo, Ti, Cu, Nb, Al, and W. The diameter of the Rowland circle amounts to 500 mm, the grating has 2,700 l/mm. Table 17.5 shows the list of lines installed in the spectrometer. The entrance slit of the spectrometer can be shifted laterally by a servo motor. Detecting measuring signals of the photomultipliers as a function of the entrance slit position yields line profiles used to define the correct position of the entrance slit in order to move the line maxima to the positions of the respective exit slits (cf. Fig. 14.61). By this, drifts of the adjustment of the spectrometer can be compensated. The PMT signals are transmitted via I/U-converters to a multichannel integrator electronics. The signals are integrated in a defined time window with respect to the laser pulse. Typical parameters for the delay and integration time are $t_{delay} = 250\,\text{ns}$, $t_{int} = 10\,\mu\text{s}$. The same integration window is used for all PMT channels.

The measurement time amounts to 2 s comprising $N_{pp} = 30$ laser pulses as prepulses to ablate surface contaminations – as, e.g., oil residues from preceding processing steps – and the spectral signals generated by the subsequent $N_{mp} = 70$ laser pulses are evaluated for material identification. During the irradiation of the laser pulses, the interaction area is flushed with filtered pressurized air. For the classification of the measured data set concerned, an expert system was developed based on a hierarchical structure of rules to identify the steel grade. According to this hierarchy, the data set is tested step by step. Based on the outcome of each test, the next rule to be applied is selected. In this way, only a subset of the rules has to be evaluated and the consultation time is minimized to a few milliseconds. A calibration is necessary for the training of the rule set. During the

Table 17.5 List of lines installed in the spectrometer of LIFT

Element	λ (nm)	E_{low} (eV)	E_{upp} (eV)
Fe I	371.99	0.00	3.33
Fe II	273.07	1.08	5.62
Ni I	351.50	0.11	3.64
Ti II	337.28	0.01	3.69
Cr II	267.72	1.55	6.18
Cr II	298.92	3.74	7.89
Al I	396.15	0.01	3.14
Nb II	319.50	0.33	4.21
Cu I	324.75	0.00	3.82
W I	400.87	0.37	3.46
Mo II	281.62	1.67	6.07

calibration, several samples of each steel grade are measured. The calibration data set establishes the knowledge base of the expert system. In addition to the normal samples, five reference samples are measured during the calibration. The data of these measurements can later be used for the recalibration of the algorithm (cf. Sect. 11.4). The recalibration can be performed within 5 min. During the automatic operation, all data sets and their classification are logged.

For the identification of different steel grades, it is necessary to detect the specific differences of the element concentrations. It is well known that internal standardization of an analyte line reduces the standard deviation of the spectral signals [cf. Sect. 11.1, relation (11.2)]. To assess the effect of internal standardization of analyte lines and the capability of a clear discrimination of different material grades, an analytical resolution A is defined as follows [cf. relation (14.1)]:

$$A = \frac{\left.\frac{I_{an}}{I_{ref}}\right|_{high} - \left.\frac{I_{an}}{I_{ref}}\right|_{low}}{\sqrt{s^2\left(\left.\frac{I_{an}}{I_{ref}}\right|_{high}\right) + s^2\left(\left.\frac{I_{an}}{I_{ref}}\right|_{low}\right)}} \tag{17.1}$$

where I_{an}, I_{ref} are the intensities of the analyte line and the reference line; s is the empirical standard deviation. The indices "high" and "low" refer to a sample containing the respective analyte at high and low concentration, respectively. In case of $A \gg 1$, the chosen analyte line and internal standardization represents a significant spectral measure for the difference of analyte concentrations, whereas for $A \approx 1$ no clear discrimination is possible. For the analyte lines of Mo, Ni, and Ti, 8–10 different reference lines were studied using the analytical resolution defined in (17.1). To calculate A, five different combinations of material grades are taken with the restriction, that the "high" samples always have approximately the same analyte concentration and the internal standard has the same concentration for the "high" and the "low" sample. Within an operational period of LIFT of 12 months, 50,000 representative data sets were selected from the steel grades 304L,

Table 17.6 Average analytical resolution $\langle A\rangle$ and standard deviation ΔA for the molybdenum line 281.62 nm for different internal standards stated in the first column

Internal standard (nm)	$\langle A\rangle$	ΔA
Fe II 273.07	5.11	0.23
Cr II 298.92	5.03	0.23
Cr II 267.72	4.60	0.65
Nb II 319.50	3.86	0.23
W I 400.87	2.86	0.29
Fe I 371.99	2.71	0.51
Ni I 351.50	1.90	0.22
Ti II 337.28	1.83	1.07
Al I 396.15	1.79	0.78
Cu I 324.75	1.33	0.38

316L, Duplex, 1.4541, and 1.4571. Each data set corresponds to one inspected pipe fitting. In the considered period of 12 months, the laser power dropped by up to 65%. Table 17.6 lists the determined average analytical resolution $\langle A\rangle$ for the five material combinations and the standard deviation ΔA. Depending on the chosen reference line, the differences in the analytical resolution vary by more than a factor of three. The highest values of $\langle A\rangle$ are attained for reference lines having an upper energy level close to the one of the analyte line. Those intensity ratios are less sensitive to changes of the induced plasma state as caused, e.g., by long-term variations of the laser pulse energy or the focusing conditions [cf. Sect. 11.1, (11.3)].

The average throughput achieved by the LIFT systems amounts to 300 specimens per hour, which is an increase by a factor of 5 in comparison to the former applied inspection by spark discharge OES. A measure of the reliability of the material identification is the percentage of workpieces, which are not the correct material grade but are classified as being okay. To study this case, specimens of the wrong material are let into the lot of pipe fittings to be inspected in regular time intervals. Since the start of the operation, every wrong steel grade introduced intentionally was identified correctly. The second possible type of error refers to the case that the material inspected is okay but it is identified to be not okay. This means the specimen is rejected although it is okay. This case occurs in less than 0.1% of all inspected specimens.

For the first time, more than 1.4 million inspections corresponding to 10^8 LIBS measurements have become available giving an insight into the long-term behavior of the system performance. Figure 17.16 shows the average laser pulse energy of 1 day and the intensity of the Fe I 371.99 nm line of a pure iron sample as a function of time. The pure iron sample (99.99 m.-% iron) is used for monitoring purposes. This sample is measured nearly every day to check the adjustment of the spectrometer (so-called profiling of the spectrometer). The laser pulse energy is measured for each inspected pipe fitting. The data points shown in Fig. 17.16 for the pulse energy are averages over an operation period of 1 day. The data reveal strong variations of the laser pulse energy and the spectral line intensity over the 5 years of operation. In a first period (denoted with "I" in Fig. 17.16) the laser pulse energy is rather

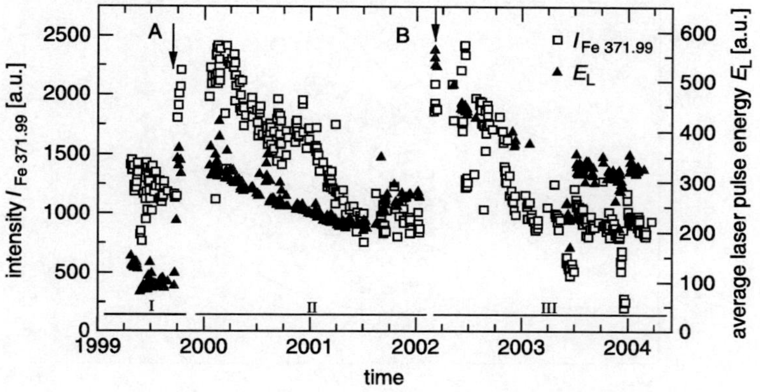

Fig. 17.16 Intensity of the Fe I 371.99 nm line of a pure iron sample and laser pulse energy as a function of time. A, new adjustment of laser; B, new laser crystal

low, which is attributed to a misalignment of an oscillator mirror of the laser head. A new adjustment of the laser head in the second half of the year 1999 (marked with "A" in Fig. 17.16) has led to a significant increase of the laser energy and of the line emission. In the following 2 years, the laser pulse energy drops continuously leading to a corresponding decrease of the line intensity (period II). After installation of a new laser crystal (marked with "B" in Fig. 17.16), the laser pulse energy rises again followed by a decay over 2 years (period III). The strong variation of the laser pulse energy is attributed to various degradation processes of the laser system (flash lamp, pump cavity, laser crystal, optical components, adjustment) occurring during the observed operation time. Figure 17.16 shows that the ratio of the iron intensity to the laser pulse energy is successively decreasing over 5 years. Comparing, e.g., the last quarter of phase III with the last quarter of phase II shows that the same level of iron intensity is achieved with a significantly higher pulse energy. This may be attributed to a gradual decrease of the transmittance of the beam path between the exit of the laser source and the specimen (cf. Fig. 17.15) and/or a gradual reduction of the spectral transmittance of the beam path of the measuring radiation from the specimen to the detectors of the spectrometer. From the available experimental data, it is not possible to decide which is the dominant underlying process. Further investigations are necessary to clarify in detail the reasons for this behavior.

Figure 17.17 shows the intensity of the iron line Fe I 371.99 nm as a function of the laser pulse energy using the same data set as in Fig. 17.16. The general trend is that higher laser pulse energies lead to higher line intensities of the iron line. Three clusters can be identified indicating different states of the measurement system in the three periods I–III (cf. ellipses marked with I–III in Fig. 17.17). Actions at the laser system as, e.g., readjustment or change of the laser crystal have changed the response characteristics of the measurement system. The laser pulse energy measured close to the laser system (cf. Fig. 17.15) is obviously not a sufficient monitoring information. The influence of the changing laser pulse energy on the

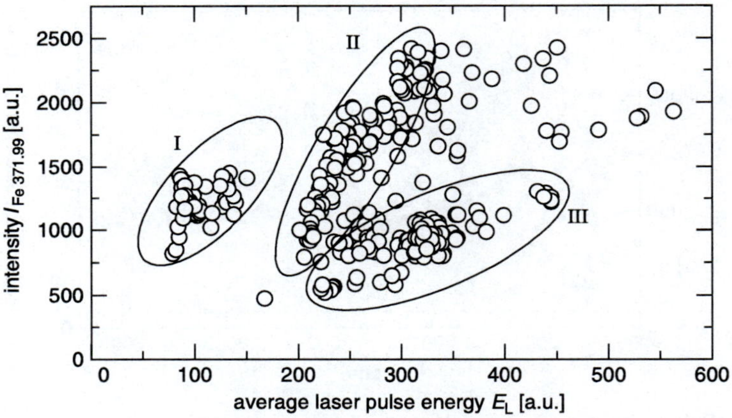

Fig. 17.17 Line intensity of the Fe I 371.99 nm line as a function of the laser pulse energy (same data set as in Fig. 17.16)

line intensity ratios used to identify the steel grades is less pronounced with relative variation of about 50%. However, such variations can only be handled by periodic recalibration of the measuring system to assure the correct identification of the steel grades to be inspected (cf. Sect. 11.4).

An approach for further improvement is to monitor additionally the laser pulse energy close to the interaction region of the laser beam with the specimen. By this, the transmittance of the laser pulse energy from the laser to the specimen can be assessed to recognize, e.g., misalignments or contaminations of the optical path between laser and specimen. Furthermore, monitor samples should be measured in regular time intervals to survey the long-term behavior of the line intensity of several emission lines having different excitation energies to detect changes of the plasma state or changes of the spectral characteristics of the receiving optics (e.g., windows, fiber optics), spectrometer and detectors (e.g., photocathodes, CCD).

The triangulation sensor measures the sample position for each inspected pipe fitting, cf. Fig. 17.15. Figure 17.18 shows the measured position d averaged for each day of operation over a period of 5 years. In principle, this distance should be constant, since it is defined by the applied fixture for the pipe fittings installed on the circular-type conveyor. Figure 17.18 shows that most of the specimens are clustered around a sample position of 232 mm.

Whereas the standard deviation for one data point – shown as error bars in Fig. 17.18 – is of the order of 1–2 mm, the overall spread of positions is up to 20 mm. This is largely due to the adjustments of the fixtures for the specimens on the circular conveyor. For different pipe-fitting geometries to be inspected, the operator has to change and to mount the respective fixtures for the pipe fitting on the circular conveyor (depending on the size and diameter of the pipe fittings). Because these actions were done manually, it is possible that the wrong fixture is taken or the

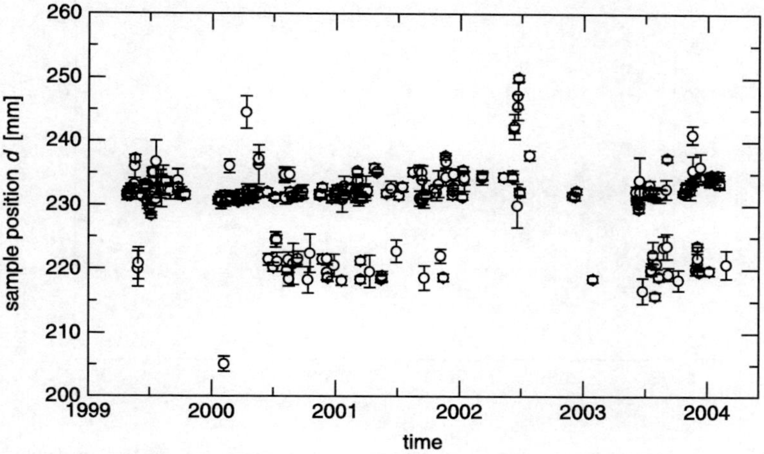

Fig. 17.18 Sample position *d* as a function of time

correct fixture is not adjusted sufficiently. In this case, different sample positions occur. Potential solutions for this problem is the use of an automated positioning system for the specimens and/or an autofocus system for the laser focusing.

The temperature measured inside the cabinet of LIFT and the position of the entrance slit of the spectrometer after the profiling is shown for a time of 5 years in Fig. 17.19. The figure shows clearly that the temperature correlates with the position of the entrance slit. Changing temperatures inside the cabinet have an influence on the mechanical structure of the spectrometer. They cause slight changes of the spatial positions of entrance slit and exit slits. By the profiling action, the optimum entrance slit position is readjusted. Figure 17.19 shows as well that the installation of a new air condition for the cabinet has led to a shift of the temperature level with a corresponding change in the position of the entrance slit. Outliers are attributed to an opening of the door of the cabinet of LIFT, which is necessary for maintenance and adjustment actions. The scattering of entrance slit positions can be further reduced by installation of a temperature stabilization for the spectrometer.

17.3 Mobile Systems

Mobile systems are defined here as LIBS instruments designed for use at different locations in an industrial environment. The equipment can be easily transported to different places of use. To this category of LIBS systems belong the so-called portable systems that can be ported by a user to the place of application [17.9, 17.10, 17.11, 17.12, 17.13, 17.14, 17.15]. There are several companies offering portable LIBS systems [17.16]. The requirements on such systems can be summarized as

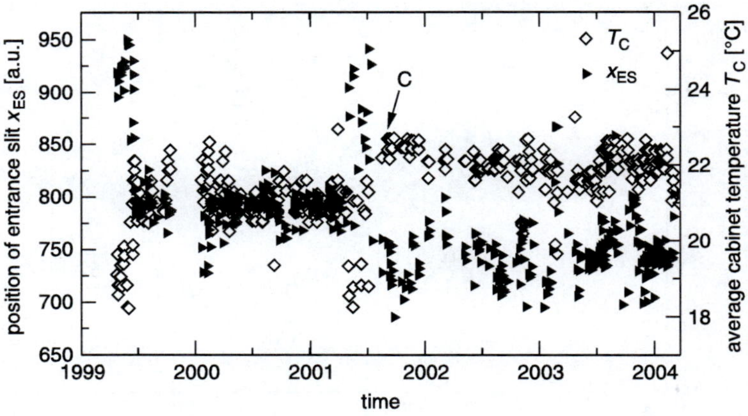

Fig. 17.19 Position of the entrance slit of the spectrometer and the temperature inside the cabinet as a function of time. C, installation of a new air condition

Fig. 17.20 Analysis of metallic samples with the compact measuring system FML

follows: compact design, capability to work under harsh industrial conditions, long-term stability and reliability, user-friendly operation and maintenance.

As an example of a mobile system designed for industrial use the instrument FML (*f*lexible *m*easuring system based on *L*IBS) will be described [17.17]. Figure 17.20 shows a photograph of FML. The spectrometer that is integrated into FML is a Paschen–Runge system with a focal length of 150 mm. It is equipped with a holographic grating and three CCD detectors that allow to measure the whole spectral range from 278 to 560 nm simultaneously with a resolution of approx. 45 pm. The signal electronic of the spectrometer is modified in such a way that it can be used to trigger the laser system and allows to record single pulse spectra.

A compact and robust Nd:YAG laser is used for the setup of FML. This laser is a flashlamp excited, Q-switched solid state laser system delivering laser pulses with < 8 ns pulse duration and a maximum pulse energy of 50 mJ at a repetition rate of up to 20 Hz. The weight of the laser head itself amounts to only about 1 kg, which makes it possible to integrate the laser directly into the measuring probe.

FML is equipped with a handheld measuring pistol, which is connected to the spectrometer and the laser power supply via a 2-m long umbilical. The plasma radiation that is generated by the laser pulses is guided to the spectrometer through a fused silica fiber within this umbilical. Additionally, argon can be delivered to the measuring probe in order to purge the measuring volume. The flexible connection between the probe and the other components of the instrument allows for a versatile use of the system even for measuring objects which are not easily accessible.

Typical measuring times of the FML range from less than a second to several 10 s, depending on the desired accuracy and precision of the measurement. For the investigations described in the following, FML was operated in a mode in which for each laser shot, the whole spectrum of the first CCD detector – ranging from 278 to 384 nm – was recorded.

One potential application for a portable measuring system is the sort identification of workpieces, half-finished products or of individual scrap particles (cf. also Sects. 13.3, 18.1, and 18.4). Metallic materials can also be analyzed by spark emission spectrometry, provided that the sample surface is suitably prepared. As the laser is capable to penetrate surface coatings, the sample preparation and the analysis can both be performed with a laser emission spectrometry system (cf. Sects. 13.1.2 and 17.1). In this case, acceptable measuring times and analytical results are only achieved if the surface layers are suitably thin and if the sample composition close to the surface is representative for the bulk.

In order to evaluate the capabilities of FML for such an application, calibration curves were recorded for alloyed aluminum samples. For aluminum, the distinction between cast aluminum and the different types of wrought aluminum is a precondition for a high-quality recycling of scrap particles.

Figure 17.21 shows an excerpt of the LIBS spectra that were measured with FML for a pure aluminum sample and for an alloyed aluminum sample containing a small amount of copper. From the comparison of the two spectra, the spectral lines of Cu at 324.8 and 327.4 nm can be clearly identified. The spectral background, given by the signal intensity at a spectral position at which no element-specific line emission can be observed, differs for the two samples. The reason for this different background was not yet investigated in detail.

However, it is important to correct the line intensities that should be correlated to the analyte concentrations for the differing background. Therefore, a spectral background position was defined for each spectral line, and the signal intensity at this background position is subtracted from the signal intensity at the peak position of the spectral line in order to calculate a background-corrected line intensity.

For the calibration, the ratio of this line intensity and the line intensity of a spectral line of the main matrix element is used (cf. Sect. 11.1). In this case, the aluminum line at 309.4 nm was selected for internal standardization.

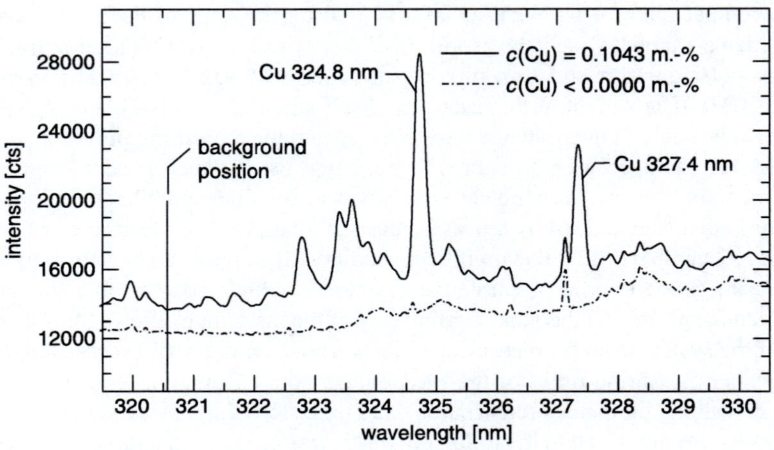

Fig. 17.21 LIBS spectra of a pure and an alloyed aluminum sample measured with FML

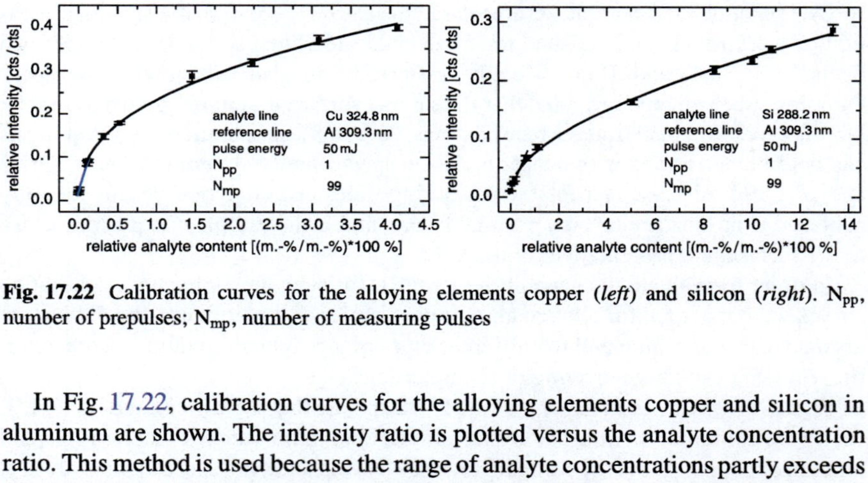

Fig. 17.22 Calibration curves for the alloying elements copper (*left*) and silicon (*right*). N_{pp}, number of prepulses; N_{mp}, number of measuring pulses

In Fig. 17.22, calibration curves for the alloying elements copper and silicon in aluminum are shown. The intensity ratio is plotted versus the analyte concentration ratio. This method is used because the range of analyte concentrations partly exceeds 10 m.-%. In consequence, this means that the concentration of the matrix element aluminum is significantly reduced and, thus, the measured intensity of the spectral line used as the internal standard decreases as well.

The calibration curves shown in Fig. 17.22 show a saturation behavior. This means that for the high concentration range, it might be better to use other, less-sensitive lines for the calibration. The measured data is fitted by an analytical function, and it was found that the saturation behavior can be described well by a power function of the type $Q^{i,r} = a + bc_{\text{rel}}^{\alpha}$ with the exponent α in the interval $0 < \alpha \leq 1$.

Additionally, a linear function was fitted to the data of the four samples with low analyte concentrations. The slope of this function gives a reasonable approximation of the slope of the calibration curve for small concentrations and was used to

Table 17.7 Analytical figures of merit for calibration curves of aluminum alloying elements measured with FML in the wavelength range from 278 to 384 nm

Element	Spectral line (nm)	LOD_{3s} (%)	r^2
Cu	324.8	0.013	0.9962
Si	288.2	0.015	0.9975
Mg	279.6	0.001	0.9943
Mg	383.8	0.057	0.9921
Fe	373.8	0.024	0.9982
Mn	294.9	0.019	0.9827
Zn	334.5	0.040	0.9859
Zr	357.6	0.005	0.9876
Ti	334.9	0.003	0.9947

calculate the limit of detection LOD_{3s} according to the 3 s-criterion, where s is the empirical standard deviation of the signals of five measurements with the pure aluminum sample performed under repeatability conditions (cf. Sect. 11.2).

Table 17.7 summarizes the LOD and the coefficient of determination r^2 of nine spectral lines that were evaluated for the determination of the eight most important alloying elements in aluminum. The data is based upon five measurements of 10 samples of certified reference material (CRM), with prepulse and 100 measuring pulses at one sample position for each of the measurements.

The average limit of detection in terms of an analyte concentration ratio amounts to 0.020 (m.-%/m.-% $\times$ 100%) with an average coefficient of determination of 0.9921. These mean values give an impression of the system performance.

For many applications, especially for sort identification or even sorting of scrap particles, the rapid availability of the measurement result is essential. For LIBS, this measuring time is directly connected to the number of laser pulses that has to be applied to the measuring object in order to achieve a significant result.

In order to investigate this influence of the number of laser pulses, different numbers of single shot spectra were accumulated numerically and the standard deviation of the signal intensity ratios was calculated. Firstly, this calculation was performed for all laser pulses applied to the sample from the first one on, but secondly it was also performed for the spectra of all laser pulses but the first. The latter case practically means that the first laser pulse is regarded as a surface-cleaning prepulse and that its spectrum is discarded. The mean RSD averaged over the 10 CRM samples and the nine investigated spectral lines is shown in Fig. 17.23 as a function of the number of measuring pulses and shows – as expected – the more measuring pulses are used, the smaller the RSD is.

Most notably, the RSD is improved by more than a factor of 10, if a single prepulse ($N_{pp} = 1$) is used and up to ten measuring pulses are evaluated. Furthermore, the improvement of the RSD by using more than ten measuring pulses is almost negligible if a single prepulse is taken into account. Due to these results, a single prepulse was always used for the data evaluation for the calibration curves that had been shown above.

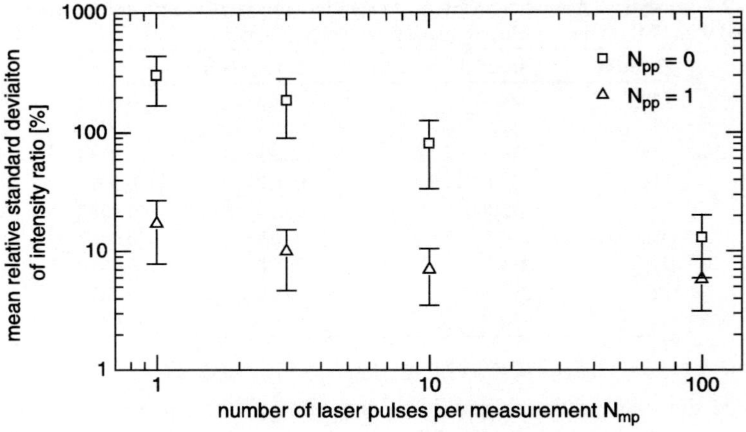

Fig. 17.23 Influence of the number of measuring pulses N_{mp} on the RSD of the intensity ratio. Pulse energy: 50 mJ; number of samples: 10; measurements per sample: 5; evaluated spectral lines: 9

In order to validate the accuracy of measurements of aluminum pieces based on the measured calibration curves, a number of aluminum-based test workpieces was measured and their analyte concentrations were calculated from the measured spectra and analysis functions.

Exemplarily, the result of these measurements is shown in Fig. 17.24 for copper. In this diagram, the analyte concentration ratio of copper to aluminum determined by LIBS is plotted versus the analyte concentration ratio of the reference measurement. Whereas the error of the measurements with FML is also shown, it has to be noted that the error of the reference analysis is unknown.

The calibration that was used for the calculation is based upon 100 measuring pulses per sample position. Only ten $(= N_{pp} + N_{mp})$ laser pulses were applied to the test workpieces, and the analyte concentration was calculated for different numbers of numerically accumulated spectra. Figure 17.24 shows the result for a single prepulse and nine measuring pulses and it can be seen that the analyte content determined with FML matches the reference analyte content within the experimental errors. The accuracy, calculated as the mean of the deviation of the measured analyte content from the reference analyte content, amounts to 0.226 (m.-%/m.-% × 100%). If three measuring pulses are evaluated, the measured Cu content matches the reference content negligibly worse, with an average accuracy of 0.244(m.-%/m.-% × 100%). Even if only two laser pulses – one prepulse and one measuring pulse – are applied to the sample at five different positions on the sample surface, the accuracy increases only slightly to 0.338(m.-%/m.-% × 100%). These values show that, depending on the specific demands of the application, the FML system can be used for a rapid assessment of alloyed aluminum measuring objects.

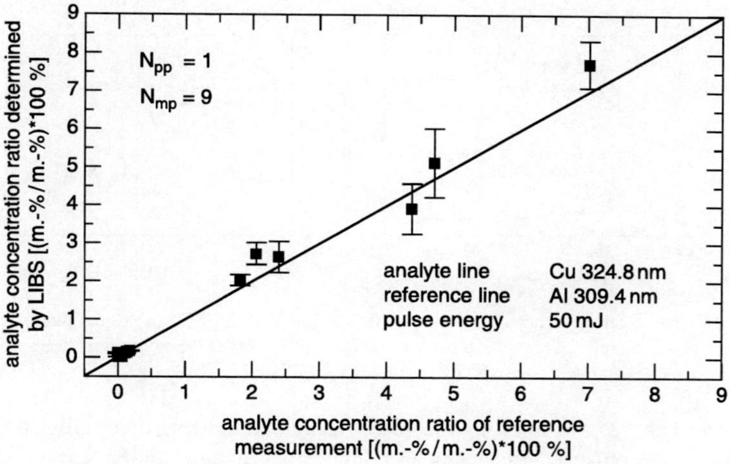

Fig. 17.24 Analyte concentration ratio determined by LIBS versus the analyte concentration ratio of the reference measurement for copper in aluminum

Another potential application for a LIBS measuring system with a flexible and compact measuring probe is the rapid depth profile analysis of coatings on sheet products. Due to the demands of the customers of sheet metal manufacturers, these surface-refined products represent an increasing fraction of the sheet product market. On the other hand, the production and the processing of these coated products requires fast measuring systems to monitor the surface coating characteristics for process control and quality assurance purposes.

Due to the restricted pulse energy of the laser system, the investigations are based on repetitive measurements at one position of a metallic sample. In a first step, a hot-dip zinc-coating with a thickness of 20 μm was analyzed, and the single-shot spectra for 200 laser pulses of 50 mJ pulse energy were recorded.

Figure 17.25a shows the signal intensity that was measured for the Fe line 373.8 nm and the Zn line 307.2 nm as a function of the laser pulse number. Even from the raw intensities, the Zn/Fe-transition can be clearly identified.

In order to determine the penetration depth per laser pulse, the derivative of the signal intensity was calculated numerically. This first derivative, which is depicted in Fig. 17.25b, shows a maximum for the Fe line and a minimum for the Zn line at approximately 50 laser pulses.

From the zinc-coating thickness of 20 μm, which had been measured using glow-discharge optical emission spectrometry (GD-OES), the average ablation rate for Zn can be estimated as 0.4 μm/pulse.

Assuming that the Zn/Fe-transition of the hot-dip coated sheet metal is very sharp with respect to the thickness of the coating, the width of the maximum of the first derivative in Fig. 17.25b can be taken as a measure for the depth resolution that was achieved with these measuring parameters. In this case, the restricted depth

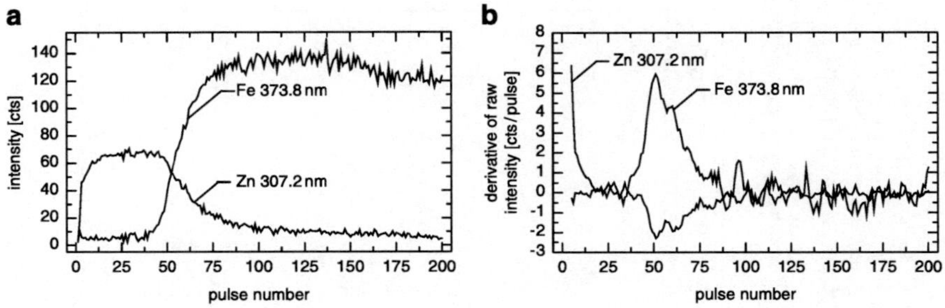

Fig. 17.25 Investigation of hot-dip zinc-coated sheet metal with FML. (**a**) signal intensity of zinc and iron, and (**b**) change of signal intensity of zinc and iron versus the number of laser pulses

resolution can be attributed to the laser beam parameters. Especially, the beam profile and the pulse energy result in cone-shaped craters on the sample surface, which in turn means that with each laser pulse, sample material from different depth regions contribute to the LIBS signal (cf. Sects. 7.1, 16.1).

The mobile system FML is a first step toward compact-measuring probes that can be integrated into automated measuring systems for industrial applications. The flexible coupling of a measuring probe to a spectrometer might be extended to a multiprobe for parallel inline inspection tasks at moving products.

17.4 Remote Systems

Several laser spectroscopic methods and systems are described in the literature for remote analysis in terms of measurements over distances of several meters and kilometers [17.18]. The applied methods are laser fluorometers, LIDAR, resonance Raman spectroscopy, and LIBS. One of the first remote LIBS systems was studied for space exploration [17.19]. A Galilean telescope was used to focus the laser beam onto soil samples at a distance of 19 m in a low pressure CO_2 atmosphere. Stainless steel samples were measured at distances of up to 45 m showing the capability to identify steel grades [17.20]. A rover-based LIBS system was described for mineralogical and elemental identification of rock and soil targets at distances of 2–6 m [17.21]. Aluminum and steel samples at high temperatures were studied using a Newton telescope of 200 mm aperture to collect the plasma emission [17.22]. A remote LIBS system was tested for real-time monitoring of high-temperature corrosion in stainless steels at temperatures up to 1,200°C [17.23]. A coaxial optical design was used for open-path remote LIBS measurements up to distances of 300 m [17.24]. However, no quantitative analytical data were presented. Laboratory-scale melts of stainless steel were measured over a distance of 7.5 m with a Czerny–Turner spectrometer having a spectral range of 30 nm and an estimated resolution of about 160 pm [17.25]. Analyte signal levels as a function of temperature increase by factors of 5–10 for temperatures up to 1,420°C. LOD were determined for Cr and Ni: 1,190 and 540 μg/g. Compared to measurements performed at smaller distances in an inert gas atmosphere as described in Sect. 13.1.3 (cf. Table 13.8), these values are worse by more than an order of magnitude, which can be attributed to the longer measuring distance and the less-defined measuring conditions as caused, e.g., by the oxidizing atmosphere. A remote LIBS system was described for elemental analysis of samples of environmental interest such as bark, leaves, soil rock, and stones [17.26]. The measuring distance was 12 m. A LIBS system operating with a frequency tripled Nd:YAG laser and measuring distances of 60 m was tested for remote imaging and remote cultural heritage ablative cleaning [17.27]. The primary mirror of the receiving Newton telescope amounts 400 mm, the spectral range detected covers 360–800 nm with a resolution of 2.2 nm. Steel slag samples at temperatures of 850°C were studied with a remote LIBS system at distances of 12 m demonstrating the capability to obtain basicity values in the range 0.9–2 with good correlation to XRF-based basicity indices [17.28]. Detection of explosive residues at distances of 45 m were studied with a remote LIBS system equipped with a Herschelian telescope [17.29].

In the following, the remote LIBS system TeleLis will be described offering a broad spectral range of more than 340 nm with a high spectral resolution of 25 pm and single as well as double pulse operation [17.30, 17.31]. The system design is tailored to industrial applications in the metal industry allowing for quantitative measurements under harsh conditions.

Figure 17.26 shows the optical setup. TeleLis comprises a pulsed Nd:YAG laser source with up to 20 Hz repetition rate and a pulse energy of up to 300 mJ. The laser system has double pulse capability with a freely choosable interpulse separation between 2 and 200 μs. The laser beam diameter is enlarged to approx. 80 mm by a Galilean telescope before being focused by an achromatic lens onto the sample. The focusing system is designed for measuring distances between 3 and 12 m. An integrated range finder – LD in Fig. 17.26 – allows to measure the distance to the sample and to adjust the focus automatically onto the sample. The plasma radiation is collected by a Newton telescope with a main mirror diameter of 300 mm. The plasma is imaged onto a fiber optic guiding the measuring light to the spectrometer entrance slit. The spectrometer is a multi-CCD Paschen–Runge system with 16 CCD detectors. It covers a wavelength range of 176–520 nm with an average spectral resolution of approximately 25 pm. The design of the laser beam guidance, the plasma radiation collection system, and the range finder optics is collinear and freely turnable within an angular range of ±45° vertically and ±90° horizontally. This allows to target a new sample within seconds without a readjustment of the optics.

TeleLis is designed as a transportable device. Figure 17.27, left shows the system ready for measurement. The laser source is mounted on the base platform together with the beam-shaping optics. The beam is then guided within the vertical arm and coupled into the telescope. The Paschen–Runge spectrometer is mounted on top of the vertical arm just above the telescope allowing for short fiber optic connection.

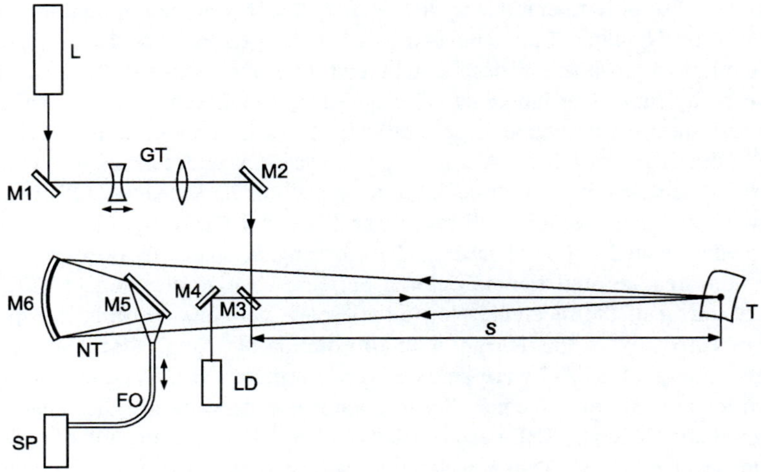

Fig. 17.26 Optical setup of the remote LIBS system TeleLis. L = laser source, GT = Galilean telescope, M1–M5 = plane mirrors, NT = Newton telescope, M6 = primary mirror of NT, T = measuring object, *s* = distance to T, LD = laser distance measurement, FO = fiber optics, SP = spectrometer

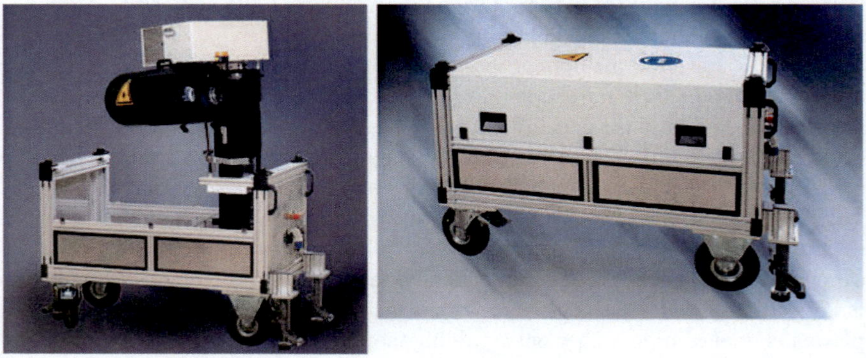

Fig. 17.27 *Left*: Remote LIBS system TeleLis with the optics in the position ready for measurement. *Right*: TeleLis system ready for transport

For transport purposes, the vertical arm with telescope and spectrometer can be folded into the transport container, which is mounted on wheels, see Fig. 17.27, right. It is not necessary to disassemble any components to transport the system. After arriving at the new measuring position, it is sufficient to unfold the optics, to repower and to reinitialize the system to be ready for measurement.

The spectrometer performance was tested by guiding the light of a mercury lamp by a fiber optic to the spectrometer. Figures 17.28 and 17.29 show the Hg I line at 289.359 nm and the Hg II line at 194.230 nm, respectively. The spectral resolution is determined to be 25 pm in both spectral regions. This resolution is a

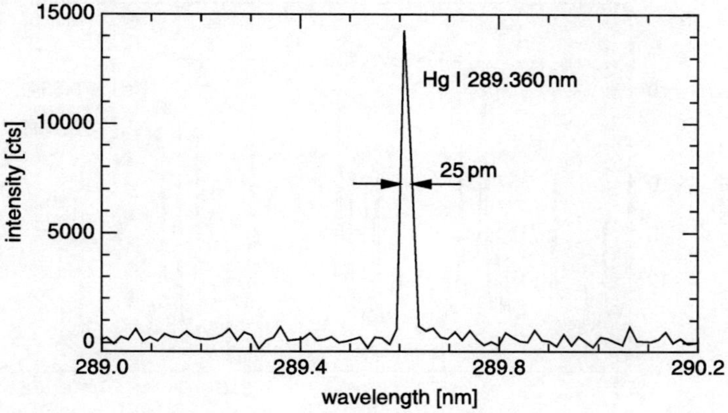

Fig. 17.28 Hg I line of a mercury lamp at 289.360 nm

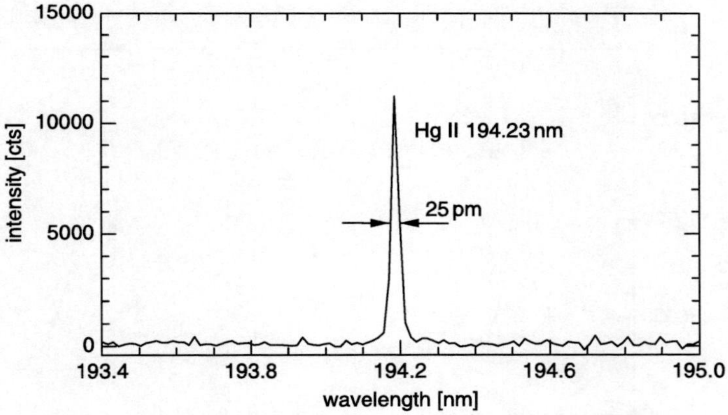

Fig. 17.29 Hg II line of a mercury lamp at 194.230 nm

necessary prerequisite for quantitative measurements of metal samples, since those samples show very line rich spectra. As an example, Figs. 17.30 and 17.31 show LIBS spectra of a steel and an aluminum sample, respectively, taken at a measuring distance of 5 m. The laser is running in double pulse mode with a pulse separation of $\Delta t = 3\,\mu s$ and a burst energy of 100 mJ. The pulse energy ratio amounts to $E_1 : E_2 = 1 : 5$. The spectra shown are gained by averaging of 50 pulses.

Figure 17.30 shows well separated closely spaced iron lines allowing to reduce interferences of analyte lines to a large extent. The spectrometer is capable to resolve self-reversal features of resonance lines as shown exemplarily in Fig. 17.31 for two aluminum resonance lines. The spectral resolution achieved is comparable to the one of a high-performance echelle system, cf. Sect. 10.3, Fig. 10.6.

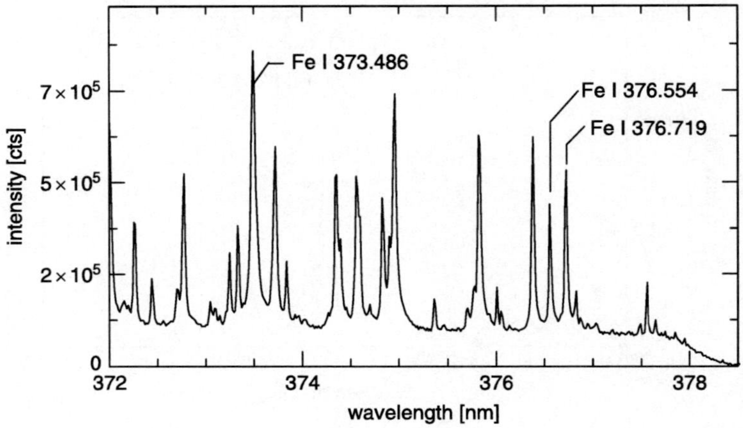

Fig. 17.30 LIBS spectrum of a steel sample, taken at a measuring distance of 5 m. For measuring parameters see text

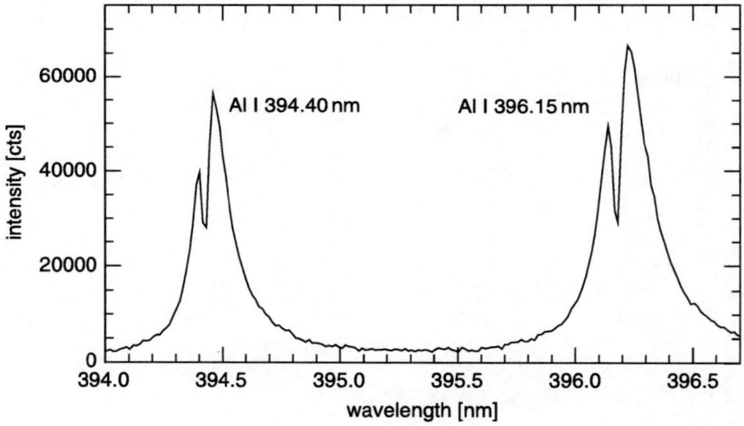

Fig. 17.31 LIBS spectrum of an aluminum sample, taken at a measuring distance of 5 m. The laser parameters are the same as those used for Fig. 17.30

The effect of double pulse mode versus single pulse excitation for remote measurements was studied for a set of analytes of the high-alloy steel sample NCS HS 23702–1 (the composition of this sample is given in Sect. 13.2, Table 13.9). Figure 17.32 shows the intensities and standard deviations of analyte lines of 12 species on a logarithmic scale. All species are measured simultaneously at a distance of 12 m. The laser is running in single and double pulse mode, respectively. The settings for double pulses are the same as stated above. The pulse energy is 190 mJ in single pulse mode. For all 13 species, the double pulse mode gives higher intensities than the single pulse mode, although the total burst energy of the double pulse is

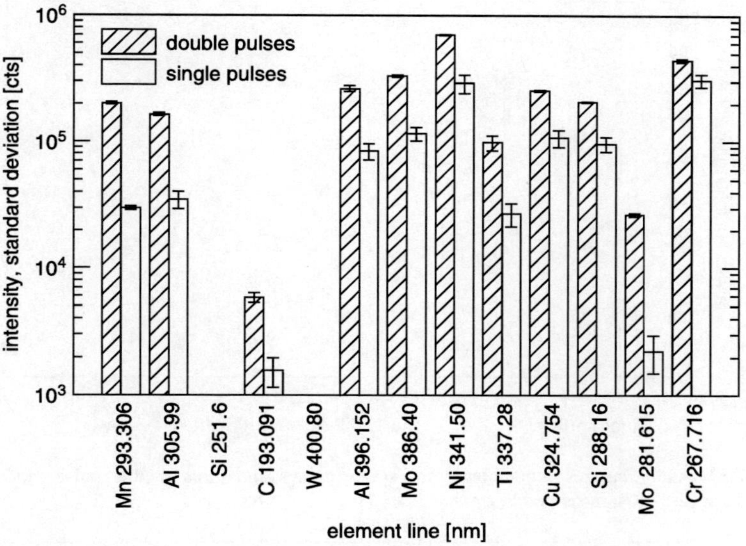

Fig. 17.32 Intensity of 12 species in single and double pulse operation of the laser source, measured simultaneously at a distance of 12 m of the high-alloy steel sample NCS HS 23702–1

about half of that used for the single pulse. At the same time, the standard deviations are significantly reduced in double pulse mode.

A measure of the capability to discriminate different analyte concentrations in a specimen is the analytical resolving power $A = (I_1 - I_2)/\sqrt{s_1^2 + s_2^2}$ [cf. definition in Sect. 14.1.2, relation (14.1)], where I_1 is the intensity of one particular spectral line of sample 1, I_2 the intensity of the same spectral line of sample 2, s_1 and s_2 are the corresponding standard deviations. Figure 17.33 shows the analytical resolving power for single pulse mode and three different double pulse modes. The double pulse modes have the same interpulse separation of $\Delta t = 5\,\mu\text{s}$, whereas the pulse ratios $E_1 : E_2$ (cf. Sect. 3.2, Fig. 3.9) are varied from 5:1, 1:1, and 1:5. The burst energy is changing slightly with different pulse ratios between $E_b = 75$ and 100 mJ. The sample distance is 5 m. The analytical resolving power is calculated from the intensity of the Si 288.2 nm line measuring two aluminum samples with different Si-content: $c_1(\text{Si}) = 25.5$ m.-%, $c_2(\text{Si}) = 1.57$ m.-%, respectively. Fifty measurement bursts are averaged ($N_{mp} = 50$) after five prebursts ($N_{pp} = 5$). In single pulse mode, the analytical resolving power reaches $A_{SP} = 7.3$. In double pulse mode, the analytical resolving power increases with increasing pulse energy of the second pulse in the burst, starting at $A_{DP}(5 : 1) = 2.3$ for a pulse ratio of 5:1, via $A_{DP}(1 : 1) = 5.4$ reaching $A_{DP}(1 : 5) = 19.2$ which is about 2.5 times higher than the corresponding value for single pulse mode. These investigations clearly show that the double pulse excitation improves the analytical performance not only at focusing

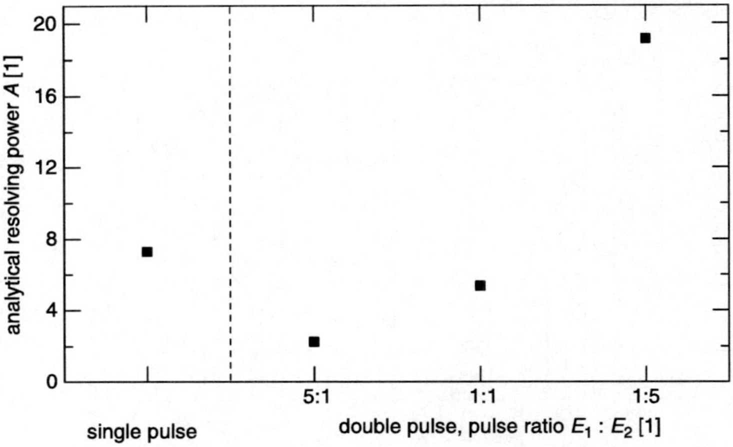

Fig. 17.33 Analytical resolving power for single pulses and three double pulse modes with different pulse ratios, measured at a distance of 5 m, $\Delta t = 5\,\mu s$

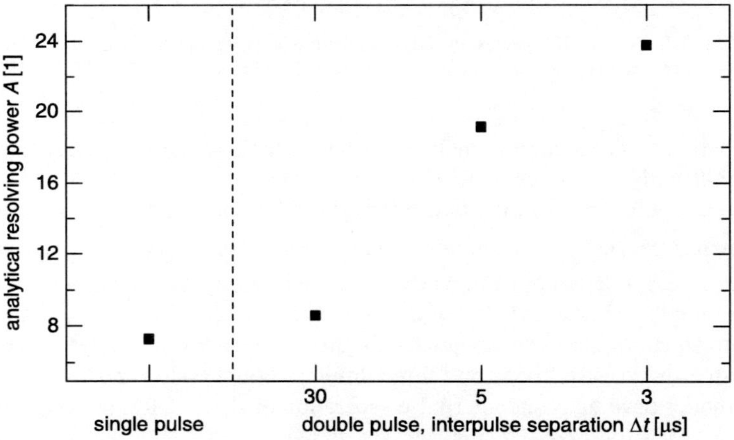

Fig. 17.34 Analytical resolving power at a measuring distance of 5 m for single and three double pulse modes with different interpulse separations. $E_1 : E_2 = 1 : 5$

distances of a few centimeters (cf. Sects. 6.1, 7.2, and 8.1) but also at distances that are orders of magnitude larger.

Figure 17.34 shows the analytical resolving power A for three different interpulse separations Δt of 30, 5, and 3 μs at a constant pulse ratio of 1:5. The total burst energy varies for the three double pulse modes in the range $E_b = 100$–140 mJ. The pulse energy for the single pulse amounts to 190 mJ. Compared to the results shown in Fig. 17.33, the analytical resolving power A is increased to 23.8 when the interpulse separation Δt is reduced to 3 μs.

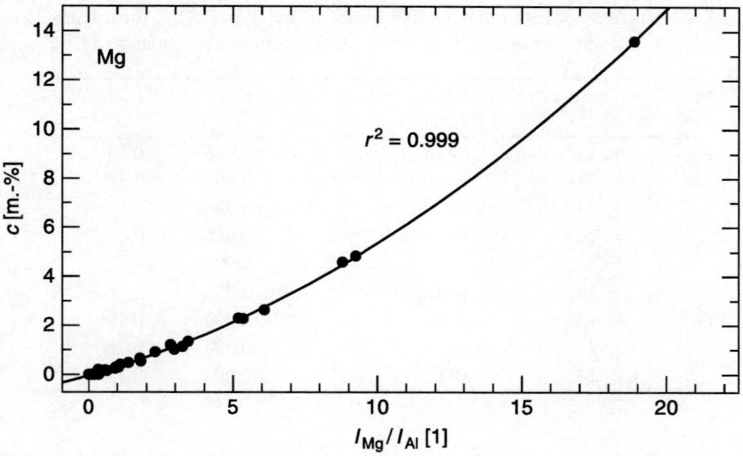

Fig. 17.35 Analysis curve for magnesium in aluminum matrix. Measuring distance 5 m, double pulse mode, interpulse separation $\Delta t = 3\,\mu s$, burst energy $E_b = 100\,mJ$, pulse ratio 1:5, $N_{pp} = 5$, $N_{mp} = 50$

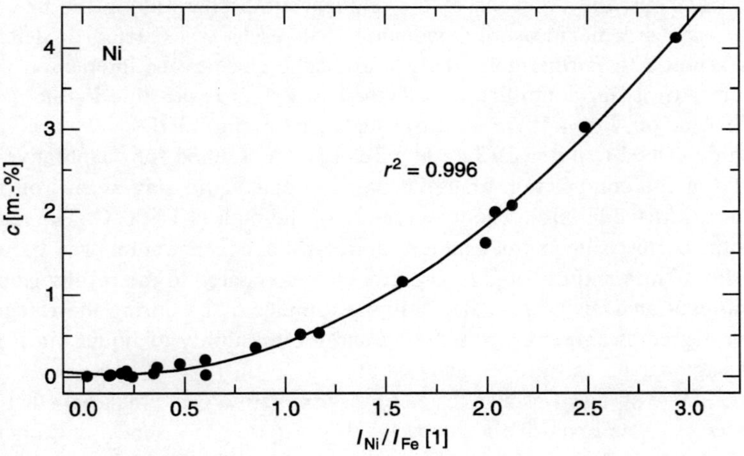

Fig. 17.36 Analysis curve for nickel in iron matrix. Measuring parameters are those given in Fig. 17.35

To study the capability for a remote quantitative analysis of analyte concentrations, analysis functions were determined using a set of aluminum and steel reference samples.

Figures 17.35 and 17.36 exemplify analysis curves for Mg in aluminum matrix and for Ni in iron matrix.

Table 17.8 summarizes the results obtained for the investigated analytes. Coefficients of determination greater than 0.99 were found for all analytes except for Mn

Table 17.8 Coefficients of determination and limits of detection (LOD) for different analytes in aluminum and iron matrix measured with remote LIBS system TeleLis at a measuring distance of 5 m

Matrix	Analyte	Concentration range (m.-%)	r^2	LOD (m.-%)
Al	Cu	0–14	0.996	0.13
	Mg	0–20	0.999	0.12
	Mn	0–2.5	0.984	
	Ni	0–4	0.996	0.07
	Si	0–7	0.996	
	Zn	0–1.8	0.998	
Fe	Cr	0–6	0.994	0.19
	Ni	0–3	0.996	0.16
	Mn	0–8	0.990	0.15

in aluminum. As expected, LOD are worse than those found for smaller measuring distances, see, e.g., Sect. 13.3, Table 13.16 for aluminum and Sect. 13.1, Tables 13.2 and 13.6 for steel. However, for the remote measurement no defined inert gas atmosphere can be applied as it was the case for the measurements described in Sect. 13.1. Thus, the influence of the ambient atmosphere along the beam path of the laser and the measuring radiation (5 m each) is expected to deteriorate the measurement. Furthermore, the gas exchange close to the interaction volume (cf. Sect. 4.6) is not controlled in a defined way as it is possible by the use of a sample stand (cf. Fig. 4.18) or a measuring lance (cf. Fig. 13.12).

The described remote LIBS system TeleLis was studied for quantitative inline analysis of the composition of liquid pig iron and liquid slag at an iron runner of a blast furnace [17.30]. The temperature of the melt is 1,500°C. The temporal evolution of the silicon, manganese, and carbon concentration was monitored inline for 40 min and the obtained results were compared to the results gained by conventional analysis of so-called lollipop samples taken during the test period. A high degree of agreement was found and the feasibility of liquid pig iron and liquid slag analysis was proven successfully.

The remote system was also tested to discriminate organic materials that are of relevance, e.g., as explosive residues [17.31]. Figure 17.37 shows line intensities of several species for the detection of ammonium nitrate on a substrate consisting of greased aluminum foil. The measurements were performed using an external echelle spectrometer (cf. Sect. 4.2, Table 4.4) instead of the built-in spectrometer. Single pulses with 190 mJ pulse energy were irradiated at a distance of 5 m. The measurement parameters used are delay time $t_{\text{delay}} = 1\,\mu\text{s}$, integration time $t_{\text{int}} = 10\,\mu\text{s}$, $N_{\text{mp}} = 3$ for substrate, $N_{\text{mp}} = 15$ for NH_4NO_3. Ammonium nitrate can clearly be detected even at low mass density levels of approx. $0.15\,\mu\text{g/mm}^2$ by the enhanced intensity of the hydrogen and cyanide emission.

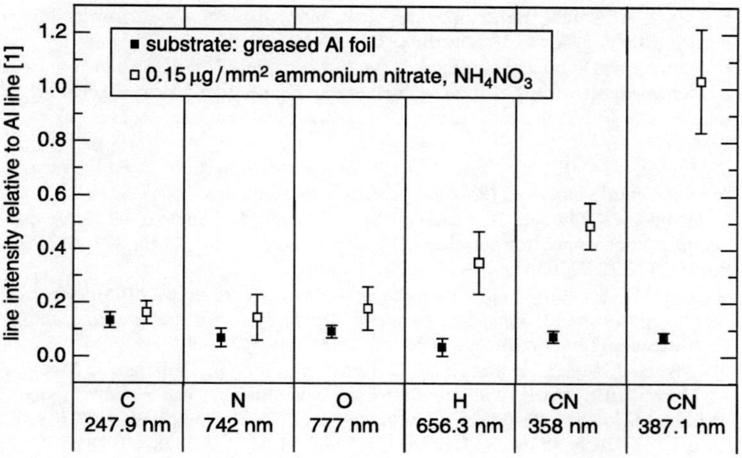

Fig. 17.37 Intensities of several species for the detection of ammonium nitrate (NH_4NO_3), measured with an echelle spectrometer at a distance of 5 m using single pulses with 190 mJ pulse energy

References

17.1. J. Vrenegor, V. Sturm, D. Eilers, R. Noll, Neuartiges Laserverfahren für die schnelle Präparation und Multielement-Analyse metallischer Prozessproben, Final Report of the Joint Research Project ATLAS funded by the Federal Ministry of Economics and Technology of Germany within the InnoNet Programm, support code IN4051, 1.1.2004–31.3.2007, 74 p

17.2. R. Noll, V. Sturm, Ü. Aydin, D. Eilers, C. Gehlen, M. Höhne, A. Lamott, J. Makowe, J. Vrenegor, Laser-induced breakdown spectroscopy – from research to industry, new frontiers for process control. Spectrochim. Acta Part B **63**, 1159–1166 (2008)

17.3. H. Balzer, M. Höhne, V. Sturm, R. Noll, Online coating thickness measurement and depth profiling of zinc coated sheet steel by laser-induced breakdown spectroscopy. Spectrochim. Acta B **60**, 1172–1178 (2005)

17.4. G. Asimellis, A. Giannoudakos, M. Kompitsas, Rapid, automated measurement of layer thickness on steel coin blanks using LIBS depth-profiling. Appl. Optics **46**, 935–942 (2007)

17.5. D. Papazoglou, V. Papadakis, D. Anglos, In situ interferometric depth and topography monitoring in LIBS elemental profiling of multi-layer structures. J. Anal. At. Spectrom. **19**, 483–488 (2004)

17.6. M. Corsi, G. Cristoforetti, M. Hidalgo, D. Iriarte, S. Legnaioli, V. Palleschi, A. Salvetti, E. Tognoni, Effect of laser-induced crater depth in laser-induced breakdown spectroscopy emission features. Appl. Spectrosc. **59**, 853–860 (2005)

17.7. Patent, DE 103 61 727, Verfahren und Vorrichtung zur Analyse fester Materialien mit der Laser-Emissionsspektrometrie, 21.7.2005

17.8. R. Noll, I. Mönch, O. Klein, A. Lamott, Concept and performance of inspection machines for industrial use based on LIBS. Spectrochim. Acta B **60**, 1070–1075 (2005)

17.9. K. Yamamoto, D. Cremers, M. Ferris, L. Foster, Detection of metals in the environment using a portable laser-induced breakdown spectroscopy instrument. Appl. Spectrosc. **50**, 222–233 (1996)

17.10. B. Castle, A. Knight, Battery powered laser-induced plasma spectrometer for elemental determinations. J. Anal. At. Spectrom. **13**, 589–595 (1998)
17.11. R. Harmon, F. De Lucia, A. Miziolek, K. McNesby, R. Walters, P. French, Laser-induced breakdown spectroscopy (LIBS) – an emerging field-portable sensor technology for real-time, in-situ geochemical and environmental analysis. Geochem. Explor. Environ. Anal. **5**, 21–28 (2005)
17.12. W. Pierce, S. Christian, *Portable LIBS Instrumentation Can Identify Trace Levels of Environmental Pollutants*, (Photonik International, Fellbach, 2006), 92–94
17.13. J. Goujon, O. Musset, A. Giakoumaki, V. Pinon, D. Anglos, E. Georgiou, A new compact laser source for portable LIBS applications. Proc. SPIE **6871**, 68712Q (2008). doi:10.1117/12.777953
17.14. A. Taffe, D. Schaurich, G. Wilsch, *Development of a portable LIBS-device for quality assurance in concrete repair*, in *Concrete Repair, Rehabilitation and Retrofitting* ed. by G. Alexander (Taylor & Francis, London, 2009), pp. 547–549
17.15. M. Myers, J. Myers, J. Sarracino, C. Hardy, B. Guo, S. Christian, J. Myers, F. Roth, A. Myers, LIBS system with compact fiber spectrometer, head mounted spectra display and hand held eye-safe erbium glass laser gun, SPIE Photonics West (2010) Solid State Lasers XIX: Technology and Devices Conference LA101, # 7578–87, 20 p
17.16. StellarNet Inc., FL; Avantes, CO; Ocean Optics, FL; Applied Photonics, UK
17.17. F. Hilbk-Kortenbruck, M. Höhne, R. Noll, M. Freit, J. Joosten, H. Falk, Compact measuring system for laser emission spectrometry of coated and uncoated metals. Proc. 9. Anwendertreffen Röntgenfluoreszenz- und Funkenemissionsspektrometrie, 59–69 (2002)
17.18. F. Hoge, *Recent advances in laser remote sensing*. Proc. SPIE **3707**, 2–9 (1999), *Laser Radar Technology and Applications IV*, ed. by G. Kamerman, C. Werner
17.19. A. Knight, N. Scherbarth, D. Cremers, M. Ferri, Characterization of laser-induced breakdown spectroscopy (LIBS) for application to space exploration. Appl. Spectrosc. **54**, 331–340 (2000)
17.20. S. Palanco, J. Baena, J. Laserna, Open-path laser-induced plasma spectroscopy for remote analytical measurements on solid surfaces. Spectrochim. Acta B **57**, 591–599 (2002)
17.21. R. Wiens, R. Arvidson, D. Cremers, M. Ferris, J. Blacic, F. Seelos, K. Deal, Combined remote mineralogical and elemental identification from rovers: field and laboratory tests using reflectance and laser-induced breakdown spectroscopy. J. Geophys. Res. **107**, 8003 (2002). doi:10.1029/2000JE001439
17.22. J. Vadillo, P. Garcia, S. Palanco, D. Romero, J. Baena, J. Laserna, Remote, real-time, on-line monitoring of high-temperature samples by noninvasive open-path laser plasma spectrometry. Anal. Bioanal. Chem. **375**, 1144–1147 (2003)
17.23. P. García, J. Vadillo, J. Laserna, Real-time monitoring of high-temperature corrosion in stainless steels by open-path laser-induced plasma spectrometry. Appl. Spectrosc. **58**, 1347–1352 (2004)
17.24. S. Palanco, J. Laserna, Remote sensing instrument for solid samples based on open-path atomic emission spectrometry. Rev. Sci. Instr. **75**, 2068–2074 (2004)
17.25. S. Palanco, S. Conesa, J. Laserna, Analytical control of liquid steel in an induction melting furnace using a remote laser-induced plasma spectrometer. J. Anal. At. Spectrom. **19**, 462–467 (2004)
17.26. C. López-Moreno, S. Palanco, J. Laserna, Remote laser-induced plasma spectrometry for elemental analysis of samples of environmental interest. J. Anal. At. Spectrom. **19**, 1479–1484 (2004)
17.27. R. Grönlund, M. Lundqvist, S. Svanberg, Remote imaging laser-induced breaddown spectroscopy and remote cultural heritage ablative cleaning. Opt. Lett. **30**, 2882–2884 (2005)
17.28. C. López-Moreno, S. Palanco, J. Laserna, Quantitative analysis of samples at high temperature with remote laser-induced breakdown spectrometry using a room-temperature calibration plot. Spectrochim. Acta B **60**, 1034–1039 (2005)

17.29. C. López-Moreno, S. Palanco, J. Laserna, F. DeLucia, J. Miziolek, J. Rose, R. Walters, A. Whitehouse, Test of a stand-off laser-induced breakdown spectroscopy sensor for the detection of explosive residues on solid surfaces. J. Anal. At. Spectrom. **21**, 55–60 (2006)
17.30. G. Mathy, B. Monfort, B. Vanderheyden, V. Tusset, Liquid steel process: advanced on line sensors under development at CRM. Metall. Anal. **30**(Suppl. 1), 6–14 (2010)
17.31. R. Noll, C. Fricke-Begemann, Stand-off detection of surface contaminations with explosives residues using laser-spectroscopic methods, in *Stand-off Detection of Suicide Bombers and Mobile Subjects*, ed. by H. Schubert, A. Rimski-Korsakov (Springer, New York, 2006), pp. 89–99

Chapter 18
Industrial Applications

This chapter describes a selection of industrial applications of LIBS systems based on methodical approaches and instrumental system designs presented in the previous chapters.

18.1 Identification Testing of High-Alloy Pipe Fittings

This application deals with the detection of material mix-ups in production lines of pipe fittings and tubes made of high-alloy steel grades. The motivation and the system design of the first generation LIBS system – called LIFT – for this task were described in Sect. 17.2. The second generation inspection system uses a flexible laser beam steering based on the optical setup shown in Fig. 13.23, where the laser beam is deflected by an *xy*-scanner to irradiate various inspection spots inside of a measuring volume of 400 mm × 400 mm × 100 mm [18.1]. Figure 18.1 shows the modules of the LIFT II system.

The spectrometer is linked via a fiber optics with the optic module. The three modules are designed in such a way that they can be integrated easily into an existing production line. Figure 18.2 shows the specimens with the laser-induced plasmas. Each specimen is measured at up to three positions to inspect the steel grades of the main body of the fitting and of the flanges.

In the production line, the test objects are placed on carriers to assure a defined orientation of the fittings. These carriers are transferred to the inspection module. After closing of a protection gate, the laser inspection is initiated. The scanner directs the laser beam sequentially to the predefined inspection positions which are transmitted automatically by the host computer where the coordinates of inspection positions for all types of pipe fittings are stored. Number and position of the inspection points can be chosen freely inside the measuring volume stated above. As soon as the scanner has reached the preset coordinates, the measurement starts. For each measurement position, the time needed amounts to about 1 s, within which up to 25 elements are detected simultaneously and the concentrations of the alloying

R. Noll, *Laser-Induced Breakdown Spectroscopy*, DOI 10.1007/978-3-642-20668-9_18,

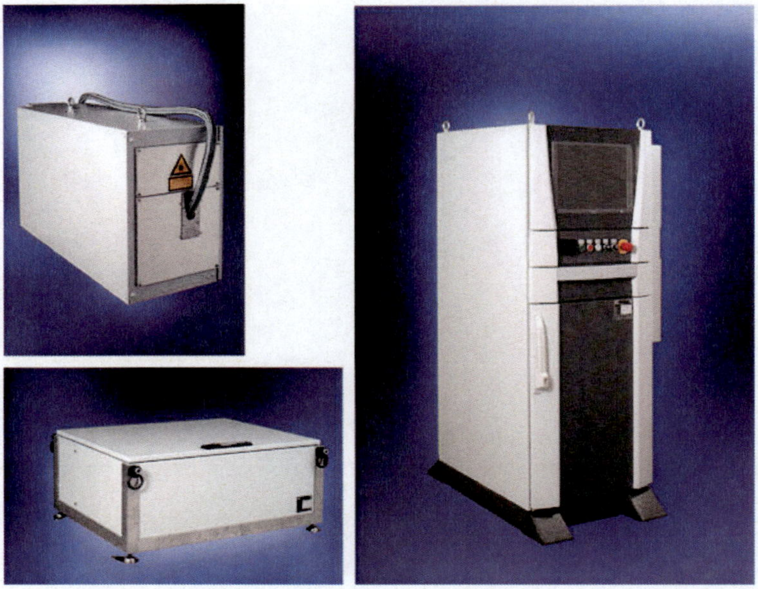

Fig. 18.1 Modules of the second generation inspection system LIFT II: *top, left*: optic module with laser source and scanner; *bottom, left*: spectrometer; *right*: electrical cabinet

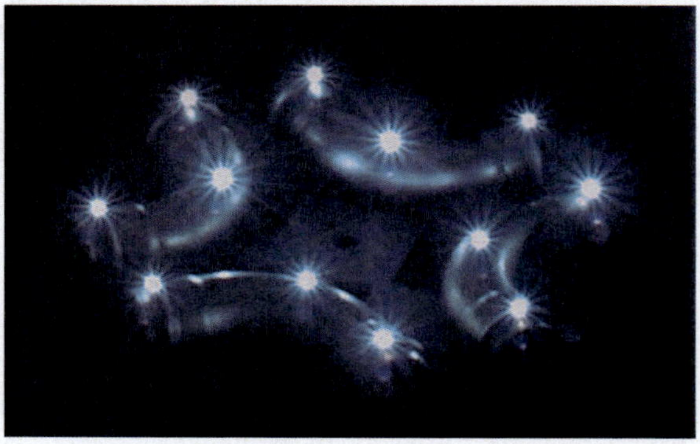

Fig. 18.2 Pipe fittings with laser-induced plasmas for mix-up detection

elements are determined automatically. If these concentrations are within the ranges specified by the manufacturer for all test objects and measurement positions, then the inspected products on the carrier are cleared for subsequent processing steps and transported onward. If a mix-up is detected, the test object concerned is identified and discharged in the next step. All measuring results for each manufactured product are documented electronically.

The inspection system LIFT II fulfills the requirements for laser class 1 according to the IEC standard [18.2]. No further protection measures are needed for the user.

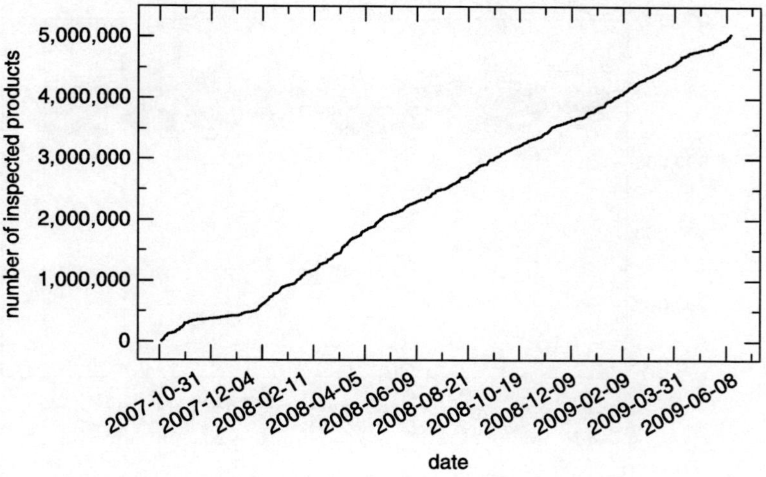

Fig. 18.3 Number of inspected products as a function of time for a second generation LIFT II system

Inspection equipment monitoring according to ISO 9001 and recalibrations are performed automatically. LIFT is equipped with a remote maintenance interface.

The described system is in use in three shift operation, 7 days a week. Figure 18.3 shows the number of inspected products over a period of about 1.5 years. More than five million products were inspected automatically in routine industrial operation. Within the shown time period more than 300 different product types and geometries were measured. The data acquired comprise more than 250 million LIBS measurements.

Figure 18.4 shows the laser power measured during each inspection over the same time period. The relative standard deviation amounts to 4.86%. A measure of the stability of the excitation process is the intensity of the zeroth order signal acquired with a separate detector installed inside the Paschen–Runge spectrometer. The zeroth order comprises all spectral components of the laser-induced plasma reflected by the grating. Figure 18.5 shows this signal over a time period of 1.5 years. The relative standard deviation is 6.51%.

The technical availability of the LIFT II system was tested according to established standards [18.3]. The availability is 99.94% demonstrating the high degree of system performance achieved.

18.2 Analysis of Slags

Motivation and LIBS setup for slag analysis were described in Sect. 14.2.2. Figure 18.6 shows the LIBS instrument with a closed cabinet ($1.0 \times 1.8 \times 1.7\,m^3$) installed $\approx$ 10 m from the vacuum degasser in the steel works. The instrument is

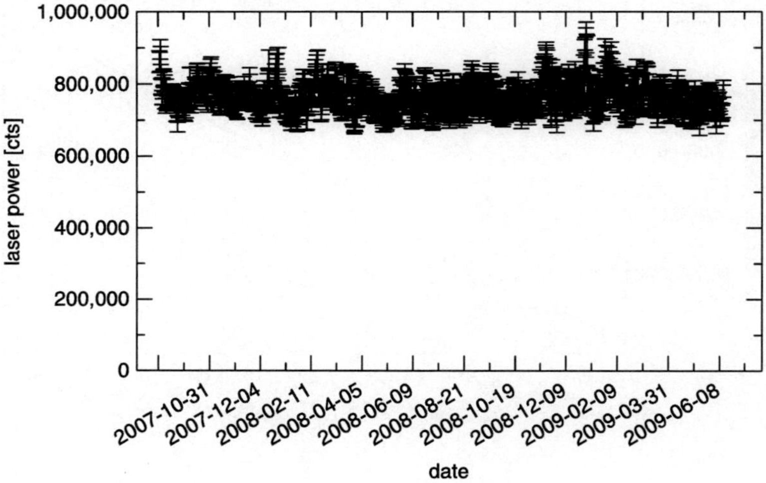

Fig. 18.4 Laser power over a time period of 1.5 years

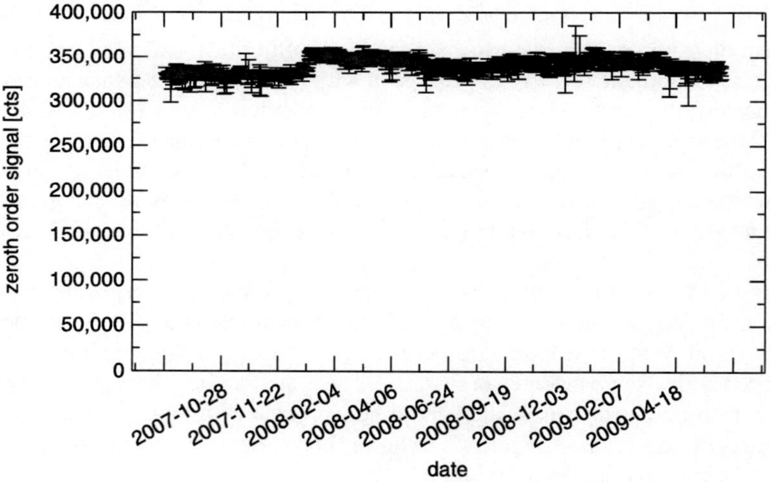

Fig. 18.5 Zeroth order signal over a time period of 1.5 years

24 h/7 d ready for operation. The routine operation is via an operator panel at the instrument. No computer keyboard or monitor is required at site in the steel works. For analysis, a button at the operator panel is pushed and the sample is put onto the sample stand. After closing the lid of the sample stand, the measurement runs automatically. The result of the analysis is transferred to the host computer (TCP/IP) of the steel works after 80 s.

Fig. 18.6 LIBS instrument installed in a steel plant (in front on the right)

The analysis of unknown samples is carried out by a multivariate data analysis of the measured LIBS signals with a partial least square regression (PLS1). The PLS1 calibration model is constructed by the measured intensities of the spectral lines I for a set of 31 production samples with BBA (beta barium borate) reference analysis. The calibration ranges given by the variation of the production samples is $w(SiO_2) \approx 0.8–10\,cg/g$, $w(CaO) \approx 50–56\,cg/g$, and $w(Al_2O_3) \approx 24–41\,cg/g$. Afterward the PLS1 calibration model is validated by the measurement of another set of 32 production samples with XRF reference mass fractions. In Fig. 18.7, the mass fractions of SiO_2 predicted by the multivariate analysis of the measured LIBS data versus the reference values are given together with the linear fitting curve [14.49]. The ideal case, i.e., identity of LIBS and reference value, is the bisecting line in this graph. The root mean square error of prediction, RMSEP, for a species i is defined as [cf. (11.19) in Chap. 11]:

$$\mathrm{RMSEP}(i) = \left[\frac{1}{N}\sum_{s=1}^{N}(w_{i,\mathrm{ref},s} - w_{i,\mathrm{LIBS},s})^2\right]^{1/2}, \qquad (18.1)$$

where $w_{i,\mathrm{LIBS},s}$ is the mass fraction predicted by multivariate analysis of the LIBS signals for analyte i and sample s, $w_{i,\mathrm{ref},s}$ is the mass fraction determined by the reference analysis, and N is the number of samples of the validation set. The RMSEP value is a measure of the average prediction error, expressed in the same units as the original values.

The corresponding values for CaO/SiO_2 are: slope 1.12, $r^2 = 0.973$, RMSEP $(CaO/SiO_2) = 1.87$ and for Al_2O_3/SiO_2: slope 1.06, $r^2 = 0.979$, RMSEP$(Al_2O_3/SiO_2) = 1.11$. The RMSEP values for CaO and Al_2O_3 directly are RMSEP(CaO) = $1.0\,cg/g$ and RMSEP$(Al_2O_3) = 1.5\,cg/g$, respectively. For comparison, RMSEP values determined by XRF analysis of homogenized vacuum slag sample (homogenized by crushing and milling to powder and afterward pressing to a solid sample)

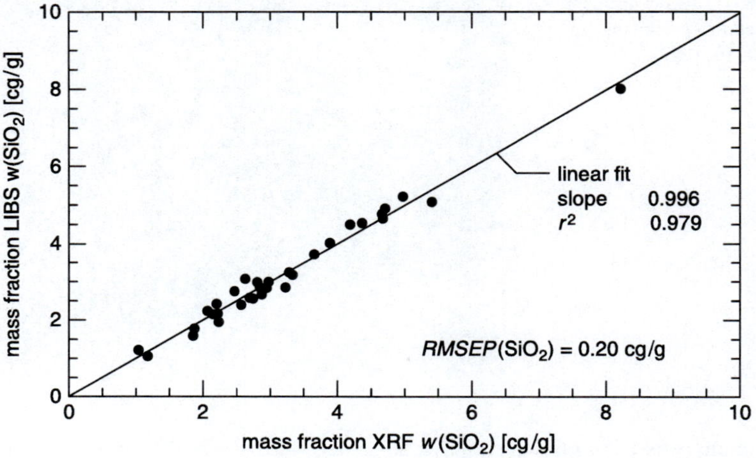

Fig. 18.7 Mass fraction of the species SiO_2 of production samples by LIBS analysis versus reference mass fraction by XRF

amount to RMSEP(CaO) = 0.85 cg/g, RMSEP(SiO_2) = 0.66 cg/g, and RMSEP (Al_2O_3) = 1.8 cg/g, respectively.

The ambient temperature at the site of the LIBS instrument in the steel works varies from $\approx$ 0 to 40° during the year. The temperature of the instrument cabinet and especially the temperature of the Paschen–Runge spectrometer is actively controlled. The resulting stability of the spectrometer temperature is $T = (33.1 \pm 0.25)$°C, see Fig. 18.8. The stable operation of the LIBS instrument in the harsh steel works environment is verified by repeated measurements of a monitor sample which is a glass-like sample remelted from slag powder of the vacuum degasser, see Fig. 18.9.

The fast analysis of vacuum slag samples by LIBS was demonstrated successfully in the steel works nearby a vacuum degasser station. The LIBS analysis takes about 80 s and a further reduction is feasible. By spatial averaging of the LIBS measurement with an optical microlens array, a representative analysis of SiO_2, CaO, and Al_2O_3 of production samples of the vacuum degasser without any further preparation was performed in spite of the sample heterogeneity and variations in color, cracks, and holes.

18.3 Characterization of Inclusions and Segregations

The concept and LIBS setup for high-speed spatially resolved microanalysis were described in Chap. 15, demonstrating the capability to gain element mappings. Figure 18.10 shows a photograph of the second generation high-speed scanning

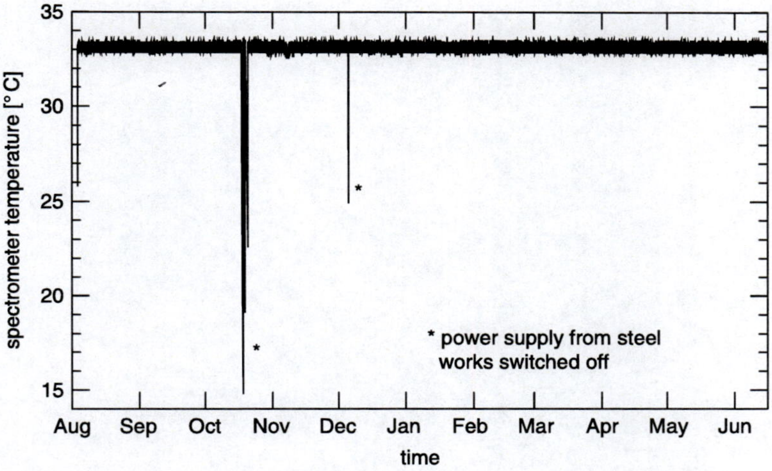

Fig. 18.8 Stability of the spectrometer temperature for a period of 10 months after installation in the steel works. The temperature is logged every 10 min (24 h/7 d)

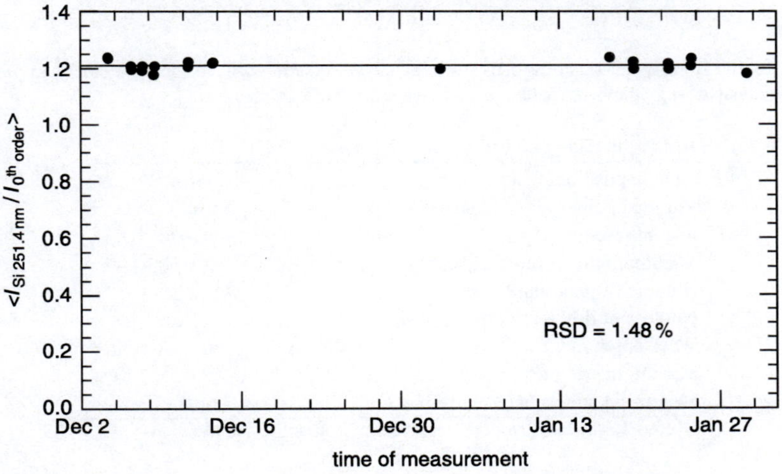

Fig. 18.9 Stability test of the LIBS instrument in the steel works by repeated measurements of a monitor sample

LIBS system SML (Scanning Microanalysis with Laser Spectrometry) developed at Fraunhofer ILT, Aachen [18.4]. The technical data of SML is given in Table 18.1.

The SML system is used to characterize inclusions and segregations in steel samples [18.5, 18.6, 18.7]. The production of high-value steel products, such as thin wires for energy-saving steel belt tires or thin sheets for lightweight packaging, requires high-quality steel grades. Inclusions reduce the steel quality by limiting

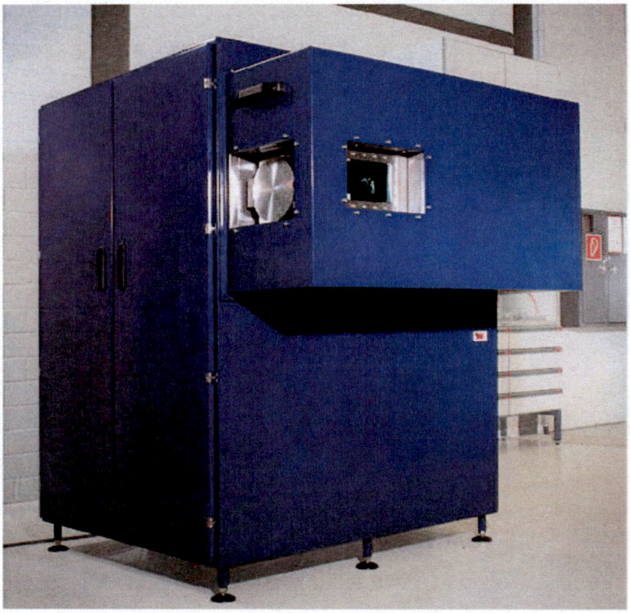

Fig. 18.10 High-speed scanning LIBS microanalysis system SML to measure steel samples for the analysis of segregations, inclusions, and decarburization zones

Table 18.1 Data of the high-speed scanning LIBS system SML

Sample positioning accuracy	$<1\,\mu m$
Minimal point to point distance	$5\,\mu m$
Spatial resolution[a]	$<20\,\mu m$
Maximum measuring frequency	1,000 Hz
Number of photomultipliers	41
Number of different detected elements	24
Wavelength range	130 − 777 nm
Maximum size of scan field	45 mm × 110 mm
Measuring atmosphere	Argon

[a]Refers to the spectroscopic analysis of a metallic sample

the minimum diameter of wires for steel belt tires or the thickness of sheets for beverage cans. Figure 18.11 illustrates these products and shows electron-microscopical images of inclusions in steel [18.8].

The development of high-quality steels requires an analyzing method to detect inclusions like Al_2O_3, aluminum nitride (AlN), TiC, SiC, CaO, ZrO_2, and others fast, reliably, and cost efficient. State of the art analyzing methods are wet-chemical analysis, spark optical emission spectrometry, micro-X-ray fluorescence analysis (μ-XRF), and scanning electron microscopical energy dispersive X-ray analysis (SEM-EDX). However, these methods have specific drawbacks. For wet-chemical

Fig. 18.11 *Left*: high-value steel products like wires for steel belt tires and thin sheets; *right*: scanning electron microscopical images of inclusions in steel; *top*: aluminum nitrite AlN; *bottom*: aluminum oxide Al_2O_3 (1), titanium nitride TiN (2), and $Ti_4C_2S_2$ (3). The scale given corresponds to 5 μm [18.8]

analysis, the inclusions are extracted chemically or electrolytically and made accessible to a subsequent analysis. The particular advantage of this approach is the high confidence level to gather all inclusions. The personnel and chemical costs are high [18.8, 18.9]. SEM-EDX requires an extensive sample preparation with diamond polishing and measuring times of a few hours. Spark optical emission spectrometry (SD-OES) has only a limited capability for a spatially resolved analysis due to the fact that the position of impact of the spark on the sample surface cannot be determined exactly (cf. Sect. 3.6, Fig. 3.17). Hence, with SD-OES, it is not possible to gain a spatially resolved analysis of inclusions with resolutions in the micrometer range. However, a statistical evaluation of single spark signals allows to estimate the steel cleanness with constraints. Especially, the light elements, oxygen and nitrogen, are partially difficult to be detected.

The methodical approach for a spatially resolved analysis with LIBS was described in Sect. 15.1. For each measuring point, only one tightly focused laser pulse is irradiated onto the sample surface. The advantages of LIBS microanalysis compared with SD-OES are electrode contamination is excluded, and the interaction zone of the laser radiation with the sample material is well defined. Compared with wet-chemical methods, consumables and time are saved. In comparison with SEM-EDX, elaborate sample preparation and long measuring times are cut down.

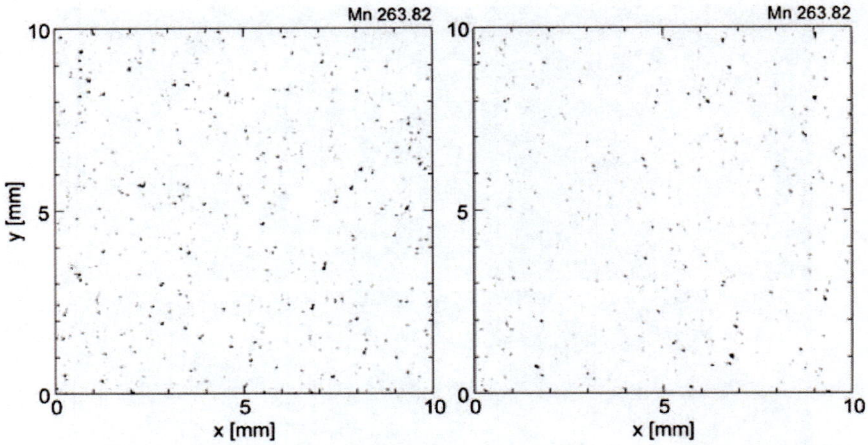

Fig. 18.12 Manganese maps of two steel samples (each 10 mm×10 mm) of different quality showing significantly different numbers of detected manganese enrichments. Measuring parameters are those given in Sect. 15.2 for Fig. 15.8

Inclusions in a steel matrix consist of, e.g., carbides, sulfides, nitrides, and oxides of the elements Al, Ca, Si, Mg, Mn, and Ti. Typical analyte lines for optical emission spectrometry lie in the wavelength range from 130 nm (O) to 777 nm (O), the highest excitation energies are 12 eV (e.g., for the nitrogen emission line at 746 nm). Since oxygen and nitrogen shall be detected, it is necessary to analyze the samples in a pure inert gas atmosphere (cf. Fig. 15.2).

Element mappings are used to identify inclusion types such as shown for manganese sulfide (MnS) in Sect. 15.2 and Fig. 15.8. An approach to obtain a measure of the cleanness of a steel sample is to count the number of measuring locations where a clear correlation of a set of elements is observed and where the intensities are greater than a predefined threshold. Figure 18.12 shows manganese maps of two steel samples of different quality [15.24]. All measuring parameters were kept constant. The number of measuring points, where the manganese intensity is greater than the average plus five times the standard deviation, amounts to 968 for the left image and 492 for the right image. An SEM-EDX measurement of these samples yields approximately the same relation for the cleanliness as the measurement with scanning LIBS.

Oxides and nitrides are critical inclusions in the steel matrix since they have a great hardness and cannot be deformed by rolling or drawing processes. Thyssen Krupp Stahl (TKS) prepared a sample of pure iron with embedded Al_2O_3 inclusions [18.10]. The sample was prepared with a grinding paper having a grain size of 80. The grains consist of 80% ZrO_2 and 20% Al_2O_3. The grinding time amounts to less than 1 min, yielding a surface roughness of below 15 μm. This simple kind of sample preparation is sufficient for scanning LIBS in contrast to SEM-EDX,

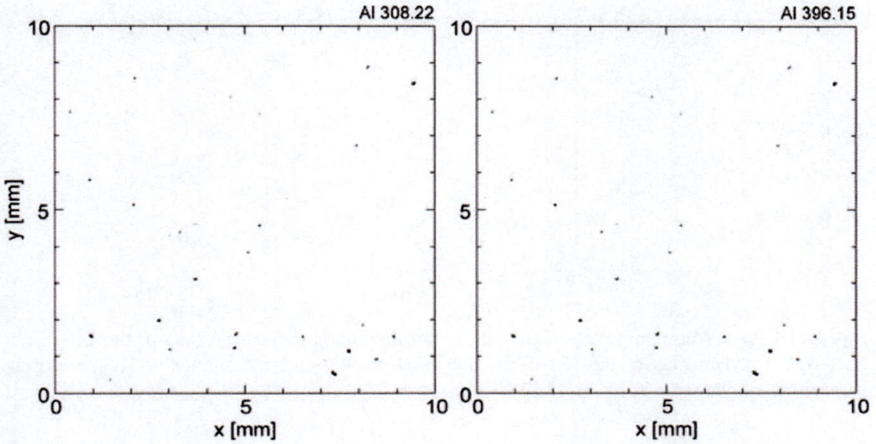

Fig. 18.13 *Left*: Aluminum map of the 308.22 nm line. *Right*: Aluminum map of the 396.15 nm line taken at the same measurement. The measuring parameters are measuring frequency 1,000 Hz, atmosphere Ar 5.0, total measuring time 11 min

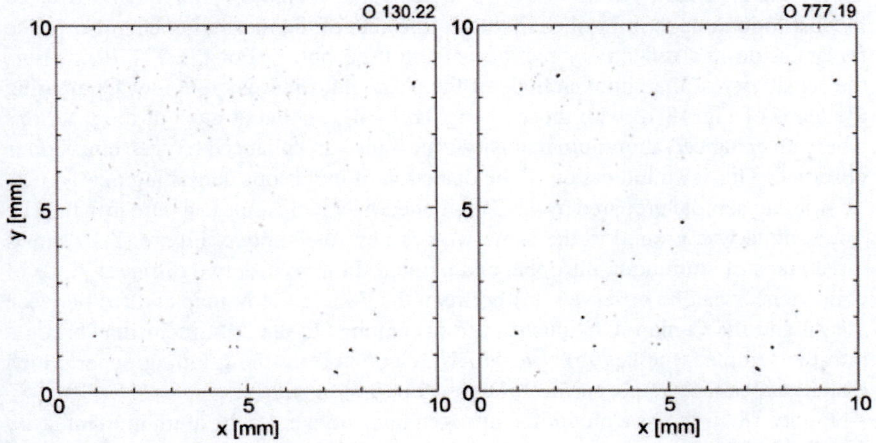

Fig. 18.14 *Left*: Oxygen map of the O I 130.22 nm line. *Right*: Oxygen map of the O I 777.19 nm line taken at the same measurement. The measuring parameters are the same as for Fig. 18.13

which requires a surface roughness of well below 1 μm and therefore a polishing of the surface for more than 20 min. Figure 18.13 shows two aluminum maps using two aluminum lines: Al I 308.22 nm and Al I 396.15 nm, having different excitation energies of 4.02 eV and 3.14 eV, respectively. Both lines were investigated simultaneously to minimize the chance of error. Most of the observed signal peaks are observed with both lines; hence, a random character of the signal peaks can be excluded.

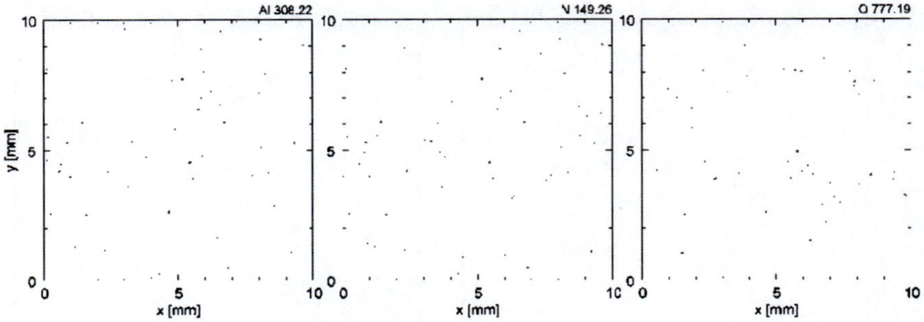

Fig. 18.15 *Left*: Aluminum map of the 308.22 nm line, *middle*: nitrogen map of the 149.26 nm line, *right*: oxygen map of the 777.19 nm line taken at the same measurement. The measuring parameters are the same as for Fig. 18.13

Figure 18.14 shows two oxygen maps using the emission lines O I 130.21 and O I 777.19 nm measured at the same time as the data for Fig. 18.13. Many of the signal peaks coincide in both maps, but not all. The signal level is for both lines comparatively small. The 130.21 nm line lies in the deep VUV and may be affected by absorption due to impurities in the atmosphere of the measuring chamber or the transmission and reflectivity features of the used optics. For the 777.19 nm line, the sensitivity of the photo cathode of the photomultiplier is quite low. Comparing the maps of Fig. 18.13 with those of Fig. 18.14 shows that at most of the positions where an enhanced aluminum intensity occurs also an enhanced oxygen emission is observed. This is an indication of the detection of inclusions consisting of Al_2O_3.

Another sample prepared by TKS contained AlN inclusions in a pure iron matrix. The sample was ground in the same way as described above. Figure 18.15 shows three maps of aluminum, nitrogen, and oxygen. In this case, two different kinds of coincidences can be observed: (a) between the Al and the N map and (b) between the Al and the O map. Obviously, case (a) belongs to the AlN inclusions brought into the sample, whereas (b) reveals Al_2O_3 grains from the grinding paper which contaminated the sample surface of the soft pure iron matrix.

Figure 18.16 shows a plot of the nitrogen intensity versus the aluminum intensity for all 250,000 data points corresponding to Fig. 18.15. Two different branches can be seen. One corresponds to the AlN inclusions, and the other one is not correlated to the nitrogen intensity and interpreted as the Al_2O_3 originating from the grinding paper. In Fig. 18.17, the oxygen intensity is plotted versus the aluminum intensity for all data points corresponding to Figs. 18.15 and 18.16. The difference between the AlN and the Al_2O_3 inclusions is difficult to detect. In order to identify the Al_2O_3 inclusions easier, in Fig. 18.18 a plot of the oxygen intensities versus the aluminum intensities is shown like in Fig. 18.17, but only those points are plotted, which belong to the lower branch in Fig. 18.16. The signal correlation between oxygen and aluminum is not very high, but a clear trend can be observed, supporting the interpretation that the lower branch in Fig. 18.16 belongs to Al_2O_3 inclusions.

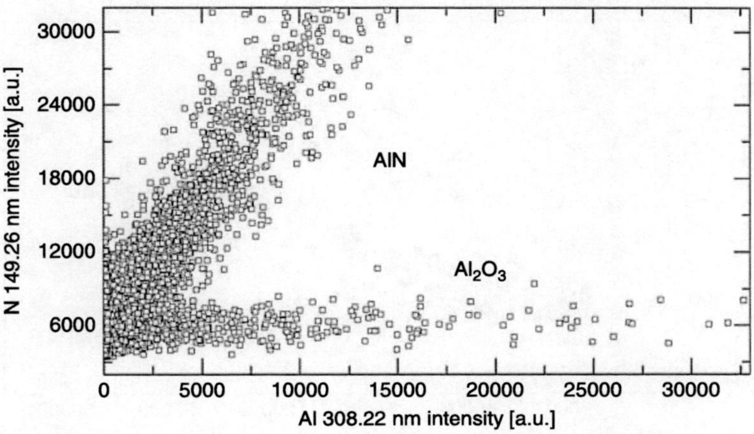

Fig. 18.16 Nitrogen intensity at 149.26 nm versus aluminum intensity at 308.22 nm for the whole measurement of 250,000 points corresponding to Fig. 18.15. The upper branch is attributed to AlN inclusions and the lower branch to Al_2O_3 inclusions (cf. Figs. 18.17 and 18.18)

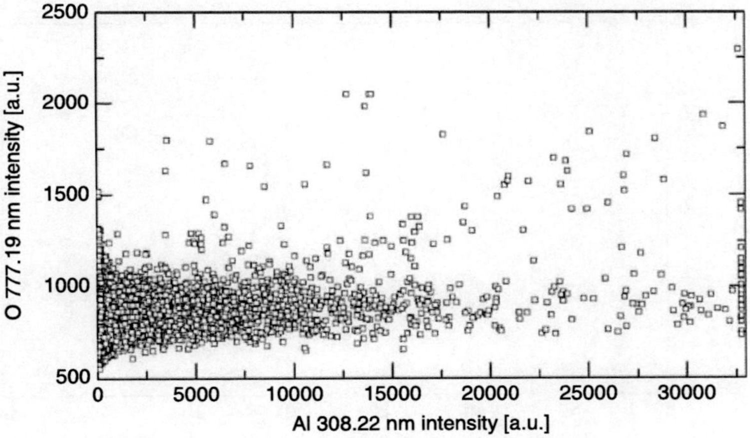

Fig. 18.17 Oxygen intensity at 777.19 nm versus aluminum intensity at 308.22 nm for the measurement corresponding to Fig. 18.15

Another approach to reveal correlations is the principal components analysis (PCA) [18.11]. Figure 18.19 shows the score plot of a PCA for the first 50,000 measuring points of the dataset of Fig. 18.15. The data points form distinctive branches. The corresponding loading plot reveals the dominant elements contributing to these branches. The AlN, copper sulfide (CuS), and SiMnCaV containing inclusions can be identified clearly.

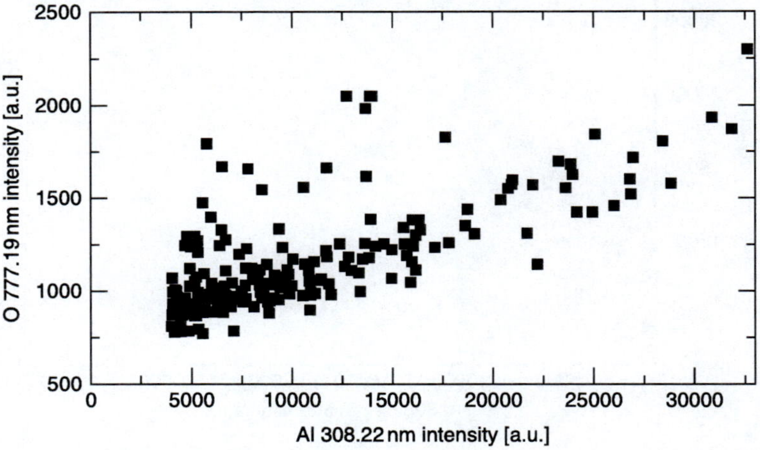

Fig. 18.18 Oxygen intensity at 777.19 nm versus aluminum intensity at 308.22 nm for the measurement corresponding to Fig. 18.15, but only the points of the lower branch of Fig. 18.16 are plotted

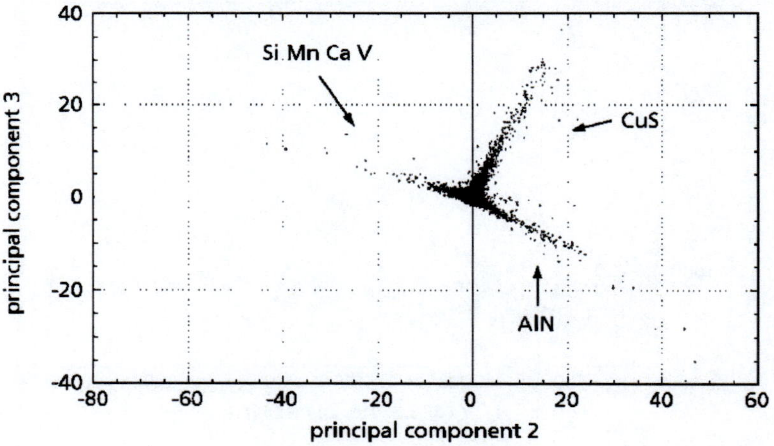

Fig. 18.19 Score plot of the principal components analysis for the first 50,000 data points of the dataset of Fig. 18.15. The principal components (PC) 2 and 3 are plotted in the x- and y-direction, respectively. The elements contributing to the distinctive branches have been identified by the corresponding loading plot

For a quantification of concentrations of inclusions, a statistical approach was studied using the skewness of the frequency distribution of the measured intensities [18.5]. Figure 18.20 shows the frequency distribution of the intensities of aluminum measured in a scan field by the SML system. The frequency distribution has an asymmetric shape as a result of the superposition of a nearly Gaussian distribution

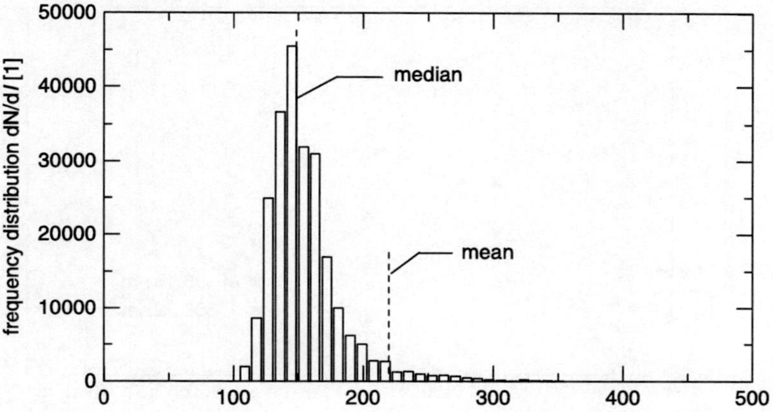

Fig. 18.20 Frequency distribution of 250,000 emission signals of an aluminum emission line. The *dashed lines* indicate the position of the median and mean value

caused by the analytes dissolved in the iron matrix and the inclusions causing outliers with significantly higher intensities. The asymmetric frequency distribution can be characterized by the skewness given by:

$$\zeta = \frac{3(M - A)}{s} \tag{18.2}$$

with M mean, A median, and s standard deviation. This skewness ζ is shown in Fig. 18.21 as a function of the aluminum oxide concentration determined conventionally. The different symbols of the data points correspond to different aluminum lines used for this evaluation. The skewness correlates with the content of aluminum oxide inclusions in the sample and hence provides a measure to quantify inclusion concentrations based on scanning microanalysis with LIBS.

Quantification of the composition of inclusions based on calibrations with Si-killed steel lollipop samples as reference materials was demonstrated successfully based on data gained with SML [18.6]. This type of sample has unusually large nonmetallic inclusions. The average chemical composition of this steel grade is 0.7% C, 0.4% Si, 1.1% Mn, 0.001% Al, 0.015% S, 0.01% P, and 0.014% Cr. The chemical composition and size of the inclusions were first characterized by SEM-EDX analysis and then by LIBS. Calibration curves were constructed using a set of inclusions identified by both methods at the same positions. Intensity data related to inclusions were extracted from the LIBS data as outlier points having an Si intensity value greater than the average of all measured Si intensities plus three times the associated standard deviation. Table 18.2 shows a comparison between LIBS and SEM-EDX values. The accuracy of the results found by LIBS is within 20% of the EDX results.

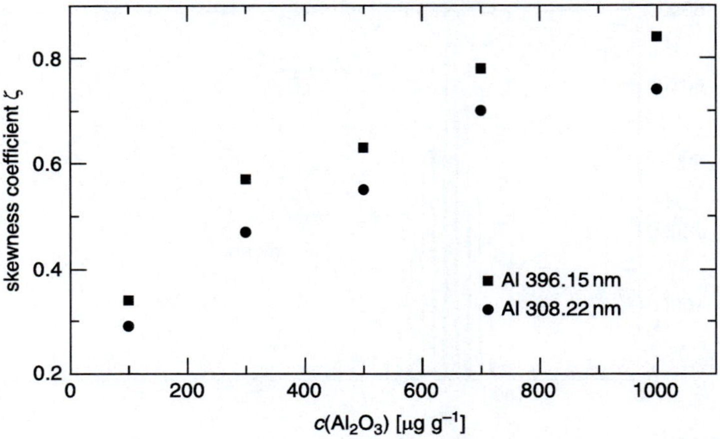

Fig. 18.21 Skewness coefficient as a function of the concentration of Al_2O_3 inclusions [18.5]

Table 18.2 Chemical quantification of nonmetallic inclusions by LIBS compared with SEM-EDX in the lollipop sample

	Ø(μm)	SiO_2	MgO	Al_2O_3	MnO	CaO	TiO_2
SEM-EDX	350	40	17	4.5	1	37	0.5
LIBS		40	17	5.7	0	37	0.4
SEM-EDX	33	47	3.9	11	9.4	28	1.1
LIBS		44	4.9	13	8.1	30	1.2
SEM-EDX	21	46	6.6	17	9.6	20	0.9
LIBS		46	6.7	21	7.7	18	1
SEM-EDX	8	47	3.1	18	10	21	1.2
LIBS		54	5.4	19	3.3	18	0.9

Values in columns 3–8 are given in m.-% [18.6]

While inclusions are characterized by high concentration gradients of certain elements with respect to the embedding material matrix, segregations have only weak concentration gradients extending over larger areas. In Fig. 18.22 sulfur, copper, and phosphorus mappings of a segregation zone in steel are shown [18.10]. An area of $40 \times 40\ mm^2$ of the steel sample was scanned with a step size of 100 μm corresponding to 400×400 measuring points. The scanning time amounts to 15 min. In this type of presentation, the element intensities at each measuring point are plotted in a gray scale representing different signal levels. The sample is a low-alloyed disc from billet. The segregation zone is clearly visible in all three mappings, see enriched element concentration along a line parallel to the y-axis at $x \approx 15$ mm. In the sulfur and copper map, the segregation shows up as a series of remaining melt zones. The phosphorus mapping additionally reveals a dendritic microstructure running to the central segregation line.

Segregation ratios can be determined, defined as ratio of analyte concentration in the segregation over the analyte concentration in the homogeneous,

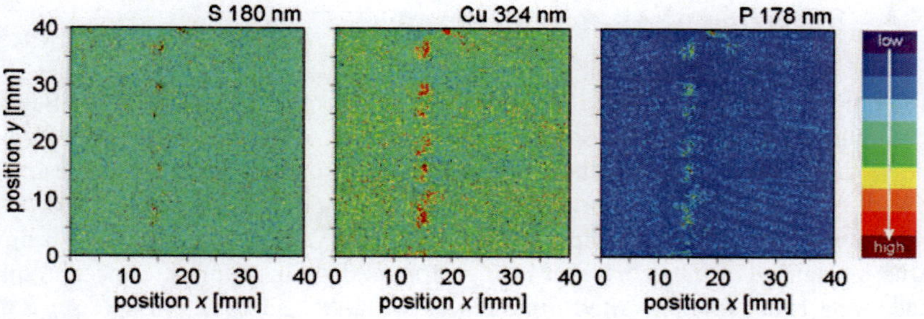

Fig. 18.22 Segregations and dendrites in steel. All mappings are gained during the same scanning measurement with SML. The sulfur, copper, and phosphorus channels show a segregation zone along the y-direction at $x \approx 15\,\mathrm{mm}$ [18.10]

Table 18.3 Segregation ratios k_s determined with ESMA and LIBS for the elements S, P, and Mn

	k_s(S)	k_s(P)	k_s(Mn)
ESMA	2.5	4.6	1.2
LIBS	2.2	4.1	1.3

segregation-free area. Table 18.3 shows a comparison of segregations ratios determined with scanning LIBS with those gained from electron scanning microanalysis (ESMA).

The capability of SML to analyze decarburization zones was demonstrated successfully [18.12]. Monitoring of decarburization layers can help controlling and improving the quality of the final steel product. Conventional methods are based on optical observation of a polished and etched surface, but this method lacks precision and consistency. Decarburization occurs when the steel starts solidifying during casting. By reaction with the ambient oxygen and moisture carbon is lost in a layer close to the surface. While pearlite is the crystal structure formed in 0.8% C solid steel, the loss of C leads to areas of pure ferrite and mixtures of ferrite and pearlite. A billet sample surface of $4 \times 1\,\mathrm{mm}^2$ was studied in a first step by optical microscopy and subsequently the same area was scanned with SML [18.12]. Both images – the conventional optical and the chemical image of the carbon gained with LIBS – correlate largely. The ferritic zones and the ferritic grain boundaries show up clearly in the carbon element mapping corresponding to low or even vanishing carbon concentrations. Carbon concentration as a function of the distance from the surface to depths of up to 10 mm were determined by averaging the carbon concentration gained in the C mapping over the width of the scan field. The results were compared with those detected by electron probe microscopy (EPMA) of the same sample showing a very high degree of correlation. The EPMA measurement required 2 h, whereas the LIBS measurement took 4 min only.

The achieved results show that scanning LIBS is applicable not only to detect and quantify inclusions, but also to visualize and quantify extended segregation as well as decarburization zones.

18.4 Recycling of Aluminum Scrap

Motivation and concept of a LIBS setup with scanner mirrors for the identification of aluminum alloys were described in Sect. 13.3. Aluminum is – after steel – one of the most important metals. Production of secondary aluminum saves up to 95% of the amount of energy needed for primary aluminum production. However, secondary production of aluminum wrought alloys (WA) requires melting of high-value primary aluminum or a presorting of shredded aluminum pieces. The main processing route for nonsorted mixtures of wrought and cast alloys (CA) is the production of CA. Float sink plants used to separate residues from old vehicles yield only a mixed aluminum shredder scrap consisting of a mixture of CA and WA. Up to now no large-scale, commercially available method exists, which would allow for an automatic separation of aluminum alloys.

LIBS offers a new approach for the high-speed characterization and identification of scrap pieces. Figure 18.23 illustrates the concept (the feeding vibrating chute is not shown) [18.13]. Scrap pieces are singularized and accelerated on a belt conveyor up to velocities of 3 m/s passing two sensor systems. In a first step, the position and size of the parts are detected by a 3D sensor. With this data the beam of the LIBS module is directed onto the object to determine the composition (cf. Fig. 13.23). Depending on the measuring result, the part is ejected by an array of pressurized valves to one of up to four sorting fractions (module ejection in Fig. 18.23). Table 18.4 shows the technical data of the developed laser sorting system for aluminum alloys. The volume depicted by dashed lines is the measuring

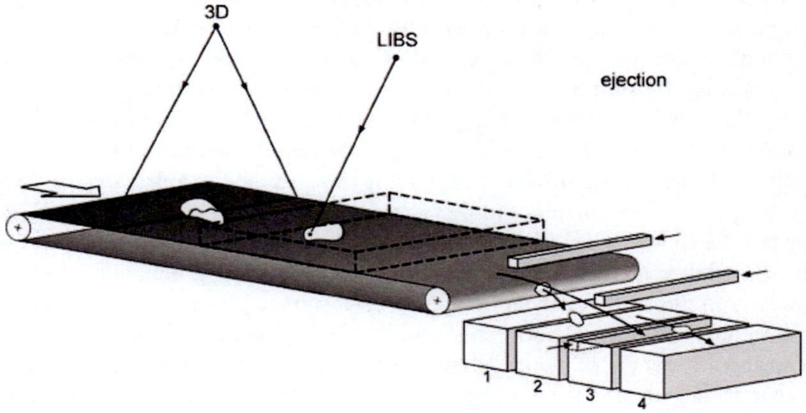

Fig. 18.23 Setup of a laser sorting system with the three main modules. *3D* = laser line section sensor to detect the position and shape of the pieces, *LIBS* = LIBS module with *xy*-scanner (cf. Fig. 13.23), *ejection* = array of pressurized valves for automatic ejection. 1–4 = boxes for the sorted fractions. The volume depicted by *dashed lines* is the measuring volume within which the laser beam of the LIBS module irradiates the pieces passing by

Table 18.4 Technical data of laser sorting system for aluminum alloys

Characteristics	Data
Laser	Nd:YAG
Simultaneously detected elements	Al, Si, Fe, Zn, Cu, Mg, Mn, C, Cd, Pb, Ti, Sn, Cr, Zr, Ni, Ca
Speed of belt conveyor	3 m/s
Sorting rate	40 pieces per second

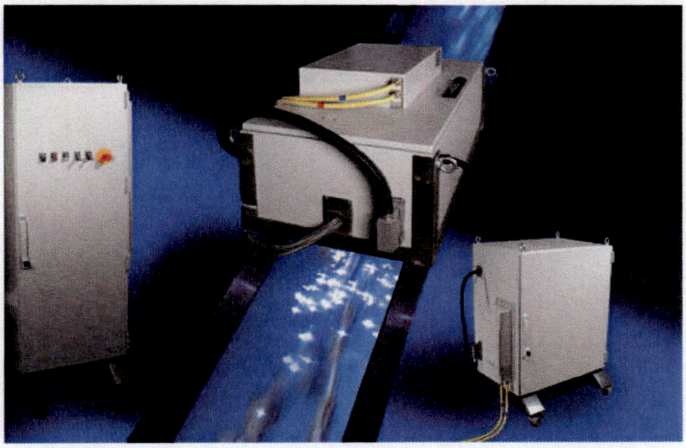

Fig. 18.24 System for the identification of scrap pieces on belt conveyors. *Left*: electric control cabinet, *middle*: laser module mounted above the belt conveyor with scrap pieces and laser-induced plasmas (time exposure showing a multitude of laser-induced plasmas), *right*: spectrometer module

volume within which the laser beam of the LIBS module irradiates the pieces passing by.

Figure 18.24 shows a laser sorting system with the laser module, which is mounted above a belt conveyor [18.1]. With a single laser pulse, the spectral signals of all elements stated in Table 18.4 are detected simultaneously.

In a first step, operating figures are defined to assess the performance of the laser sorting system shown in Fig. 18.23. In case of two material classes, there is always a finite probability of error. The symbol Y states that the material class Y is existent. The symbol X means that the material is classified – in the case described here by means of the LIBS measurement – as material class X. Table 18.5 lists, e.g., the different possible classification scenarios for the two material classes WA and CA with the corresponding conditional probabilities. The correctness of an identification result can be described by the conditional probabilities $C_{\mathrm{WA}} = p(\mathrm{WA}|\mathrm{WA})$ and $C_{\mathrm{CA}} = p(\mathrm{CA}|\mathrm{CA})$. The identification yields G_{WA} and G_{CA} are defined as ratio of the number of measuring results from which an allocation to a material class WA or CA is deduced to the number of all measuring results gained with a pure scrap charge consisting of WA and CA, respectively.

Table 18.5 Classification scenarios for two possible states and corresponding conditional probabilities

X	Y	$p = (X\|Y)$	Result
WA	WA	$1-\alpha$	Correct
CA	WA	A	Type 1 error
WA	CA	B	Type 2 error
CA	CA	$1-\beta$	Correct

WA = wrought alloy, CA = cast alloy, for further explanations see text

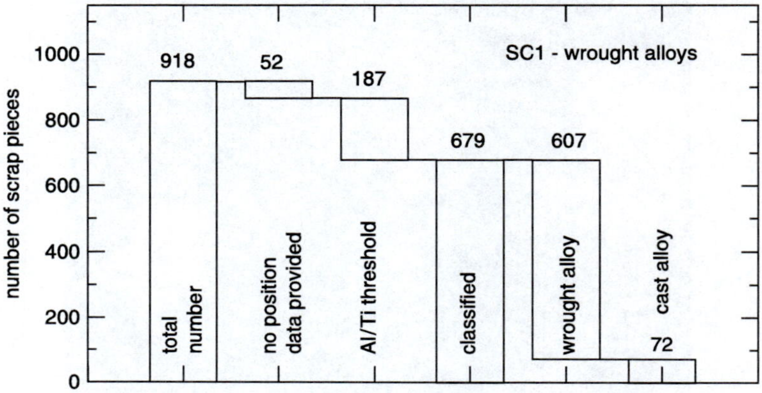

Fig. 18.25 LIBS classification result for the scrap charge SC1 consisting of wrought aluminum alloys only

The components 3D sensor and LIBS module of the system were tested using such known scrap charges SC1 and SC2, where SC1 comprises 918 scrap pieces of WA and SC2 255 scrap pieces of CA [18.15]. In order to exclude the influence of the feeding and ejection module, the pieces were put by hand on the belt conveyor. During the measurement, the pieces are moving with a speed of 3 m/s. Analyte signals of aluminum, titanium, and silicon were used to identify the material classes wrought and cast aluminum. Test runs were performed to determine the correctness and yield of the laser measurement. Figure 18.25 shows the classification results for SC1. For 52 pieces, no position data were provided by the 3D sensor (cf. Fig. 18.23), which is attributed to complex geometries of the scrap pieces which could not be measured reliably by the laser line section sensor. Without geometry data, the scanner system of the laser module cannot be controlled and therefore no LIBS results are available for these pieces. From the numbers given at the bars shown in Fig. 18.25, the identification correctness and the yield are determined (values given in brackets refer to the mass): $C_{\mathrm{WA}} = 89.4\%(91.3\,\mathrm{m.\text{-}\%})$, $G_{\mathrm{WA}} = 78.4\%(79.3\,\mathrm{m.\text{-}\%})$. The type 1 error amounts to $\alpha = 10.6\%$.

Figure 18.26 shows the classification results for SC2. Analogous to the SC1 case, the following numbers are deduced for the correctness and yield: $C_{\mathrm{CA}} = 94.0\%$

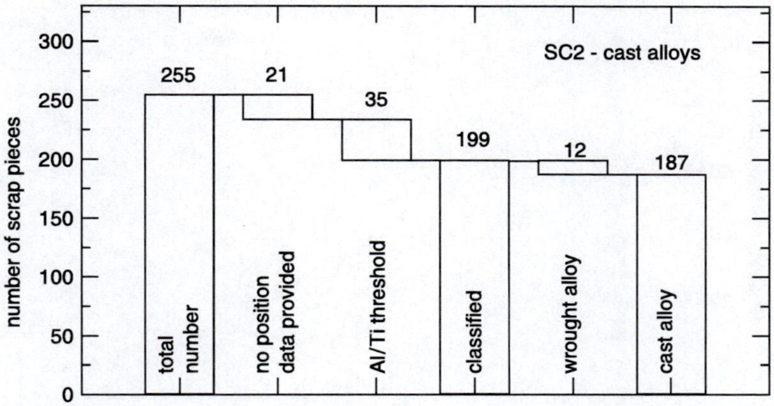

Fig. 18.26 LIBS classification result for the scrap charge SC2 consisting of cast aluminum alloys only

Table 18.6 Results of five repetitive measurements with the scrap charges SC1, wrought alloys and SC2, cast alloys [18.15]

No.	Scrap charge	G (m.-%)	C_{WA} (m.-%)	C_{CA} (m.-%)
1	SC1	74.3	91.2	
2	SC1	74.9	91.0	
3	SC1	71.3	92.0	
4	SC1	77.3	92.1	
5	SC1	73.0	92.4	
Av.		74.2	91.7	
1	SC2	70.6		94.3
2	SC2	72.4		94.4
3	SC2	73.9		93.8
4	SC2	74.5		89.2
5	SC2	77.3		96.5
Av.		73.7		93.6

Yield G and correctness C_{WA}, C_{CA} refer to the whole system including 3D sensor, LIBS module, and ejection, cf. Fig. 18.23. Av., arithmetic mean

(91.1 m.-%), $G_{CA} = 85.0\%(82.0\,\text{m.-\%})$. The type 2 error amounts to $\beta = 6.0\%$. With the average mass of the scrap pieces of the charges SC1, SC2 and a laser source with 40 Hz repetition rate, the extrapolated mass throughput per hour is 2.2 t/h for WA and 7.3 t/h for CA.

Further test runs were performed including the vibrating chute feeding the scrap pieces to the belt conveyor and the ejection module (cf. Fig. 18.23). Table 18.6 shows a summary of the results gained for five repetitive measurements with each scrap charge SC1 and SC2, respectively.

The correctnesses determined are close to those found in the first test scenario. The yield is reduced which is attributed to a partly too dense occupancy of scrap pieces on the belt conveyor and/or errors occurring in the ejection step.

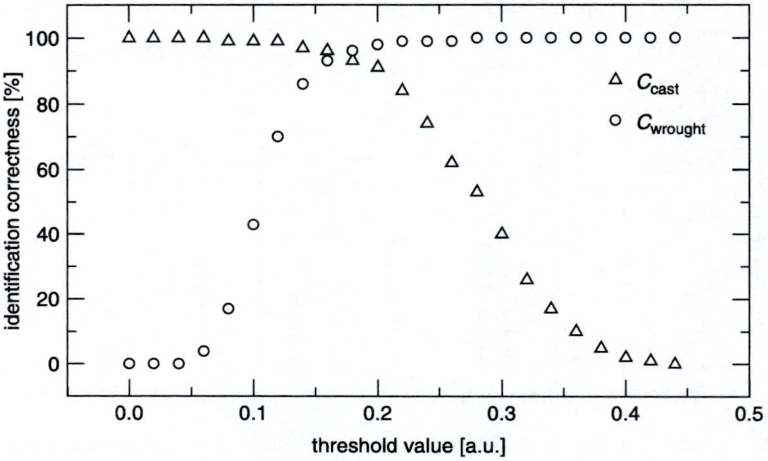

Fig. 18.27 Identification correctness for cast and wrought aluminum alloys as a function of a decision threshold. At a threshold value of 0.17 correctnesses above 90% are achieved for cast and wrought simultaneously

The identification correctness was determined by use of a known charge of aluminum scrap pieces. Scrap pieces were sorted in two classes: WA with $c(\mathrm{Si}) < 3.5\,\mathrm{m.\text{-}\%}$ and CA with $c(\mathrm{Si}) > 6.0\,\mathrm{m.\text{-}\%}$.

Figure 18.27 shows the measuring results. The identification correctness is shown as a function of a threshold value which describes a decision limit to distinguish between CA and WA. For a threshold value of 0.17, the identification correctness for CA and WA is greater than 90%. Next generation laser identification and sorting systems will allow for mass throughputs of about 5 t/h.

Within a running European project, the laser sorting will be extended to separate in future different alloy groups of wrought aluminum with sorting rates of 60 and 100 Hz [18.14]. By this technology, the material-specific recycling can be enlarged significantly compared with the present state of the art. Recently the identification of WA with a correctness >95% was demonstrated successfully for the first time [18.16].

Automatic laser-based sorting of alloys improves the sustainability of light metal cycles by provision of a technology, with counteracts the downward spiral of recycling processes toward low quality levels (down cycling).

References

18.1. C. Gehlen, J. Makowe, R. Noll, Automatisierte Verwechslungsprüfung von Edelstahlhalbzeugen in der Produktion. Stahl Eisen **129**, S70–S72 (2009)

18.2. International Standard, Safety of laser products – Part 1: Equipment classification and requirements, IEC 60825–1, Ed. 2.0, 2007–03, 200 p

18.3. German Standard (2002), Technical availability of machines and production lines, terms, definitions, determination of time periods and calculation, VDI 3423, January

18.4. R. Noll, V. Sturm, J. Makowe, Laser-Emissionsspektrometrie analysiert zeitnah Stahlproben. Stahl Eisen **129**, S61–S64 (2009)

18.5. H. Kuss, H. Mittelstädt, G. Müller, Inclusion mapping and estimation of inclusion contents in ferrous materials by fast scanning laser-induced optical emission spectrometry. J. Anal. At. Spectrom. **20**, 730–735 (2005)

18.6. F. Boué-Bigne, Analysis of oxide inclusions in steel by fast laser-induced breakdown spectroscopy scanning: an approach to quantification. Appl. Spectrosc. **61**, 333–337 (2007)

18.7. F. Boué-Bigne, Laser-induced breakdown spectroscopy applications in the steel industry: rapid analysis of segregation and decarburization. Spectrochim. Acta B **63**, 1122–1129 (2008)

18.8. E. Pappert, Studien zur Bestimmung der Bindungsformen von Al, Ti, Cr und Zn in Feststoffen mittels Röntgenspektrometrie für Anwendungen in der Stahlindustrie, Dissertation, Dortmund, 1998

18.9. Ç. Nazikkol, J. Flock, H. Mittelstädt, Klassische Methoden zur Analyse von Einschlüssen in Stählen, Proc. Nichtmetalle in Metallen, ed. by D. Hirschfeld, 2000, p. 77–84

18.10. H. Bette, R. Noll, G. Müller, H. Jansen, Ç. Nazikkol, H. Mittelstädt, High-speed scanning laser-induced breakdown spectroscopy at 1000 Hz with single pulse evaluation for the detection of inclusions in steel. J. Laser Appl. **17**, 183–190 (2005)

18.11. H. Martens, T. Mæs, *Multivariate Calibration* (Wiley, New York, 1998)

18.12. F. Boué-Bigne, Laser-induced breakdown spectroscopy applications in the steel industry: rapid analysis of segregation and decarburization. Spectrochim. Acta B **63**, 1122–1129 (2008)

18.13. R. Noll, C. Fricke-Begemann, P. Jander, T. Kuhlen, V. Sturm, P. Werheit, J. Makowe. *Perspektiven der Lasertechnik zur Steigerung der Ressourceneffizienz, Hrsg. U. Teipel, Rohstoffeffizienz und Rohstoffinnovation*, Fraunhofer Verlag, 2010, pp. S287–S298

18.14. European project, Pilot demonstration of high performance aluminium recycling by inline laser identification - PARILAS, ECO/09/256174/SI2.564797, Start: June 17th, 2010, project coordination: R. Noll, Fraunhofer ILT

18.15. Ü. Aydin, Materialidentifikation bewegter Einzelkörner aus Aluminium-Legierungen mit der Laser-Emissionsspektrometrie, Dissertation, Shaker Verlag, Aachen, 2009

18.16. P. Werheit, C. Fricke-Begemann, M. Gesing, R. Noll, Fast single piece identification with a 3D scanning LIBS for aluminium cast and wrought alloys recycling. J. Anal. At. Spectrom. 23 August 2011, DOI: 10.1039/c1ja10096c

Annex A

The annex starts with a section giving an estimate of the necessary energy to displace the ambient gas atmosphere caused by the expansion of the laser-induced plasma. The following sections list the abbreviations and symbols used in this book. The most relevant Stark data for the spectroscopic determination of the electron density of LIBS plasmas are compiled in Sect. A.4. Lists of wavelengths and elements are provided in Sect. A.5. Finally, Sect. A.6 describes a method to select spectral lines for Boltzmann plots.

A.1 Displacement of Ambient Atmosphere

As shown in Sect. 8.1 and Table 8.2 more than 97% of the ambient atmosphere particles are displaced by the expanding laser-induced plasma. A result which is consistent with the simulations presented in Sect. 10.3, Fig. 10.5. In the following the energy necessary for the displacement of the ambient atmosphere will be estimated.

The irradiated laser beam evaporates target material. The particles flowing off collide with particles of the ambient atmosphere. By these collisions, the ambient atmosphere is partially displaced. A part of the ambient gas particles diffuses into the material vapor (cf. Table 8.2, Fig. 10.5). For simplicity, it is assumed that the ambient atmosphere is displaced completely by the expanding material vapor.

In case of a complete displacement, the material vapor and plasma has to do work, which will be calculated taking the following assumptions (a) spherical expansion, (b) the ambient gas is collected at the front of the expanding plasma. The latter is a simplification and yields an upper estimate for the displacement energy (a more realistic description would have to consider the flow of ambient gas induced by the expanding vapor). Then the expanding plasma has to provide the energy given by:

$$E_{\mathrm{d}} = \int_0^R F \mathrm{d}r = \int_0^R \frac{\mathrm{d}}{\mathrm{d}t} (m_{\mathrm{a}} \dot{r}) \, \mathrm{d}r \tag{A.1}$$

R. Noll, *Laser-Induced Breakdown Spectroscopy*, DOI 10.1007/978-3-642-20668-9, with E_{d} displacement energy, R radius of the expanding plasma, m_{a} accumulated mass of ambient gas, $\dot{r}$ velocity of the accumulated ambient gas at the expanding plasma front.

For the accumulated mass of the ambient gas and a half sphere holds:

$$m_{\mathrm{a}}(t) = \rho_{\mathrm{a}} \frac{2\pi}{3} r(t)^3, \tag{A.2}$$

where ρ_{a} is the density of the ambient gas. Insertion of relation (A.2) into (A.1) leads to:

$$E_{\mathrm{d}} = \frac{2\pi \rho_a}{3} \int_0^R (r^3 \ddot{r} + 3r^2 \dot{r}^2) \mathrm{d}r. \tag{A.3}$$

The second integrand can be transformed as follows:

$$\int 3r^2 \dot{r}^2 \mathrm{d}r = \int \dot{r}^2 \mathrm{d}r^3 = \dot{r}^2 r^3 - \int r^3 \mathrm{d}\dot{r}^2 = \dot{r}^2 r^3 - 2 \int r^3 \ddot{r} \mathrm{d}r \tag{A.4}$$

which yields for (A.3):

$$E_{\mathrm{d}} = \frac{2\pi \rho_{\mathrm{a}}}{3} \left(r^3 \dot{r}^2 - \int r^3 \ddot{r} \mathrm{d}r \right) \Bigg|_0^R \tag{A.5}$$

The kinetic energy of the accumulated gas at the front of the material vapor is given by:

$$E_{\mathrm{kin}} = \frac{1}{2} \frac{2\pi \rho_{\mathrm{a}}}{3} R^3 \dot{R}^2. \tag{A.6}$$

Relations (A.5) and (A.6) show that for negligible acceleration of the front of the expanding material vapor, i.e., the second term in (A.5) can be neglected, the displacement work is just twice as much as the kinetic energy of the accumulated gas. This case is equivalent to a completely inelastic collision of the ambient gas at the front of the expanding vapor. In the rest frame of the expanding front, the incoming ambient gas loses its kinetic energy being transformed completely into internal energy.

Since the expanding material vapor has a high temperature, it emits visible radiation. In the following it is assumed that the position of the expanding front is identical to the position of the luminous front observed experimentally, cf. Sect. 8.1.

The height of the luminous front is observed with high-speed streak images in the time interval between 3 and 65 ns after the irradiation of the laser pulse with an energy of 40 mJ, see Fig. 8.6. Figure A.1 shows this data in a linear–linear presentation and a second degree polynomial as fit curve.

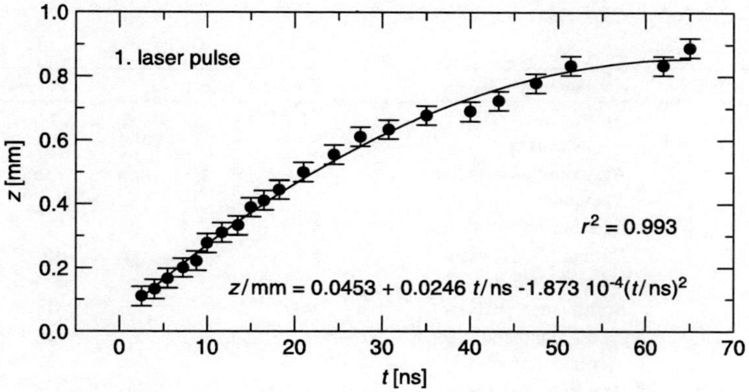

Fig. A.1 Height z of the luminous plasma front above the sample surface as a function of time after the irradiation of the first laser pulse with 40 mJ pulse energy (same data as Fig. 8.6 *top*, but linear scaling of x- and y-axis). As fit curve a second degree polynomial is taken

The general form of the polynomial is given by:

$$r = a_0 + a_1 t + a_2 t^2 \tag{A.7}$$

with (A.5) follows:

$$E_\mathrm{d} = \frac{2\pi\rho_\mathrm{a}}{3}\left(R^3\dot{R}^2 - \frac{a_2}{2}R^4\right) \tag{A.8}$$

with the fit parameters given in Fig. A.1: $a_0 = 0.0453\,\mathrm{mm}$, $a_1 = 0.0246\,\mathrm{mm/ns}$, $a_2 = -1.873\times10^{-4}\,\mathrm{mm/ns^2}$ and the density of air $1.29\,\mathrm{kg/m^3}$ relation (A.8) yields $E_\mathrm{d} = 150\,\mathrm{mJ}$ for $t = 30\,\mathrm{ns}$. This estimate of the upper limit of the displacement energy is higher than the laser pulse energy. Hence, the assumption of a complete displacement and accumulation of the ambient gas at the expanding front is a too strong simplification. Anyway, this estimate shows that the displacement energy is of the same order of magnitude as the laser pulse energy itself. An estimate of the necessary energy to heat the ablated mass of the iron target of about 100 ng (for a single laser pulse with 40 mJ, cf. Fig. 7.5) from room temperature to the melting point (average specific heat capacity of iron 0.8 J/(gK), melt temperature 1,535°C), provision of the melt enthalpy (270 J/g), heating of the melt to the evaporation temperature (assuming the same heat capacity as for solid iron [A.8]), provision of the enthalpy of vaporization (6,370 J/g), heating the vapor to 10,000 K, and ionizing 90% of the Fe particles (ionization energy 7.9 eV) yields 2.1 mJ. Therefore, the displacement energy is significantly greater than the energy necessary to produce the plasma state.

A.2 Abbreviations

Abbreviation	Explanation	Chapter(s)/Section(s)	Page(s)
AAS	Atomic absorption spectroscopy	14.1.2, 14.5	286, 338
ABS	Acrylonitrile butadiene styrene	14.1.2	286
AOM	Acousto-optic modulator	8.1	120
BAS	Bureau of Analysed Samples Ltd.	6.1, 13.1.2	86, 238
BBA	Borate bead analysis	14.2.2	316
BEC	Background equivalent concentration	11.2, 12, 13.1.3	215, 227, 250
BFR	Brominated flame retardants	14.1.2	285
BNC	Bayonet Neill Concelman	13.1.1	231
CARS	Coherent Anti-Stokes Raman spectroscopy	1	1
CCD	Charge-coupled device	4.2, 4.4, 15.1	55, 62, 388
CE	Capillary electrophoresis	2.4, 14.5	14, 338
COG	Curves-of-growths	10, 14.8.2	189, 376
CPC	Condensation particle counter	14.8.2	373
CRM	Certified reference material	2.1	11
DLVQ	Dynamic learning vector quantization	14.1.1	278
DMA	Differential mobility analyzer	14.8.2	373
DP	Double pulses	3.2, 7.3, 11.1	23, 104, 209
DPSSL	Diode-pumped solid state laser	3.1, 3.6, 4.1, 15.1	19, 35, 47, 389
EDL	Electrodeless discharge lamps	12	225
EDXRF	Energy dispersive X-ray fluorescence analysis	14.3	321
ELPI	Electrical low-pressure impactor	14.8.1	362
EOL-WEEE	End-of-life electric and electronic equipment	14.1.2	285
EPMA	Electron probe microscopy	18.3	483
ESMA	Electron scanning microanalysis	18.3	482
fb	Free–bound	8.4	161
ff	Free–free	8.4	161
FHG	Fourth harmonic generation	4.1	48
FML	Flexible measuring system based on LIBS	17.3	448

(continued)

(continued)

Abbreviation	Explanation	Chapter(s)/Section(s)	Page(s)
FWHM	Full width at half maximum	2.1, 3.1, 4.3, 6.2, 9.4, 13.1.1, 14.1.1	10, 19, 59, 92, 173, 230, 281
GC	Gas chromatography	14.5	338
GD-OES	Glow discharge optical emission spectrometry	14.1.2, 16.1, 16.2	286, 401, 416
GF	Graphite furnace	12	224
HDPE	High density polyethylene	14.1.1	278
H_γ	Hydrogen Balmer line	8.2	149
HPLC	High pressure liquid chromatography	2.4, 14.5	14, 338
HPTLC	High performance thin layer chromatography	14.5	338
HWHM	Half width at half maximum	8.2, 14.4	151, 332
ICP-AES	Inductively coupled plasma atomic emission spectrometry	14.1.2	286
ICP-MS	Inductively coupled plasma mass spectrometry	14.1.2, 14.5	286, 347
ICP-OES	Inductively coupled plasma optical emission spectrometry	12, 14.1.2	227, 302
IEC	International Electrotechnical Commission	18.1	468
ILT	Fraunhofer-Institut für Lasertechnik	Preface, 4.5	v, 66
IPDA	Intensified photodiode array	14.1.1	277
IR	Infrared	14.4, 14.5	334, 338
LA-MS	Laser ablation mass spectrometry	14.1.2	286
LAS	Laser absorption spectroscopy	1	3
LDPE	Low density polyethylene	14.1.1	278
LEAFS	Laser-excited atomic fluorescence spectrometry	12, 14.3	221, 320
LIBS	Laser-induced breakdown spectroscopy	1	2
LIDAR	Light detection and ranging	1	1
LIF	Laser-induced fluorescence	1, 12	1, 3, 221
LIFT	Laser identification of fittings and tubes	17.2, 18.1	438, 467

(continued)

(continued)

Abbreviation	Explanation	Chapter(s)/Section(s)	Page(s)
LOD	Limit of detection	11.2, 13.1.1, 13.1.2, 13.3, 14.1.2, 14.3, 17.3, 17.4	215, 234, 240, 267, 299, 323, 451, 462
LOQ	Limit of quantification	11.2	215
LTE	Local temperature equilibrium	8.4, 10	162, 187
MCI	Multichannel integrator electronics	4.5, 13.1.1	66, 231
MCP	Microchannel plate	4.2, 4.4	52, 62, 64
μ−XRF	Micro X-ray fluorescence analysis	18.3	474
MS	Mass spectrometry	14.5	338
MVDA	Multivariate data analysis	16.1	408
NBS	National Bureau of Standards	13.2	254
NCS	National Analysis Center for Iron and Steel	13.2	254
Nd:YAG	Neodymium-doped yttrium aluminum garnet	3.1	17
NIR	Near infrared	14.1.1	276
NIST	National Institute of Standards and Technology	13.1.2	238
NMR	Nuclear magnetic resonance	14.5	338
OES	Optical emission spectrometry	13.1.2	235
PA	Polyamide	3.3	27
PC	Polycarbonate,	3.3	27
	Principal component,	18.3	480
	Personal computer	17.1	435
PCA	Principal component analysis	18.3	478
PDA	Photodiode array	4.2, 15.1	53, 388
PE	Polyethylene	14.1.1	275
PET	Polyethylene terephthalate	14.1.1	275
PIN	Positive-intrinsic-negative photodiode	14.1.1	276
PLS	Partial least squares	16.1, 18.2	409, 471
PMT	Photomultiplier tube	3.5, 4.3, 4.4	30, 53, 61
PP	Polypropylene	14.1.1	275
PS	Polystyrene	14.1.1	275
PVC	Polyvinyl chloride	14.1.1	275

(continued)

(continued)

Abbreviation	Explanation	Chapter(s)/Section(s)	Page(s)
Q-switch	Variable attenuator to influence the quality factor of the laser resonator	4.1, 6	47, 83
RMSE	Root mean square error	16.1	409
RMSEP	Root mean square error of prediction	18.2	471
ROI	Region of interest	14.1.1	280
RR	Rautaruukki Oyj	14.7	358
RSD	Relative standard deviation	3.1, 14.1.2	20, 291
RSDP	Relative standard deviation of procedure	13.3	266
SBS	Stimulated Brillouin scattering	4.3	61
SD-OES	Spark discharge optical emission spectrometry	13.1, 13.1.3, 17.1, 18.3	229, 245, 430, 475
SEM	Scanning electron microscopy	1, 7.1, 15.1	3, 98, 393
SEM-EDX	Scanning electron microscopy energy dispersive X-ray fluorescence analysis	1, 18.3	3, 474
SFC	Supercritical fluid chromatography	14.5	338
SFM	Sum frequency mixing	12	223
SHG	Second harmonic generation	4.1, 12	48, 223
SML	Scanning microanalysis with laser spectrometry	15.2, 18.3	396, 473
SMPS	Scanning mobility particle sizer	14.8.2	373
SNR	Signal-to-noise ratio	14.1.2	288
SP	Single pulses	3.2, 11.1	23, 209
Spark-OES	Spark optical emission spectrometry	17.2	438
TEA	Transverse excited amplifier	14.5	350
TKS	ThyssenKrupp Steel	14.7, 16.1	358, 404
TP	Triple pulses	3.2, 7.3	25, 104
UV	Ultraviolet	3.5	33
VUV	Vacuum ultraviolet	3.5, 13.1.1	32, 230
XRF	X-ray fluorescence analysis	13.1, 13.1.2, 13.1.3, 16.1	229, 235, 245, 401

A.3 List of Symbols

Symbol	Description	Unit	Chapter(s), Section(s), Figure(s), Formula(s)	Page(s)
A	Absorptance	1	5.1, (5.1)	75
A	Dimensionless coefficient for Stark effect	1	9.4, (9.19)	173
A	Analytical resolving power, analytical resolution	1	14.1.2, (14.1), (14.4), 17.2, (17.1), Table 17.5	288, 291, 443
A	ion broadening parameter	1	A.4, Table A.1	512, 514
A_o	Overlapping area	m^2	3.6, Fig 3.20	37
A_C	Effective cross section of measuring chamber	m^2	4.6, Fig 4.17	67
A_{nm}, A_{ji}	Einstein coefficient of the transition $n \rightarrow m$, $j \rightarrow i$	s^{-1}	8.2, (8.5)	133
			9.2, (9.3), (9.8), 10.2, (10.13)	169, 170, 194
A_p	Surface area of a plasma sphere	m^2	9.2, (9.10)	171
A_s	Sampling area	m^2	3.6	38
α, $\alpha^{\nu}(\nu)$, $\alpha^{\lambda}(\lambda)$	Absorption coefficient	m^{-1}	5.3, (5.8)	79
			9.3, (9.11)	171
			9.4, (9.32)	180
α	Angle of incidence	°	4.2, Fig 4.4, (4.1)	51, 52
α	Fine structure constant	1	9.1, (9.1), 10.2, (10.15)	167, 195
a_0	Bohr radius	m	9.1, (9.1), 10.2, (10.15)	167, 195
α_O	Angle of observation	°	3.7, Fig 3.23	39
α_D	Detection angle	°	4.6, Fig 4.18, 13.1.2	68, 237
α_j	Polarizability of species j	m^{-3}	8.3, (8.12)	154
α_L	Angle of incidence of the laser beam	°	3.7, Fig 3.23	39
			3.9, Fig 3.26	42
β	Angle of diffraction	°	4.2, Fig 4.4, (4.1)	51, 52
B_{nm}	Einstein's coefficient of stimulated emission	m/kg	10.2, (10.12)	194
B_{ij}, B_{mn}	Einstein's transition probability for absorption	m/kg	9.3, (9.11)	171
			10.2, (10.12), (10.13)	194, 195
c	Vacuum speed of light	m/s	5.3, (5.8)	79
			9.2, (9.8)	170
c	Specific heat capacity	J/(kg K)	5.1, ((5.2)	75
c_j, c_a	Concentration of analyte j, a	g/g	2.1, 10	11, 192

(continued)

(continued)

Symbol	Description	Unit	Chapter(s), Section(s), Figure(s), Formula(s)	Page(s)
$\hat{c}_i$	Concentration ratio	1	11.5, (11.23), (11.24); 13.2, (13.2)	218, 254
C_s	Stark broadening parameter	Å10^{-16} cm^3	A.4, Table A.1	512, 514
χ	Adiabatic coefficient	1	8.3, (8.10)	140
d	Diameter of laser beam	m	3.3, (3.5)	28
d	Stark shift	Å	A.4, Table A.1	512, 514
d_s	Single spot diameter	m	3.6	38
d_f	Foil thickness	m	7.4, (7.1)	114
d_g	Grating period	m	4.2, (4.1)	51
d_{mc}	Diameter of material cloud	m	3.1, (3.1)	21
d_{plasma}	Diameter of the spherical plasma	m	10, (10.1)	191
dA_s	Receiving entrance aperture of the spectrometer	m^2	10.2, (10.24)	197
D	Diameter of the illuminated aperture of the focusing lens	m	3.4, (3.3); 11.1, Table 11.1	27, 209
D	Stark shift parameter	m	9.4, (9.20)	174
D_s	Stark shift parameter	Å10^{-16} cm^3	A.4, Table A.1	512, 514
D_g	Effective diameter of the grating	m	4.2, Fig 4.5, (4.5)	52
$\delta(y, z, t)$	Fringe displacement	1	8.3, (8.11)	154
δ_h	Heat penetration depth	m	3.3, Fig. 3.10, (3.2)	26
			3.6, (3.7)	34
δ_{opt}	Optical penetration depth	m	3.3, Fig. 3.10	26
			5.1, Fig. 5.1, (5.3)	77
ΔE_z^{ion}, $\Delta E_{a,z}^{ion}$	Lowering of the ionization energy, for species *a*	J	9.2, (9.7), Fig. 9.2	170
			10.2, (10.5)	193
$\Delta\lambda$	Difference between two neighboring spectral lines of equal intensity	m	4.2, (4.3)	51
$\Delta\lambda^a$	Apparative broadening given as full width at half maximum (FWHM)	m	10.1, Table 10.1, 10.2, (10.24)	192, 197
$\Delta\lambda_D$	Doppler line width (FWHM)	m	9.4, (9.18), (9.23), (9.26)	173, 174, 175
$\Delta\lambda_G^m$	Measured Gaussian part of the line width	m	9.4, (9.27), Fig 9.7	175, 178

(continued)

(continued)

Symbol	Description	Unit	Chapter(s), Section(s), Figure(s), Formula(s)	Page(s)
$\Delta\lambda_L$	Lorentz line width (FWHM)	m	9.4, (9.26)	175
$\Delta\lambda_L^a$	Apparative broadening described by a Lorentz function	m	9.4, (9.28), (9.29)	177, 178
$\Delta\lambda_L^m$	Measured Lorentzian part of the line width	m	9.4, (9.28), Fig 9.6	177
$\Delta\lambda_N$	Natural line width	m	9.4, (9.26)	175
$\Delta\lambda_{shift}$	Line shift	m	9.4, (9.20)	174
$\Delta\lambda_{Stark}$	Stark line width (FWHM)	m	9.4, (9.19), (9.22)	173, 174
$\Delta\lambda_V^a$	Total apparative broadening described by a Voigt profile	m	9.4, (9.29)	178
Δs	Distance between the beam waist location and the surface of the target	m	2.1, Fig 2.2	9
			3.4	29
			3.9, Fig 3.26; 11.1, Table 11.1	42, 210
$\Delta\tilde{\nu}$	Spectral width of the laser radiation	m^{-1}	1, Table 1.1; 4.3, Table 4.7	3, 61
$\Delta\nu_H$	Full frequency width at half maximum	s^{-1}	9.2, (9.5)	169
$\Delta\nu_V$	Frequency shift of the line maximum in relation to an unperturbed line	s^{-1}	9.2, (9.5)	169
Δt	Interpulse separation	s	3.2, Fig 3.9; 11.1, Table 11.1	25, 209
Δt_i	Interpulse separation between pulse i and pulse $i+1$	s	3.2, Fig 3.9	25
$\Delta t_{LIBS-LIF}$	Time difference between the LIBS-pulse and the LIF-pulse	s	12	224
e	Elementary charge	As	5.3, (5.9)	79
			8.4, (8.15)	161
			9.2, (9.8)	170
E_b	Burst energy, $E_b = \sum_j E_j$	J	3.2; 11.1, Table 11.1	25, 209
E_H	Ionization energy of hydrogen, Rydberg constant	eV	8.4, (8.17), (8.19); 9.1, (9.1), (9.2); 10.2, (10.16), (10.18)	163, 164, 167, 168, 195

(continued)

(continued)

Symbol	Description	Unit	Chapter(s), Section(s), Figure(s), Formula(s)	Page(s)
E_j	Energy of pulse j in a pulse burst	J	3.2, 11.1, Table 11.1	25, 209
E_L	Energy of the laser pulse	J	3.1, 3.4, (3.6)	20, 28
E_m	Specific melt enthalpy	J/kg	10, Fig 10.1	186
E_m, E_i	Energy of the lower level m, i	eV	9.3, (9.13); 9.4, (9.33); 14.5, Table 14.14	172, 180, 342
E_n, E_j	Energy of the upper level n, j	eV	8.2, Table 8.1; 9.2, (9.3); 13.1.1, Table 13.1; 13.1.2, Table 13.3	133, 169, 231, 238
E_v	Specific enthalpy of vaporization	J/kg	10, Fig 10.1	186
ε_0	Vacuum permittivity, dielectric constant	As/(Vm)	5.3, (5.9), 8.4, (8.15), 9.2, (9.8)	79, 161, 170
$\varepsilon_{a1,2}$	Weighting factor describing the proportion of an element a in the plasma in relation to its proportion in the specimen	1	10, Table 10.1	192
$\varepsilon_{d,th}$	Damage threshold radiant exposure	J/m^2	4.3, Table 4.5	59
$\varepsilon_{fb}(\lambda)$	Emission coefficient of free–bound transitions	$W/(m^3\ sr\ m)$	10.2, (10.10), (10.16)	194, 195
$\varepsilon_{fb}^{\omega,d\Omega}$	Emission coefficient for free–bound transitions	$W/(m^3\ sr\ rad\ s^{-1})$	9.1, (9.2)	168
$\varepsilon_{ff}(\lambda)$	Emission coefficient of free–free transitions	$W/(m^3\ sr\ m)$	10.2, (10.10), (10.14)	194, 195
$\varepsilon_{ff}^{\lambda}$	Emission coefficient of free–free transitions	$W/(m^3\ m)$	9.1	167
$\varepsilon_{ff}^{\omega,d\Omega}$	Emission coefficient of free–free transitions	$W/(m^3\ sr\ rad\ s^{-1})$	9.1, (9.2)	167
$\varepsilon_{ji}^{\nu,d\Omega}$	Line emission coefficient	$W/(m^3\ sr\ s^{-1})$	9.2, (9.3)	169
$\varepsilon_l(\lambda)$	Line emission coefficient	$W/(m^3\ sr\ m)$	10.2, (10.10), (10.14)	194
η	Degree of overlap	1	3.6, (3.8)	37
f	Focal length of a lens	m	3.4, (3.3)	27
f_B	Boltzmann factor	1	8.2, Table 8.2	136
f_{ij}, f_{mn}	Oscillator strength	1	9.2, (9.8); 9.4, (9.33); 13.1.2, Table 13.3; 13.2, Table 13.14	170, 180, 238, 262
$g_{a,z;n}$	Statistical weight of excited level n of species a in charge state z	1	10.2, (10.8)	193

(continued)

(continued)

Symbol	Description	Unit	Chapter(s), Section(s), Figure(s), Formula(s)	Page(s)
g_m, g_i	Statistical weight of lower level m, i	1	9.2, (9.8), 9.4, (9.33)	170, 180
g_n, g_j	Statistical weight of the upper level n, j	1	8.2, (8.5); 9.2, (9.3), (9.8)	106, 133, 169, 170
$\gamma(\lambda)$	Nonnormalized shape function	1	9.4, (9.30)	178
$\Gamma(\nu)$	Line profile as a function of the frequency	s	9.2, (9.3), (9.4)	169
$\Gamma_{nm}(\lambda)$	Line profile of the transition $n \rightarrow m$	m^{-1}	10.2, (10.12)	194
$\Gamma_L(\nu)$	Lorentz line profile	s	9.2, (9.5)	169
$\Gamma_V(\lambda)$	Voigt profile	m^{-1}	9.4, (9.23)	174
η(order)	Weighting factor between 0 and 1 describing the fraction of diffracted power for a given order of the grating spectrometer	1	10.1, Table 10.1	192
h	Planck constant	Js	8.4, (8.14)	161
h, h_c	Crater depth	m	7.1, Fig 7.1; 13.1.2, Table 13.5; 16.1, Table 16.1	98, 239, 405
$h\nu_{ji}$	Photon energy of the transition from j to i	J	9.2, (9.3)	169
$\hbar\omega$	Photon energy	J	9.1, (9.1)	167
H	Enthalpy of the plasma	J	10.4, (10.25)	203
I	Irradiance at the location of interaction	W/m^2	1, Table 1.1	3
$I(\lambda)$	Spectral radiance	$W/(m^2 sr\, m)$	10.2, (10.9)	193
$I^{acc}_{analyte}(t_{delay}, t_{int})$	Accumulated analyte signal	a.u.	14.1.2, (14.2)	291
$I_{d,th}$	Damage threshold irradiance	W/m^2	4.3, Table 4.5	59
I_f	Average irradiance at the location of the beam waist	W/m^2	3.4, (3.6)	28
$I(\lambda = \lambda_i, t)$	Spectral intensity of the plasma emission at the wavelength λ_i as a function of time	$W/(m^2 m)$	2.1, Fig 2.4	10
$I_L(R, t)$	Irradiance of the incident laser beam as a function of coordinate R and time t	W/m^2	10, Fig 10.1	186
$I^i_{\lambda L}$	Irradiance of the incident laser beam	W/m^2	2.1, Fig 2.2	9

(continued)

(continued)

Symbol	Description	Unit	Chapter(s), Section(s), Figure(s), Formula(s)	Page(s)
$I^{\mathrm{r}}_{\lambda_\mathrm{L}}$	Reflected laser irradiance	$\mathrm{W/m^2}$	2.1, Fig 2.2	9
$I^{\mathrm{s}}_{\lambda_L}$	Scattered laser irradiance	$\mathrm{W/m^2}$	2.1, Fig 2.2	9
$I_m(\lambda)$	Calculated spectral radiant flux	W/m	10.2, (10.24)	197
$I_3(\lambda)$	Calculated spectral radiant intensity at radius r_3	W/(m sr)	10.2, (10.24)	197
k_B	Boltzmann constant	$\mathrm{JK^{-1}}$	5.3, (5.13)	80
			8.4, (8.14)	161
k_s	Segregation ratio	1	18.3, Table 18.3	483
k_t	Quantity proportional to the absorption coefficient	$\mathrm{m^3}$	9.4, (9.33), Table 9.1	180, 181
κ	Thermal diffusivity	$\mathrm{m^2/s}$	3.3, (3.2)	26
			5.1, (5.2)	75
			7.4, (7.1)	114
$\ln \Lambda$	Coulomb logarithm	1	5.3, (5.13), (5.14)	80
L	Distance between the centers of two adjacent laser beam positions	m	3.6, Fig 3.20	37
$L^{\lambda,\mathrm{d}\Omega}$	Spectral radiance of a black body, Planck's law	$\mathrm{W/(m^2 sr\, m)}$	9.2, (9.9); 10.2, (10.19), (10.20)	171, 195
$L^{\lambda_{ji},\mathrm{d}\Omega}$	Spectral radiance of a black body at the wavelength of the line transition λ_{ji}	$\mathrm{W/(m^2 sr\, m)}$	9.4, (9.31)	179
λ	Wavelength of the laser radiation	m	3.4, (3.3); 11.1, Table 11.1	27, 209
λ_B	Thermal de Broglie wavelength of electrons	m	8.4, (8.14)	161
λ_exc	Wavelength of the laser used for excitation	m	1, Table 1.1	3
λ_i	Element-specific wavelength	m	2.1, Fig. 2.4	10
λ_{ji}	Wavelength of the transition from j to i	m	9.2, (9.8)	170
$\lambda_{\mathrm{ji}\,0}$	Central wavelength of the transition from j to i	m	9.4, (9.18)	173
λ_L	Landau length	m	8.4, (8.15)	161
Λ_e	Thermal de Broglie wavelength of the electrons	m	10.2, (10.5)	193

(continued)

(continued)

Symbol	Description	Unit	Chapter(s), Section(s), Figure(s), Formula(s)	Page(s)
m_e	Electron mass	kg	5.3, (5.9)	79
			8.4, (8.14), (8.19)	161, 164
M	Mass of atoms or ions	kg	8.4, (8.19)	164
M^2	Beam propagation ratio	1	3.3, (3.5), 11.1, Table 11.1	28, 209
M_a	Mass of emitting atom species a	kg	9.4, (9.18)	173
n	Diffraction order	1	4.2, (4.1), (4.3), Fig 4.4	51, 52
n	Principal quantum number	1	8.4, (8.17)	163
			9.1, (9.2)	168
n_a	Refractive index of the ambient atmosphere	1	8.3, (8.11)	154
n_e	Electron density	m^{-3}	5.3, (5.9)	79
			8.2, Table 8.2; 9.4, (9.19), (9.20)	136, 173, 174
n_{Fe}	Total iron density (atoms and ions)	m^{-3}	8.2, Table 8.2, (8.6), Table 8.3	136, 139
$n_{Fe\ I}$	Neutral iron (atoms) density	m^{-3}	8.1	175
$n_{Fe\ II}$	Density of singly ionized iron	m^{-3}	8.1	136
n_i	Density of ions	m^{-3}	5.3, (5.13)	80
n_N	Nitrogen density	m^{-3}	8.2, Table 8.2	136
n_p	Refractive index of the plasma	1	8.3, (8.11)	154
n_{ref}	Refractive index	1	4.2, (4.4)	52
N	Number of illuminated grooves of a grating	1	4.2, (4.3)	51
N	Number of samples	1	11.2, (11.10), (11.12), (11.21), (11.22)	213, 216
N_a^z	Density of ion species a with charge z	m^{-3}	9.1, (9.1)	167
			9.2, (9.3)	169
			10.2, (10.5), (10.6)	193
$N_{a\,1,2}$	Particle density of species a originating from the specimen in sphere 1 or shell 2	m^{-3}	10.1, (10.2), Table 10.1	191, 192

(continued)

(continued)

Symbol	Description	Unit	Chapter(s), Section(s), Figure(s), Formula(s)	Page(s)
$N^g_{a\,1,2}$	Particle density of species *a* originating from the ambient gas in sphere 1 or shell 2	m^{-3}	10.1, (10.3), Table 10.1	191, 192
$N^g_{\mathrm{a}}(r,t)$	Density of ambient gas of species *a*	m^{-3}	10, Fig 10.1	186
$N^g_{a\,3}$	Particle density of species *a* (atoms or molecules) of the ambient gas in shell 3	m^{-3}	10.1, (10.4), Table 10.1	191, 192
$\mathrm{N_{abl}}$, $\mathrm{N_{abl,\,3s}}$	Number of ablation bursts	1	7.3, 13.1.2	104, 238
N_{D}	Number of electrons in the Debye sphere	1	8.4, (8.16)	162
			9.4, (9.19)	173
N_{e}	Electron density	m^{-3}	8.3, (8.12)	154
			9.1, (9.1), (9.2)	167, 168
			10.1, (10.5), (10.7)	193
N_{Fe}	Number of iron particles	1	8.2, Table 8.2	136
$N_{\mathrm{Fe\,I}}$	Number of neutral iron particles	1	8.2, Table 8.2	136
$N_{\mathrm{Fe\,I}\,n}$	Number of iron atoms in the upper state *n*	1	8.2, Table 8.2	136
N_{j}	Densities of neutral atoms and ionic species	m^{-3}	8.3, (8.12)	154
$\mathrm{N_{mp}}$	Number of measuring pulses	1	3.8,11.1, Table 11.1, 13.1.1, 14.2.1	39, 210, 232, 306
$\mathrm{N_p}$	Number of simultaneously generated plasmas	1	3.6	38
$\mathrm{N_{pp}}$	Number of prepulses	1	3.8; 11.1, Table 11.1; 13.1.1, 14.2.1	39, 210, 232, 306
$\mathrm{N_{wu}}$	Number of warming-up pulses	1	3.8; 11.1, Table 11.1	39, 210
$\mathrm{NA_{ES}}$	Numerical aperture of the spectrometer seen from the entrance slit	1	4.2, (4.4)	52

(continued)

(continued)

Symbol	Description	Unit	Chapter(s), Section(s), Figure(s), Formula(s)	Page(s)
ν	Collision frequency of electrons with ions and atoms	s^{-1}	5.3, (5.8), (5.12)	79, 80
ν_0	Frequency of the line maximum for the unperturbed line	s^{-1}	9.2, (9.5)	169
ν_{ei}	Collision frequency of electrons with ions	s^{-1}	5.3, (5.12), (5.13), 8.1	80, 125
ν_{en}	Collision frequency of electrons with atoms	s^{-1}	5.3, (5.12)	80
			8.1	98
ν_{rep}	Repetition rate of the laser source	Hz	3.1, Fig. 3.1; 11.1, Table 11.1	18, 209
v_{sound}	Velocity of sound of the ambient gas	m/s	3.1, (3.1)	21
v_{th}	Thermal velocity of the atoms and ions in the plasma	m/s	10.1, (10.1)	191
ω	Angular frequency	rad/s	5.3, (5.8)	79
ω_{p}	Plasma frequency	rad/s	5.3, (5.9)	79
$p_{\text{g}}, p_{\text{g}}(r,t)$	Ambient gas pressure	N/m^2	2.1, Fig. 2.2; 10, 11.1, Table 11.1	9, 186, 211
$Q^{i,r}$	Line ratio of analyte line i to reference line r	1	11.1, (11.2); 11.3, (11.22)	208, 216
$\left\langle Q^{i,r}_{m,j}(k)\right\rangle_{N_{\text{wu}},N_{\text{pp}},N_{\text{mp}}}$	Line ratio averaged over N_{mp} measuring pulses after N_{wu} warming-up pulses and N_{pp} prepulses at a single location k	1	11.1, (11.4)	212
$\left\langle\left\langle Q^{i,r}_{m,j}(k)\right\rangle_{N_{\text{wu}},N_{\text{pp}},N_{\text{mp}}}\right\rangle_K$	As previous symbol but averaged over K locations	1	11.1, (11.5)	212
r	Degree of ionization	1	8.2, Table 8.2, Table 8.3	136, 139
r	Correlation coefficient	1	14.5, Fig 14.60	347
r^2	Coefficient of determination	1	11.5; 13.1.1, Table 13.2; 13.1.3, Table 13.8; 13.3, Table 13.16; 14.3, Table 14.12; 17.3, Table 17.7; 17.4, Table 17.8	219, 234, 251, 267, 323, 451, 462

(continued)

(continued)

Symbol	Description	Unit	Chapter(s), Section(s), Figure(s), Formula(s)	Page(s)
r_1	Radius of plasma core	m	10.1, Table 10.1	192
r_2, r_3	Radius of plasma shells	m	10.1, Table 10.1	192
r_c	Radius of the produced laser crater	m	3.6, (3.7)	34
R	Resolving power of spectrometer	1	4.2, (4.3)	51
R	Reflectivity	1	5.1, (5.3)	77
$R(i)$	Mean residual of element i	g/g	11.3, (11.21); 13.2, Tables 13.10, 13.11	216, 257, 259
R_i	Congruence of element i	g/g	13.1.2, (13.1), Table 13.7	239, 245
RMSEP(i)	Root mean square error of prediction	g/g	18.2, (18.1)	471
RSD(c)	Relative standard deviation of concentration c	%	14.2.1, Fig. 14.29, Table 14.11	304, 314
ρ	Mass density	kg/m^{-3}	5.1, (5.2)	75
ρ_0	Density of the homogeneous undisturbed ambient gas	kg/m^{-3}	8.1, (8.1)	125
$s_{\text{analyte}}^{\text{acc}}(t_{\text{delay}}, t_{\text{int}})$	Standard deviation of the accumulated analyte signal	a.u.	14.1.2, (14.3)	291
s_b	Standard deviation of the measuring values of the blank sample	1	11.2, (11.14)	214
s_1	Distance between entrance slit and mirror	m	4.2, Fig 4.5, (4.5)	52
s_{x0}, s_{c0}	Standard deviation of procedure	g/g	11.1, (11.11); 13.3, (13.4)	213, 267
s_y	Residual standard deviation	1	11.1, (11.12)	213
$S(\lambda)$	Spectrum	J/m	2.1	11
t_{delay}	Delay time between the laser pulse and the start of the integration window	s	2.1, Fig. 2.4; 11.1, Table 11.1	10, 210
t_{exc}	Pulse duration of the laser beam	s	1, Table 1.1	3
$t_{f,\alpha}$	t-Value of a one-sided test, f degrees of freedom, α level of significance	1	11.2, (11.14), (11.15), (11.17)	214, 215

(continued)

(continued)

Symbol	Description	Unit	Chapter(s), Section(s), Figure(s), Formula(s)	Page(s)
t_{fq}	Time between the ignition of the flashlamp and the Q-switching	s	3.1, Fig. 3.3	20
t_{int}	Width of the integration window	s	2.1, Fig. 2.2, Fig. 2.4; 11.1, Table 11.1	9, 10, 211
t_{rep}	Repetition time of laser pulses	s	3.1, (3.1)	21
			8.4, Fig. 8.35	161
$T_{1,2}$	Temperature in sphere 1 and shell 2	K	10.1, Table 10.1	192
T_e	Electron temperature	K	5.3, (5.13), (5.14)	80
			8.4, (8.14), (8.15)	161
T_g	Temperature of the ambient gas assumed to be stationary	K	10.1, Table 10.1	192
T_{ge}	Gas exchange time of the measuring chamber	s	4.6, (4.7)	67
τ	Decay time of plasma	s	10.1, (10.1)	191
$\tau_{e,0}$	Time for population of discrete levels according to Boltzmann statistics for neutrals	s	8.4, Fig 8.35, (8.21)	161, 165
τ_{ee}	Electron–electron relaxation time	s	8.4, Fig 8.35, (8.18)	161, 164
$\tau_{e,z-1}$	Time for population of discrete levels according to Boltzmann statistics for stage of ionization z with $z = 1$ for neutrals, $z = 2$ for singly ionized species	s	8.4, (8.21)	165
τ_{ge}	Local gas exchange time	s	3.1, (3.1)	21
			8.4, Fig 8.35	161
τ_k^0	Electron-neutral relaxation time	s	8.4, (8.19)	164
τ_k^1	Electron-ion relaxation time	s	8.4, (8.20)	164
τ_L	Temporal width (FWHM) of the laser pulse	s	3.1, Fig. 3.3; 3.4, (3.6)	20, 28, 22
			7.4, (7.1); 8.4, Fig 8.35; 11.1, Table 11.1	114, 161, 209

(continued)

(continued)

Symbol	Description	Unit	Chapter(s), Section(s), Figure(s), Formula(s)	Page(s)
τ_{Lj}	Temporal width (FWHM) of the laser pulse j in a pulse burst	s	3.2, Fig 3.9; 11.1, Table 11.1	25, 209
τ_{mod}	Time scale of multiple pulses	s	8.4, Fig 8.35	161
τ_{Plasma}	Life time of plasma	s	8.4, Fig 8.35	161
τ_s^λ	Optical thickness of a plasma layer	1	9.3, (9.16)	172
θ_0	Beam divergence angle of a Gaussian beam	rad	3.4, (3.4)	28
Θ	Beam divergence angle (full angle) of the laser beam	rad	3.4, (3.5)	28
$u_a^z(T, \Delta E_{a,z}^{ion})$	Partition function of species a with charge z as a function of temperature and reduction of ionization energy	1	9.2, (9.3), (9.6); 10.2, (10.5); 11.1, (11.3)	169, 193, 211
$u_{Fe0}, u_{Fe1}, u_{N0}, u_{N1}$	Partition function of neutral, singly ionized iron; neutral, singly ionized nitrogen	1	10.4, (10.25)	203
u^S	Partition function of the plasma state generated with SP	1	8.2, (8.7)	136
u^D	Partition function of the plasma state generated with DP	1	8.2, (8.7)	136
U_{osc}	Charging voltage of the capacitors driving the flashlamp	V	3.1, Fig 3.3	20
U_{pc1}	Bias voltage at Pockels cell	V	3.1, Fig 3.3	20
U_{PMT}	Signal of a PMT	V	3.5, Fig 3.13	30
V_C	Volume of measuring chamber	m^3	4.6, (4.7)	67
$\dot{V}_{in}$	Inlet gas flow rate	m^3/s	4.6, Fig 4.17, (4.7)	67
$\dot{V}_{mc}$	Gas flow rate in the measuring chamber	m^3/s	11.1, Table 11.1	211
V_p	Plasma volume	m^3	8.3, (8.11)	154
			9.2, (9.10)	171
V_{x0}	Relative standard deviation of procedure	%	11.1, (11.13)	213
w_0	Radius of beam waist	m	3.4, (3.3)	27

(continued)

(continued)

Symbol	Description	Unit	Chapter(s), Section(s), Figure(s), Formula(s)	Page(s)
w_b	Radius of the beam waist on the surface of the specimen	m	3.6, (3.7), (3.8)	34, 37
w_m	measured full Stark width of the line at the given wavelength	Å	A.4, Table A.1	512, 514
w_{th}	theoretically calculated full Stark width of the line at the given wavelength	Å	A.4, Table A.1	512, 514
W_e	Electron energy	eV	8.4, (8.18)	164
W_{FWHM}	Stark broadening parameter for the full width (FWHM)	m	9.4, (9.19)	173
x	Concentration of reference sample	g/g	11.2, (11.6)	212
x_{ID}	Limit of identification	g/g	11.2, (11.17)	215
x_{LOD}	Limit of detection	g/g	11.2, (11.15), (11.16), (11.17)	215
x_{LOQ}	Limit of quantification	g/g	11.2, (11.18)	215
$\hat{y}$	Prediction of measurand by regression curve	1	11.2, (11.6), (11.7), (11.8)	212
$\overline{\hat{y}'}$	Sensitivity as first derivative of the regression curve at the middle of the working range	g/g	11.2, (11.9)	213
$\bar{y}_b$	Arithmetic average of *n* measurements of the blank sample	1	11.2, (11.14)	214
y_c	Critical value	1	11.2, (11.14)	214
ξ	Parameter describing the geometry of the energy release and the shock wave expansion	1	8.1, (8.1)	125
z	Ionization stage	1	8.4, (8.17)	163
			9.1, (9.1)	167
z_R	Rayleigh length	m	3.4, Fig 3.12	29
			3.6	33
Z	Ion charge	1	5.3, (5.13)	80
ζ	Skewness	1	18.3, (18.2)	481
$\xi(\lambda)$	Spectral characteristics of the observation channel	1	10.1, Table 10.1, (10.24)	192, 197

A.4 Stark Data

Table A.1 lists a selection of Stark data for various atom and ion emission lines which are useful for diagnostic purposes of laser-induced plasmas (cf. Sect. 9.4). If several values are stated in a reference, then the average value is taken. In [A.4], Stark parameters are calculated for an electron density of 10^{16} cm^3. The calculated widths were multiplied by 2 to obtain the full widths, such as the experimental values given in, e.g., [A.1, A.3]. The experimental values taken from [A.2] and [A.9] were also multiplied by 2 to get the full width. References [A.10–A.20] give further information on Stark data as well as Table 10.2 in Sect. 10.2.

A.5 Wavelengths and Elements

The following compilation of wavelengths and elements provides an overview of those lines investigated for LIBS and described in this book. They were used, e.g., to determine calibration curves, analysis functions, detection limits, standard deviations of procedure, background equivalent concentrations, coefficients of determination, plasma temperatures, electron densities, self-absorption, temporal behavior of analyte emission. Table A.2 is arranged in ascending order of wavelengths starting with vacuum ultraviolet lines and ending in the near infrared with a ratio of the longest to the shortest wavelength of 6.4, i.e., more than two octaves. Table A.3 lists the studied elements in alphabetic order and allocates the observed emission wavelengths. In this way, both tables allow for an easy access to identify lines, their features, and usage. These tables are not intended as a comprehensive overview of emission lines studied so far for LIBS. The reader is referred to the respective literature given in the previous chapters and data bases as, e.g., [13.30, A.21]. With the help of Table A.2, these wavelengths are linked to the respective matrices investigated.

A.6 Spectral Line Selection for Boltzmann Plots

An important method to determine the temperature of laser-induced plasmas is based on the Boltzmann plot, see, e.g., Sect. 8.2, Fig 8.15 and Sect. 13.2, Fig 13.22. The choice of appropriate spectral lines is often based on experience and reference values and can even be restricted by spectroscopic or other hardware equipments. The development of spectrometers with large wavelength ranges, e.g., echelle spectrometers as described in Sect. 4.2, enables the simultaneous detection and acquisition of spectral lines with resolutions better than 20 pm, see Table 4.4 [A.22, A.23]. In the following a basic, quantitative, and objective method for the selection of spectral lines needed for temperature determinations by Boltzmann plots is

Table A.1 List of Stark data

El	IS	λ(nm)	C_s(Å 10^{-16}cm^3)	w_m, w_{th} (Å)	D_s(Å10^{-16} cm^3)	d(Å)	T(K)	n_e (10^{17} cm^{-3})	A	Ref.
Al	II	199.053	0.0440	0.044		n.i.	10,500	0.1	n.i.	[A.2]
Al	II	263.155	0.0640	0.064		n.i.	10,500	0.1	n.i.	[A.2]
Al	II	266.917	0.0062	0.0062		n.i.	10,500	0.1	n.i.	[A.2]
Al	II	281.618	0.0424	0.0424		n.i.	10,500	0.1	n.i.	[A.2]
Al	II	390.068	0.0100	0.01		n.i.	10,500	0.1	n.i.	[A.2]
Al	I	308.215	0.0398	0.51	0.0148	0.19	13,200	1.28	n.i.	[A.1]
Al	I	309.271	0.0398	0.51	0.0469	0.6	13,200	1.28	n.i.	[A.1]
Al	I	394.401	0.0336	0.840	0.0168	0.42	11,700	2.50	n.i.	[A.3]
Al	I	396.152	0.0336	0.840	0.0168	0.42	11,700	2.50	n.i.	[A.3]
Al	II	466.305	0.1206	0.121	0.0651	0.07	10,500	0.10	n.i.	[A.2]
Al	II	559.323	0.3800	0.380		n.i.	10,500	0.10		[A.2]
Al	II	704.206	0.1930	1.930	−0.0690	−0.69	18,000	1.00		[A.3]
Al	II	705.656	0.1870	1.870	−0.0710	−0.71	18,000	1.00		[A.3]
Al	II	706.362	0.1900	1.900	−0.0680	−0.68	18,000	1.00		[A.3]
Ar	II	295.539	0.0900	0.900		n.i.	28,500	1.00	n.i.	[A.3]
Ar	I	415.859	0.1952	1.210		n.i.	11,900	0.62	n.i.	[A.3]
Ar	I	425.936	0.2323	1.440		n.i.	11,900	0.62	n.i.	[A.3]
Ar	I	425,900[a]	0.2600	0.260	0.1320	0.13	10,000	0.10	0.069	[A.4]
Ar	I	427.217	0.1903	1.180		n.i.	11,900	0.62	n.i.	[A.3]
Ar	II	434.81	0.0197	0.197				1.00		[A.3]
Ar	II	457.935	0.0304	0.210	0.0087	0.06	12,000	0.69	n.i.	[A.3]
Ar	I	470.232	0.2136	1.410		n.i.	12,000	0.66	n.i.	[A.3]
Ar	II	514.531	0.5449	3.760	0.1957	1.35	12,000	0.69	n.i.	[A.3]
Ar	I	750.386	1.16					0.1		[A.5]
Ar	I	751.465	1.01					0.1		[A.5]
C	I	165.7[a]	0.0017	0.0165		n.i.	12,500	1.00	n.i.	[A.1]

C	I	165.7[a]	0.0035	0.00348	0.0021	0.0021	10,000	0.10	0.031	[A.4]
C	I	165.812	0.0036	0.060	0.0020	0.0325	14,000	1.65	n.i.	[A.1]
C	I	193.027	0.0044	0.0044	0.0027	0.0027	10,000	0.10	0.03	[A.1]
C	I	193.091	0.0050	0.0300	0.0028	0.017	11,700	0.60	n.i.	[A.1]
			0.0051	0.1120	0.0026	0.057	15,500	2.20	n.i.	[A.1]
Cl	I	725.662	0.1378	0.510		n.i.	10,000	0.37	n.i.	[A.3]
				n.i.	0.0642	0.17	9,500	0.27	n.i.	[A.3]
Cl	I	754.707	0.1486	0.550		n.i.	10,000	0.37	n.i.	[A.3]
				n.i.	0.0642	0.17	9,500	0.27	n.i.	[A.3]
Cu	I	510.554	0.0430	0.430	0.0067	0.07	10,000	1.00	n.i.	[A.3]
Fe	I	373.487	0.0090	0.090			8,000	1.00		[A.6]
Fe	I	376.554	0.0090	0.090			8,000	1.00		[A.6]
Fe	I	376.719	0.0090	0.090			8,000	1.00		[A.6]
Fe	I	378.788	0.0090	0.090			8,000	1.00		[A.6]
Fe	I	381.584	0.0140	0.140			8,000	1.00		[A.6]
Fe	I	492.050	0.1070	0.107		n.i.	8,000	0.10	n.i.	[A.1]
Fe	I	536.7	0.06–0.14				7,400–9,500			[A.7]
Fe	I	538.337	0.2120	0.212		n.i.	9,500	0.10	n.i.	[A.8]
Fe	I	538.3	0.08–0.2				6,600–10,000			[A.7]
Fe	I	540.4	0.07–0.12				7,400–9,500			[A.7]
Fe	I	542.407	0.1300	0.130		n.i.	8,000	0.10	n.i.	[A.1]
Hg	I	435.834	0.0280	0.03		n.i.	6,000	0.1	n.i.	[A.3]
K	I	766.490	0.1240	1.24	0.0275	0.28	20,000	1.0	n.i.	[A.1]

(continued)

Table A.1 (continued)

El	IS	λ(nm)	C_s(Å 10^{-16}cm^3)	w_m, w_{th} (Å)	D_s(Å 10^{-16} cm^3)	d(Å)	T(K)	n_e (10^{17} cm^{-3})	A	Ref.
Mg	I	291.545	0.0500	0.50		n.i.	10,000	1.0	n.i.	[A.1]
N	I	149.263	0.0003	0.00034		n.i.	12,350	0.1	n.i.	[A.1]
			0.0003	0.00303	0.0002	0.0016	13,000	1.0	n.i.	[A.1]
N	I	149.3[a]	0.0025	0.0025	0.0015	0.0015	10,000	0.1	0.029	[A.4]
N	I	790.45	2.4000	2.4	0.9000	0.9	10,000	0.1		[A.9]
O	I	130.4[a]	0.0016	0.00164	0.0010	0.001	10,000	0.1	0.028	[A.4]
O	I	777.3[a]	0.0630	0.0630	0.0143	0.0143	10,000	0.1	0.012	[A.4]
P	I	178.0[a]	0.0097	0.0097	0.0053	0.0053	10,000	0.1	0.043	[A.4]
Pb	I	405.781	0.0620	0.620		n.i.	11,600	1.0	n.i.	[A.1]
Si	I	288.2[a]	0.0128	0.0128	0.0079	0.0079	10,000	0.1	0.033	[A.4]

[a] Average wavelength of a multiplet

El = element, IS = ionization stage, C_s = Stark broadening parameter, w_m = measured full width of the line at the given electron density, w_{th} = theoretically calculated full width of the line at the given electron density, D_s = Stark shift parameter (calculated with d/n_e), d = Stark shift, positive if the wavelength is increased, T = temperature of the plasma, A = ion broadening parameter (cf. Sect. 9.4, (9.19), (9.20)), n.i. = no information available

Table A.2 Wavelengths and elements studied for LIBS in this book, allocation to chapters, sections, figures and tables

Wavelength (nm)	Element	Matrix	Chapter(s), Section(s), Figure(s), Table(s)	Page(s)
130.22	O I	Inclusions/steel	18.3, Fig. 18.14	477
134.72	Cl I	Cement	14.4, Fig. 14.44, Fig. 14.46, Fig. 14.49, Fig. 14.50, Fig. 14.51, Fig. 14.52	331, 332, 334, 335, 336
149.26	N I	Inclusions/steel	18.3, Fig. 18.15, Fig. 18.16	478, 479
165.81	C I	Steel	17.1, Table 17.3	438
178.28	P I	Steel, segregation/steel	3.5, Fig. 3.15; 13.1.1, Fig. 13.4, Table 13.2; 13.1.2, Table 13.6; 13.1.3, Fig. 13.15, Table 13.8; 18.3, Fig. 18.22	32, 233, 234, 240, 251, 483
180.73	S I	Steel, segregation/steel	3.5, Fig. 3.15; 13.1.1, Fig. 13.5, Table 13.2; 13.1.2, Table 13.6; 13.1.3, Table 13.8; 15.1, Fig. 15.9; 18.3, Fig. 18.22	32, 234, 240, 251, 397, 483
187.75	Fe II	Steel	3.5, Fig. 3.15; 3.8, Fig. 3.24; 13.1.1, Fig. 13.3, Fig. 13.4, Fig. 13.5; 13.1.2, Fig. 13.8, Fig. 13.9; 13.1.3, Fig. 13.14, Fig. 13.15, Fig. 13.16	32, 40, 233, 234, 241, 242, 250, 251
193.09	C I	Steel, inclusions/steel	3.8, Fig. 3.24; 13.1.1, Fig. 13.2, Fig. 13.3, Table 13.2; 13.1.2, Table 13.6, Fig. 13.8, Fig. 13.9; 13.1.3, Fig. 13.14, Table 13.8; 15.1, Fig. 15.8; 17.1, Table 17.3	40, 232, 233, 234, 240, 241, 242, 250, 251, 396, 438
194.23	Hg II	Mercury lamp	17.4, Fig. 17.29	457
212.41	Si I	Steel	6.1, Fig. 6.5; 15.1, Fig. 15.8, Fig. 15.10	91, 396, 398
213.86	Zn I	Aerosol	14.8.1, Fig. 14.84, Fig. 14.88	369, 371
225.39	Ni II	Steel, high-alloy steel	13.1.2, Table 13.6; 13.2, Table 13.10, Fig. 13.19	240, 257, 258

(continued)

Table A.2 (continued)

Wavelength (nm)	Element	Matrix	Chapter(s), Section(s), Figure(s), Table(s)	Page(s)
228.80	Cd I	Soil, polymer	12, Fig. 12.5, Fig. 12.6; 14.1.2, Table 14.6, Fig. 14.16, Fig. 14.18, Table 14.7, Table 14.8; 14.3, Table 14.12	226, 227, 295, 297, 298, 299, 303, 323
231.60	Ni II	Steel, soil	13.1.1, Table 13.2; 13.1.3, Fig. 13.16, Table 13.8; 14.3, Table 14.12	234, 251, 323
234.98	As I	Soil	14.3, Table 14.12	323
247.86	C I	Polymer, air, grease/aluminum, particulate/top gas	14.1.1, Fig. 14.2; 14.6, Fig. 14.63, Fig. 14.64, Fig. 14.65, Fig. 14.66, Fig. 14.67; 14.8.1, Fig. 14.83; 17.4, Fig. 17.37	279, 352, 353, 354, 368, 463
251.43	Si I	Slag	14.2.2, Fig. 14.33, Fig. 14.34, Fig. 14.37; 18.2, Fig. 18.9	317, 319, 473
251.61	Si I	High-alloy steel	13.2, Fig. 13.20	258
253.65	Hg I	Polymer, soil	14.1.2, Table 14.6, Table 14.7, Table 14.8; 14.3, Table 14.12	295, 299, 303, 323
257.51	Al I	Aluminum, grease/aluminum	14.8.1, Fig. 14.82, Fig. 14.83	367, 368
259.80	Sb I	Polymer	14.1.2, Table 14.6, Table 14.7, Table 14.8	295, 299, 303
259.94	Fe II	Aluminum	13.3, Table 13.15, Table 13.16, Table 13.17, Fig. 13.25	266, 267, 269, 271
263.82	Mn II	Inclusions/steel	15.2, Fig. 15.9; 18.3, Fig. 18.12	397, 476
267.72	Cr II	High-alloy steel steel, aluminum	11.5, Fig. 11.6; 13.1.1, Table 13.2; 13.2, Table 13.10, Table 13.13; 13.3, Table 13.15, Table 13.16, Table 13.17, Table 13.18; 17.2, Table 17.5, Table 17.6	219, 234, 257, 261, 266, 267, 269, 270, 443, 444
271.44	Fe II	High-alloy steel steel, particulate/top gas	11.5, Fig. 11.5; 13.1.1, Table 13.1; 13.2, Fig. 13.19, Fig. 13.20; 14.7, Fig. 14.75	218, 231, 258, 360

(continued)

Table A.2 (continued)

Wavelength (nm)	Element	Matrix	Chapter(s), Section(s), Figure(s), Table(s)	Page(s)
273.07	Fe II	Steel	3.5, Fig. 3.13; 6.1, Fig. 6.3, Fig. 6.5, Fig. 6.6; 17.1, Fig. 17.12; 17.2, Table 17.6	30, 88, 91, 92, 439, 444
273.36	Fe I	High-alloy steel	13.2, Table 13.14	262
278.81	Fe I	High-alloy steel	13.2, Table 13.14	262
279.08	Mg II	Aluminum	9.4, Fig. 9.10; 13.3, Table 13.15, Table 13.16, Table 13.17	181, 266, 267, 269
279.55	Mg II	Aluminum, water	9.4, Fig. 9.10; 14.5, Fig. 14.62; 17.3, Table 17.7	181, 349, 451
281.62	Mo II	Steel, high-alloy steel	13.1.2, Table 13.6; Fig. 13.8; 13.2, Table 13.10; 17.2, Table 17.6	240, 241, 257, 444
286.26	Cr II	High-alloy steel	11.5, Fig. 11.5; 13.2, Table 13.10, Table 13.11	218, 257, 259
288.16	Si I	Aluminum steel	11.2, Fig. 11.2; 13.1.1, Table 13.2; 13.3, Table 13.15, Table 13.16, Table 13.17, Fig. 13.24, Table 13.18; 17.3, Fig. 17.22, Table 17.7	214, 234, 266, 267, 269, 270, 450, 451
289.36	Hg I	Mercury lamp	17.4, Fig. 17.28	457
293.31	Mn II	Steel, pure iron, aluminum	6.1, Fig. 6.4, Fig. 6.6; 13.1.1, Table 13.2; 13.1.2, Table 13.6; 13.3, Table 13.15, Table 13.16, Table 13.17, Table 13.18; 15.1, Fig. 15.10; 17.1, Fig. 17.12	89, 92, 234, 240, 266, 267, 269, 270, 398, 439
294.92	Mn II	Aluminum	17.3, Table 17.7	451
296.12	Cu I	Aluminum	13.3, Table 13.15, Table 13.17, Table 13.18	266, 269, 270
296.69	Fe I	High-alloy steel	13.2, Table 13.14	262
298.92	Cr II	High-alloy steel	17.2, Table 17.6	444

(continued)

Table A.2 (continued)

Wavelength (nm)	Element	Matrix	Chapter(s), Section(s), Figure(s), Table(s)	Page(s)
305.71	Al I	Aluminum	11.2, Fig. 11.2; 13.3, Table 13.15, Fig. 13.24, Fig. 13.25	214, 266, 270, 271
307.21	Zn I	Zinc-coated steel	17.3, Fig. 17.25	454
308.22	Al I	Inclusions/steel	18.3, Fig. 18.13, Fig. 18.15, Fig. 18.16, Fig. 18.17, Fig. 18.18, Fig. 18.21	477, 478, 479, 480, 481
309.27	Al I	Slag	14.2.2, Fig. 14.37; 17.3, Fig. 17.22	319, 450
324.75	Cu I	Copper, steel, high-alloy steel, soil, water, aerosol, aluminum, segregation/steel	7.3, Fig. 7.7; 13.1.2, Table 13.3, Table 13.6; 13.2, Fig. 13.18, Table 13.10; 14.3, Fig. 14.38, Table 14.12; 14.5, Fig. 14.61; 14.8.1, Fig. 14.87; 17.3, Fig. 17.21, Fig. 17.22, Table 17.7; 18.3, Fig. 18.22	105, 238, 240, 256, 257, 322, 323, 348, 371, 450, 451, 483
327.40	Cu I	Aluminum	3.9, Fig. 3.25; 13.3, Table 13.15, Table 13.16, Table 13.17; 17.3, Fig. 17.21	42, 266, 267, 269, 450
328.23	Zn I	Zinc-coated steel	16.1, Table 16.2, Table 16.3, Fig. 16.9	410, 411, 414
334.50	Zn I	Aluminum, soil, particulate/top gas	13.3, Table 13.15; 14.3, Fig. 14.38, Table 14.12; 14.7, Fig. 14.72, Fig. 14.74; 17.3, Table 17.7	266, 322, 323, 358, 360, 451
334.94	Ti II	Aluminum	17.3, Table 17.7	451
337.28	Ti II	Aluminum	13.3, Table 13.15, Table 13.16, Table 13.17, Table 13.18	266, 267, 269, 270
341.48	Ni I	Aluminum	13.3, Table 13.15, Table 13.16, Table 13.17	266, 267, 269
351.92	Tl I	Soil	14.3, Table 14.12, Fig. 14.40, Fig. 14.41	325, 326
356.54	Fe I	High-alloy steel	13.2, Table 13.14, Fig. 13.22	262, 263
357.58	Zr I	Aluminum	17.3, Table 17.7	451
358	CN	Ammonium nitrate	17.4, Fig. 17.37	463
363.15	Fe I	High-alloy steel	13.2, Table 13.14, Fig. 13.22	262, 263

(continued)

Table A.2 (continued)

Wavelength (nm)	Element	Matrix	Chapter(s), Section(s), Figure(s), Table(s)	Page(s)
364.44	Ca I	Slag	14.2.2, Fig. 14.36	319
364.78	Fe I	High-alloy steel	13.2, Table 13.14, Fig. 13.22	262, 263
368.35	Pb I	Aerosol	14.8.1, Fig. 14.85, Fig. 14.86	369, 370
371.99	Fe I	High-alloy steel, aluminum, steel	13.2, Table 13.14, Fig. 13.22; 13.3, Table 13.15, Table 13.17; 15.1, Fig. 15.6, Fig. 15.10; 17.2, Fig. 17.16, Fig. 17.17	262, 263, 266, 269, 394, 398, 445, 446
373.83	Fe I	Aluminum	17.3, Table 17.7, Fig. 17.25	451, 454,
374.95	Fe I	High-alloy steel	13.2, Table 13.14, Fig. 13.22	262, 263
381.30	Fe I	Steel	9.4, Fig. 9.6, Fig. 9.7	177, 178
381.58	Fe I	Steel, aerosol	9.4, Fig. 9.6, Fig. 9.7; 14.8.1, Fig. 14.85, Fig. 14.86	177, 178, 369, 370
382.78	Fe I	Steel	9.4, Fig. 9.6, Fig. 9.7	177, 178
383.83	Mg I	Aluminum	17.3, Table 17.7	451
384.99	Fe I	Steel	9.4, Fig. 9.6, Fig. 9.7	177
387.1	CN	Ammonium nitrate	17.4, Fig. 17.37	463
390.55	Si I	Soil, aluminum, slag	12, Fig. 12.5; 13.3, Table 13.15, Table 13.17; 14.2.1, Fig. 14.27, Fig. 14.28; 14.2.2, Fig. 14.35	226, 266, 269, 309, 310, 318
393.37	Ca II	Water, aerosol, particulate/top gas	14.5, Fig. 14.60; 14.7, 14.8.2, Fig. 14.89, Fig. 14.91, Fig. 14.93, Fig. 14.94	347, 355, 373, 376, 378, 379
394.40	Al I	Aluminum	9.4, Fig. 9.6, Fig. 9.11; 10.2, Fig. 10.6; 17.4, Fig. 17.31	177, 182, 202, 458
396.15	Al I	Steel, zinc-coated steel, ZnAl binary alloy, aluminum, inlcusions/steel	13.1.2, Table 13.6; 16.1, Fig. 16.10; 16.2, Fig. 16.13, Fig. 16.14, Fig. 16.15, Fig. 16.16, Fig. 16.17; 17.4, Fig. 17.31; 18.3, Fig. 18.13, Fig. 18.21	240, 415, 420, 422, 423, 424, 425, 458, 477, 481
396.85	Ca II	Aerosol	14.8.2, Fig. 14.95	380
404.58	Fe I	High-alloy steel	13.2, Table 13.14, Fig. 13.22	262, 263

(continued)

Table A.2 (continued)

Wavelength (nm)	Element	Matrix	Chapter(s), Section(s), Figure(s), Table(s)	Page(s)
405.78	Pb I	Polymer, soil, particulate/top gas	14.1.2, Fig. 14.9, Fig. 14.10, Fig. 14.11, Fig. 14.12, Fig. 14.13, Table 14.6, Fig. 14.14, Fig. 14.15, Table 14.7, Fig. 14.21, Fig. 14.22, Table 14.8; 14.3, Table 14.12; 14.7, Fig. 14.72	290, 293, 294, 295, 299, 301, 303, 323, 358
406.36	Fe I	High-alloy steel	13.2, Table 13.14, Fig. 13.22	262, 263
407.17	Fe I	High-alloy steel	13.2, Table 13.14, Fig. 13.22	262, 263
410.98	Fe I	Zinc-coated steel	16.1, Table 16.3, Fig. 16.9	411, 414
411.85	Fe I	High-alloy steel	13.2, Table 13.14, Fig. 13.22	262, 263
413.47	Fe I	High-alloy steel	13.2, Table 13.14, Fig. 13.22	262, 263
419.91	Fe I	High-alloy steel	13.2, Table 13.14, Fig. 13.22	262, 263
421.94	Fe I	High-alloy steel	13.2, Table 13.14, Fig. 13.22	262, 263
422.67	Ca I	Aerosol	14.8.1, Fig. 14.84	369
425.43	Cr I	Polymer, soil	14.1.2, Table 14.6, Table 14.7, Table 14.8; 14.3, Table 14.12	295, 299, 303, 323
427.18	Fe I	Pure iron, high-alloy steel	8.2, Fig. 8.24; 13.2, Table 13.14, Fig. 13.22	150, 262, 263
430.01	Ar I	Gas	8.2, Fig. 8.24	150
430.79	Fe I	Pure iron, high-alloy steel	8.2, Fig. 8.24; 13.2, Table 13.14, Fig. 13.22	150, 262, 263
432.58	Fe I	Pure iron, high-alloy steel	8.2, Fig. 8.24, Fig. 8.26; 13.2, Table 13.14, Fig. 13.22	150, 152, 262, 263
433.12	Ar II	Gas	8.2, Fig. 8.24	150
434.52	Ar I	Gas	8.2, Fig. 8.24	150
434.81	Ar II	Gas	8.2, Fig. 8.24, Fig. 8.25; 15.1, Fig. 15.7	150, 151, 395
438.35	Fe I	Steel, high-alloy steel, zinc-coated steel	9.4, Fig. 9.5; 13.2, Table 13.14, Fig. 13.22; 16.1, Fig. 16.5, Fig. 16.6, Fig. 16.7, Table 16.2, Table 16.3, Fig. 16.8, Fig. 16.9; 16.2, Fig. 16.16, Fig. 16.17	176, 262, 263, 406, 407, 408, 410, 411, 413, 414, 424, 425

(continued)

Table A.2 (continued)

Wavelength (nm)	Element	Matrix	Chapter(s), Section(s), Figure(s), Table(s)	Page(s)
460.96	Ar II	Gas	15.1, Fig. 15.7	395
472.22	Zn I	Zinc-coated steel, ZnAl binary alloy	16.1, Fig. 16.5, Fig. 16.6, Fig. 16.7, Table. 16.2, Table. 16.3, Fig. 16.8, Fig. 16.9, Fig. 16.10; 16.2, Fig. 16.13, Fig. 16.14, Fig. 16.15, Fig. 16.16, Fig. 16.17	406, 407, 408, 410, 411, 413, 414, 415, 420, 422, 423, 424, 425
473.59	Ar II	Gas	15.1, Fig. 15.7	395
476.49	Ar II	Gas	15.1, Fig. 15.7	395
486.13	H I	Polymer	14.1.1, Fig. 14.2	279
522.72	Fe I	High-alloy steel	13.2, Table 13.14, Fig. 13.22	262, 263
531.66	Fe II	Steel	10.2, Fig. 10.5	198
532.42	Fe I	Steel	8.2, Fig. 8.13, Fig. 8.16, Table 8.3; Fig. 8.19, Fig. 8.20, Fig. 8.21; 10.2, Fig. 10.5	134, 138, 139, 142, 143, 144, 198
532.80	Fe I	Steel	6.3, Fig. 6.7; 8.2, Fig. 8.13, Fig. 8.16; 10.2, Fig. 10.5	93, 134, 138, 198
533.99	Fe I	Steel	10.2, Fig. 10.5	198
534.6	Continuum	Steel	8.2, Fig. 8.22	145
536.29	Fe II	Steel	10.2, Fig. 10.5	198
537.15	Fe I	Steel	8.2, Fig. 8.13, Fig. 8.14; 10.2, Fig. 10.5	134, 198
538.34	Fe I	Steel	8.2, Fig. 8.13, Fig. 8.14, Fig. 8.22; 8.3, Fig. 8.33; 9.1, Fig. 9.1; 9.4, Fig. 9.4, Fig. 9.5, Fig. 9.8, Fig. 9.9; 10.2, Fig. 10.5	134, 145, 158, 168, 176, 179, 180, 198
539.71	Fe I	Steel	8.2, Fig. 8.13; 10.2, Fig. 10.5	134, 198
540.58	Fe I	Steel	8.2, Fig. 8.13	134
589.00	Na I	Water, particulate/top gas, aerosol	14.5, Fig. 14.59; 14.7, Fig. 14.74; 14.8.2, Fig. 14.91, Fig. 14.92, Fig. 14.93	346, 360, 376, 377, 378
589.59	Na I	Water	14.5, Fig. 14.59	346
656.28	H I	Propane/carbon dioxide, particulate/top gas, ammonium nitrate	14.6, Fig. 14.67, Fig. 14.68; 14.7, Fig. 14.73; 17.4, Fig. 17.37	354, 355, 359, 463
725.66	Cl I	Polymer	14.1.1, Fig. 14.6	283

(continued)

Table A.2 (continued)

Wavelength (nm)	Element	Matrix	Chapter(s), Section(s), Figure(s), Table(s)	Page(s)
742.36	N I	Air, ammonium nitrate	14.6, Fig. 14.63; 17.4, Fig. 17.37	352, 463
744.23	N I	Air	14.6, Fig. 14.63	352
746.83	N I	Air, particulate/top gas, inclusions	14.6, Fig. 14.63	352
754.71	Cl I	Polymer	14.1.1, Fig. 14.6	283
766.49	K I	particulate/top gas	14.7	355
777.19	O I	Air/carbon dioxide, particulate/top gas, propane/carbon dioxide, ammonium nitrate, inclusions/steel	14.6, 14.7, Fig. 14.63, Fig. 14.65, Fig. 14.66, Fig. 14.68; 17.4, Fig. 17.37; 18.3, Fig. 18.14, Fig. 18.15, Fig. 18.17, Fig. 18.18	352, 353, 354, 355, 463, 477, 478, 479,480
777.42	O I	Air/carbon dioxide, propane/carbon dioxide	14.6, Fig. 14.63, Fig. 14.65, Fig. 14.66, Fig. 14.68	352, 353, 354, 355
777.54	O I	Air/carbon dioxide, propane carbon dioxide	14.6, Fig. 14.63, Fig. 14.65, Fig. 14.66, Fig. 14.68	352, 353, 354, 355
827.24	Br I	Polymer	14.1.2, Table 14.6, Fig. 14.17, Fig. 14.19, Table 14.7	295, 298, 299
837.59	Cl I	Cement	14.4, Fig. 14.45, Fig. 14.48, Fig. 14.50	331, 334, 335

described [A.24]. This approach implicates advantages especially in those cases where no spectral line experiences or empirical values are available for a specific diagnostic task. Additionally, even for comparatively low laser pulse energies, larger sets of spectral lines can be determined by a systematic approach, which effect a decrease in the relative error of the electron temperature to be determined. Thus, the temporal characterization of the LIBS plasma can be improved and well grounded. Low laser pulse energies are, e.g., used for microanalysis as, e.g., those described in [A.25–A.27] and Chaps. 15, 16.

The experiments are carried out with a flashlamp pumped Q-switched Nd:YAG laser at the fundamental wavelength $\lambda = 1{,}064$ nm and a repetition rate of $\nu_{\mathrm{rep}} = 20$ Hz. The laser beam diameter amounts to 6 mm. The pulse duration is about $\tau_{\mathrm{L}} = 5$ ns. The pulse energy is attenuated by a Glan-laser-prism and four different laser pulse energies, e.g., 0.2, 0.5, 1, and 2 mJ, are studied. This pulse energy regime is of interest for the design of compact LIBS instrumentation, where the use of more powerful laser sources is not possible or for LIBS in combination with laser micromachining or LIBS microanalysis [A.28, A.29]. All experiments are carried out in air at atmospheric pressure. The samples are certified C75 steel alloy

Table A.3 Elements in alphabetic order and emission wavelengths

Element	Wavelength(s) (nm)
Aluminum	257.51, 305.71, 308.22, 309.27, 394.40, 396.15
Antimony	259.80
Argon	430.01, 433.12, 434.52, 434.81, 460.96, 473.59, 476.49
Arsenic	234.98
Bromine	827.24
Cadmium	228.80
Calcium	364.44, 393.37, 396.85
Carbon	165.81, 193.09, 247.86
Chlorine	134.72, 725.66, 754.71, 837.59
Chromium	267.72, 286.26, 298.92
Copper	296.12, 324.75, 327.40
Hydrogen	486.13, 656.28
Iron	187.75, 259.94, 271.44, 273.07, 273.36, 278.81, 296.69, 356.54, 363.15, 364.78, 371.99, 373.83, 374.95, 381.30, 381.58, 382.78, 384.99, 404.58, 406.36, 407.17, 410.98, 411.85, 413.47, 419.91, 421.94, 422.67, 425.43, 427.18, 430.79, 432.58, 438.35, 522.72, 531.66, 532.42, 532.80, 533.99, 536.29, 537.15, 538.34, 539.71, 540.58
Lead	368.35, 405.78
Magnesium	279.08, 279.55, 383.83
Manganese	263.82, 293.31, 294.92
Mercury	194.23, 253.65, 289.36
Molybdenum	281.62
Nickel	225.39, 231.60, 341.48
Nitrogen	149.26, 742.36, 744.23, 746.83
Oxygen	130.22, 777.19, 777.42, 777.54
Phosphor	178.28
Silicon	212.41, 251.43, 251.61, 288.16, 390.55
Sodium	589.00, 589.59
Sulfur	180.73
Thallium	351.92
Titanium	334.94, 337.28
Zinc	213.86, 307.21, 328.23, 334.50, 472.22
Zirconium	357.58

samples (German designation 1.1750) with low concentrations of C (0.78 m.-%), Cr (0.17 m.-%), Si, (0.28 m.-%), Mn (0.7 m.-%), P (0.009 m.-%), and S (0.005 m.-%). The sample surface was put in a vertical plane and the laser beam was focused horizontally at right angle onto the sample surface by a planoconvex achromatic lens ($f = 50$ mm). The beam waist diameter is about $2w_0 = 23\,\mu$m and the average irradiance at the sample surface amounts to $I_f = 10^{10}, 2.5 \times 10^{10}, 5 \times 10^{10}$, and $10^{11}\,\text{W/cm}^2$ for the different pulse energies. The sample is moved by linear axes perpendicularly to the laser beam axis with a speed of 2 mm/min. The plasma is imaged with a reduction ratio of 1 : 0.7 using a concave mirror into an optical fiber with a core diameter of 800 μm connected to the spectrometer. The spectrometer is an echelle system with a spectral resolution of 1–22 pm in the wavelength range between 200 and 780 nm, cf. Sect. 4.2, Table 4.4. The detection system consisting of spectrometer and optical fiber uses an internal response correction function to achieve a wavelength independent response within its wavelength range. Due to the small emission intensities at laser pulse energies of only a few millijoule 100 consecutive LIBS spectra are accumulated in this wavelength range. To study the time evolution of the plasma, the line emission is investigated at nine different delay times $t_{\mathrm{delay}} = 100 - 900$ ns with an integration time of $t_{\mathrm{int}} = 100$ ns each. The upper limit for the delay time has to be chosen because of the fast decreasing spectral line intensity of the Fe II species. In the case of $E_{\mathrm{b}} = 200\,\mu$J, the $1/e$-decay time of the intensity is about 260 ns. For times up to $t_{\mathrm{delay}} = 900$ ns, a sufficient spectral line intensity can be assured for all studied pulse energies. Thus, for every pulse energy nine spectra corresponding to the nine different time windows were recorded so that in total 36 spectra were obtained for the four laser pulse energies.

Due to the comparatively low laser pulse energies used low emission line intensities are expected. In order to assure a minimum error of the determined electron temperatures, a procedure to find appropriate emission lines for the Boltzmann plots is developed. In a first step of this procedure, a preselection of spectral lines based on theoretical calculations is carried out. Afterwards, these preselected lines are assessed within an iterative optimization step using experimental data in which improper spectral lines are discarded successively. The total number of all atomic and ionic emission lines of elements contained in the used samples within the spectral range of 200–780 nm amounts to ~93,000 [A.30]. Assuming local thermal equilibrium (LTE) – see Sect. 8.4 – the electron temperature derived by the Boltzmann plot method is independent of the considered species within the plasma. The emission coefficient is proportional to the number density of the particular species in the plasma, cf. Sect. 9.2, (9.3). Thus, in the following only the emission lines of the matrix element, which in this case is Fe (>98 m.-%), are considered to determine the electron temperature using Boltzmann plots. The total number of Fe emission lines amounts to $\sim 38,000$ in the considered spectral range. Electron temperatures in LIBS plasmas are typically in the range of 5,000–16,000 K. The excitation probability decreases with the excitation energy according to the Boltzmann distribution, cf. (9.3). As on average higher ionization stages have higher excitation energies and lower occurrences in the plasma due to the Saha equation [cf. Sect. 10.1, (10.5)], and emissions of Fe lines with ionization stages higher than I and II are negligible for the expected temperature ranges. Hence, the further considered spectral lines are limited to the first two ionization stages Fe I and Fe II. The total number of remaining lines for these two species is about ~30,000. To reduce this huge number of spectral lines to only those which are clearly identifiable in the whole spectrum, it is reasonable to develop a criterion based on the emission coefficients of the spectral lines. For the remaining Fe I and Fe II spectral lines, the emission coefficients can be calculated according to the relationship (9.3). The partition function for each species can be obtained for example from the Atomic Spectral Database of the National Institute of Standards and Technology [A.21]. The electron temperature T_{e} can be estimated or derived from preliminary measurements. Here, the following temperatures were chosen:

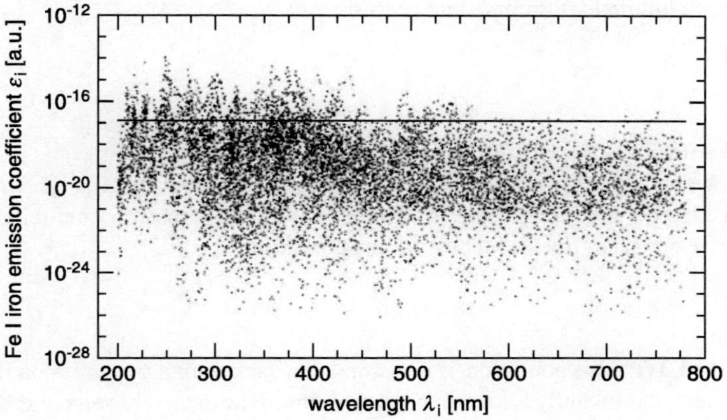

Fig. A.2 Calculated emission coefficients for all Fe I spectral lines in the considered wavelength range. The *horizontal line* represents the threshold value (mean value of all emission coefficients) for this species. The index i characterizes the individual Fe I emission line

$T_e = 8{,}900\,\mathrm{K}$ for Fe I emission lines and $T_e = 11{,}300\,\mathrm{K}$ for Fe II emission lines, respectively. In order to extract the lines with the highest emission coefficients, threshold values are calculated for each species, Fe I and Fe II. As particular threshold value the mean value of all emission lines of the regarded species is taken. The number density N can be set to a constant value, since only emission lines of the same species are compared with each other. Figure A.2 shows the half-logarithmic plot of calculated emission coefficients for Fe I emission lines versus their wavelength.

After discarding all emission lines with lower emission coefficients than the chosen threshold value in total 966 Fe I and 585 Fe II emission lines can be identified as potential candidates for the Boltzmann plots. In a second step, these remaining spectral lines have to be assessed quantitatively with respect to superposition and overlapping with other spectral lines.

Spectral lines emitted by the laser-induced plasma are broadened by a variety of mechanisms such as Stark, Van-der-Waals and Doppler broadening, cf. Sect. 9.4. Additionally, the apparative broadening of the spectrometer increases the widths of all spectral lines as well, see, e.g., Figs. 9.6 and 9.7.

Since the Boltzmann plot method relies on the measured emission line intensity, it is necessary to identify those lines which do not significantly overlap with other emission lines, e.g., due to line broadening mechanisms. Therefore, in the next step each of the remaining Fe I and Fe II lines is investigated with respect to superpositions with other emission lines which can occur in the plasma. This investigation has to cover all present species and elements in the sample even if the concentration is comparatively low. In our case for each of the remaining Fe I and Fe II emission lines a wavelength interval of 200 pm is centered at the transition wavelength λ_k. The emission coefficients ε_i of all other spectral lines within this wavelength interval are summed up according to

$$U_k = \sum_i \varepsilon_i. \tag{A.9}$$

for all i with $|\lambda_i - \lambda_k| \leq 100\,\text{pm}$ and $i \neq k$.

Afterwards, this sum of all emission coefficients in the 100 pm-neighborhood of the considered Fe emission line is compared with the emission coefficient at λ_k. Only those emission lines which fulfill the condition

$$\varepsilon_k > 3 \cdot U_k \tag{A.10}$$

are retained. With this condition, it is assured that each remaining emission line does not interfere significantly with other spectral lines within the chosen neighborhood. Even if there are still interfering emission lines, the condition (A.10) ensures that the disturbing lines have a cumulative emission coefficient of less than a third of the emission coefficient ε_k at λ_k. By this preselection for the above-described case 337 Fe I and 96 Fe II spectral lines are obtained as potential candidates for the electron temperature determination. It might happen that some of these remaining lines have the same upper energy level. In this case, only those spectral lines are retained with the highest ε_k / U_k ratio. All other lines of the same energy level are discarded, since otherwise particular energy levels would be preferred in the Boltzmann plot. By eliminating spectral lines with the same upper energy level 199 Fe I and 64 Fe II emission lines are obtained as an intermediate result after the successive preselection processes. The average ε_k / U_k ratios amount to 191.3 for Fe I and 239.7 for Fe II. As mentioned above, this result shows that even if overlapped lines are not discarded by the proposed method and used criteria, the contribution of the disturbing lines to the emission coefficient would theoretically be less than 1% and can therefore be neglected. Finally, comparing all initially regarded iron lines (9,922 Fe I, 20,162 Fe II) only 0.87% are further considered after the preselection.

After the preselection procedure, the line intensities of these emission lines have to be determined from experimentally recorded spectra in order to generate Boltzmann plots. The line profile is divided in three wavelength intervals marked with A, B, and C. Interval B contains the peak of the line and ranges in-between the closest local minima on either side of the peak. Adjacent on both sides of interval B are the intervals A and C with an extent of 10 pm each. To account for background noise of the detection system the average intensity in the intervals A and C is subtracted from the line intensity in interval B. The background corrected net line intensity can then be determined via trapezoidal integration. However, some of the preselected lines might not be detectable, visible or identifiable due to various reasons. In the case described here, the echelle spectrometer has gaps in its detectable wavelength range. Additionally, some lines may disappear in the continuum radiation emitted by the plasma which was not considered in the theoretical part as well. After eliminating emission lines which were predicted as prominent in the spectrum but not present in one of the 36 experimentally recorded spectra 167 Fe I and 48 Fe II emission lines were clearly identified and allocated.

With the spectral line intensities determined, the electron temperatures were deduced via Boltzmann plots for every time window and laser pulse energy. The relative error of the electron temperature averaged over the 36 spectra amounts to $\Delta T_e/T_e = 4.9\%$ ($r^2 = 0.72$) for Fe I lines and $\Delta T_e/T_e = 61.2\%$ ($r^2 = 0.06$) for Fe II lines. The coefficients of determination are rather poor, for iron ion lines the situation is worse compared with iron atomic lines. Some data points are far away from the regression curve. To eliminate outliers in the Boltzmann plot, an iterative method to discard spectral lines which do not match with the overall trend given by the other data points was developed. The goal of this method is to identify a set of lines for the specific time domain and pulse energy regime based on an algorithm which considers the residuum of an individual data point in the Boltzmann plot from the regression line, i.e., its distance in y-direction. For each spectral line, the following measure is calculated to ensure that the used spectral line has an emission intensity level high enough to allow detection within the entire specified time domain and pulse energy regime:

$$S_k = \sum_{x,y} D^2_{k,x,y}. \tag{A.11}$$

The indices x and y are for the different time windows and pulse energies. The index k denotes the individual spectral lines and $D_{k,x,y}$ are the deviations between a data point and the regression line. S_k corresponds to the total sum of the squared deviations for all time windows and laser pulse energies. Thus, the spectral line with the maximum value for S_k has on average, e.g., in the considered case for all four laser pulse energies and for all nine time windows, the largest deviation from the regression line. The iterative method is based on discarding this line and on recalculation of the regression function. This procedure is repeated until the coefficient of determination exceeds a threshold value. By this algebraic approach, it is ensured that data points which appear as obvious outliers in every time window are discarded. As break condition for the iterative Boltzmann plot method, an average threshold value of $r^2 = 0.98$ is taken.

This iterative method stopped after discarding 106 Fe I lines and 36 Fe II lines. By this approach, relative errors for the electron temperature of 1.8% with the remaining 61 Fe I and 4.4% with 12 Fe II emission lines averaged over all considered nine time windows and laser pulse energies were achieved. These remaining emission lines for the temperature determination are summarized in Tables A.4 and A.5.

The development of the average relative error of the electron temperature for Fe I emission lines by iteration number is shown in Fig. A.3. As can be seen the averaged relative errors are decreasing by nearly a factor of three in the case of Fe I emission lines during the iterative process. Due to a couple of outlier emission lines in the case of Fe II emission lines which are considered in the beginning of the iteration process, the averaged relative errors are reduced by even a factor of 14.

Table A.4 Spectroscopic data of selected Fe I emission lines

No.	λ (nm)	A_{nm} (1/s)	E_n(eV)	g_n
1	244.256	2.14E + 08	7.508	11
2	282.328	1.86E + 07	5.349	7
3	289.942	5.92E + 07	6.554	3
4	295.999	4.81E + 07	6.880	13
5	299.039	3.90E + 07	6.873	11
6	301.148	4.70E + 07	6.875	9
7	301.898	1.26E + 07	5.064	7
8	302.403	4.87E + 06	4.209	5
9	303.015	4.79E + 07	6.524	11
10	305.307	1.53E + 07	6.484	5
11	305.526	8.63E + 06	5.615	5
12	307.572	2.94E + 07	4.988	5
13	308.374	3.01E + 07	5.010	3
14	309.158	5.41E + 07	5.021	1
15	328.026	5.38E + 07	7.080	11
16	331.474	6.89E + 07	7.041	7
17	336.955	2.42E + 07	6.406	9
18	337.078	3.29E + 07	6.370	11
19	338.011	2.35E + 07	6.426	7
20	341.313	3.57E + 07	5.830	7
21	342.266	2.79E + 07	5.845	5
22	342.712	5.49E + 07	5.793	9
23	347.545	7.01E + 06	3.654	5
24	349.057	4.47E + 06	3.603	7
25	349.529	9.46E + 06	6.106	7
26	351.382	3.40E + 06	4.387	11
27	360.668	8.24E + 07	6.129	13
28	364.951	4.22E + 07	6.089	9
29	365.147	6.23E + 07	6.153	9
30	367.763	8.01E + 07	6.129	5
31	368.411	3.36E + 07	6.092	7
32	369.073	2.74E + 07	6.932	11
33	370.557	3.21E + 06	3.397	7
34	372.256	4.97E + 06	3.417	5
35	372.438	1.25E + 07	5.607	7
36	376.554	9.69E + 07	6.529	15
37	376.719	6.39E + 07	4.302	3
38	379.955	7.31E + 06	4.221	9
39	380.535	9.81E + 07	6.559	11
40	382.782	1.05E + 08	4.796	5
41	384.326	4.67E + 07	6.272	7
42	384.997	6.05E + 07	4.231	1
43	385.637	4.64E + 06	3.266	5
44	386.552	1.55E + 07	4.218	3
45	386.722	3.39E + 07	6.223	5
46	387.376	8.01E + 06	5.633	9
47	392.026	2.60E + 06	3.283	3

(continued)

Table A.4 (continued)

No.	λ (nm)	A_{nm} (1/s)	E_n(eV)	g_n
48	392.291	1.08E + 06	3.211	9
49	395.116	3.56E + 07	6.411	5
50	397.774	7.01E + 06	5.314	5
51	401.453	2.37E + 07	6.661	11
52	404.581	8.62E + 07	4.549	9
53	406.359	6.78E + 07	4.608	7
54	407.174	7.65E + 07	4.652	5
55	410.749	2.45E + 07	5.850	3
56	421.936	3.80E + 07	6.511	13
57	430.790	3.40E + 07	4.435	9
58	432.576	4.97E + 07	4.474	7
59	436.977	7.22E + 06	5.884	9
60	438.354	5.00E + 07	4.313	11
61	440.475	2.75E + 07	4.372	9

Table A.5 Spectroscopic data of selected Fe II emission lines

No.	λ (nm)	A_{nm} (1/s)	E_n (eV)	g_n
1	224.550	2.61E + 08	10.289	10
2	233.131	2.92E + 07	5.549	8
3	234.534	8.46E + 07	7.920	12
4	236.860	5.93E + 07	5.585	4
5	237.519	9.81E + 07	5.605	2
6	238.076	3.14E + 07	5.289	8
7	257.792	1.26E + 08	5.905	2
8	259.154	5.11E + 07	5.824	6
9	261.383	1.99E + 08	4.849	2
10	262.829	8.56E + 07	4.837	4
11	263.957	1.07E + 08	8.036	2
12	271.441	5.48E + 07	5.553	6

Figure A.4 shows the development of the averaged coefficient of determination for Fe II emission lines.

The coefficient of determination in iteration step 37 exceeds the threshold value of $r^2 = 0.98$ so that in total 36 Fe II emission lines are discarded in this case. The corresponding Boltzmann plots are shown in Figs. A.5 and A.6. The relative temperature errors are reduced in this particular case to $\Delta T_e / T_e = 1.7\%$ for Fe I and $\Delta T_e / T_e = 5.9\%$ for Fe II emission lines. This effect is due to the fact that on average a certain coefficient of determination is required which in turn is responsible for the exclusion of emission lines with comparatively high residue to the regression line. The higher relative error for Fe II emission lines can be related to the lower number of remaining spectral lines compared with those for Fe I. This result is a consequence of the fact that ionized species in the plasma relax continuously to lower ionization stages.

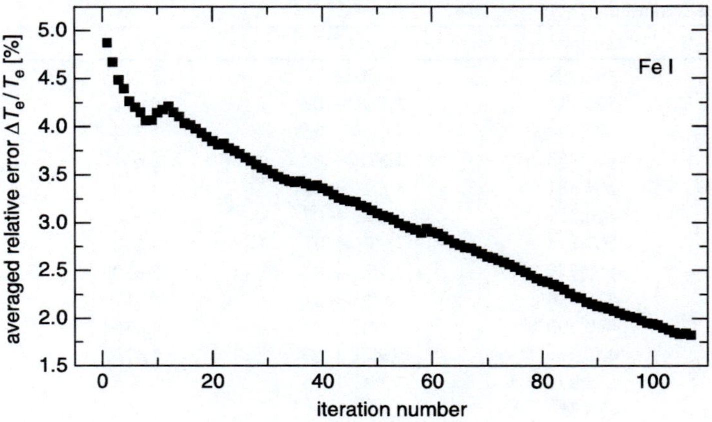

Fig. A.3 Development of the averaged relative errors of electron temperature derived with Fe I emission lines

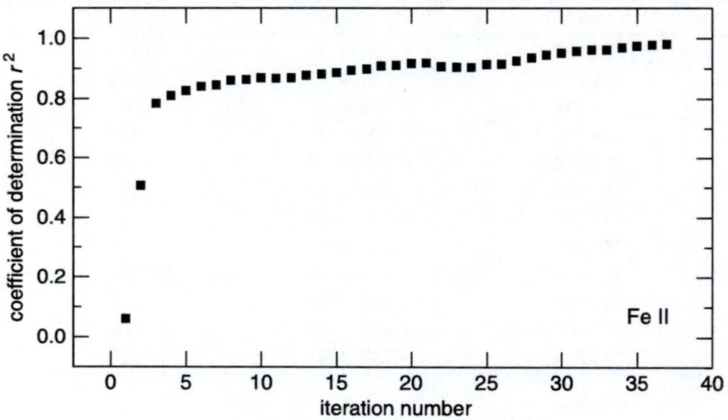

Fig. A.4 Development of the averaged coefficient of determination for the considered 36 Boltzmann plots vs. the iteration number for Fe II emission lines

In Figs. A.7 and A.8, the calculated electron temperatures are shown for the investigated pulse energies and delay times t_{delay}. For Fe I and Fe II, two different electron temperatures exist within the plasma. Starting from $t_{\mathrm{delay}} = 100$ ns to $t_{\mathrm{delay}} = 900$ ns for both ionization stages and most pulse energies the temperatures are decreasing. Comparing those values with the estimated electron temperatures at the beginning of the theoretical line preselection, the average electron temperature $T_e \sim 8{,}300$ K of the atomic lines for all investigated times and pulse energies is $\sim$600 K below the chosen temperature at the beginning of the line selection process. In the case of the ionic lines, the average temperature $T_e \sim 12{,}200$ K is about 900 K

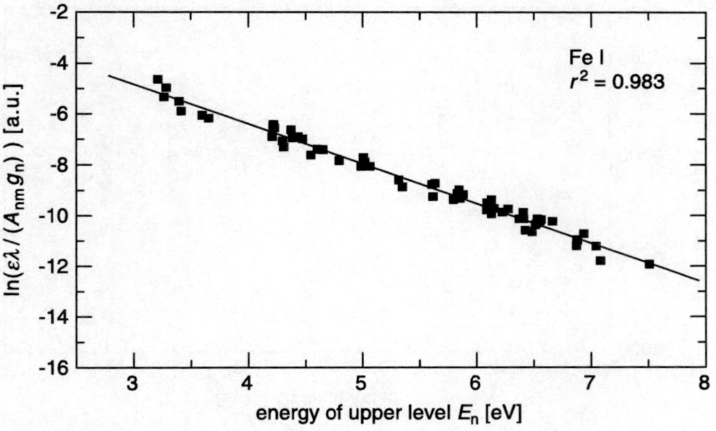

Fig. A.5 Boltzmann plot with 61 Fe I emission lines for an exemplarily chosen time window $t_{delay} = 900$ ns, $t_{int} = 100$ ns, $E_b = 0.5$ mJ. The relative temperature errors are $\Delta T_e / T_e = 1.7\%$ for this time window

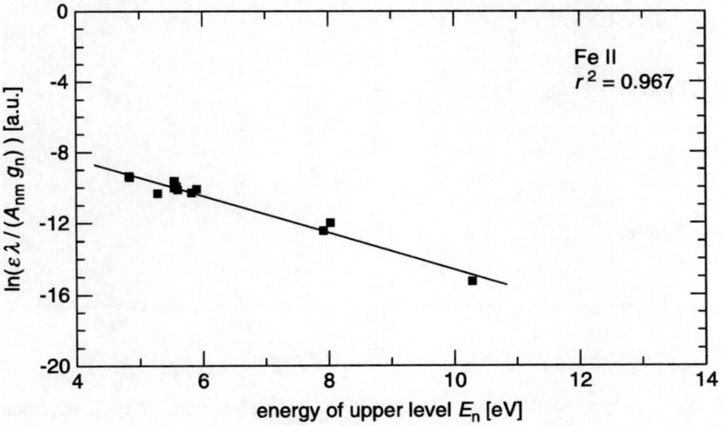

Fig. A.6 Boltzmann plot with 12 Fe II emission lines, other data as in Fig. A.5. The relative temperature error is $\Delta T_e / T_e = 5.9\%$

higher than the initial one. These average temperatures are in the same regime like those that were reported by Detalle et al. where electron temperatures between 7,000 and 10,000 K were detected using similar laser irradiances [A.31]. To determine the electron temperature, six Fe I spectral lines in the wavelength range between 370 and 380 nm were chosen. Only delay times between 100 and 400 ns were investigated, with an error of 37.5% for $t_{delay} = 400$ ns.

It should be noticed that the above-described iterative Boltzmann plot procedure is a systematic approach to improve the precision of the electron temperature

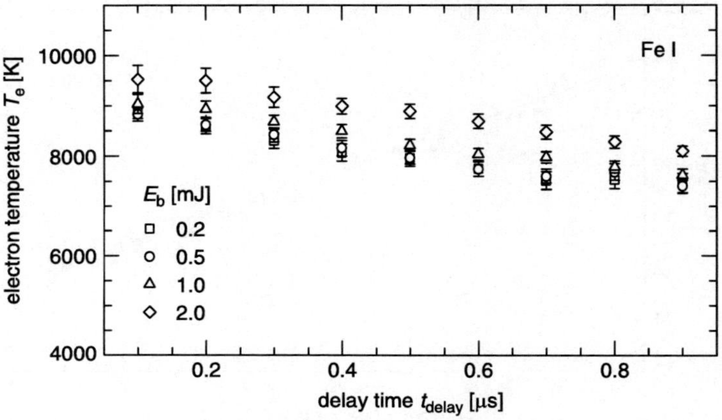

Fig. A.7 Electron temperatures determined by Boltzmann plots with 61 Fe I emission lines at different delay times for pulse energies between 0.2 and 2 mJ

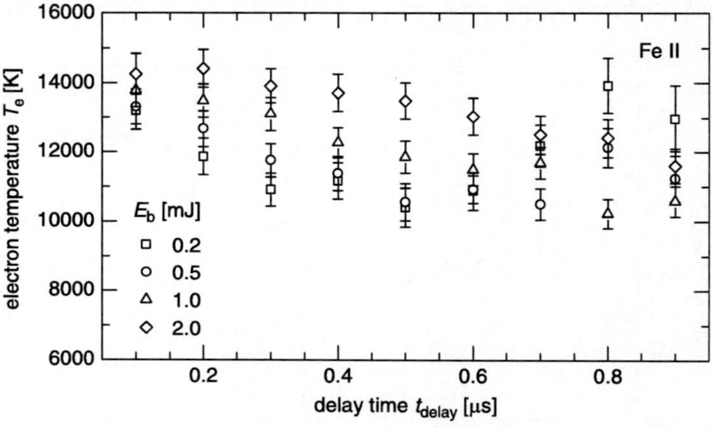

Fig. A.8 Electron temperatures determined by Boltzmann plots with 12 Fe II emission lines at different delay times for pulse energies between 0.2 and 2 mJ

determination. However, the reduction of the relative temperature errors does not necessarily imply an improvement regarding the accuracy. The assessment of accuracy would require a comparison between independent techniques for the determination of the electron temperature.

The results show that even at comparatively low laser pulse energies ($E_b \leq 2$ mJ) up to 61 atomic emission lines can be identified within a typical spectral range for temperature measurements. It should be mentioned that the chosen criteria within the proposed procedure were applied to achieve pragmatic but quantitatively well-grounded results for the experiment. In fact, this result improves the precision of the electron temperature determination. The three main advantages of the described method are: the transferability to other problems, e.g., samples with arbitrary chemical composition; the quantitative validation between theoretical predictions and experimental data; the systematic approach to find appropriate spectral lines for LIBS applications.

References

A.1. N. Konjevic, M. Dimitrijevic, W. Wiese, Experimental Stark widths and shifts for spectral lines of neutral and ionized atoms, J. Phys. Chem. Ref. Data, **13**, 619–647 (1984)

A.2. C. Colon, G. Hatem, E. Verdugo, P. Ruiz, J. Campos, Measurement of the Stark broadening and shift parameters for several ultraviolet lines of singly ionized aluminum, J. Appl. Phys. **73**, 4752–4758 (1993)

A.3. N. Konjevic, W. Wiese, Experimental Stark widths and shifts for spectral lines of neutral and ionized atoms, J. Phys. Chem. Ref. Data, **19**, 1307–1385 (1990)

A.4. H. Griem, *Spectral Line Broadening by Plasmas*, (Academic, London, 1974), Appendix IV, p. 320

A.5. Z. Szymanski, J. Kurzyna, W. Kalita, The spectroscopy of the plasma plume induced during welding of stainless steel and titanium, J. Phys. D. Appl. Phys. **30**, 3153–3162 (1997)

A.6. J. Aguilera, C: Aragon, Curves of growth of spectral lines emitted by a laser-induced plasma: influence of the temporal evolution and spatial inhomogeneity of the plasma, Spectrochim. Acta B 58 221–237 (2003)

A.7. A. Lesage, J. Lebrun, J. Richou, Temperature dependence of Stark parameters for Fe I lines, Astrophys. J. **360**, 737–740 (1990)

A.8. M. Rösner-Kuhn, D. Matson, K. Drewes, U. Thiedemann, G. Kuppermann, M. Flemings, M. Frohberg, Enthalpies and heat capacities of liquid Fe-Cr-Ni alloys with the focus on pure liquid chromium, Thermochim. Acta **314**, 123–129 (1998)

A.9. A. Bartecka, T. Wujec, J. Halenka, J. Musielok, Experimental Stark-broadening studies of the N I multiplet at 7904.5 A, Eur. Phys. J. D **29**, 265–271 (2004)

A.10. S. Freudenstein, J. Cooper, Stark broadening of Fe I 5383 A, Astron. Astrophys. **71**, 283–288 (1979)

A.11. V. Bakshi, R. Kearney, Measurement of Stark width of some Ar I transitions in a d.c. argon plasma jet at atmospheric pressure, J. Quant. Spectrosc. Radiat. Transf. **42**, 405–413 (1989)

A.12. M. Dimitrijevic, S. Sahal-Brechot, Stark broadening of Ca II spectral lines, J. Quant. Spectrosc. Radiat. Transf. **49**, 157–164 (1993)

A.13. J. Puric, S. Djenize, A. Sreckovic, S. Bukvic, S. Pivalica, J. Labat, Stark widths of singly-ionized iron spectral lines, Astron. Astrophys. Suppl. Ser. **102**, 607–609 (1993)

A.14. M. Dimitrijevic, S. Sahal-Brechot, Stark broadening of Al I spectral lines, Physica Scripta **49**, 34–38 (1994)

A.15. K. Dzierzega, K. Musiol, Stark broadening and shift for Ar II lines, J. Quant. Spectrosc. Radiat. Transf. **52**, 747–754 (1994)

A.16. M. Dimitrijevic, S. Sahal-Brechot, Stark broadening of Mg I spectral lines, Physica Scripta **52**, 41–51 (1995)

A.17. E. Sarandaev, M. Salakhov, Regularities in the Stark widths and shifts of spectral lines of singly-ionized aluminium, J. Quant. Spectrosc. Radiat. Transf. **56**, 399–407 (1996)

A.18. J. Knauer, M. Kock, Experimental Stark broadening constants for Ar I, Ar and II Kr I resonance lines in the vacuum ultraviolet region, J. Quant. Spectrosc. Radiat. Transf. **56**, 563–572 (1996)

A.19. N. Nessib, Z. Lakhdar, Stark broadening of neutral oxygen lines in the impact and quasistatic approximations, Physica Scripta **54**, 608–613 (1996)

A.20. L. Popovic, M. Dimitrijevic, Stark broadening of heavy ion lines: As II, Br II, Sb II and I II, Physica Scripta **53**, 325–331 (1996)
A.21. NIST Atomic Spectra Database, Vers. 4, Lines, http://physics.nist.gov/PhysRefData/ASD/lines_form.html
A.22. V. Detalle, R. Héon, M. Sabsabi, L. St-Onge, An evaluation of a commercial echelle spectrometer with intensified charge-coupled device detector for materials analysis by laser-induced plasma spectroscopy, Spectrochim. Acta B **56**, 1011–1025 (2001)
A.23. M. Sabsabi, V. Detalle, M. A. Harith, W. Tawfik, H. Imam, Comparative study of two new commercial echelle spectrometers equipped with intensified CCD for analysis of laser-induced breakdown spectroscopy, Appl. Opt. **42**, 6094–6096 (2003)
A.24. Ü. Aydin, P. Roth, C. Gehlen, R. Noll, Spectral line selection for time-resolved investigations of LIBS plasmas by an iterative Boltzmann plot method, Spectrochim. Acta B **63**, 1060–1065 (2008)
A.25. P. Fichet, J.-L. Lacour, D. Menut, P. Mauchien, A. Rivoallan, C. Fabre, J. Dubessy, M.-C. Boiron, in *Micro LIBS Technique*, ed. by A.W. Miziolek, V. Palleschi, I. Schechter. Laser-Induced Breakdown Spectroscopy (LIBS) – Fundamentals and Applications (Cambridge University Press, Cambridge, 2006), ISBN-13: 9780511243042
A.26. C. Geertsen, J.-L. Lacour, P. Mauchien, L. Pierrard, Evaluation of laser ablation optical emission spectrometry for microanalysis in aluminium samples, Spectrochim. Acta B **51**, 1403–1416 (1996)
A.27. V. Detalle, J.-L. Lacour, P. Mauchien, A. Semerok, Investigation of laser plasma for solid element composition microanalysis, Appl. Surf. Sci. **138**, 299–301 (1999)
A.28. C. Hartmann, A. Gillner, Investigaion on Laser Micro Ablation of Steel Using ps-IR Pulse Bursts, *LIA Conference Proceedings 2007*, (LIA, Orlando, 2007), pp. 38–44
A.29. H. Bette, R. Noll, in *High-Speed, High-Resolution LIBS Using Diode-Pumped Solid State Lasers*, ed. by A. Miziolek, V. Palleschi, I. Schechter. Laser-Induced Breakdown Spectroscopy, Chap. 14 (Cambridge University Press, Cambridge, 2006), pp. 490–515
A.30. R. Kurucz, B. Bell, 1995 *Atomic Line Data, Kurucz CD-ROM No. 23* (Smithsonian Astrophysical Observatory, Cambridge, MA)
A.31. V. Detalle, J. Lacour, P. Mauchien, A. Semerok, Investigation of laser plasma for solid element composition microanalysis, Appl. Surf. Sci. **138–139**, 299–301 (1999)
A.32. S. Brym, R. Ciurylo, E. Lisicki, R. Trawinski, Pressure broadening and shift of the 326.1 nm Cd line perturbed by argon, Physica Scripta **53**, 541–544 (1996)
A.33. Z. Szymanski, J. Kurzyna, W. Kalita, The spectroscopy of the plasma plume induced during welding of stainless steel and titanium, J. Phys. D. Appl. Phys. **30**, 3153–3162 (1997)
A.34. S. Pellerin, K. Musiol, J. Chapelle, Measurement of atomic parameters of singly ionized argon lines – III. STARK broadening parameters, J. Quant. Spectrosc. Radiat. Transf. **57**, 377–393 (1997)
A.35. J. Val, J. Aparicio, S. Mar, Experimental Stark widths and shifts of several Ne I spectral lines, Astrophys. J. **513**, 535–541 (1999)
A.36. N. Konjevic, Plasma broadening and shifting of non-hydrogenic spectral lnines: present status and applications, Phys. Rep. **316**, 339–401 (1999)
A.37. N. Konjevic, A. Lesage, J. Fuhr, W. Wiese, Experimental Stark widths and shifts for spectral lines of neutral and ionized atoms, J. Phys. Chem. Ref. Data **31**, 819–927 (2002)
A.38. A. Bartecka, T. Wujec, J. Halenka, J. Musielok, Experimental Stark-broadening studies of the N I multiplet at 7904.5 A, Eur. Phys. J. D **29**, 265–271 (2004)
A.39. J. Hoffman, Z. Szymanski, Time-dependent spectroscopy of plasma plume under laser welding conditions, J. Phys. D. Appl. Phys. **37**, 1792–1799 (2004)
A.40. S. Bukvic, A. Sreckovic, S. Djenize, Mg II h and k lines Stark parameters, New Astron. **9**, 629–633 (2004)

Index

R. Noll, *Laser-Induced Breakdown Spectroscopy*, DOI 10.1007/978-3-642-20668-9,